Lecture Notes in Electrical Engineering 345

More information about this series at http://www.springer.com/series/7818

Jengnan Juang
Editor

Proceedings of the 3rd International Conference on Intelligent Technologies and Engineering Systems (ICITES2014)

Springer

Editor
Jengnan Juang
School of Engineering
Mercer University
Macon, GA, USA

ISSN 1876-1100 ISSN 1876-1119 (electronic)
Lecture Notes in Electrical Engineering
ISBN 978-3-319-17313-9 ISBN 978-3-319-17314-6 (eBook)
DOI 10.1007/978-3-319-17314-6

Library of Congress Control Number: 2015953832

Springer Cham Heidelberg New York Dordrecht London

Printed on acid-free paper

Springer International Publishing AG Switzerland is part of Springer Science+Business Media (www.springer.com)

Foreword

On behalf of the organizing committee, we are pleased to extend our warmest welcome to all the participants of International Conference on Intelligent Technologies and Engineering System (ICITES 2014) at the International Convention Center Kaohsiung in Taiwan. The organizing committee has put forth an excellent technical program covering various aspects of Intelligent Technologies and Engineering Systems.

The primary objective of this conference is to bring together on a common platform, academicians, researchers, application engineers, industry personnel, and the users of emerging intelligent technologies and engineering systems. Delegates and experts are invited to participate and present their work and the conference will be of interest to a wide spectrum of professionals at all levels. The conference is intended to encourage and facilitate knowledge sharing, interactions, and discussions on future evolutions of new challenges in the twenty-first century. There will be keynote speeches, invited presentations, technical sessions, panel discussions, and local tours.

We have accepted 80 papers consisting of plenary sessions and technical sessions in which over 120 authors from several countries around the world will give their presentations. The organizing committee has tried its best to make the conference as comfortable as possible to all the participants.

We are grateful to the participants—in particular, those who organized sessions and presented papers. We would like to extend our sincere thanks to all the invited speakers, members of the organizing committee, review committee, local arrangement committee, and the persons from industries. We also would like to thank all the graduate and undergraduate students who are presenting their research work as well as volunteering to help in organizing this conference. We understand what it takes to put together an event such as this and would like to express our sincere thanks to all involved in the preparation of this conference. We also want to thank all of the colleagues and staff who worked on the conference technical program.

The success of this conference is due to those of you who planned and designed the technical program, participated in reviewing the manuscripts, coordinated the exhibit, and volunteered to moderate technical sessions.

We look forward to seeing you all again in the next conference.

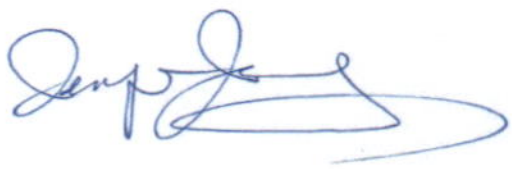

Macon, GA, USA Jengnan Juang

Preface

The papers presented at the International Conference on Intelligent Technologies and Engineering Systems (ICITES 2014) held at the International Convention Center Kaohsiung in Taiwan during December 19–21, 2014 are compiled in this volume. The primary objective of this conference is to bring together on a common platform, academicians, researchers, application engineers, industry personnel, and the users of emerging intelligent technologies and engineering systems. The conference is intended to encourage and facilitate knowledge sharing, interactions, and discussions on future evolutions of new challenges in the twenty-first century.

Over 250 participants attended the conference, made technical presentations, and discussed about various aspects of intelligent technologies and engineering systems. The number of papers published in this volume and the number of unpublished presentations at the conference indicate the evidence of growing interest in the areas of emerging intelligent technologies and engineering systems.

The papers published in this volume include but not limited to the following areas of interest: intelligent decision models, intelligent computing systems and applications, smart sensor network and applications, networking and signal processing, signal processing and computing technologies, systems control and applications, software engineering, intelligent electronic circuits and systems, innovative motor electronics and computer applications, mechanical and materials engineering, materials and communications, functional materials and radio frequency applications, and applied technology for engineering materials.

We would like to appreciate the efficient work done by Springer-Verlag in publishing this volume.

Macon, GA, USA Jengnan Juang

Conference Theme

The Third International Conference on Intelligent Technologies and Engineering Systems (ICITES 2014) was held on 19–21 December 2014, at International Convention Center Kaoshiung, Kaohsiung, Taiwan. This conference offered a great podium for scientists, engineers, and practitioners from a variety of background to present and discuss the latest research results, ideas, developments, and applications in intelligent technologies and engineering systems. The major areas of activity included networking, signal processing, artificial intelligence, software engineering, intelligent electronic circuits and systems, power system, communications, material and mechanical engineering and advance material.

Committees

General Chair

Jengnan Juang, Professor, Department of Electrical and Computer Engineering, Mercer University, USA

General Co-chairs

Hans Georg Bock, University of Heidelberg, Germany
Richard W. Longman, Columbia University, USA
Wen-Chung Chang, Southern Taiwan University of Science and Technology, Taiwan
Hsin-Nin Hsieh, World Invention Property Associations, Taipei, Taiwan

International Program Committee (Chair: Jengnan Juang)

Jer-Nan Juang, National Cheng Kung University, Taiwan
Chih-Chiang Hua, National Yunlin University of Science and Technology, Taiwan
J. Walton, The University of Edinburgh, UK
Fuh-Shyang Juang, National Formosa University, Yunlin, Taiwan
F. Frank Chen, The University of Texas-San Antonio, Texas, USA
Ekaterina Kostina, Marburg University, Marburg, Germany
Man-Li Hsieh, World Invention Property Associations, Taipei, Taiwan
Yi-Cheng Huang, National Changhua University of Education, Changhua, Taiwan
Benjamas Panomruttanarug, King Mongkut's University of Technology, Thailandthai
Jeh Won Lee, Yeungnam, University in Gyeongbuk, Korea
Wen-Yuan Chen, National Chin-Yi University of Technology

Jengnan Juang, Mercer University, USA
Richard W. Longman, Columbia University, USA
Hans Georg Bock, University of Heidelberg, Germany
Yoko Amano, Nihon University, Japan
Shujiro Dohta, Okayama University of Science, Japan
Win-Get Luo, National Chin-Yi University of Technology
R. Radharamanan, Mercer University, Georgia, USA
Tetsuya Akagi, Okayama University of Science, Japan
Wei-Ching Chuang, National Formosa University, Yunlin, Taiwan
Andrew J. Tickle, Coventry University, UK

Technical Program Committee (Chair: Wei-Ching Chuang)
Minh Q. Phan, Dartmouth College, New Hampshire, USA
R. Radharamanan, Mercer University, Georgia, USA
Richard W. Longman, Columbia University, USA
Hans Georg Bock, University of Heidelberg, Germany
Jeh Won Lee, Yeungnam, University in Gyeongbuk, Korea
Robert Howard, ATA, USA
Jengnan Juang, Mercer University, USA
Ghadir Radman, Tennessee Technological University, USA
Andrew J. Tickle, Coventry University, UK
Soo cheol Lee, Daegu University, Korea
F. Frank Chen, The University of Texas—San Antonio, USA
Ekaterina Kostina, Marburg University, Germany
Yung-Chiang Ting, Far East University, Tainan, Taiwan
Cheng-Fu Yang, National University of Kaohsiung, Taiwan
Cheng-Yi Chen, Cheng Shiu University, Kaohsiung, Taiwan
Yi-Cheng Huang, National Changhua University of Education Changhua, Taiwan
Wing-Kwong Wong, National Yunlin University Scene, Yunlin, Taiwan
Wen-Chung Chang, Southern Taiwan University, Tainan, Taiwan
Tetsuya Akagi, Okayama University of Science, Japan
Wei-Ching Chuang, National Formosa University, Yunlin, Taiwan

National Advisory Committee (Chair: Fuh-Shyang Juang)
Wing-Kwong Wong, National Yunlin University of Science & Technology, Taiwan
Cheng-Fu Yang, Department of Chemical and Materials Engineering, National University of Kaohsiung, Taiwan
Cheng-Yi Chen, Department of Electrical Engineering, Cheng Shiu University, Taiwan
Wen-Chung Chang, Department of Electronic Engineering, Southern Taiwan University of Science and Technology, Taiwan
Fuh-Shyang Juang, Department of Electro-Optical Engineering & Graduate Institute of Electro-Optical and Materials Science, National Formosa University, Taiwan

Keynote Speakers

Date: Friday, 19 December 2014
Time: 10:20–11:10
Richard W. Longman
Fellow AIAA
Fellow AAS
Professor of Mechanical Engineering
Columbia University in New York City, USA
Title: Laser Communication Between Spacecraft—A New Frontier for High Precision Control Systems

Abstract

The bandwidth for interplanetary communication is limited. It is not possible now to send high definition TV images from Mars to Earth in real time. Communication using lasers, LaserCom, addresses issues of power limitation. LaserCom experiments will be performed from the International Space Station and have recently been performed with a satellite and the moon. But LaserCom introduces the need for high precision pointing together with spacecraft jitter cancellation. Experimental comparisons of control law approaches to address these needs are presented, including some fundamental limitations resulting from the Bode Integral Theory—or the waterbed effect.

Date: Friday, 19 December 2014
Time: 11:10–12:00
Ramachandran Radharamanan, Ph.D.
Professor of Industrial Engineering
Director of Mercer Engineering Entrepreneurship Education Program (MEEEP)
Director of Mercer Center for Innovation and Entrepreneurship (MCIE)
Fellow, International Society for Productivity Enhancement (ISPE)
Mercer University School of Engineering

Title: Additive Manufacturing in Engineering Education

Abstract

Additive Manufacturing (AM) is any of various processes for making three-dimensional object of almost any shape from a 3D model or any other electronic data source primarily through additive processes in which successive layers of material are laid down under computer control. In this paper, the importance of incorporating AM as part of manufacturing curriculum in engineering education is emphasized. The difference between traditional manufacturing processes (subtractive manufacturing, SM) and AM are highlighted. Recent advances in the AM technologies that specialize in rapid prototyping of three-dimensional objects such as photopolymerization, powder bed fusion, extrusion, beam deposition, and 3D printing processes are discussed. Some of the AM student projects completed using 3D Scanning and 3D Printing at Mercer University School of Engineering are presented and discussed.

Date: Friday, 19 December 2014
Time: 13:10–14:00

Wen Jauh Chen
Professor
Dean of College of Humanities and Applied Sciences
National Yunlin University of Science and Technology, Yunlin, Taiwan, R.O.C.

Title: Study of Pt catalyst on graphene and its application to electrochemical biosensor

Abstract

A nonenzyme glucose biosensor was fabricated based on Pt nanoparticles on graphene (PtNPs/GN). The PtNPs/GN was formed from graphite by modified Hummers and Offeman's method and polyol synthesis method. The graphene oxide (GO) sheets were synthesized by modified Hummers and Offeman's method. The graphene is to obtain by reduction of GO films by sodium tetrahydridoborate. Then, the graphene deposited with Pt particles were put in ethylene glycol for a reflux at 130, 150, and 170 °C for 6 h and 18 h, respectively. The Taguchi method was used to find the optimum parameters for optimizing the growth of Pt/graphene nanocomposites. The electrocatalytic abilities of Pt/graphene nanocomposites were evaluated by the electrochemically active surface area (ECSA). The nanocomposites (Pt/graphene) were characterized by X-ray diffraction (XRD), scanning electron microscope (SEM), and transmission electron microscopy (TEM). The electrocatalytic characteristics of the Pt/graphene nanocomposites were evaluated for the nonenzymatic oxidation of hydrogen peroxide (H_2O_2) and glucose by using cyclic voltammetry and amperometry.

The results show that the PtNPs/GN composites modified GCE shows a wide linear range and high sensitivity for H_2O_2 and glucose detection. The PtNPs/GN composites also exhibit superior electrochemically active surface area (ECSA) and the increasing ECSA increased the oxidation peak currents of glucose. This reveal that the PtNPs/GN can be used as promising electrocatalyst supports for electrochemical sensors and biosensors.

Date: Saturday, 20 December 2014
Time: 11:10–12:00
Hans Georg Bock
Co-Chairman, Strategic Committee for Mathematical Modeling, Simulation and Optimization (KoMSO) at the Federal Ministry of Education and Science, Interdisciplinary Center for Scientific Computing IWR, University of Heidelberg, Germany

Title: Intelligent Optimization of Engineering Problems in Fluid Dynamics: Shape Optimization of Turbine and Compressor Blades

Abstract

The shape optimization of turbine and compressor blades is a crucial step within the design cycle of a whole turbomachine, where already minimal variations have a drastic influence on the efficiency of the turbine. In addition, many important manufacturing restrictions have to be taken into account. This paper is a report on a joint project between academia and the turbo machinery industry leading to efficient solution software for this problem to be used in the daily work of the designing engineers. The lecture analyses the special properties of the optimization problem, which aims to minimize the loss of pressure subject to a strongly nonlinear PDE boundary value problem as a constraint and additional geometry and manufacturing constraints. Since already the "forward" simulation problem requires the numerical solution of a complex fluid dynamics boundary value problem (BVP), optimization methods based on repeated solution of the simulation problem are extremely time consuming. We present a new "simultaneous" optimization approach, in which the optimization process is performed simultaneously with the solution of the BVP. This way, the optimization effort is only a small multiple of the effort needed for the simulation only. The solution method can be described as a parallel, multiple set points, reduced Sequential Quadratic Programming method that takes the special structure due to the discretized fluid dynamics problem into account. Convergence properties are discussed and numerical results are presented.

Date: Saturday, 20 December 2014
Time: 13:10–14:00
Jin-Tsong Jeng
IET Fellow
Professor of Department of Computer Science and Information Engineering, National Formosa University
Dean of College of Electrical and Computer Engineering, National Formosa University, Taiwan

Title: Intelligent Symbolic Data Fuzzy Clustering on Smart Phone

Abstract

Intelligent fuzzy clustering algorithms have been widely used such as pattern recognition, data analysis, machine learning, fuzzy modeling, electrical engineering, etc. In this presentation, we focus on the partitioning-based intelligent fuzzy clustering approach. About partitioning-based clustering Bezdek (1980) firstly proposed an improved K-means clustering algorithm; namely, fuzzy c-means (FCM) clustering algorithm for the single-valued data. On the other hand, interval fuzzy c-means (IFCM) clustering method is one common clustering approach proposed for symbolic interval-valued data. That is, Carvalho (2007) firstly extended the FCM to IFCM clustering. However, it still has noisy and outliers' problems. For the extension on the outliers' problem, Jeng et al., (2010) proposed robust interval competitive agglomeration clustering algorithm to overcome outliers. At the same time, Chuang et al. (2013) proposed interval PCM (IPCM) clustering algorithm and interval FPCM (IFPCM) clustering algorithm to extend IFCM. That is, we will show our intelligent symbolic data clustering algorithm to overcome the conventional symbolic data clustering algorithm for the symbolic interval data clustering in noisy and outlier environments. Besides, we also provide some further study directions in this topic. For the further App's development with intelligent symbolic data fuzzy clustering, we firstly use C# language to develop the improved FCM clustering algorithm windows mobile software on the Windows Mobile systems with .NET Compact Framework. From the implement results, the intelligent fuzzy symbolic data clustering algorithms have the fast convergence on the Windows Mobile systems. That is, we successfully integrate symbolic clustering algorithm on smartphone.

Session List

Friday (19 Dec 2014)
14.00–15.30
Advance Material
Oral room (A) 211, 213, 221, 233 (Chair: Dr. Jui-Che Huang)
Oral room (B) 272, 210, 212, 219, 225, 206 (Chair: Dr. Ramachandran Radharamanan)
Oral room (C) 208, 248, 259, 235, 257 (Chair: Dr. Tetsuya Akagi)
Poster session 207, 223, 240, 241, 281, 250, 269, 275, 284, 282, 266

16.00–17.30
Artificial Intelligence
Oral room (A) 265, 215, 217, 224, 244 (Chair: Dr. Richard Longman)
Communication
Oral room (B) 218, 263, 261, 285, 286, 254, 280 (Chair: Dr. Wei-Ching Chuang)
Oral room (C) 239, 267, 216, 227 (Chair: Dr. Shujiro Dohta)
Microwave Communication
Oral room (D) 262, 270, 271, 246, 268 (Chair: Dr. Behnam Kamali)
Saturday (20 Dec 2014)

14.00–15.30
Control and Software Engineering
Oral room (A) 256, 278, 264, 232 (Chair: Dr. Anthony Choi)
Oral room (B) 204, 209, 229, 231, 234 (Chair: Hans Georg Bock)
Oral room (C) 237, 238, 242, 260, 283 (Chair: Dr. Ramachandran Radharamanan
Intelligent Electronic Circuits and Systems
Oral room (D) 220, 228, 236, 279 (Chair: Dr. Fuh-Shyang Juang)

16.00–17.30
Intelligent Electronic Circuits and Systems
Oral room (A) 203, 243, 247, 253, 258 (Chair: Dr. Win-Jet Luo)
Material and Mechanical Engineering
Oral room (B) 273, 276, 245 (Chair: Dr. Richard Longman)
Microwave Communication
Oral room (C) 214, 226, 277, 251, 274, 252 (Chair: Jun-Ing Ker)

Program Schedule

December 19, 2014 (Friday)

International Conference Hall (Administration building)				
09:00–10:00	*Registration and Welcome Reception*			
10:00–10:20	Opening Ceremony, General Chair Prof. Jengnan Juang			
10:20–11:10	Keynote Speech 1 **Prof. Richard W. Longman**			
11:10–12:00	Keynote Speech 2 **Prof. Ramachandran Radharamanan, Ph.D.**			
12:00–13:10	*Lunch*			
13:10–14:00	Keynote Speech 3 **Prof. Wen Jauh Chen**			
14:00–15:30	Oral room A	Oral room B	Oral room C	Poster Session
15:30–16:00	*Tea Time (Break)*			
16:00–17:30	Oral room A	Oral room B	Oral room C	Oral room D

December 20, 2014 (Saturday)

International Conference Hall (Administration building)				
09:30–10:00	*Registration*			
10:00–11:00	Committees Conference			
11:00–12:00	Keynote Speech 4 **Prof. Hans Georg Bock**			
12:00–13:10	*Lunch*			
13:10–14:00	Keynote Speech 5 **Prof. Jin-Tsong Jeng**			
14:00–15:30	Oral room A	Oral room B	Oral room C	Oral room D
15:30–16:00	*Tea Time (Break)*			
16:00–17:30	Oral room A	Oral room B	Oral room C	Oral room D
18:00–20:00	*Banquet*			

December 21, 2014 (Sunday)

International Conference Hall (Administration building)	
09:30–10:00	*Registration*
10:00–12:00	Committees meeting
15:30–19:30	Conference Tour (Surcharge)

Paper Abstract

ICITES2014-203

Active Closed-Loop Gap Control for Aerostatic Bearing

Jyh-Chyang Renn[1*] and Yaw-Ming Chiou[2]

[1]Department of Mechanical Engineering, National Yunlin University of Science and Technology, Douliou 640, Taiwan

[2]Department of Mechanical Engineering, National Yunlin University of Science and Technology, Douliou 640, Taiwan

Abstract. Advantages of aerostatic bearing are clean, no pollution, and very low friction force. Therefore, it can easily be found in many high-precision machine tools and measurement equipments. However, owing to the compressibility of air, the stiffness of aerostatic bearing is relatively low compared to traditional ball bearing or hydrostatic bearing. In order to improve this fault, an active closed-loop gap control for aerostatic bearing with higher stiffness is proposed. First of all, an aerostatic bearing is designed and manufactured according to a previous report. Next, an experimental test rig based on LabVIEW software for the proposed active closed-loop gap control is constructed. Finally, it is proved that the gap between the aerostatic bearing and workpiece can be held at a stable and constant value even when the bearing is subjected to variable external disturbances.

Keywords: Aerostatic bearing, Active control, Closed-loop gap control

ICITES2014-204

Firefly Algorithm for Power Economic Emission Dispatch

Chao-Lung Chiang

Department of Electronic Engineering, Nan Kai University of Technology, Nan-Tou 542, Taiwan

Abstract. This work proposes a firefly algorithm for the optimal economic emission dispatch (EED) of the hydrothermal power system (HPS), considering non-smooth fuel cost and emission level functions. The firefly algorithm (FA) can efficiently search and actively explore solutions. The multiplier updating (MU) is introduced to handle the equality and inequality constraints of the HPS, and the å-constraint technique is employed to manage the multi-objective problem. To show the advantages of the proposed algorithm, one example addressing the best compromise is applied to test the EED problem of the HPS. The proposed approach integrates the FA, the MU, and the å-constraint technique, revealing that the proposed approach has the following merits—ease of implementation; applicability to non-smooth fuel cost and emission level functions; better effectiveness than the previous method, and the requirement for only a small population in applying the optimal EED problem of the HPS.

Keywords: Firefly algorithm, Multiplier updating, Economic emission dispatch

ICITES2014-206

Application of the Taguchi Method to the Warpage for Light Guide Plate

Po-Jen Cheng[1] and Chin-Hsing Cheng[2*]

[1]Department of Electrical Engineering, Nan Jeon University of Science and Technology, Tainan 73746, Taiwan

[2]Department of Electrical Engineering, Feng Chia University, Taichung 40724, Taiwan

Abstract. In recent years, the LED backlight module has become the main development of the thin-film transistor liquid crystal display (TFT-LCD) module. The light guide plate warpage is the major source of ripple phenomenon found in the TFT-LCD module. In this paper, different molding technological parameters are used for the experiment by the Taguchi method in the manufacturing process to find the factors of light guide plate warpage. The results of this manufacturing process are used to identify the optimization for production processes.

Keywords: Thin-film transistor liquid crystal display (TFT-LCD), Taguchi method, Light guide plate warpage

ICITES2014-207

Diagnosis of Muscle Properties in Low-Back-Pain with Electrical Bioimpedance Methodology

Chin-Sung Hsiao[1*], Yung-Tsung Yang[1] and Tai-Shin Ching[2]

[1]Department of Photonics and Communications Engineering, Asia University, Taichung 41354, Taiwan

[2]Department of Electrical Engineering National Chi Nan University, Nantou, Taiwan, R.O.C.

Abstract. An approach to detect low-back-pain (LBP) is developed in this study. In this work, the electrical bioimpedance methodology (EBM) was proposed to measure the impedance and phase angle of the low-back muscle at various frequency bands for the healthy group and the patients suffering from low-back-pain. After normalization of the measured impedances and phase angle, characteristic analyses of muscle property for low-back-pain patients and the healthy using statistical software SPSS with paired sample *t*-test and independent sample *t*-test is feasible to evaluate the dissimilarity of the two groups. This proposed methodology is enabled to have the capability of offering a fast and accurate technique for some muscle-related diseases.

Keywords: Low-back-pain, Electrical bioimpedance methodology, PSS Paired sample *t*-test, Muscle-related disease

ICITES2014-208

Low-Cost Wearable Control Valves with No Mechanical Sliding Parts in Valves

Tetsuya Akagi[1*], Shujiro Dohta[1], Ayumu Ono[1], and Abdul Nasir[1]
[1]Department of Intelligent Mechanical Engineering, Okayama University of Science, 1-1, Ridai-cho, Kita-ku, Okayama 700-0005, Japan

Abstract. The wearable driving system requires the pneumatic soft actuators, valves, and a controller. The complex inner configuration of the valve requiring relatively high precision prevents to fabricate a low-cost driving system. The cost of valves occupies most of the total cost. In this paper, two types of low-cost control valves with no mechanical sliding parts in valve are described. One is an on/off type valve using vibration motor, another is a servo valve using buckled tube. The operating principle and performance of both valves are also introduced.

Keywords: Valve using vibration motor, Servo valve using buckled tube

ICITES2014-209

Estimation of Residual Traveling Distance for Power Wheelchair Using Neural Network

Pei-Chung Chen[1*], Xiao-Qin Li[2], and Si-HanHaung[1]
[1]Department of Mechanical Engineering, Southern Taiwan University of Science and Technology, Tainan 71005, Taiwan
[2]Department of Electronic of Information Engineering, Ningbo Polytechnic, Zhejiang, Ningbo 315800, China

Abstract. The residual traveling distance of a power wheelchair is difficult to estimate due to the unknown factors of user manipulation behavior and journey characteristics of wheelchair. A virtual residual energy estimation system for power wheelchair based on neural network is proposed to estimate virtual residual energy which could be transformed into residual traveling distance. Two types of estimation systems with three training processes are presented. The estimated results are provided and compared. The results indicate that type-A estimation system with adaptive learning rate is a feasible solution based on economic factor and estimated performance.

Keywords: Residual traveling distance, Residual energy, Power wheelchair

ICITES2014-210

Residual Whitening Method for Identification of Induction Motor System

Chien-Hsun Kuo[1*] and D.-M. Yang[1]

[1]Department of Mechanical and Automation Engineering, Kao Yuan University, Kaohsiung 812, Taiwan

Abstract. This paper identifies the induction motor system by residual whitening method. Through residual whitening, the optimal properties of the Kalman filter could be enforced for a finite set of data. This technique uses AutoRegressive Moving Average with eXogeneous input (ARMAX) model which is combined of ARX (AutoRegressive with eXogeneous input) and MA (Moving-Average) models. Numerical and experimental results are shown for the identified induction motor system.

Keywords: Residual whitening, ARMAX model, Eigen system realization algorithm (ERA)

ICITES2014-211

Analysis and Simulation of Small-sized Quasi-servo Valve Using Tiny On/off Control Valve

So Shimooka[1*], Shujiro Dohta[1], Tetsuya Akagi[1], and Yoshinori Moriwake[1]

[1]Department of Intelligent Mechanical Engineering, Okayama University of Science, Okayama 700-0005, Japan

Abstract. Today, the care and welfare pneumatic equipment to support a nursing care and a self-reliance of the elderly and the disabled are actively researched and developed by many researchers. These wearable devices require many servo valves for multidegrees of freedom and precise control performance of the wearable actuator. The total weight of the wearable devices increases according to the degree of freedom. In our previous study, a small-sized and light-weight pressure control type quasi-servo valve was developed. The valve consists of two on/off control valves and an embedded controller. In this study, the quasi-servo valve composing of much smaller-sized (40 % in mass, 42 % in volume) on/off valves is proposed and tested. The analytical model of the tested valve is proposed and the system parameters are identified. As a result of the comparison between experimental results and simulated, it was confirmed that the proposed analytical model and the identified system parameters were valid.

Keywords: Quasi-servo valve, Small-sized control valve, Embedded controller

ICITES2014-212

Analysis of Flexible Thin Actuator Using Gas-Liquid Phase-Change of Low Boiling Point Liquid

Yasuyuki Tsuji[1*], Shujiro Dohta[1], Tetsuya Akagi[1], and Yuto Fujiwara[1]

[1]Department of Intelligent Mechanical Engineering, Okayama University of Science, Okayama 700-0005, Japan

Abstract. In our previous study, the flexible thin actuator using gas-liquid phase-change of a low boiling point liquid that can generate large force was proposed and tested. The tested actuator is an envelope-type actuator that is made of laminating plastic sheets, low boiling point liquid, and a flexible heater. In this paper, the analytical model of the flexible thin actuator was proposed and tested. The system parameters of the actuator were also identified. As a result, it was confirmed that the proposed analytical model can predict the behavior of the real actuator.

Keywords: Analysis, Flexible thin actuator, Gas-liquid phase-change, Low boiling point liquid, Portable rehabilitation device, Silent movement

ICITES2014-213

Development of Flexible Pneumatic Cylinder with String Type Displacement Sensor for Flexible Spherical Actuator

Yasuko Matsui[1], Tetsuya Akagi[1], Shujiro Dohta[1], and Shinsaku Fujimoto[1]
[1]Department of Intelligent Mechanical Engineering, Okayama University of Science, Okayama 700-0005, Japan

Abstract. This study aims at developing a potable rehabilitation device which can be safe to use while holding it. In our previous study, a novel flexible pneumatic cylinder that can be used even if it is deformed by external force has been developed. A portable rehabilitation device using the flexible spherical actuator that consists of two ring-shaped flexible pneumatic cylinders was proposed and tested. The attitude control system using a tiny embedded controller, four small-sized quasi-servo valves and two accelerometers was also proposed and constructed. The attitude control of the device was executed. In the next step, it is necessary to recognize the relative position between both stages to prevent both hands to contact each other. In this study, the low-cost flexible displacement sensor using the nylon string coated with carbon is proposed and tested. As a result, it is confirmed that the tested sensor can measure the displacement of the cylinder.

Keywords: Portable rehabilitation device, Flexible pneumatic cylinder, Flexible spherical actuator, Nylon string coated with carbon, String type displacement sensor

ICITES2014-214

Evaluation of Computing Reliability in Internet Computing

Shin-Guang Chen
Department of Industrial Management, Tungnan University, New Taipei City 222, Taiwan

Abstract. Internet computing is now a blooming technology which refers to multiple independent computing units working together over the Internet. The computing or calculating units are processor-memory pairs which distribute over one or more computers networked by information links over the Internet. It is important to maintain a highly efficient computing environment, namely, computing reliability. However, very few literature concerns this issue. This paper proposes an evaluation method for the computing reliability in Internet computing. A numerical example is presented for exploring the proposed method. The results show that the proposed method is helpful in the evaluation of computing reliability for Internet computing applications.

Keywords: Internet computing, Computing reliability, Processor–memory pair, Flow network, Minimal path

ICITES2014-215

An Effective Method for Classification of White Rice Grains Using Various Image Processing Techniques

Suchart Yammen[1] and Chokcharat Rityen[2]

[1]Department of Electrical and Computer Engineering, Naresuan University, Phisanulok 65000, Thailand

[2]Department of Electrical Engineering, Rajamangala University of Technology Lanna, Tak 63000, Thailand

Email: [1]sucharty@nu.ac.th, [2]chok.kpg@gmail.com

Abstract. This paper presents an algorithm for classifying grains of white rice by using image processing. Each image size is acquired via a digital camera. The resolution is 720 × 480 pixels. The algorithm begins with improving grain images, converting these images into binary images by using Otsu's method, removing noise from the binary images by applying the morphological method with square structural elements, detecting each grain boundary by using the Canny operator, and determining the length of each grain by using the Euclidean method. Next, the grain length is used for classifying the rice grains according to the Rice Standards of Thailand. The testing results from processing 500 grain images; one grain per image, the algorithm provides good performance with the mean absolute error of 0.01 mm in length. For 300 grain images with some grains per image, the algorithm provides good classification with an average accuracy of 99.33 %.

Keywords: Image processing, Morphological, Classification, White rice grains

ICITES2014-216

Applying ZigBee Wireless Sensors and Photovoltaic System to Plant Factory

Shun-Peng Hsu[1], Yi-Nung Chung[1], Chih-Chung Yu[2], and Young-Chi Hsu[1]

[1]Department of Electrical Engineering, National Changhua University of Education, Changhua 500, Taiwan

[2]Department of Electrical Engineering, Da-yeh University, Changhua 515, Taiwan

Abstract. This study proposes to apply wireless monitor and photovoltaic (PV) system to plant factory. The major structures include solar energy system, ZigBee wireless sensor network, and plant factory. In this system, the photovoltaic panels produce energy and the charge controller is used to manage the energy consumption of the load. In order to monitor the plant factory, a ZigBee wireless sensor system is applied. It will monitor the environment situations of the plant factory and supply the information of temperature, humidity, and illumination via the computer network. The system will adjust the environment conditions to enhance the plant growth. According to the experimental results, the system proposed in this paper can enhance the growth of the plants and reach the goal of saving energy also.

Keywords: Plant factory, Solar energy power, ZigBee wireless sensor network

ICITES2014-217

Applying Particle Filter Technology to Object Tracking

Tun-Chang Lu, Shun-Peng Hsu, Yu-Xian Huang, Yi-Nung Chung, and Shi-Ming Chen

Department of Electrical Engineering, National Changhua University of Education, Changhua 500, Taiwan

Abstract. This study proposes an approach to track moving object and to predict the observed targets based on the particle filter. This system includes three parts which are the foreground segmentation, the partial filtering, and the particle filter for tracking objects. In order to estimate the location of next state and track the moving objects, it applies the prior and current state based on the particle filter technology. Experimental result shows that this method can track objects accurately.

Keywords: Particle filter, Foreground segmentation, Track objects

ICITES2014-218

Measurement of Thickness and Refractive Index of Optical Samples in FD-OCT with Two Orthogonal Polarized Lights

Ya-Fen Chang, Yu-An Chen, Hsu-Chih Cheng*

Department of Electro-Optical Engineering, National Formosa University, Yunlin 632, Taiwan

Abstract. This paper proposed an improved structure of Frequency-Domain optical coherence tomography (FD-OCT) which can measure the thickness and the refractive index of unknown sample simultaneously. In the conventional FD-OCT system, the mirror signal and autocorrelation terms will be generated after the inverse Fourier Transform and result in measurement range is restricted. Therefore, this study presents an improved method using the theory of phase-shifting algorithm with two orthogonal polarized lights to increase measurement range and eliminate unnecessary noise. By phase-shifting algorithms, eliminating unnecessary noise become possible and the measurement range is doubled. In other words, the structure of FD-OCT achieves simultaneous measurement of thickness and refractive index of optical samples based on full-range measurement.

Keywords: Optical coherence tomography (OCT), Optical imaging, Orthogonal polarized lights

ICITES2014-219

Fabrication of a Peristaltic Micropump with UV Curable Adhesive

Yi-Chu Hsu*, Jeffrey Levin, and Hsiao-Wei Lee
Department of Mechanical Engineering, Southern Taiwan University of Science and Technology, Tainan 71005, Taiwan

Abstract. This paper describes a method to fabricate a peristaltic micropump using UV curable adhesive (NOA81). This study utilized replication method to fabricate UV curable adhesive-based structure. NOA81 was also used as a bonding material to fabricate micropump's main body, which consists of two PMMA plates and UV structure. In this study, NOA81 reached a fully cured state in 10 min and the highest bonding strength is 10.18 J/m^2. Partially cured NOA81 mold would produce bad replication effect, because during partially cured period, a slight amount of pressure will change the geometry. Under confocal microscope, the geometrical errors from fully cured and post-cured NOA81 structure mold are below the maximum tolerance (10 %). The highest membrane displacement and flow rate of the micropump are 0.824 μm and 21.87 μL/min, respectively, while the maximum backpressure is 4.42 Pa.

Keywords: NOA81, UV curable adhesive, Peristaltic micropump

ICITES2014-220

Detecting Object Edges by Xtion Pro and Open Sources

Cheng-Tiao Hsieh[1*]
[1]Department of Industrial Design, Ming Chi University of Technology, New Taipei City 24301, Taiwan

Abstract. Kinect is a popular device used in many applications such as interactive PC games, robot, virtual reality, 3D sensing, and etc. Especially in 3D sensing, many researchers and engineers utilized Kinect-based devices to develop a cheap 3D scanners. 3D scanners usually provide a better and quicker way to support the needs of 3D printing industry. This advantage is really helpful to speed up the development of 3D printing industry. Regarding this fact, this paper attempts to propose an approach to track object edges. The approach is capable of detecting edges of a given object by Xtion pro. Many existed edge detection methods have been developed based on color information. However, this approach relies on spatial information—depth to detect object edges. The approach had also been developed successfully by open sources like OpenNI, OpenCV, and Point Cloud Library. Developing a cheap 3D scanner for 3D printing market becomes possible.

ICITES2014-221

Temperature Control of a Baking System for an Ultra-high-vacuum Insertion Device

Jui-Che Huang*, Yu-Yung Lin, Chin-Kang Yang, Yung-Teng Yu, Cheng-Hasing Chang, and Ching-Shang Hwang
Magnet Group, National Synchrotron Radiation Research Center, Hsinchu, Taiwan

Abstract. Baking an undulator artificially accelerates outgassing, and is one of the most important and difficult factors in constructing an in-vacuum undulator. The total duration of baking was 76 h and each component in the undulator required a separate temperature for baking. Automatic procedures and devices for temperature control during baking were constructed and continually applied to achieve the required ultra-high vacuum (pressure $P < 2 \times 10^{-10}$ Torr). This paper describes details of the baking and the devices used and constructed in National Synchrotron Radiation Research Center.

Keywords: Ultra-high vacuum, Insertion device, Vacuum bake-out

ICITES2014-223

Microarray Data Analysis with Support Vector Machine

Si-Hao Du[1], Jin-Tsong Jeng[2*], Shun-Feng Su[1], and Sheng-Chieh Chang[3]

[1]Department of Electrical Engineering, National Taiwan University of Science and Technology, Taipei, Taiwan

[2]Department of Computer Science and Information Engineering, National Formosa University, Yunlin County, Taiwan (*tsong@nfu.edu.tw)

[3]Aeronautical Systems Research Division, National Chung-Shan Institute of Science and Technology, Taichung, Taiwan

Abstract. Microarray data analysis approach has become a widely used tool for disease detection. It uses tens of thousands of genes as input dimension that would be a huge computational problem for data analysis. In this paper, the proposed approach deals with selection of feature genes and classification of microarray data under support vector machine (SVM) approach. Feature genes can be found out according to the adjustable epsilon-support vector regression (epsilon-SVR) and then to select high ranked genes after all microarray data. Moreover, multi-class support vector classification (multi-class SVC) and cross-validation methods applies to acquire great prediction classification accuracy and less computing time.

Keywords: Support vector machine, Support vector regression, Multi-class support vector classification, Feature genes, Microarray data analysis

ICITES2014-224

Feature Selection Algorithm for Motor Quality Types Using Weighted Principal Component Analysis

Yun-Chi Yeh, Liuh-Chii Lin*, Mei-Chen Liu, and Tsui-Shiun Chu

Department of Electronic Engineering, Chien Hsin University of Science and Technology, Zhongli 320, Taiwan, R.O.C.

Abstract. This paper proposes a qualitative feature selection for motor quality types using Weighted Principal Component Analysis (WPCA) method. The WPCA includes two processes, one is the Procedure-FFV (find the final weights) process and the other is the Procedure-DPC (determine the principal components) process. The input variables of the WPCA are nine original features and the output variables are six qualitative features. Experimental results indicate that the proposed WPCA provides an efficient, simple, and fast method for feature selection on motor's current waveforms.

ICITES2014-225

Feasibility Test of Range of Motion Exercises for Ankle Joints Rehabilitation Using Pneumatic Soft Actuators

Hironari Taniguchi[1*], Noriko Tsutsui[2], and Yoshiaki Takano[1]
[1]Department of Electronics and Control Engineering, National Institute of Technology, Tsuyama College, Tsuyama-shi, Okayama 708-8509, Japan
[2]Advanced Mechanical and Control System Engineering, National Institute of Technology, Tsuyama College, Tsuyama-shi, Okayama 708-8509, Japan

Abstract. Patients with movement disabilities have been increasing steadily by accidents and diseases. If the condition does not improve, contractures may occur in some joints and muscles. The contractures often tend to occur in ankle joints which are the most important body parts for everyday living. Thus, it is important to start rehabilitation therapy shortly after an accident and a disease. The purpose of this study is to develop a rehabilitation device for ankle joints. In our previous study, we proposed a pneumatic actuator with soft material. The actuator has many advantages such as low mass, flexibility, safety, and user-friendliness. Therefore, we focused on the actuator as a drive source of the rehabilitation equipment and the actuator used for the range of motion (ROM) exercises. We measured the ROM needed for ankle joint. As a result, we confirmed that the actuator is able to provide several ROM exercises.

ICITES2014-226

3D Motion Editing Through B-Spline Fitting with Constraints

Mankyu Sung
Department of Game & Mobile Contents, Keimyung University, Korea

Abstract. This paper proposes a novel motion editing algorithm that uses human motion capture data for animating 3D characters. First, the algorithm fits the 2D root joint trajectory with the cubic B-Spline through least-square minimization. In this process, it finds the optimal number of control points based on an error threshold. Once it gets control points, users are allowed to change the positions of control points, which is able to create a new trajectory. The new trajectory is then fed into the original motion so that it is modified to reflect the new trajectory. To find exact parameter of the spline curve representing root joint position, the algorithm performed the arc-length parameterization on the curve. Since the motions are forced to change the root joint positions, the result may violate the fidelity of the original motions, which may cause some artifacts such as foot skating. To fix them, the IK (Inverse Kinematics) solver is applied to motions to change the limb orientation. Although the IK solver can change the orientation of original motions, if the differences between the modified trajectory and original trajectory are too big, then the result motion produces awkward poses over times. In order to prevent them, our algorithm puts constraints on the control points of curve automatically so that users are able to edit the trajectory freely without considering whether it produces natural motions or not.

Keywords: Motion capture, 3D Motion editing, Computer animation

ICITES2014-227

An ARAR-Tree-Based Diagnosis Mechanism for Rule Anomalies Among Internet Firewalls

Chi-Shih Chao

Department of Communications Engineering, Feng Chia University, Taichung 40725, Taiwan

Abstract. While configuring firewalls, firewall rule ordering and distribution must be done cautiously on each of cooperative firewalls. However, network operators are prone to incorrectly configuring firewalls because there are commonly hundreds of thousands of filtering rules (i.e., rules in the Access Control List file; or ACL for short) which could be set up in a firewall, not mention these rules among firewalls could affect mutually. To speed up the crucial but laboring inspection of rule configuration on firewalls, this paper describes our developed diagnosis mechanism which can speedily figure out rule anomalies among firewalls with an innovative data structure—Adaptive Rule Anomaly Relationship tree (or ARAR tree). With the aid of this data structure and associated algorithms, significant improvements have been made in the field.

Keywords: Defense in depth, Firewall rule anomalies, ARAR tree, Diagnosis reuse

ICITES2014-228

Development of Active Orthosis for Lumbago Relief

Shinsaku Fujimoto[1], Tetsuya Akagi[1], and Feifei Zhao[2]

[1]Department of Intelligent Mechanical Engineering, Okayama University of Science, 1-1 Ridaicho, Kita-ku Okayama, Japan

[2]Tsuyama National College of Technology, 624-1, Numa, Tsuyama, Okayama, Japan

Abstract. It is important to develop the orthosis which improves the Quality of Life (QOL) and maintains health conditions. As one of the treatment methods done to lumbago, the waist fixation method with the spinal brace or the orthosis is performed. A waist active orthosis implemented with pneumatic flexible actuators is developed and pressure control method of pneumatic flexible actuators is established. Orthosis control valve (on/off valve) system is modeled and the reliability of exhaust and supply model has been validated through experiment. The effectiveness of the proposed control method has been validated through the control experiment using the orthosis control valve.

Keywords: Pneumatic actuator, Orthosis, Lumbago, Control system design

ICITES2014-229

Controller Design by Time-Domain Objective Functions Using Cuckoo Search

Huey-Yang Horng
Department of Electronic Engineering, I-Shou University, Kaohsiung 84001, Taiwan

Abstract. In this research, a new optimization algorithm, called the cuckoo search algorithm, is introduced for design controller. A large proportion of industrial systems are represented by linear time-invariant transfer functions. The proportional-integral-derivative (PID) controller is one of the most widely used functions. The lead-lag controller is a more practical alternative. Traditionally, time-domain or frequency-domain methods have been used to design a lead-lag controller to design specifications. This paper focused on the design of controller both PID and lead-lag controller, by optimization of the time-domain objective function. The proposed objective function includes time-domain specifications, including the rise time, peak time, maximum overshoot, setting time, and steady-state error. In the paper, Cuckoo Search algorithm is chosen to finding the optimal solutions. Cuckoo Search is meta-heuristic optimization method recently developed. That is a type of population-based algorithm inspired by the behavior of some Cuckoo species in combination with the Lévy flight behavior. Given that the plant is modeled according to a linear time-invariant transfer function, the proposed method designs the controller capable of approaching the specifications.

Keywords: PID controller, Lead-lag controller, Cuckoo search, PSO

ICITES2014-231

Behavior Network-based Risk Recognition Method

Jeonghoon Kwak[1], Suhyun Gong[2], and Yunsick Sung[1]
[1]Department of Game Mobile Contents, Keimyung University, Daegu, South Korea
[2]Department of Computer Engineering, Keimyung University, Daegu, South Korea
{jeonghoon, suhyun.gong, yunsick}@kmu.ac.kr
Corresponding Author: yunsick@kmu.ac.kr

Abstract. This paper proposes a two-behavior networks-based method to automatically detect whether a situation is risky or not. Behavior network is used to analyze measured values from the sensors of a smartphone. Bayesian probability is also used for implementing such a behavior network. An experiment was conducted to validate that the speed of recognizing risk situations was improved by learning such situations iteratively through behavior networks with Bayesian probability.

Keywords: Behavior network, Bayesian probability, Situation classification

ICITES2014-232

Reinforcement Learning with Multiple Actions

Riku Nishiyama[1*] and Satoshi Yamada[1]
[1]Department of Intelligent Mechanical Engineering, Okayama University of Science, Okayama 700-0005, Japan

Abstract. The effectiveness of the reinforcement learning with multiple actions was investigated. The multiple actions consist of a group of simple actions and the termination conditions. To evaluate the learning performance of the reinforcement learning with multiple actions, it was compared with the reinforcement learning with simple actions. To develop more general learning algorithm, the reinforcement learning with multiple actions using INGnet was also investigated.

Keywords: Reinforcement learning, Multiple actions, INGnet, Khepera

ICITES2014-233

Wind Turbine Blade Load Alleviation Performance Employing Individual Pitch Control

Chin-Fan Chen[1], Chi-Jo Wang[2], Alireza Maheri[3], and Terrence Macquart[4]

[1]Department of Electrical Engineering, Southern Taiwan University of Science and Technology, Tainan 71005, Taiwan

[2]Department of Electrical Engineering, Southern Taiwan University of Science and Technology, Tainan 71005, Taiwan

[3]Faculty of Engineering and Environment, Northumbria University, Newcastle Upon Tyne NE1 8ST, UK

[4]School of Computing, Engineering and Information Sciences, Northumbria University, Newcastle Upon Tyne NE1 8ST, UK

Abstract. Nowadays, the usage of renewable energy has become popular. For the wind energy, as the wind turbine getting larger, more wind energy can be extracted. However, wind turbine components also became bulkier and hence increased the cost of installation. Furthermore, loads on the larger rotor also increased and made its life span shorter. As a result, the aim of the project is to investigate the implementation of individual pitch control (IPC) for alleviating wind turbine blade loads and then increase the cost-effectiveness. This study covers the operation of variable wind speed with different controllers, namely pitch control, torque control, and individual pitch control. The result will corporate the implementation of proportional-integral-derivative controller (PID) into the torque and pitch control as well as the direct-quadrature (d-q) transformation for individual pitch control.

ICITES2014-234

Planet Editing Method Using Leap Motions

Ji Won Kim, Phil Young Kim, and Yunsick Sung

Department of Game Mobile Contents, Keimyung University, Daegu, South Korea {jiwon1219, kimpy1111, yunsick}@kmu.ac.kr, Corresponding Author: yunsick@kmu.ac.kr

Abstract. Universal courseware using motion recognition devices is very useful for students to learn universal science because it enables interactive learning. However, given that students change the radius of planets by the distance between their hands, it is nearly impossible to denote larger radii with their hands within fixed hand recognition area of the motion recognition devices by a single expression.

This paper proposes a method that controls planets using motion recognition devices for courseware contents. The distance between hands is expressed by the sum of the difference of each position of the user's hands. In the experiment, we validated the change of the radius of planets according to the difference of each recognized hand.

Keywords: Leap motion, Universal courseware, Motion recognition

ICITES2014-235

Development of Flexible Haptic Robot Arm Using Flexible Pneumatic Cylinders with Backdrivability for Bilateral Control

Takafumi Morimoto[1], Mohd Aliff[1], Tetsuya Akagi[1], and Shujiro Dohta[1]

[1]Department of Intelligent Mechanical Engineering, Okayama University of Science, Okayama 700-0005, Japan

Abstract. In the remote controlled rehabilitation device, the physical therapists must recognize the situation of the patient. Therefore, a haptic device that the therapists can feel the reaction force is required. In our previous study, to realize the haptic device driven by pneumatic pressure, the flexible pneumatic cylinder with back drivability was proposed and tested. In this paper, the flexible haptic robot arm using the cylinders for human wrist rehabilitation is proposed and tested. The low-cost bilateral control system using two embedded controller and the tested cylinders for bilateral control is also proposed and tested. The bilateral control using the tested robot arm is carried out.

Keywords: Flexible pneumatic cylinder with back drivability, Flexible haptic robot arm, Rehabilitation device, Bilateral control

ICITES2014-236

Integration Method of Proxy and Producer–Consumer Patterns

Jeonghoon Kwak, Jaehak Uam, and Yunsick Sung

Department of Game Mobile Contents, Keimyung University, Daegu, South Korea {jeonghoon, jaehak, yunsick}@kmu.ac.kr

Corresponding Author: yunsick@kmu.ac.kr

Abstract. This paper proposes a novel design pattern that combines proxy and producer–consumer patterns to solve problems of a scenario generator for verifications of sensors in smart space. In addition, how to apply the novel design pattern to scenario generator is introduced. Proxy pattern hides the complexity of the process of generating scenarios and solves the storage problem of a scenario generator. By applying the proposed Proxy–Producer pattern including multiple threads, the generation time of the huge number of scenarios was reduced.

Keywords: Internet of things, Scenario generation, Bayesian probability, Design pattern, Proxy pattern, Producer–consumer pattern

ICITES2014-237

Bayesian Probability and User Experience-Based Smart UI Design Method

Junhyuck Son and Yunsick Sung
Department of Game Mobile Contents, Keimyung University, Daegu, South Korea
{junhyuck12, yunsick}@kmu.ac.kr
Corresponding Author: yunsick@kmu.ac.kr

Abstract. This paper proposes a method that controls User Interfaces by changing properties of UIs after analyzing the patters of events invoked by users. The events and criteria are defined and handled to determine the properties of UIs. In the experiment, our proposed method was applied to an education authoring tool. The changes of UIs after the analysis of user events were validated.

Keywords: User interface, User experience, Bayesian probability

ICITES2014-238

Bayesian Probability-Based Hand Property Control Method

Phil Young Kim, Ji Won Kim, and Yunsick Sung
Department of Game Mobile Contents, Keimyung University, Daegu, South Korea
{kimpy1111, jiwon1219, yunsick}@kmu.ac.kr
Corresponding Author: Yunsick Sung, yunsick@kmu.ac.kr

Abstract. Deficiencies in low-priced motion recognition devices lead to diverse kinds of errors in recognizing palms and hands. To utilize lower-priced devices better, the recognition rate of the properties of hands should be improved. This paper proposes a method that revises recognition errors in properties of hands. By calculating the Bayesian probability of the directions of a recognized palm, the directions were revised.

Keywords: Leap motion, Motion recognition device, Bayesian probability

ICITES2014-239

Implementation of a Delta–Sigma Analog-to-Digital Converter

Chin-Fa Hsieh[1*], Tsung-Han Tsai[1], Chun-Sheng Chen[2], and Yu-Hao Hsieh[2]
[1]Department of Electrical Engineering, National Central University, Chung-Li 32001, Taiwan
[2]Department of Electronic Engineering, China University of Science and Technology, Taipei 11581, Taiwan

Abstract. The sigma–delta analog-to-digital converter (ADC) has less consumption of circuit power and can achieve higher resolution. In this paper, a sigma–delta ADC which contains a second-order sigma–delta modulator is presented. The modulator architecture is first designed by using the behavioral simulation of MATLAB, and then the TSMC 0.18 μm single-poly six-metal process. Layout of each analog block has been shown. Simulation results show that, with an input of a −6 dB 1 kHz sine, the delta–sigma analog-to-digital converter can achieve an SNR

of 87.2 dB. The core size is 0.6456 mm × 0.3340 mm. With a 16-bit resolution, it is suitable for audio applications.
Keywords: ADC, Sigma–Delta, Modulation

ICITES2014-240

An Experimental Investigation of Effect on Engine Performance by Controlling the Temperature of the Fuel

Ming-Hsien Hsueh
Department of Industrial Engineering and Management, National Kaohsiung University of Applied Sciences, Kaohsiung 807, Taiwan

Abstract. The advantages of controlling the fuel temperature in the engine system are presented in this paper. By using the thermoelectric material, engine fuel can be cooled or heated to control the temperature of the inlet mixture. In this investigation, the preheating time of the engine can be reduced by heating the fuel. The fuel consumption, the concentration of oxides of nitrogen can be decreased by cooling the fuel. At the same time, the power of the engine also can be increased by cooling the fuel.

Keywords: Fuel temperature, Thermoelectric material, Oxides of nitrogen, Power

ICITES2014-241

Modeling of an AlGaAs-Based VCSEL with Bragg Mirrors

Shu-Hui Liao
Department of Electronic Engineering, Chung Chou University of Science and Technology, Changhua County 51003, Taiwan

Abstract. This paper discusses methods of modeling Bragg mirrors for vertical cavity surface emitting lasers (VCSEL). The propagation matrix method was used to present the design and performance of a Vertical Cavity Surface Emitting Laser) with AlAs/AlGaAs Bragg mirrors for electromagnetic radiation centered at 980 nm. We adopted and compared with AlAs/AlGaAs periodic dielectric layer stacks consisting of 20 and 40 identical layer-pairs. The result indicates that the reflectivity of the Bragg mirror is close to unity over a wavelength band width centered at wavelength of 980 nm. It was found that the results achieved from the simulation showed a high correlation with predicted results.

Keywords: VCSEL, Bragg mirror, propagation matrix method

ICITES2014-242

Risk Decision Method Based on Sensory Values of Smart Devices

Taejun Son[1], Jeonghoon Kwak[1], Suhyun Gong[2], and Yunsick Sung[1]
[1]Department of Game Mobile Contents in Keimyung University, Daegu, South Korea
[2]Department of Computer Engineering in Keimyung University, Daegu, South Korea
{jason.son,jeonghoon,suhyun.gong,yunsick}@kmu.ac.kr
Corresponding Author: yunsick@kmu.ac.kr

Abstract. Smartphone sensors can be useful in recognizing risk situations faced by users. However, the use of diverse kinds of sensors for risk estimation increases the complexity of such analyses. This paper proposes a method to recognize the risk quotient in user situations by measuring and managing the data that smartphone sensors provide.

Keywords: Risk recognition, Normalization, School-zone, Internet of things

ICITES2014-243

A Flywheel Energy Storage System Suspended by Active Magnetic Bearings Using a Fuzzy Control with Radial Basis Function Neural Network

Van-Sum Nguyen[1*], Hong-Van Tran[2], Lai-Khac Lai[3], Nguyen Thi-Hoai Nam[4], Tran Huu-Chau Giang[5], and Le-Phuong Truong[6]

[1]Department of Electrical Engineering, Da-Yeh University, Changhua 51591, Taiwan

[2]Department of Automation Technology, Thu Duc College of Technology, Ho Chi Minh, Vietnam

[3]Department of Science Technology & Environment, Thai Nguyen University, Vietnam

[4]Department of Electrical Engineering, Hue Industrial College, Hue City, Vietnam

[5]Department of Training, Hue Industrial College, Hue City, Vietnam

[6]Department of Electromechanical and Electronics, LacHong University, Bien Hoa City, Vietnam

Abstract. A flywheel energy storage system (FESS) is an effective energy-saving device. It works by accelerating a rotor flywheel disc at a very high speed and maintaining the energy in the system as rotational energy. Active magnetic bearings (AMBs) are ideally suited for use at high-speed and are so used in FESSs. This work develops a mathematical model of the levitation force and rotational torque of a flywheel. The system for controlling the position of the flywheel is designed based on a neural fuzzy controller. A mathematical model of an AMB system comprises identification followed by collection of information from this system. A fuzzy logic controller (FLC), the parameters of which are adjusted using a radial basis function neural network, is applied to the unbalanced vibration in FESS. The results obtained concerning the FESS indicate that the system exhibited satisfactory control performance including transient and steady-state responses under various operating conditions.

Keywords: Flywheel, Energy storage system, AMB, FLC, Neural network

ICITES2014-244

A Matrix Robot for Block Stacking

Yu-Ching Su, Ming-Chang Chen, and Wing-Kwong Wong

Department of Electronic Engineering, National Yunlin University of Science and Technology, Yunlin 64002, Taiwan

Abstract. This research studies how to control a robot to stack pieces of block according to the block structure shown in a given image. In this project, a LabVIEW program is designed and run on a Lego Mindstorms NXT, which in turn controls a robot built with components of Matrix Robotics. There are two operation modes for the block stacking robot. In the first mode, a user uses an Android smartphone to control the robot with Bluetooth to move around, pick up blocks, and stack them. In the second mode, the robot identifies the block structure in a given image with OpenCV and picks up the needed blocks to stack them to build the block structure. The robot was tested with two images of simple block structures in the second mode. The success rates ranged from 70 to 80 %. Future work should further identify the weaknesses of the robot and improve its performance.

Keywords: Android, LabVIEW, Matrix Robotics, NXT, OpenCV

ICITES2014-245

A Study of Regular Transmission Delay in Bluetooth Communications

Komang Oka Saputra, Wei-Chung Teng*, Pin-Yen Chou, and Tien-Ruey Hsiang
Department of Computer Science and Information Engineering, National Taiwan University of Science and Technology, Taipei City 106, Taiwan

Abstract. This paper studies a special case of transmission delay when two devices communicate by Bluetooth technology. Transmission delays of packets are usually distributed randomly over some range, or the delay jitter, in wireless or wired communication. However, it is observed that under certain conditions, the transmission delays of consecutive packets may form into parallel dotted lines, and the intervals between a line and its next one are almost the same. The characteristics of the dotted-line delays, like the lifetime of one dotted line, are deduced to help develop an algorithm for detecting the period of this phenomenon. Experiments are further conducted to reveal how factors like operating system, packet sending period, and Bluetooth chips may affect the pattern of regular transmission delays.

Keywords: Transmission delay, Bluetooth, Raining

ICITES2014-246

IEEE 802.16j-Based Multihop Relays for Future Performance Enhancement of Aeronautical Mobile Airport Communications Systems (AeroMACS)

Behnam Kamali
Department of Electrical and Computer Engineering, Mercer University, Macon, GA 31207, USA

Abstract. The Aeronautical Mobile Airport Communication Systems (AeroMACS) has already been deployed in nine major US airports. This technology is used to support fixed and mobile ground to ground applications and services. In this article, it is demonstrated that IEEE 802.16j-amendment-based WiMAX is most feasible for future enhancement of AeroMACS networks. Perhaps the most important benefit of application of multihop relays in AeroMACS is the flexible, power efficient, and cost-effective radio range extension into severely shadowed airport areas. Multihop relay operational modes are discussed. The key concept of "*multihop gain,*" which explains how multihop relays enable performance enhancement in AeroMACS networks, is introduced. Under a reasonable set of assumptions, multihop gain is quantified in the form of an equation that provides a raw measure of this gain in Decibel.

Keywords: AeroMACS, IEEE 802.16, Multihop relay, WiMAX

ICITES2014-247

Explore the Influential Factors on Maritime Accidents by Regression Approach

Chien-Chang Chou[1,2], Cheng-Yi Chen[3], Jeng-Ming Yih[4], Kur-Eng Chang[5], and Chung-Ping Wu[6]

[1]Department of Shipping Technology, National Kaohsiung Marine University, Kaohsiung, Taiwan

[2]Chou's Science Research Center, Kaohsiung, Taiwan

[3]Department of Electrical Engineering, Cheng Shiu University, Kaohsiung, Taiwan

[4]Center of General Education, Min-Hwei College of Health Care Management, Tainan, Taiwan

[5]Department of Marine Information and Technology, National Kaohsiung Marine University, Kaohsiung, Taiwan

[6]Department of Shipping Technology, National Kaohsiung Marine University, Kaohsiung, Taiwan

Abstract. This paper investigates the important factors on the maritime accidents in the harbors and waters surrounding Taiwan. Based on the collected data including wind, wave, tide, current, and maritime accidents, a regression model is used to discover the influential factors on the maritime accidents. It is found that various ports in Taiwan have different influential factors on the maritime accidents. Finally, some useful suggestions are given to the managers of harbors and the decision makers of governmental maritime departments to increase the navigation safety of ships in the harbors and waters surrounding Taiwan.

Keywords: Regression analysis, Navigation safety, Safety management

ICITES2014-248

Analytical Model of Pipe Inspection Robot Using Flexible Pneumatic Cylinder

Haojun Qiu[1*], Shujiro Dohta[1], Tetsuya Akagi[1], So Shimooka[1], and Shinsaku Fujimoto[1]

[1]Department of Intelligent Mechanical Engineering, Okayama University of Science, 1-1 Ridai-cho, Kita-ku, Okayama 700-0005, Japan

Abstract. A pipe inspection robot is useful to reduce the inspection cost. In our previous study, a novel pipe inspection robot using a flexible pneumatic cylinder that can be driven even if it bends has been proposed and tested. The built-in pneumatic driving system using a tiny embedded controller and small-sized valves has been also proposed and tested to decrease the mass of the robot for increasing inspection area. In this paper, in order to find out the optimal driving pattern and length of the robot, an analytical model of the pipe inspection robot is proposed. The model consists of two on/off control valves and a sliding mechanism which is composed of a flexible pneumatic cylinder.

Keywords: Pipe inspection robot, Flexible pneumatic cylinder, Analytical model

ICITES2014-250

Thermal Resistance Characterization of SiGe-Based HBTs on Thick-Film and Thin-Film SOI

Shu-Hui Liao
Department of Electronic Engineering Chung Chou University of Science and Technology, Changhua County 51003, Taiwan

Abstract. The nonlinear behavior of thermal resistance (RTh) of SiGe HBTs on both thick-film and thin-film SOI substrates were investigated at different levels of dissipated power densities by the ISE-TCAD simulator in this paper. We examined both of the buried-oxide thickness and the silicon thickness effects on the device characteristics for thermal resistance. DESSIS-ISE was used to compute the temperature distribution from the given power. Simulation results suggest that the amount of self-heating is strongly dependent on device structure. Owing to the thick-film SiGe on SOI works like a bulk SiGe HBT, the enhanced silicon thickness could degrade the thermal resistance. Thus, it needs to be carefully considered in device design. The thermal resistance characteristics revealed for SiGe HBTs on thick-film and thin-film SOI may help us to establish more accurate thermal models for reliability of circuit design and device technology optimization.

Keywords: SOI, HBT, SiGe, Thermal resistance

ICITES2014-251

Improved Item Relational Structure Theory Based on Liu's Item Ordering Consistence Property

Hsiang-Chuan Liu[1] and Jing-Ming Ju[2,3*]
[1]Department of Biomedical Informatics, Asia University, Taichung 41354, Taiwan
[2]Department of Early Childhood Education, Asia University, Taichung 41354, Taiwan
[3]Department of Psychology, Asia University, Taichung 41354, Taiwan
[1]lhc@asia.edu.tw; [2]jjm3222@gmail.com
*Corresponding Author

Abstract. The well-known Ordering Theory (OT) does not consider the nonindependence property of ordered items. For improving this drawback, the Item Relational Structure Theory (IRST) and Liu's Ordering Theory (LOT) were proposed. Although IRST is more sensitive than LOT and OT, the IRST does not satisfy Liu's consistency property of ordering relationship as LOT. In this paper, for overcoming the abovementioned drawback by mathematical analysis, an improved item relational structure theory (IIRST) was proposed; it satisfies item ordering consistency property, and it is more sensitive than LOT and OT.

Keywords: Ordering theory, Item relational structure theory, Nonindependence, Ordering consistency property

ICITES2014-252

Enhancement of Digitized Old Film Using Inpainting-Based Method

Chung-Ming Li and Day-Fann Shen
Department of Computer, National Yunlin University of Science and Technology, Yunlin 64002, Taiwan

Abstract: In this paper, our goal is to remove blotch noises as well as straight line scratches commonly seen in old film movies using digital image processing techniques. We proposed an adaptive blotch detection method based on our blotches characteristic analysis, where we calculate the size of each connected region using Connected Component methods to distinguish blotches and non-blotches, as a result, the proposed method can accurately mark the blotches noises. We also proposed a method to avoid misclassification of small objects as blotches using, improving the accuracy of marked blotches. We adopt and improve Inpainting algorithm to repair the blotches noise. Inpainting algorithm is highly dependent on the order in which the filling proceeds; we also extend this approach to color images. In addition, we proposed criterion for performance evaluation of the proposed method.

Keywords: Old film, Image restoration, Scratch, Blotch, Connected component method, Inpainting

ICITES2014-253

Environmental Controlling System for Orchid Growth Utilizing Thermoelectric Cooling Modules

Kao-Feng Yarn[1], M.-H. Shih[2], Ming-Chu Hsieh[3], Yu-Jen Pan[4], and Win-Jet Luo[5*]
[1]Department of Electronic Engineering, Far East University, Tainan 744, Taiwan
[2]Department of Mechanical Engineering, Chinese Military Academy, Fengshan 83059, Taiwan
[3]Department of Mechanics, National Chin-Yi University of Technology, Taichung 411, Taiwan
[4]Department of Digital Content and Technology, Ta Hwa University of Science and Technology, Hsinchu 307, Taiwan
[5]Graduate Institute of Precision Manufacturing, National Chin-Yi University of Technology, Taichung City 41170, Taiwan
PO Box Graduate Institute of Precision Manufacturing, National Chin-Yi University of Technology, Taichung City 41170, Taiwan
Tel: 886-4-23924505 ext 5110 (office) E-mail: wjluo@ncut.edu.tw
*Corresponding Author

Abstract. For the development of delicate agriculture, in addition to technical problems, some environmental and energy-saving issues have to be concerned. In this study, a thermal electric cooling module was applied to a small-scale environmental controlling system instead of traditional vapor compressed cooling modules in order to light the weight of the plant cultivating system. In the cultivating system, the environmental temperature can be maintained in the range of 22–26 °C, and the relative humidity can also be controlled in the range of RH = 70 % to RH = 75 %. Light-emitting diode and automatic water system were applied to supply sufficient light intensity and water for photosynthesis of the cultivated plants. Through the plant cultivating system, suitable environmental conditions can be generated automatically for plant growth even without artificial charge for a long period. Utilizing environmental controlling technology to cultivate organic vegetables and plants can promote the plant quality which cannot be affected by the environmental climate change as well.

Keywords: Pant cultivation, Photosynthesis, Thermoelectric cooling, Environmental control

ICITES2014-254
Study of Nickel Catalysts Deposited by Using the Electroless Plating Method and Growth the Multiwall Carbon Nanotubes

Chih-Yi Lin[1], Jian-Liang Pan[2], Chia-Ching Wu[3*], and Wen-Chung Chang[1]
[1]Southern Taiwan University of Science and Technology, Tainan, Taiwan, R.O.C.
[2]Department of Applied Cosmetology, Kao Yuan University, Kaohsiung, Taiwan, R.O.C.
[3]Department of Electronic Engineering, Kao Yuan University, Kaohsiung, Taiwan, R.O.C.

Abstract. In this study, the electroless plating has been successfully applied for nickel catalyst layer and the multiwall carbon nanotubes (MWNTs) grown by chemical vapor deposition (CVD). Sulfuric acid solution was used as buffer to adjust and maintain pH value of electroless plating solution on 4.5. The structural, element, and quality of MWNTs were investigated with field emission scanning electron microscope (FE-SEM), X-ray diffraction patterns (XRD), energy dispersive spectrometer (EDS), and Raman spectrometer. From the FE-SEM image, it showed that the density of MWNTs increased as the deposition time of nickel catalyst layer increased. This result caused by the formation of nickel nucleation becomes rich as the immersion of the substrate in electroless plating solution was longer, and this benefited the growth of carbon nanotubes. The Raman analysis demonstrated that the ID/IG ratio of MWNTs decreases as the deposition times of nickel catalyst layer increases, indicating that more graphene MWNTs structures were formed.

Keywords: Multiwall carbon nanotubes, Thermal chemical vapor deposition, Electroless plating method, Nickel catalyst

ICITES2014-256

Photovoltaic Power Generation System Modeling Using an Artificial Neural Network

Cheng-Ting Hsu[1], Roman Korimara[1], Lian-Jou Tsai[1], and Tsun-Jen Cheng[1*]
[1]Department of Electrical Engineering, Southern Taiwan University of Science and Technology, Tainan 71005, Taiwan
*Corresponding Author: chengtj@mail.stust.edu.tw

Abstract. This paper presents a solar power modeling method using an application of the Levenberg–Marquardt (L–M) algorithm. This L–M algorithm has been adopted and incorporated into back propagation learning algorithm for training a feed forward neural network. With this model, the photovoltaic power generation can be approximated. Meteorological data and the historical output power data of the Taiwan Chimei Island photovoltaic plant were selected for this study. The proposed model is evaluated by comparing the simulated results with the actual measured values and is found to be in good agreement.

Keywords: Artificial neural networks, PV model, Chimei Island

ICITES2014-257

Thermal Stress Analysis of Layer-Wise Functionally Graded Material Beam Considering Neutral Plane

Young-Hoon Lee[1*] and Ji-Hwan Kim[2]
[1]Department of Mechanical and Aerospace Engineering, College of Engineering, Seoul National University, Seoul 151-744, Republic of Korea
[2]Institute of Advanced Aerospace Technology, Department of Mechanical and Aerospace Engineering, College of Engineering, Seoul National University, Seoul 151-744, Republic of Korea

Abstract. Stress analysis of layer-wise Functionally Graded Materials (FGMs) beam is investigated in this work. Layer-wise theory is a distributed model instead of the continuous model for the FGMs, and the material properties are not isotropic. Therefore, neutral surface of FGMs is different from the isotropic material. And the surface is determined by using the first moment with respect to temperature-dependent Young's modulus. In the numerical results, axial and shear stress distribution are calculated by considering neutral plane according to the number of layers of FGMs.

Keywords: Layer-wise, Functionally graded materials, Neutral plane

ICITES2014-258

A Smart Home Automation System

Chien-Yuan Liu
Department of Computer Science and Information Engineering, Cheng-Shiu University, Kaohsiung 83347, Taiwan

Abstract. Internet of things enabled by the information and communication technologies is driving the promotion of smart living applications around the world.

Convenient control with energy-saving consideration is one of the most important functions for smart homes. This paper presents the research work for a smart home automation system linked via wireless communications. The system is composed of a smartphone running an APP for control operations, a communication gateway, and multiple remote control devices. The functions of the system are correctly performed as the expected requirements of power socket control and environmental weather detection.

Keywords: Smart APP, Automation system, Smart home

ICITES2014-259

Material Structure Selection of Solution Blue OLEDs Using a Design of Experiment

Yu-Sheng Tsai[1], Apisit Chittawanij[1*], Lin-Ann Hong[1], Fuh-Shyang Juang[1], and Van-Huong Tran[2]

[1]Institute of Electro-Optical and Materials Science, National Formosa University, Huwei, Yunlin 63208, Taiwan

[2]Department of Power Mechanical Engineering, National Formosa University, Huwei, Yunlin 63208, Taiwan

*Tel: 886-5-6315029, Fax: 886-5-6331211, E-mail: abcd_omo@hotmail.com

Abstract. A blue small-molecular organic light-emitting diode (SM-OLED) based on a solution-process is investigated in this study. Design of experiment (DOE) with response surface methodology (RSM) was applied to optimize the driving voltage and current efficiency of blue SM-OLED devices. The spin-coating speed of the PEDOT: PSS as hole injection layer and the 26DCzPPy: FIrpic as emitting layer were chosen as two main process input factors. Analysis of variance (ANOVA) was adopted to identify significant factors before regression models were obtained. The optimal material structure was determined by minimizing and maximizing a desirability function relating to selected critical quality characteristics including the driving voltage and current efficiency, respectively.

Keywords: Solution-process, Blue OLEDs, Design of experiments, Stamping process

ICITES2014-261

A Design of a Light Guide for Highly Uniform Illumination

Kun-Yi Lee[1*], Chun-Han Chou[2], Wei-Sung Weng[3], Wei-Ching Chuang[3], Wei-Yu Lee[1], and Allen Jong-Woei Whang[2]

[1]Department of Electrical Engineering, China University of Science and Technology, Taipei City 115, Taiwan

[2]Department of Electronic Engineering, National Taiwan University of Science and Technology, Taipei City 106, Taiwan

[3]Department of Electro-Optical Engineering, National Formosa University, Yunlin County 632, Taiwan

Abstract. In the paper, we propose a design of a light guide for highly uniform illumination. The design is based on refractive equation theory and conservation of energy theory. We used above methods to design the secondary optics lens. The methods can quick design a lens with high uniformity. Finally, we get a Fresnel lens for the illumination of uniformity near to 92 %.

Keywords: Secondary optics lens, Fresnel lens

ICITES2014-262

A Study on the Performance Characteristics of a Synchronous Elastic FIR Filter

Ren-Der Chen and Sheng-Yu Kao

Department of Computer Science and Information Engineering, National Changhua University of Education, Changhua 500, Taiwan

Abstract. Elastic systems provide tolerance to the unpredictable timing variations in computation and communication delays. For a synchronous elastic circuit, the handshaking mechanism to synchronize the data communication between sender and receiver is produced at the level of cycle in which the events are synchronized with the clock. In this paper, the performance characteristics, i.e., area, delay, and power, of a synchronous elastic circuit are studied by the implementation of a finite impulse response (FIR) filter. The filter is designed as a two-stage pipeline, and to be compared with its nonelastic counterpart, both ASIC and FPGA implementations have been made for various orders of the filter.

Keywords: Elastic, FIR, Filter

ICITES2014-263

A Design of Cavity Filters Based on Photonic Crystal Slab Waveguide with Liquid Crystal

Wei-Sung Weng[1], Hui-Chun Lin[1], Kun-Yi Lee[2*], Li-Ling Chu[1], Hsin-Jung Lee[2], and Wei-Ching Chuang[1]

[1]Department of Electro-Optical Engineering, National Formosa University, Yunlin County 632, Taiwan

[2]Department of Electrical Engineering, China University of Science and Technology, Taipei City 115, Taiwan

Abstract. In this paper, an optical filter based on photonic crystal structure with liquid crystal is proposed. Tunable spectral characteristics of the filters are demonstrated by changing the indexes of refraction of liquid via the external electric fields. The simulation results show the filters are promising in the optical communications.

Keywords: Photonic crystals, Optical filters, Liquid crystals

ICITES2014-264

A Self-Focus High-Precision Scheme Applied to HCPV Solar Center Tracker

Yiing-Yuh Lin[1*] and Fu-Mao Jhuang[1]

[1]Department of Aeronautics and Astronautics, National Cheng Kung University, Tainan 70101, Taiwan

Abstract. An image processor with a self-focus scheme to estimate the center of the solar disc is presented in this paper. It can be utilized in the automatic tracking module on an HCPV power generation unit to closely follow the solar center with high precision under various weather conditions. To start the proposed procedures, a photo containing the solar image is taken first by a digital-sensing device. Then, it is sent to the algorithm and goes through a series of steps to identify and to isolate the solar image. Finally, the solar center is located by the three-point center method. The proposed scheme, based on a weighted brightness factor of the image, evaluates the image center by comparing the estimated solar diameter to the reference one and iteratively modifies the thresholds in the image isolation step. The results show total processing time can be a fraction of a second and the estimation accuracy reaches to less than an arc second, depending on the pixels in the photo taken.

Keywords: Solar tracking, Image processing, Threshold, Solar center, HCPV

ICITES2014-265
Controlling Flock Through Normalized Radial Basis Function Interpolation
Mankyu Sung
Department of Game & Mobile Contents, Keimyung University, Korea

Abstract. This paper introduces a controllable real-time flocking simulation framework through a vector field based on normalized radial basis function. During the design process, the framework subdivides the entire simulating environment into small cells, the so-called grid structure, and then assigns a vector per each cell, which represents a 2D vector field. The vectors of the field are automatically calculated by specifying a set of *control vectors* which are used for interpolating all vectors on the field. The interpolation scheme is based on normalized radial basis function. Once the construction of vector field is done, at the low level, flocks are simulated by following the vector field in the grid structure. Throughout this process, the position of individual agents is updated and collisions between the flock and the static obstacles are avoided by emitting a repulsive vector around the obstacles on the field. Interindividual collisions are also handled through fast lattice-bin method which can minimize the number of comparison for detecting collisions.

Keywords: Flocking, Computer animation, Radial basis functions

ICITES2014-266
Novel AF Relay Design for Optimizing an MIMO Relay Network Under Backward Non-flat Fading Channels
Chun-Hsien Wu*
Department of Photonics and Communication Engineering, Asia University, Taichung 41354, Taiwan

Abstract. This paper presents a novel AF relay design for an MIMO relay network optimization under received power constraint. With no CSI at the source, the optimum multiple relay precoders of an MIMO relay network are devised for combating the backward non-flat fading channels. Simulation cases in terms of BER performance validate the proposed relay design and justify the proposed approach in its ability to pursue an optimum delay transmitted block for detection at the receiver.

Keywords: Amplify-and-forward, AF, Oblique projection, Relay precoder

ICITES2014-267

Development of Wearable Power Assist Wear Using Pneumatic Actuators

Feifei Cho[1], Xiangpan Li[2], and Toshiro Noritsugu[1]

[1]National Institute of Technology, Tsuyama College, Tsuyama, Okayama 708-8025, Japan

[2]Henan University of Science and Technology, Kaiyuan Road, Luoyang, Henan 471023, China

Abstract. This research focuses on developing a safe, lightweight, power-assisted device that can be worn by people during lifting or static holding tasks to prevent low-back-pain (LBP). In consideration of flexibility, light weight, and large force to weight ratio, two types of pneumatic actuators are employed in assisting low back movement for safety and comfort. The device can be worn directly on the body like normal clothing. Because there is no rigid exoskeleton frame structure, it is lightweight and user friendly.

Keywords: Wearable, Assist wear, Pneumatic actuator

ICITES2014-268

Cost Model of Physical Activity Monitoring Systems

Jen-Liang Cheng[1*], Chien-Chih Chen[2], and Ling-Hsuan Chen[3]

[1]Department of Medical Informatics, Tzu-Chi University, Hua-Lien City 97004, Taiwan

[2]Institute of Medical Sciences, Tzu-Chi University, Hua-Lien City 97004, Taiwan

[3]Department of Medical Informatics, Tzu-Chi University, Hua-Lien City 97004, Taiwan

Abstract. A physical activity (PA) monitoring system comprises a number of wearable devices based on a certain measuring technique and a way to collect and record data on the devices. Recent studies showed that the system is an effective intervention tool to increase participants' PA level. Given the trend of physical inactivity among students, it is desired to widely deploy the system in schools to decelerate the trend. Based on the need of school, health educators must assess capability and cost of monitoring systems associated with various measuring techniques. The assessment, however, is difficult due to the lack of cost model for a PA monitoring system to be run in real setting. In this article, we propose the cost model so that cost of a PA monitoring system can be estimated according to the selected measuring technique. The model explains the infeasibility of pedometer- or accelerometer-based PA program in population level and implies the preference of an automated measurement system.

Keywords: Physical activity (PA) promotion, School-based PA program, Pedometer, Objective measurement, Population-level setting

ICITES2014-269
The BCI Control Applied to the Interactive Autonomous Robot with the Function of Meal Assistance

Shih-Chung Chen*, Chih-Hung Hsu, Hsuan-Chia Kuo, and Ilham A.E. Zaeni
Department of Electrical Engineering, Southern Taiwan University of Science and Technology, Tainan 71005, Taiwan

Abstract. A brain–computer interface (BCI) system is proposed to control an interactive autonomous robot with a function to assist with feeding meals. The subject's electroencephalogram (EEG), regarded as the control command, can be utilized to combine with system integration technologies to establish a BCI control robot system with an automatic feeding function. At present, the integrated technologies of the automatic feeding robot encompasses image recognition, voice recognition, the robot's mechanism design, the gripper, tactile sensor design, etc. The automatic feeding robot can be controlled by steady-state visual-evoked potential (SSVEP)-based BCI to use the gripper grasping a utensil to ladle food to the subject's mouth successfully. The signal processing algorithm adopted for the SSVEP-based BCI is magnitude squared coherence (MSC). Ten subjects participated in the BCI test for choosing the food on the plate. The average of MSC values for different visual stimulation frequencies were calculated and compared.

Keywords: Brain–computer interface (BCI), Electroencephalogram (EEG), Steady-state visual evoked potential (SSVEP), Magnitude squared coherence (MSC)

ICITES2014-270
Design of a Transparent Pipeline-Based Multiplier

Ren-Der Chen and Xiang-Chih Kuo
Department of Computer Science and Information Engineering, National Changhua University of Education, Changhua 500, Taiwan

Abstract. This paper implements an 8×8 multiplier based on the transparent pipeline architecture. A transparent pipeline can lower the power consumption by reducing the number of clock pulses required for data latch controlling. The efficiency of power saving is evaluated here by applying the multiplier to the multiplication of two sparse matrices. It can be seen from the experimental results that, when compared with the traditional synchronous multiplier using flip-flops as storage elements, the improvement in power consumption is obvious only when the sparsity of the matrix reaches a certain amount.

Keywords: Transparent pipeline, Multiplier, Sparse matrix

ICITES2014-271

An IP-Based Design to Achieve Power Reduction

Chin-Fa Hsieh[1*], Tsung-Han Tsai[1], and Chih-Hung Lai[1]

[1]Department of Electrical Engineering, National Central University, Chung-Li 32001, Taiwan

Abstract. In this paper, an IP-based design for power reduction on a one-dimension, lifting-based discrete wavelet transform (DWT) is presented. The prototype architecture is coded in VerilogHDL and simulated using Quartus-II to verify the function. Based on this prototype architecture, a low-power operator (adder and subtractor) IP, which is designed based on a full-custom design methodology, plays a role in replacing the main operations. The simulation result shows that power consumption can be reduced by 16.31 %. The architecture can be used as an independent IP core of a wavelet-based application.

Keywords: IP, Discrete wavelet transform, VerilogHDL

ICITES2014-272

Crystalline Indium-Doped Zinc Oxide Thin Films Prepared by RF Magnetron-Reactive Sputtering

Chien-Chen Diao[1], Chia-Ching Wu[1], and Cheng-Fu Yang[2*]

[1]Department of Electronic Engineering, Kao Yuan University, Kaohsiung, Taiwan, R.O.C.

[2]Department of Chemical and Materials Engineering, National University of Kaohsiung, Kaohsiung, Taiwan, R.O.C.

Abstract. The characteristic of indium-doped zinc oxide (IZO) thin films are closely related with the composition of the target, the deposition technique, and various process parameters such as the substrate temperature, the deposition pressure, the distance from target to substrate, and so on. In this study, IZO thin films have been deposited onto glass substrates with different deposition pressures by using the radio frequency magnetron-reactive sputtering method. The structural, optical, and resistivity properties of IZO thin films were investigated using a field emission scanning electron microscope (FE-SEM), X-ray diffraction patterns (XRD), UV-visible spectroscopy, and Hall-effect analysis. XRD analysis on IZO thin films showed that only the (002) diffraction peak was observable, indicating that the IZO films showed a good c-axis orientation perpendicular to the glass substrates. As the deposition pressure of IZO thin films change from 5×10^{-3} to 5×10^{-2} Torr, the thickness decreased from 220 to 72 nm, the grain size increased from 74 to 23 nm, and the resistivity increased from 2.03×10^{-3} to 1.65×10^{-1} Ω cm. The lowest resistivity value of 2.03×10^{-3} Ω m was obtained for a deposition pressure of 5×10^{-3} Torr.

Keywords: Transparent conductive oxide, Indium-doped zinc oxide, RF magnetron sputtering, Pressure

ICITES2014-273

The Construction and Efficiency Validation of an Educational Module of Pneumatics Robotics Arm Controlled by Programmable Logic Controller (PLC)

Pornpirom Fak-orn[1], Atjanapon Yomkurd[1], Natawat Punfong[1], and Mana Thanaon[1*]

[1]Department of Electrical Engineering, Rajamangala University of Technology Lanna, Muang, Tak 63000, Thailand

Abstract. This paper presents construction and efficiency validation of an educational module of pneumatics robotics arm controlled by programmable logic controller (PLC). This module was developed from a student project in the Bachelor of Science in technical education in field of electrical engineering. This is also expected to be used in teaching a subject, PLC, to obtain a high vocational certificate or diploma in the electrical power curriculum at Rajamangala University of Technology Lanna (RMUTL). The module consists of three main systems, pneumatics, a robotic arm, and PLC, integrated as a laboratory work station. Among various types of PLC, Omron CPM2A model was selected to play as controller here. At least three lab sheets were set and tested. The verification and validation have been done on three topics, body, usage and experiment through 5 experts and 20 students in second year of diploma. The reports from all experts and all topics show the average *t*-score of 4.52 and the average score from students of 4.535.

Keywords: Pneumatics, Robotics arm, Programmable logic controller (PLC)

ICITES2014-274

GPU Computations on Hadoop Clusters for Massive Data Processing

Wenbo Chen[1], Shungou Xu[1], Hai Jiang[2], Tien-Hsiung Weng[3*], Mario Donato Marino[4], Yi-Siang Chen[3], and Kuan-Ching Li[3]

[1]School of Information & Technology of Lanzhou University, China

[2]Department of Computer Science, Arkansas State University, AR, USA

[3]Department of Computer Science and Information Engineering, Providence University, Taiwan

[4]Independent Researcher, Italy

{chenwb, xushg12}@lzu.edu.cn, hjiang@astate.edu, {thweng, kuancli}@pu.edu.tw *Corresponding Author

Abstract. Hadoop is well-designed approach for handling massive amount of data. Comprised at the core of the Hadoop File System and MapReduce, it schedules the processing by orchestrating the distributed servers, providing redundancy and fault tolerance. In terms of performance, Hadoop is still behind high performance capacity due to CPUs' limited parallelism, though. GPU-accelerated computing involves the use of a GPU together with a CPU to accelerate applications to data processing on GPU cluster toward higher efficiency. However, GPU cluster has low-level data storage capacity. In this paper, we exploit hybrid model of GPU and Hadoop to make best use of both capabilities, and the design and implementation of application using Hadoop and CUDA is presented through two interfaces: Hadoop Streaming and Hadoop Pipes. Experimental results on K-means algorithm are presented as well discussed their performance.

Keywords: Hadoop, GPU, HPC, Massive data processing

ICITES2014-275
An HHT-Based Music Synthesizer

I-Hao Hsiao, Chun-Tang Chao, and Chi-Jo Wang
Department of Electrical Engineering, Southern Taiwan University of Science and Technology, Tainan, Taiwan

Abstract. Synthesizing musical sound plays an important role in modern music composition. Composers nowadays can easily take advantage of powerful and user-friendly personal computers to produce the desired musical sound with a good music synthesis method. In this paper, the Hilbert-Huang Transform (HHT) time-frequency analysis method is employed, in an attempt to implement a new efficient music synthesizer. By applying the HHT technique, the original varying-pitch music signals can be decomposed into several intrinsic mode functions (IMF) based on the empirical mode decomposition (EMD). The instantaneous amplitude and frequency of IMFs can be further obtained by Hilbert transform. By extracting the main spectrum coefficients of the instantaneous amplitude and frequency of the IMFs, the original musical signal can be reconstructed with little error. Experimental results indicate the feasibility of the proposed method.

Keywords: Music synthesis, Hilbert-Huang transform (HHT), Empirical mode decomposition (EMD), Intrinsic mode function (IMF)

ICITES2014-276
Decentralized Robust Estimation of Interconnected Systems with State-Dependent Impulse Disturbances

Cheng-Fa Cheng and Shih-Jyun Lin
Department of Communications, Navigation and Control Engineering, National Taiwan Ocean University, Keelung 20224, Taiwan

Abstract. The problem of local observer synthesis for uncertain interconnected systems subjected to nonlinear interaction functions, bounded uncertainties, and state-dependent impulse disturbances is investigated. The observation scheme is only based on certain functional properties of the uncertainty and interaction bounds and sufficient conditions will be provided such that the observation error of the uncertain state-jump interconnected system will be practically stable by the proposed local observer scheme. Furthermore, within the derived lower bound of the impulse interval, the boundedness of the observation errors of the uncertain interconnected system with the equidistant impulse disturbance will be guaranteed and the radius of the attraction ball can be measured. Finally, an example with simulations is given to illustrate the application of our results

Keywords: Interconnected systems, Robust, Local observer, State-dependent impulse disturbance, Practical stability

ICITES2014-277

A Lean Analysis to Automate the Draw Operations at a Tube Manufacturing Company

Jun-Ing Ker[1*], Chandra Mani Shrestha[1], and Yichuan Wang[2]
[1]Industrial Engineering, Louisiana Tech University, Ruston, LA 71272, USA
[2]Department of Aviation and Supply Chain Management, Auburn University, Auburn, AL 36830, USA

Abstract. The objective of this research is to conduct a lean analysis of the Draw Cell operations at a manufacturer that fabricates stainless steel tubes and to implement Lean concepts for automation. Videotaping operations, interviewing operators, and studying the engineered layout of the operation were carried out to understand the existing operations. Industrial engineering knowledge of Lean, 8 wastes, Kaizen, Single Minute Exchange of Die, and Value Stream Map were applied to improve the setup of workstations and material handling procedures. These principles helped make the current draw cell to operate more effectively and economically without major renovations.

Keywords: Tube drawing, Draw operation automation, Lean analysis, Kaizen, Single minute exchange of die

ICITES2014-278

Classification of the ECG Signal Using Artificial Neural Network

Andrew Weems[1*], Mike Harding[1], and Anthony Choi[2]
[1]Department of Biomedical Engineering, Mercer University, Macon, GA 31207, USA
[2]Department of Electrical and Computer Engineering, Mercer University, Macon, GA 31207, USA

Abstract. Recording of electrocardiogram (ECG) signals and the correlation to cardiovascular diseases are major problems in today's society. A common abnormality is arrhythmia, which is unexpected variation in cardiac rhythm. The goal of this study is to analyze these types of signals and find a more efficient way to classify these signals. Currently, medical devices for detecting ECG signals are at least 85 % accurate in analyzing the data. Neural networks have progressed quickly over the past few years and have the capability of recognizing many types of variation in these signals. The pattern recognition power of Artificial Neural Networks (ANNs) is a valuable tool when classifying ECG signals in cardiac patients. Data obtained from the PhysioBank ATM was used to analyze the structure of an ANN and the effect that it has on pattern recognition. The results show that only one misclassification occurred resulting in an accuracy of 96 %.

Keywords: Artificial neural network, ECG, MATLAB, Signal classification, Cardiac abnormalities, Cardiac arrhythmias

ICITES2014-279

Integration Method of Composite Pattern for Solving Structure Problems of Visitor Pattern

Jaehak Uam, Jeonghoon Kwak, and Yunsick Sung

Department of Game Mobile Contents, Keimyung University, Daegu, South Korea
{jaehak, jeonghoon, yunsick}@kmu.ac.kr

Corresponding Author: yunsick@kmu.ac.kr

Abstract. This paper proposes a method in which the Visitor pattern is used to apply the Composite pattern for solving the structure problems of the scenario generator, based on Bayesian probability. The traditional Visitor pattern is not ideal for adding new member functions due to its structure problems. By applying the Composite pattern, our approach provides flexibility to interfaces for adding and deleting new member functions.

Keywords: Internet of things, Scenario generator, Design pattern, Visitor pattern, Composite pattern

ICITES2014-280

A Circuit Design of Dimming T8 LED Tube for High Power Factor Correction

Manh Tran Van[1], Wei-Sung Weng[1], Chun-Shan Liu[2], Shen-Yuar Chen[1], and Wei-Ching Chuang[1*]

[1]Department of Electro-Optical Engineering, National Formosa University, Yunlin 632, Taiwan

[2]Department of Electrical Engineering, National Formosa University, Yunlin 632, Taiwan

Abstract. An advanced method to raise the power factor (PF) with fewer components than the traditional PFC circuits is proposed. It also provides an efficient solution to supply power for light-emitting diodes (LEDs). The structure of driver and control circuit for T8-LED tube is designed and verified as well.

Keywords: Pulse width modulation (PWM), Fly-back converter, Power factor correction

ICITES2014-281

Anti-ESD Improvement by the Bulk-FOX Structure in HV nLDMOS Devices

Shen-Li Chen*, Shawn Chang, Yu-Ting Huang, and Shun-Bao Chang

Department of Electronic Engineering, National United University, MiaoLi City 36003, Taiwan

Abstract. This work is focused on a 0.25 μm 60 V high-voltage nLDMOS devices which will be integrated with an FOX structure in the bulk region, and evaluate the impacts on its anti-ESD protection ability. It is found that as an FOX structure adding, and as the FOX area ratio is increased, I_{t2} value will be enhanced too. When the FOX area ratio is about 83.5 %, I_{t2} value has a maximum value ~6 A. However, as the FOX area ratio is increased, the Ron value will be declined. From the experimental data, it is revealed that V_{t1} (V_h) value decreased more than 16.9 % (35.6 %), anti-ESD ability increased more than 170.2 %, and Ron decreased more than 81.2 % as compared with the Ref. DUT.

Keywords: Electrostatic discharge (ESD), Field oxide (FOX), Holding voltage (V_h), n-channel lateral-diffused MOS (nLDMOS), On resistance (Ron), Secondary breakdown current (I_{t2}), Trigger voltage (V_{t1})

ICITES2014-282

N+ Extended-Distribution Influences on Anti-ESD Ability in the 60-V pLDMOS-SCR (NPN Arranged-Type)

Shen-Li Chen*, Yu-Ting Huang, Shawn Chang, and Shun-Bao Chang
Department of Electronic Engineering, National United University, MiaoLi City 36003, Taiwan

Abstract. In order to effectively improve the reliability capability, a p-channel lateral-diffused MOSFETs with an embedded SCR which is formed by implanting N+ doses in both sides of the drain end, and this structure called as the "*npn*" arranged-type of pLDMOS-SCR in this paper (diffusion region of drain side is N+-P+-N+). Then, changing the layout manner of N+ implants in both side of a drain-side P+ region is investigated in this paper by a 0.25-μm 60-V BCD process. In this planning idea, the layout types of N+ region are continuously extended into the drain-side. From the experimental results, due to all of their secondary breakdown current (I_{t2}) values are so good reached above 7 A, it can be found that the layout manner of continuous extended types in the drain-side have a little impact on the ESD capability. However, the major repercussion is the Vh value will be decreased about 10.8–49.5 %.

Keywords: Electrical over stress (EOS), Electrostatic discharge (ESD), Latchup (LU), p-Channel lateral-diffused MOSFET (pLDMOS), Secondary breakdown current (I_{t2}), Transmission-line pulse (TLP), Trigger voltage (V_{t1})

ICITES2014-283

WristEye: An Elderly Computer Learning Assistant System with Wrist-Wearable Devices

Wan-Jung Chang[1*], Yi-You Hou[2], Rung-Shiang Cheng[3], and Ming-Che Cheng[4]

[1]Department of Electronics Engineering, Southern Taiwan University of Science and Technology, 71005 Tainan, Taiwan (R.O.C.)
[2]Department of Electrical Engineering, Far East University, Tainan 74448, Taiwan (R.O.C.)
[3]Department of Computer and Communication, Kun Shan University, Tainan 71070, Taiwan (R.O.C.)
[4]Department of Electrical Engineering, National Cheng Kung University, Tainan 701, Taiwan (R.O.C.)

Abstract. This paper develops an elderly computer learning assistant system with wrist-wearable devices, designated as WristEye, which can be used to analyze the computer learning attitudes, reactions, and behaviors of elderly individuals whilst in computer learning classes. WristEye is equipped with a kinematic sensor to effectively detect the changes in the orientation and vertical acceleration of the elderly wrist and to determine the corresponding operations in learning computer, i.e., moving mouse, hitting keyboard, idle, and swing mouse. Furthermore, a remote backend server receives the detected signal from the wearable unit via a Wireless Sensor Network (WSN) and then identifies the corresponding computer learning effectiveness. The experimental results show that WristEye has a classification accuracy to recognize computer learning status of elderly individuals.

Keywords: Wearable, Elderly, Learning

ICITES2014-284

A Study of Ultrafast Laser-Based X-Ray Sources for In-Line Phase-Contrast Biomedical Imaging

Mario do Nascimento[1*] and Jengnan Juang[2]
[1]Department of Electrical and Computer Engineering, Mercer University, Macon, GA 31207, USA
[2]School of Engineering, Mercer University, Macon, GA, USA

Abstract. Ever since the Coolidge tube was created in 1913, the same design has been used for X-ray imaging devices. As modern medicine advances, the Coolidge tube is reaching its limited full potential. For this reason, this research analyses two of the modern alternatives for X-ray imaging: laser-based X-ray sources and phase-contrast imaging. According to this study, lasers are able to provide significantly better focal spot sizes, much better temporal resolution, and a high repetition rate. As for integrating phase-contrast imaging, this research depicts how phase-contrast imaging is overwhelmingly more sensitive to elements found in soft tissue and refraction is far more likely to occur than absorption, which has the potential to decrease exposure time. Although these technologies have been found to be more advantageous than the current technologies, there are still essential improvements and research that need to be accomplished before clinical use.
Keywords: Laser X-rays, Phase-contrast imaging, Biomedical imaging

ICITES2014-285

PC-USB-Based Real-Time Control Systems Using Quadratic Optimal Control Method

Dershyang Ker
Program of Game Design, China University of Science and Technology, Taipei 115, Taiwan

Abstract. The goal of this paper is to implement the quadratic optimal control algorithm in PC-USB-Based real-time control systems by using mathematic model blocks in VisSim/USB software package environment and a USB-based module. PC is used as a controller to apply quadratic optimal control algorithm in PC-USB-Based real-time control systems. First, a design example based on the quadratic optimal control algorithm is given. The theoretical state feedback controller can be obtained. An alternative approach using VisSim's OptimizePro to automatically calculate optimal state feedback controller has been verified to be a very powerful method to avoid tedious theoretical design. Second, the model of the real control system, FB-33 control system, can be set up in VisSim. The quadratic optimal controller can be easily obtained without the theoretical design. Finally, the VisSim/USB is used to implement the quadratic optimal control and USB-based module to control the FB-33 servo control system. The satisfied results are shown in this paper.

Keywords: PC-USB-based real-time control system, Quadratic optimal control, USB-based module

ICITES2014-286

A Bibliometric Analysis on Data Mining Using Bradford's Law

Jiann-Min Yang[1], Shu-Feng Tseng[2], and Yu-Ling Won[3*]
[1]Department of MIS, National Chengchi University, Taipei 116, Taiwan
[2]Department of MIS, National Chengchi University, Taipei 116, Taiwan
[3]Department of MIS, National Chengchi University, Taipei 116, Taiwan

Abstract. The subject-specific Bibliometric Analysis on Data Mining complies with the Bradford's Law to show the $1{:}n{:}n^2$ relationship and the zone distribution of research journals. The overall count of articles in each zone is about the same. The top 3 author countries contribute to more than 50 % of total journal articles. However, big differences in productivity exist among them. Most of the high ranking author organizations are academic institutions and many of them belong to the top 3 author countries. The total article count grows year by year except a sudden drop in 2007 but grows again in 2008 and afterwards. The top 3 research areas are quite stable. There are some variations in other research areas. Some drop down the article counts and some others grow gradually.

Keywords: Data mining, Bradford's law, Bibliometric analysis

Contents

Chapter 1
Active Closed-Loop Gap Control for Aerostatic Bearing

Jyh-Chyang Renn and Yaw-Ming Chiou

Abstract Advantages of aerostatic bearing are clean, no pollution and very low friction force. Therefore, it can easily be found in many high-precision machine tools and measurement equipments. However, owing to the compressibility of air, the stiffness of aerostatic bearing is relatively low compared to traditional ball bearing or hydrostatic bearing. In order to improve this fault, an active closed-loop gap control for aerostatic bearing with higher stiffness is proposed. First of all, an aerostatic bearing is designed and manufactured according to a previous report. Next, an experimental test rig based on LabVIEW software for the proposed active closed-loop gap control is constructed. Finally, it is proved that the gap between the aerostatic bearing and workpiece can be held at a stable and constant value even when the bearing is subjected to variable external disturbances.

Keywords Aerostatic bearing • Active control • Closed-loop gap control

1.1 Introduction

Aerostatic bearings have been widely used for measuring instruments, machine tools, dental drills, jet engines, and computer peripheral devices because of their substantially low friction loss and heat generation [1]. Like other fluid lubricated bearings, aerostatic bearings serve two purposes. One is to support an external load and the other is to lubricate a pair of surfaces. Nowadays, the aerostatic bearings are commonly used in the field of precision engineering. Figure 1.1 shows the schematic sectional views of five commonly used aerostatic bearings with different types of restrictors [1–3]. The gas from an external source is fed into the clearance space through flow restrictor, and escapes continuously to the atmosphere from the outside edges of the bearing. Thus, a thin film of air is established to support the external load. However, such aerostatic bearings belong to passive control schemes.

J.-C. Renn (✉) • Y.-M. Chiou
Department of Mechanical Engineering, National Yunlin University of Science and Technology, Douliou 640, Taiwan
e-mail: rennjc@yuntech.edu.tw

J. Juang (ed.), *Proceedings of the 3rd International Conference on Intelligent Technologies and Engineering Systems (ICITES2014)*, Lecture Notes in Electrical Engineering 345, DOI 10.1007/978-3-319-17314-6_1

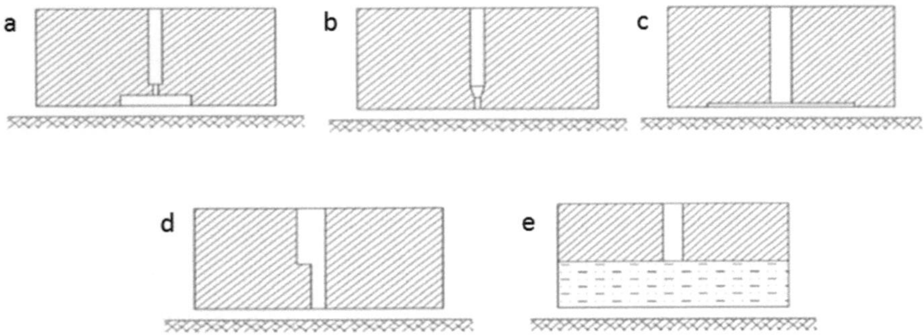

Fig. 1.1 Basic types of inlet restrictors. (**a**) Orifice. (**b**) Inherent. (**c**) Surface. (**d**) Slot. (**e**) Porous

Lower stiffness is inevitably its major fault. In addition, if the external load is quite large, the thin air-film may no longer exist due to the compressibility of air [5, 6]. Consequently, real contact between the aerostatic bearing and workpiece may occur which will damage the surface of the workpiece. In this chapter, therefore, an active closed-loop gap control scheme for aerostatic bearing is proposed. Using closed-loop control, it is expected that an aerostatic bearing with higher stiffness can be implemented.

1.2 The Design of Aerostatic Bearing

Generally speaking, due to the design of a thin cylindrical air chamber at the outlet of the orifice restrictor, the orifice-type aerostatic bearing possesses highest stiffness among five different types of restrictors shown in Fig. 1.1. However, improper design of the orifice-type restrictor may also give rise to the pneumatic hammer instability [1, 2]. In this study, therefore, the inherent type of restrictor is chosen to design the aerostatic bearing. Figure 1.2a shows the geometry and dimension of the developed aerostatic bearing. It is worth mentioning that, to increase the overall stiffness, three holes of diameter ψ 0.4 mm together with three pairs of radial and circular grooves of depth 0.04 mm are designed and manufactured [4]. The real picture of the developed aerostatic bearing is shown in Fig. 1.2b.

1.3 Experimental Test Rig for Closed-Loop Gap Control

In this chapter, a simple but effective test rig to evaluate the performance of the closed-loop gap control is proposed and realized. The schematic layout of the test rig is depicted in Fig. 1.3a. In addition, the real picture of this test device is shown in Fig. 1.3b. The description of the developed test rig is briefly summarized as follows.

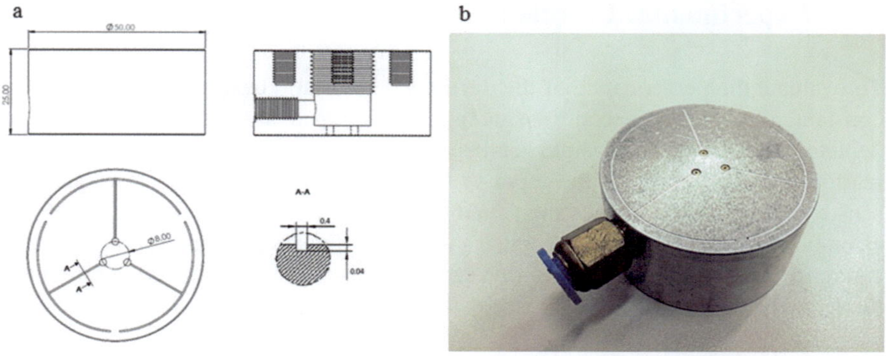

Fig. 1.2 The developed aerostatic bearing, (**a**) geometry and dimension, (**b**) real picture

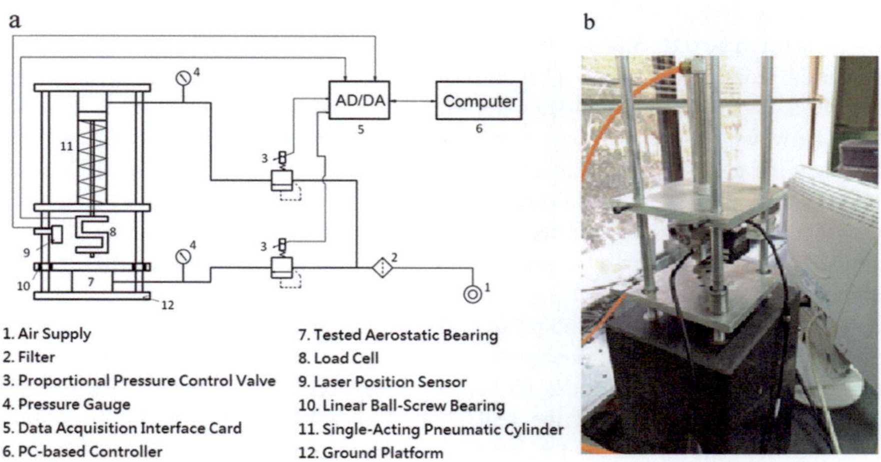

Fig. 1.3 The test rig, (**a**) schematic layout, (**b**) real picture

To accomplish the closed-loop air-film gap control, it is necessary to detect and measure the real actual gap between the aerostatic bearing and the ground platform. In this study, a laser position sensor (Keyence IL series) with maximal resolution of 1 μm is employed. In addition, to generate variable external loads or disturbances, a single-acting pneumatic cylinder (Mindman, MCMI series) at the top and a proportional pressure control valve (FESTO, VPPM series) are utilized. Setting different input pressure to the cylinder can generate various external loads to the aerostatic bearing. Besides, a second proportional pressure control valve of the same kind is also used to electrically adjust the input pressure to the aerostatic bearing. Other necessary sensors, like the S-type load cell (Transcell, BSA series) and pressure gauges, etc. are also built into the test rig. Finally, the control, monitoring and data acquisition are all integrated into a PC-based LabVIEW software controller.

1.4 Experimental Results and Discussions

To evaluate the performance of the developed aerostatic bearing with closed-loop constant gap control, a control strategy is developed and its block diagram is depicted in Fig. 1.4. The desired gap between the aerostatic bearing and the ground platform, R_{GAP}, serves as the input to the control system and the laser position sensor detects and feedbacks the actual gap signal, X, to the controller. Thus, the error between the desired and actual gap drives the proportional pressure valve trying to compensate and reduce this error. In this study, the utilized controller is the simple PID controller described by the following (1.1).

$$u(t) = K_{\mathrm{p}}e(t) + K_I\int_0^\tau e(\tau)d\tau + K_D\frac{de(\tau)}{d\tau}. \tag{1.1}$$

From the previous report [4], the optimal gap corresponding to largest stiffness was found to be around 0.01 mm. Therefore, in the real experiments, the desired gap, R_{GAP}, is set to be 0.01 mm and the magnitude of variable external square-wave load is varied from 10 to 95 N. The corresponding experimental result is shown in Fig. 1.5a. Clearly, the steady-state performance of constant gap control is satisfactory. However, in the transient response, it is observed that there is real contact between the aerostatic bearing and the ground platform due to the larger external load as well as the inevitable compressibility of air. To avoid this real contact, a modified PID controller is further proposed in this chapter. In details, if the detected actual gap is larger than a preset value (0.006 mm), then the normal PID control algorithm described by (1.1) will be executed. However, if the actual gap is detected to be smaller than the preset value, a large positive voltage (7 V) will be sent directly to the proportional pressure valve and results in the maximal pressure output. This maximal pressure (6 bar) then supports the aerostatic bearing and pushes it upwards to avoid the real contact. The flow chart of this switching mechanism is shown in Fig. 1.6. Finally, the corresponding experimental result is depicted in Fig. 1.5b. It is observed that there is no more real contact between the aerostatic bearing and the ground platform.

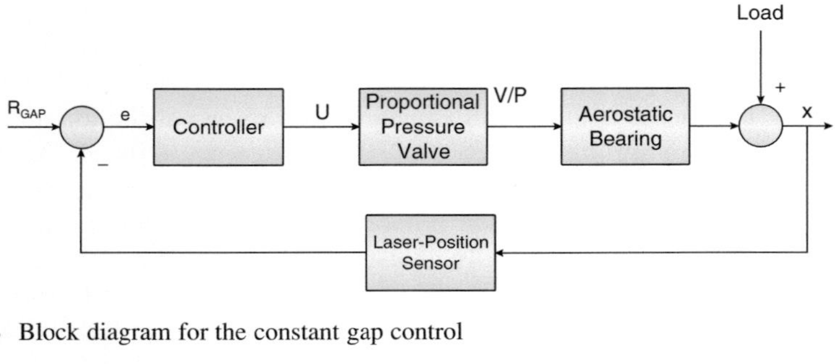

Fig. 1.4 Block diagram for the constant gap control

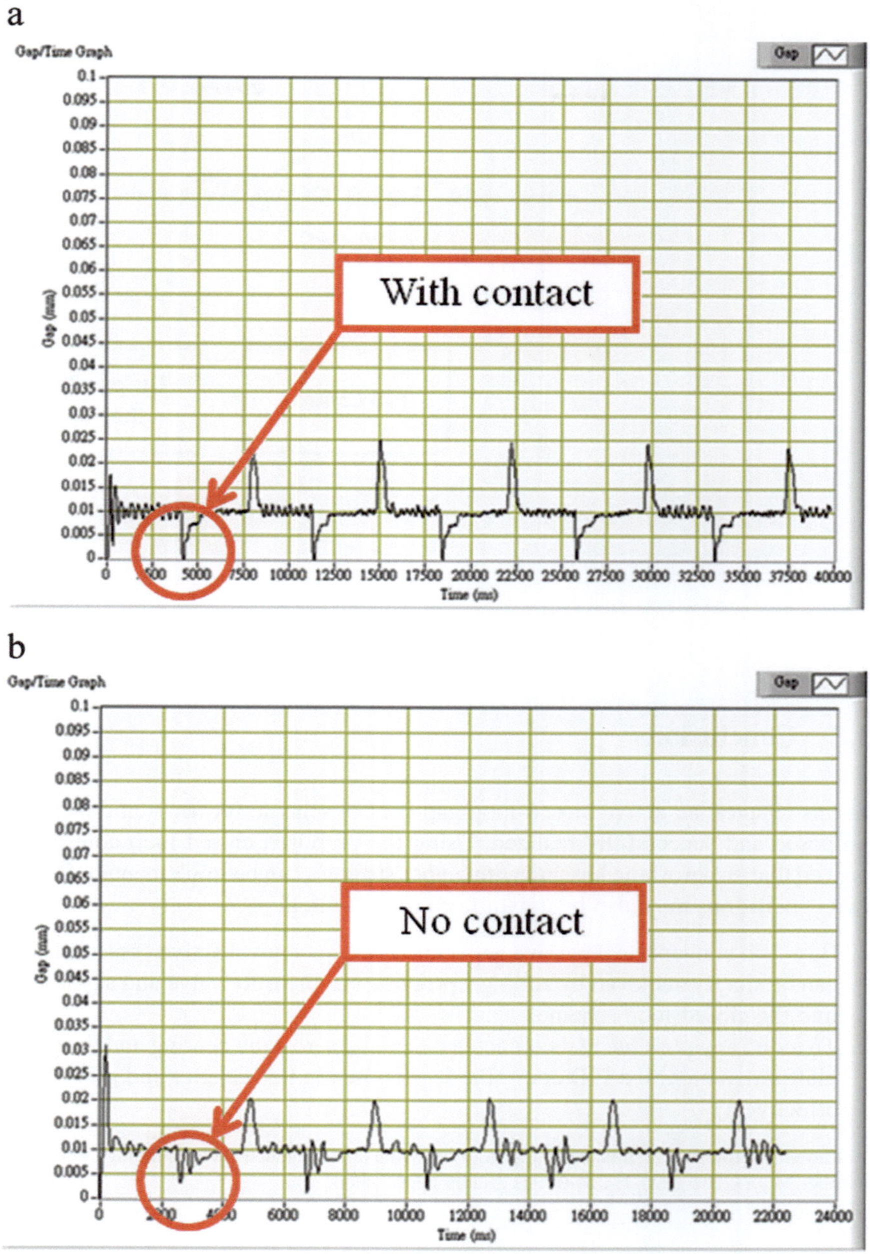

Fig. 1.5 Experimental results of closed-loop constant gap control ($R_{GAP} = 0.01$ mm), (**a**) using normal PID control, (**b**) using modified PID control

Fig. 1.6 Flow chart of controller switching mechanism

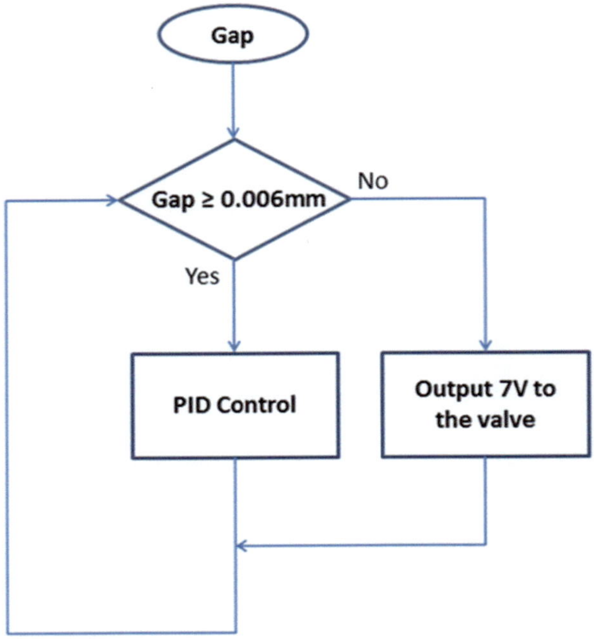

1.5 Conclusions

In this chapter, an active closed-loop gap control scheme for aerostatic bearing is proposed and successfully realized. Using the proposed closed-loop control, it is proved that an aerostatic bearing with higher stiffness can be implemented. Besides, two conclusions may also be drawn from this research.

1. Based on conventional passive aerostatic bearing, a novel active control strategy can be simply achieved by adding a proportional pressure valve and a gap sensor into the closed-loop scheme.
2. To avoid any possible real contact between the aerostatic bearing and the ground platform, a modified PID controller is proposed in this chapter and proved to be effective.

Acknowledgement The financial support of the Ministry of Science and Technology under grant number MOST 103-2221-E-224-047 is greatly appreciated.

References

1. O'Donoghue, J.P., Rowe, W.B.: Hydrostatic bearing design. Tribol. Int. **2**, 25–71 (1969)
2. Rowe, W.B.: Hydrostatic, Aerostatic and Hybrid Bearing Design. Butterworths, London (1983)
3. Renn, J.C., Hsiao, C.H.: Experimental and CFD study on the mass flow-rate characteristic of gas through orifice-type restrictor in aerostatic bearings. Tribol. Int. **37**, 309–315 (2004)
4. Chen, M.F., Lin, Y.T.: Static behavior and dynamic stability analysis of grooved rectangular aerostatic thrust bearings by modified resistance network method. Tribol. Int. **35**, 329–338 (2002)
5. Bassani, R., Piccigallo, B.: Hydrostatic Lubrication. Elsevier, Amsterdam (1992)
6. Canbazoğlu, S., Canbulut, F.: A note on the flow coefficients of capillary tube and small orifice restrictors exposed to very low Reynolds number flow. Ind. Lubr. Tribol. **57**(3), 116–120 (2005)

Chapter 2
Firefly Algorithm for Power Economic Emission Dispatch

Chao-Lung Chiang

Abstract This work proposes a firefly algorithm for the optimal economic emission dispatch (EED) of the hydrothermal power system (HPS), considering non-smooth fuel cost and emission level functions. The firefly algorithm (FA) can efficiently search and actively explore solutions. The multiplier updating (MU) is introduced to handle the equality and inequality constraints of the HPS, and the ε-constraint technique is employed to manage the multi-objective problem. To show the advantages of the proposed algorithm, one example addressing the best compromise is applied to test the EED problem of the HPS. The proposed approach integrates the FA, the MU, and the ε-constraint technique, revealing that the proposed approach has the following merits—ease of implementation; applicability to non-smooth fuel cost and emission level functions; better effectiveness than the previous method, and the requirement for only a small population in applying the optimal EED problem of the HPS.

Keywords Firefly algorithm • Multiplier updating • Economic emission dispatch

2.1 Introduction

Traditionally, in the short-term scheduling of a fixed water head, the variation of the net head can be ignored only for relatively large reservoirs, in which case power generation depends only on the discharge of water [1]. Recently, Basu [2] modeled the HPS problem as a multi-objective problem and solved it using a weighted combination. Nevertheless, the weighting method linearly combined the objectives as a weighted sum. The objective function thus formed may lose significance because the various multiple noncommensurable factors are incorporated into a single function. This study employs the ε-constraint technique [3] to handle the multi-objective problem.

C.-L. Chiang (✉)
Department of Electronic Engineering, Nan Kai University
of Technology, Nan-Tou 542, Taiwan
e-mail: t129@nkut.edu.tw

J. Juang (ed.), *Proceedings of the 3rd International Conference on Intelligent Technologies and Engineering Systems (ICITES2014)*, Lecture Notes in Electrical Engineering 345, DOI 10.1007/978-3-319-17314-6_2

Firefly algorithm (FA) was developed by Yang [4]. It is a new metaheuristic nature-inspired algorithm, based on the flashing light of fireflies has been successfully applied to solve different engineering problems [5–7]. This chapter throws a light on how well the firefly algorithm is utilized to solve the EED problems. The EED problem is very difficult to be solved by direct approach and thus creates prominent damage to the power system operation and planning in the existing scenario, so a metaheuristic approach such as FA is generally preferred for optimal EED solutions.

2.2 Problem Formulation

The following objectives and constraints of the HPS with N_i thermal units and N_h hydro plants over M time subintervals are considered.

2.2.1 *Economic Objective* F$_1$

The fuel cost function of each thermal unit considering the valve-point loadings is realistically expressed as the superposition of a quadratic function and a sinusoidal function. The total fuel cost can be accurately denoted in terms of real power output as a non-smooth cost function:

$$F_1 = \sum_{m=1}^{M} \sum_{i=1}^{N_i} t_m \left[a_i + b_i P_{mi} + c_i P_{mi}^2 + \left| e_i \sin\left\{ f_i \left(P_i^{\min} - P_{mi} \right) \right\} \right| \right] \tag{2.1}$$

where F_1 is the total cost of generation; P_{mi} is the generation of the ith thermal unit in the mth subinterval; a_i, b_i, and c_i are coefficients of the cost curve of the ith generator; e_i and f_i are fuel cost coefficients of the ith unit with valve-point loadings, and t_m is the generating duration.

2.2.2 *Emission Objective* F$_2$

Fossil-based generating stations are the primary sources of nitrogen oxides, so the Environmental Protection Agency has strongly urged them to reduce their emissions. In this study, the amount of emitted nitrogen oxides is taken as the selected index from the perspective of environmental conservation. The emission from each generator is given as a function of its output, which is the sum of a quadratic and exponential functions. The emission objective can be mathematically modeled as [1]:

$$F_2 = \sum_{m=1}^{M}\sum_{i=1}^{N_i} t_m\left[\alpha_i + \beta_i P_{mi} + \gamma_i P_{mi}^2 + \xi_i e^{(\zeta_i P_{mi})}\right] \tag{2.2}$$

where α_i, β_i, γ_i, ξ_i, and ζ_i are coefficients of generator emission characteristics.

2.2.3 System Constraints

The power balance and water availability equality constraints of the HPS are imposed.

2.2.3.1 Power Balance Equality Constraints

$$\sum_{i=1}^{N_i} P_{mi} + \sum_{h=1}^{N_h} P_{mh} - P_{mD} - P_{Lm} = 0, \quad m = 1, \ldots, M \tag{2.3}$$

The power balance constraints (2.3) are M equalities, where P_{mh} is the generation of the hth hydro plant in the mth subinterval, and P_{mD} is the total demand in the mth subinterval. The P_{Lm} is the real power loss of the transmission lines in the mth subinterval, and is given as follows:

$$P_{Lm} = \sum_{i=1}^{N_i+N_h}\sum_{j=1}^{N_i+N_h} P_{mi} B_{ij} P_{mj} \tag{2.4}$$

2.2.3.2 Water Availability Equality Constraints

$$\sum_{m=1}^{M} t_m\left(a_{0h} + a_{1h} P_{mh} + a_{2h} P_{mh}^2\right) - W_h = 0, \quad h = 1, \ldots, N_h \tag{2.5}$$

The water availability constraints (2.5) are N_h equalities; a_{0h}, a_{1h}, and a_{2h} are characteristic coefficients of the hth hydro unit, and W_h is the water availability of the hth hydro unit.

System limits. The inequality constraints of the HPS imposed on unit output are (2.6) and (2.7), respectively:

$$P_i^{\min} \le P_i \le P_i^{\max} \tag{2.6}$$

$$P_h^{\min} \le P_h \le P_h^{\max} \tag{2.7}$$

where $P_i^{\min}$ and $P_i^{\max}$ are the minimum and maximum limits of the ith thermal generator, and $P_h^{\min}$ and $P_h^{\max}$ are the minimum and maximum bounds of the hth hydro unit.

2.3 The Proposed Algorithm

2.3.1 The ε-Constraint Technique

The ε-constraint technique [3] is used to generate pareto-optimal solutions for the multi-objective problem. To proceed, one of the objective functions constitutes the primary objective function and all other objectives act as constraints. To be more specific, this procedure is implemented by replacing one objective in the EED problem with one constraint. Reformulate the problem as follows:

$$\begin{aligned}
&\min_{P_{mi}(m=1,\ldots,M \text{ and } i=1,\ldots,N_i+N_h)} && F_j(P_{mi}), \quad j=1 \text{ or } 2 \\
&\text{Subject to} && F_k(P_{mi}) \le \varepsilon_k; \quad k=1 \text{ or } 2, \text{and } k \ne j \\
& && \sum_{i=1}^{N_i} P_{mi} + \sum_{h=1}^{N_h} P_{mh} - P_{mD} - P_{Lm} = 0 \\
& && \sum_{m=1}^{M} t_m\left(a_{0h} + a_{1h}P_{mh} + a_{2h}P_{mh}^2\right) - W_h = 0 \\
& && P_i^{\min} \le P_i \le P_i^{\max}, \quad i=1,\ldots,N_i \\
& && P_h^{\min} \le P_h \le P_h^{\max}, \quad h=1,\ldots,N_h
\end{aligned} \tag{2.8}$$

where $F_j(P_{mi})$ and $F_k(P_{mi})$ are the objective functions to be minimized over the set of admissible decision vector P_{mi}. Where ε_k is the maximum tolerable objective level. The value of ε_k is chosen for which the objective constraints in problem (2.8) are binding at the optimal solution. The level of ε_k is varied parametrically to evaluate the impact on the single objective function $F_j(P_{mi})$.

2.3.2 The FA

FA is naturally inspired from flashing light of fireflies. For a given optimization problem, the objective function of FA is affiliated to the intensity of light. This assists the swarm of fireflies to relocate to more brighter and attractive places for efficient optimal solutions. Although, the FA has various resemblances with other swarm intelligence algorithms, namely Artificial Bee Colony (ABC), Ant Colony, Cuckoo Search, and Particle Swarm optimization (PSO), but its simplicity both in conceptualwise and implementation makes it distinct from other algorithms. According to [4] the characteristic feature of the FA is the fact that it simulates a parallel independent run strategy, where in every iteration, a swarm of n fireflies has generated n solutions. Each firefly works almost independently and as a result the algorithm will converge very quickly with the fireflies aggregating closely to the optimal solution [5–7].

2.3.3 The MU

Herein, the MU [8] is introduced to handle this constrained optimization problem. Such a technique can overcome the ill-conditioned property of the objective function.

Considering the nonlinear problem with general constraints as follows:

$$\begin{aligned} & \min_x F(x) \\ \text{subject to} \quad & h_k(x) = 0, \quad k = 1, \ldots, m_e \\ & g_k(x) \leq 0, \quad k = 1, \ldots, m_i \end{aligned} \tag{2.9}$$

where $h_k(x)$ and $g_k(x)$ stand for equality and inequality constraints, respectively.

The augmented Lagrange function (ALF) [6] for constrained optimization problems is defined as:

$$\begin{aligned} L_a(x, \nu, \upsilon) = f(x) &+ \sum_{k=1}^{m_e} \alpha_k \left\{ [h_k(x) + \nu_k]^2 - \nu_k^2 \right\} \\ &+ \sum_{k=1}^{m_i} \beta_k \left\{ \langle g_k(x) + \upsilon_k \rangle_+^2 - \upsilon_k^2 \right\} \end{aligned} \tag{2.10}$$

where α_k and β_k are the positive penalty parameters, and the corresponding Lagrange multipliers $\nu = (\nu_1, \ldots, \nu_{m_e})$ and $\upsilon = (\upsilon_1, \ldots, \upsilon_{m_i}) \geq 0$ are associated with equality and inequality constraints, respectively.

The contour of the ALF does not change shape between generations while constraints are linear. Therefore, the contour of the ALF is simply shifted or biased in relation to the original objective function, *f*(*x*). Consequently, small penalty parameters can be used in the MU. However, the shape of contour of L_a is changed by penalty parameters while the constraints are nonlinear, demonstrating that large penalty parameters still create computational difficulties. Adaptive penalty parameters of the MU are employed to alleviate the above difficulties. More details of the MU are found in [8].

2.4 System Simulations

An HPS was employed to demonstrate the effectiveness of the proposed approach, as determined by the quality of the solutions obtained. This test system includes two hydro plants and four thermal generators whose characteristics are the same as those in [2]. The short-term scheduling of this HPS is divided into four subintervals and involves four subinterval demands. For the purpose of comparing the previous method [2] with the same situations, the duration of each subinterval is 12 h. The transmission loss (P_{Lm}) in each subinterval was represented using *B*-coefficient method. The proposed algorithm was compared with Non-dominated Sorting Genetic Algorithm-II (NSGA-II) [2], Strength Pareto Evolutionary Algorithm-2 (SPEA2) [2], and Multi-objective Differential Evolution (MODE) [2] in the best compromise. The computation was implemented on a personal computer (P5-3.0 GHz) in FORTRAN-90. Setting factors utilized in this case were as follows: the population size *Np* was set to 5, and iteration numbers of the outer loop and inner loop were set to (outer, inner) as (50, 5,000) for the proposed FA-MU. The implementation of this example can be described as follows:

$$L_a(x,\nu,\upsilon) = f(x) + \sum_{k=1}^{4} \alpha_k \left\{ [h_k(x)+\nu_k]^2 - \nu_k^2 \right\} + \sum_{k=1}^{3} \beta_k \left\{ \langle g_k(x)+\upsilon_k \rangle_+^2 - \upsilon_k^2 \right\} \tag{2.11}$$

$$\mathrm{SCV} = \sum_{k=1}^{4} |h_k| + \sum_{k=1}^{2} \max\{g_k, 0.0\} \tag{2.12}$$

where

$$F_1 = \sum_{m=1}^{4}\sum_{i=1}^{4} t_m \left[a_i + b_i P_{mi} + c_i P_{mi}^2 + \left| e_i \sin\left\{ f_i\left(P_i^{\min} - P_{mi}\right)\right\}\right| \right] (\$) \tag{2.13}$$

$$F_2 = \sum_{m=1}^{4}\sum_{i=1}^{4} t_m \left[\alpha_i + \beta_i P_{mi} + \gamma_i P_{mi}^2 + \xi_i e^{(\zeta_i P_{mi})} \right] (\mathrm{lb}) \tag{2.14}$$

And subject to

$$h_1 \sim h_4 : \sum_{i=1}^{4} P_{mi} + \sum_{h=1}^{2} P_{mh} - P_{mD} - P_{Lm} = 0 \tag{2.15}$$

$$g_1, g_2 : \sum_{m=1}^{4} t_m \left(a_{0h} + a_{1h} P_{mh} + a_{2h} P_{mh}^2 \right) - W_h \le 0 \tag{2.16}$$

$$g_3 : F_2 - E_{\text{lim}} \le 0 \tag{2.17}$$

This scheduling of the best compromise includes the prime function (2.11) with 24 variables (P_{11}, …, P_{16}, P_{21}, …, P_{26}, P_{31}, …, P_{36}, P_{41}, …, P_{46}), 4 equality constraints (h_1, …, h_4), and 3 inequality constraints (g_1, g_2, g_3). The g_3 stands the violation of emission criterion for the expected ε_2. For comparison, the sum of the equality and inequality constraint violations defined as $\text{SCV} = \sum_{k=1}^{4} |h_k| + \sum_{k=1}^{2} \max\{g_k, 0.0\}$ is used to evaluate the effect of the equality and inequality constraints on the final solutions. SCV doesn't take g_3 into account for the purpose of directly using results obtained from the previous algorithms.

Table 2.1 lists the compared results of the best compromise obtained by NSGA-II [2], SPEA2 [2], MODE [2], and the proposed FA-MU. The cost (F_1) obtained by the proposed approach is satisfactory, in relation to those obtained by NSGA-II [2], SPEA2 [2], and MODE [2]. The proposed FA-MU completely meets the system constraints (SCV = 0.00). It is superior to NSGA-II [2], SPEA2 [2], and MODE [2] in the quality of solutions. Results in this case, with SCV are 3.89 and 5.84, obtained by NSGA-II [2] and SPEA2 [2], respectively. There are infeasible solutions. Consequently, the proposed FA-MU is more effective and efficient than the previous methods.

Table 2.1 Compared results of the previous methods and FA-MU

Method item	NSGA-II [2]	SPEA2 [2]	MODE [2]	FA-MU
h_1	0.00	0.00	0.00	0.00
h_2	0.00	0.00	0.00	0.00
h_3	0.00	0.00	0.00	0.00
h_4	0.00	0.00	0.00	0.00
g_1	1.90	3.03	−3.18	0.00
g_2	1.99	2.81	−3.84	0.00
g_3	–	–	–	0.00
SCV	**3.89**	**5.84**	0.00	0.00
F_1 ($)	68,332.9417[a]	68,392.3888[a]	68,388.1897	67,027.0135
F_2 (lb)	25,278.2860	26,005.7492	25,759.3182	25,278.2308
CPU_time (s)	–	–	–	9.47

[a]Infeasible solution

2.5 Conclusions

The proposed FA-MU yields optimal values, taking into account different objectives, and the pareto-optimal set represents the trade-off between the objectives. The proposed approach integrates the FA, the MU and the ε-constraint technique, showing that the proposed algorithm has the following merits—(1) ease of implementation; (2) applicability to non-smooth fuel cost and emission level functions; (3) better effectiveness than the previous method, and (4) the need for only a small population. System simulations have shown that the proposed approach has the advantages mentioned above for solving optimal EED problems of the HPS.

References

1. Zaghlool, M.F., Trutt, F.C.: Efficient methods for optimal scheduling of fixed head hydrothermal power systems. IEEE Trans. Power Syst. **3**(1), 24–30 (1988)
2. Basu, M.: Economic environmental dispatch of fixed head hydrothermal power systems using nondominated sorting GA-II. Appl. Soft Comput. **11**, 3046–3055 (2011)
3. Lin, J.G.: Multi-objective problems: pareto-optimal solutions by method of proper equality constraints. IEEE Trans. Autom. Control **21**(5), 641–650 (1976)
4. Yang, X.S.: Nature-Inspired Metaheuristic Algorithms, 1st edn. Lunver Press, Beckington (2008)
5. Yang, X.S., Hosseini, S.S.S., Gandomic, A.H.: Firefly algorithm for solving non-convex economic dispatch problems with valve loading effect. Appl. Soft Comput. **12**, 1180–1186 (2012)
6. Sahu, R.K., Panda, S., Padhan, S.: A hybrid firefly algorithm and pattern search technique for automatic generation control of multi area power systems. Int. J. Electr. Power Energy Syst. **64**, 9–23 (2015)
7. Naidu, K., Mokhlis, H., Bakar, A.H.A., Terzija, V., Illias, H.A.: Application of firefly algorithm with online wavelet filter in automatic generation control of an interconnected reheat thermal power system. Int. J. Electr. Power Energy Syst. **63**, 401–413 (2014)
8. Chiang, C.L.: Improved genetic algorithm for power economic dispatch of units with valve-point effects and multiple fuels. IEEE Trans. Power Syst. **20**(4), 1690–1699 (2005)

Chapter 3
Application of the Taguchi Method to the Warpage for Light Guide Plate

Po-Jen Cheng and Chin-Hsing Cheng

Abstract In recent years, the LED backlight module has become the main development of the thin film transistor liquid crystal display (TFT-LCD) module. The light guide plate warpage is the major source of ripple phenomenon found in the TFT-LCD module. In this chapter, different molding technological parameters are used for the experiment by the Taguchi method in the manufacturing process to find the factors of light guide plate warpage. The results of this manufacturing process are used to identify the optimization for production processes.

Keywords Thin film transistor liquid crystal display (TFT-LCD) • Taguchi method • Light guide plate warpage

3.1 Introduction

Light-emitting diodes (LEDs) have many favorable characteristics, including a small form factor, lower power consumption, instant power-on capacity, and an absence of mercury or any other form of toxic substance. As a result, they have attracted considerable attention from liquid crystal display (LCD) manufacturers as a potential light source for backlight units.

The light guide plate (LGP) is the prime component of LCD. Its function is to heighten the brightness of light and control light uniformity by guiding the beam. Molding components may reach a precise tolerance, resulting in improved mechanical and machine quality. Injection parameters are used to fit molding components

P.-J. Cheng
Department of Electrical Engineering, Nan Jeon University of Science and Technology, Tainan 73746, Taiwan

C.-H. Cheng (✉)
Department of Electrical Engineering, Feng Chia University, Taichung 40724, Taiwan
e-mail: chcheng@fcu.edu.tw

J. Juang (ed.), *Proceedings of the 3rd International Conference on Intelligent Technologies and Engineering Systems (ICITES2014)*, Lecture Notes in Electrical Engineering 345, DOI 10.1007/978-3-319-17314-6_3

on thicker LGPs. The LGP warpage is the major source of ripple phenomenon found in the thin film transistor liquid crystal display (TFT-LCD) module. Therefore, studying the factors that influence the molding precision of the LGP is crucial to producing high-quality and thin plates [1].

This chapter utilizes the Taguchi design method to optimize the design parameters of a LED backlight unit for LCD. The Taguchi method, a popular experimental design method in the industry, can overcome the shortcomings of full factorial design when implementing fractional factorial design. The latter approach optimizes parameter design, but with fewer experiments. Traditional experimental design is used to improve the mean level of a process such as the arithmetic mean of a sample. In modern quality engineering, experimental design work is performed to develop robust designs to improve the quality of the product. Taguchi's parameter design is intended to yield robust quality by reducing the effects of environmental conditions and variations due to the deterioration of certain components. This high quality is achieved by the selection of various design alternatives or by varying the levels of the design parameters for component parts or system elements. Taguchi's parameter design can optimize performance characteristics by setting design parameters and reducing the sensitivity of the system performance to sources of variation [2–4].

3.2 Light Guide Plate

Figure 3.1 presents a schematic illustration of the backlight unit. As shown in Fig. 3.1, the backlight unit comprises a color-mixing zone containing a reflector and two rows of LEDs and an extractor zone containing an arrangement of optical microstructures and a LGP. The light emitted by the LEDs is directed into the color-mixing zone by the reflector and then enters the extractor zone, where it is reorientated and redistributed in such a way as to achieve a uniform illumination of the overlying LCD panel. Different molding technological parameters are used for the experiment by the Taguchi method in the manufacturing process to determine the factors of LGP warpage.

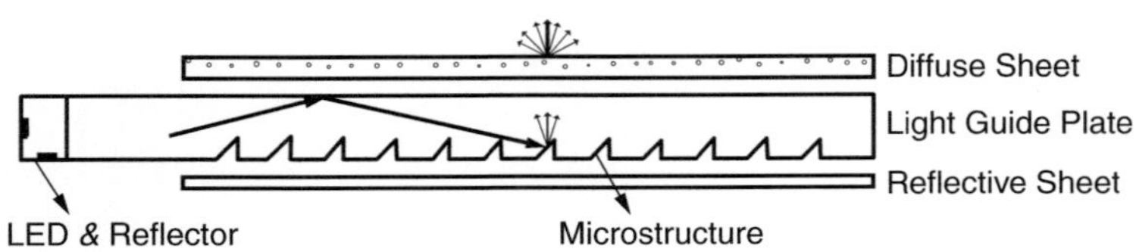

Fig. 3.1 Schematic illustration of backlight unit [2]

3.3 Experimental Methodology

Essentially, traditional experimental design procedures are too complicated and not easy to use. A large number of experimental works must be carried out when the number of process parameters increases. To solve this problem, the Taguchi method uses a special design of orthogonal arrays to study the entire parameter space with only a small number of experiments. Orthogonal arrays have been widely utilized in engineering analysis and consist of a plan of experiments with the objective of acquiring data in a controlled way, in order to obtain information about the behavior of a given process. The greatest advantages of this method are experimental time and cost savings, as well as quick determination of significant factors [5–7].

Based on the average output value at each parameter level, main effect analysis is performed. Furthermore, a statistical analysis of variance (ANOVA) is performed to see which process parameters are statistically significant. With the main effect and ANOVA analysis, the optimal combination of the process parameters can be predicted. Finally, a confirmation test is conducted to verify the optimal process parameters obtained from the parameter design [8].

3.4 Experimental Procedure

3.4.1 Establishment of Orthogonal Array

The injection-molding products are influenced by a number of process parameters. The Taguchi experimental of L_{27} orthogonal test was adopted to identify and optimize the process parameters on the various manufacturing processes. Those factors are temperature of core, temperature of cavity, injection speed, holding pressure, angle of LGP, drying time, cooling time, cooling pressure, and number of jigs. A product's warpage is regarded as the quality index. Three levels for every factor are shown in Table 3.1 [9].

3.4.2 Conducting of the Experiment

There are 27 experiments required to determine the optimum combination of the levels of these factors. In the Taguchi method, the term "signal" represents the desirable value for the output characteristic and the term "noise" represents the undesired value for the output characteristic. Therefore, the *S*/*N* ratio is the ratio of the desirable value to the undesired value. Taguchi used the *S*/*N* ratio to measure the quality characteristic deviating from the desired value. There are several *S*/*N* ratios available depending on the type of characteristic: smaller-the-better,

Table 3.1 The levels of effective factors for experimental variance

Factors/levels		1	2	3
Core temperature (°C)	A	75	85	95
Cavity temperature (°C)	B	73	83	93
Injection speed (IP/IV/mm)	C	50	150	250
Holding pressure (HP/s)	D	100	200	300
Angle of LGP (°)	E	60	90	180
Drying time (h)	F	2	4	6
Cooling time (h)	G	24	48	72
Cooling pressure (g)	H	1,000	2,000	3,000
Number of jig	I	2	4	6

Table 3.2 Response table of *S/N* ratios for all levels of all factors

Settings/ factors	A	B	C	D	E	F	G	H	I
1	−5.50	−4.56	−3.87	−1.98	−3.44	−4.69	−2.42	−5.18	−5.42
2	−2.96	−3.56	−3.21	−4.79	−3.52	−3.84	−4.70	−2.70	−3.71
3	−3.48	−3.83	−4.86	−5.17	−4.99	−3.42	−4.83	−4.07	−2.82

nominal-the-best, or larger-the-better. Smaller-the-better is used in our experiments. The smaller-the-better quality characteristic can be explained as follows:

$$S/N = -10 \cdot \log_{10}\left[\frac{1}{n}\sum_{i=1}^{n} y_i^2\right]. \tag{3.1}$$

S/N ratio values are calculated by taking into consideration (3.1). Five experiments are carried out on the same machine for reducing the variances between different machines. The responses obtained for all levels of factors are shown in Table 3.2.

3.4.3 Analysis of Variance

ANOVA is a computational technique used to estimate the relative significance of each process parameter in terms of percent contribution to the overall response. ANOVA is also required for estimating the variance of error for the effects and confidence interval of the prediction error. The ANOVA table contains the degrees of freedom, sum of squares, mean square, and percentage contribution. The parameters with higher percentage contribution are ranked higher in terms of

Table 3.3 Variance analysis table

Factors	Degree of freedom	Sum of squares	Mean square	*F* ratio	*P*-value	Percentage contribution (%)
A	2	1.189	0.5945	3.39	0.086	11.80
B	2	0.2315	0.1158	0.66	0.543	2.30
C	2	0.5145	0.2572	1.47	0.287	5.11
D	2	2.1381	1.0691	6.09	0.025	21.22
E	2	0.6255	0.3128	1.78	0.229	6.21
F	2	0.117	0.0585	0.33	0.726	1.16
G	2	1.343	0.6715	3.83	0.068	13.33
H	2	0.8818	0.4409	2.51	0.142	8.75
I	2	1.6294	0.8147	4.64	0.046	16.17
Error	8	1.4041	0.1755			0
Total	26	10.0739				100

importance in the experiment and also have significant effects in controlling the overall response [10, 11].

To conduct ANOVA, the sum of squares (SS) is calculated first. It is a measure of the deviation of the experimental data from the mean value of the data. The sum of squares due to various factor A (SSF_A) can be calculated as

$$SSF_A = \sum_{i=1}^{27} (m_{Ai} - m_{\text{all A}})^2 \tag{3.2}$$

SSF_B, SSF_C, SSF_D, SSF_E, SSF_F, SSF_G, SSF_H, and SSF_I can be obtained in the same way. The variance analysis results are shown in Table 3.3. Analysis of the test data reveals that the core temperature, holding pressure, cooling time, and number of jigs have great impact on the warpage level.

3.4.4 Design Optimization

It is noted in Table 3.2 that the best combination of design parameters is (A_2, B_2, C_2, D_1, E_1, F_3, G_1, H_2, I_3). Table 3.4 shows the various optimized factors, and the optimized results are shown in Table 3.5. The *S/N* ratio improvement from −5.35 to −5.22 is obtained from the optimized results.

Table 3.4 Various optimized factors

Factors		Levels	Conditions
Core temperature (°C)	A	2	85
Cavity temperature (°C)	B	2	83
Injection speed (IP/IV/mm)	C	2	150
Holding pressure (HP/s)	D	1	100
Angle of LGP (°)	E	1	60
Drying time (h)	F	3	6
Cooling time (h)	G	1	24
Cooling pressure (g)	H	2	2,000
Number of jig	I	3	6

Table 3.5 Optimized results

Experiments	Ex. 1	Ex. 2	Ex. 3	Ex. 4	Ex. 5	*S/N*
Optimized parameters	0.64	0.64	0.39	0.57	0.45	−5.22

3.5 Conclusions

By the way of orthogonal testing, this chapter studies different molding technology parameters (temperature of core and cavity, injection speed, holding pressure, angle of LGP, drying time, cooling time, cooling pressure, and number of jigs), which have different levels of impacts on LPG warpage. From the data of variance analysis results, the most significant factors are core temperature, holding pressure, cooling time, and number of jigs. The results of this manufacturing process can be used to identify the optimization for production processes. Thus, the ripple phenomenon found in the TFT-LCD module due to LPG warpage can be avoided.

Acknowledgments This work was supported by the Ministry of Science and Technology of Taiwan, Republic of China, under Grant number MOST 103-2221-E-035-039.

References

1. Yin, S., Wang, J., Ruan, F.: Impact on PC light guide plate warpage caused by injection parameters. In: 2010 International Conference on Digital Manufacturing & Automation, pp. 823–829 (2010)
2. Lin, C.F., Wu, C.C., Yang, P.H., Kuo, T.Y.: Application of Taguchi method in light-emitting diode backlight design for wide color gamut displays. J Disp. Technol. **5**, 323–330 (2009)
3. Lin, H.L., Chou, C.P.: Modeling and optimizing of Nd:YAG laser micro-weld process using Taguchi method and a neural network. Int. J. Adv. Manuf. Technol. **37**, 513–522 (2008)
4. Taguchi, G.: Taguchi Methods: Design of Experiments. American Supplier Institute, Inc., Dearhorn, MI (1993)

5. Bagci, E., Aykut, S.: A study of Taguchi optimization method for identifying optimum surface roughness in CNC face milling of cobalt-based alloy (satellite 6). Int. J. Adv. Manuf. Technol. **29**, 940–947 (2006)
6. Yang, W.H., Tang, Y.S.: Design optimization of cutting parameters for turning operations based on the Taguchi method. J. Mater. Proc. Technol. **84**, 122–129 (1998)
7. Taguchi, G.: Introduction to Quality Engineering. Asian Productivity Organization, Tokyo (1990)
8. Badkar, D.S., Pandey, K.S., Buvanashekaran, G.: Parameter optimization of laser transformation hardening by using Taguchi method and utility concept. Int. J. Adv. Manuf. Technol. **52**, 1067–1077 (2010)
9. Roy, R.K.: Design of Experiments Using the Taguchi Approach. Wiley, New York (2001)
10. Hunping, L., Guoqun, Z., Shanting, N., Yiguo, L.: Optimization of technology parameters for the plane-strain component in the process of gas quenching. Appl. Math. Model. **32**, 860–872 (2008)
11. Jeang, A.: Robust cutting parameters optimization for production time via computer experiment. Appl. Math. Model. **35**, 1354–1362 (2011)

Chapter 4
Diagnosis of Muscle Properties in Low-Back-Pain with Electrical Bioimpedance Methodology

Chin-Sung Hsiao, Yung-Tsung Yang, and Tai-Shin Ching

Abstract An approach to detect low-back-pain (LBP) is developed in this study. In this work, the electrical bioimpedance methodology (EBM) was proposed to measure the impedance and phase angle of the low-back muscle at various frequency bands for the healthy group and the patients suffering from LBP. After normalization of the measured impedances and phase angle, characteristic analyses of muscle property for LBP patients and the healthy using statistical software SPSS with paired sample *t*-test and independent sample *t*-test is feasible to evaluate the dissimilarity of the two groups. This proposed methodology enables to have the capability of offering a fast and accurate technique for some muscle-related diseases.

Keywords Low-back-pain • Electrical bioimpedance methodology • SPSS • Paired sample *t*-test • Muscle-related disease

4.1 Introduction

Low-back-pain (LBP) is a common ailment that degrades the quality of living of human being. Around 60–80 % adults in the world [1, 2] have been affected by this symptom. The prevalence of LBP has existed for decades. In Taiwan over 10 % of adults have suffered from the LBP and spent about three billion for medical treatment every year. Additionally, near 70 % of adults in the United States and United Kingdom have also been influenced by LBP [3, 4] and contacted professional medical doctors for appropriated treatment.

C.-S. Hsiao (✉) • Y.-T. Yang
Department of Photonics and Communications Engineering,
Asia University, Taichung 41354, Taiwan
e-mail: cshsiao@asia.edu.tw

T.-S. Ching
Department of Electrical Engineering, National Chi Nan University, Nantou, Taiwan, ROC

J. Juang (ed.), *Proceedings of the 3rd International Conference on Intelligent Technologies and Engineering Systems (ICITES2014)*, Lecture Notes in Electrical Engineering 345, DOI 10.1007/978-3-319-17314-6_4

Up to now, many methods to diagnose LBP have been proposed so as to reach an effective cure. These methods are computed tomography scan (CT) [5], discography, electromyography (EMG) [6], nerve conduction studies (NCS), magnetic resonance imaging (MRI) [7, 8], myelography, and X-ray. Computed tomography scan and X-ray are popular and proper schemes for detecting the bone problems while these two methods seem unsuitable for the diagnosis of the LBP problems. MRI is a good way enabling accurately diagnose all muscle-related problems but it is too expensive for most people to accept it. EMG is a medical technology that has the capability of detecting some nerve-based muscle problems. Myelography and discography are both invasive ways by which a radiopaque is injected to spinal subarachnoid for diagnosis of bone problems, while some side-effects such as nausea, vomiting, and headache might be induced. Besides, the devices for these measuring technologies mentioned above are costly and the operation of those devices must be dealt with by professional doctors or by the trained technicians of hospital. Based on these reasons mentioned above, a methodology for fast and accurate detecting LBP is proposed and developed.

The aim of this study is to provide an approach that has the potential for analyzing the dissimilarity of the muscle characteristics for the patients with LBP and the healthy from the bands of low frequency till high frequency. In addition, this present method could offer a promising technique for accurate assessment of the LBP problems in terms of the normalized impedance and phase angle of the muscle Z value as well as supply a low-cost, real-time, noninvasive manner of diagnosis.

4.2 Method and Testing Procedure

Electrical bioimpedance methodology (EBM) has widely been used for decades to study the human physical composition because of the appearance of medical instrument in market since mid-1980s. Method proposed for electrical bioimpedance measurement is generally equipped with the bipolar-electrode technique as shown in Fig. 4.1a. As seen from Fig. 4.1a, the two electrodes are applied as the input of the current signals, while the voltage signal is accessed at the cross terminal in which the electrode–electrolyte interface impedance is in series with the sample impedance and the parasitic impedances are large enough that can lead to disturb the measurements, especially at low frequencies range. To eliminate this inaccurate measurement problem, we here proposed a tetrapolar (four-electrode) method to enhance the performance of the measurement as shown in Fig. 4.1b.

Sixteen patients with LBP ailment with equal number of gender, age at 45 ± 7, the 16 healthy people has the same number in gender without LBP within 6 months, age at 48 ± 10, participated in this work. EBM was conducted by means of the measurement of the impedance and phase angle by which the LBP patients and the healthy lied in the probe bed. Figure 4.2 show a schematic diagram of electrode position attached for measuring the electrical bioimpedance of the patients with

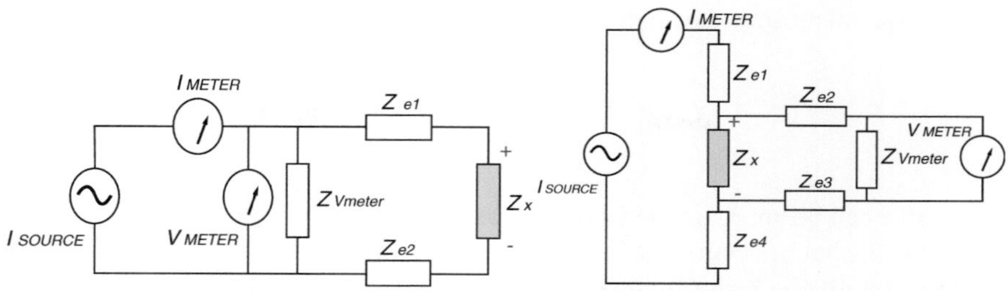

Fig. 4.1 (**a**) A bipolar-electrode technique, (**b**) tetrapolar technique for impedance measurement

Fig. 4.2 Schematic diagram of the position of electrodes. The upper two electrodes are placed 5 cm below the fifth lumbar vertebra and is 5 cm separated with each other, the other two electrodes are placed 5 cm below the upper two electrodes

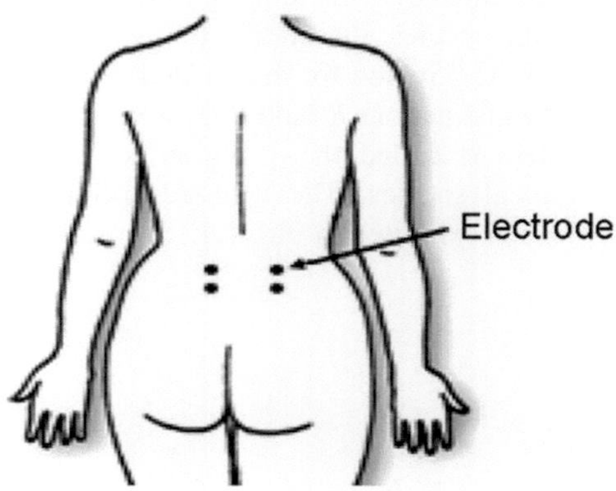

Fig. 4.3 The equivalent circuit of a single cell

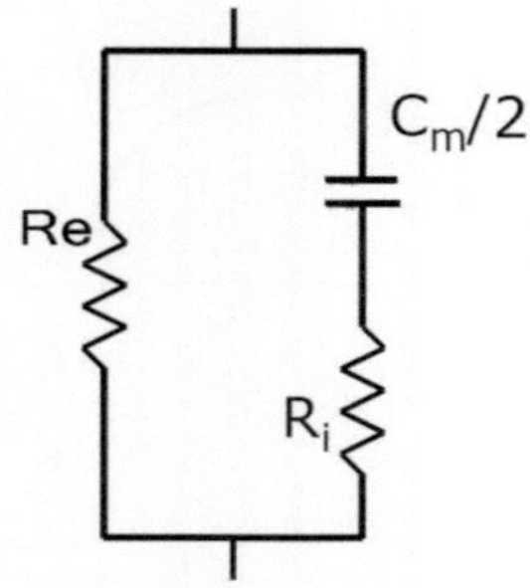

LBP and the healthy. As observed from the diagram, a pair of electrodes are placed 5 cm apart from the fifth lumbar vertebra and 5 cm apart between the right and the left sides. Prior to attaching an electrode to the proper positions of the low back, alcohol was used to clean skin for removing containment or oil thoroughly. An equivalent circuit for single cell of muscle is shown in Fig. 4.3, in which R_e and R_i are defined as the extracellular resistance and the intracellular resistance, respectively, and C_m stands the cell membrane capacitor.

The equivalent impedance of the single cell can be written as

$$Z = \frac{1}{\left(\frac{1}{R_e}\right) + \frac{1}{\left(\frac{1}{2\pi f C_m}\right) + R_i}} \tag{4.1}$$

the z value can be approximated as $z = R_e$ at low frequency with the cell membrane capacitor C_m being deemed as open circuit and approximated as $z = R_e/R_i$ at high frequency with C_m being thought as short circuit based on the electrical theory. Generally, the measurement of dielectric property for biological materials is progressed with a triple-frequency method in which the named α region is located at low frequency band ranging from 10 Hz to 10 kHz for the diffusing detection of the ionic species, the middle frequency band ranging from 10 kHz to 10 MHz is termed the β region for the dielectric property measurement of the cell membrane and the interactions R_e and R_i as well, while the highest frequency band higher than 10 MHz is called the γ region for the aqueous content measurement of the biological species. The detailed spectrum distribution for the triple-frequency regions is exhibited in Fig. 4.4.

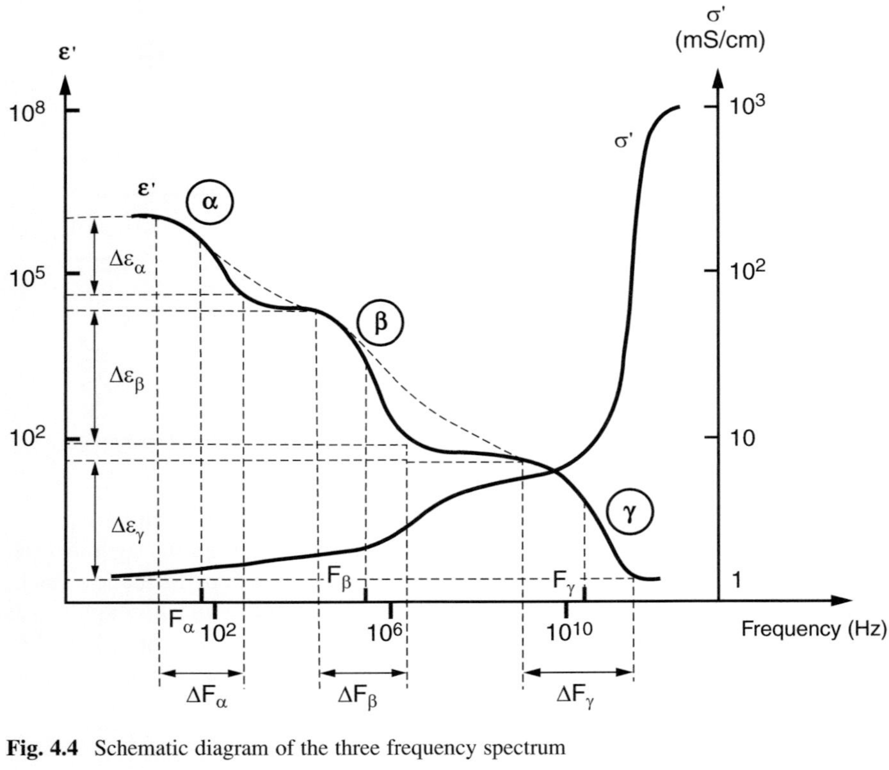

Fig. 4.4 Schematic diagram of the three frequency spectrum

EBM is performed by placing two-electrode probe with the amplitude of voltage uplimited 200 mv and connecting the electrodes to the LBP patients and the healthy for 2 min. Seven frequency bands range at 0.02, 25.02, 50.02, 200.02, 800.02, 3,200, and 10,000 kHz is randomly adopted for the whole measurement. The measured current value is read out from the current meter and the impedance value Z ($Z = V/I$) could be reached by dividing the current value from the settled voltage V based on the Ohm's law. All the impedances acquired at each frequency mentioned above were then normalized for statistical analyses.

4.3 Normalization and Statistical Analysis

It is essential to normalize the impedance values obtained from the EBM measure for the healthy and the LBP patients in statistical analysis. In the present work, the normalized impedance (phase angle) is defined as the value obtained with the muscle impedance (phase angle) divided by the spine impedance (phase angle) at the test frequency.

In order to verify the performance of the proposed EBM is feasible, statistical analyses using SPSS with margin $p = 0.05$ is carried out to evaluate the normalized impedance (phase angle) that were expressed as the mean values and the standard deviation. The paired-samples *t*-test (within group) and the independent sample *t*-test (between groups) were adopted to assess the dissimilarity of normalized impedance (phase angle) between the LBP patients and the healthy.

4.4 Results

Thirty-two subjects including the healthy and the LBP patients with the same number of gender participated in this measurement. Impedance analyzer 6420c from Wayne Co. was used to measure all participators; 7 frequency bands of 0.02, 25.02, 50.02, 200.02, 800.02, 3,200, and 10,000 kHz were randomly chosen for characteristic analysis of normalized impedance and phase angle. Figure 4.5 shows the normalized impedance spectrum of the LBP patients and the healthy

It can be seen from Fig. 4.5 that the normalized impedances Z both for the LBP patients and the healthy keep increasing along with the testing frequency until at middle frequency $f = 800$ kHz. The normalized impedance Z of the LBP patients is higher ($p < 0.05$) than the healthy at low frequency $f = 20$ Hz. On the contrary, this situation is changed for testing frequency above 20 Hz in which the LBP patients are with smaller ($p < 0.05$) values with respect to the healthy. Table 4.1 details the independent-samples *t*-test for the impedance normalized for the two groups.

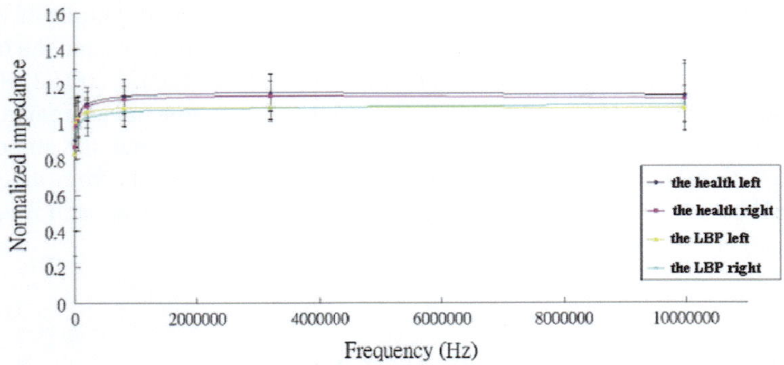

Fig. 4.5 The normalized impedance spectrum of the paired sample *t*-test for the LBP patient and the healthy

Normalization of phase angle is also carried out for accurate assessment of the proposed EBM. The normalized phase angle spectrum for the LBP patients and the healthy is detailed in Fig. 4.6. It can be seen from this figure that the normalized phase angle of the LBP patients keeps increasing along with the testing frequency until $f = 3{,}200$ kHz, while this situation is on the opposite side for the healthy. The normalized phase angle of the LBP patients is higher than the healthy from the frequency $f = 800$–$10{,}000$ kHz that is well matched the test ($p > 0.05$) of SPSS as shown in Fig. 4.7. Table 4.2 details the independent-samples *t*-test for the impedance normalized for the two groups.

4.5 Conclusions

In conclusion, the proposed EBM for the characterization of the muscle electrical properties in LBP has been confirmed by SPSS analyses in terms of the paired sample *t*-test and the independent sample *t*-test with the margin value of $p < 0.05$. Additionally, this present methodology has the potentiality of providing low-cost, reliable, noninvasive and real-time diagnostics of low-back pain.

Acknowledgment This work was supported by grants 102-Asia-46 from Asia University.

Table 4.1 Details of the independent-samples *t*-test for the impedance normalized for the two groups

		Independent samples test								
		Leven's test for equality of variances		*t*-Test for equality of means						
									95 % confidence interval of the difference	
		F	Sig.	*t*	df	Sig. (2-tailed)	Mean difference	Std. error difference	Lower	Upper
20 Hz	Equal variances assumed	0.214	0.646	0.102	34	0.919	0.01554	0.15204	−0.29345	0.32452
	Equal variances not assumed			0.102	33.780	0.919	0.01554	0.15204	−0.29352	0.32460
25 KHz	Equal variances assumed	1.843	0.184	0.168	34	0.867	0.00738	0.04392	−0.08187	0.09664
	Equal variances not assumed			0.168	32.903	0.867	0.00738	0.04392	−0.08198	0.09674
50 KHz	Equal variances assumed	2.133	0.153	0.600	34	0.553	0.02232	0.03722	−0.05332	0.09796
	Equal variances not assumed			0.600	32.706	0.553	0.02232	0.03722	−0.05343	0.09807
200 KHz	Equal variances assumed	1.222	0.277	1.803	34	0.080	0.05157	0.02860	−0.00654	0.10968
	Equal variances not assumed			1.803	33.767	0.080	0.05157	0.02860	−0.00656	0.10970
800 KHz	Equal variances assumed	0.019	0.893	2.518	34	0.017	0.06597	0.02620	0.01273	0.11921
	Equal variances not assumed			2.518	33.231	0.017	0.06597	0.02620	0.01269	0.11926
3.2 MHz	Equal variances assumed	0.785	0.382	2.744	34	0.010	0.07249	0.02642	0.01879	0.12619
	Equal variances not assumed			2.744	31.486	0.010	0.07249	0.02642	0.01864	0.12635
10 MHz	Equal variances assumed	5.717	0.022	1.113	34	0.273	0.05328	0.04786	−0.04399	0.15054
	Equal variances not assumed			1.113	24.320	0.277	0.05328	0.04786	−0.04543	0.15199

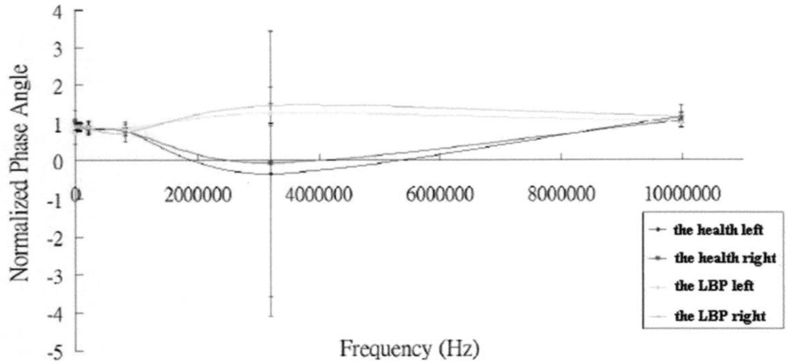

Fig. 4.6 The normalized phase angle spectrum of the paired sample *t*-test for the LBP patients and the healthy

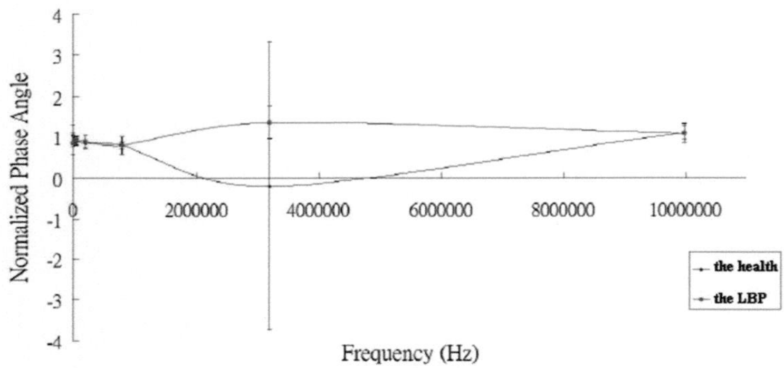

Fig. 4.7 The phase angle spectrum of the independent sample *t*-test for the LBP patients and the healthy

Table 4.2 Details of the independent-samples *t*-test for the impedance normalized for the two groups

		Independent samples test								
		Leven's test for equality of variances		*t*-Test for equality of means						
									95 % confidence interval of the difference	
		F	Sig.	*t*	df	Sig. (2-tailed)	Mean difference	Std. error difference	Lower	Upper
20 Hz	Equal variances assumed	0.175	0.678	2.201	34	0.035	0.19877	0.09030	0.01526	0.38228
	Equal variances not assumed			2.201	33.896	0.035	0.19877	0.09030	0.01524	0.38230
25 KHz	Equal variances assumed	5.040	0.031	−0.369	34	0.714	−0.00911	0.02467	−0.05924	0.04102
	Equal variances not assumed			−0.369	30.175	0.715	−0.00911	0.02467	−0.05947	0.04125
50 KHz	Equal variances assumed	3.873	0.057	−0.840	34	0.407	−0.02590	0.03082	−0.08854	0.03674
	Equal variances not assumed			−0.840	29.694	0.407	−0.02590	0.03082	−0.08887	0.03707
200 KHz	Equal variances assumed	8.010	0.008	−1.077	34	0.289	−0.04606	0.04278	−0.13299	0.04087
	Equal variances not assumed			−1.077	26.267	0.291	−0.04606	0.04278	−0.13394	0.04183
800 KHz	Equal variances assumed	12.296	0.001	−0.213	34	0.833	−0.01165	0.05469	−0.12279	0.09949
	Equal variances not assumed			−0.213	22.361	0.833	−0.01165	0.05469	−0.12496	0.10166
3.2 MHz	Equal variances assumed	13.621	0.001	−1.864	34	0.071	−1.55871	0.83626	−3.25819	0.14078
	Equal variances not assumed			−1.864	17.446	0.079	−1.55871	0.83626	−3.31963	0.20222
10 MHz	Equal variances assumed	1.041	0.315	0.126	34	0.900	0.00863	0.06834	−0.13025	0.14751
	Equal variances not assumed			0.126	29.990	0.900	0.00863	0.06834	−0.13094	0.14820

References

1. Heliovaara, M., Makela, M., Knekt, P., Impivaara, O., Aroma, A.: A determination of sciatica and low back pain. Spine **16**, 608–614 (1991)
2. Nurminen, M.: Reanalysis of the occurrence of back pain among construction works: modeling for the interdependent effects of heavy physical work, earlier back accident and aging. Occup. Environ. Med. **54**, 807–811 (1997)
3. Frymoyer, J.W., Cats-Baril, W.L.: An overview of the incidences and costs of low back pain. Orthop. Clin. North Am. **22**, 263–271 (1991)
4. Walsh, K., Cruddas, M., Coggon, D.: Low back pain in eight areas of Britain. J. Epidemiol. Community Health **46**, 227–230 (1992)
5. Ohashi, K., El-Khoury, G.Y., Musculoskeletal, C.T.: Recent advances and current clinical applications. Radiol. Clin. North Am. **47**, 387–409 (2009)
6. Pandey, B., Mishra, R.B.: An integrated intelligent computing model for the interpretation of EMG based neuromuscular diseases. Expert Syst. Appl. **36**, 9201–9213 (2009)
7. Fayad, F., Colau, M.M.L., Drapé, J.L., Feydy, A., Chemla, N., Quintéro, N., Rannou, F., Fermanian, S.P.J., Revel, M.: Reliability of a modified Modic classification of bone marrow changes in lumbar spine MRI. Joint Bone Spine **76**, 286–289 (2009)
8. Kamath, S., Jain, N., Goyal, N., Mansour, R., Mukherjee, K.: Incidental findings on MRI of the spine. Clin. Radiol. **64**, 353–361 (2009)

Chapter 5
Low-Cost Wearable Control Valves with No Mechanical Sliding Parts in Valves

Tetsuya Akagi, Shujiro Dohta, Ayumu Ono, and Abdul Nasir

Abstract The wearable driving system is required the pneumatic soft actuators, valves, and a controller. The complex inner configuration of the valve requiring relatively high precision prevents to fabricate a low-cost driving system. The cost for valve occupies most of the total cost. In this chapter, two types of low-cost control valves with no mechanical sliding parts in valve are described. One is an on/off type valve using vibration motor, another is a servo valve using buckled tube. The operating principle and performance of both valves are also introduced.

Keywords Valve using vibration motor • Servo valve using buckled tube

5.1 Introduction

Recently, wearable driving systems using pneumatic soft actuators for power assisted nursing care and rehabilitation have received much attention and many active researches have been carried out [1–3]. In such a system, an actuator and a control valve are mounted on the human body [4–7]. In order to support the multidegrees of freedom of human motion, many wearable pneumatic actuators and valves need to be used. Since the wearable actuator can be made of lightweight and low-cost elastic materials, the size and mass of the valve is more affected in the total volume of the wearable control system. Especially, the typical electromagnetic solenoid valve drives its spool using a solenoid to open the flow passage as shown in Fig. 5.1. The solenoid valves also have a sliding mechanism requiring relatively high precision to keep a seal while the spool is moving. This complex inner configuration prevents to fabricate a low-cost valve. Since a cheap microcomputer has been used as a controller, the cost of the valve occupies most of the total cost of wearable pneumatic driving system.

The purpose of our study is to develop a low-cost, small-sized, and lightweight control valve that can construct inexpensive wearable pneumatic driving system.

T. Akagi (✉) • S. Dohta • A. Ono • A. Nasir
Department of Intelligent Mechanical Engineering, Okayama University of Science, 1-1, Ridai-cho, Kita-ku, Okayama 700-0005, Japan
e-mail: akagi@are.ous.ac.jp

J. Juang (ed.), *Proceedings of the 3rd International Conference on Intelligent Technologies and Engineering Systems (ICITES2014)*, Lecture Notes in Electrical Engineering 345, DOI 10.1007/978-3-319-17314-6_5

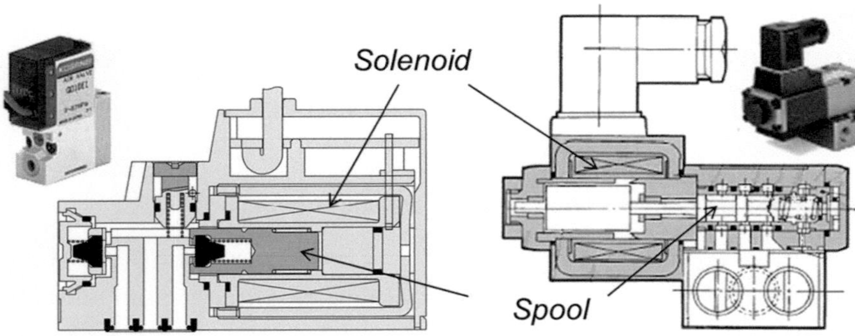

Fig. 5.1 Inner construction of the typical electromagnetic on/off valve and servo valve

We think that the simplification of the inner construction of the valve is one of solution for decreasing the cost. To simplify the inner construction, we proposed several valves with no mechanical sliding parts [8–10]. In this chapter, two low-cost control valves proposed in our previous study are introduced. One is an on/off type control valve using the vibration motor [8]. Another is a servo type control valve using buckled tube [9, 10]. The construction and operating principle of both valves will be also described. The performance of both valves is also introduced.

5.2 On/Off Valve Using Vibration Motor

Figure 5.2a, b shows the operational image of a typical electromagnetic on/off valve and the proposed valve, respectively. In both valves, the steel ball is always applied by the upper force according to the differential pressure between the inlet and the outlet of the orifice as shown in Fig. 5.2. To open the typical valve, it needs a larger longitudinal pulling force to overcome the pushing force generated by the supplied pressure. Therefore, the typical valve needs a larger solenoid. In the proposed valve as shown in Fig. 5.2b, the horizontal force to the steel ball is applied to move it by using a smaller force.

Figure 5.3a, b shows the photograph and schematic diagram of the tested on/off type valve, respectively. The valve consists of a flexible tube with the inner diameter of 2.5 mm and the outer diameter of 4 mm, an acrylic orifice with inner diameter of 0.5 mm, a steel ball with outer diameter of 2 mm and a vibration motor (Shicoh Co. Ltd. SE-4C-1E). The acrylic orifice and the steel ball are inserted into the flexible tube. The vibration motor is set on the outer side of the flexible tube by an acrylic connector. There are no mechanical sliding parts in the tube. The volume of the tested valve including the vibration motor is about 1 cm^3, that is the valve has a length of 20 mm, a width of 5 mm, and a height of 10 mm. The mass of the tested valve is very small, only 2 g. The valve can be fabricated at a lower cost because of

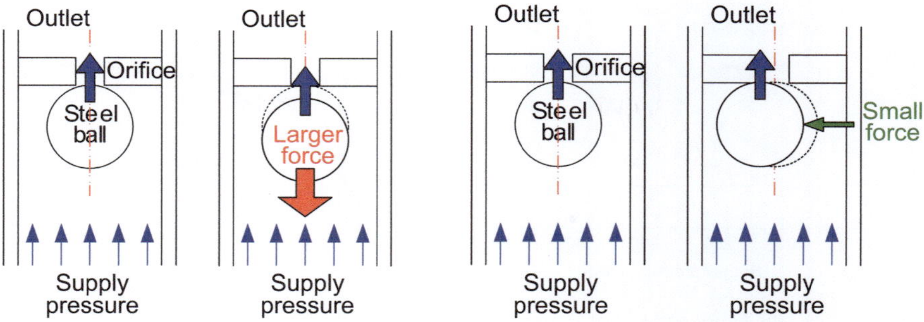

Fig. 5.2 Fundamental concept of the tested valve: (**a**) *left figure* shows the case of the typical electromagnetic valve. (**b**) *Right figure* shows the case of the proposed valve

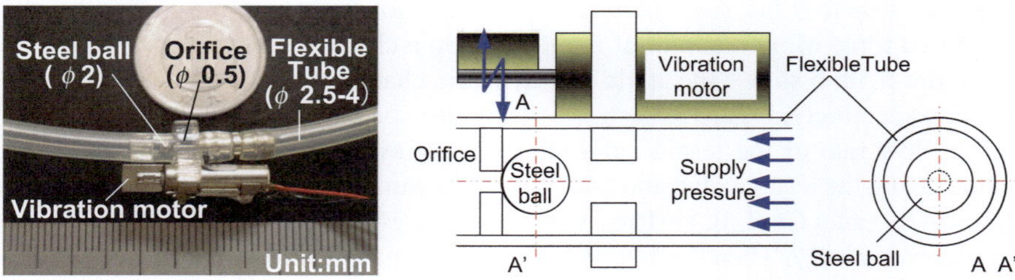

Fig. 5.3 The on/off valve using vibration motor. (**a**) *Left photograph* shows the view of the tested valve. (**b**) *Right schematic diagram* shows the inner construction of the valve

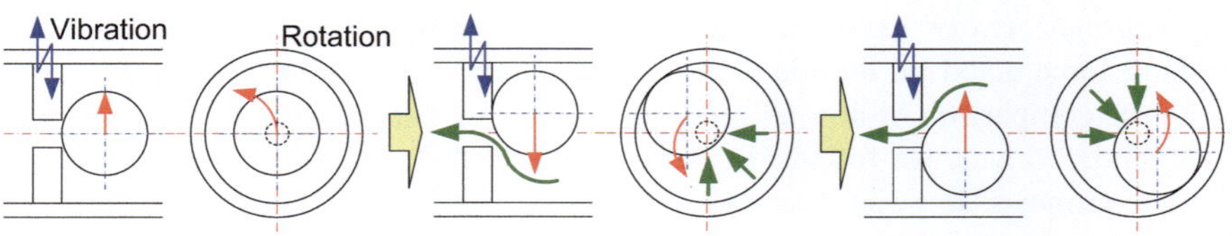

Fig. 5.4 Operating principle of the on/off valve using vibration motor

simple construction inside of the valve. The estimated cost of the valve is about 2 US dollars.

Figure 5.4 shows the schematic diagram in the operation of the valve. The operating principle of the valve is as follows. When the vibration motor is driven, the tube connected to the vibration motor is oscillated. From the law of inertia, the inner steel ball stays in a certain position of the tube. The inner steel ball contacts the inner wall of the tube. The inner ball starts to move and rotate along to the inner wall of the tube. By applying continuous vibration, the ball rotates around the inner wall of the tube. It should be noted that the opening of the orifice keeps a constant sectional area while the vibration motor is being driven. It means that the tested valve generates a stable flow rate while the valve is working. When the vibration

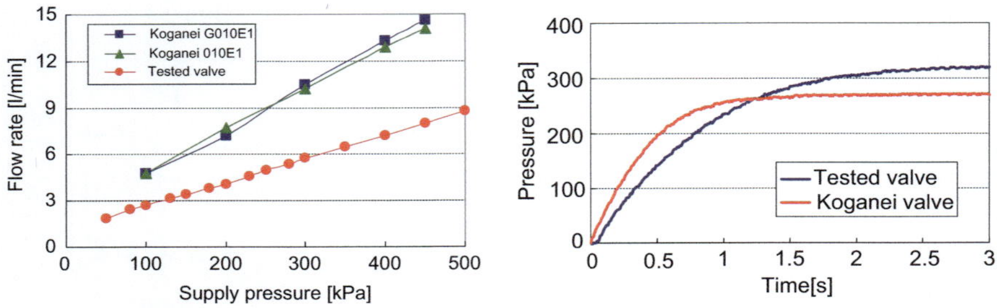

Fig. 5.5 (**a**) *Left figure* shows the statics of the valve, that is, the relation between the supply pressure and the output flow rate of the valve. (**b**) *Right figure* shows the dynamics of the valve, that is, the transient response of the output pressure from the valves when the stepwise input while the valves connected to the McKibben actuator with an initial volume of about 47 cm^3

motor is stopped, the steel ball automatically moves toward the orifice by the generated force of momentum of the flow, as in a check valve.

Figure 5.5a, b shows the static and dynamic characteristics of the tested on/off valve, respectively. Figure 5.5a shows the relation between the supply pressure and output flow rate of the tested valve. In Fig. 5.5, symbols ●, ▲, and ■ show the results using the tested valve and two types of pneumatic on/off control valve on the market (Koganei Co. Ltd. G010E1 with a mass of 16.8 g and 010E1 with a mass of 20 g), respectively. Then the size of both commercial valves is about 9 cm^3. These commercial pneumatic control valves are relatively small-sized valves available in the market. The supply pressure of these valves was changed from 50 to 500 kPa. From Fig. 5.5a, it can be seen that the output flow rate of the tested valve has a linear relationship between the supply pressure and the output flow rate. It can be seen that the sectional area of the orifice in the tested valve does not change even if the pushing force acted on the inner ball increases according to the supply pressure. From the comparison of normalized output flow rate between the tested valve and commercial valves, we found that the normalized flow rate of the tested valve is about ten times as large as those using the commercial valves. The normalized flow rate by the mass of the tested valve is about six times larger than the commercial valves.

Figure 5.5b shows the transient response of the pressure in the McKibben actuator with the initial volume of about 47 cm^3 using the tested valve and the commercially available on/off valve (Koganei Co. Ltd. G010E1) when we apply a stepwise operating voltage change. In Fig. 5.5b, the blue line shows the result using the tested valve. The red line shows the result in the case using the commercial valve. From Fig. 5.5b, it can be seen that the pressure response using the tested valve is a little slower than in the case using the commercial valve because of the difference of the maximum output flow rate. However, it can be seen that the response of the tested valve is stable even if the valve is operated by vibration. We find that it needs about 0.04 s as a corresponding time of the time delay. This time delay is time until the steel ball starts to rotate. This value is not so large compared with the commercially available valve, that is from 0.01 to 0.03 s.

5.3 Servo Valve Using Buckled Tube

Figure 5.6a, b shows the construction and the operating principle of a servo valve using buckled tubes, respectively [9, 10]. The valve consists of two buckled soft polyurethane tubes (SMC Co. Ltd. TUS0425: inner diameter of 2.5 mm, outer diameter of 4.0 mm), a small-sized RC servo motor (GWS Co. Ltd. PICO/STD/F rotational speed: 500 °/s), an acrylic valve holder, and an acrylic tube holder disk. Two buckled tubes are used for supply and exhaust. In the initial condition, the buckling tubes are held by the acrylic rotary disk so that each of buckled angle is 63° in clockwise and counterclockwise from the longitudinal axis of fixed tubes, respectively. Then, each end of the supply and exhaust port are fixed at the tube holder by passing through the smaller hole set on the position with the radius of 20 mm from the motor shaft. The tubes are possible to slide along the suppressed plates on the rotary disk. This arrangement of both tubes that includes an initial buckling angle of 63° and buckling point with radius of about 4 mm from the motor shaft could be obtained from the experiment. The alternative arrangement of two buckled tubes helps to decrease the reaction torque for the motor. It means that the reaction torques from two tubes are balanced in the initial condition.

The operating principle of the valve is as follows. When the servo motor rotates clockwise, the buckled angle of exhaust tube is decreased and at the same time the buckled angle of supply tube is increased. Then, it causes the increasing of the sectional area in the supply tube by releasing the bending force at the buckling point, while the bending force acted on the buckling point of the exhaust tube is increasing. By increasing the buckled force, the exhaust tube closed surely. The mass of the valve is 22 g. The mass and the size of this valve are suitable for a wearable device. The mass of this proposed valve is smaller compared with the typical small-sized servo valve such as FESTO MPYE-5-1/8-HF-010 B that has a mass of 300 g. The estimated cost of the valve is low, that is about 9 US dollars.

Figure 5.7a shows a relation between the incremental rotational angle of the RC servo motor and the output flow rate of the tested valve. In Fig. 5.7a, the solid line shows the result in the case of increasing the rotational angle from −30° to 30°. The broken line shows the result in the case of decreasing in the rotational angle from 30° to −30°. From Fig. 5.7a, it can be also seen that this relation is almost linear

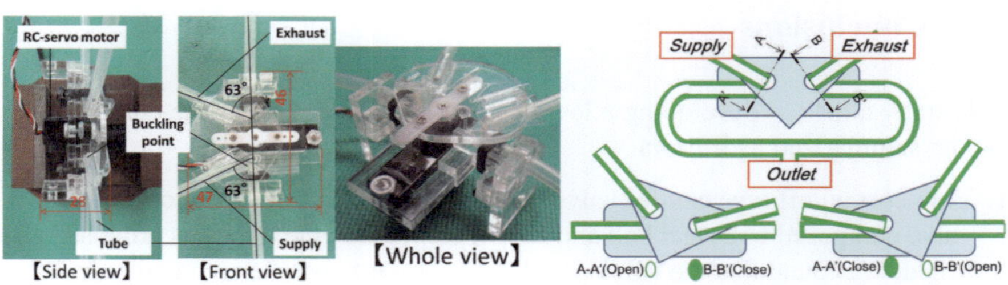

Fig. 5.6 Servo valve using buckled tube. (**a**) *Left figure* shows the view of the valve. (**b**) *Right figure* shows the operating principle of the valve

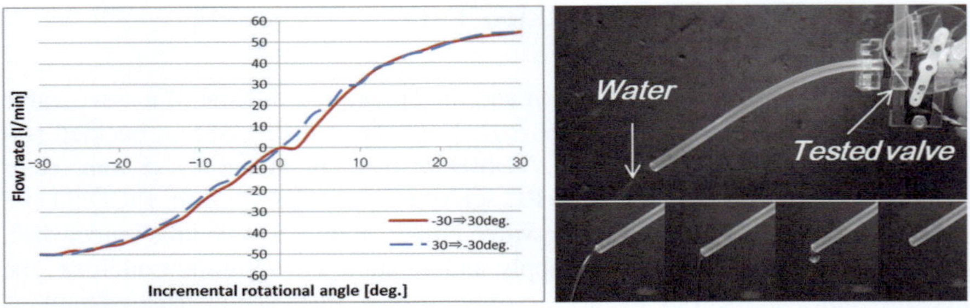

Fig. 5.7 Characteristics of the servo valve using buckled tube. (**a**) *Left figure* shows the relation between the rotational angle of servo motor and output flow rate of the valve. (**b**) *Right figure* shows transient view of flow rate control of tap water using the valve

with little hysteresis. We can also see that the valve has an overlap between $-1°$ and $2°$ that can hold the output pressure. This function is useful to control the pneumatic actuator from the view point of lower energy consumption. The tested servo valve has some advantages that are small-sized, light-weight, and low-cost. The valve also has no mechanical sliding parts in adjusting mechanism of sectional area where the working fluid contacts. In addition, the working fluid in the valve is quite isolated from dirty elements such as a lubricating oil, metal chips, and static electricity. It means that the tested valve can be used as a flow rate control valve of a medicine, flammable liquid, and gas. From the fundamental operational principle of the valve, it is possible to control the flow rate even if either liquid or gas is used as a working fluid. In order to confirm the advantage, the flow control of the valve using tap water was carried out. Figure 5.7b shows the transient view of the flow control using tap water. In the experiment, the output flow rate of tap water is changed from the maximum to zero and from zero to maximum. From Fig. 5.7b, it can be seen that the tested valve can control the liquid as a same manner of operating the gas. We confirm that the proposed valve has an ability to apply various fields such as a medical treatment, food production, and water hydraulic system while keeping the production costs low.

5.4 Conclusions

This study aimed at developing a low-cost and small-sized wearable control valve can be summarized as follows.

1. Low-cost on/off type control valve that consists of a vibration motor and a check valve composed of a steel ball and an orifice in a flexible tube was proposed and tested. We found that the valve could hold constant opening area while the inner steel ball rotated around the inner wall of the tube by giving vibration. We

confirm that the estimated cost of the valve is extremely low, that is only 2 US dollars.

2. The low-cost wearable servo valve that consists of two buckled tubes driven by the inexpensive RC servo motor was proposed and tested. The output flow rate of the tested valve for various rotational angle of the motor was investigated. As a result, it was found that the relation between the rotational angle and the output flow rate of the valve is almost linear with less hysteresis. The low-cost servo valve that the estimated cost is about 9 US dollars was realized.
3. From the flow rate control of tap water, we confirmed the tested servo valve has many abilities to apply various fields such as a medical treatment, food production, and water hydraulic system while keeping the production cost and mass extremely low compared with the commercially available water hydraulic valve that is made of stainless steel.

Acknowledgments Finally, we express thanks that this work was supported in part by the Ministry of Education, Culture, Sports, Science and Technology of Japan through a QOL Innovative Research Program (2012-) and Grant-in-Aid for Scientific Research (C) (Subject No. 24560315).

References

1. Yamamoto, K., et.al.: Development of wearable power assisting suit. In: Proceedings of the FLUCOME 2003, 113pdf, pp. 1–6 (2003)
2. Kobayashi, H., et al.: Realization of all 7 motions for the upper limb by a muscle suit. J. Robot. Mechatron. **16**(5), 504–512 (2004)
3. Noritsugu, T., et al.: Development of power assist wear using pneumatic rubber artificial muscles. J. Robot. Mechatron. **21**(5), 607–613 (2009)
4. Nagata, Y. (ed.): Soft Actuators—Forefront of Development, NTS Ltd. (2004) (in Japanese)
5. Akagi, T., et al.: Development of wearable pneumatic actuator and multiport pressure control valve. J. Robot. Mechatron. **17**(5), 529–536 (2005)
6. Uehara, S., et al.: Unconstrained vibrational pneumatic valves for miniaturized proportional control devices. In: Proceedings of 9th International Conference Mechatronics Technology (2005)
7. Nishioka, Y, et al.: A new pneumatic control system using multiplex pneumatic transmission. In: Proceedings of the 7th JFPS International Symposium on Fluid Power 2008, pp. 439–442 (2008)
8. Akagi, T., et al.: Development of small-sized flexible pneumatic valve using vibration motor and its application for wearable actuator. J. Comput. Appl. Technol. **39**(1/2/3), 86–92 (2010)
9. Nasir, A., et al.: Development of low-cost wearable servo valve using buckled tube. Appl. Mech. Mater. **393**, 532–537 (2013)
10. Nasir, A., et al.: Development of small-sized servo valve controlled by using buckled tube and its application. J. JSME **7**(4), 516–527 (2013)

Chapter 6
Estimation of Residual Traveling Distance for Power Wheelchair Using Neural Network

Pei-Chung Chen, Xiao-Qin Li, and Yong-Fa Koh

Abstract The residual traveling distance of a power wheelchair is difficult to estimate due to the unknown factors of user manipulation behavior and journey characteristics of wheelchair. A virtual residual energy estimation system for power wheelchair based on neural network is proposed to estimate virtual residual energy which could be transformed into residual traveling distance. Two types of estimation systems with three training processes are presented. The estimated results are provided and compared. The results indicate that type-A estimation system with adaptive learning rate is a feasible solution based on economic factor and estimated performance.

Keywords Residual traveling distance • Residual energy • Power wheelchair

6.1 Introduction

Power wheelchair is one of the most commonly used mobility-assisted devices for elderly or disabled people. However, the residual traveling distance of power wheelchair is one of concerning topics for the users. The residual traveling distance of power wheelchair will be affected by the residual capacity of battery, user weight, user's manipulation behavior, traveling speed and journey characteristics of power wheelchair. Due to the above unknown factors, user cannot exactly know the residual traveling distance of power wheelchair from the residual capacity of battery.

Neural network, having the better fault tolerance and self learning capability, is often applied to solve nonlinear problems or achieve approximate system model without reprogramming or other interference in the program itself [1, 2]. Therefore,

P.-C. Chen (✉) • Y.-F. Koh
Department of Mechanical Engineering, Southern Taiwan University of Science and Technology, Tainan 71005, Taiwan
e-mail: chenpc@mail.stust.edu

X.-Q. Li
Department of Electronic & Information Engineering, Ningbo Polytechnic, Ningbo, ZheJiang 315800, China

J. Juang (ed.), *Proceedings of the 3rd International Conference on Intelligent Technologies and Engineering Systems (ICITES2014)*, Lecture Notes in Electrical Engineering 345, DOI 10.1007/978-3-319-17314-6_6

it is widely applied to estimate the battery state of charge (SOC), such as the estimation of lithium battery SOC using adaptive neural network [3], the estimation of battery SOC of electric scooters by fuzzy neural network [4]. Fewer researches discuss the residual traveling distance estimation of electric scooter [5] or electric wheelchair [6].

In this chapter, a neural-network-based virtual residual energy estimation system for power wheelchair is proposed. The estimation system has four inputs and one output called virtual residual energy. The physical residual traveling distance of wheelchair could be achieved by transforming the virtual residual energy into residual traveling distance. The details are discussed in the following sections.

6.2 Residual Traveling Distance Estimation System

Due to the factors of user manipulation behavior and wheelchair journey, characteristics are not measurable, neural network approach is introduced to estimate the virtual residual energy of power wheelchair and then transforms it into the physical residual traveling distance of power wheelchair. Figure 6.1 shows the proposed neural network with p inputs, q outputs, and two hidden layers.

The output of the jth neuron of the nth layer could be expressed as

$$y_j^n = f\left(\mathrm{net}_j^n\right) \tag{6.1}$$

The jth neuron's input of the nth layer could be expressed as

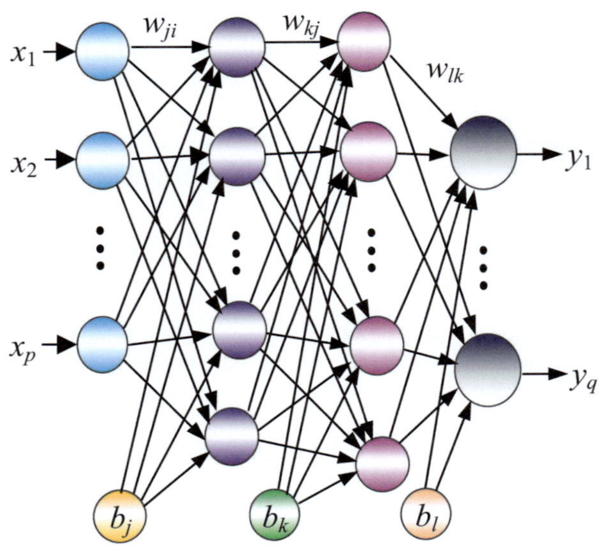

Fig. 6.1 The proposed neural network has p inputs, q outputs, and two hidden layers

$$\mathrm{net}_j^n = \sum_i w_{ji}^n y_i^{n-1} + b_j^n \tag{6.2}$$

where y_j^n and b_j^n denote the output and bias of the jth neuron of the nth layer, respectively. $f(\cdot)$ is the activation function, w_{ji}^n is a weight between the jth neuron of the nth layer and the ith neuron of the $(n-1)$th layer.

The activation function is defined as follows.

$$f(x) = \begin{cases} 0 & x \le 0 \\ x & 0 < x \le 1 \\ 1 & x > 1 \end{cases} \tag{6.3}$$

For economic consideration, virtual frictional force and virtual residual energy are defined, and a virtual residual energy estimation system for power wheelchair based on neural network is proposed to estimate the virtual residual energy. The proposed estimation system has four inputs called lithium battery SOC, normalized output current IL_{n}, normalized wheelchair traveling speed $\mathrm{Spd_n}$, and virtual frictional force VFr, and one output called virtual residual energy y_{v}. Because the orders of input variables are not same, in order to increase the learning performance and estimation accuracy, the physical datum of power wheelchair are normalized first. The normalization process is defined as follows.

$$IL_{\mathrm{n}} = \frac{I_{\mathrm{L}}}{I_{\mathrm{Lmax}}},\quad IV_{\mathrm{n}} = \frac{I_{\mathrm{L}}V_{\mathrm{B}}}{IV_{\mathrm{max}}},\quad S\,\mathrm{pd_n} = \frac{\mathrm{Speed}}{\mathrm{Speed_{max}}},\quad \mathrm{Rtd_n} = \frac{\mathrm{RTD}}{\mathrm{RTD_{max}}}$$

where I_{L} and V_{B} are output current and voltage of lithium battery, respectively. Speed and RTD are traveling speed and residual traveling distance of wheelchair, respectively. $\mathrm{Rtd_n}$ is the normalized traveling speed of wheelchair. The parameters: IV_{max} is 280 W, $\mathrm{Speed_{max}}$ is 10 km/h, I_{Lmax} is 10 A, and $\mathrm{RTD_{max}}$ is 30 km, respectively. Two virtual variables are defined as follows.

Virtual frictional force

$$\mathrm{VFr} = \frac{IV_{\mathrm{n}}}{S\,\mathrm{pd_n}} \tag{6.4}$$

Virtual residual energy

$$y_{\mathrm{v}} = \mathrm{VFr} \times \mathrm{Rtd_n} \tag{6.5}$$

By the $\mathrm{Rtd_n}$ definition and (6.5), the estimated residual traveling distance of power wheelchair is defined as

$$y_{\mathrm{r}} = \frac{\mathrm{RTD}_{\max}}{\mathrm{VFr}} \times y_{\mathrm{v}} \tag{6.6}$$

6.3 Training of Virtual Residual Energy Estimation System

Two types of estimation systems are proposed with outputs defined as follows.

(a) Type-A estimation system output

$$y_{\mathrm{v}}^{\mathrm{A}} = \sum_{k} w_{lk} y_k \tag{6.7}$$

(b) Type-B estimation system output

$$y_{\mathrm{v}}^{\mathrm{B}} = f\left(\sum_{k} w_{lk} y_k + b_l\right) \tag{6.8}$$

In order to achieve the optimal weights and biases of proposed estimation system, back-propagation algorithm and steepest descent method are introduced to gradually decrease the error between the network output and desired output. The objective function is defined as

$$E = \frac{1}{2}\left(y_{\mathrm{v}}^{*} - y_{\mathrm{v}}\right)^2 \tag{6.9}$$

where y_{v}^{*} denotes the desired output and y_v denotes $y_{\mathrm{v}}^{\mathrm{A}}$ or $y_{\mathrm{v}}^{\mathrm{B}}$ defined in (6.7) or (6.8). To decrease the error between the output and desired output of the proposed neural network, the weights and biases are adjusted based on the steepest descent method by submitting a set of training data. By the steepest descent method, the tuning laws of weights and biases for type-A estimation system could be expressed as

$$w_{ji}(t+1) = w_{ji}(t) + \alpha\left[\sum_{k} \delta_k^3 w_{kj}\right] g\left(\mathrm{net}_j\right) y_i^1 \tag{6.10}$$

$$b_j(t+1) = b_j(t) + \beta\left[\sum_{k} \delta_k^3 w_{kj}\right] \tag{6.11}$$

where learning rates $\alpha > 0$, $\beta > 0$, $\delta_k^n = -\frac{\partial E}{\partial y_j^n}$ and $g(x) = \begin{cases} 1 & 0 < x \leq 1 \\ 0 & \text{otherwise} \end{cases}$. Other tuning laws similar to the above equations are not presented in this chapter.

6.4 Experiment Results

Figure 6.2 is the experiment device with 24 V/10 Ah lithium battery as the power supply. The experiment datum including wheelchair speed, current, and voltage of lithium battery are collected in 250 ms by a 16 bits embedded system SPEC061A. In order to raise the data correctness and reduce the noise, take the average of 240 records of data as a training data (or testing data), that is, each record of training data (or testing data) is 1 min.

Five testing samples with randomly selected 30 testing points are employed to estimate the residual traveling distance of power wheelchair. The parameters for testing samples are summarized in Table 6.1.

The root mean square error of residual energy and of residual traveling distance are $\mathrm{RMSE_v} = \left[\frac{1}{M}\sum_{i=1}^{M} \mathrm{Error}_\mathrm{v}^2(i)\right]^{1/2}$ and $\mathrm{RMSE_r} = \left[\frac{1}{M}\sum_{i=1}^{M} \mathrm{Error}_\mathrm{r}^2(i)\right]^{1/2}$, respectively.

$\mathrm{Error_v} = y_\mathrm{v}^* - y_\mathrm{v}$, $\mathrm{Error_r} = y_\mathrm{r}^* - y_\mathrm{r}$, and y_r^* is the physical residual traveling distance of wheelchair, y_v denotes y_v^A or y_v^B and M is the number of evaluating point.

The adaptive laws for learning rates α and β are defined as

$$\alpha = \begin{cases} \alpha \times \mathrm{Lr_{inc}} & \mathrm{RMSE_v} < k_\mathrm{inc}\mathrm{RMSE_{pre}} \\ \alpha \times \mathrm{Lr_{dec}} & \mathrm{RMSE_v} > k_\mathrm{dec}\mathrm{RMSE_{pre}} \end{cases} \quad (6.12)$$

$$\beta = \begin{cases} \beta \times \mathrm{Lr_{inc}} & \mathrm{RMSE_v} < k_\mathrm{inc}\mathrm{RMSE_{pre}} \\ \beta \times \mathrm{Lr_{dec}} & \mathrm{RMSE_v} > k_\mathrm{dec}\mathrm{RMSE_{pre}} \end{cases} \quad (6.13)$$

where $\mathrm{Lr_{inc}}$ and $\mathrm{Lr_{dec}}$ are 1.05 and 0.7, respectively, k_inc and k_dec are 0.995 and 1.0, respectively, and $\mathrm{RMSE_{pre}}$ is the last value of RMSE_v.

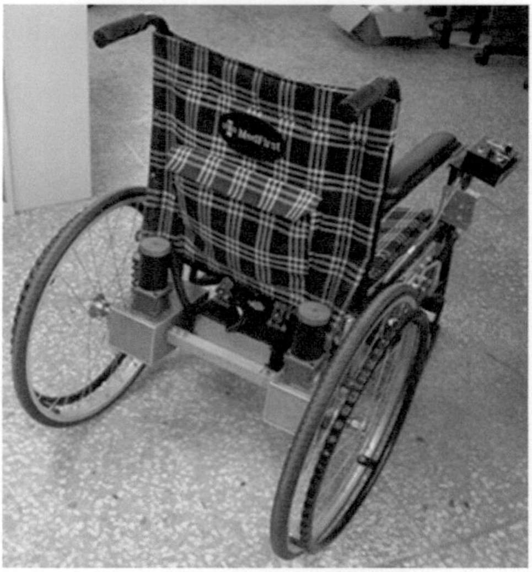

Fig. 6.2 Physical model of power wheelchair used in this chapter

Table 6.1 Summary of the characteristic parameters for testing samples

Testing sample	User weigh (kg)	Average traveling speed (km/h)	Traveling distance (km)
T0717	61	4.3	21.8
T0809	69	4.6	18.8
T0824	49	4.7	24.7
T0825	49	5.5	26.0
T0118	56	5.7	20.6

Type-A estimation systems without and with adaptive learning rate and type-B estimation system with adaptive learning rate and momentum [1] are considered and evaluated. Each estimation system has five neurons in the first hidden layer and four neurons in the second hidden layer.

The learning epoch of estimation system is 30,000 times. The $RMSE_v$ of training results for type-A estimation system without and with adaptive learning rate are 0.0169 and 0.013, respectively, while type-B estimation system is 0.0082. Figure 6.3 compares the estimated results of residual traveling distance for testing sample T0809 using the three proposed virtual residual energy estimation systems. Red "+" denotes the physical residual traveling distance, blue "*" is the estimated results of type-A estimation system with $RMSE_r$ is 1.919 km, black "○" denotes the estimated results of type-A estimation system with adaptive learning rate and its $RMSE_r$ is 1.018 km, while green "△" represents type-B estimation system with $RMSE_r$ is 1.235 km. The best estimated result of the testing sample T0809 is type-A estimation system with adaptive learning rate. Other estimated results for different testing samples are summarized in Table 6.2.

Though the best estimated result shown in Table 6.2 is testing sample T0118 using type-B estimation system, the estimated performance is not necessarily the best based on the estimated results of testing samples T0809 and T0824. The results indicate that type-A estimation system with adaptive learning rate is a feasible solution based on economic factor and estimated performance.

6.5 Conclusions

A virtual residual energy estimation system for power wheelchair is proposed. The estimated virtual residual energy is then transformed into physical residual traveling distance of power wheelchair. Two types of estimation systems with three different training processes, without and with adaptive learning rates, and with adaptive learning rates and momentum, the training results and their estimated results are compared. Though type-B estimation system has the best training performance ($RMSE_v$ is 0.0082), the estimated performances are not necessarily the best of the three estimation systems. Considering the economic factor and estimated performance, type-A estimation system with adaptive learning rate is a feasible solution.

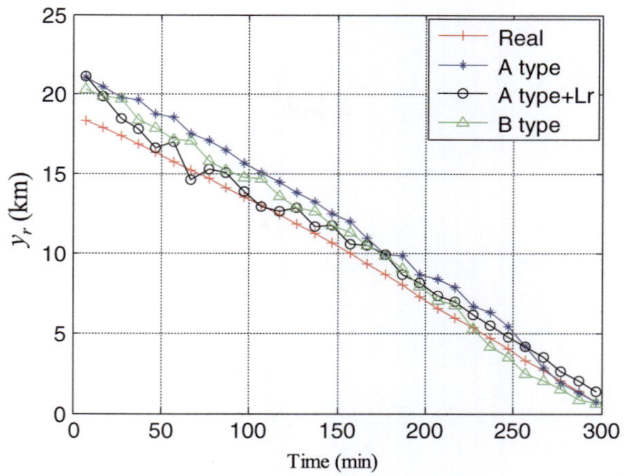

Fig. 6.3 Estimated results of residual traveling distance for testing sample T0809 using the three proposed virtual residual energy estimation systems

Table 6.2 Summary of $RMSE_r$s of estimated results for testing samples

Testing sample	$RMSE_r$ for type-A estimation system (km)	$RMSE_r$ for type-A estimation system with adaptive learning rate (km)	$RMSE_r$ for type-B estimation system (km)
T0717	0.573	0.750	0.584
T0809	1.919	1.018	1.235
T0824	1.830	1.134	1.234
T0825	1.607	0.568	0.453
T0118	1.354	0.610	0.368

Acknowledgments This chapter is supported by Ministry of Science and Technology of the Republic of China under the grant number MOST 103-2221-E-218-013.

References

1. Haykin, S.: Neural Networks—A Comprehensive Foundation, 2nd edn. Prentice-Hall, Inc., Upped Saddle River, NJ (1998)
2. Ham, F.M., Kostanic, I.: Principles of Neurocomputing for Science and Engineering. McGraw-Hill, New York (2000)
3. Grewal, S., Grant, D.A.: A novel technique for modeling the state of charge of lithium ion batteries using artificial neural networks. In: INTELEC 2001, pp. 174–179 (2001)
4. Lee, D.T., Shiah, S.J., Lee, C.M., Wang, Y.C.: State of charge estimation for electric scooters by using learning mechanisms. IEEE Trans. Veh. Technol. **56**(2), 544–556 (2007)
5. Ting, C.H., Tsai, C.S., Fang, Y.L.: Estimating the residual travel distance of an electrical scooter. In: 2009 IEEE/ASME International Conference on Advanced Intelligent Mechatronics, pp. 1348–1352 (2009)
6. Chen, P.C., Koh, Y.F.: Residual traveling distance estimation of an electric wheelchair. In: BMEI 2012, pp. 790–794 (2012)

Chapter 7
Residual Whitening Method for Identification of Induction Motor System

Chien-Hsun Kuo and D.-M. Yang

Abstract This chapter identifies the induction motor system by residual whitening method. Through residual whitening, the optimal properties of the Kalman filter could be enforced for a finite set of data. This technique uses AutoRegressive Moving Average with eXogeneous input (ARMAX) model which is combined of ARX (AutoRegressive with eXogeneous input) and MA (Moving-Average) models. Numerical and experimental results are shown for the identified induction motor system.

Keywords Residual whitening • ARMAX model • Eigensystem Realization Algorithm (ERA)

7.1 Introduction

System identification or modeling is the process of building mathematical models of dynamical systems based on the available input and output data from the systems. In theory, when one identifies a system from input–output data in a time domain, it is assumed that the data length is long enough and the ARX model order is sufficiently large. Then the identified observer tends to be the optimal Kalman filter gain in the presence of process and measurement noise. Under these conditions, the resultant residual of the filter is minimized, uncorrelated with the input and output data, and also white. However, in practice, since only a finite set of data is available and choosing the large ARX model order is limited by the computation constraint, the identified system model and observer may not be optimal. Shan et al. [1] proposed an algorithm which balances a data fidelity term with a norm induced by the set of single pole filters to identify the unknown system from noisy linear measurements. Smith [2] used a nuclear norm minimization-based method for frequency domain subspace identification. In [3], the residual whitening method

C.-H. Kuo (✉) • D.-M. Yang
Department of Mechanical and Automation Engineering,
Kao Yuan University, Kaohsiung 812, Taiwan
e-mail: t30081@cc.kyu.edu.tw

J. Juang (ed.), *Proceedings of the 3rd International Conference on Intelligent Technologies and Engineering Systems (ICITES2014)*, Lecture Notes in Electrical Engineering 345, DOI 10.1007/978-3-319-17314-6_7

is performed to improve observer/Kalman identification (OKID). Through residual whitening, the optimal properties of the Kalman filter could be enforced for a finite set of data. This means that for a given set of finite data, one can identify the system and observer whose residual is minimized. And one can also relax the requirement of the model order to reduce the computation burden, especially for several input and output systems. In this chapter, the induction motor system is identified by residual whitening method.

7.2 System Identification Algorithm

7.2.1 Model Structures

A finite-dimensional, linear, discrete-time, time invariant stochastic system can be expressed as:

$$x_{k+1} = Ax_k + Bu_k + w_k, \quad y_k = Cx_k + v_k, \tag{7.1}$$

where $x \in R^{n\times 1}$, $u \in R^{ni\times 1}$, $y \in R^{no\times 1}$ are state, input, and output vectors, respectively; [A,B,C] are the state-space system matrices. The sequences of process noise $w \in R^{n\times 1}$ and measurement noise $v \in R^{no\times 1}$ are assumed white, Gaussian, zero mean. The noises w_k and v_k are also assumed uncorrelated with covariance Q and R, respectively.

If one defines the error between the actual output y_k and the estimated output $C\hat{x}_k$ as residual ε_k, one can have

$$\hat{x}_{k+1} = A\hat{x}_k + Bu_k + AK\varepsilon_k, \quad y_k = C\hat{x}_k + \varepsilon_k, \tag{7.2}$$

where $\hat{x}$ the estimated state vector and $K \in R^{n\times no}$ is the steady-state Kalman filter gain [4]. In a Kalman filter sense, (7.2) is also called a filter-innovation model [5] which is the best description of a stochastic system.

Next, a modified model structure is formulated to reduce the requirement of model order. The term My_k will be added and subtracted to the right hand side of the state equation in (7.2) to yield

$$\begin{aligned} \hat{x}_{k+1} &= A\hat{x}_k + Bu_k + AK\varepsilon_k + My_k - My_k \\ &= (A + MC)\hat{x}_k + (AK + M)\varepsilon_k + Bu_k - My_k \\ y_k &= C\hat{x}_k + \varepsilon_k \end{aligned} \tag{7.3}$$

Now the new relationship of reference input and output becomes

$$y_k = \sum_{i=1}^{\infty} C\widetilde{A}^{i-1}(-M)y_{k-i} + \sum_{i=1}^{\infty} C\widetilde{A}^{i-1}Bu + \sum_{i=1}^{\infty} C\widetilde{A}^{i-1}\overline{M}\varepsilon_{k-i} + \varepsilon_k, \tag{7.4}$$

where

$$\widetilde{A} = A + MC, \quad \overline{M} = M + AK. \tag{7.5}$$

Make $\tilde{A}$ asymptotically stable, $\widetilde{A}^{i-1} \approx 0$ if $i \geq p$ for a sufficient large number p, (7.4) becomes

$$y_k = \sum_{i=1}^{p} h_i y_{k-i} + \sum_{i=1}^{p} t_i u_{k-i} + \sum_{i=1}^{p} s_i \varepsilon_{k-i} + \varepsilon_k, \tag{7.6}$$

where

$$h_i = C\widetilde{A}^{i-1}(-M), \quad t_i = C\widetilde{A}^{i-1}B, \quad s_i = C\widetilde{A}^{i-1}\overline{M}, \quad i = 1, 2, \ldots, p. \tag{7.7}$$

Equation (7.6) is an ARMAX model containing the dynamics of residual. As stated in (7.5), $\widetilde{A} = A + MC$, the matrix M is used to make the new system (7.3) more stable than the original one in (7.2), especially, since it can be used to reduce the requirement of ARMAX model order p.

7.2.2 Residual Whitening

Suppose that there are N data points of y_k and u_k, $k = 0, 1, \ldots, N-1$, are given. Define

$$Y = \begin{bmatrix} y_0 & y_1 & \ldots & y_q & \ldots & y_{N-1} \end{bmatrix}, \quad R = \begin{bmatrix} \varepsilon_0 & \varepsilon_1 & \ldots & \varepsilon_{N-1} \end{bmatrix},$$

$$\theta = \lfloor t_1 \quad h_1 \quad \ldots \quad t_p \quad h_p \rfloor, \quad \Psi = \lfloor s_1 \quad s_2 \quad \ldots \quad s_p \rfloor,$$

$$V = \begin{bmatrix} 0 & u_0 & \ldots & u_{q-1} & \ldots & u_{N-2} \\ 0 & y_0 & \ldots & y_{q-1} & \ldots & y_{N-2} \\ \ldots & \ldots & \ldots & \ldots & \ldots & \ldots \\ 0 & 0 & 0 & u_0 & \ldots & u_{N-q-1} \\ 0 & 0 & 0 & y_0 & \ldots & y_{N-q-1} \end{bmatrix},$$

$$W = \begin{bmatrix} 0 & \varepsilon_0 & \varepsilon_1 & \ldots & \varepsilon_{p-1} & \ldots & \varepsilon_{N-2} \\ 0 & 0 & \varepsilon_0 & \ldots & \varepsilon_{p-2} & \ldots & \varepsilon_{N-3} \\ \ldots & \ldots & \ldots & \ldots & \ldots & \ldots & \ldots \\ 0 & 0 & 0 & \ldots & \varepsilon_o & \ldots & \varepsilon_{N-p-1} \end{bmatrix}$$

Then (7.6) could be written in a matrix form

$$Y = \theta V + \Psi W + R. \tag{7.8}$$

The least squares solution of the parameters $\hat{\theta}$ and $\hat{\Psi}$ that minimize the cost function $J = \sum_{k=0}^{N-1} \varepsilon_k^T \varepsilon_k = \mathrm{tr}\left(RR^T\right)$ is

$$\begin{bmatrix} \hat{\theta} & \hat{\Psi} \end{bmatrix} = Y \begin{bmatrix} V \\ W \end{bmatrix}^{+} = Y \begin{bmatrix} V^T & W^T \end{bmatrix} \cdot \begin{bmatrix} VV^T & VW^T \\ WV^T & WW^T \end{bmatrix}^{-1}.$$

The estimated parameters could also be expressed as

$$\begin{aligned} &\hat{\theta} = Y\Lambda - \hat{\Psi} W \Lambda, \quad \hat{\Psi} = Y(I - \Lambda V) W^T \lambda^{-1}, \\ &\text{where } \Lambda = V^T \left(VV^T\right)^{-1}, \quad \lambda = W(I - \Lambda V) W^T. \end{aligned} \tag{7.9}$$

Examining the first term of $\hat{\theta}$ in (7.9) which is the ordinary least squares solution in open-loop system; and in the second term one may consider it as a bias term. One can write, hence,

$$\begin{aligned} &\hat{\theta} = \theta^{\mathrm{LS}} - \theta^{\mathrm{bias}}, \\ &\text{where } \theta^{\mathrm{LS}} = Y\Lambda = YV^T\left(VV^T\right)^{-1}, \quad \theta^{\mathrm{bias}} = \hat{\Psi} W \Lambda = \hat{\Psi} W V^T \left(VV^T\right)^{-1}. \end{aligned} \tag{7.10}$$

The parameter matrix, $\hat{\Psi}$, in (7.9) could also be written as

$$\begin{aligned} \hat{\Psi} &= Y(I - \Lambda V) W^T \lambda^{-1} = \left(Y - \theta^{\mathrm{LS}} V\right) W^T \left[WW^T - WV^T\left(VV^T\right)^{-1} VW^T\right]^{-1} \\ &= \hat{e}^{\mathrm{LS}} W^T \left[WW^T - WV^T\left(VV^T\right)^{-1} VW^T\right]^{-1}. \end{aligned} \tag{7.11}$$

where $\hat{e}^{\mathrm{LS}}$ denotes the colored residual between the measurement Y and the least squares estimation $\theta^{\mathrm{LS}} V$.

7.2.3 Iterative Procedure for Identification

1. An initial estimate of the parameter θ, denoted by $\hat{\theta}^{\mathrm{LS}}$, is computed from the ordinary least squares solution. And assuming that an estimate of the parameter matrix, Ψ, denoted by $\hat{\Psi}_0$ is zero.

$$\hat{\theta}^{\mathrm{LS}} = YV^T\left(VV^T\right)^{-1}, \quad \hat{\Psi}_0 = 0.$$

2. Calculate the colored residual sequence $\hat{e}^{\mathrm{LS}}$ which corresponds to the initial estimate ARMAX model parameter $\hat{\theta}^{\mathrm{LS}}$. $\hat{e}^{\mathrm{LS}} = Y - \hat{\theta}^{\mathrm{LS}}V$.This is also an initial estimate of the white residual sequence denoted by $\hat{e}_0 = \hat{R}_0$.
3. Construct the residual matrix $\hat{W}_0$ and then the parameter $\hat{\Psi}$ could be updated by (7.11).

$$\hat{\Psi}_1 = \hat{e}^{\mathrm{LS}}\hat{W}_0^T\left[\hat{W}_0\hat{W}_0^T - \hat{W}_0V^T\left(VV^T\right)^{-1}V\hat{W}_0^T\right]^{-1}.$$

The updated parameter $\hat{\Psi}_1$ is used to correct the initial bias as follows

$$\hat{\theta}_1^{\mathrm{bias}} = \hat{\Psi}_1\hat{W}_0\Lambda = \hat{\Psi}_1\hat{W}_0V^T\left(VV^T\right)^{-1}, \quad \hat{\theta}_1 = \hat{\theta}^{LS} - \hat{\theta}_1^{bias}.$$

4. Compute the new whitened residual sequence $\hat{R}_1$ by using the estimated parameters $\hat{\theta}_1$ and $\hat{\Psi}_1$ as follows: $\hat{R}_1 = Y - \hat{\theta}_1V - \hat{\Psi}_1\hat{W}_0$.
5. Iterate the procedure from step 3 to step 4 by generating the new residual matrix $\hat{W}_1$ and using the updated parameters $\hat{\theta}_1$ and $\hat{\Psi}_1$. The next cycle is to calculate $\hat{\Psi}_2$, $\hat{\theta}_2^{\mathrm{bias}}$, and $\hat{\theta}_2$.

After having obtained the ARMAX model estimated parameters θ, and Ψ, one can use the estimated coefficients to construct the open-loop system Markov parameter $Y_s(k) = CA^{k-1}B$, Kalman filter Markov parameters $Y_k(k) = CA^{k-1}AK$, and $Y_m(k) = CA^{k-1}M$, $k = 1, 2, \ldots, p, p+1, \ldots$ as following:

$$Y_s(k) = t_k + \sum_{i=1}^{k} h_iY_s(k-i), \quad Y_m(k) = -h_k + \sum_{i=1}^{k-1} h_iY_m(k-i),$$

$$Y_k(k) = h_k + s_k + \sum_{i=1}^{k-1} h_iY_k(k-i).$$

$\{A \quad B \quad C \quad M \quad K\}$ can be realized by the Eigensystem Realization Algorithm (ERA) [6].

7.3 Experimental Setup

The experimental apparatus used in this study consists of a 3 hp, 1,800 rpm (i.e., $fr = 30$ Hz), 4-pole three-phase induction machine driving a 5 kW DC generator via a flexible coupling, as shown in Fig. 7.1. The generator is used to absorb the energy generated by the motor. A piezo-electric accelerometer is mounted on the housing of the induction electrical machine to measure the vibration signal. Three current sensors were used to record the three phase current signals from the inverter. All measured signals were sampled at 6 kHz via a real-time data acquisition device

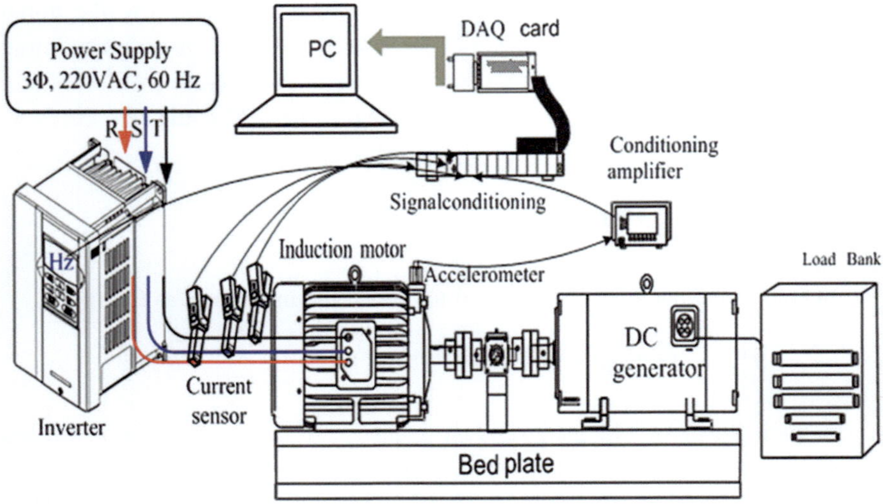

Fig. 7.1 Experimental setup

(Type NI 6062E) under the motor no-load condition. The frequency of inverter varies from 1 to 60 Hz.

7.4 Numerical Validation

The inputs consist of three currents into the motor and the output is the vibration signal from the accelerometer. Each data contains 122,880 points. Two sets of input–output data are recorded, one set is implemented for system identification, and the other is used for comparison between the predicted output and the true one. Figure 7.2 shows new test data (solid) and ARMAX predicted output (dashed) from 8,000 to 8,500 points. Figure 7.3 shows the frequency plot of the real (solid) and predicted output (dashed) from Fourier Transform.

7.5 Conclusions

In this chapter, the residual whitening technique has been used to identify the induction motor system. The experimental results show that the predicted output is very close to the test data. The effectiveness of the proposed approach is verified. Actually, for the high level of noise cases, the most significant error is from the original estimated parameters which have been corrupted by the disturbance.

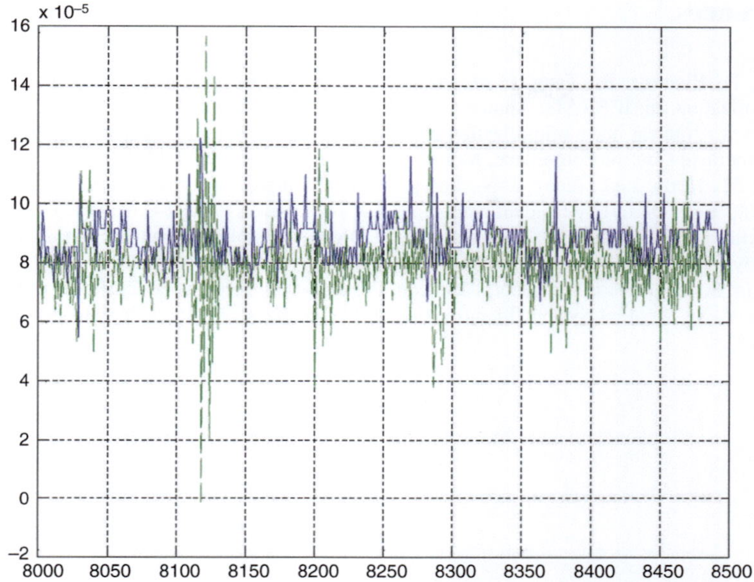

Fig. 7.2 New test data (*solid*) and ARMAX predicted output (*dashed*) from 8,000 to 8,500 points

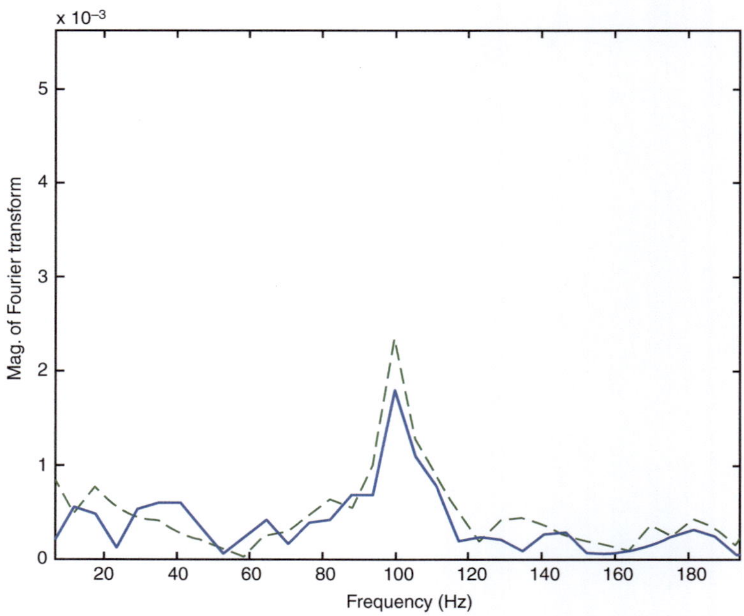

Fig. 7.3 The frequency plot of the real (*solid*) and predicted output (*dashed*) from Fourier transform

References

1. Shan, P., Bhaskar, B., Tang, G., Recht, B.: Linear system identification via atomic norm regularization. In: IEEE 51st Annual Conference on Decision and Control (2012)
2. Smith, R.: Nuclear norm minimization methods for frequency domain subspace identification. In: American Control Conference, Montreal, Canada, pp. 2689–2694 (2012)
3. Phan, M., Horta, L.G., Juang, J.-N., Longman, R.W.: Improvement of observer/Kalman filter identification (OKID) by residual whitening. J. Vib. Acoust. **117**, 232–239 (1995)
4. Franklin, G.F., Powell, J.D., Workman, M.L.: Digital control of dynamic systems. Addison-Wesley Publishing Company, Reading, MA (1990)
5. Haykin, S.: Adaptive Filter Theory, 2nd edn. Prentice-Hall, Englewood Cliffs, NJ (1991)
6. Juang, J.-N.: Applied System Identification. PRT Prentice-Hall, Inc., Englewood Cliffs, NJ (1994)

Chapter 8
Analysis and Simulation of Small-Sized Quasi-servo Valve Using Tiny On/Off Control Valve

So Shimooka, Shujiro Dohta, Tetsuya Akagi, and Yoshinori Moriwake

Abstract Today, the care and welfare pneumatic equipment to support a nursing care and a self-reliance of the elderly and the disabled are actively researched and developed by many researchers. These wearable devices require many servo valves for multidegrees of freedom and precise control performance of the wearable actuator. The total weight of the wearable devices increases according to the degree of freedom. In our previous study, a small-sized and lightweight pressure control-type quasi-servo valve was developed *t*. The valve consists of two on/off control valves and an embedded controller. In this study, the quasi-servo valve composing of much smaller-sized (40 % in mass, 42 % in volume) on/off valves is proposed and tested. The analytical model of the tested valve is proposed and the system parameters are identified. As the result of the comparison between experimental results and simulated, it was confirmed that the proposed analytical model and the identified system parameters were valid.

Keywords Quasi-servo valve • Small-sized control valve • Embedded controller

8.1 Introduction

Today, the care and welfare pneumatic equipment to support a nursing care and a self-reliance of the elderly and the disabled are actively researched and developed by many researchers [1–3]. These wearable devices require many servo valves for multidegrees of freedom and precise control performance of the wearable actuator [4]. The total weight load of the wearable devices increases according to the degree of freedom. Therefore, we aim to develop a small-sized, lightweight, and low-cost quasi-servo valve using on/off control valves to reduce the burden of the user instead of expensive and bulky electropneumatic servo valves. In our previous

S. Shimooka (✉) • S. Dohta • T. Akagi • Y. Moriwake
Department of Intelligent Mechanical Engineering, Okayama University of Science, Okayama 700-0005, Japan
e-mail: t14rm01ss@ous.jp

J. Juang (ed.), *Proceedings of the 3rd International Conference on Intelligent Technologies and Engineering Systems (ICITES2014)*, Lecture Notes in Electrical Engineering 345, DOI 10.1007/978-3-319-17314-6_8

studies [5–7], an inexpensive pressure control-type quasi-servo valve using a low-cost embedded controller and a pressure transducer was proposed and tested. In this chapter, a quasi-servo valve by using the much smaller, lighter, and inexpensive on/off control valves was produced and tested. The precise analytical model of the tested valve was also proposed in order to utilize an optimal design. And the values of system parameters are identified by measurement and simulation. The simulated transient response of the output pressure in the tracking control was also tested and compared with experimental one.

8.2 Quasi-servo Valve and Its Analytical Model

Figures 8.1 and 8.2 show the schematic diagram and the view of the quasi-servo valve, respectively. The valve consists of two on/off type control valves (SMC Co. Ltd., S070C-SDG-32) whose both output ports are connected to each other. One valve is used as a switching valve to supply or exhaust, and the other is used as a PWM control valve that can adjust output flow rate like a variable fluid resistance [5]. The size of the on/off valve is $12 \times 33 \times 7$ mm, and the mass is only 6 g. The previous on/off valve (Koganei Co. Ltd., G010HE-1) was $33 \times 20 \times 10$ mm, the mass was 15 g [5]. The new valve is 58 % smaller and 60 % lighter than the previous valve. The maximum output flow rate is 15 L/min at the supply pressure of 500 kPa.

Figure 8.3 shows the analytical model of the tested three-port-type on/off valve. The valve consists of a solenoid, a popet (armature with a spring) and three ports. Figure 8.4 shows the analytical model of the tested quasi-servo valve. The generated electromagnetic force F_e is given by

$$F_\mathrm{e} = K_\mathrm{c} i, \tag{8.1}$$

where i is a current and K_c is a conversion factor from the current to the force. The electrical circuit of the solenoid is given by

$$R_v i + L_v \frac{di}{dt} = e_v, \tag{8.2}$$

where R_v, e_v, and L_v are the electrical resistance, an input voltage of the solenoid coil, and an inductance of the coil, respectively. The equation of the motion of the armature is given by

$$M_v \frac{dx^2}{dt^2} + C_v \frac{dx}{dt} + K_v x = F_\mathrm{e}, \tag{8.3}$$

where M_v, C_v, and K_v are the mass of the armature, the damping coefficient, and the spring constant, respectively.

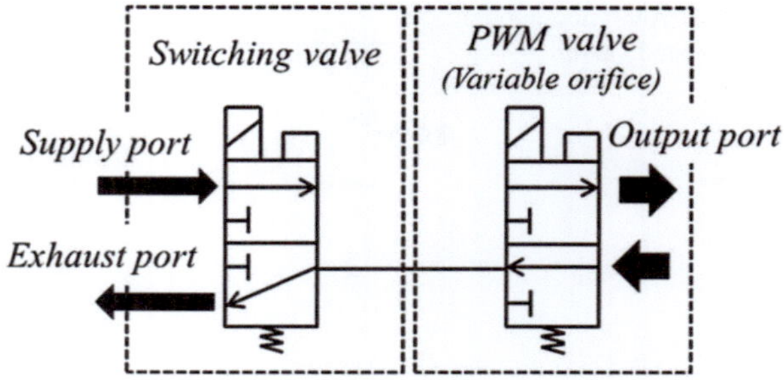

Fig. 8.1 Schematic diagram of quasi-servo valve

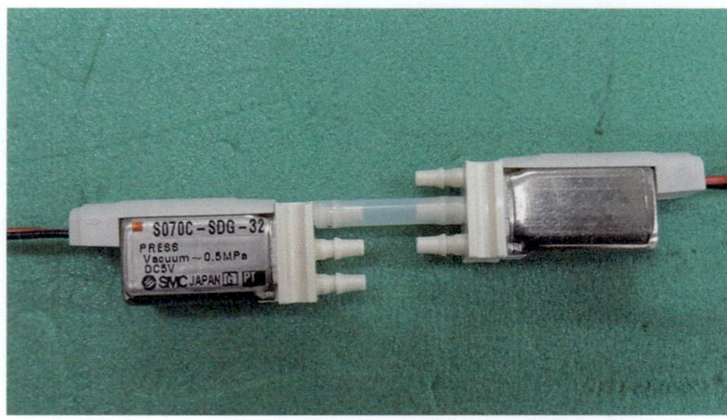

Fig. 8.2 View of quasi-servo valve

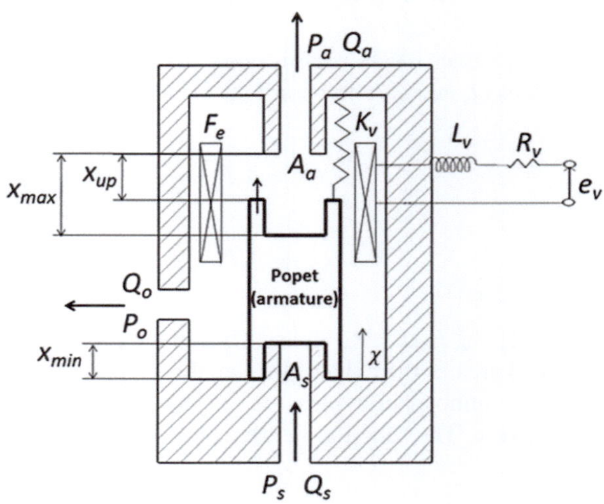

Fig. 8.3 Analytical model of 3-port on/off valve

The sectional area of the valve will change based on the displacement of the armature x. The sectional area of the supply port A_s and the area of exhaust port A_a start to change at the point of a certain displacement x_{min} (=0.12 mm) of the armature. Then both of supply port and exhaust port open when $x < x_{up}$. Where $x_{up} = x_{max} - x_{min}$. When $x = x_{up}$, the exhaust port is closed and the armature stops at

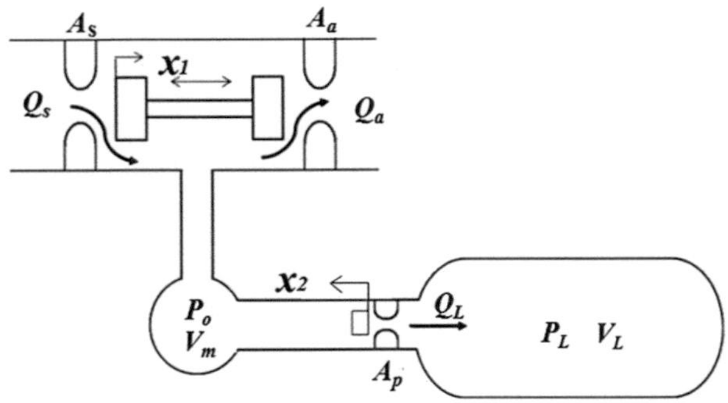

Fig. 8.4 Analytical model of the tested valve

the maximum displacement $x_{\max}$. Therefore, both sectional areas A_s and A_a are given by (8.4a)–(8.4c). The sectional area must be chosen as the narrowest area of the flow field. Therefore, in the case if the cylindrical sectional area of the valve seat (orifice) is smaller than the sectional area of the orifice that is maximum value A_{a0} or A_{s0}, the effective sectional area is changed as the linear function of the displacement x.

$$A_s = 0, \quad A_a = \sqrt{4\pi A_{a0}}\left(x_{up} - x\right) \quad (0 < x < x_{\min}) \tag{8.4a}$$

$$A_s = \sqrt{4\pi A_{s0}}(x - x_{\min}), \quad A_a = \sqrt{4\pi A_{a0}}\left(x_{up} - x\right) \quad \left(x_{\min} < x < x_{up}\right) \tag{8.4b}$$

$$A_s = \sqrt{4\pi A_{s0}}(x - x_{\min}), \quad A_a = 0 \quad \left(x_{up} < x < x_{\max}\right). \tag{8.4c}$$

In 2-port valve (PWM valve), same equations as (8.4a) and (8.4c) are used. Mass flow rates Q_s and Q_a shown in Fig. 8.4 are given as follows.

$$Q_s = A_s P_s \sqrt{\frac{2}{RT}} f(z), \quad z = \frac{P_o}{P_s} \tag{8.5a}$$

$$Q_a = A_a P_o \sqrt{\frac{2}{RT}} f(z), \quad z = \frac{P_a}{P_o}, \tag{8.5b}$$

where, A_s, A_a, P_s, P_o, P_a, R, and T mean the sectional area of the supply orifice, the sectional area of the exhaust orifice, the supply pressure, pressure in the connected chamber, atmospheric pressure, a gas constant, and an absolute temperature, respectively. The function $f(z)$ that expresses the state of flow is given as follows.

$$f(z) = \sqrt{\frac{\kappa}{\kappa - 1}\left(z^{\frac{2}{\kappa}} - z^{\frac{(\kappa+1)}{\kappa}}\right)} \quad (0.528 < z \leqq 1) \tag{8.6a}$$

$$f(z) = \sqrt{\frac{\kappa}{\kappa+1}\left(\frac{2}{\kappa+1}\right)^{\frac{2}{(\kappa-1)}}} \quad (0 \leqq z \leqq 0.528), \tag{8.6b}$$

where κ means a specific heat ratio (=1.4).

If the pressure change in the connected chamber volume V_m is assumed as an adiabatic change, the relation between P_o and the output flow rate Q_o is given by

$$\frac{dP_o}{dt} = \frac{\kappa RT}{V_m} Q_o \tag{8.7}$$

$$Q_o = Q_s - Q_a - Q_L, \tag{8.8}$$

where Q_L is the flow rate through the orifice in the PWM valve and is expressed by

$$Q_L = A_p P_o \sqrt{\frac{2}{RT}} f(z), \quad z = \frac{P_L}{P_o} \quad (P_o \geqq P_L) \tag{8.9a}$$

$$Q_L = -A_p P_L \sqrt{\frac{2}{RT}} f(z), \quad z = \frac{P_o}{P_L} \quad (P_o < P_L), \tag{8.9b}$$

where A_p is the sectional area of the orifice in the PWM valve (=A_s). The relation between pressure P_L in the volume and Q_L is given by

$$\frac{dP_L}{dt} = \frac{\kappa RT}{V_L} Q_L. \tag{8.10}$$

By using (8.1)–(8.10), the behaviour of the tested valve can be predicted.

8.3 Simulated Results and Discussion

The identified system parameters of the on/off valve are shown in Table 8.1. As the method of identification, the parameters that could not be measured directly will be identified so as to agree the calculated value with the experimental value by using the on/off valve. The inductance L was calculated from the outer diameter and the number of the turns of the coil. The resistance R_v, the mass of the armature M, and the spring constant of the coil spring K_v were measured directly by using the disassembled parts of the valve. The sectional area of the supply port and the exhaust port were determined by using the value in the product catalogue and were adjusted by a simulation.

Figure 8.5 shows the relation between the duty ratio and the mean output flow rate in the chamber. In the experiment, the switching valve is always open and the input duty ratio of the PWM valve is changed. The output flow rate is measured by the digital flow meter (SMC Co. Ltd., PF2A710-01-27: measuring range: 1–10 L/

Table 8.1 Identified system parameters

Parameter	Value	Unit
A_{s0}	0.202×10^{-6}	m^2
A_{a0}	0.23×10^{-6}	m^2
C_v	0.05	N s/m
K_c	0.4	N/A
K_v	311	N/m
L_v	0.00778	H
M_v	0.563×10^{-3}	kg
R_v	51.1	Ω
x_{min}	0.12×10^{-3}	m
x_{max}	0.32×10^{-3}	m

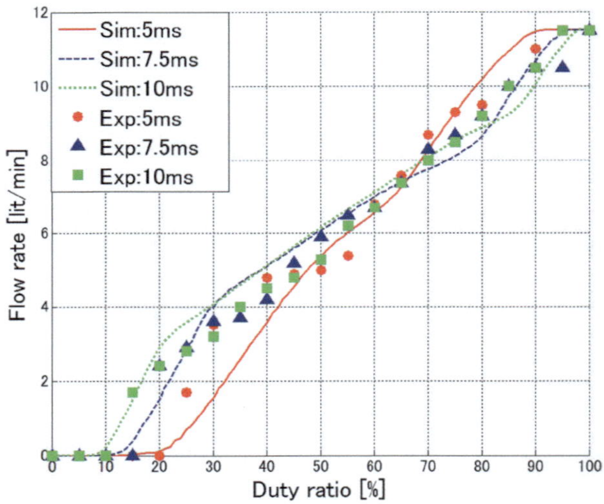

Fig. 8.5 Relation between duty ratio and flow rate

min). The supply pressure is 500 kPa. The periods of 5, 7.5, and 10 ms are selected as the PWM period by considering the dynamics of the on/off valve with the time delay of 3 ms. In Fig. 8.5, each colored symbol shows the experimental result using various PWM period. Each colored line shows the calculated results using the proposed model and the identified parameters. From Fig. 8.5, it can be seen that the calculated results using the proposed model agree well with the experimental results. The theoretical result can predict the nonlinear characteristics of the tested valve such as the dead zone of output flow rate in lower duty ratio and the saturation in higher duty ratio. In addition, the model can predict the overall relation between the input duty ratio and the output flow rate of the valve. From the result, we can conclude that the calculated results using the proposed model and the identified parameters are valuable and reliable to estimate the performance of the valve.

In order to investigate the dynamics of the tested valve, the pressure control was carried out by using the control system shown in Fig. 8.6. The control system

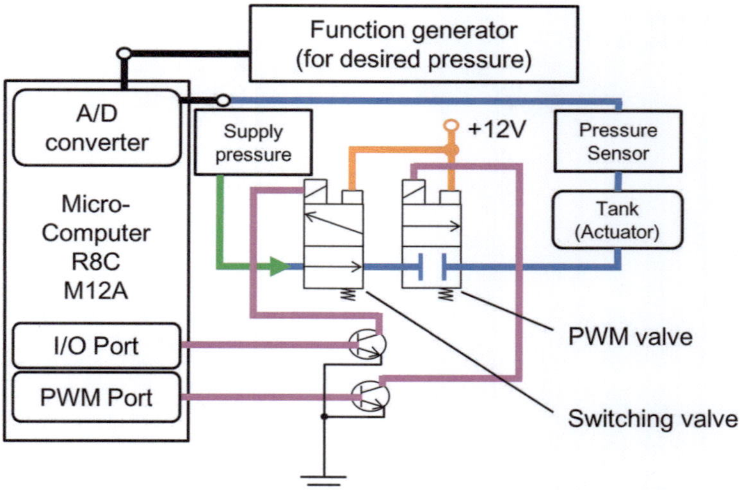

Fig. 8.6 Schematic diagram of control system

consists of the quasi-servo valve, a pressure transducer (Matsushita Electronics Co. Ltd., ADP5160), and an inexpensive embedded controller (Renesas Co. Ltd. R8C12M). The pressure control is done as follows. First, the embedded controller gets the sensor output voltage and the reference voltage through an inner 10 bit A/D converter. Then, the error between sensor output and the reference voltage is calculated by a controller, and the control input is calculated based on a control scheme. Finally, the control input is applied to the quasi-servo valve. P control scheme is expressed by the following equation.

$$u = K_{\mathrm{p}} e_c + 20.0 \tag{8.11}$$

Zwhere u, K_{p} (=1.5 %/kPa) and e_c mean the input duty ratio, the proportional gain and the error from the reference pressure, respectively. Because the valve has a dead zone as shown in Fig. 8.5, the input duty ratio of the PWM valve is always added by 20.0 %. In the simulation, the model of the embedded controller includes a 10 bit A/D converter and PWM port with PWM period of 5 ms. The sampling period of calculation is 0.01 ms and the sampling period of control is 3.2 ms. Figure 8.7 shows the simulated frequency response of the pressure for reference frequency of 1 Hz. From Fig. 8.7, it can be seen that the calculated result agrees with the experimental.

Fig. 8.7 Example of frequency response at 1 Hz

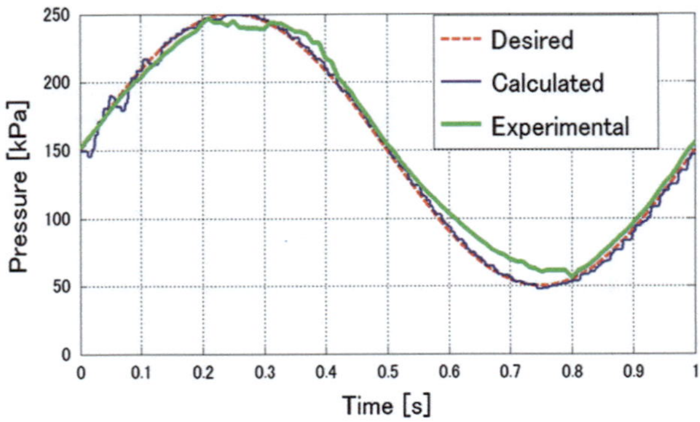

8.4 Conclusions

This study can be summarized as follows. The new small-sized quasi-servo valve using a tiny on/off valve which is 58 % smaller and 60 % lighter than previous one was proposed and tested. The analytical model of the tested quasi-servo valve was proposed and the system parameters were identified. The statics and the pressure control of the valve were investigated. As a result, it was confirmed that the proposed analytical model and the identified system parameters were valid.

Acknowledgments This study was supported by the MEXT of Japan through a Financial Assistance Program for QOL Innovative Research (2012-).

References

1. Noritsugu, T., Takaiwa, M., Sasaki, D.: Development of power assist wear using pneumatic rubber artificial muscles. J. Robot. Mechatron. **21**, 607–613 (2009)
2. Kobayashi, H., Shiban, T., Ishida, Y.: Realization of all 7 motions for the upper limb by a muscle suit. J. Robot. Mechatron. **16**, 504–512 (2004)
3. Ishii, M., Yamamoto, K., Hyodo, K.: Stand-alone wearable power suit—development and availability. J. Robot. Mechatron. **17**(5), 575–583 (2005)
4. Nagata, Y. (ed).: Soft Actuators—Forefront of Development, NTS Ltd., pp. 291–335 (2004)
5. Zhao, F., Dohta, S., Akagi, T.: Development and analysis of small-sized quasi-servo valves for flexible bending actuator. Trans. Robot. Autom. JSME (C) **76**(772), 3665–3671 (2010)
6. Moriwake, Y., Akagi, T., Dohta, S., Zhao, F.: Development of low-cost pressure control type quasi-servo valve using embedded controller. J. Procedia Eng. **41**, 493–500 (2012)
7. Shimooka, S., Dohta, S., Akagi, T., Moriwake, Y., Zhao, F.: Estimation of pressure control performance in low-cost quasi-servo valve using embedded controller. In: Lecture Notes in Electrical Engineering, vol. 293, pp. 359–366. Springer, Switzerland (2014)

Chapter 9
Analysis of Flexible Thin Actuator Using Gas–Liquid Phase-Change of Low Boiling Point Liquid

Yasuyuki Tsuji, Shujiro Dohta, Tetsuya Akagi, and Yuto Fujiwara

Abstract In our previous study, the flexible thin actuator using gas–liquid phase-change of a low boiling point liquid that can generate large force was proposed and tested. The tested actuator is an envelope-type actuator that is made of laminating plastic sheets, low boiling point liquid, and a flexible heater. In this chapter, the analytical model of the flexible thin actuator was proposed and tested. The system parameters of the actuator were also identified. As a result, it was confirmed that the proposed analytical model can predict the behavior of the real actuator.

Keywords Analysis • Flexible thin actuator • Gas–liquid phase-change • Low boiling point liquid • Portable rehabilitation device • Silent movement

9.1 Introduction

Due to the aging of Japanese society and the decreasing birth rate, an important problem to keep a quality of life (QOL) for the elderly has occurred [1]. Especially, the decrease in physical ability of the elderly will be concerned to increase in nursing care task. Therefore, the national budget for social welfare will be increased. In order to solve this problem, that is, to improve a quality of life (QOL) for the elderly and the disabled and to decrease the budget for nursing care, a simple rehabilitation device to keep their physical ability might be one of solutions. The actuator used in such a rehabilitation device requires lightweight, softness, small size and larger generated force that can support human body [2–4]. Therefore, many researchers are studying wearable actuators such as a McKibben artificial muscle [5–7]. The McKibben actuator can generate larger force that is more than 300 N by using the actuator whose diameter of 10 mm in the initial condition. However, the actuator cannot work with long stroke. Maximum stroke is

Y. Tsuji (✉) • S. Dohta • T. Akagi • Y. Fujiwara
Department of Intelligent Mechanical Engineering, Okayama University of Science, Okayama 700-0005, Japan
e-mail: t13rm03ty@ous.jp

J. Juang (ed.), *Proceedings of the 3rd International Conference on Intelligent Technologies and Engineering Systems (ICITES2014)*, Lecture Notes in Electrical Engineering 345, DOI 10.1007/978-3-319-17314-6_9

one fourth of the original length of the actuator. To get larger force by using other pneumatic actuator, larger sectional area of the actuator or higher pressure supplied to the actuator is needed. In ideal, a thin actuator that can generate larger force and larger displacement is required. Therefore, in our previous study [8], the envelope-type actuator driven by pneumatic pressure was proposed and tested as a flexible thin actuator. The actuator can be used as a rehabilitation device for shoulder by being sandwiched in armpit. In addition, the flexible thin actuator using gas–liquid phase-change of low boiling point liquid was also proposed and tested [9]. It was also confirmed that the tested actuator could be driven silently. In the next step, it is necessary to develop the method to estimate the dynamics of the actuator to design it so as to get the suitable behavior. In this chapter, an analytical model of the tested actuator is proposed. The identified parameters and the comparison of the dynamic characteristics between the theoretical and the experimental results of the actuator will be also described.

9.2 Flexible Thin Actuator Driven by Low Boiling Point Liquid

In our pervious study, the thinner and lighter wearable actuator was proposed and developed. Figure 9.1 shows the construction and the view of flexible thin actuator that is an envelope-type actuator. The actuator consists of four sheets of paper and plastic laminate films as shown in Fig. 9.1. The left figure in Fig. 9.1 shows the shape of paper A and B. The middle figure shows the sectional structure of the actuator. The paper B works as a gusset in the actuator. The size of plastic laminate film is larger than the size of paper A and B. The plastic laminate films are bonded to each other at the margin of the films with the exception of paper area. Then, the length of L_A, L_B, and H in paper A and B are 90, 110, and 40 mm, respectively. From Fig. 9.1 on the right, it can be seen that the thickness of the actuator was greatly deformed about 39 times from the initial thickness of 1.5 mm with no supplied pressure to maximum thickness of 58.4 mm. By using typical laminate

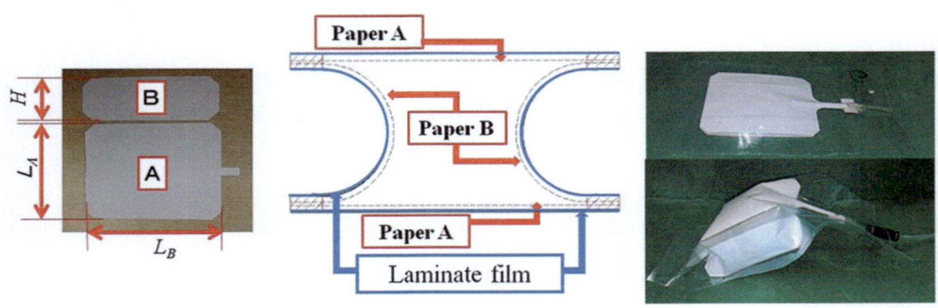

Fig. 9.1 Contraction and view of the flexible thin actuator that is envelope-type actuator

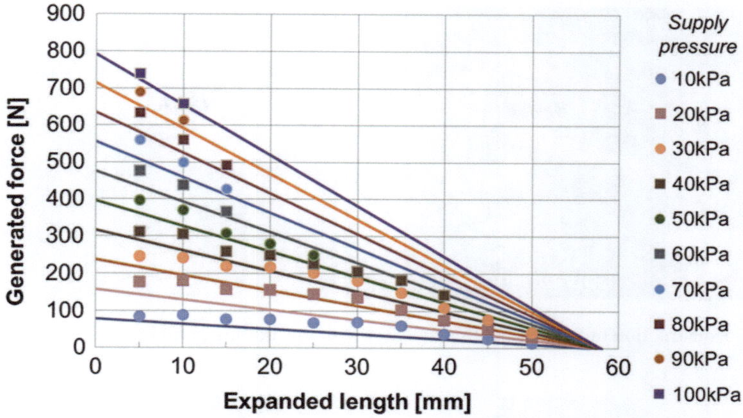

Fig. 9.2 Relation between the expanded displacement and the generated force of the actuator

films and paper that can be easily got on the market as a stationery, the actuator can be quickly and easily fabricated. The time for production of the actuator is several minutes.

Figure 9.2 shows the relation between the expanded length (displacement) and the generated force of the actuator. From Fig. 9.2, it can be seen that the maximum generated force of 657 N could be obtained in the case of supplied pressure of 100 kPa. From these results, the following experimental relation between the displacement and the generated force of the actuator can be obtained.

$$F = A_0 P (1 - X/X_M) \tag{9.1}$$

where, F, A_0, P, X, and X_M mean the generated force, equivalent sectional area, supplied pressure, displacement, and maximum displacement of the actuator, respectively. In Fig. 9.2, the solid lines show calculated results using (9.1) with $A_0 = 0.8 L_A L_B = 0.00792\ \mathrm{m}^2$ and $X_M = 2.9 L_C = 0.058$ mm. From Fig. 9.2, it can be seen that the proposed experimental relation expressed by (9.1) about the generated force and displacement of the actuator can predict well the static characteristics of the tested actuator.

Figure 9.3 shows the construction of the flexible thin actuator using gas–liquid phase-change of low boiling point liquid and a flexible heater. The flexible heater consists of a copper tape as an electrode, a functional paint (Future Carbon GmbH Co. Ltd., Carbon e-Therm PUR-120) [10] and plastic laminate films. The size of the heater as shown in Fig. 9.3 on the left is 60 × 65 mm. The thickness of the heater is 0.5 mm. The heater has flexibility. The flexible heater is installed into the envelope-type actuator with a low boiling point liquid. As a low boiling point liquid, Functional liquid was used as a low boiling point liquid (3MTM Co. Ltd., Novec™7000) [10]. The liquid is used as a cleaning liquid for the electronic circuit board. The boiling point is 34 °C. The liquid is nonflammable and nonpoisonous

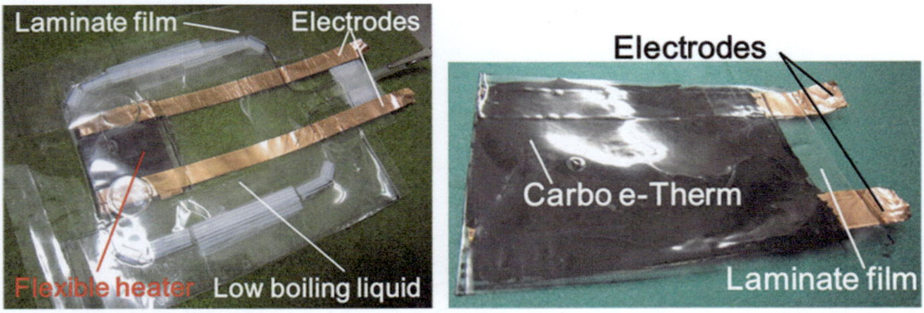

Fig. 9.3 Construction of the flexible thin actuator with low boiling point liquid and flexible heater

Fig. 9.4 Transient view of movement of the tested actuator when the input voltage of 10 V is applied (Time: 5, 10, 15 s)

fluid virtually. The high volume resistivity shows that the liquid can contact the electric heater directly during the high applied current. The low boiling point liquid with volume of about 50 cm^3 is also poured into the actuator. To apply the voltage from the outside of the actuator, two copper tapes were connected with the heater. The size of the actuator is 110 mm in length, 90 mm in width, and 6 mm in thickness. The total mass of the actuator including the liquid is 80 g. Figure 9.4 shows the transient view of the movement of the tested actuator when the input voltage of 10 V was applied. Each figure shows the view of the actuator when 5, 10, 15 s from the beginning of starting to apply the voltage. It can be seen that the actuator reached at maximum displacement within 15 s. After stopping the supply of the voltage, the actuator was contracted according to the decrease of the temperature of the liquid by natural cooling. It was confirmed that the actuator did not make any sound while being driven.

9.3 Analytical Model of the Actuator

Figure 9.5 shows a simple analytical model of the tested actuator to estimate the generated force of the actuator. From the experimental results and geometric configuration of the actuator as shown in Figs. 9.1 and 9.2, the following relationships can be obtained. $L_A = 4.5L_C$, $L_B = 5.5L_C$, $A_0 = 0.8L_AL_B$, $X_M = 2.9L_C$. Where

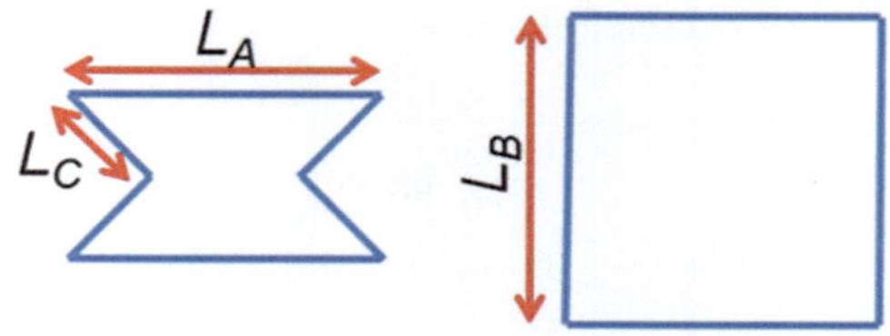

Fig. 9.5 Basic shape of the envelope-type actuator

L_C is a half length of the gusset as shown in Fig. 9.5. From (9.1) and these relations, the following equation can be obtained.

$$F = 19.8PL_C^2 - 6.82PXL_C \tag{9.2}$$

In order to estimate the dynamics of the actuator, it is necessary to make an analytical model of gas–liquid phase change of low boiling point liquid. Therefore, the following energy equilibrium between the inner thermal energy of the actuator and the given energy by the heater and the atmosphere is considered.

$$m_G\{(T_b - T_0)\alpha + L\} = k_w\int W_{in}dt - \int A_hC_h(T_b - T_a)dt \tag{9.3}$$

where, A_h, C_h, k_w, L, m_G, W_{in}, and α are the equivalent contacting area with the atmosphere (=0.0336 m^2), the convective heat conductivity, the efficiency of the heater, the latent heat of vaporization (=142 kJ/kg), the mass of the gas, the heating value of the heater (=41.6 W), and the specific heat (=1,300 J/kg K) of the low boiling point liquid, respectively. T_a, T_b, and T_0 mean the absolute temperature of the atmosphere (=296 K), the boiling point and the initial temperature of the liquid, respectively. Then, the left side term shows the energy required to make the liquid boiled. The right side terms express the energy from the heater and the air cooling, respectively. From Clausius–Clapeyron equation, the boiling point T_b of the liquid in the actuator is changed according to the inner pressure (=101.3 kPa) of the actuator P_i and can be given by the following equation.

$$T_b = \frac{T_{ba}L}{L - RT_{ba}\ln(P_i/P_a)} \tag{9.4}$$

where P_a, R, and T_{ba} show the atmospheric pressure, the gas constant (=287 J/(kg K)), and the boiling point at the atmospheric pressure (=307 K).

Then, it is assumed that the pressure of the gas is equal to the atmospheric pressure by changing the inner volume of the actuator. The volume V_{Ga} of the gas that is vaporized from the liquid is given as follows.

$$V_{Ga} = \frac{m_G}{M_{La}}M_{Ga} \tag{9.5}$$

where M_{La} and M_{Ga} show the molar mass of the liquid (=188 kg/mol) and gas (=22.4 × 10^{-3} m^3/mol) at the atmospheric pressure. From the geometric

Table 9.1 Identified parameters of the flexible thin actuator using the low boiling point liquid

Parameter	value	Parameter	value	Parameter	value
A_h (m^2)	0.0336	L_C (m)	0.02	T_0 (K)	296
α (J/kg K)	1,300	M_G (kg)	Valiable	T_a (K)	296
C_h (W/m^2 K)	6	M_{Ga} (m^3/mol)	22.4×10^{-3}	T_b (K)	Valiable
K_w	0.5	M_{LA} (kg/mol)	188.054	T_{ba} (K)	307
L (kJ/kg)	142	P_a (kPa)	101.3	V_{Ga} (m^3)	Valiable
L_A (m)	0.09	P_i (kPa)	101.3	W_{in} (W)	41.6
L_B (m)	0.11	R (J/(kg K))	287	x (m)	Valiable

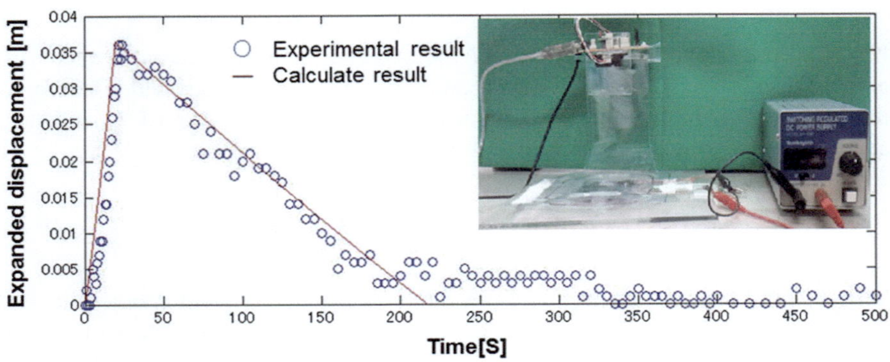

Fig. 9.6 Transient response of the displacement of the actuator

relationship of the actuator, the relation between the volume and the displacement of the actuator x can be obtained as follows.

$$V_{Ga} = x\left(L_A - \sqrt{L_C^2 - (x/2)^2}\right)(L_B - x) \quad (9.6)$$

By solving (9.3)–(9.6), the dynamic behavior of the actuator can be calculated.

Table 9.1 shows the identified parameters of the actuator. Most parameters are measured and investigated from the specification. The parameters of C_h and k_w are identified so that the calculated result of the behavior of the actuator agrees with the experimental result.

Figure 9.6 shows the transient response of the displacement of the actuator. In the experiment, the displacement of the actuator was measured by the infrared displacement sensor (Sharp Co. Ltd. GP2Y0A21YK0F) when the input voltage of 10 V was applied for 120 s. After 120 s, the input voltage was stopped. The actuator was gradually deformed to the original shape by cooling of natural heat dissipation. In Fig. 9.6, symbols and line show the experimental and calculated results using the proposed analytical model and identified parameters, respectively. It can be seen that the calculated result agrees with the experimental results up to about 700 s from the beginning. The difference between the experimental and calculated results was also found. It is considered that it is caused by the stiffness of the gusset in the actuator. Therefore, it is necessary to improve the analytical model as our further work.

9.4 Conclusions

This study that aims to develop and analyze the wearable thin actuator driven by low boiling point liquid for portable rehabilitation device can be summarized as follows.

The analytical model of the actuator that includes the gas–liquid phase-change and the energy equilibrium between the inner thermal energy of the actuator and the given energy by the heater and the atmosphere was proposed and tested. The system parameters of the actuator were also identified. As a result, the calculated displacement of the actuator using the proposed model and the identified parameters can predict the behavior of the tested actuator well. It is confirmed that the proposed model is valid to estimate the dynamics of the tested actuator theoretically.

References

1. Ministry of Internal Affairs and Communications. Statistics Bureau, Director-General for Policy Planning (Statistical Standards) & Statistical Research and Training Institute, Statistical handbook of Japan 2013 [Online]
2. Akagi, T., Dohta, S.: Development of wearable pneumatic actuator and multiport pressure control valve. J. Robot. Mechatron. **17**(5), 529–536 (2005)
3. Akagi, T., Dohta, S., Ihara, M.: Improvement of McKibben artificial muscle with long stroke motion and its application. J. Syst. Des. Dyn. **4**(4), 538–551 (2010)
4. Nagata, Y. (ed.): Soft Actuators—Forefront of Development, NTS Ltd., pp. 291–335 (2004)
5. Yamamoto, K., Ishii, M., Hyodo, K., Yoshimitsu, T., Matsuo, T.: Development of wearable power assisting suit. In: Proceedings of the International Symposium on Fluid Control, Measurement and Visualization 2003, 113pdf, pp. 1–6 (2003)
6. Kobayashi, H., Shiban, T., Ishida, Y.: Realization of all 7 motions for the upper limb by a muscle suit. J. Robot. Mechatron. **16**(5), 504–512 (2004)
7. Noritsugu, T., Takaiwa, M., Sasaki, D.: Development of power assist wear using pneumatic rubber artificial muscles. J. Robot. Mechatron. **21**(5), 607–613 (2009)
8. Tsuji, Y., Dohta, S., Akagi, T., Noguchi, S.: Development of envelope-type pneumatic thin actuator. In: Proceedings of the FLUCOME 2013, OS4-03-2pdf, pp. 1–7 (2013)
9. 3M™ Novec™ 7000 Engineered Fluid (Catalog) [Online]
10. Akagi, T., Dohta, S., Fujimoto, S., Tsuji, Y., Fujiwara, Y.: Development of flexible thin actuator driven by low boiling point liquid. Int. J. Mater. Sci. Eng. **3**(1), 55–58 (2015)

Chapter 10
Development of Flexible Pneumatic Cylinder with String-Type Displacement Sensor for Flexible Spherical Actuator

Yasuko Matsui, Tetsuya Akagi, Shujiro Dohta, and Shinsaku Fujimoto

Abstract This study aims at developing a potable rehabilitation device which can be safe to use while holding it. In our previous study, a novel flexible pneumatic cylinder that can be used even if it is deformed by external force has been developed. A portable rehabilitation device using the flexible spherical actuator that consists of two ring-shaped flexible pneumatic cylinders was proposed and tested. The attitude control system using a tiny embedded controller, four small-sized quasi-servo valves, and two accelerometers was also proposed and constructed. The attitude control of the device was executed. In the next step, it is necessary to recognize the relative position between both stages to prevent both hands to contact each other. In this study, the low-cost flexible displacement sensor using the nylon string coated with carbon is proposed and tested. As a result, it is confirmed that the tested sensor can measure the displacement of the cylinder.

Keywords Portable rehabilitation device • Flexible pneumatic cylinder • Flexible spherical actuator • Nylon string coated with carbon • String-type displacement sensor

10.1 Introduction

In an aging society, it is required to develop a system to aid in nursing care [1] and to support the activities of daily life for the elderly and the disabled [2]. In addition, the rehabilitation devices help the elderly who was injured temporally to recover their physical ability for keeping Quality of Life (QOL). The actuators used in such a system need to be flexible so as not to injure the human body [3]. The purpose of this study is to develop and improve a portable rehabilitation device that can be safe enough to use it while handling it with hands. In our previous study, a novel flexible

Y. Matsui (✉) • T. Akagi • S. Dohta • S. Fujimoto
Department of Intelligent Mechanical Engineering, Okayama University of Science, Okayama 700-0005, Japan
e-mail: akagi@are.ous.ac.jp

J. Juang (ed.), *Proceedings of the 3rd International Conference on Intelligent Technologies and Engineering Systems (ICITES2014)*, Lecture Notes in Electrical Engineering 345, DOI 10.1007/978-3-319-17314-6_10

pneumatic cylinder that can be used even if the cylinder is deformed by external forces has been proposed and tested [4]. The portable rehabilitation device using the flexible spherical actuator that consists of two flexible pneumatic cylinders was proposed and tested [5]. The inexpensive attitude measuring system using an embedded controller and accelerometers was also proposed and tested [5]. An attitude control system of the portable rehabilitation device using an embedded controller and four quasi-servo valves [6] composed of inexpensive on/off control valves was also constructed and several kinds of attitude control were executed. In the next step, it is necessary to recognize the relative position between both stages to prevent both hands to contact each other. In order to know the relative position, a displacement sensor of the flexible pneumatic cylinder is required. Furthermore, it is ideal to construct the sensor at low-cost. In this study, a novel low-cost flexible displacement sensor is proposed and tested. The performance of the united sensor with the cylinder is investigated.

10.2 Flexible Spherical Actuator

Figure 10.1 shows the construction of a rodless-type flexible pneumatic cylinder developed by us [4]. The cylinder consists of a flexible tube as a cylinder and gasket, one steel ball as a cylinder head and a slide stage that can move along the outside of the cylinder tube. The steel ball in the tube is pinched by two pairs of brass rollers from both sides of the ball. The operating principle of the cylinder is as follows. When the supply pressure is applied to one side of the cylinder, the inner steel ball is pushed. At the same time, the steel ball pushes the brass rollers and then the slide stage moves toward opposite side of the pressurized while it deforms the tube.

Figure 10.2 shows the appearance and the detailed construction of the handling stage of the tested spherical actuator using flexible pneumatic cylinders. The actuator could give the rehabilitation motion for their shoulders and arms while the patient is holding both handling stages with hands. The spherical actuator consists of two ring-shaped flexible pneumatic cylinders which are intersected at right angle and each slide stage of the flexible cylinder is fixed on each handling stage as shown in Fig. 10.2 on the right. The size of the actuator is 260 mm in width and 270 mm in height. The total mass of the device is 310 g. In order to measure the attitude angle of each handling stage, two accelerometers are used as angular sensors.

Figure 10.3 shows the transient view and the response of the stage angle of the spherical actuator. In the experiment, the sequential on/off operation of the control valve every 0.8 s was done. The supply pressure of 450 kPa is applied. From Fig. 10.3, it can be seen that the actuator can create the different attitudes easily. It was found that the measuring device could also observe the inclined angle between both handling stages in the actuator. However, the relative position between both stages cannot be recognized because of lack of information about the displacement between both stages. In the next step, it is necessary to recognize the relative

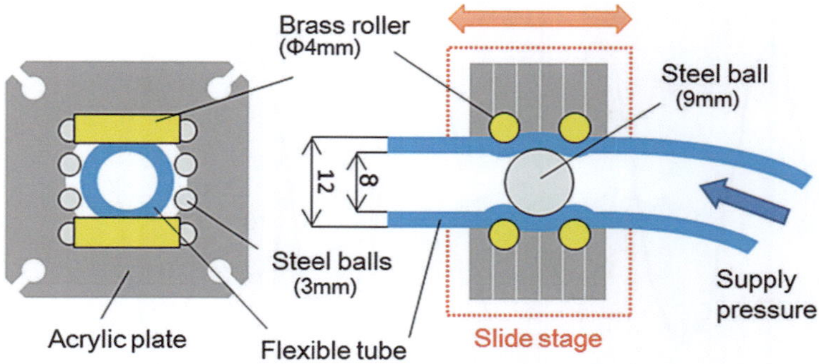

Fig. 10.1 Construction of the flexible cylinder

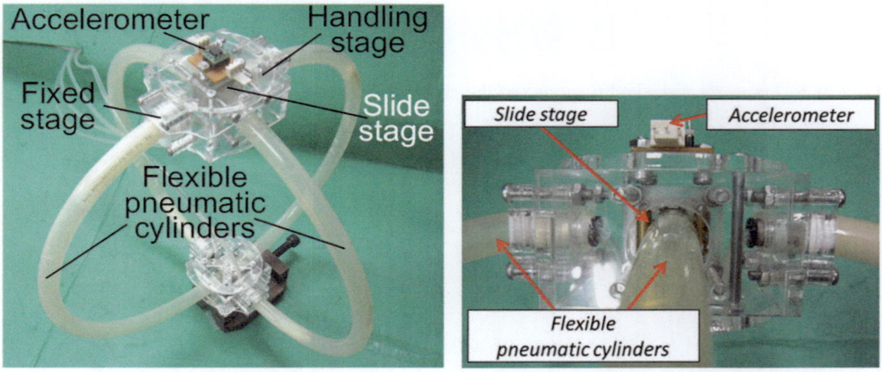

Fig. 10.2 Appearance and the detailed construction of the handling stage of the spherical actuator

position between both stages to prevent both hands to contact each other. In order to get the relative position, a displacement sensor of the flexible pneumatic cylinder is required.

10.3 Flexible Linear Encoder

As a displacement sensor of the flexible pneumatic cylinder, the flexible linear encoder as shown in Fig. 10.4 was proposed and tested in our previous study [7]. The linear encoder consists of the yellow-colored flexible tube with many circumferential slits on the surface of the tube and a sensor unit with four photoreflectors (Kodenshi Co. Ltd., SG-105). Slits are arranged with a distance of 2 mm and a width of 1 mm. Figure 10.4 on the right shows the inner construction of the sensor unit. The sensor consists of two pairs of the detecting units (a–a′ and b–b′) with each unit having two photoreflectors. In the detecting unit, two photoreflectors are set facing each other through the flexible tube. Each detecting unit is set in parallel with a

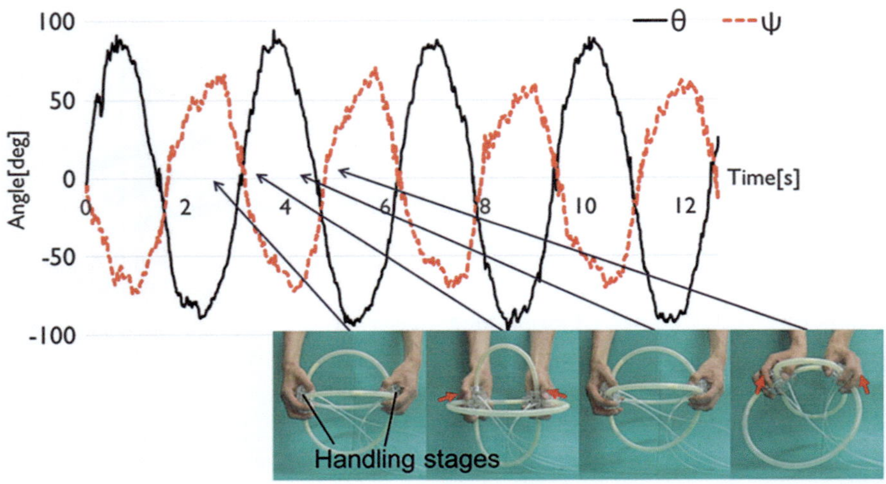

Fig. 10.3 Transient view and response of the stage angle of the spherical actuator

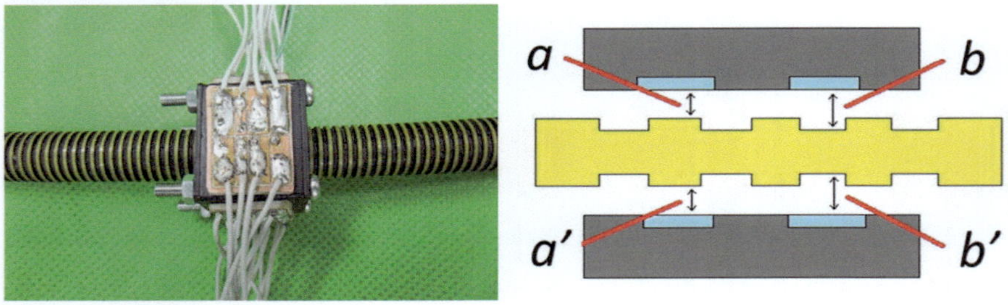

Fig. 10.4 Appearance of the flexible linear encoder and the inner construction of the sensor unit

distance of 6.1 mm to make two kinds of signals, Phase A and Phase B, shifted by 90°. By measuring this distance from each detecting unit and using the programmed up/down counter constructed on the embedded controller. By this method, the linear encoder can measure the displacement of the flexible tube even if the tube bends.

Figure 10.5 shows the transient view of the movement of the one-ring-shaped flexible cylinder with diameter of 280 mm in tracking position control using the linear encoder. In the experiment, the displacement of the cylinder can be controlled according to the desired position given by the potentiometer by the embedded controller (Renesas Co. Ltd., H8/3364) and on/off valves (Koganei Co. Ltd., G010E1). It can be seen that the slide stage could be controlled of its position by using the linear encoder. However, in the case when the larger deformation was given to the ring-shaped cylinder so that it changes to the elliptic shape, the sensor unit was caught by the step of the slit tube. Therefore, it will be needed to develop the other flexible sensor that can be used when the larger deformation is occurred on the cylinder.

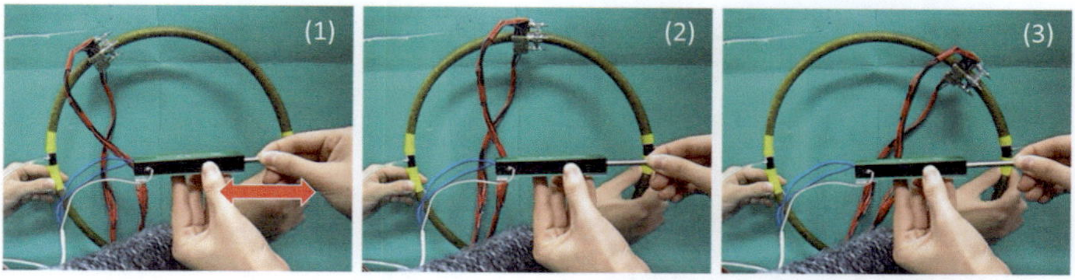

Fig. 10.5 Transient view of the movement of the one-ring-shaped flexible cylinder with diameter of 280 mm in tracking position control using the tested linear encoder

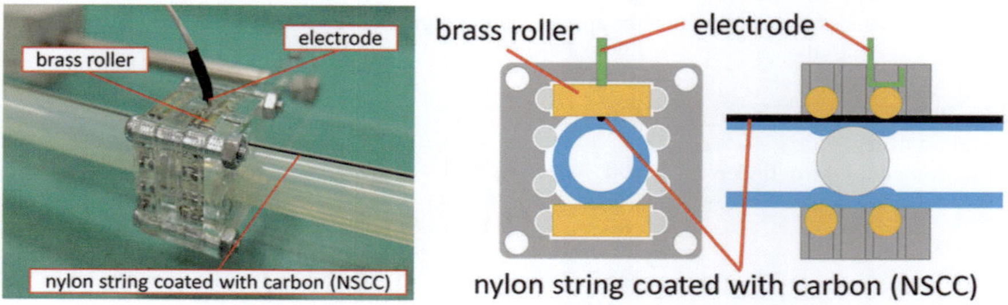

Fig. 10.6 Appearance of the flexible pneumatic cylinder with string-type displacement sensor and the inner construction of the slide stage with sensor

10.4 String-Type Displacement Sensor

Figure 10.6 shows the construction of a tested string-type flexible displacement sensor (it is called "STDS" for short) and the inner construction of the slide stage with sensor. The novel tested cylinder consists of flexible pneumatic cylinder and a nylon string coated with carbon (it is called "NSCC" for short). The diameter of the NSCC is about 0.6 mm. In order to set the NSCC on the flexible tube, the tube has a groove with depth of 0.4 mm along the longitudinal axis of the tube. It works as a flexible potentiometer by sliding the brass roller along NSCC while keeping the electrical contact. The estimated cost of the STDS is extremely low, that is less than 1 US dollars.

Figure 10.7 shows the experimental setup using the STDS and the relation between the displacement of the cylinder measured by a potentiometer and the output signal of the STDS (A/D value). The experimental setup consists of a STDS on the flexible pneumatic cylinder, a potentiometer that is set parallel to the cylinder and a microcomputer (Renesas Co. Ltd., SH7125). The microcomputer gets the output voltages from the STDS and the potentiometer through A/D converter. From Fig. 10.7 on the right, it can be seen that the relation between the output from the STDS and displacement is linear. It means that the tested STDS is useful to apply it as a displacement sensor.

Figure 10.8 shows the transient response of the displacement of the cylinder. In Fig. 10.8, broken line and solid lines show the measured displacement using the

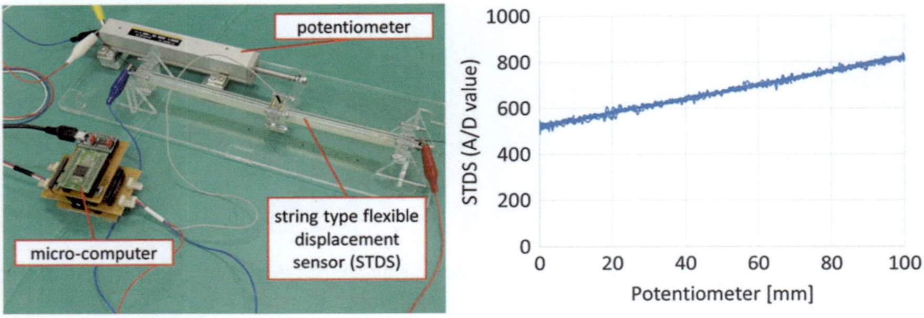

Fig. 10.7 Experimental setup using the STDS and the relation between the displacement of the cylinder measured by a potentiometer and the output signal of the STDS (A/D value)

Fig. 10.8 Transient response of the displacement of the cylinder

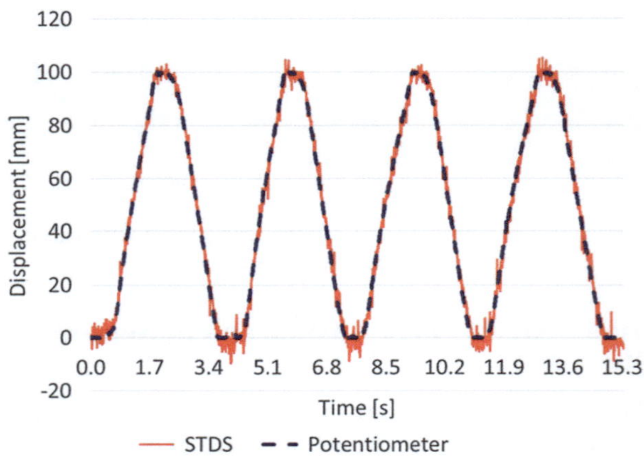

potentiometer and the tested STDS, respectively. It can be seen that the displacement measured by the tested sensor (STDS) agrees well with the displacement measured by the potentiometer. It is confirmed that the STDS is useful to apply it as a flexible potentiometer because of its extremely low-cost and compact sensor configuration.

10.5 Conclusions

This study that aims to develop the displacement sensor for the portable rehabilitation device can be summarized as follows.

1. As a displacement sensor of the flexible pneumatic cylinder, the developed flexible linear encoder developed before was applied to the device. As a result, it can be seen that the position of the slide stage could be controlled by using the

linear encoder. However, it was also concluded that the sensor unit was caught by the step of the slit tube in the case of larger deformation of the tube.

2. The string-type flexible displacement sensor (STDS) using a NSCC that can be used when the larger deformation is occurred on the cylinder was proposed and tested. As a result, it is confirmed that the STDS is useful because of its extremely low-cost and compact sensor configuration.

Acknowledgments This study was supported by the Education Ministry of Japan through a Financial Assistance Program for QOL Innovative Research (2012-) and a Grant-in-Aid for Scientific Research (C) (Subject No. 24560315).

References

1. Noritsugu, T., Takaiwa, M., Sasaki, D.: Development of power assist wear using pneumatic rubber artificial muscles. J. Robot. Mechatron. **21**, 607–613 (2009)
2. Kobayashi, H., Shiban, T., Ishida, Y.: Realization of all 7 motions for the upper limb by a muscle suit. J. Robot. Mechatron. **16**, 504–512 (2004)
3. Nagata, Y. (ed.): Soft Actuators—Forefront of Development, NTS Ltd., pp. 291–335 (2004)
4. Akagi, T., Dohta, S.: Development of a rodless type flexible pneumatic cylinder and its application. Trans. JSME, Ser. C **73**, 2108–2114 (2007)
5. Matsui, Y., Akagi, T., Dohta, S., Aliff, M., Liu, C.: Development of portable rehabilitation device using flexible spherical actuator and embedded controller. In: Proceedings of the 2nd International Conference on Intelligent Technologies and Engineering Systems (ICITES2013). Lecture Notes in Electrical Engineering, vol. 293, pp. 231–238. Springer, Heidelberg (2014)
6. Moriwake, Y., Akagi, T., Dohta, S., Zhao, F.: Development of low-cost pressure control type quasi-servo valve using embedded controller. J. Procedia Eng. **41**, 493–500 (2012)
7. Akagi, T., Dohta, S., Matsushita, H., Fukuhara, A.: Development of flexible pneumatic cylinder with built-in flexible linear encoder and flexible bending sensor. J. Syst. Des. Dyn. **6**(4), 359–372 (2012)

Chapter 11
Evaluation of Computing Reliability in Internet Computing

Shin-Guang Chen

Abstract Internet computing is now a blooming technology which refers to multiple independent computing units working together over the Internet. The computing or calculating units are processor–memory pairs which distribute over one or more computers networked by information links over the Internet. It is important to maintain a highly efficient computing environment, namely, computing reliability. However, very few literature concerns this issue. This chapter proposes an evaluation method for the computing reliability in Internet computing. A numerical example is presented for exploring the proposed method. The results show that the proposed method is helpful in the evaluation of computing reliability for Internet computing applications.

Keywords Internet computing • Computing reliability • Processor–memory pair • Flow network • Minimal path

11.1 Introduction

Internet computing is a technology which refers to multiple independent computing units working together over the Internet. The computing units are processor–memory pairs which distribute over one or more computers networked by information links. Such linkage is important to maintain a highly efficient computing environment, namely, computing reliability. A popular way of improving the reliability of Internet computing is to construct a reliable architecture of the network [1]. Regarding the evaluation of computing reliability in a given network, very few literature concerns this issue [2]. However, on telecommunication networks, there is a vast amount of literature regarding such evaluation of network reliabilities.

Traditionally, two-terminal network reliability, all-pair network reliability, and k-out-of-n network reliability [3] are deeply investigated. Moreover, Boubour

S.-G. Chen (✉)
Department of Industrial Management, Tungnan University, New Taipei City 222, Taiwan
e-mail: bobchen@mail.tnu.edu.tw

J. Juang (ed.), *Proceedings of the 3rd International Conference on Intelligent Technologies and Engineering Systems (ICITES2014)*, Lecture Notes in Electrical Engineering 345, DOI 10.1007/978-3-319-17314-6_11

et al. [4] discussed a new use of safe Petri nets in the field of distributed discrete event systems with application to telecommunication network management. Tomasgard et al. [5] formally described new optimization models for telecommunication networks with distributed processing around the shift of focus towards processing, the new technological aspects, and how to utilize flexibility to cope with uncertainty. Ke and Wang [6] examined two compensating methods that account for imperfect nodes and perfect nodes, and presented a revised Torrieri method to evaluate the bi-state network reliability in terms of network partitioning method.

This chapter proposes a method based on network theory [7] for the evaluation of computing reliability in Internet computing. This method involves two parts: measurement and integration. At first, the information links are regarded as arcs, and the processor–memory pairs are regarded as nodes. Each node transmits information to any other nodes and vice versa. So, the first step is to measure the reliability of each arc. Then, the overall computing reliability is obtained by the integration of all individual reliabilities, and is formally defined by the probability that the maximal information flow of the entire network is not less than the required information volume *d*. Usually, the minimal path (MP) method [8] is used for the integration of such probability. The remainder of the work is described as follows: The assumptions are addressed in Sect. 11.2. The Internet computing model is described in Sect. 11.3. Section 11.4 gives the proposed method. Then, a numerical example is demonstrated in Sect. 11.5. Section 11.6 draws the conclusion and discussions of the chapter.

11.2 Assumptions

Let (A, N, M) be a network, where $A = \{a_i | 1 \leq i \leq n\}$ is the set of arcs representing the information linkages, $N = \{b_j | 1 \leq j \leq s\}$ is the set of nodes representing the processor–memory pairs, and $M = (m_1, m_2, \ldots, m_n)$ is a vector with m_i (an integer) being the maximal capacity of a_i. Such a network is assumed to satisfy the following assumptions.

1. Each node can transmit/receive information to/from the other nodes.
2. The capacity of each arc a_i is an integer-valued random variable which takes values from the set $\{0, 1, 2, \ldots, m_i\}$ according to the observed empirical distribution.
3. The nodes are perfect. That is, the nodes are excluded from the reliability integration.
4. Flow in the network satisfies the flow-conservation law [7].
5. The arcs are statistically independent of each other.

11.3 The Internet Computing Model

Suppose $\rho_1, \rho_2, \ldots, \rho_z$ are totally the MPs from one node to the other nodes. Thus, the network model can be described in terms of two vectors: the capacity vector $X = (x_1, x_2, \ldots, x_n)$ and the flow vector $F = (f_1, f_2, \ldots, f_z)$ where x_i denotes the current capacity of a_i and f_j denotes the current flow on ρ_j. Then such a vector F is feasible iff

$$\sum_{j=1}^{z} \left\{ f_j \middle| a_i \in \rho_j \right\} \le m_i, \qquad \text{for each } i = 1, 2, \ldots, n. \tag{11.1}$$

Constraint (11.1) describes that the total flow through a_i cannot exceed the maximal capacity of a_i. Such set of F is denoted as $U_M \equiv \{F|F \text{ is feasible under } M\}$. Similarly, F is feasible under $X = (x_1, x_2, \ldots, x_n)$ iff

$$\sum_{j=1}^{z} \left\{ f_j \middle| a_i \in \rho_j \right\} \le x_i, \qquad \text{for each } i = 1, 2, \ldots, n. \tag{11.2}$$

For clarity, let $U_X = \{F|F \text{ is feasible under } X\}$. The maximal flow under X is defined as $V_X \equiv \max\left\{ \sum_{j=1}^{z} f_j | F \in U_X \right\}$.

11.3.1 The Computing Reliability Evaluation

Given a required information volume d, the computing reliability denoted by R_d is the probability that the maximal information flow of the network is no less than d, i.e., $R_d \equiv \Pr\{X|V_X \ge d\}$. To calculate R_d, it is advantageous to find the minimal vector in the set $\{X|V_X \ge d\}$. A minimal vector X is said to be a d-MP for d iff (1) $V_X \ge d$ and (2) $V_Y < d$ for any other vector Y such that $Y < X$, in which $Y \le X$ iff $y_j \le x_j$ for each $j = 1, 2, \ldots, n$ and $Y < X$ iff $Y \le X$ and $y_j < x_j$ for at least one j. Suppose there are totally t d-MPs for d: $X_1, X_2, \ldots, X_t$, the computing reliability is equal to

$$R_d = \Pr\left\{ \bigcup_{k=1}^{t} \{X|X \ge X_k\} \right\}, \tag{11.3}$$

which can be calculated by inclusion–exclusion principle or RSDP method [9].

11.3.2 The Generation of All **d**-*MPs for* **d**

At first, we find the flow vector $F \in U_M$ such that the total flow of F equals d. It is defined in the following demand constraint,

$$\sum_{j=1}^{z} f_j = d. \tag{11.4}$$

Then, let $\mathbf{F} = \{F | F \in U_M \text{ and satisfies (11.4)}\}$. If X is a d-MP for d, then there is an $F \in \mathbf{F}$ such that

$$x_i = \sum_{j=1}^{z} \left\{ f_j | a_i \in \rho_j \right\}, \quad \text{for each } i = 1, 2, \ldots, n. \tag{11.5}$$

This is a necessary condition for a d-MP. To explain that, consider a d-MP X and an $F \in U_X$ feasible under $\mathbf{F}$. It is known that $\sum_{j=1}^{z} \left\{ f_j | a_i \in \rho_j \right\} \leq x_i$, $\forall i$. Suppose there is a k such that $x_k > \sum_{j=1}^{z} \left\{ f_j | a_k \in \rho_j \right\}$. Set $Y = (y_1, y_2, \ldots, y_{k-1}, y_k, y_{k+1}, \ldots, y_n) = (x_1, x_2, \ldots, x_{k-1}, x_k - 1, x_{k+1}, \ldots, x_n)$. Hence $Y < X$ and $F \in U_Y$ (since $\sum_{j=1}^{z} \left\{ f_j | a_i \in \rho_j \right\} \leq y_i$, $\forall i$), which indicates that $V_Y \geq d$ and contradicts to that X is a d-MP for d. Thus $x_i = \sum_{j=1}^{z} \left\{ f_j | a_i \in \rho_j \right\}$, $\forall i$. Given $F \in \mathbf{F}$, we generate a capacity vector $X_F = (x_1, x_2, \ldots, x_n)$ via (11.5). The set $\Lambda = \{X_F | F \in \mathbf{F}\}$ is built. Let $\Lambda_{\min} = \{X | X \text{ is a minimal vector in } \Lambda\}$. Then, $\Lambda_{\min}$ is the set of d-MPs for d.

11.4 The Proposed Method

11.4.1 The Measurement

At first, to build up the reliability data of each individual arc, a tool for counting the transferred information volume is required. Figure 11.1 gives an example of counting the information packets between computers. Then, we can make the statistics for the observation during a period of time. Suppose Table 11.1 is the results of observation. By statistics, we can immediately obtain the derived empirical distribution in Table 11.2.

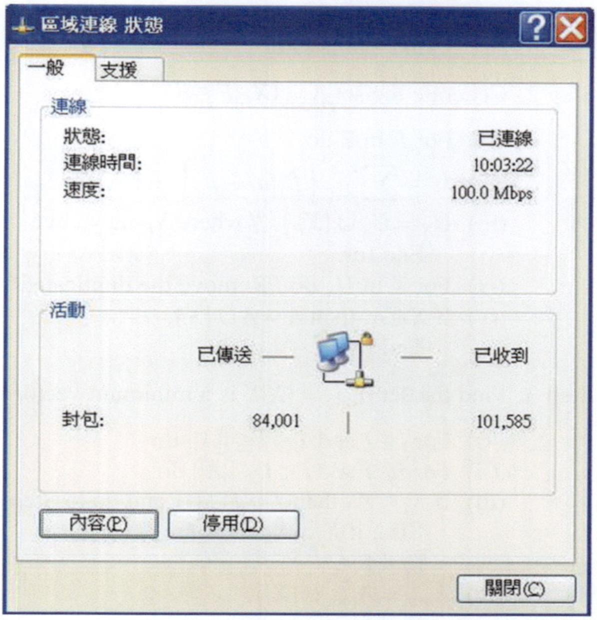

Fig. 11.1 A tool for counting the information packets between computers

Table 11.1 The observations of arc a_1

	The # of transmitted packets						
Items	0	1	2	3	4	5	6
The occurrences	50	66	78	99	120	82	45

Table 11.2 The empirical distribution of arc a_1

	The # of transmitted packets						
Items	0	1	2	3	4	5	6
The probabilities	0.09	0.12	0.14	0.18	0.22	0.15	0.08

11.4.2 The Algorithm

The following algorithm is proposed to search for all *d*-MPs for *d* in the network.

Algorithm 1: Searching for all *d*-MPs for *d*.

Step 1. Find the feasible flow vector $F = (f_1, f_2, \ldots, f_z)$ satisfying both capacity and demand constraints.

(i) Enumerate f_j for $1 \le j \le z$, $0 \le f_j \le \min\{m_i | a_i \in \rho_j\}$ do

(ii) If f_j satisfies the following constraints

$$\sum_{j=1}^{z} \left\{ f_j | a_i \in \rho_j \right\} \le m_i \quad \text{and} \quad \sum_{j=1}^{z} f_j = d, \quad \text{for} \quad 1 \le i \le n,$$

then $\mathbf{F} = \mathbf{F} \cup (F)$.
End enumerate.

Step 2. Generate the set $\Lambda = \{X_F | F \in \mathbf{F}\}$.

(i) For F in $\mathbf{F}$ do
(ii) $x_i = \sum_{j=1}^{z} \left\{ f_j | a_i \in \rho_j \right\}$, for $i = 1, 2, \ldots, n$.
(iii) $U_X = U_X \cup \{X_F\}$. // where X_F may have duplicates.
End for.
(iv) For X in U_X do //Remove the duplicated vectors.
(v) If $X \notin \Lambda$, then $\Lambda = \Lambda \cup \{X\}$.
End for.

Step 3. Find the set $\Lambda_{\min} = \{X | X$ is a minimum vector in $\Lambda\}$. Let $J = \{j | X_j \notin \Lambda_{\min}\}$.

(i) For $i \notin J$ and $1 \leq i \leq ||\Lambda||$ do
(ii) For $j \notin J$ and $i < j \leq ||\Lambda||$ do
(iii) If $X_j \leq X_i$, then $J = J \cup \{i\}$ and go to Step 3(i).
Else, if $X_j > X_i$, then $J = J \cup \{j\}$.
End for.
(iv) $\Lambda_{\min} = \Lambda_{\min} \cup \{X_i\}$.
End for.
(v) Return $\Lambda_{\min}$.

Step 1 indicates that according to the MPs, the feasible F under Constraints (11.1) and (11.4) was enumerated into $\mathbf{F}$. Then, the candidate vector set Λ for d-MPs can be derived from $\mathbf{F}$ under (11.5) at Step 2. Finally, the set of d-MPs was created by Step 3.

11.5 Numerical Example

Figure 11.2 gives an example of Internet computing with four processor–memory pairs. There are 12 MPs: $\rho_1 = \{b_1, a_1, b_4\}$, $\rho_2 = \{b_1, a_2, b_2\}$, $\rho_3 = \{b_1, a_6, b_3\}$, $\rho_4 = \{b_2, a_3, b_3\}$, $\rho_5 = \{b_2, a_5, b_4\}$, $\rho_6 = \{b_2, a_2, b_1\}$, $\rho_7 = \{b_3, a_3, b_2\}$, $\rho_8 = \{b_3, a_4, b_4\}$, $\rho_9 = \{b_3, a_6, b_1\}$, $\rho_{10} = \{b_4, a_4, b_3\}$, $\rho_{11} = \{b_4, a_1, b_1\}$, $\rho_{12} = \{b_4, a_5, b_2\}$. Table 11.3 gives the empirical distributions of six arcs of the network. The required information volume is four packets per second.

After calculation, one gets the computing reliability $R_4 = 0.9604$ by a RSDP program coded with PROLOG running on a PC with four CPUs of Intel Core i5 @ 2.8 GHz, 2.79 GHz, and 2 GB RAM.

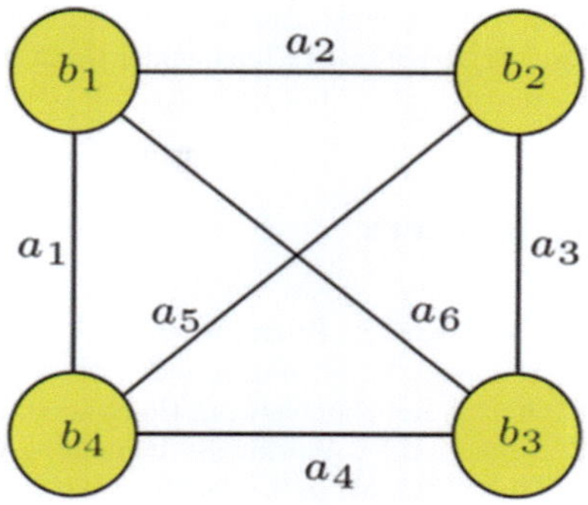

Fig. 11.2 An example of Internet computing with four processor–memory pairs

Table 11.3 The empirical distributions of six arcs of the network

	The # of transmitted packets						
Arcs	0	1	2	3	4	5	6
a_1	0.09	0.12	0.14	0.18	0.22	0.15	0.08
a_2	0.07	0.11	0.15	0.19	0.21	0.17	0.10
a_3	0.05	0.08	0.16	0.22	0.21	0.27	0.01
a_4	0.02	0.13	0.18	0.27	0.18	0.21	0.01
a_5	0.10	0.09	0.20	0.09	0.16	0.23	0.13
a_6	0.11	0.10	0.24	0.10	0.14	0.10	0.21

11.6 Conclusion and Discussions

This chapter proposes an evaluation method for the computing reliability in Internet computing. Internet computing is a technology which refers to multiple independent computing units working together over a network. The computing units are processor–memory pairs which distribute over one or more computers networked by information links. Such linkage is important to maintain a highly efficient computing environment. This method involves two parts: measurement and integration. At first, the information links are regarded as arcs, and the processor–memory pairs are regarded as nodes. Each node transmits information to any other nodes and vice versa. So, the first step is to measure the reliability of each arc. Then, the computing reliability of the entire network is obtained by the integration of all individual reliabilities, and is formally defined by the probability that the maximal information flow of the network is no less than the required information volume d.

This chapter also presents an example of Internet computing with four processor–memory pairs. Each unit can communicate with any other units freely. After observing the computing reliability for each individual arc, the overall computing reliability is obtained by integrating all individual reliabilities. The results show that the proposed method is helpful in the evaluation of computing reliability for Internet computing applications.

Acknowledgment This work was supported in part by the Ministry of Science and Technology, Taiwan, Republic of China, under Grant No. MOST 103-2221-E-236-005.

References

1. Nikoletseas, S., Palem, K., Spirakis, P., Yung, M.: Automata, languages and programming. In: Abiteboul, S., Shamir, E. (eds.) Short Vertex Disjoint Paths and Multiconnectivity in Random Graphs: Reliable Network Computing. LNCS, vol. 820, pp. 508–519. Springer, Berlin (1994)
2. Birman, K.P.: Worldwide computing and its applications. In: Masuda, T., Masunaga, Y., Tsukamoto, M. (eds.) Building Secure and Reliable Network Applications. LNCS, vol. 1274, pp. 15–28. Springer, Berlin (1997)
3. Chao, M.T., Fu, J.C., Koutras, M.V.: Survey of reliability studies of consecutive-k-out-of-n:F and related systems. IEEE Trans. Reliab. **44**, 120–127 (1995)
4. Boubour, R., Jard, C., Aghasaryan, A., Fabre, E., Benveniste, A.: A Petri net approach to fault detection and diagnosis in distributed systems. Part I: Application to telecommunication networks, motivations, and modelling. In: Proceedings of the 36th IEEE Conference on Decision and Control, San Diego, CA, USA, vol. 1, pp. 720–725 (1997)
5. Tomasgard, A., Audestad, J.A., Dye, S., Stougie, L., van der Vlerk, M.H., Wallace, S.W.: Modelling aspects of distributed processing in telecommunication networks. Ann. Oper. Res. **82**, 161–185 (1998)
6. Ke, W.J., Wang, S.D.: Reliability evaluation for distributed computing networks with imperfect nodes. IEEE Trans. Reliab. **46**, 342–349 (2002)
7. Ford, L.R., Fulkerson, D.R.: Flows in networks. Princeton University Press, Princeton (1962)
8. Chen, S.G., Lin, Y.K.: Search for all minimal paths in a general large flow network. IEEE Trans. Reliab. **61**, 949–956 (2012)
9. Zuo, M.J., Tian, Z., Huang, H.Z.: An efficient method for reliability evaluation of multistate networks given all minimal path vectors. IIE Trans. **39**, 811–817 (2007)

Chapter 12
An Effective Method for Classification of White Rice Grains Using Various Image Processing Techniques

Suchart Yammen and Chokcharat Rityen

Abstract This chapter presents an algorithm for classifying grains of white rice by using image processing. Each image size is acquired via a digital camera. The resolution is 720 × 480 pixels. The algorithm begins with improving grain images, converting these images into binary images by using Otsu's method, removing noise from the binary images by applying the morphological method with square structural elements, detecting each grain boundary by using the Canny operator, and determining the length of each grain by using the Euclidean method. Next, the grain length is used for classifying the rice grains according to the Rice Standards of Thailand. The testing results from processing 500 grain images; one grain per image, the algorithm provides good performance with the mean absolute error of 0.01 mm in length. For 300 grain images with some grains per image, the algorithm provides good classification with an average accuracy of 99.33 %.

Keywords Image processing • Morphological • Classification • White rice grains

12.1 Introduction

According to Ministry of Commerce's B.E. 2540 Thai rice standard, there are procedures and classification of different Thai rice species. However, the judgment of the grain quality used in here is based on physical examination of white rice grain considering only rice classification shown in Table 12.1.

S. Yammen (✉)
Department of Electrical and Computer Engineering, Naresuan University, Phisanulok 65000, Thailand
e-mail: sucharty@nu.ac.th

C. Rityen
Department of Electrical Engineering, Rajamangala University of Technology Lanna, Tak 63000, Thailand
e-mail: chok.kpg@gmail.com

J. Juang (ed.), *Proceedings of the 3rd International Conference on Intelligent Technologies and Engineering Systems (ICITES2014)*, Lecture Notes in Electrical Engineering 345, DOI 10.1007/978-3-319-17314-6_12

Table 12.1 Classification standard for white rice grain

Grades	Grain classification (%)			
	Long grain			Short grain ($\leq$6.2 mm)
	Class 1 ($\geq$7.0 mm)	Class 2 ($\geq$6.6–7.0 mm)	Class 3 (>6.2–6.6 mm)	
100 % Grade A	$\geq$70.00	–	$\leq$5.00	0.00
100 % Grade B	$\geq$40.00	–	–	$\leq$5.00
100 % Grade C	$\geq$30.00	–	–	$\leq$5.00
5 %	$\geq$20.00	–	–	$\leq$10.00
10 %	$\geq$10.00	–	–	$\leq$15.00
15 %	$\geq$5.00	–	–	$\leq$30.00

Nowadays, owners of rice mills in Phitsanulok of Thailand have experts to classify the rice grain using their two naked eyes. This often causes price argument for inspecting the grain quality of white rice between farmers as a seller and mill owners as a buyer. From the mentioned argument, the farmers need an effective tool for grain classification in order to increase accuracy and reliability in verifying the rice grain quality.

Many researchers have recently developed various tools for different types of object classification. Ajmal and Hussain [1] presents the detection and classification of vehicle type via aerial photography camera using an image processing method. This process is necessary to determine an edge and set a threshold value with morphological processing for vehicle detection. The result shows that types of vehicles can be appropriately classified with an accuracy rate of 85 %. Then, Lin et al. [2] presents statistical techniques from the popular histogram with determining the adaptive threshold value for inspecting defects that occur on a railway. Moreover, Sim et al. [3] presents a Contrast Enhancement Bilateral Closing Top-hat Otsu Thresholding (CEBICTOT) for the crack detection in structure. This method has five steps. The first step is to increase the contrast of an input image, and the input image is then converted to a gray-scale image. The second step is to bring the gray-scale image through a bilateral filter. The third step is to delete a background of the filtered image with opening and closing top-hat methods. The fourth step is to remove the image background using Otsu's method. The fifth step is to reduce noise from the obtained image for giving clear crack of the structure.

In this chapter, an effective method for grain classification of white rice is developed by using various techniques in image processing. The used techniques in order are the correlation method [4], the Otsu method [5, 6], the connected component labeling method [7], the morphological method with structural elements [8, 12], and the canny operator [9–11]. The chapter is organized as follows. The proposed method for grain classification of white rice is given in Sect. 12.2. Experimental results and discussions are thoroughly provided in Sect. 12.3. Finally, conclusion is presented in Sect. 12.4.

12.2 The Proposed Method for Grain Classification of White Rice

The procedure of the proposed method for grain classification of white rice is divided into nine following steps.

Step 1. Prepare input image data. Testing images is divided into two sets. The first set is composed of 500 images of white rice; one grain per image, as shown in Fig. 12.1. The second set consists of 300 grain images with some grains per image, as shown in Fig. 12.2. Each image of both datasets is 720 pixels × 480 pixels, and has a reference paper whose length and width are 20 mm and 1 mm, respectively. The reference paper is used to be an actual scale for determining the length of the rice grain.

Step 2. Convert input color images to gray-scale images. Each input color obtained image from Step 1 is converted to gray-scale image using the relationship (12.1):

$$\text{Grayscale} = [(R + G + B)/3] \tag{12.1}$$

The three values of R, G, and B are the three intensities of the red light, the green light, and the blue light, respectively. Symbol [•] is the operator that converts a real number to an integer. Thus, grayscale is an interger between 0 and 255.

Step 3. Convert gray-level images into binary images. Each obtained image from Step 2 is converted to a binary image using the Otsu method with automatic threshold t. The t value is determined from the optimization problem by applying the minimum variance between the groups ($\sigma_b^2[t]$) as the relationship (12.2):

$$\max_{t \in \{0,1,2,\ldots,255\}} \sigma_b^2[t] = \max_{t \in \{0,1,2,\ldots,255\}} \frac{\{\mu_T \omega[t] - \mu[t]\}^2}{\omega[t](1-\omega)[t]} = \sigma_b^2[t^*] \tag{12.2}$$

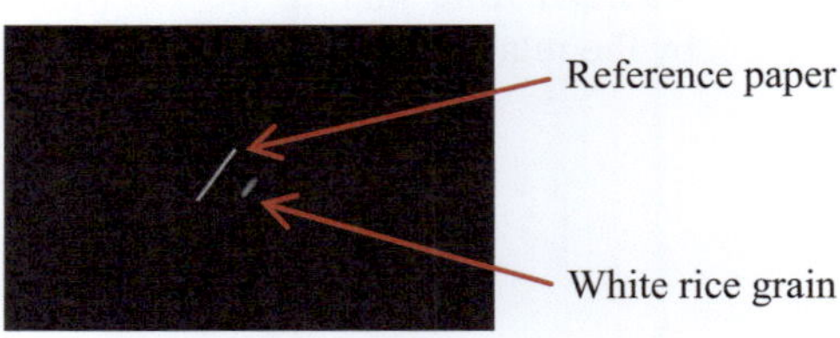

Fig. 12.1 Image of one white rice grain

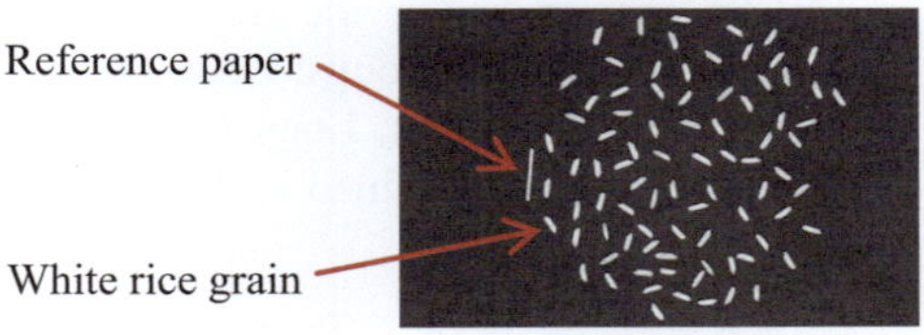

Fig. 12.2 Image of some grains

The ($\omega[t]$) value is the sum of the probability when the gray level is less than t. Both (μ_T) and ($\mu[t]$) values are the gray-level average and the first moment of the histogram whose the gray-level ranges from 0 to $(t-1)$. Then, the optimum threshold (t^*) value is used for converting the gray-scale image ($w[x, y]$) to the binary image ($b[x, y]$) by using the relationship (12.3):

$$b[x,y] = \begin{cases} 1 & \text{if} \quad w[x,y] \geq t^* \\ 0 & \text{if} \quad w[x,y] < t^* \end{cases} \tag{12.3}$$

Step 4. Reduce noise by using the morphological method with structural elements.
Small objects (or noise) of each image obtained from Step 3 are removed by using the morphological opening method with structure elements of six sizes of 3×3, 5×5, 7×7, 9×9, 11×11, and 13×13 pixels in order to get complete grains in terms of shape. The opening operator of image A and structure element B denoted by the symbol $A \circ B$ is defined as:

$$A \circ B = (A \Theta B) \oplus B \tag{12.4}$$

The symbols of Θ and $\oplus$ are the erosion operator and the dilation operator, respectively.

Step 5. Detect grain edges of white rice. Each binary image obtained from Step 4 is operated by using the Canny operator to get the image edge of the rice grains.

Step 6. Find the length of each grain in the image obtained from Step 5. The distance of any pair of coordinates (x_n, y_n) for $n = 1, 2, 3, \ldots, N$ on the edge can be computed by using the relationship (12.5):

$$d[n,k] = \sqrt{(x_n - x_{n+k})^2 + (y_n - y_{n+k})^2} \tag{12.5}$$

where $k = 1, 2, 3, \ldots, N - n$. Then, the longest distance ($d[n^o, k^o]$) is obtained from finding the solution of the optimization problem as given by the relationship (12.6):

$$\max_{\substack{n \in \{1,2,3,\ldots,N\} \wedge \\ k \in \{1,2,3,\ldots,N-n\}}} d[n,k] = d[n^o, k^o] \tag{12.6}$$

Step 7. Classify the length of the rice grain according to the RST. Each length of the rice grain on the image obtained from Step 6 is divided into the following four groups: Long Grain Class 1 (≥7.0 mm), Long Grain Class 2 (≥6.6–7.0 mm), Long Grain Class 3 (>6.2–6.6 mm), and Short Grain (≤6.2 mm) as defined in Table 12.1.

Step 8. Specify grades of the rice grain. Count the number of the rice grains in each group of the image obtained from Step 7 to find proportion of the group for classifying the grade of the rice grains as defined in Table 12.1.

Step 9. Display the results indicating the obtained grade of the rice grain in each image through the computer monitor.

12.3 Results and Discussions

To test the efficiency of the developed algorithm, each length of the rice grain on 500 images is calculated by using the algorithm as mentioned in Sect. 12.2. The obtained length of the grain is compared with that of the grain, which is measured with a vernier caliper, as shown in Fig. 12.3. The algorithm also provides good efficiency with the mean absolute error of 0.01 mm.

In addition, 300 grain images with some grains per image are used to test the effectiveness of the grain classification by using the developed algorithm with structural elements of six sizes. In Table 12.2, the rice grains can be divided into

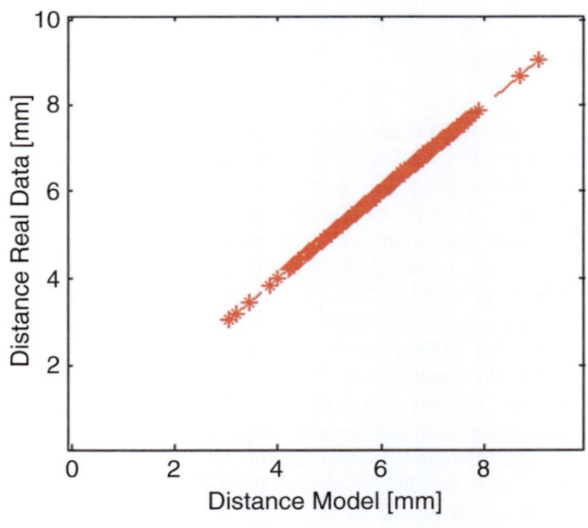

Fig. 12.3 A plot of the grain length obtained from the algorithm and the vernier caliper

Table 12.2 The classification results

	White rice 100 % Grade A			White rice 100 % Grade B			White rice 10 %		
Pixels	True	False	Sum	True	False	Sum	True	False	Sum
3 × 3	94	6	100	99	1	100	100	0	100
5 × 5	99	1	100	99	1	100	100	0	100
7 × 7	99	1	100	99	1	100	99	1	100
9 × 9	99	1	100	99	1	100	98	2	100
11 × 11	100	0	100	97	3	100	95	5	100
13 × 13	100	0	100	49	51	100	2	98	100

three grades: 100 % Grade A, 100 % Grade B, and 10 % for each case of the structure elements. It is clear that the 5 × 5 structural element provides the highest accuracy of grain classification for the three grades. The accuracy for 10 % Grade is 100 %, and the accuracy for 100 % Grade A or 100 % Grade B is 99 %. The dataset is shown in Table 12.2.

12.4 Conclusions

The developed algorithm can be applied for the grain classification of white rice with the average accuracy of 99.33 % for the three grades: 10 % and 100 % Grade A and B. The algorithm also provides highly accurate length of the grain length with the mean absolute error of 0.01 mm.

Acknowledgment This project is financially supported by Hands-on Research Rajamangala University of Technology Lanna.

References

1. Ajmal, A., Hussain, I.M.: Vehicle detection using morphological image processing technique. Paper presented at the international conference on multimedia computing and information technology (MCIT), 2–4 March 2010
2. Lin, J., Luo, S., Li, Q., Zhang, H., Ren, S.: Real-time rail head surface defect detection: a geometrical approach. Paper presented at the IEEE international symposium on industrial electronics, ISIE 2009, 5–8 July 2009
3. Sim, K.S., Kho, Y.Y., Tso, C.P.: Application of contrast enhancement bilateral closing top-hat Otsu thresholding (CEBICTOT) technique on crack images. Paper presented at the 7th IEEE international conference on cybernetic intelligent systems, CIS 2008, 9–10 September 2008
4. Liang, P., Zhiwei, X., Jiguang, D.: Fast normalized cross-correlation image matching based on multiscale edge information. Paper presented at the 2010 international conference on computer application and system modeling (ICCASM), 22–24 October 2010
5. Alasdair, M.: Introduction to Digital Image Processing with Matlab. Thomson Course Technology, Boston (2004)
6. Otsu, N.: A threshold selection method from gray-level histograms. IEEE Trans. Syst. Man Cybern. **9**(1), 62–66 (1979)
7. Baraghimian, G.A.: Connected component labeling using self-organizing feature maps. Paper presented at the proceedings of the 13th annual international conference on computer software and applications conference, COMPSAC 89, 20–22 September 1989
8. Soille, P.: Morphological Image Analysis: Principles and Applications. Springer, Berlin (2003)
9. Bing, W., ShaoSheng, F.: An improved canny edge detection algorithm. Paper presented at the second international workshop on computer science and engineering, WCSE'09, 28–30 October 2009

10. Han, J., Kamber, M.: Data Mining Concepts and Techniques. Elsevier/Morgan Kaufmann, Amsterdam/San Francisco (2006)
11. Canny, J.: A computational approach to edge detection. [Online]. http://ieeexplore.ieee.org/stamp/Stamp.jsp?tp=&arnumber=4767851 (2010)
12. Gonzalez, R.C., Woods, R.E.: Digital image processing. Prentice Hall, Upper Saddle River (2002)

Chapter 13
Applying ZigBee Wireless Sensors and Photovoltaic System to Plant Factory

Shun-Peng Hsu, Yi-Nung Chung, Chih-Chung Yu, and Young-Chi Hsu

Abstract This study proposes to apply wireless monitor and photovoltaic (PV) system to plant factory. The major structures include solar energy system, ZigBee wireless sensor network, and plant factory. In this system, the photovoltaic panels produce energy and the charge controller is used to manage the energy consumption of the load. In order to monitor the plant factory, a ZigBee wireless sensor system is applied. It will monitor the environment situations of the plant factory and supply the information of temperature, humidity, and illumination via the computer network. The system will adjust the environment conditions to enhance the plant growth. According to the experimental results, the system proposed in this chapter can enhance the growth of the plants and reach the goal of saving energy also.

Keywords Plant factory • Solar energy power • ZigBee wireless sensor network

13.1 Introduction

Due to human overexploitation of energy resources, the demand of energy is increased significantly in recent years. Because of large amounts of factories and automobiles produce CO_2 which will enhance the greenhouse effect. It causes global warming and climate change. Therefore, the renewable energy has played an important role today. In this chapter, a solar power system [1, 2] is applied for the plant factory to reduce the CO_2 concentration of atmosphere. The solar cell can convert light energy to electrical energy by two effects which are the

S.-P. Hsu • Y.-N. Chung (✉) • Y.-C. Hsu
Department of Electrical Engineering, National Changhua University of Education, Changhua 500, Taiwan
e-mail: ynchung@cc.ncue.edu.tw

C.-C. Yu
Department of Electrical Engineering, Da-Yeh University, Changhua 515, Taiwan

J. Juang (ed.), *Proceedings of the 3rd International Conference on Intelligent Technologies and Engineering Systems (ICITES2014)*, Lecture Notes in Electrical Engineering 345, DOI 10.1007/978-3-319-17314-6_13

photoconductive effect and the internal electric field. The solar power transformation is affected by two major factors which are sun radiation and temperature. In general, the transformation efficiency of solar panel is low, so the maximum power point (MPP) tracking (MPPT) algorithm [3, 4] must be applied to enhance the solar energy.

On the other hand, traditional agriculture is seriously affected by the environment, so the concept of plant factory getting more and more attention in recent years. Many companies and researchers have invested in this research area and developed greenhouses or plant factories [5]. There are some factors that will affect the growth of plants, which are the water supply, strength of the light, concentration of CO_2, and temperature. In a general greenhouse, it usually uses soil tillage. Even though the environment conditions of greenhouse can be controlled and managed, but the soil nutrients and moisture content will change day by day. The values of electrical conductivity (EC) and PH will change also. To control and measure the EC and PH values are not easy. The soilless cultivation is one of the solutions. In a particular environment such as a plant factory [5], these factors can be controlled and can enhance the growth rate of plants. In order to control these factors, one wireless monitoring system is adopted in this system also. In this research, one plant factory is developed for soilless cultivation. This plant factory combines wireless transmission technology using ZigBee sensor [6], therefore the administrators can control and monitor the plant factory in real time.

ZigBee is a wireless technology, which is used for short distance wireless transmission. ZigBee has its own unique wireless standard, so it can transmit signals between many different sensors and communicate with each other. Because it is short distance signal transmission equipment, if you want to spread far distance, you should use the deliver way to get to the final PC for analysis and access. This study uses ZigBee wireless sensor network to monitor various environment information and to control the plant factory situations. ZigBee has the following characteristics which are power saving, high reliability message transmission, and expansion capability. If on the data transmission side there is no confirmation message received, it will send a message to ensure the reliability of data transmission. ZigBee can also be made with other existing internet connections, even via the internet to control two ZigBee devices in different places.

13.2 System Design

In this research, one plant factory is developed for soilless cultivation. The light source uses LED light and the EC and PH values are controlled also. In order to monitor the plant factory situations, the ZigBee technology is applied to monitor the lightness, temperature, etc. The major parts of this system include solar power system, ZigBee wireless sensor network, and plant growth equipment, which are described as follows. The solar power system is consisted of an array of solar photovoltaic (PV) panels to collect solar energy, and through a charge controller to control the power management based on sunshine situations. If the current of PV array is under a certain threshold value, the controller switches to discharge mode, then the electric energy will be supplied by the storage battery or grid power system.

The output power of a solar energy system is depended on solar irradiation and temperature conditions. In order to obtain the maximum output power, the MPPT algorithm is adopted in a solar energy system. In this chapter, an improved method based on Perturbation and Observation method (P&O) [3, 4] is developed. The basic concept of the P&O method, the current operating point and the subsequent perturbation point, then observes whether the output power should move up towards the MPP within one sampling period. After that, the comparison process will be repeated periodically in every sampling. Such algorithm results in more perturbation loss. An improved tracking algorithm for a solar energy generation system is designed in this chapter. Figure 13.1 shows the flow chart of an improved

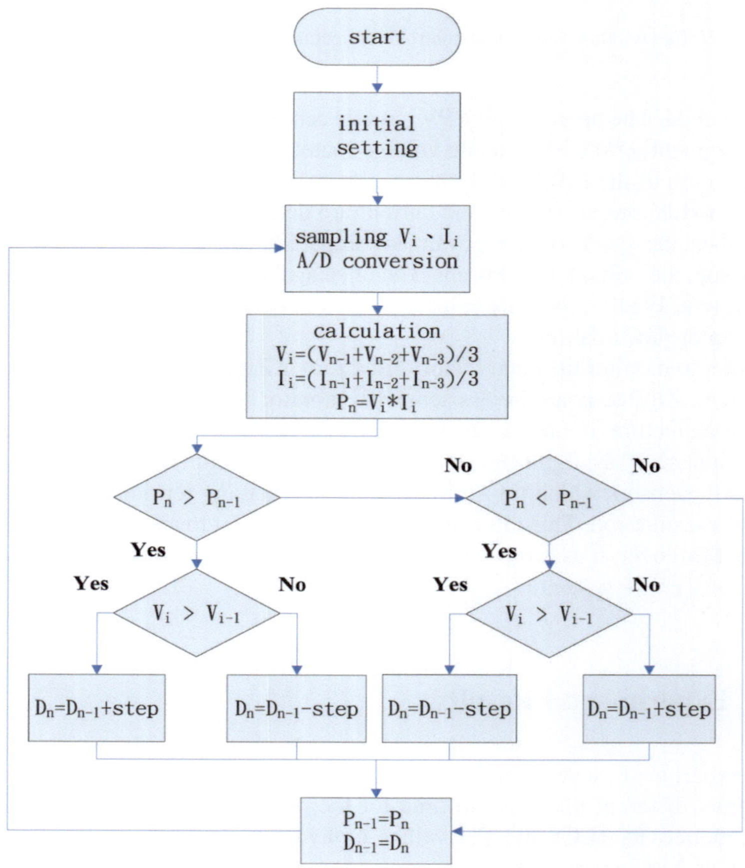

Fig. 13.1 Improved P&O tracking algorithm

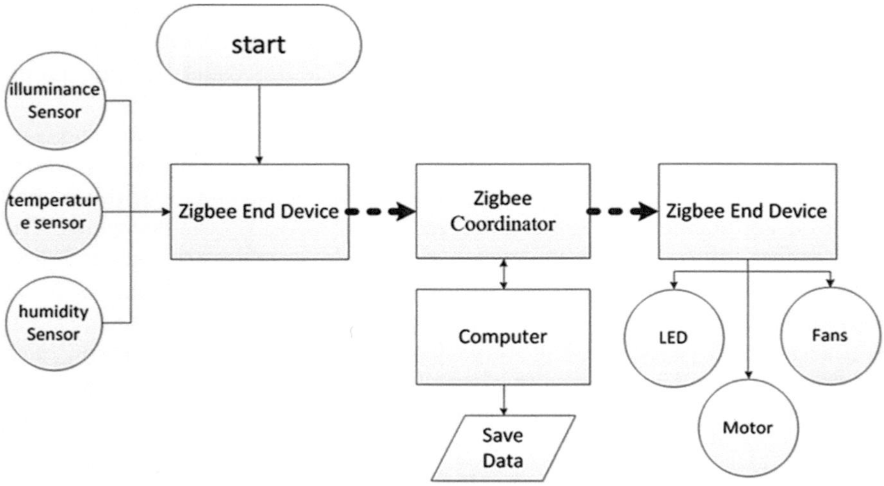

Fig. 13.2 ZigBee wireless sensor and control architecture

MPPT method. The operation of a PV module can be divided into two characteristic areas. Augmenting the PV module voltage increases the power when operating on the left region of the MPP and decreases the power when on the right of the MPP. The PV module output voltage and current are detected first to calculate its output power, then the given output power is compared with the preceding one. After comparison, the voltage value is checked once again to avoid power loss caused by misjudgment. Finally, the duty value is shifted to determine the subsequent step of scaling up or down the load.

In order to monitor the plant factory situations, a ZigBee technology is applied in this system. ZigBee is a wireless sensor to monitor and control the entire system whose architecture is shown as Fig. 13.2. It monitors important environment information such as lightness, temperature, and humidity. The information is detected by sensors which transmit them to the receiving terminal device through wireless transmission. This information can also transmit to a computer and present to users. Moreover, it uses the software to access this information and to the control functions of entire system.

13.3 Experimental Results

In the experiment, several different situations are considered. Based on soilless cultivation, different illumination time for the plant growth is adopted. The electrical conductivity (EC) and PH values of hydroponic solution are to maintain constant in first experiment. In second experiment, both EC and PH values are inconstant. The entire system is monitored by ZigBee system. According to the

Table 13.1 Plant net weight change during the experimental period

Illumination time (h)	No. of plant	Day 1	Day 15	Net weight change	Average weight (g)	EC and PH values
12	1	4.16	26.78	22.62	20.46	Constant
	2	3.7	22.53	18.83		
	3	3.99	23.93	19.94		
12	1	3.39	17.8	14.41	16.15	Inconstant
	2	3.48	17.23	13.75		
	3	3.38	23.67	20.29		
18	1	1.81	29.77	27.96	24.82	Constant
	2	2.47	30.78	28.31		
	3	2.04	20.22	18.18		
18	1	1.49	32.39	30.9	23.43	Inconstant
	2	2.44	13.14	10.7		
	3	1.73	30.42	28.69		
24	1	4.80	53.89	49.09	43.37	Constant
	2	5.25	43.15	37.9		
	3	4.34	47.47	43.13		
24	1	3.97	37.6	33.63	42.75	Inconstant
	2	5.07	55.45	50.38		
	3	4.0	48.25	44.25		

experimental procedure described as above, the net weight change of plants after the period of fifteen days of cultivation is shown in Table 13.1. Based on the experimental results, the plant average net weight is heavier if the illumination time is longer. Moreover, the plant average net weight is heavier when both EC and PH values are kept constant.

13.4 Conclusions

In this research, one plant factory is developed for soilless cultivation. The LED light source is used and the EC and PH values are under control also. In order to monitor the plant factory situations, the ZigBee technology is applied to monitor the lightness, temperature, and humidity. Based on the experimental results, the plant average net weight is heavier if the illumination time is longer. Moreover, the plant average net weight is heavier when both EC and PH values are kept constant. We convince that the system proposed in this chapter can enhance the growth of the plants and reach the goal of saving energy also.

Acknowledgement This work was supported by the National Science Council under Grant NSC 102-2221-E-018-018-.

References

1. Huang, B.J., Hsu, P.C., Wu, M.S., Chen, K.Y.: A high-performance stand-alone solar PV power system for LED lighting. In: Proceeding of IEEE 35th Photovoltaic Specialists Conference (2010)
2. Wahab, H.A., Dukea, M., Carson, J.K., Anderson, T.: Studies of control strategies for building integrated solar energy system. In: Proceeding of IEEE First Conference on Clean Energy and Technology CET (2011)
3. Wang, F., Wu, X., Lee, F.C., Wang, Z., Kong, P., Zhuo, F.: Analysis of unified output MPPT control in subpanel PV converter system. IEEE Trans. Power Electron. **29**(3), 1275–1284 (2013)
4. Gules, R., De Pellegrin Pacheco, J., Hey, H.L., Imhoff, J.: A maximum power point tracking system with parallel connection for PV stand-alone applications. IEEE Trans. Ind. Electron. **55**(7), 2674–2683 (2008)
5. Chuang, K.-C., Juang, D.-J., Liu, W.-C., Chung, Y.-N., Hsu, C.-H., Hu, Y.-N.: A dynamic load variation circuit design of perturbation and observation method for a solar power system. ICIC Express Lett. B Appl. **2**(4), 765–769 (2011)
6. Faheem, I., Siddiqui, A.A., Byung Kwan, I., Chankil, L.: Remote management and control system for LED based plant factory using ZigBee and Internet. In: Advanced Communication Technology (ICACT) (2012)

Chapter 14
Applying Particle Filter Technology to Object Tracking

Tun-Chang Lu, Shun-Peng Hsu, Yu-Xian Huang, Yi-Nung Chung, and Shi-Ming Chen

Abstract This study proposes an approach to track moving objects and to predict the observed targets based on the particle filter. This system includes three parts which are the foreground segmentation, the partial filtering, and the particle filter for tracking objects. In order to estimate the location of next state and track the moving objects, it applies the prior and current state based on the particle filter technology. Experimental result shows that this method can track objects accurately.

Keywords Particle filter • Foreground segmentation • Track objects

14.1 Introduction

In this chapter, the image processing and tracking techniques are applied to track the moving objects. The preliminary processing is adopted before the image processing. Many researchers usually use the optical flow and the background subtraction methods to filter out the interest moving object for following process. The optical flow method will suffer high computational cost and fail due to the ambient illumination changes in the environment [1]. The background subtraction method is easy to implement and has satisfied low computable cost, nevertheless, it will be influenced by the surrounding illumination and the environment conditions [2]. This chapter proposes to use the temporal differencing method which uses properly in the dynamical background and don't often modify the reference background [3]. That also has lower computational time and less influence by the ambient illumination.

T.-C. Lu • S.-P. Hsu • Y.-X. Huang (✉) • Y.-N. Chung • S.-M. Chen
Department of Electrical Engineering, National Changhua University of Education, Changhua 500, Taiwan
e-mail: d0052005@mail.ncue.edu.tw

J. Juang (ed.), *Proceedings of the 3rd International Conference on Intelligent Technologies and Engineering Systems (ICITES2014)*, Lecture Notes in Electrical Engineering 345, DOI 10.1007/978-3-319-17314-6_14

The National Television Standards Committee (NTSC) system for color space conversion is used in image segmentation that can lower the shadow effect which affects the tracking correctness. Tracking the moving objects is the major operation after preliminary image processing. The algorithm of particle filter is applied to trace object in this chapter. The particle filter is applied to the non-Gaussian background which represented by state spatial model in the nonlinear dynamical system. It extends the traditional Kalman filter features and don't have any restrictions to the system noise and noise measurement, then, is properly used in the nonlinear system and have more precise-to-close prediction and reach the real-time tracking [4].

14.2 Image Processing

The input image should be identified in the moving object from background before the tracking process. This chapter proposes to use the temporal differencing method to extract the moving objects. This method subtracts current picture with previous picture in the cluster of image to obtain the moving object continuously. The object tracking will be affected by the ambient illumination change and shadows that cause some interference to get miscalculation. So, a proper color conversion must be done for the preprocessing. This chapter uses the NTSC system to convert the RGB image because it can provide the luminance, hue, and color saturation of input image. The luminance presents the gray level information in which the hue and saturation present the color information [5].

In order to filter out noise, this chapter uses an adaptive median filter (AMF) [6] to decrease noise that can prevent the general median filter causing the original image to be blurred and distorted. The algorithm of AMF has two main definitions. The output value will be the original intensity if the median value of gray level intensity and the intensity of current processing position are between the maximal and minimal value in the mask. If the above conditions are not met, the mask size will be increased and repeat this examination, otherwise the output is median value. In the process, it should apply the binary transformation with a proper threshold to the previous processed images for clearly obtaining the moving object in image. The binary images may have some fractures or unfilling situation. This chapter applies the morphological closing to optimize the binary image.

14.3 Particle Filter

In this chapter, a particle filter algorithm is applied to track targets. Particle filter that uses recursive computing to update the random discrete probability density function offers a probabilistic framework for dynamic state estimation. The tracking conception of particle filter is to maintain a probability distribution over the state of the traced object. Particle filter uses a set of weighted samples or particles to exhibit this distribution. So, each particle represents a forecast which is one possible location of the traced object. The set of particles will have more weight at positions that the traced object is possible to be there. This weighted distribution is propagated through time by using the Bayesian filtering equations, hence, which can determine the object trajectory by taking the particle with the highest weight at each time step.

At initial state, first it creates a group of particles $\{x_0^i\}_{i=1}^{N_s}$ from the prior probability $P(x_0)$ and $\{\omega_k^i, i = 0, \ldots, N_s\}$ is associated weight for $\{x_{0:k}^i, i = 0, \ldots, N_s\}$. The particle weights update to current state and posterior probability of object state at k time can be discretely weighted.

$$p(x_{0:k}|z_{1:k}) \approx \sum_{i=1}^{N_s} \omega_k^i \delta(x_{0:k} - x_{0:k}^i) \tag{14.1}$$

The set of particle $\{x_{0:k}^i\}_{i=1}^{N_s}$ can be obtained from importance density function $q(x_{0:k-1}|z_{1:k-1})$ and then

$$\omega_k^i \propto \frac{p(x_{0:k}^i|z_{1:k})}{q(x_{0:k}^i|z_{1:k})} \tag{14.2}$$

where

$$q(x_{0:k}|z_{1:k}) = q(x_k|x_{0:k-1}, z_{1:k})q(x_{0:k-1}|z_{1:k-1}) \tag{14.3}$$

The posterior probability density function can be represented by

$$\begin{aligned} p(x_{0:k}|z_{1:k}) &= \frac{p(z_k|x_{0:k}, z_{1:k-1})\,p(x_{0:k-1}|z_{1:k-1})}{p(z_k|z_{1:k-1})} \\ &= \frac{p(z_k|x_k)\,p(x_k|x_{k-1})}{p(z_k|z_{1:k-1})}\,p(x_{0:k-1}|z_{1:k-1}) \\ &\propto p(z_k|x_k)\,p(x_k|x_{k-1})\,p(x_{0:k-1}|z_{1:k-1}) \end{aligned} \tag{14.4}$$

That can obtain the update formula of importance weights by using the formulae (14.3) and (14.4) to substitute in the formula (14.2). Then the formula of importance weights is

$$\begin{aligned} \omega_k^i &\propto \frac{p(z_k|x_k^i)\,p(x_k^i|x_{k-1}^i)\,p(x_{0:k-1}^i|z_{1:k-1})}{q(x_k^i|x_{0:k-1}^i, z_{1:k})q(x_{0:k-1}^i|z_{1:k-1})} \\ &= \omega_{k-1}^i \frac{p(z_k|x_k^i)\,p(x_k^i|x_{k-1}^i)}{q(x_k^i|x_{0:k-1}^i, z_{1:k})} \end{aligned} \tag{14.5}$$

Each importance weight ω_k^i represents the quality to the associated input series of x_k^i, therefore, the samples with low weight mean being not the major part of posterior distribution that tend to useless samples and seem helpless for final estimation. Too many useless samples cause the filter low computing and the degeneracy problem. To decide amount of degeneracy, that needs to define the standard measurement approximation N_{eff} for associated effective samples. The definition is

$$N_{\text{eff}} = N_s \Big/ \left(1 + \text{var}_{q\left(\cdot|z_{1:k}\right)}\left(\omega_k^i\right)\right) \leq N_s \tag{14.6}$$

The var(ω_k^i) is the squared error of ω_k^i. Usually, N_{eff} cannot be calculated accurately, but it may obtain approximation from the following formula.

$$\hat{N}_{\text{eff}} = \frac{1}{\sum_{i=1}^{N_s} \left(\omega_k^i\right)^2} \tag{14.7}$$

From this formula, that means the degeneracy problems are serious if the N_{eff} is lower. Hence, first to set the thresholds $N_{\text{threshold}}$ of effective samples, it should resample when $N_{\text{eff}} = 1/\sum_{i=1}^{N_s} \left(\omega_k^i\right)^2 < N_{\text{threshold}}$. That needs to eliminate particles with low weight and choose more particles in more probable regions which replicate the effective sample with higher weight to make up the discarded samples previously. Figure 14.1 is the diagram of resampling.

The above circles represent particles before resampling and the radius is directly proportioned to associated weight value. The below circles represent particles after resampling and own the same weight $1/N_s$. That will increase the sampled probability after larger particles (higher weight) are divided into more particles and also solve the particle shortage and degeneracy problems.

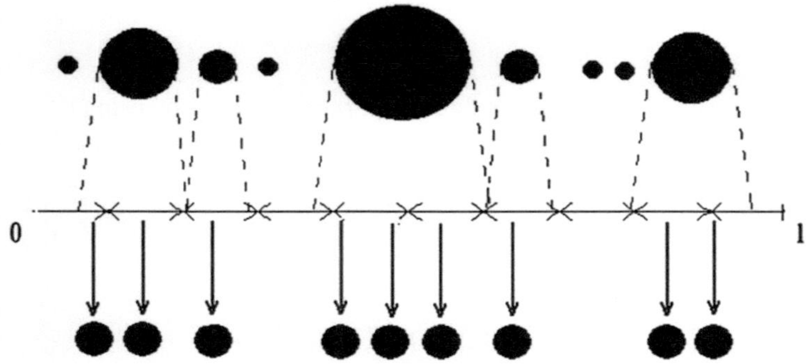

Fig. 14.1 Diagram of resampling

After obtaining the importance density function $q(x_{0:k}|z_{1:k})$ and normalizing the weights, then, that can figure out the following estimation and trace the object trajectory. The estimation of state and estimation of square error are shown as follows.

$$\hat{x}_x = \sum_{i=1}^{N} \omega_k^i x_k^i \tag{14.8}$$

$$P_k = \sum_{i=1}^{N} \omega_k^i \left(x_k^i - \hat{x}_k\right)\left(x_k^i - \hat{x}_k\right)^{\mathrm{T}} \tag{14.9}$$

If the tracked object is still in image read from the input films, the scheme should go back to step 2 for keeping on the recursive computing and object tracing. That should stop the particle computing and cease the object tracking if the tracked object leaves the image or disappears which is determined when no object present in the input image. It will not waste computing resource to calculate during no-objects-exist state.

14.4 Experimental Results

This chapter chooses two different scenarios for experiment which have indoor and outdoor situations that include various interference factors to verify the proposed method. Figure 14.2a is an indoor scenario without illumination change. At initialize step, the particles have been spread with uniform distribution in image which sets the total particle number to 500. That should begin to track and circle the object when it appears in image. This situation can be traced accurately which is shown in Figs. 14.2b and c, respectively.

In this scenario, the moving object will be interfered by the vivid illumination change and accompanied shadow, but the tracking is not influenced as shown in Fig. 14.3.

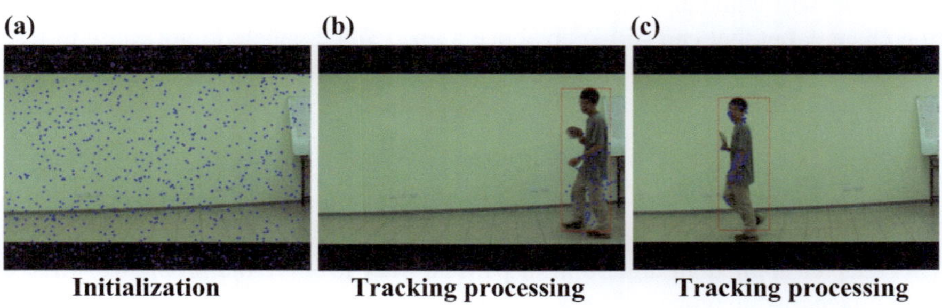

Fig. 14.2 (**a**) Initialization, (**b**) tracking processing, and (**c**) tracking processing

Fig. 14.3 Object tracking for illumination change and shadow interference

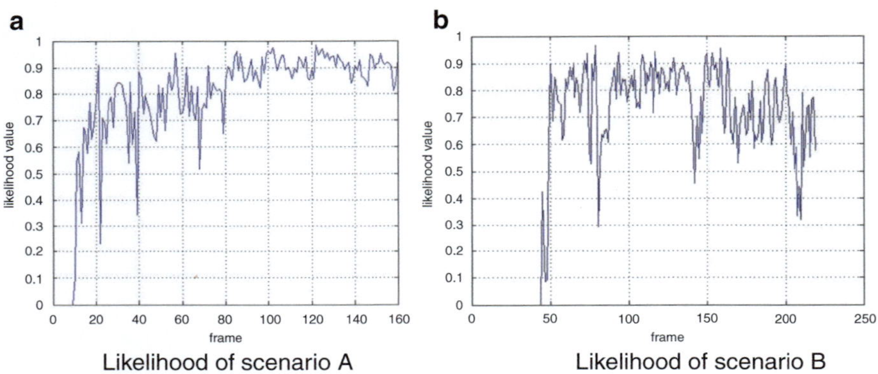

Fig. 14.4 (**a**) Likelihood of scenario A and (**b**) likelihood of scenario B

This chapter uses likelihood level to analyze the experiments. The design method put in 500 particles to track object at the initialized state. That will capture the first 10 % particles of higher likelihood value (50 particles) to examine the likelihood level of every frame. To inspect scenario A, the observed object appears at 0.5 s (12th frame), so the prior frames' likelihood value is zero as shown in Fig. 14.4a. The average of likelihood value for later frames is high and proven to have good tracking. In the scenario B, the likelihood value that is shown in Fig. 14.4b will undulate obviously owing to be interfered by the violent illumination change, but that can still keep above a good value which do not cause to lose the tracked object.

14.5 Conclusions

This chapter applied an algorithm of particle filter to solve the problems suffered from image tracing process. The scenarios of image tracking will be varied by many factors such as shadow interference, object sheltered, and illumination change. That can lead to the erroneous system operation and affect the tracking result. This chapter simulates these various interference conditions for experiment. According to the experimental result, it can prove the image processing technology proposed by this chapter which may eliminate these interference factors in the tracking process and cause this system better to obtain the object tracking accurately.

Acknowledgment This work was supported by the National Science Council under Grant MOST 103-2221-E-018-017.

References

1. Kim, W., Kim, C.: Background subtraction for dynamic texture scenes using fuzzy color histograms. IEEE Signal Process. Lett. **19**(3), 127–130 (2012)
2. Han, B., Davis, L.S.: Density-based multi-feature background subtraction with support vector machine. IEEE Trans. Pattern Anal. Mach. Intell. **34**(5), 1017–1023 (2012)
3. Dianat, R., Kasaei, S.: Change detection in optical remote sensing images using difference-based methods and spatial information. IEEE Geosci. Remote Sens. Lett. **7**(1), 215–219 (2010)
4. Thomas, V., Ray, A.K.: Fuzzy particle filter for video surveillance. IEEE Trans. Fuzzy Syst. **19**(5), 937–945 (2011)
5. Yoon, I., Kim, S., Kim, D., Hayes, M.H., Paik, J.: Adaptive defogging with color correction in the HSV color space for consumer surveillance system. IEEE Trans. Consum. Electron. **58**(1), 111–116 (2012)
6. Huang, Y.-X., Jhao, G.-W., Chung, Y.-N., Hsu, C.-H.: Application of image processing technology in detection and tracking of vehicles on the road. ICIC Exp. Lett. **8**(5), 1301–1305 (2014)

Chapter 15
Measurement of Thickness and Refractive Index of Optical Samples in FD-OCT with Two Orthogonal Polarized Lights

Ya-Fen Chang, Yu-An Chen, and Hsu-Chih Cheng

Abstract This chapter proposed an improved structure of frequency-domain optical coherence tomography (FD-OCT) which can measure the thickness and the refractive index of unknown sample simultaneously. In the conventional FD-OCT system, the mirror signal and autocorrelation terms will be generated after the inverse Fourier Transform and result in measurement range is restricted. Therefore, this study presents an improved method using the theory of phase-shifting algorithm with two orthogonal polarized lights to increase measurement range and eliminate unnecessary noise. By phase-shifting algorithms, eliminating unnecessary noise becomes possible and the measurement range is doubled. In other words, the structure of FD-OCT achieves simultaneous measurement of thickness and refractive index of optical samples based on full-range measurement.

Keywords Optical coherence tomography (OCT) • Optical imaging • Orthogonal polarized lights

15.1 Introduction

Optical coherence tomography is a novel imaging technique. In recent years, with the expansion of the diagnosis region and required higher resolution [1], optical coherence tomography has continued in the development. Optical coherence tomography technology becomes more sophisticated, it can measure the internal structure of biological tissue [2], perform internal transverse cross-sectional images with high-resolution, high-speed image capture rate, high detection sensitivity, and three-dimensional imaging technology [3]. It is also a common application in ophthalmology, for example detects retinal imaging and nerve fiber thickness [4].

Y.-F. Chang • Y.-A. Chen • H.-C. Cheng (✉)
Department of Electro-Optical Engineering, National Formosa University, Yunlin 632, Taiwan
e-mail: janua2905@gmail.com

J. Juang (ed.), *Proceedings of the 3rd International Conference on Intelligent Technologies and Engineering Systems (ICITES2014)*, Lecture Notes in Electrical Engineering 345, DOI 10.1007/978-3-319-17314-6_15

A phase-retrieval algorithm which is an optical signal-processing method in frequency-domain optical coherence tomography (FD-OCT) has been proposed [5]. However, this method is tedious and time-consuming because standard phase-retrieval algorithms need five interferograms. Then a phase-shifting method is applied in the FD-OCT, the method can be used to eliminate the coherent noise terms and the dc peak only need two recorded interferograms [6]. It utilized phase-shifting algorithm to achieve high-speed and long-depth-range probing by FD-OCT. Adding polarization plates and quarter-wave plates with appropriate angles into the OCT structure so that a $\pi/2$ phase shift is obtained between two orthogonally polarized lights [7]. It can achieve a full-range measure of the thickness of unknown sample. In this chapter, the improved structure of FD-OCT to simultaneously measure thickness and refractive index of unknown samples have been presented. By phase-shifting algorithms over two orthogonally polarized lights, eliminating those unnecessary noise terms becomes possible, and the measurement range is doubled.

15.2 Theory and Schematic of Proposed FDOCT Scheme

As shown in Fig. 15.1, the proposed FD-OCT setup consists of a low-coherence light source, one polarizer, a quarter-wave plate, two beam splitters, three mirrors, and an optical spectrum analyzer (OSA). In architecture, the light source SLD is divided into three main optical paths by two beam splitters. The SLD light source is divided into three optical paths (M_R-arm, M_L-arm, M_S-arm) by beam splitters BS1 and BS2. The path M_R-arm is a reference arm used to detect, M_L-arm is another reference arm, and M_S-arm is a sample arm. The light source is ASLD155-200-B-FC with a power of 10.6 mW, central wavelength of 1,550 nm, and spectral bandwidth of 40 nm.

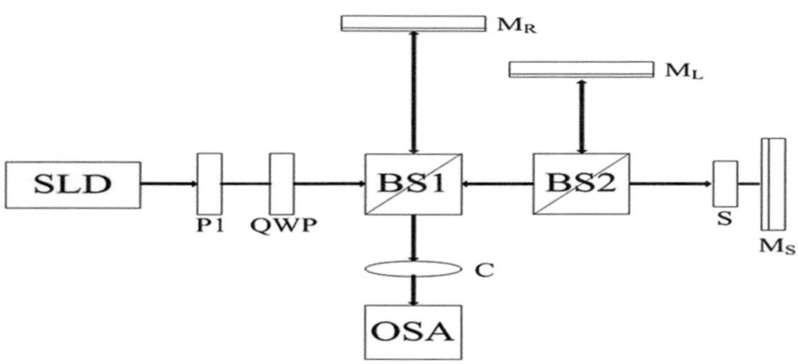

Fig. 15.1 Schematic illustration of proposed FD-OCT scheme (*SLD* super luminescent diode, P_1 polarizer, *QWP* quarter-wave plate, *BS1* and *BS2* beam splitter, *S* sample, M_R, M_L, and M_S mirror, *C* collimator, *OSA* optical spectrum analyzer)

To obtain the first interferogram, the light source passing through the polarizer P_1 placing 0° and then through a quarter-wave plate (QWP) placing 0° and divided into three main optical paths by two beam splitters. Then the reflected light from these three mirrors and sample is obtained in the OSA. It generated the first interferogram and records it in the OSA. The second interferogram is obtained from the polarizer P_1 placing 90°. It can generate a $\pi/2$ phase difference interferogram with the first interferogram. The two interferograms (0° and 90° phase difference) are received by the OSA. After using phase-shifting algorithm [6, 7], the information such as thickness and refractive index of the unknown sample can be obtained. It utilizes the interference signals from M_R, M_L, and M_S arms to measure the thickness and the refractive index of the unknown sample. Before measuring the unknown sample, it is necessary to calibrate the optical path length of the mirror M_L the same as the mirror M_S when the sample has not been placed in M_S arm.

As shown in Fig. 15.2 A, B, D, and C are the surfaces of sample (front-surface and back-surface), M_S, and M_L, respectively. P_1 is the optical path length of reflected light from the front-surface of the unknown sample, P_2 is the optical path length of reflected light from the back-surface of the unknown sample, P_3 is the optical path length of reflected light from the mirror M_S, and P_4 is the optical path length of reflected light from the mirror M_L. Due to addition of the unknown sample in M_S arm thus increasing the optical length, P_3 and P_4 are not the same optical path lengths. The relationships between P_1, P_2, P_3, and P_4 are shown below:

$$d_1 = P_2 - P_1 = n_s \times L \tag{15.1}$$

$$d_2 = P_3 - P_4 = (n_s - n_a) \times L \tag{15.2}$$

From the above (15.1) and (15.2), the following relationship can be derived:

$$n_s = n_a \times \frac{P_2 - P_1}{(P_2 - P_1) - (P_3 - P_4)} = n_a \times \frac{d_1}{d_1 - d_2} \tag{15.3}$$

where n_s is the refractive index of samples, n_a is the refractive index of air, L is the thickness of the unknown sample. In the above formulas, the thickness L and the refractive index n_s of the unknown sample can be obtained.

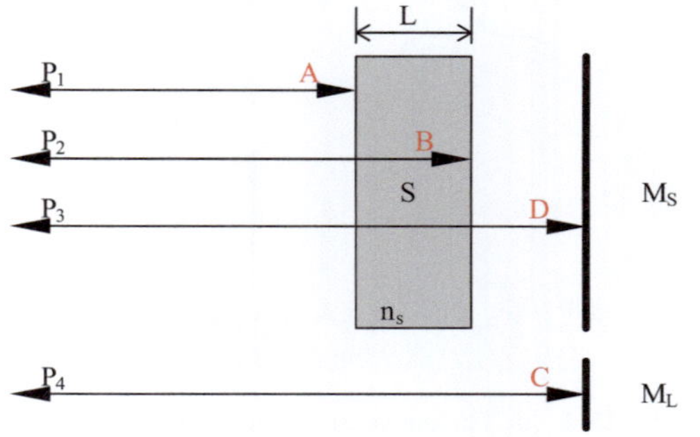

Fig. 15.2 The optical path length of each surface by the M_L and M_S arm

15.3 Experimental Results

Before measuring the unknown sample, the calibration of the optical path difference between the M_L-arm and the M_S-arm is needed. Then, the next step needs to insert a glass plate with refractive index 1.5 and thickness 1.09 mm as the unknown sample to measure. In this article, the measurement position is a short distance in front of the sample.

Figure 15.3 shows that the two interferograms (0° and 90° phase difference) are obtained from the OSA. These two interferograms can use the phase algorithm to obtain the final result as shown in Fig. 15.4. Figure 15.4a is the conventional IFFT transfer of Fig. 15.3a. Obviously, the result shows the four main peaks of A, B, C, and D

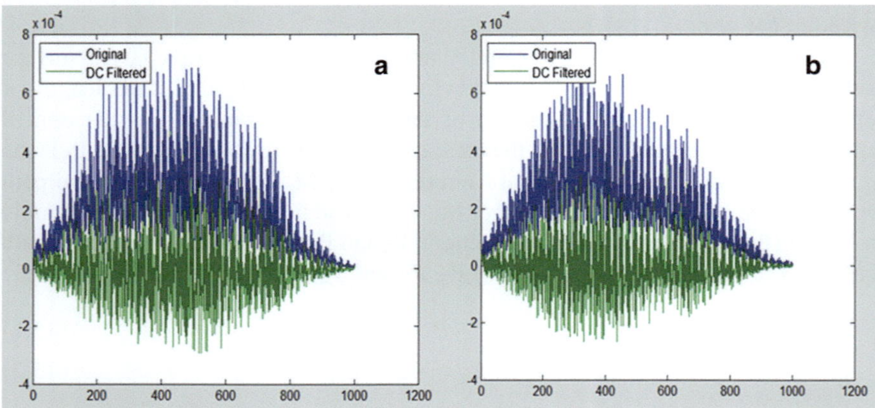

Fig. 15.3 (**a**) The first interferogram with (*green line*) and without (*blue line*) DC filtering; (**b**) The second interferogram with (*green line*) and without (*blue line*) DC filtering

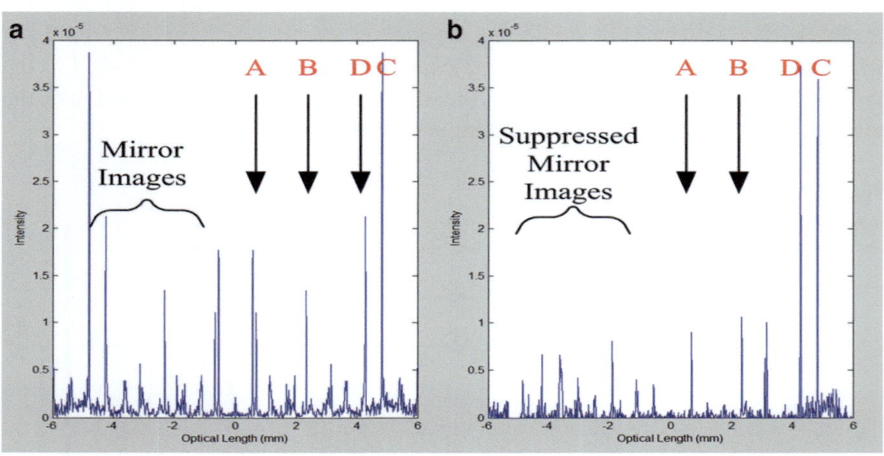

Fig. 15.4 (**a**) The conventional IFFT result of FD-OCT interferogram and (**b**) the IFFT result of two interferograms by phase-shifting algorithm

in the right part and their mirror images in the left part. After using the phase algorithm, the mirror images of A, B, C, and D are suppressed as shown in Fig. 15.4b.

After using (15.1)–(15.3), it can be obtained, the measured value of the thickness and the refractive index. For the measurement in the phase-shifting FD-OCT, the measured values of thickness and the refractive index are 1.0802 mm (deviation is about 0.899 %) and 1.522 (error of 1.47 %) respectively at this measurement position.

15.4 Conclusions

This study presents a new method to eliminate unnecessary noise and simultaneous measurement of thickness and refractive index of optical samples. Using $\pi/2$ phase-shifting on two orthogonally polarized light and adding a new reference arm make this system achieve full-range measurement. In the measurement, the unnecessary signals that cannot be completely eliminated. The reason is that the polarizer's or the quarter-wave plate's angle are not accurate enough. The other reason is that the intensity received by two OSA is not equal. However, the phase-shifting FD-OCT has a disadvantage that the system can't measure the birefringence sample. It also needs double measuring time in this system.

Acknowledgments The current authors gratefully acknowledge the financial support provided to this study by the Ministry of Science and Technology, Taiwan under the grant MOST 103-2622-E-150-013-CC3.

References

1. Haruna, M., Ohmi, M.: Simultaneous measurement of the phase and group indices and the thickness of transparent plates by low-coherence interferometry. Opt. Lett. **23**(12), 966–968 (1998)
2. Huang, D., Swanson, E.A., Lin, C.P., Schuman, J.S., Stinson, W.G., Chang, W., Hee, M.R., Flotte, T., Gregory, K., Puliafito, C.A., Fujimoto, J.G.: Optical coherence tomography. Science **254**, 1178–1181 (1991)
3. Vabre, L., Dubois, A., Boccara, A.C.: Thermal-light full-field optical coherence tomography. Opt. Lett. **27**, 530–532 (2002)
4. Wojtkowski, M., Leitgeb, R., Kowalczyk, A., Bajraszewski, T., Fercher, A.F.: In vivo human retinal imaging by Fourier domain optical coherence tomography. J. Biomed. Opt. **7**(3), 457–463 (2002)
5. Wojtkowski, M., Kowalczyk, A., Leitgeb, R., Fercher, A.F.: Full range complex spectral optical coherence tomography technique in eye imaging. Opt. Lett. **27**, 1415–1417 (2002)
6. Leitgeb, R.A., Hitzenberger, C.K., Fercher, A.F., Bajraszewski, T.: Phase-shifting algorithm to achieve highspeed long-depth-range probing by frequency-domain optical coherence tomography. Opt. Lett. **28**, 2201–2203 (2003)
7. Cheng, H.C., Shiu, M.S.: Experimental demonstration of high-speed full-range Fourier domain optical coherence tomography imaging using orthogonally polarized light and a phase-shifting algorithm. Appl. Opt. **51**(36), 8762–8768 (2012)

Chapter 16
Fabrication of a Peristaltic Micropump with UV Curable Adhesive

Yi-Chu Hsu, Jeffrey Levin, and Hsiao-Wei Lee

Abstract This chapter describes a method to fabricate a peristaltic micropump using UV curable adhesive (NOA81). This study utilized replication method to fabricate UV curable adhesive-based structure. NOA81 was also used as a bonding material to fabricate micropump's main body, which consists of two PMMA plates and UV structure. In this study, NOA81 reached a fully cured state in 10 min and the highest bonding strength is 10.18 J/m^2. Partially cured NOA81 mold would produce bad replication effect, because during partially cured period, a slight amount of pressure will change the geometry. Under confocal microscope, the geometrical errors from fully cured and post-cured NOA81 structure mold are below the maximum tolerance (10 %). The highest membrane displacement and flow rate of the micropump are 0.824 μm and 21.87 μL/min, respectively, while the maximum backpressure is 4.42 Pa.

Keywords NOA81 • UV curable adhesive • Peristaltic micropump

16.1 Introduction

The majority of the micropumps reported are diaphragm micropumps [1]. An example of a diaphragm micropump is peristaltic micropump. In a peristaltic micropump, sequential actuations of actuating diaphragm in a desired fashion can generate fluid flow rates in a controlled direction. Piezoelectric actuation is the most popular method applied for peristaltic micropumps, because it has short response time, a high output force at a lower voltage and commercially available. As a result, piezoelectric actuation can be easily integrated in micropump system.

The high demand for micropumps is driving a need for a low cost fabrication for mass productions. New material such as UV curable adhesive is gaining recognition

Y.-C. Hsu (✉) • J. Levin • H.-W. Lee
Department of Mechanical Engineering, Southern Taiwan University of Science and Technology, Tainan 71005, Taiwan
e-mail: yichu@mail.stust.edu.tw

J. Juang (ed.), *Proceedings of the 3rd International Conference on Intelligent Technologies and Engineering Systems (ICITES2014)*, Lecture Notes in Electrical Engineering 345, DOI 10.1007/978-3-319-17314-6_16

in microfluidic field. Several researchers have published results in relation to the fabrication of microfluidic devices from UV curable adhesive products, such as NOA81 [2–4], NOA63 [5], NOA68, and NOA74 [6]. From these NOA types, NOA74 (80–95 cps) and NOA81 (300 cps) have low viscosity than NOA63 and NOA68 [7]. The lower the viscosity, the easier it will be to deposit the UV adhesive on top of a substrate. Between NOA74 and NOA81, NOA81 (90-Shore D) has higher hardness value rather than NOA74 (30-Shore D) [7]. In this work, PZT actuation was used as a driving mechanism and this actuation type has a high output force. The micropump's structure (which was made by UV adhesive) should be able to withstand the high output force. Based on these reasons, NOA81 was chosen to become the micropump's structure.

Bonding is also an important aspect of microfluidic, which is still quite challenging. The most common techniques are anodic bonding, chemical bonding, ultrasonic bonding, and glue adhesive bonding. These techniques have some challenges in relation to cleanliness of the wafer, high temperature, and oxidation problems. Glue adhesive bonding, due to its capillary effects, there is a chance that the glue flows into the channels, thus blocking them. One of the advantages from UV curable adhesives is that it can turn from liquid to solid state with UV light in a short amount of time.

Even though UV curable adhesives have been utilized in microfluidic field, there is no publication regarding the use of this material in the fabrication of a micropump. In this chapter, the first workable peristaltic micropump ever made from UV curable adhesive was fabricated. The peristaltic micropump will undergo several experiments to determine the performance from this UV adhesive-based peristaltic micropump. Razor blade test was performed to determine the bonding strength of the NOA81. In this chapter, the NOA81 transfer characteristics were investigated. A confocal microscope was utilized to measure the geometrical changes of the micropump's structure between PMMA mold and NOA81 structure mold. Several conditions were investigated for NOA81 structure mold, such as partially cured, fully cured, and post-cured mold.

16.2 NOA81 Transfer Characteristics

To investigate the replication effects between partially cured, fully cured, and post-cured NOA81 mold, a confocal microscopy (KEYENCE, VK-X100/X200) was used to measure the geometry change between PMMA mold and NOA81 mold. The target areas of this investigation consist of 4 spots, which are the three channel's centers and one chamber (Fig. 16.1). The tolerance of the error from replication effects is set to be 10 %. The geometrical error can be calculated using (16.1):

$$\text{Error} = \left(\frac{\text{NOA81}_{\text{depth}} - \text{PMMA}_{\text{depth}}}{\text{PMMA}_{\text{depth}}}\right) \times 100\ \% \tag{16.1}$$

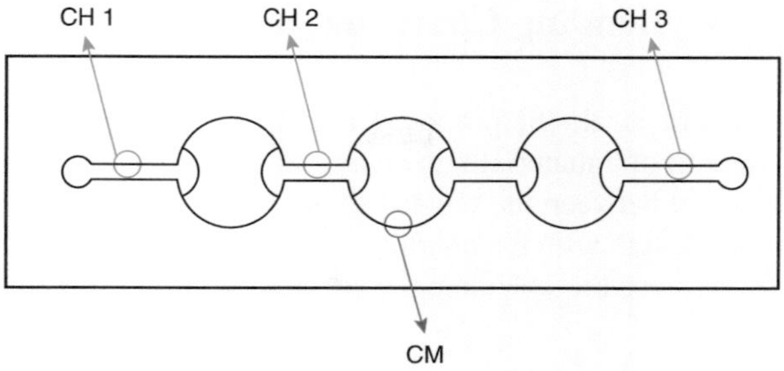

Fig. 16.1 The four spots of investigations under a confocal microscope

16.2.1 Experimental Setup of NOA81 Transfer Characteristics

Poured PDMS mold (Sylgard 184, Dow Corning) on top of PMMA master mold and then put it in an oven and heated at 95 °C for 1 h. After 1 h, peeled off the PDMS mold from PMMA master mold and deposited NOA81 on top of PDMS mold around 0.25 cm^3. PMMA plate was placed on top of NOA81. Cured the substrate with UV light (UVA-201PO, 40 mW/cm). There are three conditions for this experiment. *Partially cured*; three different curing times are 1, 5, and 10 min. *Consistency of the NOA81 mold*; the depths of five NOA81 mold were measured in order to know the reliability of the NOA81 to replicate microstructure. *Post cured*; after fully cured the NOA81 molds were post-cured for 60 min to investigate change in geometry.

16.2.2 Results and Discussions of NOA81 Transfer Characteristics

One-minute and 5-min samples have error above 10 %. One-minute sample has 11.13 % error on CH 2 spot, while 5-min sample has 17.6 % error on CH 2 and 15.94 % on CM. Meanwhile, 10-min sample does not show any errors that exceed the maximum tolerance. During partially cured period, a slight amount of pressure during peeling can affect the geometry of the microstructure. In this case, the NOA81 will reach the fully cured state in 10 min of curing. From five fully cured samples, the geometrical errors are all below 10 %. It shows the consistency of the NOA81 to replicate a microstructure, thus NOA81 is a reliable material to be used in a replication method. The maximum error from post-cured NOA81 mold is −4.29 % (minus means that post-cured mold has shallower depth than fully cured mold). This result shows that post curing does not affect the structure's geometry on NOA81 mold.

16.3 Bonding Characterization

Maszara et al. [8] proposed a simple and elegant way to measure the bonding strength of a material known as razor blade method. A blade with known thickness inserted between the bonded samples and a formula was developed to correlate the crack length with the bonding strength in terms of surface energy (γ). Equation 16.2 [8] was used to calculate the surface energy.

$$\gamma = \frac{3Et_w^2 t_b^2}{32L^4} \tag{16.2}$$

where E is the Young's modulus, t_w is the wafer thickness, L is the crack length, and t_b is the blade thickness.

16.3.1 Experimental Setup of Razor Blade Test

After the NOA81 transfer process, the NOA81 structure is sandwiched between PMMA plates and clamped together to align the inlet, outlet, and chamber areas. NOA81 was injected into the substrates using a syringe for six times. After each injection, the NOA81 was partially cured for 1 min in order to prevent it from blocking the channels. After the sixth injection, the curing times were varied. Finally, the razor blade was inserted into the bonded sample and let the crack to stabilize for a day.

There were two conditions for this experiment, each condition with five samples (1, 3, 5, 7, and 10 min). These time variations were made during the sixth injection. *First condition*, razor blades were directly inserted after curing. *Second condition*, after curing, samples were heated on a hotplate at 50 °C for 12 h, then the razor blades were inserted (symbolized with "H" to differentiate between the first condition).

16.3.2 Results and Discussions of Razor Blade Test

The parameters used to calculate surface energy using (16.2) are as follows; $E = 3.1$ GPa, $t_w = 1.2$ mm, and $t_b = 0.45$ mm. The highest surface energy for normal and heated samples is 3.56 and 10.16 J/m^2. The heated sample surface energy value is almost four times bigger than the normal sample. During fabrication process of a peristaltic micropump, after bonding process the micropump's main body was heated for 12 h at 50 °C on a hotplate to achieve the highest surface energy. Typical values of adhesive surface energy for non-toughened adhesives are of the order of 10–100 J/m^2, with 1-mm adhesive layer thickness [9]. This razor blade test proves that the experimental results are reasonable.

16.4 Fabrication Process of Peristaltic Micropump

The design of the peristaltic micropump used in this research is demonstrated in Fig. 16.2. The microchannels depth and width are expected to be 200 μm and 80 μm, respectively. While the chambers depth is 15 μm. Hot embossing technique was utilized to transfer the geometry on PMMA substrate from a master mold, made from Unsaturated Polyester (UP) [10].

A mixture of silicon elastomer and a curing agent mixed at 10:1 ratio. The mixture was stirred carefully and placed in an oven to be vacuumed in order to remove the air bubbles. Placed the PMMA mold onto a flat surface of a glass petri dish. Degassed PDMS was then poured on the PMMA mold. Placed the petri dish in an oven and cured at 90 °C for 1 h. After 1 h, the PDMS mold was cut using a lancet. NOA81 was used to create the UV mold. Poured NOA81 onto PDMS mold using a pipette (±0.25 cm^3) and placed PMMA plate on top PDMS mold and UV adhesive. Fully cured the UV adhesive with UV exposure for 10 min using a commercial UV light (UVA 201-PO, 40 mW/cm). PMMA top plate (which was cut to create a linear array in rectangle shape) was brought into contact with UV structure and clamped together. NOA81 was injected using a syringe on six sides of the clamped PMMA plates. After every injection, NOA81 has to be partially cured for 1 min to prevent it from blocking the microchannels. On the sixth injection, the adhesive was fully cured for 10 min. After the exposure, place the substrates on a hotplate at 50 °C for 12 h.

Three 6-mm square piezoelectric plates were attached to glass substrates using silver epoxy to create piezoelectric actuators for three micropump chambers. Actuators with known thickness of 191 μm were attached to UV-based chambers by epoxy.

16.4.1 Experimental Setup of Micropump Performance

The peristaltic micropump was actuated using a function generator and power amplifier. Signals from function generator and power amplifier were displayed by digital oscilloscope. The membrane displacement was measured using a fiber optic measurement system (MTI Instruments, MTI 2000 Fotonic Sensor). The fotonic sensor probe was positioned above middle PZT actuator and aligned above the

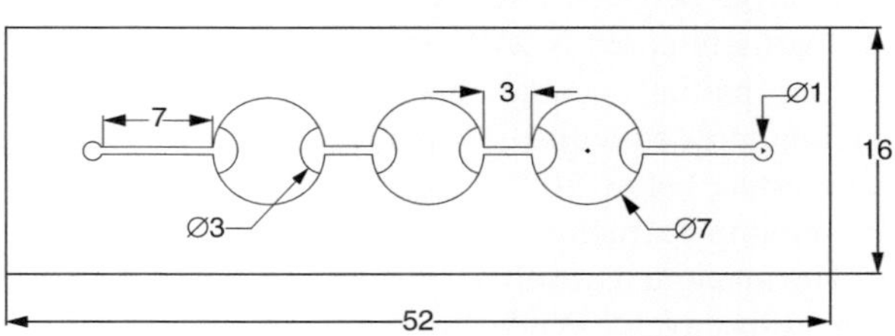

Fig. 16.2 Geometry of the peristaltic micropump

center of the actuator. In this research, the flow rate was estimated by using a microbalance to measure the change in weight of the fluid over a specified amount of time. The pumping fluid for all experiments was DI water. The experimental processes commenced at a constant driving voltage of $100V_{pp}$ and frequencies ranging from 25 to 400 Hz.

The backpressure of a micropump provides an indication of its pumping power. The maximum backpressure of a micropump can be measured by gradually raising the height of the output relative to the micropump and measuring the height of the pumping fluid at which, the flow rate reduces to zero. In this work, the micropump was driven by a constant $100V_{pp}$ and an excitation frequency of 37.5 Hz.

16.4.2 Results and Discussions of Micropump Performance

The maximum membrane displacement and flow rate are 0.824 μm and 21.87 μL/min, respectively. Both were achieved at 37.5 Hz excitation frequency. The highest backpressure is 4.42 Pa. Previous work from Cheng [10], the flow rate and backpressure is 133.6 μL/min at 400 Hz and 11.8 Pa. The biggest membrane displacement is 0.98 μm at 37.5 Hz. There are no significant differences in membrane displacement and backpressure. Cheng's micropump has deeper channel depth. Theoretically, shallower depth produces bigger fluidic resistance in microchannel. The bigger the fluidic resistance, the smaller the flow rate will be.

The NOA81 has lower Young's modulus (1.38 GPa) [7] rather than PMMA's (3.1 GPa). Probably during the actuation process, the deflections from PZT actuators create a deformation on NOA81 structure, thus made the flow rate even smaller.

16.5 Conclusions

A fabrication method for a UV adhesive-based peristaltic micropump was developed. UV curable adhesive has been served both as a bonding material and part of a micropump main body.

In this study, the NOA81 reached the fully cured state in 10 min time. The highest bonding strength is 10.16 J/m^2. Under a confocal microscope, the geometrical errors between PMMA mold and NOA81 structure mold were investigated. For partially cured NOA81 mold, 1- and 5-min samples have errors above 10 %. During partially cured state, a slight amount of pressure can change the micropump's structure geometry. For fully and post-cured mold, all of the samples have errors below 10 %. This proved the consistency of the NOA81 to replicate micropump's structure.

The maximum membrane displacement and flow rate are 0.824 μm and 21.87 μL/min, respectively. While the biggest backpressure of the micropump is 4.42 Pa.

Shallower channel produce bigger fluidic resistance, as a result the micropump's flow rate is smaller. NOA81 structure is softer than PMMA, this happened because the NOA81 structure undergoes a deformation during actuation process, making the flow rate even smaller.

References

1. Laser, D.J., Santiago, J.G.: A review of micropumps. J. Micromech. Microeng. **14**, 35–64 (2004)
2. Sollier, E., Murray, C., Maoddi, P., Di Carlo, D.: Rapid prototyping polymers for microfluidic devices and high pressure injections. Lab Chip **11**, 3752–3765 (2011)
3. Bartolo, D., Degre, G., Nghe, P., Studer, V.: Microfluidic stickers. Lab Chip **8**, 274–279 (2008)
4. Wagli, P., Homsy, A., de Rooij, N.F.: NOA81 for fabrication of microfluidic devices with adjustable surface properties and high chemical resistance against IR-transparent organic solvents. In: Eurosensors XXIV Conference, Austria (2010)
5. Dupont, E.P., Luisier, R., Gijs, M.A.M.: NOA 63 as a UV-curable material for fabrication of microfluidic channels with native hydrophilicity. Microelectron. Eng. **87**, 1253–1255 (2010)
6. Mokkapati, V.R.S.S., Bethge O., Hainberger, R., Brueckl, H.: In: Microfluidic Chips Fabrication from UV Curable Adhesives for Heterogeneous Integration. IEEE, San Diego (2012)
7. Norland Optical Adhesive Information. http://www.norlandprod.com/UVdefault.html
8. Maszara, W.P., Goetz, G., Caviglia, A., McKitterick, J.B.: Bonding of silicon wafer for silicon-on-insulator. J. Appl. Phys. **64**, 4943–4950 (1988)
9. Kinloch, A.J.: Adhesion and Adhesives Science and Technology. Chapman and Hall, New York (1987)
10. Cheng, P.Y.: Unsaturated polyesters as stamps for hot embossing of micropumps. Master thesis, Southern Taiwan University of Science and Technology (2014)

Chapter 17
Detecting Object Edges by Xtion Pro and Open Sources

Cheng-Tiao Hsieh

Abstract Kinect is a popular device used in many applications such as interactive PC games, robot, virtual reality, and 3D sensing. Especially in 3D sensing, many researchers and engineers utilized Kinect-based devices to develop cheap 3D scanners. 3D scanners usually provide a better and quicker way to support the needs of 3D printing industry. This advantage is really helpful to speed up the development of 3D printing industry. Regarding this fact, this chapter attempts to propose an approach to track object edges. The approach is capable of detecting edges of a given object by Xtion pro. Many existed edge detection methods have been developed based on color information. However, this approach relies on spatial information-depth to detect object edges. The approach had also been developed successfully by open sources like OpenNI, OpenCV, and Point Cloud Library. Developing a cheap 3D scanner for 3D printing market becomes possible.

Keywords Xtion pro • OpenNI • OpenCV • Point Cloud Library • Object edge tracking

17.1 Introduction

The first generation of Kinect was developed by Microsoft and applied in Xbox 360 for supporting the needs of a natural interaction PC games by the end of 2010. By means of its functions, users are able to control PC games intuitively through human body gestures. Its successes have grabbed people's attentions and encouraged researchers and engineers to further explore its possible applications in various areas, such as robot, 3D sensing, virtual reality, etc. In robot industry, engineers had successfully installed a Kinect on the top of a Quadrotor for detecting

C.-T. Hsieh (✉)
Department of Industrial Design, Ming Chi University of Technology,
New Taipei City 24301, Taiwan
e-mail: cthsieh@maiil.mcut.edu.tw

J. Juang (ed.), *Proceedings of the 3rd International Conference on Intelligent Technologies and Engineering Systems (ICITES2014)*, Lecture Notes in Electrical Engineering 345, DOI 10.1007/978-3-319-17314-6_17

obstacles [1]. In medical area, researchers utilized a Kinect to capture and recognize hand gestures and then to control a DaVinci robotic surgical system to operate a surgery. In 3D sensing field, a Kinect had been implemented as a key component of a 3D scanner, such as ReconstructMe and KinFu [2].

The price of the Kinect-based scanners is much cheaper than other scanners in current market. This advantage does open the gate of 3D scanning-related applications, such as Reverse Engineering, computer animation, 3D printing, etc. Especially in 3D printing industry, a Kinect-based scanner is capable of creating available models to support a building task of 3D printers. According to this advantage, this chapter attempts to develop an approach for implementing an Xtion Pro capable of detecting edges of objects automatically. Most of these methods of tracking object edges depend on image processing. However, this study proposed an object tracking approach based on the depth map captured by Xtion pro. After object edges are detected by this approach, the objects can be easily extracted from a raw depth map. In the future, those processes will be integrated to a completed 3D Kinect-based scanning system.

17.2 Motion Sensing Device

Microsoft Kinect consists of a set of motion sensing devices including multi-array microphone, RBG camera, and depth sensor [3]. The multi-array microphone is an array of four microphones which provide users to control application by voice. The RGB camera is used to capture a color RGB image. By an appropriate 3D registration, a color image can be mapped into a correspondent depth map for creating a color point cloud. About the depth sensor, it consists of an infrared projector and a monochrome CMOS. It is based on the principle of the structured light to capture depths. Figure 17.1a shows an Xtion Pro, a motion sensor developed by ASUS. It only generates a depth map. A projected infrared pattern as shown in Fig. 17.1b is used to find correspondences and further to generate a depth map. The SDK of Xtion Pro is OpenNI, an open source of Kinect framework [4]. In the end of 2013,

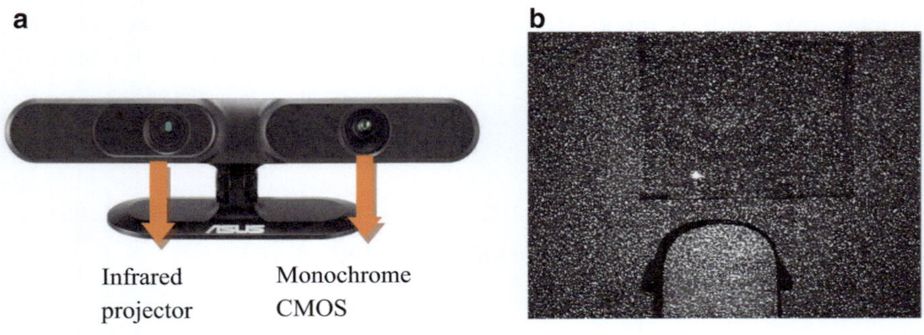

Fig. 17.1 (**a**) ASUS's Xtion Pro motion sensing device and (**b**) infrared pattern image

PrimeSense was bought by Apple Inc. The official OpenNI website had been closed in April 2014, but OpenNI users still can obtain the last edition of OpenNI from the Structure Sensor official website.

17.2.1 Accuracy Issue of an Xtion Pro

The official ASUS's specification indicates that the available working range of Xtion Pro is in the interval from 0.8 to 3.5 m. As we know, long working distance is not appropriate for small and detail feature scanning. According to this problem, Xtion Pro does support a near field mode for small and detail feature scanning. The working range under a near mode had been upgraded to the interval from 0.4 to 3.5 m. The scanning resolution of Xtion Pro is dependent on the resolution of its infrared monochrome CMOS. Based on the ASUS's specification, its maximum resolution is 640×480 in 30 fps video ratio. This notes that the amount of depths captured by Xtion Pro must be less than 307,200. This study only focused on the 3D sensing applications. In order to obtain accurate depths from Xtion Pro for advanced 3D sensing applications, this study utilized the maximum setting parameters to explore the accuracy of depths received by Xtion Pro. To do that, this study set up an experiment. An Xtion Pro was installed in front of a blank board to receive a set of sequent depth maps with a constant video ratio, 30 fps as shown in Fig. 17.2a. The distance between Xtion Pro and blank board is about 1,415 mm. Figure 17.2b shows the obtained depths based on the first video frame.

In order to analyze the accuracy of the depths from Xtion pro, this study utilized Xtion Pro to collect several pairs of depth maps captured by sequential Xtion Pro frames. Since the Xtion Pro was fixed, theoretically the deviations of correspondent depths within each pair of sequential depth maps should be indicated. Unfortunately, the obtained results are not as we had expected. Figure 17.3 shows the

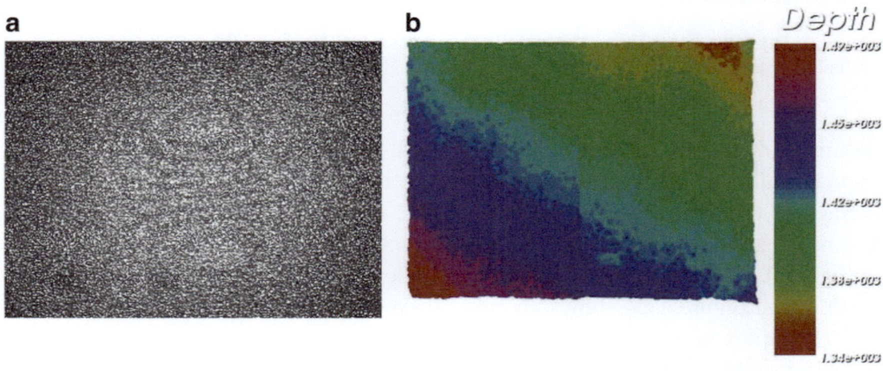

Fig. 17.2 (**a**) An infrared capturing scene of Xtion Pro for an accuracy analysis process and (**b**) a depth map received by Xtion Pro with a constant video ratio 30 fps

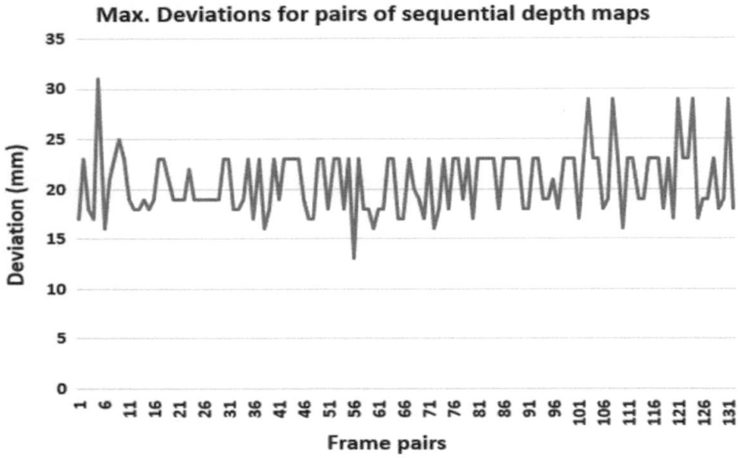

Fig. 17.3 The maximum deviations of pairs of sequential depth maps

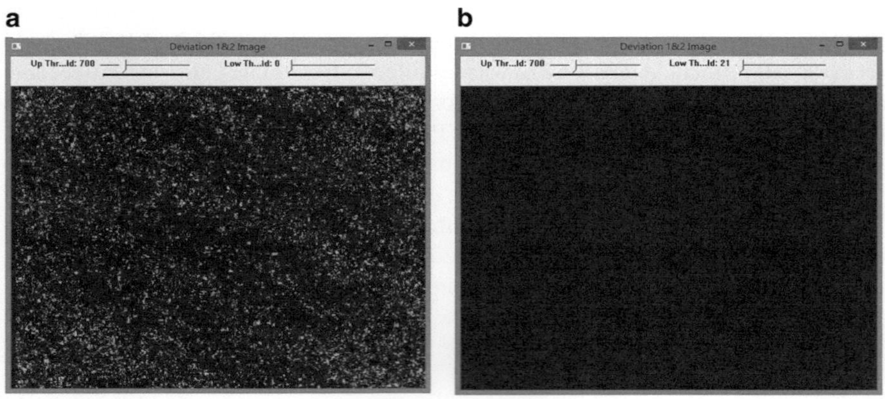

Fig. 17.4 (**a**) A deviation map based on the threshold value 0 mm and (**b**) a deviation map based on the lower threshold value 21 mm

maximum deviations of several pairs of depth maps. From the figure, it illustrates the maximum deviations disturbed around the interval between 13 and 31 mm. The average of the maximum deviation is 20.712 mm. According to this result, a top threshold used to remove depth errors can be set up to 20.712 mm.

Figure 17.4 shows the deviation maps calculated based on pairs of sequential depth maps. Figure 17.4 shows an error filters with various lower thresholds. The filter is similar to a pass-through filter which only removes the errors less than a given threshold called a lower threshold. Based on the result as Figs. 17.3 and 17.4b indicate that Xtion Pro errors can be removed from a depth map when the lower threshold is larger than the average of the maximum deviations.

17.3 Approach Description

The main task of this approach is to track object edges by an Xtion Pro. Most of the existing methods for tracking object edges are dependent on color or gray intensity [5]. It means that those existing methods require a RGB or B/W camera to capture color or gray intensity information from camera scene for further object edge tracking. Unfortunately, Xtion Pro doesn't support image information and it can only generate depth map. Due to this limitation, this study attempted to utilize depth information to detect object edges. In 2013, author had presented a novel approach to extract one or multiple objects from the depth maps of Xtion Pro [6, 7]. Actually, the approach was developed based on the principle of the Background subtraction. The approach requires users to take a background depth map first. This requirement may be inconvenient to users. According to this limitation, author tried to seek an alternative approach which is capable of performing object extraction without the prerequirement.

The approach had been successfully completed to detect object edges based on a set of spatial information—depths. By means of the approach, object edges can be detected and its results can further support to construct a friendly object extraction method. The flow chart as shown in Fig. 17.5 illustrates the workflow of the approach. In order to make an Xtion pro ready to perform specific tasks, the OpenNI library was introduced to configure and control Xtion Pro. The configurations include setting video frame ratio, video resolution, and output data type. Starting device and closing device during a task can also be done by OpenNI [4]. The second step is to receive a sequence of depth maps. The third step is to calculate deviations of correspondent depths in each pair of sequential depth maps. Since the Xtion pro is fixed in the entire process, a correspondent depth of a given depth can be defined easily without using algorithm. Once deviations are calculated, those deviations also can be collected into a deviation map. The forth step is using a filter to remove the defined errors. The intensities of the errors must be less than a given threshold. The filter had been developed by the Point Cloud Library [8]. The final step is to visualize the results. This part had been developed by the OpenCV library [9].

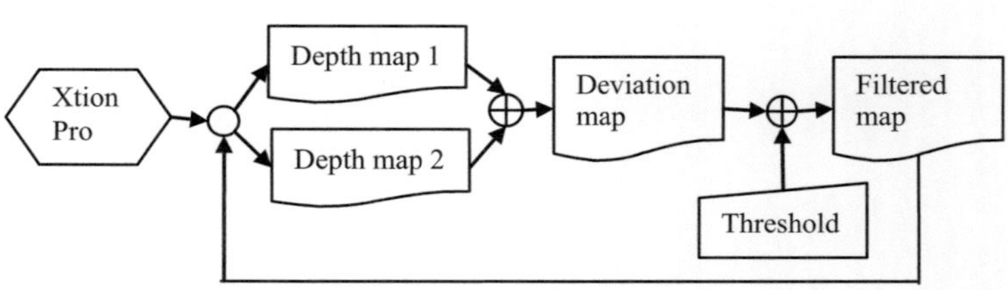

Fig. 17.5 The flowchart of the object edge detection approach

17.4 Results

The above approach is capable of supporting visual information to users to observe the deviations of a sequence of depth maps. Theoretically, the deviations should be a blank scene as shown in Fig. 17.4b, but unfortunately it is not. Especially, a given environment contains many objects and objects are separated randomly as shown in Fig. 17.6a. Figure 17.6b shows the deviations of two sequential depth maps associated with a threshold interval between 0 and 8,000. The maximum deviations in this given setups and conditions were distributed within the interval between 4,522 and 4,983 mm.

The above result didn't show a good result of tracking object edges, because there are many unpredictable errors caused by object edges. Therefore, it is needed to adjust the lower and upper thresholds to obtain a better result of object edge tracking as shown in Fig. 17.7.

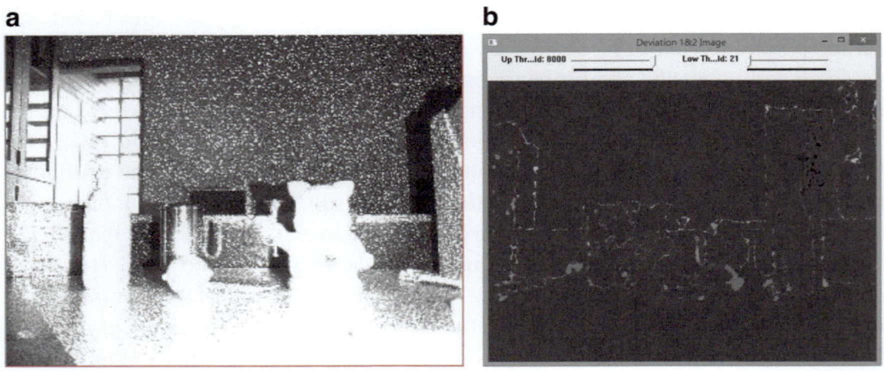

Fig. 17.6 (**a**) An infrared scene showing environment for object edge tracking process and (**b**) A deviation map associated with a lower threshold value = 21

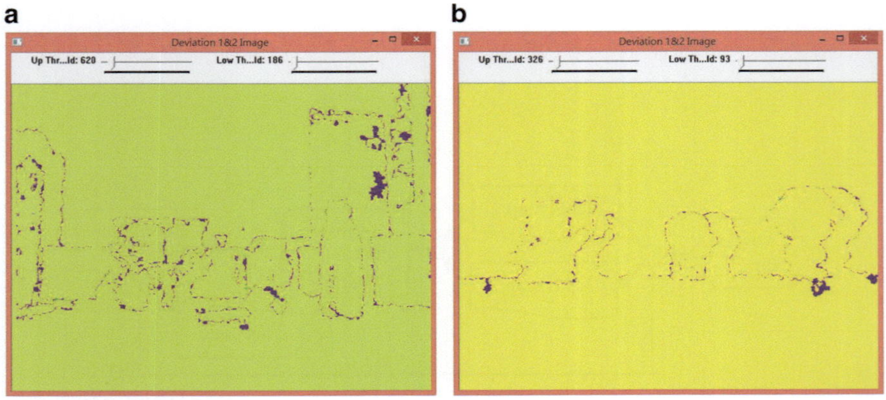

Fig. 17.7 (**a**) A better result of object edges tracking by filtering the deviation map with an up threshold = 620 and a low threshold = 186 and (**b**) A better example with less and isolated objects

17.5 Conclusions

From the previous section, it clearly illustrates that the areas near to edges of objects usually have larger deviations. As we know, Xtion Pro relies on the principle of structured light that projects an infrared pattern to measure depths from hit objects. The projected patterns near to the areas of object edges are usually unstable to be measured precisely. It is why to cause larger deviations near those areas. By means of this event, this chapter developed an approach to detect object edges based on larger deviations. This study had utilized open sources to code the approach. In the future, this developed approach will be continuously updated and improved for making objects to be extracted easily as well as satisfying the needs and requirements of 3D printing.

References

1. Huang, A.S., et al.: Visual odometry and mapping for autonomous flight using an RGB-D camera. In: International Symposium on Robotics Research (ISRR) (2011)
2. Newcombe, R.A., et al.: KinectFusion: real-time dense surface mapping and tracking. In: 10th IEEE International Symposium on Mixed and Augmented Reality (Ismar) (2011)
3. MicroSoft, Human Interface Guidelines—Kinect for Windows 1.5 ed. (2012)
4. OpenNI_group OpenNI User Guide (2013)
5. Sonka, M., Hlavac, V., Boyle, R.: Image Processing, Analysis, and Machine Vision. Thomson-Engineering, Toronto (2007)
6. Hsieh, C.-T.: An efficient scanning approach for multiple targets. In: Juang, J. (ed.) Lecture Notes in Electrical Engineering, pp. 991–997. Springer, Cham (2013)
7. Hsieh, C.-T.: Extracting a scanned target from a raw scanned data set produced by Kinect-based 3D scanner. In: 2013 International Conference on Applied Technology, Sun Moon Lake, Taiwan (2013)
8. Rusu, R.B.: 3D is here: Point Cloud Library (PCL). In: 2011 I.E. International Conference on Robotics and Automation (ICRA), pp. 1–4. IEEE Xplor, Shanghai, China
9. OpenCV, Open Source Computer Vision Library—Reference Manual 2001: Intel

Chapter 18
Temperature Control of a Baking System for an Ultrahigh-Vacuum Insertion Device

Jui-Che Huang, Yu-Yung Lin, Chin-Kang Yang, Yung-Teng Yu, Cheng-Hasing Chang, and Ching-Shang Hwang

Abstract Baking an undulator artificially accelerates outgassing, and is one of the most important and difficult factors in constructing an in-vacuum undulator. The total duration of baking was 76 h and each component in the undulator required a separate temperature for baking. Automatic procedures and devices for temperature control during baking were constructed and continually applied to achieve the required ultrahigh vacuum (pressure $P < 2 \times 10^{-10}$ Torr). This chapter describes details of the baking and the devices used and constructed in National Synchrotron Radiation Research Center.

Keywords Ultrahigh vacuum • Insertion device • Vacuum bake-out

18.1 Introduction

In-vacuum undulators are widely used in a contemporary synchrotron radiation facility, such as SPring-8, ESRF, SOLEIL, and SACLA. In plan phase one, Taiwan Photon Source (TPS) will install seven in-vacuum undulators to serve five beamlines. To attain a long beam lifetime, the pressure in the accelerator must be maintained at an ultrahigh vacuum (UHV) (pressure $P < 2 \times 10^{-10}$ Torr). As an in-vacuum undulator is directly connected to the storage-ring vacuum chamber, the same vacuum requirement is necessary. The vacuum system of a TPS in-vacuum undulator includes four NEG pumps (SAES MK500-SP8, 500L/s), two ion-pumps (Canon Anelva, PIC-052NP 125L/s), two BA vacuum gauges (Canon Anelva, NIG-2TF), and two residual-gas analyzers (Inficon, Transpector H100M, 100 amu) [1, 2], but even such a great pumping capacity is inadequate to achieve UHV in a short period of time. Baking an undulator artificially accelerates the outgassing, and is one of the most important processes in constructing and

J.-C. Huang (✉) • Y.-Y. Lin • C.-K. Yang • Y.-T. Yu • C.-H. Chang • C.-S. Hwang
Magnet Group, National Synchrotron Radiation Research Center, Hsinchu, Taiwan
e-mail: huang.juiche@nsrrc.org.tw

J. Juang (ed.), *Proceedings of the 3rd International Conference on Intelligent Technologies and Engineering Systems (ICITES2014)*, Lecture Notes in Electrical Engineering 345, DOI 10.1007/978-3-319-17314-6_18

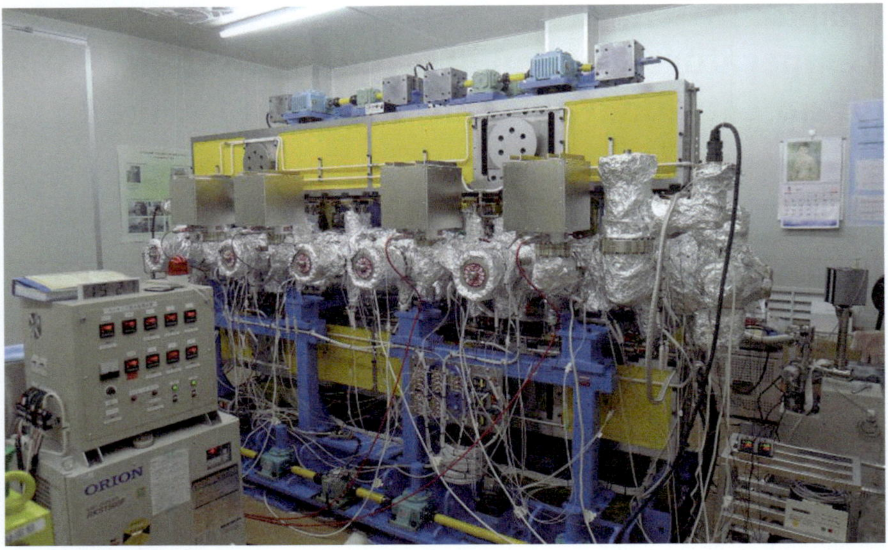

Fig. 18.1 TPS in-vacuum undulator bake-out in the laboratory

commissioning an in-vacuum undulator. The total duration of baking a TPS undulator was 76 h: the first 8 h involved raising the temperature, during the next 60 h the temperature was in a stable state, and the temperature decreased during the last 8 h. Figure 18.1 shows a TPS in-vacuum undulator that was baked in the laboratory.

18.2 Temperature Control

An in-vacuum undulator involves assembly of magnets into vacuum chambers. To achieve a UHV condition, high-capacity vacuum pumps, special surface treatments, and baking of the undulator are essential. Baking an undulator requires a separate temperature for each component, as listed in Table 18.1. The SUS304 vacuum chamber must be heated above 180 °C to remove hydrogen, but, to take into account the thermal deformation, the maximum temperature must be limited to 210 °C. The undulator magnet (model NX-38EH, NEOMAX Engineering) has great coercivity (2,760 kA/m) and was annealed at temperature 145 °C before assembly into the undulator. The magnets can thus withstand a temperature at least 125 °C without demagnetization. Without a cooling system for the magnet array, the temperature of the magnet might easily exceed 145 °C from the radiation in the vacuum chamber, thereby risking demagnetization.

A RF-transition taper is made of beryllium copper (BeCu) to possess spring-like characteristics, according to the Larson-Miller experimental formula (Eq. (18.1))

Table 18.1 Temperature control for UHV undulator

Component	Material	Controlled temperature (°C)
Main chamber	SUS 304	180–210
Magnets	NdFeB	<125
End chamber	SUS304	<130
RF-transition taper	BeCu	<135
Photon absorber	OHFC 1010	<140
Flange	With copper gasket	<200
Ion-pump	With magnet	<140

[3], One can estimate the remaining stress under a combination of temperature and exposure duration necessary for the design,

$$P = \left(\frac{9}{5}T + 492\right)(20 + \log t)10^{-3} \tag{18.1}$$

in which P denotes Larson-Miller's parameter corresponding to a given remaining stress; T denotes the temperature (°C) and t denotes exposure duration (h).

The baking temperature of the RF transition must be limited to 135 °C to ensure that the degradation of the spring is less than 3 % under baking for 72 h. The photon absorber is designed to protect downstream components, so to avoid heating from synchrotron radiation, so water-cooling pipe is brazed (welded) onto the photon absorber. High temperature baking results in oxidation at the location of brazing and small air holes increase the possibility of water leakage. From experience at SPring8 [4], the baking temperature of the photon absorber must be controlled below 140 °C. For the flange of the vacuum chamber, a temperature above 200 °C creates a high risk of vacuum leakage; the flange must thus not be covered sometimes with aluminum foil.

Four ion-pumps are assembled in an undulator; as each ion-pump consists of magnets to deflect the ionized particles, the ion-pump must be baked below 140 °C to avoid magnet demagnetization. As many components are baked at varied temperatures, the baking must be automated to avoid error of human control.

18.3 Power Control and Distributed Panel

An undulator lacks a regular shape; as each component is baked at a separate temperature, various heaters are necessary. Table 18.2 shows the number and type of heater used for a 3-m undulator. As a BA gauge must be baked at a temperature above 300 °C, special 300-W heaters are used.

To decrease the work loading, heaters of the same type are collected and made to one connection. We constructed a power control and distribution panel for 380/220 V mains power supply and the total current assumption is 30 A. For precise

Table 18.2 Various heaters for baking a 3-m in-vacuum undulator

Heater	Heater power (W)	Number	Current (A)
Main chamber heater	200	4	5.4
End chamber heater	200	2	1.8
Bellows heater	100	16	1.0
Flange heater	100	10	0.7
Taper heater	100	10	0.7
NEG pump chamber heater	200	8	1.0
Ion-pump heater	200	4	5.1
Ion-pump chamber heater	200	4	1.0
BA gauge chamber heater	300	2	2.6

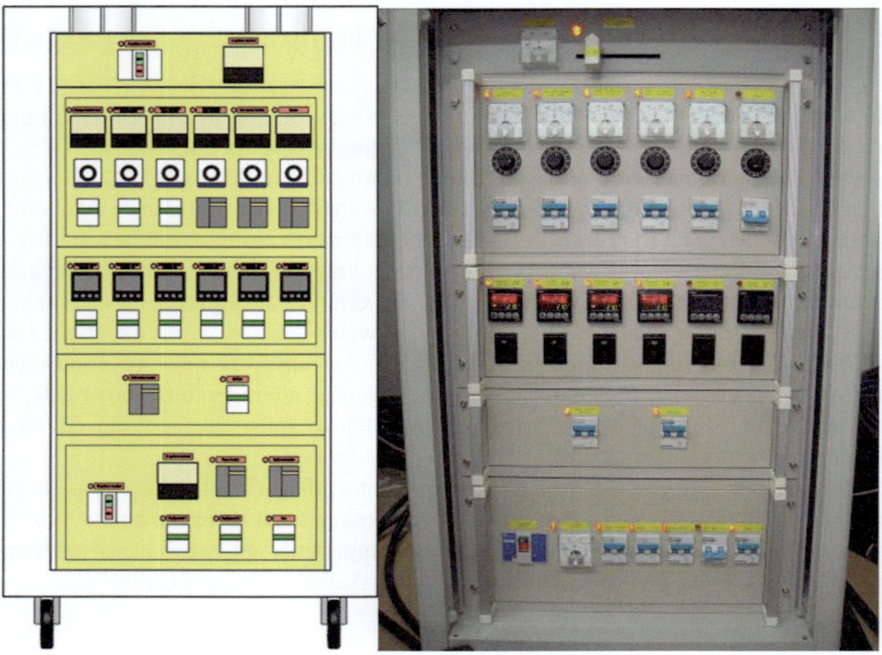

Fig. 18.2 Design and manufacture of power control and distribution panel

control of the temperature, K-type thermocouples serve as temperature feedback and are connected to temperate controllers. Six temperature controllers are used to stabilize the main- and end-chamber temperatures, so that thermal deformation and heating power are controlled. In some locations in which the deformation or temperature stability is less important, an open-loop heating control is used. Figure 18.2 shows the NSRRC power control panel that supplies the current for various heaters; in the second rack, a temperature-feedback system is implemented.

18.4 Cooling and Heating System for Magnet Arrays

A system for cooling or heating a magnet array was designed at SPring8 [5], and is shown in Fig. 18.3; hot water circulates in the cooling channel of the magnet array. During increasing temperature, the water heats the magnets. When the temperature of the magnet array exceeds 120 °C, the machine provides pressurized water at 110 °C for cooling. As the water must remain in the liquid phase at 110 °C, a hot-water pump provides an additional pressure, 2 MPa. The thermocouples in contact with the magnet send signals to the cooling machine; six electrical thermostat-control valves automatically decide the duration of the cooling water. The magnets are slowly heated by radiation from the vacuum chamber; the temperature of the upper magnet array increases more rapidly than that of the lower magnet array because hot air floats outside the vacuum chamber. Once the electrical thermostat-control valves open, the temperature of a magnet can rapidly decrease. The variation of the temperature of the magnet array is maintained about ±1.5 °C in the state of stable temperature.

Fig. 18.3 Circulating pressurized hot-water machine for cooling or heating the magnet arrays

18.5 Thermal Expansion

Thermal expansion is a critical issue during baking of an undulator; a large thermal expansion might result in component failure, requiring reassembly. As the magnet is mounted on a copper I-beam and as the vacuum chamber is made of SUS 304, the thermal expansion and temperature rise vary greatly. At the stage of increasing temperature, the rates of temperature increase of the magnet array and vacuum chamber are kept at 2.5 °C/5 min and 5 °C/5 min, respectively, to avoid a large difference of thermal expansion between the vacuum chamber and the magnet array beam. If the temperature of the vacuum chamber increases too rapidly, the bellows shaft might tilt at a stage of increasing temperature.

During the state of stable temperature, the thermal expansions of the SUS vacuum chamber and the magnet array are calculated to be 2.3 and 4.5 mm, respectively. The new sliding-type design of RF transition (connecting the SUS vacuum chamber and the magnet array) is implemented in a TPS in-vacuum undulator, and allows a large difference of thermal expansion between the chamber and the magnet beam [4].

18.6 Temperature and Vacuum Pressure

In the state of stable temperature, the temperature of the upper vacuum chamber is 2–5 °C greater than that of the lower vacuum chamber, because hot air rises. The temperature of the magnet array and the RF transition oscillated with a regular period because the hot-water machine controlled the temperature with pressurized hot water. Figure 18.4a shows the pressure and temperature of the magnet array and the vacuum chamber.

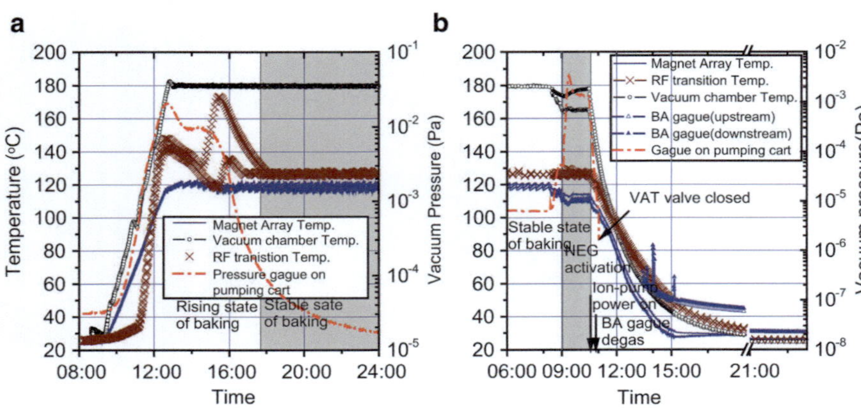

Fig. 18.4 State of temperature (**a**) rising and (**b**) decreasing during baking

After 60 h in a state of stable temperature, the NEG activation began at 450 °C for 1 h, followed by power to the ion-pump (Fig. 18.4b). The magnet temperature must be kept lower by 10 °C during the NEG activation. In the final step before the system temperature decreases, the degassing of two BA, one extractor gauge and two RGA occurs. In the state of temperature decreasing, the NEG and IP are at full pumping speed. The temperature decreases 2.5 °C/5 min for the magnet array and 5 °C/5 min for the vacuum chamber. 12 h after the baking is finished; the pressure of the in-vacuum undulator can attain 9.5×10^{-11} Torr.

18.7 Conclusions

Baking of an in-vacuum undulator has been successfully performed several times in NSRRC. Circulating hot water and the power control and distribution panel were used to control the baking temperature within ±1.5 °C. Semiautomatic procedures improved the efficiency of vacuum baking.

1. Detailed baking of complex vacuum system is discussed in this chapter and these procedures and experiences can be implemented in other vacuum systems.
2. Setting the baking temperature depends strongly on the knowledge of material science, the manufacturing process, and thermal expansion.
3. The high-temperature baking process shorten the time to achieve an UHV ($P < 2 \times 10^{-10}$ Torr).

References

1. Huang, J.C., et al.: Performance of NSRRC in-vacuum undulator. IEEE Trans. Appl. Supercond. **24**(3), 503704 (2014)
2. Wu, L.H., Huang, J.C., et al.: Baking test for in-vacuum undulator. In: Fourth International Particle Accelerator Conference, THPFI075, Shanghai, China (2013)
3. Beryllium copper. http://www.ngk.co.jp/english/products/electronics/berylliumcopper/strip/strip_data_b03.html
4. Private communication with Dr. H. Kitamura in SPring8 (2013)
5. Manual of hot-water-circulating machine, Asahi Co., Japan (2012)

Chapter 19
Microarray Data Analysis with Support Vector Machine

Si-Hao Du, Jin-Tsong Jeng, Shun-Feng Su, and Sheng-Chieh Chang

Abstract Microarray data analysis approach has became a widely used tool for disease detection. It uses tens of thousands of genes as input dimension that would be a huge computational problem for data analysis. In this chapter, the proposed approach deals with selection of feature genes and classification of microarray data under support vector machine (SVM) approach. Feature genes can be finding out according to the adjustable epsilon-support vector regression (epsilon-SVR) and then to select high ranked genes after all microarray data. Moreover, multi-class support vector classification (multi-class SVC) and cross-validation methods apply to acquire great prediction classification accuracy and less computing time.

Keywords Support vector machine • Support vector regression • Multi-class support vector classification • Feature genes • Microarray data analysis

19.1 Introduction

Machine learning is one of artificial intelligence, which has the ability to learn. Machine learning techniques have been successfully applied to cancer classification for microarray data [1]. In machine learning approach, one of popular approaches is support vector machine (SVM) that can deal with classification under support vector classification (SVC) and regression analysis under support vector regression (SVR) [2]. In recent years, "feature selection" became a popular topic. It means used some methods to find feature genes from original genes. In general, cost will

S.-H. Du • S.-F. Su
Department of Electrical Engineering, National Taiwan University of Science and Technology, Taipei, Taiwan

J.-T. Jeng (✉)
Department of Computer Science and Information Engineering, National Formosa University, Yunlin County, Taiwan
e-mail: tsong@nfu.edu.tw

S.-C. Chang
Aeronautical Systems Research Division, National Chung-Shan Institute of Science and Technology, Taichung, Taiwan

J. Juang (ed.), *Proceedings of the 3rd International Conference on Intelligent Technologies and Engineering Systems (ICITES2014)*, Lecture Notes in Electrical Engineering 345, DOI 10.1007/978-3-319-17314-6_19

increase under the number of genes in disease detection. Besides, many studies focused on combined with feature selection and SVM to deal with that reduce gene number and classification [3–7]. Zhang et al. [3] proposed r-test methods convert gene ranking results into position p-values to evaluate the significance of genes. Tang et al. [4] purposed a new two-stage SVM-recursive feature elimination (SVM-RFE) algorithm what overcomes the instability problem of the SVM-RFE to achieve better algorithm utility. And then have demonstrated that the two-stage SVM-RFE is significantly more accurate and more reliable than the SVM-RFE. Kung and Mak [5] purposed a fusion strategy to integrate the diversified information embedded in the symmetric doubly supervised (SDS) formulation. However, simulation studies on protein sequence data for subcellular localization confirm that the prediction can be significantly improved by combining vector-index-adaptive SVM (VIA-SVM) with relevance scores (e.g., Signal-to-Noise Ratio (SNR)) and redundancy metrics (e.g., Euclidean distance). In Leung and Hung [6], a multiple-filter-multiple-wrapper (MFMW) approach is proposed that makes use of multiple filters and multiple wrappers to improve the accuracy and robustness of the classification, and to identify potential biomarker genes. Lee and Leu [7] purposed a novel hybrid method for feature selection in microarray data analysis. The method first uses a genetic algorithm with dynamic parameter setting (GADP) to generate a number of subsets of genes and to rank the genes according to their occurrence frequencies in the gene subsets. Second applies the χ^2-test for homogeneity to select a proper number of the top-ranked genes for data analysis. Finally, they use the classification of SVM to verify the efficiency of the selected genes. Based on the above description, there are many studies focused on the topic of feature selection, and get good experiment results for prediction. For the SVM, the most of results are used SVC to do classification. In this chapter, we apply SVM for the microarray data analysis, the process in feature selection with SVR and in classification with multi-class SVC. That is, the analysis of microarray, selection of feature gene, and classification all use SVM in this chapter.

19.2 Characteristic of Ovarian Microarray Data

There are 41 samples and each sample is a piece of the microarray. These microarray samples are divided into four classes; namely, normal ovaries class, benign ovarian tumors (OVT) class, ovarian cancers at stage I (OVCAI) class, and ovarian cancers at stage III (OVCAIII) class in the ovarian cancer microarray data. Tissues applied in this study included 6 normal ovaries class, 13 OVT class, 7 OVCAI class, and 15 OVCAIII class in Table 19.1. All ovarian cancer microarray procedures

Table 19.1 Category of ovarian microarray

	Normal	OVT	OVCAI	OVCAIII	Total
Sample number	6	13	7	15	41

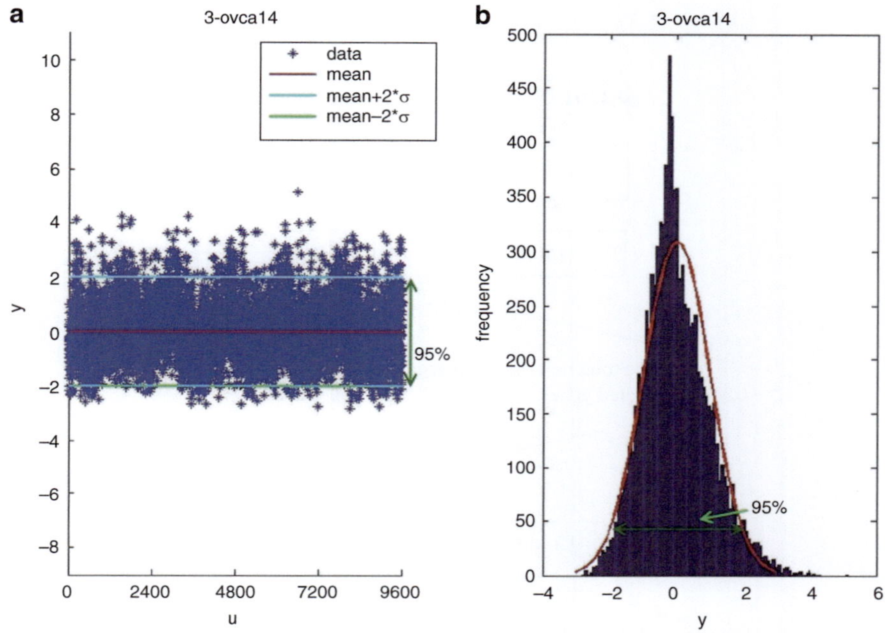

Fig. 19.1 Feature genes selection based on statistics theory

were performed in a dust/climate control laboratory at China Medical University. A sequence-verified human cDNA library containing 9,600 human cDNA clones was a kind gift from the National Health Research Institute of Taiwan [8].

Figure 19.1a shows the microarray data information where u is the number of genes in microarray and y is log(based 2) of R/G normalized ratio. R is magnitude of Cy5 and G is magnitude of Cy3. Traditionally, biologists found out feature genes based on statistics theory. The method is calculated p-value, based on mean and standard deviation. They usually use another 5 % genes to be feature genes for disease detection (see Fig. 19.1b). In general, the character of microarray data has a wave property from Fig. 19.1a. Therefore, the nature of microarray data is nonlinear. Hence, we proposed SVM that can deal with nonlinear problem to improve statistical method.

19.3 The Proposed Approach

SVM is a new classification and regression technique that was proposed by Vapnik, and successfully applied to many different fields [9]. The concept of SVM is that separate different high-dimensional labeled data according to optimal hyperplane. Besides, in SVM the data applies kernel function to map input data into another

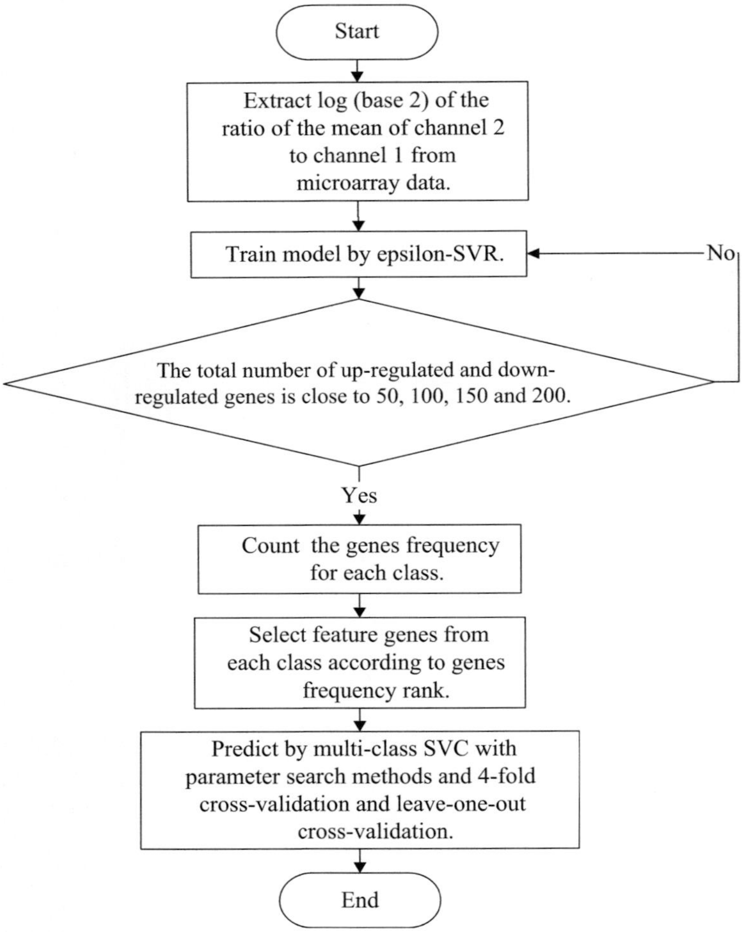

Fig. 19.2 The flowchart of the proposed approach

space. This chapter chooses radial basic function in multi-class SVC and epsilon-SVR. The flowchart of the purposed approach is shown in Fig. 19.2.

In general, the fluorescent dyes Cy3 (green) and Cy5 (red) are most often used to prepare labeled cDNA for microarray hybridizations. In this chapter, we only consider the magnitude of Cy5 and Cy3 in microarray data. Firstly, log(base 2) of the R/G ratio of the mean of channel 2 to channel 1 from microarray data is used. The genes expression data had been recorded in column named log 2 ratio normalized *R*/*G* mean as follows:

$$\text{Log(base2) of } R/G \text{ Normalized Ratio (Mean)} = \log_2 \frac{\text{Cy5}}{\text{Cy3}}. \tag{19.1}$$

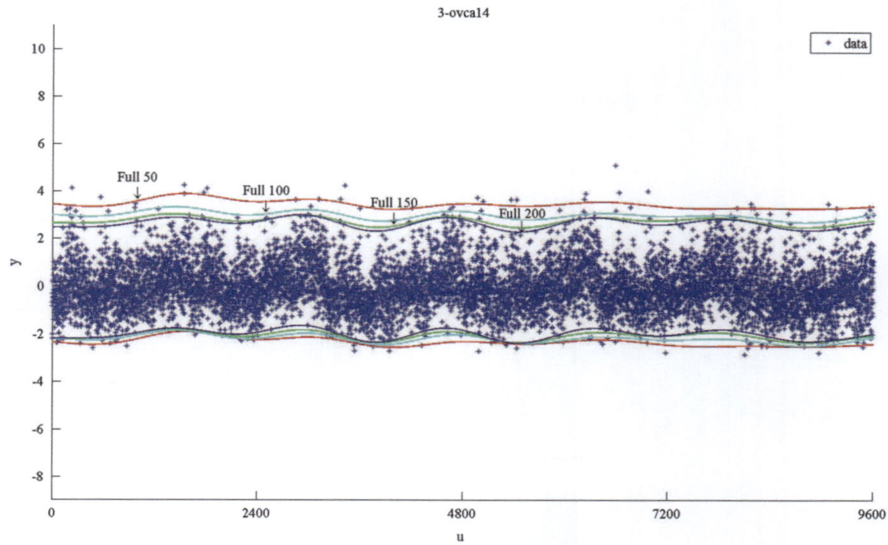

Fig. 19.3 The feature genes, finding out via the proposed SVR with different ε values

Secondly, based on adjust epsilon-SVR to build a smooth curve that can find out the total number of upregulated and downregulated genes is close to 50, 100, 150, and 200 that is shown in Fig. 19.3. The main concept of ε-SVR is proposed to find out the feature gene as in Fig. 19.4 under certain ε in SVR. If ε increased as red arrow then the total number of upregulated and downregulated would be reduced. The parameter ε could control how many genes would be filtered. Hence, using ε-SVR to filter out four classes of microarray data and record upregulated and downregulated genes is close to 50, 100, 150, and 200.

Thirdly, count and record the genes frequency for each class. For example: the gene named A, it was filtered five times in class 1, ten times in class 2, and three times in class 3 and recorded it into gene sets like "Full 50," "Full 100," "Full 150," and "Full 200". "Full 50" means a gene set that finds out the total number of upregulated and downregulated genes close to 50 with each sample from original genes. Fourth, select feature genes from each class according to genes frequency rank (from high to low, and if had existed then selected minor). Finally, use multi-class SVC with parameter search methods to classification microarray data according to fourfold cross-validation and leave-one-out (LOO) cross-validation.

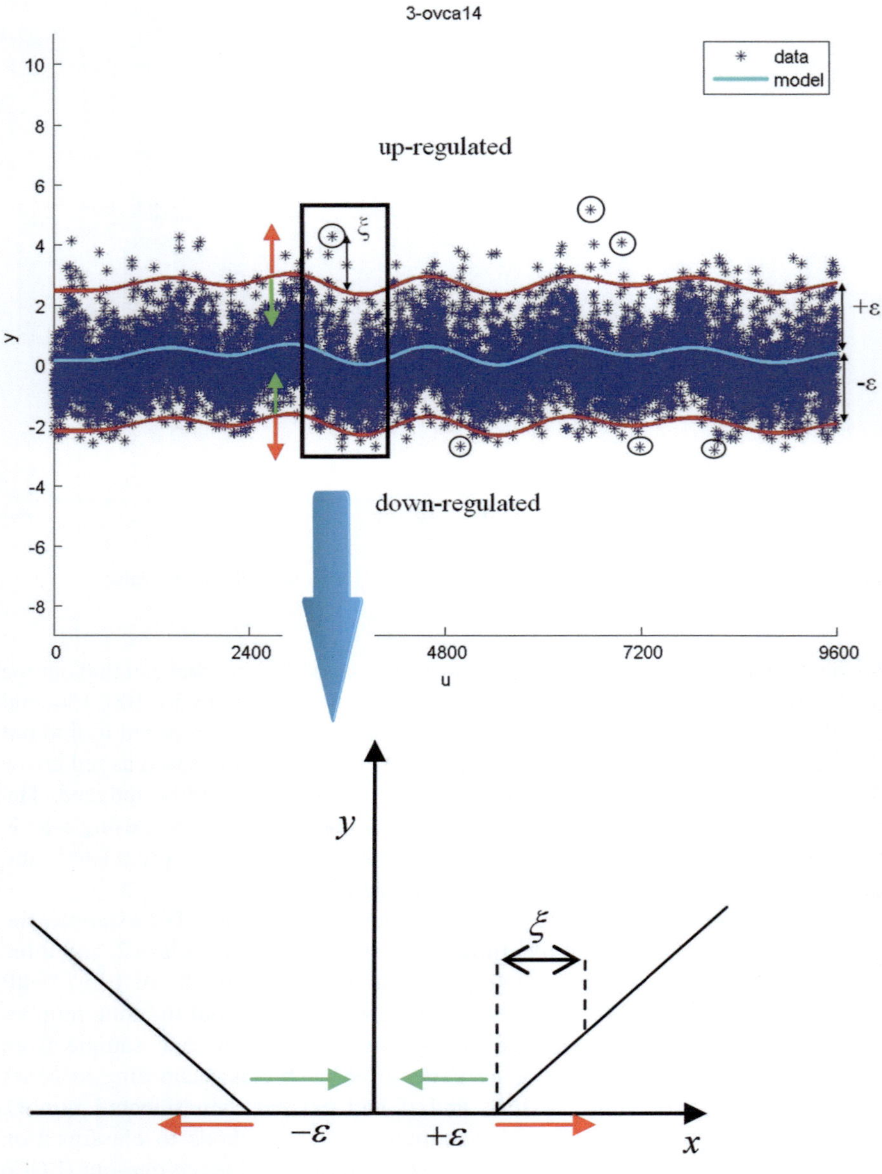

Fig. 19.4 The soft margin loss setting for SVM

19.4 Experiment Results

In this study, the number of experiment sample is 41 and the number of genes is 9,600. Tissues were applied in the current study that included 6 normal, 13 benign OVT, 7 OVCAI, and 15 OVCAIII. Figure 19.5a and b shows the microarray dataset used the proposed approach with different feature genes under fourfold and LOO cross-validations, respectively.

In general, more genes in microarray don't guarantee to get greater prediction classification accuracy as Fig. 19.5. Besides, the prediction classification accuracy hadn't linear relationship with genes number absolute. In Fig. 19.5b, the best prediction classification accuracy with LOO had used two or three genes. Table 19.2 shows the results of experiments that got greater predictive classification accuracy with less than the original gene number, whether fourfold and LOO cross-validation in this chapter.

From the above results, in this chapter we successfully apply SVM for microarray data analysis.

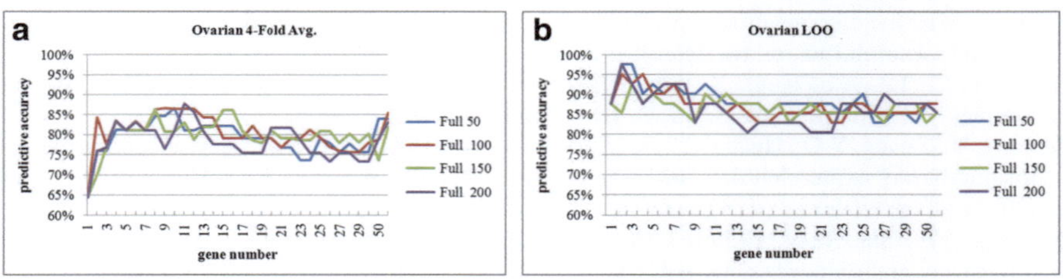

Fig. 19.5 Shown the proposed approach for the ovarian microarray data with different feature genes under (**a**) fourfold cross-validation and (**b**) LOO cross-validation

Table 19.2 The best prediction of classification accuracy of ovarian cancers under different feature genes with fourfold and LOO cross-validation

	Ovarian microarray data			
Gene set	Gene number	Fourfold average (%)	Gene number	LOO (%)
Original	9,600	83.16	9,600	82.93
Full 50	10	86.70	2	97.56
Full 100	9	86.70	2	95.12
Full 150	8	86.29	3	92.68
Full 200	11	87.71	2	97.56

19.5 Conclusions

It is difficult to find out the feature genes for cancer research. Additionally, the cost of disease detection has a relation with the number of genes in microarray. Hence, in this chapter, the proposed approach can reduce the number of genes after epsilon-SVR analysis. Also the simulation results revealed that higher prediction accuracy with less than the original gene number. That means the proposed approach can be effectively applied to selecting feature genes and prediction from microarray data with lower cost.

Acknowledgment The authors wish to thank that this work was supported by National Science Council Under Grant NSC 95-2221-E-150-085, NSC 101-2221-E-150-048-MY2.

References

1. Peterson, C., Ringnér, M.: Analyzing tumor gene expression profiles. Artif. Intell. Med. **28**, 59–74 (2003)
2. Furey, T.S., Cristianini, N., Duffy, N., Bednarski, D.W., Schummer, M., Haussler, D.: Support vector machine classification and validation of cancer tissue samples using microarray expression data. Bioinformatics **16**, 906–914 (2000)
3. Zhang, C., Lu, X., Zhang, X.: Significance of gene ranking for classification of microarray samples. IEEE/ACM Trans. Comput. Biol. Bioinf. **3**, 31–320 (2006)
4. Tang, Y., Zhang, Y.Q., Huang, Z.: Development of two-stage SVM-RFE gene selection strategy for microarray expression data analysis. IEEE/ACM Trans. Comput. Biol. Bioinf. **4**, 365–381 (2007)
5. Kung, S.Y., Mak, M.W.: Feature selection for self-supervised classification with applications to microarray and sequence data. IEEE J. Sel. Top. Signal Process. **2**, 297–309 (2008)
6. Leung, Y., Hung, Y.: A multiple-filter-multiple-wrapper approach to gene selection and microarray data classification. IEEE/ACM Trans. Comput. Biol. Bioinf. **7**, 108–117 (2010)
7. Lee, C.P., Leu, Y.: A novel hybrid feature selection method for microarray data analysis. Appl. Soft Comput. **11**, 20–213 (2011)
8. Jeng, J.T., Lee, T.T., Lee Y.C.: Classification of ovarian cancer based on intelligent systems with microarray data. In: IEEE International Conference on Systems, Man and Cybernetics, pp. 1053–1058 (2005)
9. Vapnik, V.N.: The Nature of Statistical Learning Theory. Springer, New York (1995)

Chapter 20
Feature Selection Algorithm for Motor Quality Types Using Weighted Principal Component Analysis

Yun-Chi Yeh, Liuh-Chii Lin, Mei-Chen Liu, and Tsui-Shiun Chu

Abstract This chapter proposes a qualitative feature selection for motor quality types using Weighted Principal Component Analysis (WPCA) method. The WPCA includes two processes, one is the Procedure-FFV (find the final weights) process and the other is the Procedure-DPC (determine the principal components) process. The input variables of the WPCA are nine original features and the output variables are six qualitative features. Experimental results indicate that the proposed WPCA provides an efficient, simple, and fast method for feature selection on motor's current waveforms.

Keywords Weighted Principal Component Analysis (WPCA) • Feature selection • DC motor

20.1 Introduction

Feature selection [1] is an extensively adopted dimensionality reduction technique, and has been the focus of much research in pattern recognition, machine learning, and data mining. There have been several investigations dealing with the feature selection for analog signals. For instance, principal component analysis (PCA) [2] and Fisher's linear discriminate analysis (Fisher's LDA) [3] have been extensively adopted in pattern recognition and general feature selection problems. These methods find a mapping between the original feature space and a low-dimensional feature space. All the above methods require some complicated mathematical calculations to achieve their aims, as is well known. However, fast, reliable, and efficient algorithms become much important if the analysis of analog

Y.-C. Yeh (✉) • L.-C. Lin • M.-C. Liu • T.-S. Chu
Department of Electronic Engineering, Chien Hsin University of Science and Technology, Zhongli 320, Taiwan, ROC
e-mail: yunchi@uch.edu.tw

J. Juang (ed.), *Proceedings of the 3rd International Conference on Intelligent Technologies and Engineering Systems (ICITES2014)*, Lecture Notes in Electrical Engineering 345, DOI 10.1007/978-3-319-17314-6_20

signals for feature selection systems is applied to areas involving limited energy consumption. Hence, this study proposes a simple, fast, and reliable method called Weighted Principal Component Analysis (WPCA) for effective feature selection for motor quality types [4].

20.2 The Proposed WPCA

The features in the motor's current waveform are the location, duration, amplitudes, and shapes of the waves. These features can be recognized by the experienced operators for recognizing whether the motor is good or not. If it has defects, the kind of defects is also determined. Table 20.1 and Fig. 20.1 list ten important test points of the motor's current waveform. Table 20.2 lists nine original features based on many experiments [5].

After providing the nine original features of the current waveform of motor, the next important task is the qualitative feature selection. In this study, the qualitative feature selection for determining motor's quality types using the WPCA is introduced. The WPCA includes two processes, one is the Procedure-FFV process and the other is the Procedure-DPC process [4].

20.2.1 Procedure-FFV: Find the Final Weights for the jth Index W_j

Assume each sample vector has nine original features (see Table 20.2). The procedure to find the final weights for the *j*th index values of W_j is briefly depicted as follows [6].

Table 20.1 Definition of ten important test points of the motor's current waveform [5]

Serial no. of test point	Symbol	Test point description
1	Pole-up$_1$	The commuted-points (up) of each cycle
2	Pole-down	The commuted-points (down) of each cycle
3	max$_1$	The maximum value point of waveform 1 of each cycle
4	min$_1$	The minimum value point of waveform 1 of each cycle
5	max$_2$	The maximum value point of waveform 2 of each cycle
6	min$_2$	The minimum value point of waveform 2 of each cycle
7	pole-up$_2$	The second point after the pole-up$_1$
8	max$_{22}$	The second point after the max$_2$
9	min$_{22}$	The third point before the min$_2$
10	min$_{23}$	The third point after the min$_2$

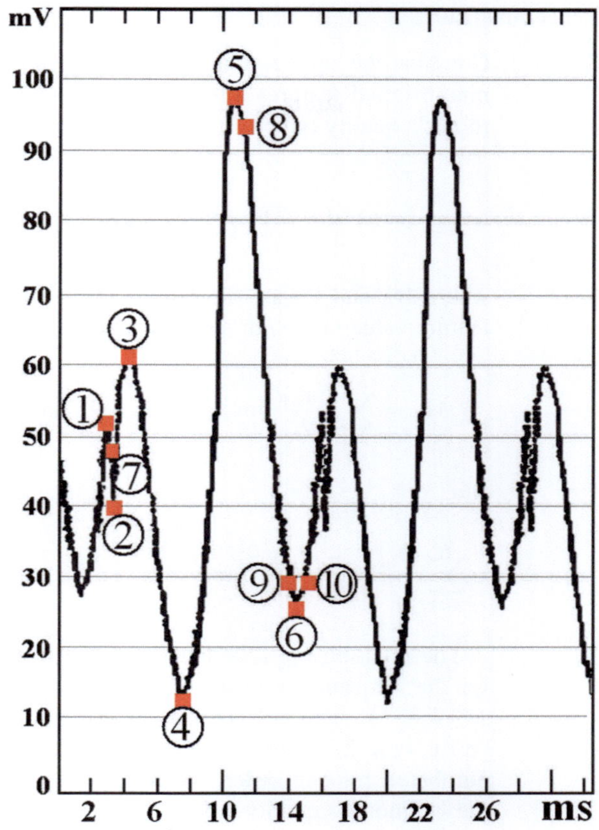

Fig. 20.1 Ten important test points of the motor's current waveform (good DC motor)

Table 20.2 Definition of the nine original features [5]

Feature's serial no.	Feature's symbol	Feature description	Unit
$F_{1k,n}$	$V_{1,2}$	The amplitude between point 1 and point 2 in Fig. 20.1	mV
$F_{2k,n}$	$V_{3,4}$	The amplitude between point 3 and point 4 in Fig. 20.1	mV
$F_{3k,n}$	$V_{1,7}$	The amplitude between point 1 and point 7 in Fig. 20.1	mV
$F_{4k,n}$	$V_{6,9}$	The amplitude between point 6 and point 9 in Fig. 20.1	mV
$F_{5k,n}$	$V_{6,10}$	The amplitude between point 6 and point 10 in Fig. 20.1	mV
$F_{6k,n}$	$V_{1,5}$	The amplitude between point 1 and point 5 in Fig. 20.1	mV
$F_{7k,n}$	$V_{4,6}$	The amplitude between point 4 and point 6 in Fig. 20.1	mV
$F_{8k,n}$	$V_{3,5}$	The amplitude between point 3 and point 5 in Fig. 20.1	mV
$F_{9k,n}$	$V_{5,8}$	The amplitude between point 5 and point 8 in Fig. 20.1	mV

Procedure-FFV:

Step 1-1. Combine the square value of sample mean and sample variance of the nine original features into a set of variables $\{T_{jk}\}$, $j = 1, \ldots, 18$, and motor's quality type $k = 1, 2, \ldots, 11$,

$$T_{jk} = \begin{cases} M_{ik}^2; & j \leq 9, \; j = i \\ V_{ik}; & j > 9, \; j = i + 9 \end{cases} \tag{20.1}$$

where M_{ik} and V_{ik} are the sample mean and sample variance of the ith feature value in the kth motor's quality type, respectively, defined as

$$M_{ik} = \sum_{n=1}^{N_k} \frac{F_{ik,\,n}}{N_k} \quad \text{for } i = 1, \ldots, 9, \text{ and } k = 1, \ldots, 11 \tag{20.2}$$

and

$$V_{ik} = \sum_{n=1}^{N_k} \frac{(F_{ik,\,n} - M_{ik})^2}{N_k} \quad \text{for } i = 1, \ldots, 9, \text{ and } k = 1, \ldots, 11 \tag{20.3}$$

The notation $F_{ik,\,n}$ is defined as the feature value of the ith feature for the nth samples in the kth motor's quality type, where $1 \leq i \leq 9$, $1 \leq k \leq 11$, $1 \leq n \leq N_k$, and 11 motor's quality type, "error type 1," "error type 2," "error type 3," …, "error type 10," and "good," are numbered by k in order, where N_k denotes the number of samples of the kth motor's quality type.

Step 1-2. The elements of $T_{jk} = \{T_{j1}, T_{j2}, \ldots, T_{j11}\}$ are sorted in reverse numerical order. The sorted T_{jk} becomes $\{T_{jr} \geq T_{js} \geq \cdots \geq T_{jt}\}$ for $j = 1, \ldots, 18$, and $r, s, t \in \{1, \ldots, 11\}$.

Step 1-3. Decide the weight pointer P_{jr}. For the kth feature, the weight pointer P_{jr} is decided as follows. If $\frac{T_{jr}}{T_{js}} \geq \gamma$, then $P_{jr} = 1$, else $P_{jr} = 0$. The threshold value γ is selected based on a lot of experimental results for different motor's current waveforms. For the kth feature, γ must be satisfied that the number of $P_{jr} = 1$ is larger than 1. When $P_{jr} = 1$, the corresponding T_{jr} contains the extremely important information about the motor's current waveforms.

Step 1-4. Decide the weight W_j. Each W_j $(1 \leq j \leq 18)$ is dependent on P_{jr}. If all P_{jr} $(1 \leq j \leq 18)$ are 0s, then all W_j $(1 \leq j \leq 18)$ are assigned with 0s. Otherwise, each W_j $(1 \leq j \leq 18)$ is assigned with a random value.

Step 1-5. Calculate the final value of W_j. First, calculate S_k for the kth motor's quality type as follows.

$$S_k = \sum\nolimits_{j=1}^{18} T_{jk} W_j; \quad k = 1, \ldots, 11 \tag{20.4}$$

Then, the following conditions must be satisfied:

$$\frac{|S_r - S_t|}{D} \leq 0.1, \quad r, t \in \{1, \ldots, 11\}, \ r \neq t, \tag{20.5}$$

and

$$\frac{\sum_{j=1}^{18} T_{jk} W_j P_{jk}}{S_k} \geq \varepsilon, \quad k = 1, \ldots, 11 \tag{20.6}$$

where $D = \frac{\sum_{k=1}^{11} S_k}{11}$, and ε is a constant value. It is noted that (20.4) is used to give the normalized values of S_k for $k = 1, \ldots, 11$. As discussed further below, the cumulative proportion of eigenvalues typically selected by WPCA method is 90 %. Therefore, $\varepsilon = 100\ \% - 90\ \% = 0.1$ in (20.5) is suggested. If computed $\{S_1, S_2, \ldots, S_{11}\}$ can satisfy (20.4) and (20.5), thus W_j is the final weight. Otherwise, go to Step 1-4. After the final weights for the *j*th index W_j $(1 \leq j \leq 18)$ is obtained, the next task is to determine the number of principal components for the *k*th motor's quality type.

20.2.2 Procedure-DPC: Determine the Principal Components (DPC)

The procedure for performing this task is depicted as follows [4, 6].

Procedure-DPC:

Step 2-1. Find the significant features related to $W_j \neq 0$. In the above procedure, the total number of M_{ik} and V_{ik} corresponding to $W_j \neq 0$ and $W_{i+9} \neq 0$ are L_1 and L_2, respectively, where $i \in 1, \ldots, 9$. Two significant features are given, one is the corresponding feature $F_{\alpha_i k,\ n}$, $(\alpha_1, \alpha_2, \ldots, \alpha_{L_1} \in \{1, \ldots, 9\})$, and the other is $F_{\beta_i k,\ n}$, $\beta_1, \beta_2, \ldots, \beta_{L_2} \in \{1, \ldots, 9\}$.

Step 2-2. To obtain the sample covariance matrix, the proposed method firstly sets column vector $\overline{g}_{k,\ n}$ as

$$\overline{g}_{k,\ n} = \begin{bmatrix} F_{\alpha_1 k,\ n} \\ \vdots \\ F_{\alpha_{L_1} k,\ n} \\ \hline F_{\beta_1 k,\ n} - M_{\beta_1 k} \\ \vdots \\ F_{\beta_{L_2} k,\ n} - M_{\beta_{L_2} k} \end{bmatrix} \in R^{L \times 1}, \tag{20.7}$$

where $\alpha_1, \alpha_2, \ldots, \alpha_{L_1} \in \{1, \ldots, 9\}$, $\beta_1, \beta_2, \ldots, \beta_{L_2} \in \{1, \ldots, 9\}$, and $L = L_1 + L_2$. The total number of the significance means and variances is reduced from 18 to L. It will be discussed in the next section.

Step 2-3. Find the sample covariance matrix $\overline{G}_k$ of the kth motor's quality type. First, the indexes with $W_j = 0$ in Step 1-4 of Procedure-FFV are removed. Let elements of the column vector $\overline{w}$, which is formed by $W_j \neq 0$, be related to elements of $\overline{g}_{k,\ n}$ in (20.6). Second, the sample covariance matrix of the kth heartbeat case is then defined as follows.

$$\overline{G}_k = \sum_{n=1}^{N_k} \frac{1}{N_k} \mathrm{diag}[\overline{w}]^{0.5}\ \overline{g}_{k,n}\ \overline{g}_{k,n}^T \mathrm{diag}[\overline{w}]^{0.5}, \quad k = 1, \ldots, 11 \tag{20.8}$$

Step 2-4. For each k, find the eigenvalues and eigenvectors of the matrix $\overline{G}_k$ for the kth heartbeat case. After eigen decomposition of $\overline{G}_k$, (20.7) can be rewritten as

$$\overline{G}_k = [\overline{e}_{1k}, \overline{e}_{2k}, \ldots, \overline{e}_{Lk}] \begin{bmatrix} \lambda_{1k} & & & 0 \\ & \lambda_{2k} & & \\ & & \ddots & \\ 0 & & & \lambda_{Lk} \end{bmatrix} \begin{bmatrix} \overline{e}_{1k}^{\mathrm{T}} \\ \overline{e}_{2k}^{\mathrm{T}} \\ \vdots \\ \overline{e}_{Lk}^{\mathrm{T}} \end{bmatrix} \tag{20.9}$$

$$k = 1, \ldots, 11$$

where eigenvalues $\lambda_{1k} \geq \lambda_{2k} \geq \cdots \geq \lambda_{Uk} \geq \cdots \geq \lambda_{Lk}$ are in reverse numerical order and $\bar{e}_{1k}, \bar{e}_{2k}, \ldots, \bar{e}_{Lk}$ are their corresponding eigenvectors. Since $|\overline{e}_{ik}| = 1$ $(i = 1, 2, \ldots, L)$, λ_{ik} is also called the power of the $\bar{e}_{ik}$. The next step retains only U principal eigenvectors of $\overline{G}_k$ for further processing using the determination procedure for the motor's quality types.

Step 2-5. The first U eigenvectors in (20.9) are used as principle components if the cumulative proportion CP_k exceeds a certain threshold. A useful compromise criterion is a value of 90 %. The cumulative proportion CP_k of the eigenvalues is defined as follows [2].

$$\mathrm{CP}_k = \frac{\lambda_{1k} + \lambda_{2k} + \cdots + \lambda_{Uk}}{\lambda_{1k} + \lambda_{2k} + \cdots + \lambda_{Uk} + \cdots + \lambda_{Lk}} \geq 90\ \% \tag{20.10}$$

The purpose of Procedure-DPC is to provide eigenvectors of U principle components for motor's quality type determination of the following subsection. After performing the Procedure-DPC, six qualitative features $V_{1,2}$, $V_{3,4}$, $V_{1,7}$, $V_{1,5}$, $V_{4,6}$, and $V_{3,5}$ are selected (see Table 20.2) [2, 4, 6].

20.3 Evaluation and Results

This section introduces a fuzzy logic system for the diagnosis of the motor's quality type. The fuzzy logic consists of four parts: fuzzy sets definition, fuzzy rule base establishment, fuzzy inference engine design, and defuzzification [7]. The input variables of the fuzzy rule base are six qualitative features (after performing the Procedure-DPC), namely $V_{1,2}$, $V_{3,4}$, $V_{1,7}$, $V_{1,5}$, $V_{4,6}$, and $V_{3,5}$ (see Table 20.2). The output variable is "motor's quality type," which has 11 motor's quality types of "error type 1," "error type 2," "error type 3," . . ., "error type 10," and "good." In the experiment, the total classification accuracy was approximately 99.705 % by fuzzy logic [5]. Results of the experiment indicate that the six qualitative features are the best choice for qualitative feature selection for motor quality types using WPCA method.

20.4 Conclusions

This study proposes a qualitative feature selection for motor quality types using WPCA method. The proposed WPCA has several advantages: (1) good detection results: the total classification accuracy was approximately 99.705 %; (2) simplicity: complicated mathematical computations such as cross-correlation and Fourier transformation are unnecessary; (3) high speed: It can recognize good or defective motors as well as defect types in less than 0.5 s. Experimental results indicate that the proposed WPCA provides an efficient, simple, and fast method for feature selection on motor's current waveforms.

Acknowledgement The work was supported by the Ministry of Science and Technology of Republic of China under Grant MOST 103-2221-E-231-019.

References

1. Dash, M., Liu, H.: Feature selection for classification. Intell. Data Anal. **1**, 131–156 (1997)
2. Jolliffe, I.T.: Principal Component Analysis. Springer, New York (1986)
3. Ren, H., Chang, Y.L.: Feature extraction with modified Fisher's linear discriminant analysis. Proc. SPIE **5995**, 56–62 (2005)
4. Yue, H.H., Tomoyasu, M.: Weighted principal component analysis and its applications to improve FDC performance. In: International Conference on Decision and Control, ICDC-2004, Bahamas, pp. 4262–4267 (2004)
5. Yeh, Y.C.: Fuzzy logic method for motor quality types on current waveforms. Measurement **46**, 1682–1691 (2013)
6. Yeh, Y.C., Chu, Y., Chang, W.T.: Cardiac Arrhythmia diagnosis on ECG signals using weighted principal component analysis. In: ICME, pp. 357–362 (2014)
7. Zadeh, L.A.: Fuzzy sets. Inf. Control **8**, 338–353 (1965)

Chapter 21
Feasibility Test of Range of Motion Exercises for Ankle Joints Rehabilitation Using Pneumatic Soft Actuators

Hironari Taniguchi, Noriko Tsutsui, and Yoshiaki Takano

Abstract Patients with movement disabilities have been increasing steadily by accidents and diseases. If the condition does not improve, contractures may occur in some joints and muscles. The contractures often tend to occur in ankle joints which are the most important body parts for everyday living. Thus it is important to start rehabilitation therapy shortly after an accident and a disease. The purpose of this study is to develop a rehabilitation device for ankle joints. In our previous study, we proposed a pneumatic actuator with soft material. The actuator has many advantages such as low mass, flexibility, safety, and user-friendliness. Therefore, we focused on the actuator as a drive source of the rehabilitation equipment and the actuator used for the range of motion (ROM) exercises. We measured the ROM needed for ankle joint. As the result, we confirmed that the actuator is able to provide several ROM exercises.

Keywords Pneumatic soft actuator • Rehabilitation device • Ankle joints

21.1 Introduction

Patients with impairment of motor functions have been increasing steadily by several accidents and diseases. For example, strokes threaten the health of many people in the world; one in six people worldwide will have a stroke in their lifetime [1]. In Japan, there were 1.37 million stroke patients in 2005 [2] and they are increasing annually. The onset of a stroke often causes movement disabilities like partial paralysis. If the condition does not improve, contractures may occur in some joints and

---

H. Taniguchi (✉) • Y. Takano
Department of Electronics and Control Engineering, National Institute of Technology, Tsuyama College, Tsuyama-shi, Okayama 708-8509, Japan
e-mail: taniguti@tsuyama-ct.ac.jp

N. Tsutsui
Advanced Mechanical and Control System Engineering, National Institute of Technology, Tsuyama College, Tsuyama-shi, Okayama 708-8509, Japan

J. Juang (ed.), *Proceedings of the 3rd International Conference on Intelligent Technologies and Engineering Systems (ICITES2014)*, Lecture Notes in Electrical Engineering 345, DOI 10.1007/978-3-319-17314-6_21

muscles. Therefore, it makes daily life difficult for the patients. Furthermore, if the contract is associated with severe disease, a rehabilitation therapy causes acute pain to the patients. These mean that it becomes especially difficult to control the body, thus it is very important to start the rehabilitation therapy before developing the contracture.

Normally, range of motion (ROM) exercise for ankle joints is prescribed by a physical therapist to prevent the joints contracture and improve the patient's symptoms. However, the patients cannot receive enough rehabilitation, because the time of therapy from the therapist is limited. In addition, the therapists will be generally less numerous than the patients, they will expect to increase workloads.

Many researchers have developed mechanical ankle joints rehabilitation devices to improve movement disabilities [3–5]. However, these devices are able to exercise only range of ankle joint motion. Therefore the ROM exercise device was developed for ankle joints which provides multiple degree of freedom ROM exercises. Firstly, we focused on a pneumatic soft actuator as the drive source of the device. This actuator has many advantages such as lightweight, flexibility, safety, and with a high affinity for people. The rehabilitation device is more likely to be popular with users since safety is essential for them.

In this chapter, we report the methods of ROM exercises for ankle joints using the pneumatic soft actuators. We fabricated an experimental device with mechanism elements and pneumatic equipment. We also confirmed the performance of our proposal through experiments.

21.2 Concept of Rehabilitation Device

We investigated methods of ankle joints rehabilitation performed by a physical therapist from the Department of Rehabilitation of Tsuyama Central Hospital. ROM exercises for ankle joints have six motions which are plantar flexion/dorsiflexion, eversion/inversion, and abduction/adduction motions. Table 21.1 shows the normal ROM for ankle joint. If a contracture occurs around the ankle, the ROM is reduced relatively. Therefore, it is necessary to improve the function to move the ankle joint between the all of normal ROM exercises. Our investigation revealed that ROM exercises as the Table 21.1 shows are needed for use in ankle rehabilitation device. In this study, we devised a mechanism for ankle movement using pneumatic soft actuators.

Table 21.1 The normal ROM for ankle joints

Name	ROM (deg)
Plantar flexion	0–45
Dorsiflexion	0–20
Adduction	0–20
Abduction	0–30
Inversion	0–30
Eversion	0–20

21.3 Pneumatic Soft Actuator

Figure 21.1a shows configuration of a "pressure bag". The pressure bag is an integral component of proposed pneumatic soft actuator and consists of an aluminum pouch, a canvas sheet, and a polyurethane air tube. The aluminum pouch is formed by sealing the edge of aluminum sheet stacked in pairs. The pouch is coverd with the canvas sheet. It is also used to prevent the burst of aluminum pouch. Furthermore, the canvas sheet covers to control the movement of pressure bag. The actuator shown in Fig. 21.1b is used for plantar flexion/dorsiflexion and eversion/inversion movements. It is assembled from the pressure bag of two different sizes; pressure bag A is 140 mm in width and 160 mm in length, and pressure bag B is 80 mm in width and 160 mm in length. The other actuator shown in Fig. 21.1c is used for abduction/adduction movements. It is also assembled from two, pressure bag A.

The actuator acts on forepart of the foot directly, so that it is almost the same size as the average Japanese man's foot. The actuators were designed by using database information about body dimensions from National Institute of Advanced Industrial Science and Technology (AIST) [6]. The displacement and generated force were measured by limiting air pressure from 0.01 to 0.07 MPa when changing at every 0.01 MPa. As the results, the maximum displacement value for plantar flexion/dorsiflexion movements, eversion/inversion movements, and abduction/adduction movements were 175 mm, 35 mm, and 120 mm, respectively. The generated force was obtained more than 30 N at both actuators. We therefore confirmed that the characteristics were enough to rehabilitate ankle joints of patient.

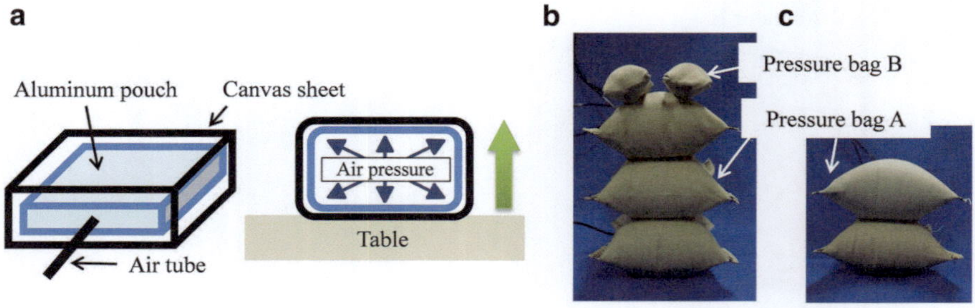

Fig. 21.1 (**a**) Schematic of the pressure bag, (**b**) pneumatic soft actuator for plantar flexion/dorsiflexion and eversion/inversion movements and (**c**) pneumatic soft actuator for abduction/adduction movements

21.4 Experimental Device of ROM Exercise for Ankle Joints Rehabilitation

21.4.1 Design and Driving Principle of the Device

A configuration of the experimental device for ankle joints rehabilitation is shown in Fig. 21.2a. The device consists mainly of a robust acrylic chassis and the pneumatic soft actuators. The actuator for plantar flexion/dorsiflexion and eversion/inversion movements was fixed on the acrylic board at the upper part of the top of the foot and at the posterior part of the bottom of the foot, respectively. The actuator for abduction/adduction movements was also fixed on the acrylic board at the inside and outside of the foot.

Subsequently, we describe the drive principle for ROM exercise of the ankle joint. In demonstrating the abduction motion of right foot, the foot is first held in place by both the actuators as shown in Fig. 21.2b. Next, when the compressed air flowed into the actuator for abduction motion, the actuator is expanded and then moves the foot such as shown in Fig. 21.2c.

21.4.2 Experimental Method

ROM exercises for the ankle joints require the ankle to be moved to its maximum ROM. In addition, the joints of patients without contracture can be considered the same as the joints of healthy individuals. Therefore, to verify the problems with the rehabilitation device related to ankle joints with unlimited ROM, we tested by an

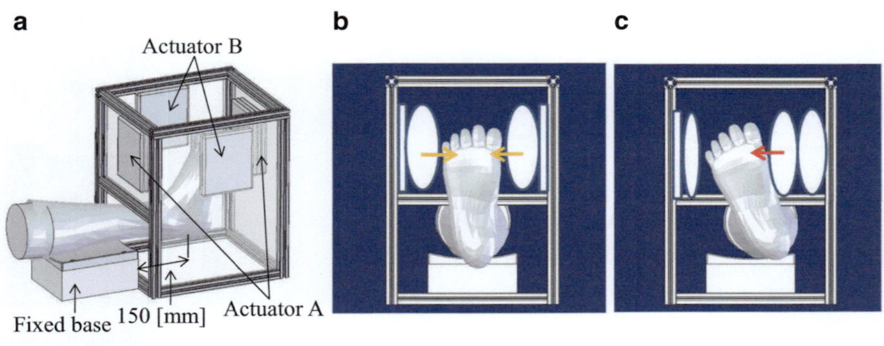

Fig. 21.2 (**a**) The experimental device for ankle joints rehabilitation, (**b**) initial state and (**c**) the drive principle for abduction motion of right foot. Actuator A means the pneumatic soft actuator for plantar flexion/dorsiflexion and eversion/inversion movements and also Actuator B means for abduction/adduction movements

Table 21.2 Subject data of healthy nine individuals

No.	1	2	3	4	5	6	7	8	9	IA	JA
Foot length	225	225	267	250	250	270	268	236	240	248	244
Foot breadth	100	98	105	98	105	100	110	100	100	102	98
Foot circumference	245	235	255	250	240	250	245	245	257	247	241
Lateral malleolus height	80	75	80	73	75	75	70	70	70	74	68

IA and JA mean average value of nine individuals and average value of approximately 500 Japanese people, respectively

experimental device with healthy nine individuals from 19 to 38 years old. Detailed information about their foot was shown in Table 21.2. We also asked a physical therapist to test the device and obtained their assessment.

The ROM was measured as follows. First, the ankle was strapped on a fixed base 150 mm away from the acrylic chassis such as shown Fig. 21.2a and reflective markers were attached to surface of the foot. Compressed air, which was adjusted by an electro-pneumatic regulator, flowed into the actuators at 0.05 MPa. Ankle motion during operation of the experimental device was captured with a high-speed camera (HAS-L1, DITECT Co., Ltd., Tokyo, Japan). The movement of the markers was then analyzed from the captured footage using motion analysis software (Dipp-Motion PRO, DITECT Co., Ltd.). This allowed us to calculate six motion patterns in the ROM exercise.

21.4.3 Experimental Results and Discussion

The results of ROM exercises test are shown in Fig. 21.3. We measured the same ROM motion three times. A target area is illustrated with a broken line. The measurement values within broken line frame indicate that the ankle joint could be moved to its maximum ROM.

In the dorsiflexion and abduction motion, the target angle was achieved at least once for all research participants. In the plantar flexion and adduction motion, the measurement for most of the participants has a tendency to fall below the target angle. However, all measurement angles did not achieve the target angles in the inversion and eversion motion. We considered due to the fact that the actuator did not properly operate to the foot sometimes. For example, we confirmed that the actuator deviates to the upper side of the instep at the plantar flexion motion.

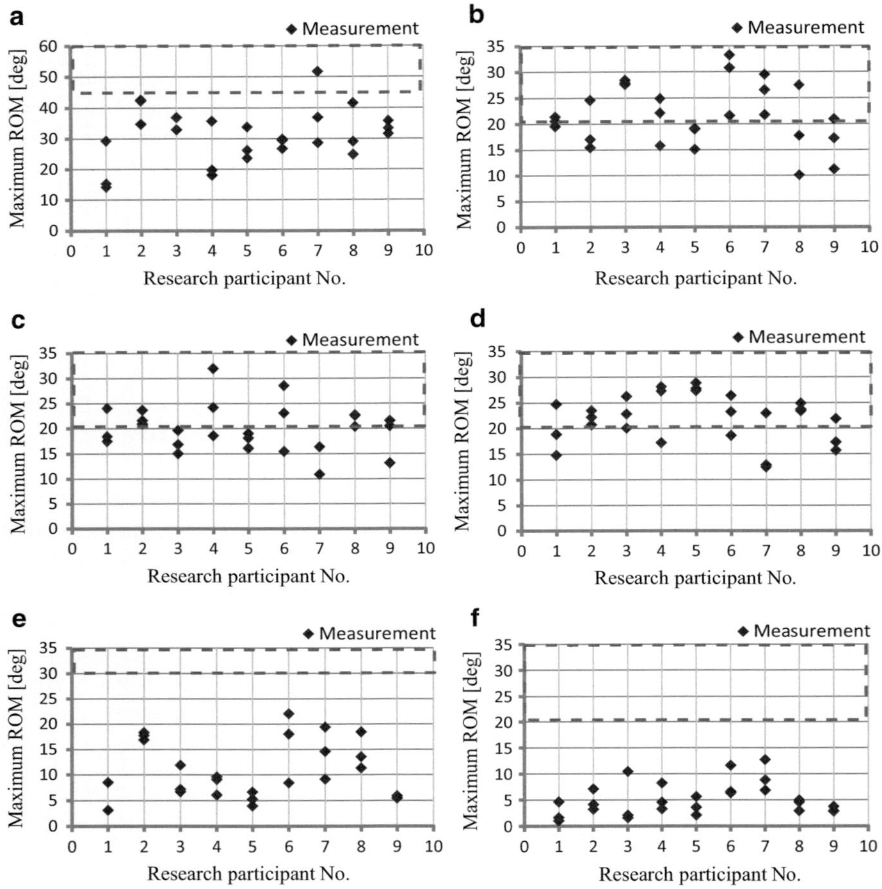

Fig. 21.3 The measurement results of the (**a**) plantar flexion motion, (**b**) dorsiflexion motion, (**c**) adduction motion, (**d**) abduction motion, (**e**) inversion motion, and (**f**) eversion motion

21.5 Conclusion

In this chapter, we designed a novel pneumatic soft actuator and proposed a method of ROM exercise for ankle joints using its actuator. The pneumatic soft actuator is assembled from a pressure bag of two different sizes. We then prototyped the experimental device with the actuators and investigated the ankle movement with healthy nine individuals. The device was fabricated so that the pneumatic soft actuators directly acted on the foot.

The experimental results of the ROM exercises showed the device can achieve a target angle in three of the six motions. Especially, the ankle joint could not be moved to its maximum ROM in the inversion and eversion motion. Of course, there were some problems with the actuators and these faults should be improved. However, the experimental results indicate the feasibility of the actuator as an ankle rehabilitation device.

Acknowledgments This work was supported by JKA and its promotion funds from KEIRIN RACE. That also funded by Fluid Power Technology Promotion Foundation and a grant of the Ministry of Education, Culture, Sports, Science and Technology (MEXT) in Japan.

References

1. World Stroke Campaign (WSC), http://www.worldstrokecampaign.org/2011/Pages/Home.aspx
2. Ministry of Health, Labour and welfare, http://www.mhlw.go.jp/toukei/saikin/hw/kanja/05/05.html
3. Petrosenko, R.D., Vandervoort, A.A., Chesworth, B.M., Porter, M.M., Campbell, G.J.: Development of a home ankle exerciser. Med. Eng. Phys. **18**, 314–319 (1996)
4. Jamwal, P.K., Xie, S.Q., Aw, K.C.: Kinematic design optimization of a parallel ankle rehabilitation robot using modified genetic algorithm. Robot. Auton. Syst. **57**, 1018–1027 (2009)
5. Jamwal, P.K., Xie, S.Q., Toshi, Y.H., Aw, K.C.: Forward kinematics modeling of a parallel ankle rehabilitation robot using modified fuzzy inference. Mech. Mach. Theory **45**, 1527–1554 (2010)
6. AIST Japanese Body Dimension Data, 1991–92, https://www.dh.aist.go.jp/database/91-92/

Chapter 22
3D Motion Editing Through B-Spline Fitting with Constraints

Mankyu Sung

Abstract This paper proposes a novel motion editing algorithm that uses human motion capture data for animating 3D characters. First, the algorithm fits the 2D root joint trajectory with the cubic B-Spline through least-square minimization. In this process, it finds the optimal number of control points based on an error threshold. Once it gets n control points, users are allowed to change the positions of control points, which is able to create a new trajectory. The new trajectory is then fed into the original motion so that it is modified to reflect the new trajectory. To find exact parameter t of the spline curve representing root joint position, the algorithm performed the arc-length parameterization on the curve. Since the motions are forced to change the root joint positions, the result may violate the fidelity of the original motions, which may cause some artifacts such as foot skating. To fix them, the IK (Inverse Kinematics) solver is applied to motions to change the limb orientation. Although the IK solver can change the orientation of original motions, if the differences between the modified trajectory and original trajectory are too big, then the result motion produces awkward poses over times. In order to prevent them, our algorithm puts constraints on the control points of curve automatically so that users are able to edit the trajectory freely without considering whether it produces natural motions or not.

Keywords Motion capture • 3D motion editing • Computer animation

22.1 Introduction

Data-driven motion synthesis methods have been a popular way for animating 3D characters for the last decades [1]. However, editing motion data still poses a lot of huddles to overcome. First, because motion data are nothing more than just 3D signal over time, it must consider the temporal coherence to maintain the smoothness. Second, from the original 3D motion data, a high-level mathematical model is

M. Sung (✉)
Department of Game & Mobile Contents, Keimyung University, Daegu, South Korea
e-mail: mksung@kmu.ac.kr

J. Juang (ed.), *Proceedings of the 3rd International Conference on Intelligent Technologies and Engineering Systems (ICITES2014)*, Lecture Notes in Electrical Engineering 345, DOI 10.1007/978-3-319-17314-6_22

needed to be constructed to manipulate the data without changing values at every single frame. Third, since manual editing on motion data may produce unnatural movement of 3D characters, it needs an automatic way of detecting whether or not the current editing produces bad motions and then alerting users to that symptom immediately. This paper proposed a novel spline-based motion editing technique with enforcing constraints. Among all kinds of different motion types, the proposed method targets on the most popular locomotion such as walking, running, and standing motions. The method first models the 3D root joint trajectory as a 2D curve, which represents global positions of whole characters over time. And then, by using the optimal least-square cubic B-spline-fitting technique, the algorithm finds m piecewise spline-curve segments that consist of n control points. Once it finds the B-spline curve, users are allowed to change positions of control points. Those changes produce a new shape of curve. In this process, to minimize the visual discrepancy from the original motion, the arc-length parameterized spline curve is adopted. The new shape of the curve, which becomes the new root joint trajectory of a character, is then reflected back to the original motion automatically. In this process, the positions and orientations of root joints are adjusted to the position and tangent of the edited curve. Since the motions are forced to change the original root trajectory, some artifacts such as foot skating may occur. To fix the problem, the IK (Inverse Kinematics) solver is applied on the edited automatically. The IK solver adjusts the limb configuration, which corresponds to the knee and ankle joints, so that feet are maintained on the ground. Although users are allowed to change the position of control points of the curve, if the positional difference from the original control points is too big, then the resulting motion does have some visual artifacts such as unnatural stretch of limbs. In this case, the algorithm puts the constraint on the control points so that they are always within a specific circle. The radiuses of circles are computed through the curvatures of the trajectory.

22.2 Related Works

Many data-driven motion synthesis techniques have been proposed by computer animation researchers for more than decades. Signal processing-based techniques were the first attempt to modify the original motion by applying several filters on the motion signal [2]. These techniques were simple to implement but hard to edit because they considered each joint datum individually. Rearrangement-based motion editing techniques tried to build a high-level data structure by finding similar poses of the motions and connecting them [3, 4]. Their methods produced good quality of motions provided that there were many input motion data. A similar technique to what this chapter is presenting was proposed by Chapter [5]. This chapter also was using spline curve to model the root joint trajectory for editing and applying constrain solution to the reestablished geometric features of the motion. However, it did not specify how to obtain the optimal spline curve from the original root trajectory, and did not have a safe-net feature that prevented users from changing arbitrary path that resulted in unnatural motions. Our method fits the motion trajectory to the optimal spline curve based on the error estimation, and puts the constraints on the position of control points. In editing the curve, users are not able to change the control points beyond some range limit, thanks to the constraints enforced on the control points.

22.3 Algorithms

A human motion capture data can be considered as a set of postures over time. The posture is sampled at regular time rate and each posture at a particular time is called a *Frame*. Let a motion M be given as follows:

$$M = \{F_1, F_2, \ldots, F_n\}^T \tag{22.1}$$

where $F_i, 1 \leq i \leq n$ is the posture at the frame i. For a human character with m number of joints, a frame F_i can be represented as a vector that includes root joint position and orientations of m joints.

$$F_i = \left\{p_i, q_i^1, q_i^2, \ldots, q_i^m\right\} \tag{22.2}$$

where $p_i \in \mathbb{R}^3$ is the global position of the root joint and $q_i^j \in \mathbb{S}^3, 1 \leq j \leq m$ is jth orientation of joint. In particular, the orientation q_i^1 is the global orientation of root joint. From frames $F_i, 1 \leq i \leq n$, the root trajectory T can be represented as a vector as follows:

$$T = \{p_1, p_2, \ldots, p_n\} \tag{22.3}$$

The 3D position of p_i, then can be projected onto the ground so that y coordinates can be eliminated. Those 2D points represent a trajectory $\hat{T}$.

$$\hat{T} = \{\hat{p}_1, \hat{p}_2, \ldots, \hat{p}_n\} \tag{22.4}$$

where $\hat{p}_i \in \mathbb{R}^3$.

The proposed approach fits the trajectory $\hat{T}$ with the uniform cubic B-Spline curve by least-square manner. The planar B-Spline curve is defined as $f(t) = \sum_{j=1}^{k} \mathrm{B}_j(t)Q_j$, where $\mathrm{B}_j(t)$ are the B-Spline basis functions of a fixed order and knots, Q_i is the control points and t is the parameter. In the case of Cubic B-spline, the basis functions can be found in Chapter [6]. Our goal is to find the Q_i by solving linear equation of form $Ax = B$ where B is $n \times 1$ matrix of $\hat{T}$ and A is $n \times k$ matrix in which k is the number of control points and x corresponds to Q_j. From this linear equation, the algorithm is trying to find x, which is the position of control points, that minimizes $\|Ax - b\|^2$. This equation then can be represented as normal equation $A^T A_x = A^T b$ where the matrix A is sparse matrix. This normal equation can be solved using Cholesky factorizations method. The proposed method applies LAPACK software to solve this equation [7].

One significant step for solving the least-square-based B-spline fitting is to determine the number of control points. This number of control points must be entered as an input value before fitting. As we increase the number of control points, we can fit the $\hat{T}$ more tightly. Yet, it takes long time to find the positions of control points through least-square minimization, and it is also quite inconvenient to edit all the control points afterward. However, if the number of control points is not sufficient, there is big discrepancy between $\hat{T}$ and the spline curve, which does not produce good quality of motions. This approach iterates the spline-fitting process as it increases the number of control points, denoted as k_i, from minimum $k_{\min}$ to maximum number $k_{\max}$. The first step of this iteration is to find the parameter t where $0 \leq t \leq k - 3$ in the curve that has the exact offset vector of original motion trajectory. That is, if the length of the offset vector is defined as $d_j = \|\hat{p}_j - \hat{p}_{j-1}\|$, then, it needs to find parameter $\hat{t}_j$ from t that matches the d_j as close as possible on the fitted spline curve with k control points. This reparameterization of curve parameter $\hat{t}_j$ based on the arc-length is called *arc-length parameterization*. More details can be found in [6]. By using parameter $\hat{t}_j$ on the curve with k_i control points, the algorithm keeps checking whether the error summation $E(k_i)$ is less than predefined threshold value θ. Mathematically, the $E(k_i)$ is defined in the (22.5).

$$E(k_i) = \sum_{j=0}^{n} \left\| f\left(\hat{t}_j\right) - \hat{p}_j \right\| \tag{22.5}$$

where j is the frame number, n is the total number of frames, $\hat{p}_j$ is the 2D root position of root joint, and $f\left(\hat{t}_j\right)$ is the 2D position of curve at arc-length parameter at parameter $\hat{t}_j$.

If $E(k_i) = \theta$, the algorithm stops the iteration and uses the k_i as the final number of control points. The threshold value θ is automatically set as follows:

$$\theta = \frac{\sum_{j=0}^{n-1} \left\| \hat{p}_j - \hat{p}_{j-1} \right\|}{h} \tag{22.6}$$

where h is the height of 3D characters. Basically, the threshold is set to be proportional to the length of motions while being inversely proportional to character size. Figure 22.1 shows the spline-fitting results with different number of control points k_i In this figure, gray lines indicate the original $\hat{T}$, and red lines the fitted spline curve $f\left(\hat{t}_j\right)$. The control points are represented as yellow dots. Note that as the number of control points increases, the motion trajectory fits more tightly.

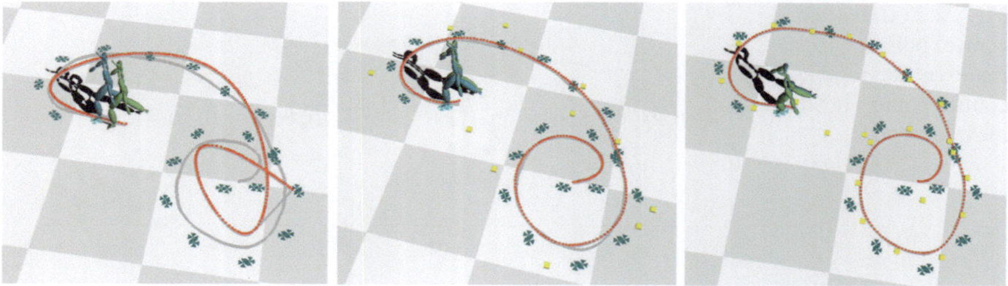

Fig. 22.1 Comparison of spline-curve fitting, using different numbers of control points *Left*: 5 control points are used. *Middle*: 10 control points are used. *Right*: 20 control points are used

After the spline fitting is finished, users are allowed to change the positions of control points. For example, in Fig. 22.1, users just pick one control point with the mouse and drag it to move the control point to another place. That repositioning of control points changes the shape of the curve automatically. However, due to the locality property of piecewise B-spline, those changes only affect corresponding segments of the curve, which is a useful property for editing motions.

And then, the newly edited curve must be fed into the original motion for further processing. Let's say the $f'(\hat{t}_j)$ is the newly edited curve, then, from the $f'(\hat{t}_j)$, original $\hat{T}$ is needed to be modified to $\hat{T}'$. In order to do that, the algorithm first needs to construct a *displacement map*, say $m_i \in \mathbb{S}^3$, which is the orientation difference between the global orientation of root joint, q_i^1, at frame i and global tangent orientation of *unedited* spline curve at t, denoted as $r(t_i) \in \mathbb{S}^3$. Mathematically, let $\alpha(t_i)$ be the tangent angle at parameter t_i of the curve. Then, $r(t_i)$ can be calculated as $r(t_i) = \left(\cos\left(\alpha(t_j)/2\right), 0, \sin\left(\alpha(t_j)/2\right), 0\right)$, in which four values in the parentheses represent (w, x, y, z) of a quaternion respectively. After that, the displacement map m_i can be computed as follows:

$$m_i = q_i^1 \cdot r(t_i)^{-1}, \quad 0 \leq i \leq n \tag{22.7}$$

where both q_i^1 and $r(t_i)$ are quaternions.

The displacement map represents the relationship between root joint orientation and tangent of spline curve. This relation must be preserved even after the spline curve is edited. That is, on the edited curve $f'(\hat{t}_i)$, the new orientation $\hat{q}_i^1$ and position $\hat{p}_i$ of root joint can be estimated as follows:

$$\begin{cases} \hat{q}_i^1 = r(\hat{t}_i) \cdot m_i & 0 \leq i \leq n \\ \hat{p}_i = f'(\hat{t}_i) \end{cases} \tag{22.8}$$

The equation of (22.8) only sets the root joint. Other joints must be adjusted as well to satisfy important geometrical constraints. One of the important geometrical constraints is the *foot plant constraint*, meaning that one of the feet must be planted

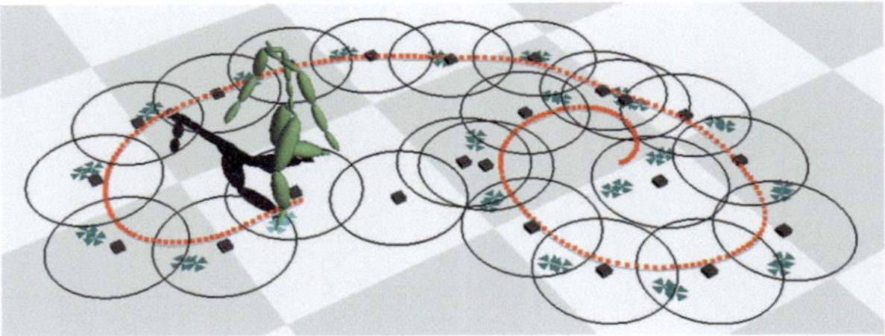

Fig. 22.2 Constrained control points. *Circles* represent the regions that control points are allowed to move around

on the ground for some amount of frames. If one foot has such a plant constraint over a particular range of frames in the original motion, then the same constraint must be preserved on the edited motion. Since the motions are forced to change the root joint position and orientation through (22.8), the *foot skating* artifact, which is an error that causes some feet to float on the ground, may occur. In order to fix this problem, limb configuration must be modified. The algorithm applies the IK solver that is proposed by [8]. This method performs the per-frame-based numerical IK solver to get the knee angle for given foot positions, and then filter the motion afterward for smoothing.

Although this method finds the limb configuration most of the time, it may cause unnatural stretch of legs to satisfy the foot plant constraint. This occurs when the edited motion trajectory is too much different from the original motion trajectory, especially when the motion path has a high curvature. To prevent this, the proposed method puts the constraints on the position of control points. As shown in Fig. 22.2, circles representing the *safe* regions for each control points are put. If the control point is inside the circle, the motion won't have any artifact such as leg stretch. The radius of circle is automatically calculated through the curvature of trajectory. That is, if the curvature at a particular control point is small, then the algorithm sets a big circle around it, otherwise, a small circle is set automatically. Due to these constraints, when users are changing the position of control points outside the circle, the algorithm forces it back inside the circle, which guarantees good quality of motions.

22.4 Experiments

I performed several experiments to validate our proposed algorithm. I used OpenGL for 3D rendering and used BVH (BioVision Hierarchy) motion capture files for tests. The Fig. 22.3 compared the original motion and the edited motion applied proposed algorithms. On the left side of Fig. 22.3, the original was just straight walking motion. Therefore, if I just used this motion, the 3D character had to pass through the obstacle in the middle. However, if I applied our algorithm, the motion can be edited so easily that the 3D character can make a detour. On the right side of Fig. 22.3, from the same straight walking motion, I made a curved motion. In

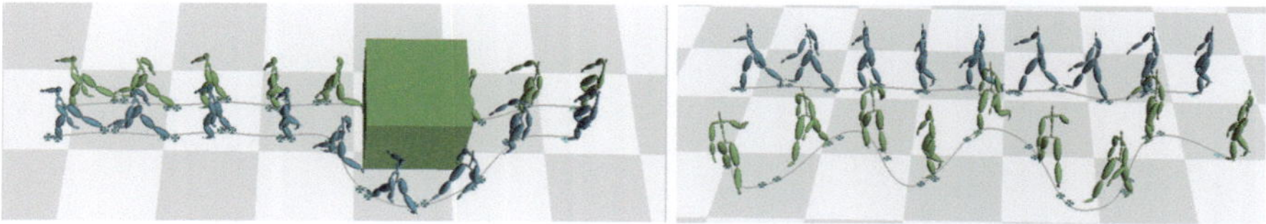

Fig. 22.3 *Left*: The proposed algorithm is able to modify the motion so that the 3D character can detour the obstacle. *Right*: From the original straight walking motion, the proposed algorithm made a zig-zag motion

both cases, the control points were constrained. Therefore, control points couldn't move beyond the limit. From these experiments, I found that the motions were modified in real-time rate as I changed the shape of the curve.

22.5 Conclusion

In this paper, I introduce a novel algorithm for editing human motion capture data through optimal spline fitting based on least-square minimization with capability of putting constraints on the control points. This algorithm is able to edit the motion in real-time without causing any visual artifact. As a future plan, I would like to extend our algorithm to more general motions such as dancing or ballet dancing in which motion paths are more complicated and very hard to model with a single spline curve.

References

1. Wang, X., Chen, Q., Wang, W.: 3D human motion editing and synthesis: a survey. Comput. Math. Meth. Med. **2014**, 11 Article ID 104535 (2014)
2. Bruderlin, A., Williams, L.: Motion signal processing. In: Proceedings of the 22nd annual conference on computer graphics and interactive techniques, pp. 97–104 (1995)
3. Kovar, L., Gleicher, M., Pighin, F.: Motion graphs. ACM Trans. Graph. **21**(3), 473–482 (2002)
4. Jehee, L., Jinxiang, C., Reitsma, P., Hodgins, J., Pollard, N.: Interactive control of avatars animated with human motion data. ACM Trans. Graphics **21**(3), 491–500 (2002)
5. Gleicher, M.: Motion: path editing. In: Proceedings of 2001 ACM symposium. Interactive 3D Graph. ACM, Mar. (2001)
6. Shirley, P., Marschner, S.: Fundamentals of computer graphics. CRC, Boca Raton (2009)
7. Linear Algebra Package (LAPACK), http://www.netlib.org/lapack/
8. Kovar, L., Schreiner, J., Gleicher, M.: Footskate cleanup for motion capture editing. SCA: Proceedings of 2002 ACM SIGGRAPH/eurographics symposium on computer animation, New York, pp. 97–104 (2002)

Chapter 23
An ARAR-Tree-Based Diagnosis Mechanism for Rule Anomalies Among Internet Firewalls

Chi-Shih Chao

Abstract While configuring firewalls, firewall rule ordering and distribution must be done cautiously on each of the cooperative firewalls. However, network operators are prone to incorrectly configuring firewalls because there are commonly hundreds of thousands of filtering rules (i.e., rules in the Access Control List file; or ACL for short) which could be setup in a firewall, not mentioning these rules among firewalls could affect mutually. To speed up the crucial but laboring inspection of rule configuration on firewalls, this chapter describes our developed diagnosis mechanism which can speedily figure out rule anomalies among firewalls with an innovative data structure—Adaptive Rule Anomaly Relationship tree (or ARAR tree). With the aid of this data structure and associated algorithms, significant improvements have been made in the field.

Keywords Defense in depth • Firewall rule anomalies • ARAR tree • Diagnosis reuse

23.1 Introduction

Network firewalls and their associated filtering rules should be discreetly deployed and configured for cooperative, integrated, and in-depth network security protection. However, in a large and complex network equipped with numbers of firewalls, it is very likely for a network manager to get trouble while setting the firewall rules (i.e., ACL rules) since maintaining the security consistency between firewall rule configuration and the demands of network security policies is always time-consuming, laboring, and error-prone.

The security inconsistency typically can be revealed by either the occurrence of anomalies between the firewall rules or demand-mismatching of network security policies. Al-Shaer et al. formally define an anomaly as a duplicate or multiple rule-

C.-S. Chao (✉)
Department of Communications Engineering, Feng Chia University, Taichung 40725, Taiwan
e-mail: cschao@fcu.edu.tw

J. Juang (ed.), *Proceedings of the 3rd International Conference on Intelligent Technologies and Engineering Systems (ICITES2014)*, Lecture Notes in Electrical Engineering 345, DOI 10.1007/978-3-319-17314-6_23

matching for a packet in a rule set. Based on the concept, they further define several different intra-/inter-ACL anomalies among the firewall rules [1]. Nevertheless, because a Finite-State-Machine (or FSM)-based comparison between each pair of rules should be conducted for anomaly checking, their anomaly diagnosis will meet an inefficiency when the number of rules or firewalls is on the increase.

To lower the comparison times between firewall rules needed in [1], Yin et al. [2] segment the IP address space formed by the source and destination networks into blocks where each block is precisely cut out by the IP addresses in the <conditional field> of each firewall rule. Utilizing these varying-sized blocks, a SIERRA tree is built and two conflict rules would be hanged on the same branch of the tree. The network manager just needs to do the anomaly inspections/checking on rules in the same spatial block(s), instead of wasting time to conduct a comprehensive pairwise rule comparisons like [1]. Yet, this approach would lead to a fatal drawback in a networking environment with the need of frequent rule updates. A clean-slate reconstruction of the SIERRA tree is very possibly unavoidable when a rule deletion or insertion is performed. Once one rule changes, a change for the whole spatial rule relationship would occur, and the corresponding data structures could be reconstructed. This drawback also means the local diagnosis results, i.e., the intra-ACL rule diagnosis results, can hardly be reused for the diagnosis of inter-ACL rule anomalies. Thus, by the same token, it is very likely that the modification or reconfiguration of firewall rules for new demands of network security could fail to go live in time in the face of different threats.

The rest of the chapter is organized as following: In Sect. 23.2, the innovative data structure—ARAR tree with its basic operations is shown. The diagnosis mechanism for intra-ACL and inter-ACL rule anomalies is introduced in Sect. 23.3 where our performance evaluation is also included. At last, Sect. 23.4 concludes this chapter and shows some of the future trends of our system development.

23.2 ARAR Tree

To have a clean overlook of our mechanism, Fig. 23.1 network is used and built in our lab where Fig. 23.2 shows those filtering rules with port 80 which are configured in firewall H, G, and C, respectively, for the routing path from network domain D2 to domain D7 (the dotted line in Fig. 23.1). In our work, the IP address ranges of the source network domain and destination network domain of a designated routing path are employed as two axes to form a rectangle traffic plane; later, with the fields of <source_IP> and <destination_IP>, the IP address space of each ACL filtering rule can be depicted as a smaller rectangle and put on some proper place of this traffic plane (see Fig. 23.3).

Referring to the coding tree data structure widely used in data/video compression [3], the traffic plane will be split recursively as well as exponentially if a split block finds there are more than two rules within it (Fig. 23.4), rather than splitting the traffic plane into a matrix containing fixed-sized smaller blocks as done in our

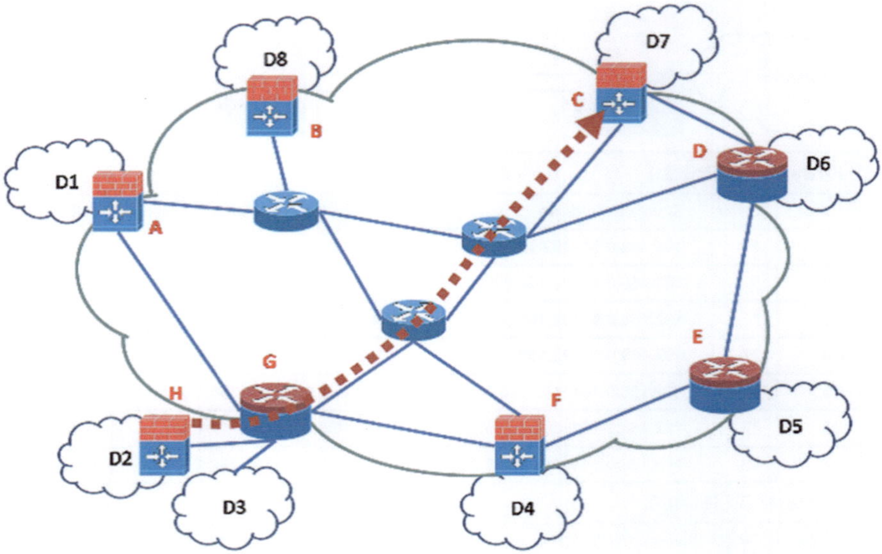

Fig. 23.1 Example network

previous work [4]. After that, the address space of a filtering rule can be recorded in our ARAR tree in the form of □–○–△, where □ contains the values of the conditional fields of the rule, ○ is used to indicate the split block(s) spanned by the address space of the rule, △ shows the label (or the order) of the rule. By dealing with each rule in this fashion, the ARAR tree depicting the structural configuration of Fig. 23.4 can be shown as Fig. 23.5.

23.3 Diagnosis with ARAR Tree

From Fig. 23.5, it can be found that there are nine branches containing more than one △ leaves, which indicates merely the IP address spaces of those rules in these branches intersect with one another and hence incur intra-ACL (or, in this case, intra-firewall) rule anomalies. So, we simply have to do the pairwise rule comparisons for anomaly checking on the rules at the same branch within these nine branches. Comparing to [1], without the ARAR tree, three times rule pairwise comparisons are required for anomaly checking.

To isolate the inter-ACL (or, here, inter-firewall) rule anomalies, in our approach, it can easily be done by simply reusing the ARAR trees built for the diagnosis of intra-ACL (or intra-firewall) rule anomalies. We can first do the intra-ACL anomaly diagnosis for rules inside two designated firewalls individually, which can lead to the construction of two ARAR trees separately for the diagnosis of intra-ACL rule anomalies. Later, to obtain the diagnosis of inter-firewall rule

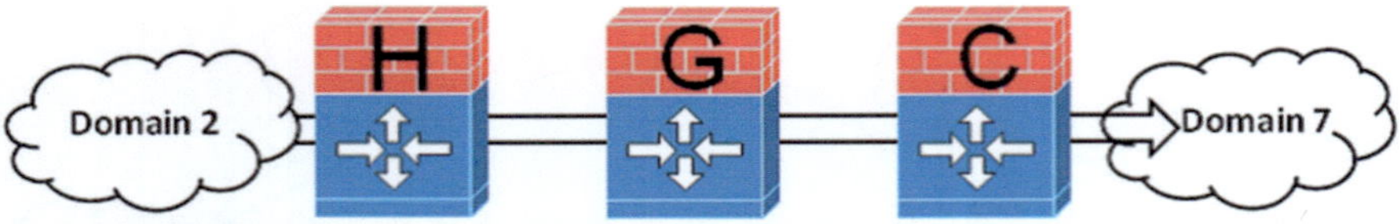

Firewall H , Port80

Name	Order	Source IP Address	Destination IP Address	Action
R1	1	192.168.0.64~192.168.0.95	192.168.1.128~192.168.1.210	accept
R2	5	192.168.0.0~192.168.0.95	192.168.1.160~192.168.1.210	deny
R3	7	192.168.0.0~192.168.0.95	192.168.1.128~192.168.1.159	deny
R4	21	192.168.0.0~192.168.0.223	192.168.1.0~192.168.1.63	accept
R5	22	192.168.0.144~192.168.0.223	192.168.1.32~192.168.1.63	deny
R6	39	192.168.0.144~192.168.0.195	192.168.1.32~192.168.1.127	deny
R7	40	192.168.0.144~192.168.0.223	192.168.1.96~192.168.1.127	accept

Firewall G , Port80

Name	Order	Source IP Address	Destination IP Address	Action
R1	3	192.168.0.192~192.168.0.223	192.168.1.32~192.168.1.127	deny
R2	8	192.168.0.144~192.168.0.223	192.168.1.64~192.168.1.95	accept
R3	10	192.168.0.32~192.168.0.95	192.168.1.64~192.168.1.95	deny
R4	12	192.168.0.0~192.168.0.95	192.168.1.0~192.168.1.95	accept
R5	15	192.168.0.144~192.168.0.192	192.168.1.159~192.168.1.200	deny

Firewall C , Port80

Name	Order	Source IP Address	Destination IP Address	Action
R1	5	192.168.0.192~192.168.0.223	192.168.1.32~192.168.1.127	deny
R2	6	192.168.0.144~192.168.0.192	192.168.1.159~192.168.1.200	deny

Fig. 23.2 Filtering rules with port 80 for the routing path from domain D2 to domain D7

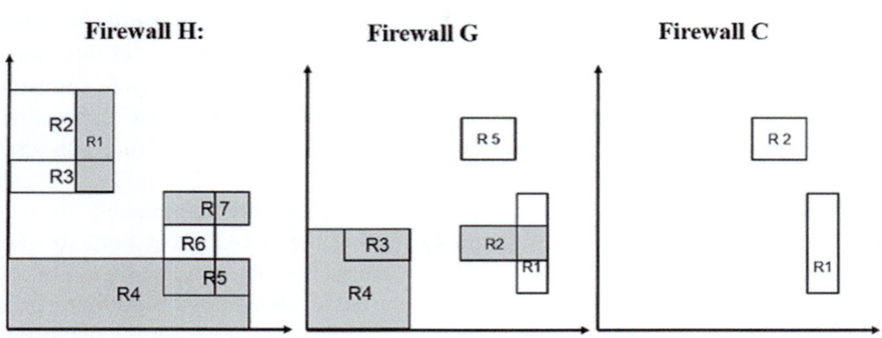

Fig. 23.3 The IP address space for each of related filtering rules in firewalls H, G, and C, respectively

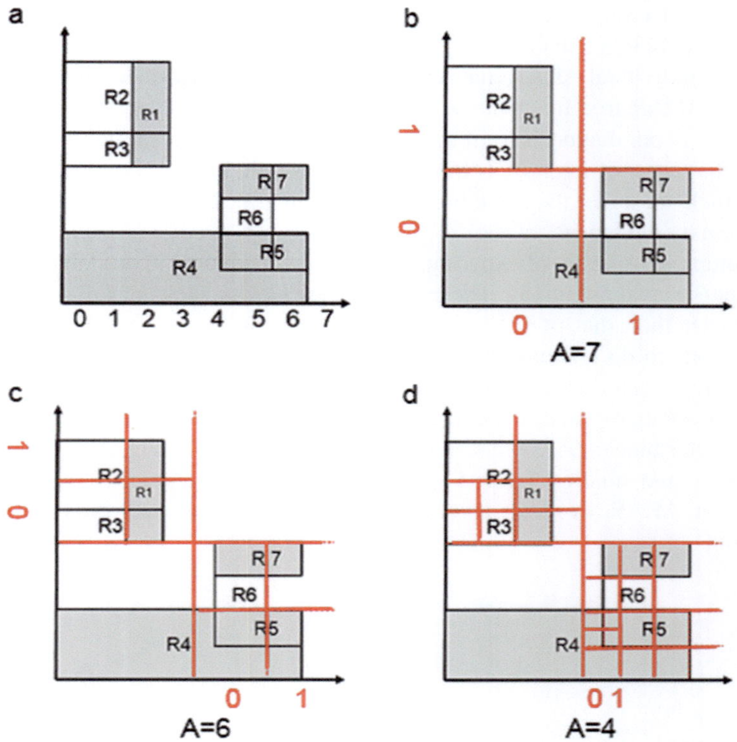

Fig. 23.4 Exponential splitting on a traffic plane

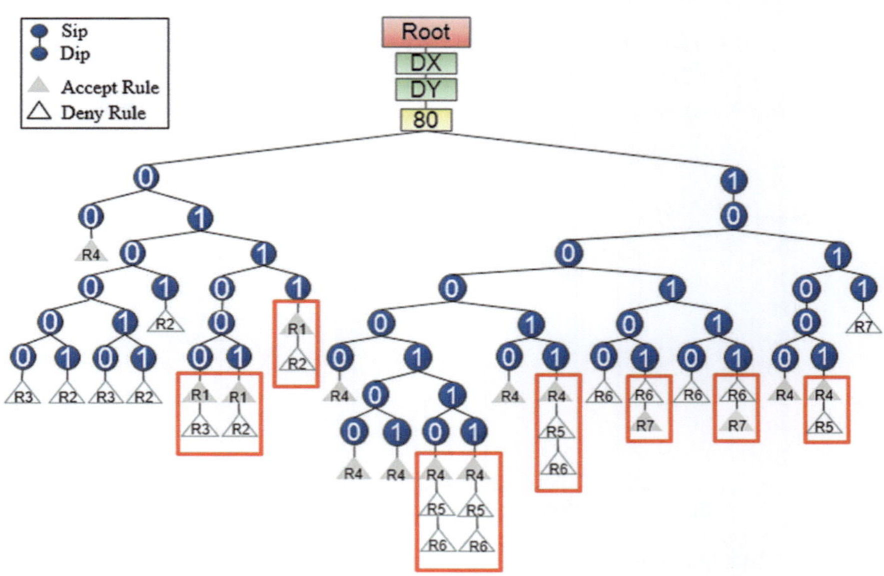

Fig. 23.5 The corresponding ARAR tree of Fig. 23.4

anomalies between these two firewalls, tree integration can be made by adjusting the tree level/height and collecting the leave △ nodes belonging to the same branch of the two individual ARAR trees and putting them together under the same branch of a new ARAR tree for inter-ACL rule anomaly diagnosis. Later, following the same logic in our diagnosis for intra-ACL rule anomalies, the pairwise comparisons for the diagnosis of inter-ACL rule anomalies would only be conducted for those rules which are under the same branch of the integrated ARAR tree for inter-ACL rule anomaly diagnosis.

A comprehensive set of experiments had been conducted in our lab to obtain the performance evaluation [5]. The experimental results show our diagnosis is ten times faster than that of [1], which needs a substantial amount of pairwise rule comparisons to do anomaly diagnosis. In addition, Figs. 23.6 and 23.7 show the performance comparisons between our ARAR-tree-based system and the previous version (RAR-tree-based) done in 2013 [4], which splits the traffic plane into fixed-sized smaller blocks (A is the size of blocks). In the case of Fig. 23.6, rules have a larger size (and smaller size in Fig. 23.7) of filtering area/space than A. It can be found that ARAR tree can give us significant improvements on rule anomaly diagnosis.

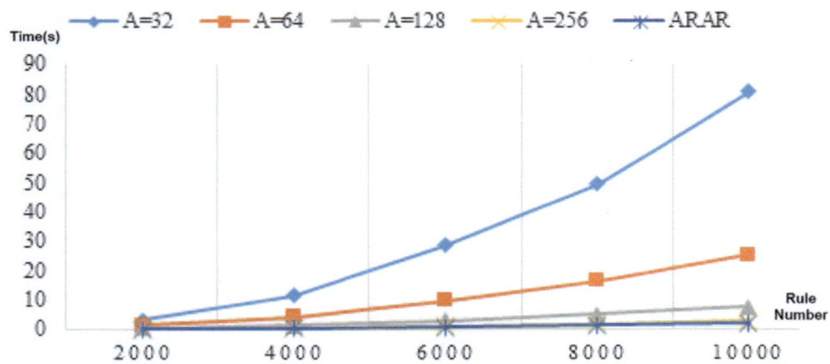

Fig. 23.6 ARAR-tree-based system vs. RAR-tree-based system with larger rule filtering area

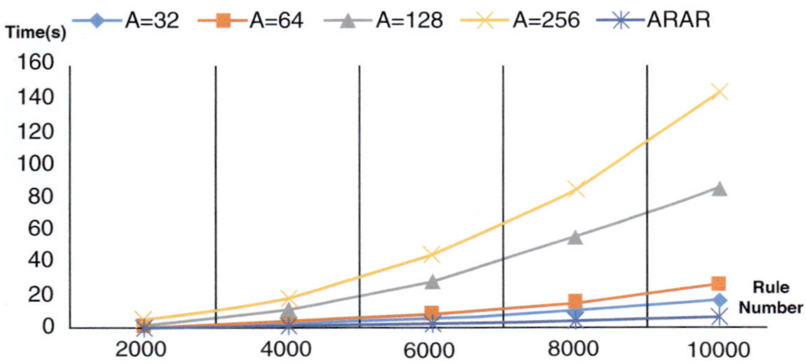

Fig. 23.7 ARAR-tree-based system vs. RAR-tree-based system with smaller rule filtering area

23.4 Conclusions

With the implementation of the ARAR tree, our diagnosis mechanism for firewall rule anomalies meets the planned requirements: Efficiency, expandability, and feasibility. Although we get a noticeable achievement on our system development, as the next steps, more novel ingredients are expected to get in to complete our diagnosis scheme, e.g., migrating the current mechanism to IPv6 networking environment or adding inspection functions for behavior mismatching between two firewalls.

Acknowledgment This work is supported by MOST, ROC, under contract MOST-2221-E-035-100.

References

1. Al-Shaer, E., Hamed, H., Boutaba, R., Hasan, M.: Conflict classification and analysis of distributed firewall policies. IEEE J. Select. Areas Commun. **23**(10), 2069–2084 (2005)
2. Thanasegaran, S., Yin, Y., Tateiwa, Y., Katayama, Y., Takahashi, N.: Topological approach to detect conflicts in firewall policies. In: 23rd IEEE international parallel and distributed processing symposium. http://SSN-1569173665-paper-3.pdf, IEEE Press (2009)
3. Dai, X.-Q.: Data Compression. ISBN 978-957-442-517-4. Flag. Taipei, Taiwan (2009)
4. Chao, C.-S.: A novel and feasible system for rule anomaly and behavior mismatching diagnosis among firewalls. Springer LNEE 234 (2013)
5. Chao, C.-S., Chiu, C.-T.: An adaptive RAR tree-based diagnosis system for rule anomalies and behavior mismatching among firewalls. In: 2013 national computer symposium (NCS2013). Session 3, No. 20 (2013)

Chapter 24
Development of Active Orthosis for Lumbago Relief

Shinsaku Fujimoto, Tetsuya Akagi, and Feifei Zhao

Abstract It is important to develop the orthosis which improves the Quality of Life (QOL) and maintains health conditions. As one of the treatment methods done to lumbago, the waist fixation method with the spinal brace or the orthosis is performed. A waist active orthosis implemented with pneumatic flexible actuators is developed and pressure control method of pneumatic flexible actuators is established. Orthosis control valve (on/off valve) system is modeled and the reliability of exhaust and supply model has been validated through experiment. The Effectiveness of the proposed control method has been validated through the control experiment using the orthosis control valve.

Keywords Pneumatic actuator • Orthosis • Lumbago • Control system design

24.1 Introduction

It is important to improve the QOL (Quality of Life) and maintain the health condition for elderly people. Our special attention is directed to lumbago (low-back pain) because many people feel a back pain in daily life. As one of the treatment methods done to lumbago, the waist fixation method with the spinal brace or the orthosis is performed. Generally, an orthosis is wrapped around the waist made of cloth and the rubber textile. However, the existing orthosis have the following difficult and important problems.

1. *Problem difficult to meet on-demand requirements*: It is difficult to change size or shape of the orthosis by daily demand actions and physical conditions of users.

S. Fujimoto (✉) • T. Akagi
Department of Intelligent Mechnical Engineering, Okayama University of Science, Ridaicho, Kita-ku, Okayama, Japan
e-mail: fuji@are.ous.ac.jp

F. Zhao
Tsuyama National College of Technology, 624-1, Numa, Tsuyama, Okayama, Japan

J. Juang (ed.), *Proceedings of the 3rd International Conference on Intelligent Technologies and Engineering Systems (ICITES2014)*, Lecture Notes in Electrical Engineering 345, DOI 10.1007/978-3-319-17314-6_24

2. *Problem to suppress blood stream*: The orthosis causes many physical problems such as pain, skin eruptions, and bloodstream suppression for a long-time use.
3. *Problem difficult to customize*: It is difficult to select size and shape of the appropriate orthosis which is best fitted to an individual body.

To solve these problems, a waist-active orthosis implemented with pneumatic flexible actuators is developed and pressure control method of pneumatic flexible actuators is established. However, the customizing problems have not been solved.

In this study, firstly, the prototype of pneumatic textile actuator (hereafter called PTA) is developed. To measure the body pressure between the belly and the orthosis, the pressure sensor is developed. However, when the capacity (volume) of the PTA is small, the orthosis control (on/off) valve is vibrated continuously. To solve the problem, the orthoses control valve system is modeled and the reliability of exhaust and supply model is validated through experiment. Finally, a new control method is proposed for the orthosis control (on/off) valve. Reliability of the proposed control method is confirmed by using the control experiment.

24.2 Prototype of Active Orthosis for Lumbago Relief

As one of the treatment methods done to lumbago, the waist fixation method with spinal braces or orthoses is performed. Therefore, this chapter describes the structure of the proposed active orthosis for a lumbago relief.

In order to provide orthosis for the waist which effectively fastens the pelvis, the proposed orthosis does not slip up even if a hard exercise is conducted during carrying. And the orthosis has effects on the support of the pelvis and pain relief in the prognosis of lumbago treatment or the recurrent prevention and prophylaxis of lumbago. McKibben-type actuator [1] is very lightweight because the main element consists of a thin rubber tube covered with fiber sleeves. The McKibben-type actuator is a pneumatic actuator with a high power-to-weight ratio and further having high affinity (physical flexibility) with a human. Therefore, the McKibben-type actuator was adopted as an actuator of active orthosis. Figure 24.1 shows the structure of PTA which was developed for the active orthosis.

The developed PTA has a structure which inwrought with a long McKibben-type actuator into two soft cloths. The operating principle of PTA is very simple. When the compressed air is inject into the supply port, the McKibben-type actuator contracts to an axial direction. As a result, the PTA shrinks by the seam constraint.

Figure 24.2 shows the prototype of the active orthosis for the lumbago relief. The proposed active orthosis consists of five PTAs (=belts) and the body pressure sensor. The size of the orthosis is $970 \times 200 \times 7$ mm, and the mass is 350 g. The active orthosis has a double structure. One structure is the body-orthosis that is configured in the main belt and the ilium belts (right and left). Another structure is two X-type belts.

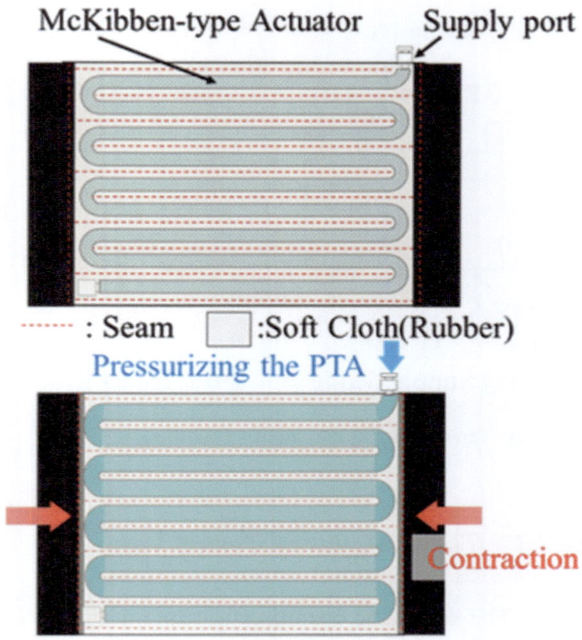

Fig. 24.1 Structure of PTA

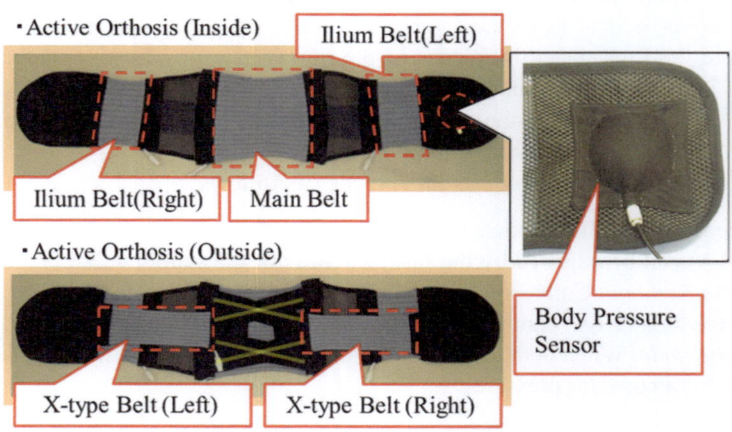

Fig. 24.2 Prototype of active orthosis

24.3 The Modeling of Orthosis Control Valve

When the capacity (volume) of the PTA (Ilium belts) is small, the orthosis control (on/off) valve is vibrated continuously. Therefore, to solve this problem, the orthoses control valve system is modeled and the reliability of exhaust and supply model is validated through experiment in this section.

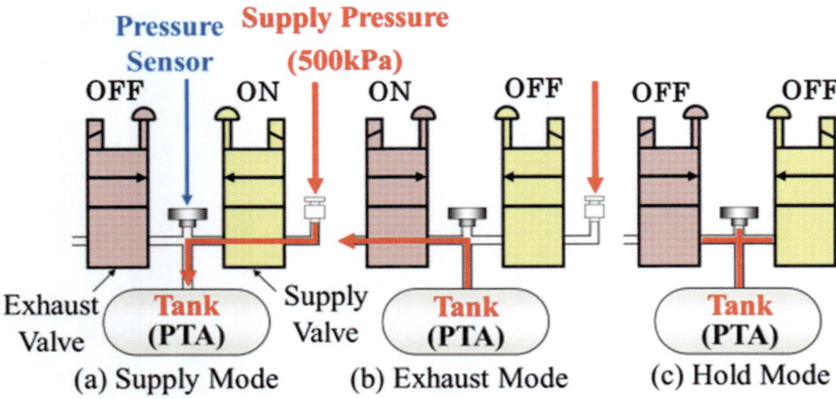

Fig. 24.3 Three modes of the orthosis control valve

24.3.1 *Three Modes of Tested Valve*

Figure 24.3 shows the operating principle of the orthosis control valve. The tested valve consists of two on/off-type control valves (Koganei Co. Ltd., G010HE-1) that both output ports are connected each other. One valve is used as the supply valve, and another is used as the exhaust valve. Two valves can adjust output flow rate like a variable fluid resistance by means of the fast switching. The size of the on/off valve is 33 × 19.6 × 10 mm, and the mass is 15.3 g. The total mass of the orthosis control valve including the controller (Microcomputer: Renesas Co. Ltd., H8/3,664 F) is very light, that is about 350 g.

The orthosis control valve has three modes as follows:

(a) *Supply mode*: The exhaust valve is the off state. And, the pressure of the tank (PTA) can be adjusted by the fast switching of the supply valve.
(b) *Exhaust mode*: Conversely, the supply valve is the off state. And, the pressure of the tank (PTA) can be adjusted by the fast switching of the exhaust valve.
(c) *Hold mode*: When both valves are the off states, the pressure of the tank (PTA) is kept a constant pressure.

24.3.2 *Modeling of On/Off Valve*

Figure 24.4 shows the analytical model of the orthosis control valve. The mass flow rate of supply valve Q_i and the exhaust valve Q_o are given as follows.

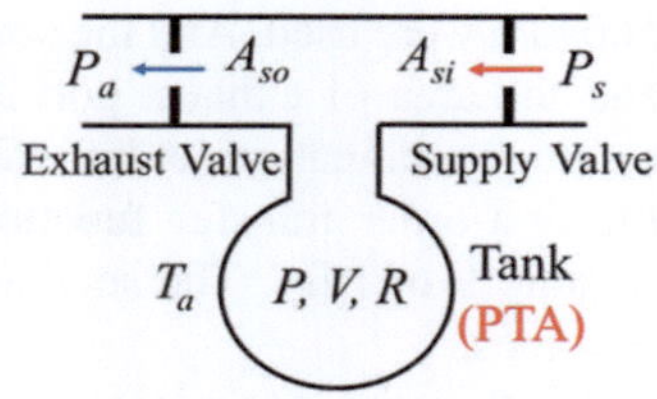

Fig. 24.4 Analytical model of valve

$$Q_i = A_{si}P_s\sqrt{\frac{2}{RT_a}}g(z_i), \quad z_i = \frac{P}{P_s}: \quad \text{Supply State} \tag{24.1}$$

$$Q_o = A_{so}P_a\sqrt{\frac{2}{RT_a}}g(z_o), \quad z_o = \frac{P_a}{P}: \quad \text{Exhaust State} \tag{24.2}$$

where R and T_a mean a gas constant and an absolute temperature, respectively. The function $g(z)$ that expresses the state of flow is given as follows.

$$g(z) = \sqrt{\frac{\kappa}{\kappa-1}\left(z^{2/\kappa} - z^{(\kappa+1)/\kappa}\right)} \quad [0.528 \le z \le 1]: \quad \text{Subsonic Flow} \tag{24.3}$$

$$g(z) = \sqrt{\frac{\kappa}{\kappa+1}\left(\frac{2}{\kappa+1}\right)^{2/(\kappa-1)}} \quad [0 \le z \le 0.528]: \quad \text{Choked Flow} \tag{24.4}$$

where κ means a specific heat ratio (=1.4). The pressure P in the volume V of PTA is given by the next equation.

$$\dot{P} = \frac{dP}{dt} = \frac{\kappa RT_a}{V}(Q_i): \quad \text{Supply State} \tag{24.5}$$

$$\dot{P} = \frac{dP}{dt} = \frac{\kappa RT_a}{V}(-Q_o): \quad \text{Exhaust State} \tag{24.6}$$

The systems of (24.5) and (24.6) are nonlinear systems with respect to the input of sectional area A_{s*}($*=$i or o). Here, the simple idea is to approximate a nonlinear system by a linear one (around the pressure point $P = P_E$ ($z = 0.528$))

$$\dot{x} = -\frac{2.70 \times 10^2 \cdot A_{so}}{V}x - \frac{4.61 \times 10^4}{V}v \quad \because x = P_E - P,\ v = A_o: \quad \text{Supply State} \tag{24.7}$$

$$\dot{x} = -\frac{2.78 \times 10^2 \cdot A_{si}}{V}x + \frac{4.68 \times 10^4}{V}v \quad \because x = P_E - P,\ v = A_i: \quad \text{Exhaust State} \tag{24.8}$$

In the linearization, the atmospheric pressure P_a of 101.3 kPa, the room temperature T_a of 298 K, the gas constant R of 287 J/kg/K, the supply pressure P_s of

500 kPa were used. And the sectional area of the supply port A_{si} (=0.45 × 10^{-6} m^2) and the area of exhaust port A_{so} (=0.60 × 10^{-6} m^2) were adopted the values of catalog specification (on/off valve). The obtained linear system can be expressed by the first-order transfer function. The pole of first-order system depends on the volume V of PTA. The smaller the volume of PTA, the faster the response speed becomes.

On the other hand, suppose that the sectional area $v = A_*$ of the on/off valve is opened (or closed) slowly. Suppose that the switching area A_* ($* =$ i or o) of valve is approximated by dead time L (=3 ms) and time constant T_o (=2 ms) of a primary delay system.

$$v(s) = A_*(s) = \frac{A_{s*}}{T_o s + 1} \cdot e^{-Ls} \cdot u(s) \quad \text{New Input} \begin{cases} u(s) = 1 : & \text{Supply Mode} \\ u(s) = -1 : & \text{Exhaust Mode} \\ u(s) = 0 : & \text{Hold Mode} \end{cases} \tag{24.9}$$

From (24.7)–(24.9), the transfer function of on/off valve system with volume of PTA is given by a second-order form and a dead time.

To validate the reliability of the proposed linear model, the verification experiment of supply (500 kPaG) and exhaust (0 kPaG) motion was performed on conditions of different volumes. Figure 24.5 shows the experimental result and

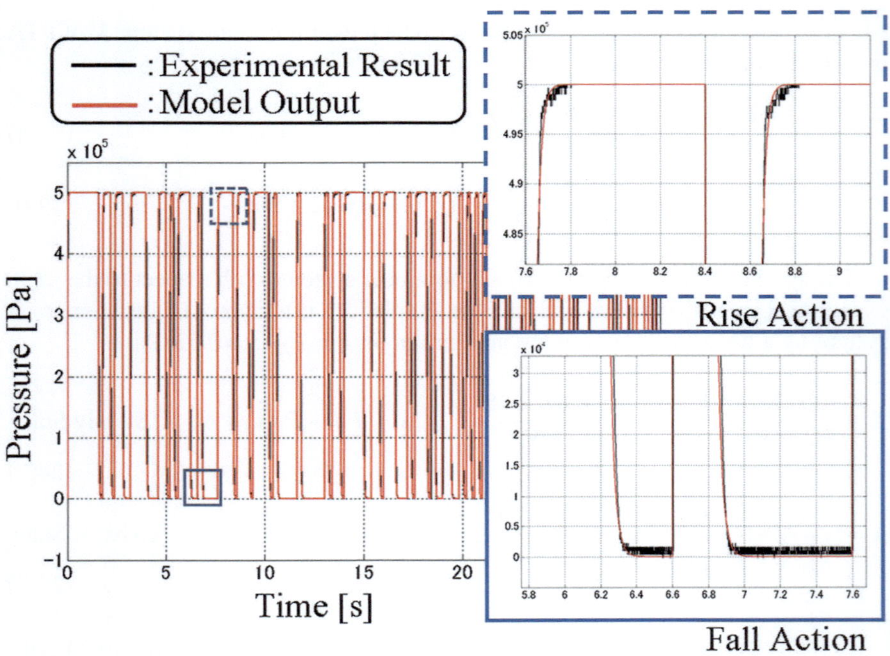

Fig. 24.5 Experimental result and model outputs

output of the proposed model at the volume $V = 3$ mL. From Fig. 24.5, it can be seen that the output results (rise action and fall action) using the proposed model agree well with the experimental result. The proposed model was a very simple model, but it could be confirmed that the actual valve system including volume can be represented by means of a second-order form with a dead time.

24.4 Control System Design for On/Off Valve

When the capacity (volume) of the PTA is small, the orthosis control (on/off) valve is vibrated continuously. Therefore, to solve this problem, the orthoses control valve system is modeled and the reliability of exhaust and supply model is validated through experiment in this chapter.

24.4.1 Previous Control Method

In general, the following control-law has been used to the previous controller [2] and [3] for the on/off valve.

$$u_v(t) = e(t) + \alpha \dot{e}(t) \quad \because \alpha = k \cdot \Delta T, \quad e = P_d - P : \text{error} \tag{24.10}$$

where ΔT is the sampling period, k is the positive control parameter and α is derivative gain. Figure 24.6 shows the previous control scheme. This control law is applied to the following three modes:

(a) Supply Mode : $u_v(\mathrm{t}) > w$ (b) Exhaust Mode : $u_v(\mathrm{t}) < -w$
(c) Hold Mode : $-w \leqq u_v(\mathrm{t}) \geqq w$ $\because w$: Control accuracy of pressure

This control law has some problems that there are no determining method of control parameter k (or α) and the control accuracy w. And the time delay is not considered in the control system design.

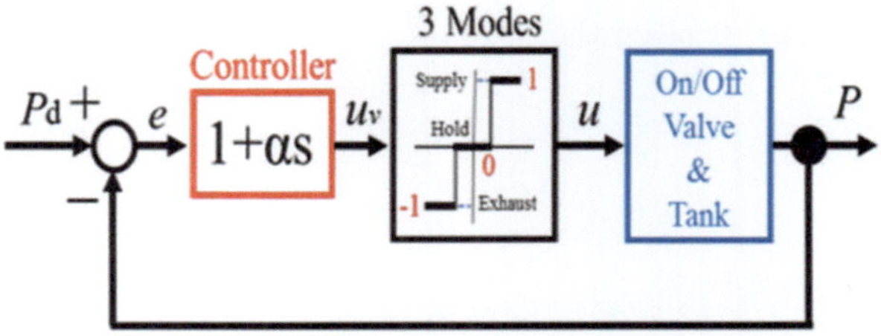

Fig. 24.6 Previous control system

24.4.2 Proposed Control Design Method

The linear time-invariant error system with the time delay L is given by

$$\begin{aligned} \dot{\boldsymbol{e}}(t) &= A\boldsymbol{e}(t) + Bu(t-L) \quad \because \boldsymbol{e} = \left[P_d - P - \dot{P}\right]^{\mathrm{T}} \\ y &= \boldsymbol{Ce}(t) \end{aligned} \tag{24.11}$$

where $\boldsymbol{e}$(t) is the state, u(t) is the control input, and P_{d} is the desired pressure.

Consider the following quadratic performance index J for the systems (24.11)

$$J = \int_0^{\infty} \left\{ \mathrm{e}^{\mathrm{T}}(t)\boldsymbol{C}^{\mathrm{T}}\boldsymbol{QCe}(t) + u(t)Su(t) \right\} \mathrm{d}t \quad \text{to minimize} \tag{24.12}$$

where the matrix Q is nonnegative definite and S is positive value.

The control law of this problem has already been solved by Ref. [4].

$$u(t) = -[K_1 \quad K_2] \left[\boldsymbol{e}(t) + \int_{-L}^{0} \Phi(\tau + L)\boldsymbol{B}u(t+\tau)\mathrm{d}t \right] \tag{24.13}$$

where K_1 and K_2 are state feedback gains, respectively. $\boldsymbol{\Phi}(t)$ is the state transition matrix. Thus, the control parameter α is obtained as the value of $-K_2/K_1$.

In the case of on/off valve, the control input is constant value during the time delay L. And, the second term on the right side (24.13) becomes the constant vector.

$$\int_{-L}^{0} \Phi(\tau + L)\boldsymbol{B}u(t+\tau)\mathrm{d}t = [e_{1\varepsilon} \quad e_{2\varepsilon}]^{\mathrm{T}} \tag{24.14}$$

where $e_{1\varepsilon}$ and $e_{2\varepsilon}$ mean the minimum accuracies of the pressure and the derivative pressure, respectively. In other words, the minimum accuracy $e_{1\varepsilon}$ of the pressure corresponds to the control accuracy w and the time delay is considered in the control system design. The effectiveness of the proposed control method was confirmed by control experiment. The Step responses of pressure were performed from 0 to 0.35 MPaG. Figure 24.7 shows the simple experimental device.

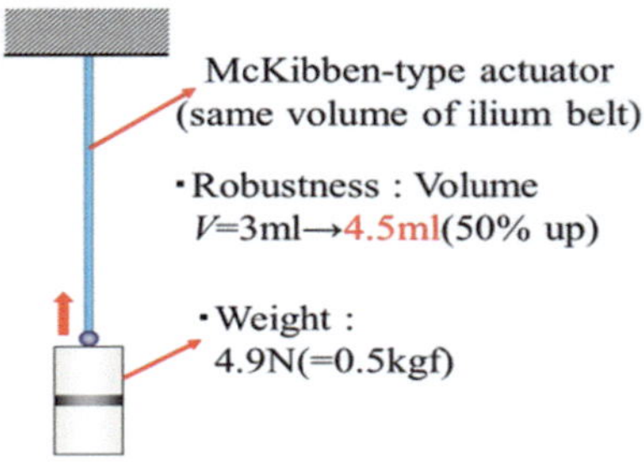

Fig. 24.7 Experimental device

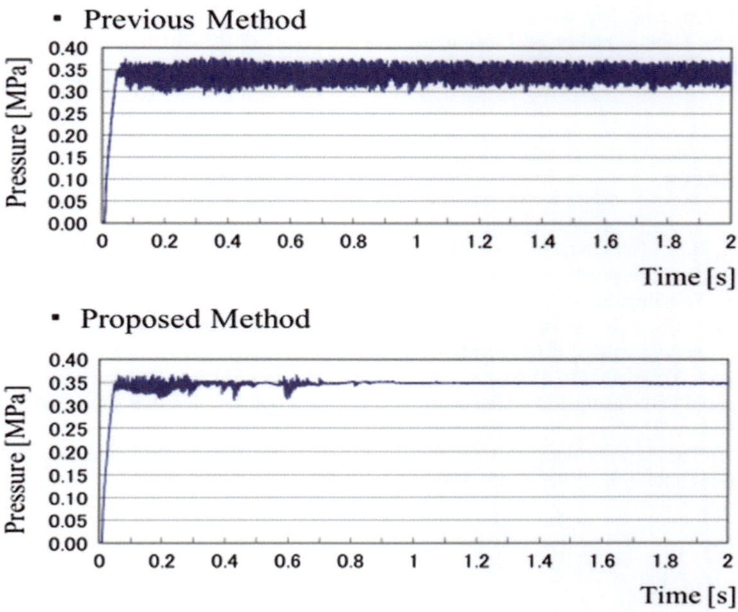

Fig. 24.8 Control performance comparison

The volume (diameter) of McKibben-type actuator is the same volume ($V = 3$ mL, Diameter: 2 mm) of the ilium belt. To verify the robustness, the volume of McKibben-type actuator is changed to the volume 4.5 mL (50 % up), and the load is increased to (4.9 N = 0.5 kgf) on the McKibben-type actuator. Compared with the previous result as shown in Fig. 24.8, it can be seen that the sustained oscillation of output pressure can be reduced in the case of the proposed method. However, the control parameter α is the same value (0.04). If the time delay is considered in the control system design, the sustained oscillation can be avoided.

24.5 Conclusions

This study is aimed to develop the active orthosis for lumbago relief and is summarized as follows:

1. The pole of first-order system depends on the volume V of PTA.
2. The smaller volume of PTA, the faster response speed becomes.
3. The on/off valve system including volume can be represented by means of a second-order form with a dead time.
4. The sustained oscillation can be avoided by considering the dead time.

Acknowledgments The authors would like to acknowledge financial support for this work provided by QOL (Quality of Life) Innovative Research (2012–2016) from the Ministry of Education, Culture, Sports, Science and Technology of Japan.

References

1. Schulte, H.F.: The characteristics of the McKibben artificial muscle. In: The application of external power in prosthetics and orthotics. National Academy of Science National Research Council, Washington, DC, pp. 94–115 (1961)
2. van Varseveld, R.B., Bone, G.M.: Accurate position control of a pneumatic actuator using On/Off solenoid valves. IEEE/ASME Trans. Mechatron. **2**(3), 195–204 (1997)
3. Moriwake, Y., Akagi, T., Dohta, S., Zhao, F.: Development of low-cost pressure control type quasi-servo valve using embedded controller. Int. Symp. Robot. Intell. Sensors **2012**, 493–500 (2012)
4. Abe, N.: The limiting optimal regulator for systems with time-delay considering disturbance decoupling. Mem. Inst. Sci. Tech. Meiji Univ. **39**(1), 1–6 (2000)

Chapter 25
Controller Design by Time-Domain Objective Functions Using Cuckoo Search

Huey-Yang Horng

Abstract In this research, a new optimization algorithm, called the cuckoo search algorithm, is introduced for design controller. A large proportion of industrial systems are represented by linear time-invariant transfer functions. The proportional-integral-derivative (PID) controller is one of the most widely used functions. The lead-lag controller is a more practical alternative. Traditionally, time-domain or frequency-domain methods have been used to design a lead-lag controller to design specifications. This chapter focused on the design of controller both PID and lead-lag controller, by optimization of the time-domain objective function. The proposed objective function includes time-domain specifications, including the rise time, peak time, maximum overshoot, setting time, and steady-state error. In the chapter, Cuckoo Search algorithm is chosen to finding the optimal solutions. Cuckoo Search is metaheuristic optimization method recently developed. That is a type of population-based algorithm inspired by the behavior of some Cuckoo species in combination with the Lévy flight behavior. Given that the plant is modeled according to a linear time-invariant transfer function, the proposed method designs the controller capable of approaching the specifications.

Keywords PID controller • Lead-lag controller • Cuckoo Search • PSO

25.1 Introduction

Most industrial plant systems can be represented by a linear time-invariant transfer function. Proportional-integral-derivative (PID) controllers are probably the most commonly used controllers in industrial applications. Numerous methods have been proposed for tuning the PID controller parameters [1–3].

Lead-lag controllers provide a more practical alternative. The design of the lead-lag controller has been studied [4–6]. Ou and Lin proposed a method based on Generic Algorithm (GA) and Particle Swarm Optimization (PSO) to design the PID

H.-Y. Horng (✉)
Department of Electronic Engineering, I-Shou University, Kaohsiung 84001, Taiwan
e-mail: hyhorng@isu.edu.tw

J. Juang (ed.), *Proceedings of the 3rd International Conference on Intelligent Technologies and Engineering Systems (ICITES2014)*, Lecture Notes in Electrical Engineering 345, DOI 10.1007/978-3-319-17314-6_25

controller, and then compared the results [7]. Horng used cuckoo search to design lead-lag controller [8]. Cuckoo search algorithm is one of the latest metaheuristic techniques, developed by Yang and Deb [9–11].

In this chapter, cuckoo search algorithm that uses a time-domain objective function to design the controller is proposed. If the plant could be modeled as a linear time-invariant transfer function, the proposed method can design a controller that approaches or meets the time-domain specifications. The objective function includes eight specifications, they are delay time, rise time, first peak time, maximum peak time, percent maximum overshoot, percent maximum undershoot, setting time, and steady-state error. This is improved by the formation of reference [8].

25.2 Time-Domain Objective Functions

Typical unit-step response of a control system illustrating the time-domain specifications are percentage maximum overshoot OS %, delay time T_d, rise time, T_r, setting time T_s, and steady-state error E_{ss} [1]. Let $y(t)$ be the unit-step response and y_{ss} the steady-state error. For a more general system, the following specifications will be considered:

1. First Peak time, T_p. The time to reach the first peak.
2. Maximum Peak time, T_m. The time to reach the maximum peak.
3. Percentage Maximum overshoot, US % is defined as

$$y_{us} = \min(y(t)),\ \ t \geq T_p,\quad \text{US } \% = \begin{cases} \dfrac{(y_{ss} - y_{us})}{y_{ss}}, & \text{if } \ y_{us} < y_{ss}, \\ 0, & \text{if } \ y_{us} \geq y_{ss}. \end{cases} \tag{25.1}$$

First define deviation ratio (DR)

$$\begin{aligned} \mathrm{DR}(\mathrm{TDS}) &= f(\mathbf{x}|\mathrm{TDS} : \mathrm{lb}, \mathrm{ub}) \\ &= \begin{cases} 0, & \text{if } 0 \leq \mathrm{lb} \leq f(\mathbf{x}|\mathrm{TDS}) \leq \mathrm{ub} \\ \dfrac{f(\mathbf{x}|\mathrm{TDS}) - \mathrm{ub}}{\mathrm{ub}}, & \text{if } f(\mathbf{x}|\mathrm{TDS}) > \mathrm{ub} \\ \dfrac{\mathrm{lb} - f(\mathbf{x}|\mathrm{TDS})}{\mathrm{lb}}, & \text{if } f(\mathbf{x}|\mathrm{TDS}) < \mathrm{lb} \end{cases} \end{aligned} \tag{25.2}$$

where *TDS* is the time-domain specification, i.e., rise time, first peak time, maximum peak time, etc. The proposed objective function is

$$\begin{aligned}\mathrm{TDOF} = \big(&w_1\mathrm{DR}(T_d) + w_2\mathrm{DR}(T_r) + w_3\mathrm{DR}\big(T_p\big) + w_4\mathrm{DR}(T_m)\\ &+w_5\mathrm{DR}(\mathrm{OS}\,\%) + w_6\mathrm{DR}(\mathrm{US}\,\%) + w_7\mathrm{DR}(T_s) + w_8\mathrm{DR}(E_{ss})\big)/\mathrm{TW}\end{aligned} \tag{25.3}$$

where $\mathrm{TW} = \sum_{i=1}^{8} w_i$. In (25.3), w_i represents weights reflecting the relative importance of the corresponding term. DR denotes the deviation ratio of desired interval, from lower bound (lb) to upper bound (ub). For second-order system, the first peak time is the maximum peak time. But, for the general system, they are not the same.

The design of controller becomes in the minimization of the TDOF for all possible parameters. Moreover, some tolerances in the time-domain specifications are allowable as in (25.2).

25.3 The Controller

Two kinds of controllers are considered, there are PID and lead-lag (or lag-lead) controller. For PID control, one typically has

$$G_c(s) = K_p + \frac{K_i}{s} + K_d\, s. \tag{25.4}$$

The transfer function of a lead-lag (or lag-lead) controller is written as:

$$G_c(s) = K\left(\frac{T_1 s + 1}{\alpha T_1 s + 1}\right) \cdot \left(\frac{T_2 s + 1}{\beta T_2 s + 1}\right), \tag{25.5}$$

where $K > 0,\ \ \alpha > 1,\ \ T_1 > 0,\ \ \beta < 1, T_2 > 0$.

25.4 Cuckoo Search Algorithm

In this chapter, a new optimization algorithm, called the cuckoo search algorithm, is introduced for design controller [9–11]. The algorithm uses a combination of a local random walk and the global random walk, controlled by a parameter p_a. This allows for proper balance between exploration and exploitation of the solution space. The local random walk can be written as

$$x_i^{t+1} = x_i^t + \alpha s \otimes H(p_a - \varepsilon) \otimes \left(x_j^t - x_k^t\right) \tag{25.6}$$

where x_i^t and x_k^t are two different solutions taken by random permutation. In (25.6), $H(u)$ is a Heaviside function, ε is a random number got from a uniform distribution, and s is the step size. On the other hand, the global random walk is figured out using Lévy flights:

$$x_i^{t+1} = x_i^t + \alpha L(s, \lambda) \tag{25.7}$$

where

$$L(s, \lambda) = \frac{\lambda \Gamma(\lambda) \sin(\pi\lambda/2)}{\pi} \cdot \frac{1}{s^{1+\lambda}}, \quad s \geq s_0 \geq 0 \tag{25.8}$$

Here, $\alpha = 0.01$ is the step size scaling factor.

In the following design procedure, set maximum generation equals to 750, and $P_a = 0.25$. Generate initial population of 15 host nests, which cause closed-loop stable (by using Routh-Hurwitz criterion).

25.5 Illustrative Examples

Example 1 A unity feedback system has the forward transfer function:

$$G_p(s) = \frac{150}{s^2 + 15s + 50}. \tag{25.9}$$

as in Fig. 25.1. The closed-loop system is stable. The system type refers to the order of the pole of $G_p(s)$ at $s = 0$. Thus, the closed-loop system having the forward-path transfer function of (25.9) is type 0. The step response of the uncompensated system is shown in Fig. 25.3. Here, the peak time is 0.249 s, the percentage overshoot is 13.3 %, and the steady-state error E_{ss} is equal to 0.273.

As shown, the uncompensated system does not satisfy the design specifications in Table 25.1. Hence, the controller is used to improve the transient response as in Fig. 25.2. All the weight w_i are set to 1. First, the PID controller will be designed for the system. When the design procedure is completed, the parameters are $K_p = 5.3161$, $K_i = 10.2197$, and $K_d = 0.2299$. Apart from delay time, most of the time-domain specifications are very close to demand. Next, the lead-lag controller will be designed. When completed, the parameters are $K = 59.6384$, $\alpha = 26.0208$, $T_1 = 0.2319$, $\beta = 0.1520$, and $T_2 = 0.0911$. All the desired specifications are fulfilled. If the deviation ratio of delay time would be reduced, the weight of delay time will increase to 6, and keep the others the same value. After the adjustments of weights, and redesign the controller again. The parameters found are, $K_p = 3.0046$, $K_i = 7.3653$, and $K_d = 0.1339$. The deviation ratio of all specifications are listed in Table 25.2. Since the specifications of both led-lag controllers are full-filed, the adjustments produce influence of no importance as shown in Fig. 25.4.

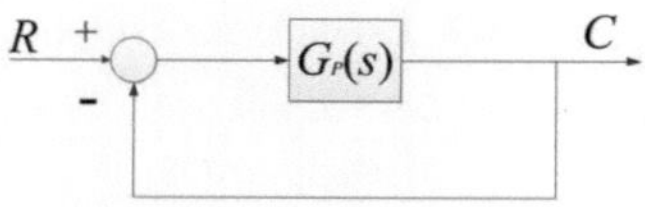

Fig. 25.1 Unity feedback system without controller

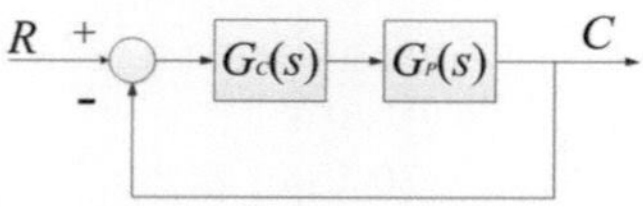

Fig. 25.2 Unity feedback system with controller

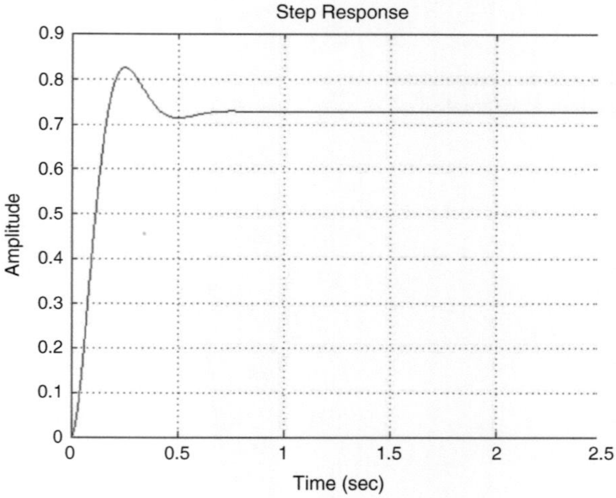

Fig. 25.3 Uncompensated system in Example 1

Table 25.1 Example 1: Controller design with $w_i = 1$, for $i = 1, 2, \ldots, 8$

Spec.	Desired interval	w_i	PID 1	PID 1 DR	Lead-lag 1	Lead-lag 1 DR
T_p	[0.0988, 0.1008]	1	0.1014	0.0063	0.1006	0
T_m	[0.0988, 0.1008]	1	0.1014	0.0063	0.1006	0
T_r	[0.0477, 0.0487]	1	0.0466	0.0223	0.0487	0
T_d	[0.0307, 0.0314]	1	0.0185	0.3970	0.0311	0
$OS\,\%$	[0.0, 0.03]	1	0.0300	0	0.0245	0
$US\,\%$	[0.0, 0.02]	1	0	0.0200	0.0021	0
T_s	[0.1014, 0.1035]	1	0.1031	0	0.1023	0
E_{ss}	[0.0, 0.22]	1	0.0326	0	0.0045	0

Example 2 A unity feedback system has the following forward transfer function:

$$G_p(s) = \frac{700}{(s+2)(s+4)(s+6)}.$$

The closed-loop system is unstable. The step response of the uncompensated system is shown in Fig. 25.5. The design specifications are listed in Table 25.3.

Table 25.2 Example 1: Controller design with $w_4 = 6$, and $w_i = 1$, for $i \neq 4$

Spec.	Desired interval	w_i	PID 2	PID 2 DR	Lead-lag 2	Lead-lag2 DR
T_p	[0.0988, 0.1008]	1	0.1542	0.5300	0.1006	0
T_m	[0.0988, 0.1008]	1	0.1542	0.5300	0.1006	0
T_r	[0.0477, 0.0487]	1	0.0749	0.5382	0.0485	0
T_d	[0.0307, 0.0314]	6	0.0307	0.0000	0.0308	0
$OS\,\%$	[0.0, 0.03]	1	0.0300	0	0.0300	0
$US\,\%$	[0.0, 0.02]	1	0.0200	0	0.0148	0
T_s	[0.1014, 0.1035]	1	0.1055	0.0197	0.1023	0
E_{ss}	[0.0, 0.22]	1	0.0453	0	0.0061	0

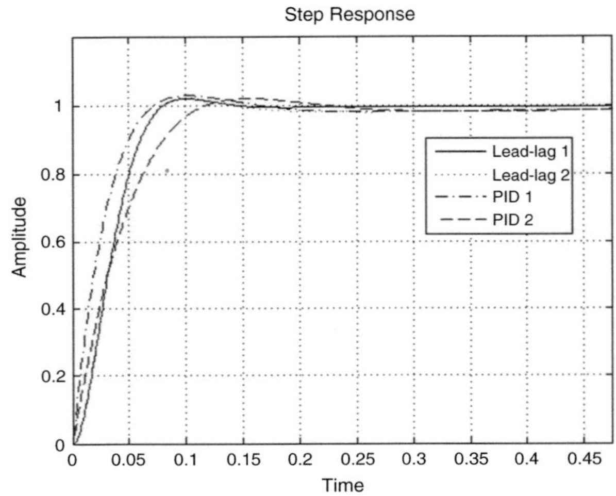

Fig. 25.4 Compensated system in Example 1

When the design procedure is finished, the PID parameters are $K_p = 0.2431$, $K_i = 0.3030$, and $K_d = 0.2299$. Furthermore, the lead-lag parameters are $K = 7.1739$, $\alpha = 67.1652$, $T_1 = 0.4054$, $\beta = 0.0010$, and $T_2 = 0.4112$. The deviation ratio of delay time has been improved (Fig. 25.6).

The weight of delay time T_d increase by a factor of 6, while keeping the others the same. Redesign the controller again; the PID parameters found are $K_p = 0.2431$, $K_i = 0.2840$, and $K_d = 0.0523$ (Table 25.4).

25.6 Conclusions

A large part of an industrial plant system may be represented by the linear time-invariant transfer function. A simple procedure is used to design the controller to meet or approach the specification with cuckoo search algorithm has been proposed

Fig. 25.5 Uncompensated system in Example 3

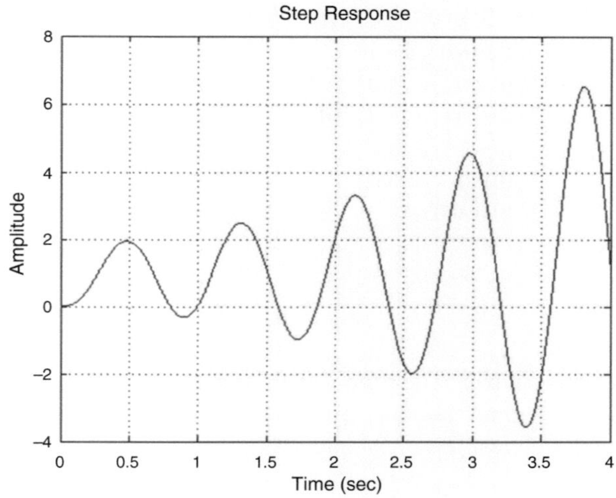

Table 25.3 Example 2: Controller design with $w_i = 1$, for $i = 1, 2, \ldots, 8$

Spec.	Desired interval	w_i	PID 2	PID 2 DR	Lead-lag 2	Lead-lag 2 DR
T_p	[0.7411, 0.7561]	1	0.6434	0.1319	0.7423	0
T_m	[0.7411, 0.7561]	1	0.6434	0.1319	0.7423	0
T_r	[0.3581, 0.3653]	1	0.3251	0.0921	0.3711	0.0159
T_d	[0.2305, 0.2352]	1	0.2229	0.0329	0.2514	0.0687
$OS\,\%$	[0.0, 0.03]	1	0.0300	0	0.0300	0
$US\,\%$	[0.0, 0.02]	1	0.0114	0	0.0085	0
T_s	[0.7606, 0.7760]	1	0.6557	0.1379	0.0107	0
E_{ss}	[0.0, 0.22]	1	0.2263	0.0286	0.0058	0

Fig. 25.6 Compensated system in Example 3

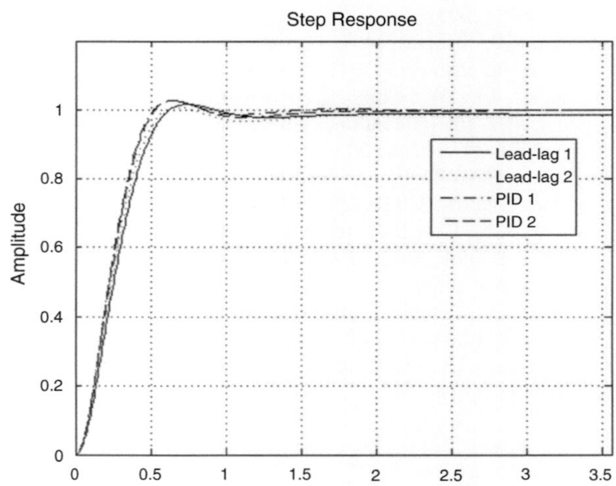

Table 25.4 Example 2: controller design with $w_3 = 6$, and $w_i = 1$, for $i \neq 3$

Spec.	Desired interval	w_i	PID 2	PID 2 DR	Lead-lag 2	Lead-lag 2 DR
T_p	[0.7411, 0.7561]	1	0.6495	0.1235	0.6805	0.0818
T_m	[0.7411, 0.7561]	1	0.6495	0.1235	0.6805	0.0818
T_r	[0.3581, 0.3653]	1	0.3306	0.0767	0.3490	0.0254
T_d	[0.2305, 0.2352]	6	0.2277	0.0119	0.2361	0.0040
$OS\,\%$	[0.0, 0.03]	1	0.0300	0.0304	0.0141	0
$US\,\%$	[0.0, 0.02]	1	0	0	0.0200	0
T_s	[0.7606, 0.7760]	1	0.1016	0	0.6619	0.1297
E_{ss}	[0.0, 0.22]	1	0.0089	0	0.2414	0.0974

in this chapter. The proposed time-domain objective function is expressed in terms of peak time, maximum overshoot, maximum undertow, and setting time of the unit-step response. Computer simulations support the usefulness of the method.

References

1. Golnaraghi, F., Kuo, B.C.: Automatic control systems, 9th edn. Wiley, Hoboken (2011)
2. Okada1, Y., Yamakawa, Y., Yamazaki, T., Kurosu, S.: Tuning method of PID controller for desired damping coefficient. In: SICE anal conference, Kagawa University, Japan, Sept. 17–20, pp. 795–799. (2007)
3. Panda, S., Sahu, B.K., Mohanty, P.K.: Design and performance analysis of PID controller for an automatic voltage regulator system using simplified particle swarm optimization. J. Franklin Inst. **349**(8), 2609–2625 (2012)
4. Yang, J., Chen, C.F., Chen, C.S., Xu, Y.S.: An approach to automatic tuning of phase-lead and phase-lag compensators. In: Proceedings of the 30th IEEE conference on decision and control, Brighton, England, Vol. 3. pp. 2944–2945. (1991)
5. Tan, N.: Computation of stabilizing lag/lead controller parameters. Comput. Electr. Eng. **29**, 835–849 (2003)
6. Sree, R.P., Srinivas, M.N., Chidambaram, M.: A simple method of tuning PID controllers for stable and unstable FOPTD systems. Comput. Chem. Eng. **28**, 2201–2218 (2004)
7. Ou, C., Lin, W.: Comparison between PSO and GA for parameters optimization of PID controller. In: International conference on mechatronics and automation. Luoyang, China, pp. 2471–2475. (June 2006)
8. Horng, H.Y.: Design of lead-lag controller via time-domain objective function by using cuckoo search, (ICITES2013). Lect. Notes Electr. Eng. **293**, 1083–1091 (2014)
9. Yang, X.S., Deb, S.: Cuckoo search via Lévy flights. In: Proceedings of world congress on nature & biologically inspired computing, India. IEEE Publications, USA, pp. 210–214. (December 2009)
10. Yang, X.S., Deb, S.: Engineering optimisation by cuckoo search. Int. J. Math. Modell. Numer. Optim. **1**(4), 330–343 (2010)
11. Yang, X.S., Deb, S.: Cuckoo search: recent advances and applications. Neural Comput. Appl. **24**, 169–174 (2014)

Chapter 26
Behavior Network-Based Risk Recognition Method

Jeonghoon Kwak, Suhyun Gong, and Yunsick Sung

Abstract This chapter proposes a two-behavior networks-based method to automatically detect whether a situation is risky or not. Behavior network is used to analyze measured values from the sensors of a smart phone. Bayesian probability is also used for implementing such a behavior network. An experiment was conducted to validate that the speed of recognizing risk situations was improved by learning such situations iteratively through behavior networks with Bayesian probability.

Keywords Behavior network • Bayesian probability • Situation classification

26.1 Introduction

Many studies have been conducted to recognize risk situations. Among these, few studies use Bayesian probability-based behavior network to recognize such situations [1]. The risk situations faced by school children can be recognized by measured sensor values of smart phones. These values are utilized to calculate the Bayesian probability. The Bayesian probability is then utilized by the behavior network that recognizes risk situations through five states: Sensing, Bayesian, critical, process, and learning states.

The above behavior network-based risk situation recognition approach had problems because it could not estimate the level of riskiness of a particular situation. Therefore, risk situations should be expressed in terms of the level of riskiness and determine a way to handle the situations accordingly. This chapter proposes a method that uses and connects Bayesian probability-based two-behavior networks to denote the level of riskiness of situations. By denoting two levels, each behavior network of the corresponding level can handle its risk situations by executing corresponding processes.

J. Kwak • S. Gong • Y. Sung (✉)
Department of Game Mobile Contents, Keimyung University, Daegu, South Korea
e-mail: jeonghoon@kmu.ac.kr; suhyun.gong@kmu.ac.kr; yunsick@kmu.ac.kr

J. Juang (ed.), *Proceedings of the 3rd International Conference on Intelligent Technologies and Engineering Systems (ICITES2014)*, Lecture Notes in Electrical Engineering 345, DOI 10.1007/978-3-319-17314-6_26

The rest of the chapter is organized as follows. Section 26.2 describes research-conducted to recognize various risk situations. In Sect. 26.3, a behavior network-based two-level situation recognition method is described. Section 26.4 introduces the implementation of our proposed method. Finally, we conclude in Sect. 26.5.

26.2 Related Work

Development of smart devices and improved Internet speeds facilitate the provision of diverse Internet-based services. In particular, various sensors are embedded on smart phones that allow users to recognize and evade risk situations. This section describes various studies conducted to determine risk situations faced by users.

There has been research to design a system that notifies risk situations faced by infants to their parents by analyzing various situations in home network environments [2]. The information related to infants can be acquired based on Open Service Gateway initiative (OSGi), utilizing various sensors and RFIDs. Ontology-based database and reasoning engines are used to recognize risk situations accurately by analyzing the gathered information.

Research on a system that notifies guardians and medical institutions about the situation of elderly people along with real-time monitoring has been conducted [3]. RFIDs are used to analyze the amount of activity in wards and infrared sensors are used to sense proximity distance. The system may be used to accurately check the health status of elderly people and notify guardians in case of emergent situations.

A study that proposes a system to facilitate pedestrians who use headphones to recognize risk situations through responses based on external sounds [4]. This system uses sound to calculate movement and direction of risk elements. If a risk element is detected, the music automatically stops.

This chapter proposes a method to recognize and learn the levels of riskiness of situations utilizing measured values of various sensors and applying behavior networks. This will allow users to determine the levels of riskiness of situations automatically, and provide risk situation notification services.

26.3 Four-Level-Based Risk Recognition Process

This chapter proposes a method that defines two risk levels, lower level and upper level, by using Bayesian probability-based two-behavior networks and determines a way to handle the risk situations with the behavior networks. When two levels are defined, upper level is more risky than lower level situation.

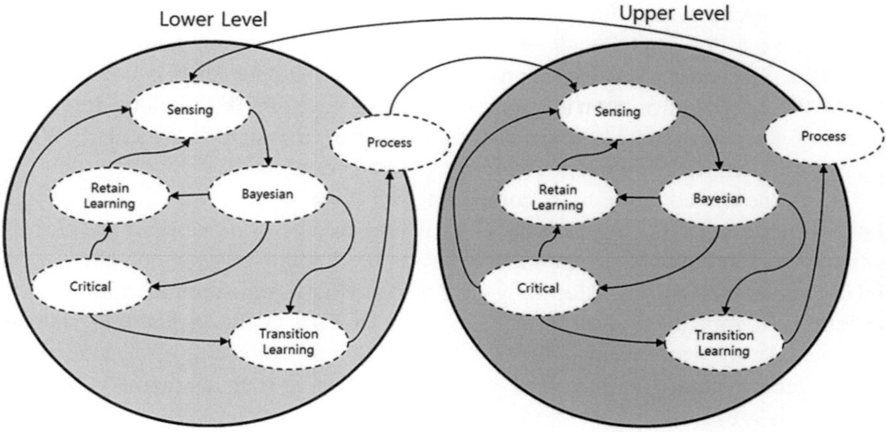

Fig. 26.1 Proposed two-level-based behavior network

It is assumed that the proposed system in this chapter recognizes risk situations by converting multiple sensor values into a single mixed value m_t at situation s_t, where s_t is defined based on all sensor values. The states defined in each behavior network, i.e., sensing, Bayesian, critical, transition learning, retention learning, and process states, are shown in Fig. 26.1.

First, situation s_t and mixed value m_t are received during the sensing state in a behavior network. Once both are received, Bayesian state is executed. These two values are unchanged until the sensing state is executed again.

Second, Bayesian probability p_{st} of the behavior network corresponding to the situation s_t is calculated by the level transition count (LTC) q_{st} and the level retention count (LRC) r_{st} as shown in (26.1), where LTC q_{st} is the number of the occurrences of the transition learning state and LRC r_{st} is the number of the occurrences of retention learning state. First, whether the sum of LTC q_{st} and LRC r_{st} is 0 or not is checked. If the sum is 0, critical state is executed. If the sum is greater than 0, Bayesian probability p_{st} is obtained by dividing LTC q_{st} by the summation of LTC q_{st} with LRC r_{st}.

$$p_{s_t} = \frac{q_{s_t}}{q_{s_t} + r_{s_t}} \tag{26.1}$$

The calculated Bayesian probability p_{st} of the lower level is compared with the upper critical probability (UCP) β. The calculated Bayesian probability p_{st} of the upper level is compared with lower critical probability (LCP) α. If Bayesian probability p_{st} is greater than the UCP β or is smaller than the LCP α, user confirmation is required. If the user concludes that the current situation is risky, transition learning state is executed. If not, retention learning state is executed. When a user does not decide within fixed period, transition learning state is executed. Similarly when Bayesian probability p_{st} of upper level is smaller than the LCP α, user confirmation is required and in the case of risk situations, transition learning state is executed, otherwise ratention learning state is executed. When there is no confirmation, ratention learning state is executed.

Third, the mixed value m_t is compared with upper critical value (UCV) δ and lower critical value (LCV) ζ during critical state to determine whether the current situation is risky or not. In the case of lower level, the UCP δ is compared with mixed value m_t. If mixed value m_t is greater than UCP δ, transition learning state is executed. If mixed value m_t is smaller than UCP δ, ratention learning state is executed. In the case of upper level, UCP δ is compared with mixed value m_t. If mixed value m_t is smaller than UCP δ, transition learning state is executed. If mixed value m_t is greater than UCP δ, ratention learning state is executed.

Fourth, LTC q_{st} and LRC r_{st} are updated considering the situation s_t during transition learning state and ratention learning state. Transition learning state increases LTC q_{st} by 1 and changes the levels. It is then transferred to process state. Retention learning state increases LRC r_{st} by 1 if LRC r_{st} is not 0 and maintains the levels.

Fifth, process state handles risk situations by performing corresponding tasks and then is transferred to upper level or lower level. We enter sensing state when either an upper level or a lower level is reached.

In the lower and upper levels, Bayesian state compares Bayesian probability with the critical probabilities. Critical state then compares the mixed value with the critical values to decide whether upper level is transferred or not.

26.4 Experiment

In our experiment, we determine whether two level-based behavior networks could recognize the levels of riskiness of situations when school children walk on streets. The values obtained from the sensors of the smartphone, Galaxy S4, used by the school children were measured. Then, corresponding risk situations were recognized. In this experiment, four levels were defined namely, safety level, attention level, warning level, and risk level. Our approach was applied only to warning and risk levels. To validate our approach, the behavior networks utilizing only critical values were used with our approach.

The proposed method recognized all safe situations correctly. In the case of using critical values, some risk situations were recognized as safe situations according to the sensor values as show in Fig. 26.2.

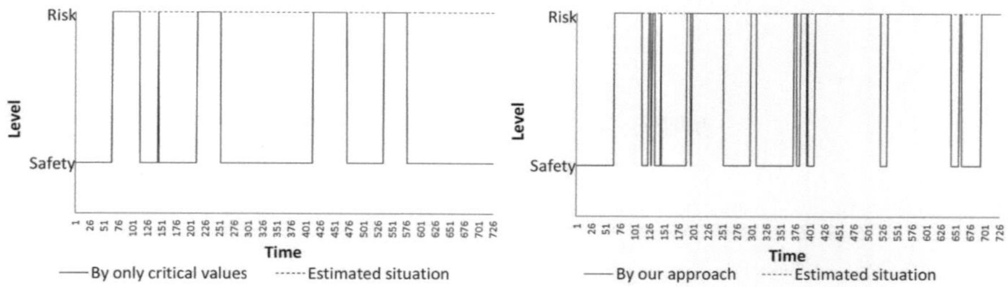

Fig. 26.2 Levels of risky situations

26.5 Conclusions

This chapter proposed a method to recognize risk situations based on two behavior networks. Each behavior network handles its risk situations. The proposed two-level-based behavior networks were applied to the smartphones of school children. Some of the risk situations were recognized only by our proposed method.

Acknowledgement This work was supported by the Korea Foundation for the Advancement of Science & Creativity (KOFAC), and funded by the Korean Government (MOE).

References

1. Kwak, J., Sung, Y.: Bayesian probability-based behavior network design for recognizing dangerous situations of school children. In: 2014 KIPS fall conference, Busan, South Korea, November 7–8. (2014)
2. Jeon, H.-K., Park, Y.-J., Lee, J.-H.: Design of infant danger notification system using context-information on the home network environment. In: Proceeding of KIIT summer conference, Gyeongbuk Gumi, Kumoh National Institute of Technology, pp. 324–329. (2007)
3. Cha, S.-H., Kim, D.-Y., Choi, J.-H., Lee, J.-E., Kim, K.-H., Cho, K.-H.: Design and implementation of an OSGI-based old age patient care system in embedded programming on RFIDs and infrared sensors. J. Korea Inform. Commun. Soc. (J-KICS) **33**(11), 1005–1012 (2008)
4. Yoon, S.: A study on the automatic detection method of urgent traffic situation and direction for headphone. Andong National University, Master thesis. (2012)

Chapter 27
Reinforcement Learning with Multiple Actions

Riku Nishiyama and Satoshi Yamada

Abstract The effectiveness of the reinforcement learning with multiple actions was investigated. The multiple actions consist of a group of simple actions and the termination conditions. To evaluate the learning performance of the reinforcement learning with multiple actions, it was compared with the reinforcement learning with simple actions. To develop more general learning algorithm, the reinforcement learning with multiple actions using INGnet was also investigated.

Keywords Reinforcement learning • Multiple actions • INGnet • Khepera

27.1 Introduction

The reinforcement learning (RL) learns the behavior through trial-and-error interactions with a dynamic environment [1, 2]. Since RL does not need the precise information on the desired output or desired action, it is considered to be applicable to many types of control problems. However, it has the disadvantage to take a large learning time to train the network for the complex control problems. In order to apply RL to the practical control problems, more effective learning algorithm is required. To tackle this, the RL with multiple actions [3] was investigated. In the usual RL, the output of RL is a simple action, e.g., turn right or turn left in the navigation control. So the learning agent must learn the input–output relations of the huge number of combinations of input and simple actions. The multiple actions consist of several simple actions and the termination conditions, e.g., avoid walls in the navigation control. By using multiple actions as output of RL system, the learning agent selects appropriate action sets for the situation, and is expected to decrease the learning time.

In animal brain, motor controls are performed in the primary motor cortex, the cerebellum, and the basal ganglia. It is considered that cerebellum learns the smooth and precise control. The basal ganglia are considered to perform RL. In

R. Nishiyama • S. Yamada (✉)
Department of Intelligent Mechanical Engineering, Okayama University of Science, Okayama 700-0005, Japan
e-mail: yamada@are.ous.ac.jp

J. Juang (ed.), *Proceedings of the 3rd International Conference on Intelligent Technologies and Engineering Systems (ICITES2014)*, Lecture Notes in Electrical Engineering 345, DOI 10.1007/978-3-319-17314-6_27

the brain, it is considered that the RL does not learn the selection of each movement of each muscle, but the selection of behaviors. RL with multiple actions is expected to be a model of RL in the brain.

In this study, RL with multiple actions is applied to the control problems in which the combination of two types of sensors (AND, OR, and XOR) must be discriminated. In future, the multiple actions should be created autonomously by the learning agent. In this study, they were made manually. The learning performance of RL with multiple actions was compared with that of the modular RL with simple actions [4]. As the method to process the sensor information, CMAC [4] and the incremental normalized Gaussian networks (INGnet) [5] were compared.

27.2 Reinforcement Learning with Multiple Actions

Multiple actions consist of several simple actions, which are used for the intermediate purpose, and the termination conditions. If the multiple actions are performed, the state of the control object is changed deterministically. Figure 27.1 shows the difference between the control by using simple actions (a) and that by using multiple actions (b). In the case of RL with simple actions, the learning agent must select the actions many times to avoid walls. On the other hand, in the case of that with multiple actions, it just selects the "avoiding walls" multiple action once. In RL, Q-values are defined for the combination of the state and output (simple actions). So the learning agent must learn the huge number of input–output combinations. By using multiple actions, the number of the combination of the state and output (multiple actions) is expected to be decreased.

The following nine multiple actions were made manually by using three simple actions (turn left, turn right, and headway) for this study's task (see Sect. 27.3).

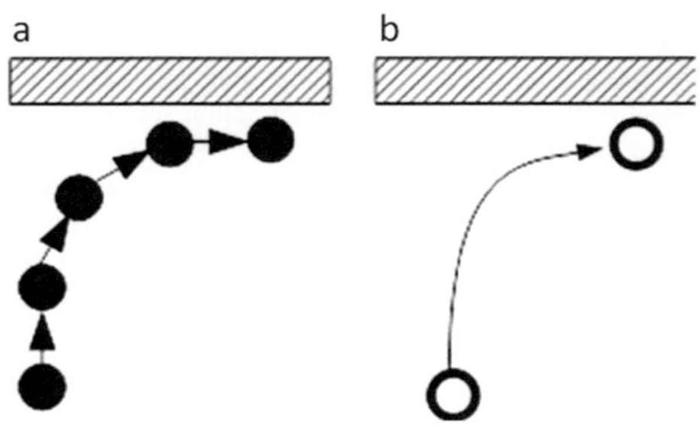

Fig. 27.1 The trajectory of simple actions (**a**) and multiple actions (**b**)

1. GoStraight(): Go straight until distance sensors or light sensors detect walls or objects.
2. TurnRightSensD(N): Turn right until *N*th distance sensor detects walls or objects.
3. TurnLeftSensD(N): Turn left until *N*th distance sensor detects walls or objects.
4. TurnRightSensO(N): Turn right until *N*th light sensor detects lamps.
5. TurnLeftSensO(N): Turn left until *N*th light sensor detects lamps.
6. SearchLamp(): Turn until light sensor detects lamps.
7. ApproachBar(): Approach bars by using the information obtained from the linear CCD array.
8. ApproachLamp(): Approach lamps by using light sensors.
9. SearchBar(): Turn until the linear CCD array detects bars.

27.2.1 Learning Algorithm

In this study, RL with multiple actions are trained by Sarsa (λ) using CMAC [4] or by Q-learning using INGnet [5].

Sarsa (λ) *using CMAC*. The Q-values for the action a in the state s are calculated as follows:

$$Q(s,a) = \sum_{i=1}^{n} w_a(i)\phi_s(i) \tag{27.1}$$

where $w_a(i)$ and ϕ_s denote the parameter vector for the action a and the feature vector for the state s, respectively. The action a is selected according to $Q(s,a)$ which produces maximal values in the state s. The parameter vector is updated as follows:

$$\begin{aligned} w_a(i) &= w_a(i) + \alpha\hat{r}(t)e_a(i) \\ \hat{r}(t) &= r(t+1) + \gamma Q(s_{t+1}, a_{t+1}) - Q(s_t, a_t) \\ e_a(i) &= \begin{cases} 1 & a = a_t, \phi_s = \phi_{s_t} \\ 0 & a \neq a_t, \phi_s = \phi_{s_t} \\ \lambda e_a(i) & \text{otherwise} \end{cases} \end{aligned} \tag{27.2}$$

where α, $\hat{r}(t)$, $e_a(i)$, $r(t)$, γ, λ denote the learning rate, TD errors, the eligibility trace, the reinforcement signal, the discount factor, and the decay rate, respectively.

Q Learning using INGnet. The Gaussian activation function is calculated as follows:

$$a_k(X) = \mathrm{e}^{-0.5\|M_k(X-c_k)\|^2} \tag{27.3}$$

where $a_k(X)$, X, M_k, c_k denote the activation function, the n-dimensional input vector, the matrix describing the activation function's feature, and the center of the activation function, respectively. The basis function, $b_k(X)$, is a normalized activation function, such that the summation of the basis function at each point is 1. In the INGnet, a new processing unit (a new basis function), whose center is the current input vector and the matrix is $M_k = \mathrm{diag}(\mu_i)$, is added if the following conditions are satisfied:

$$|\hat{r}\,(t)| > e_{\max} \text{ and } \max_k\, a_k(X) < a_{\min} \tag{27.4}$$

where $e_{\max}$, $a_{\min}$, μ_i denote the threshold for errors, the threshold for the activation, and the inverse of the radius of the activation function, respectively. The action a is selected proportionally to the probability $p(a)$, which is calculated as follows:

$$p(a) = \frac{\exp(Q(s,a)/T)}{\sum_b \exp(Q(s,b)/T)} \tag{27.5}$$

where T is a parameter. Q values are calculated and the parameter vectors are updated similarly to the Sarsa (λ).

27.3 Control Task

The reinforcement learning is applied to the control of Khepera robot. The Khepera is a cylindrical robot of 55 mm in diameter and has two DC motors, IR sensors (Fig. 27.2) and several options. The sensors of Khepera used in this study are optical sensor mode (light S) and distance sensor mode (distance S) of IR sensors and a

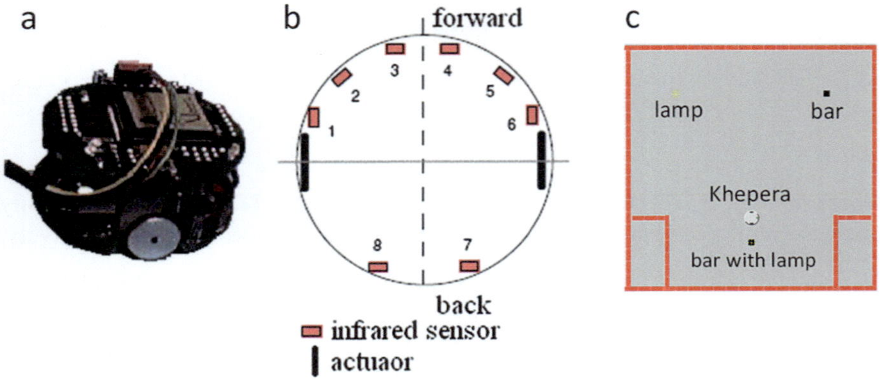

Fig. 27.2 Khepera and the environment for the control task. (**a**) Photograph of Khepera, (**b**) IR sensor position of Khepera, (**c**) the environment for the control task

Table 27.1 Rewards for approaching objects in the task

Task	Lamp	Bar	Bar with lamp
AND	−1	−1	+1
OR	+1	+1	+1
XOR	+1	+1	−1

linear CCD array (CCD S). The distance S was used to avoid walls, and the light S and CCD S were used to discriminate objects. All results were obtained by the computer simulation which used the modified "Khepera Simulator".

There are three objects, a lamp (upper left), a bar (upper right), and a bar with a lamp (lower center), in the field (Fig. 27.2c). The object of the task is to approach the targets with avoiding walls and other objects. The target is determined dependent on the task, which is shown in Table 27.1. The reward of +1 is given when the robot approaches the target. The reward of −1 is given when the robot approaches the other objects and walls. In this study, RL was applied to the control problem in which the combination of the light S and CCD S (AND, OR, and XOR) must be discriminated.

27.4 Results and Discussions

The learning performance of RL with multiple actions was compared with that of simple actions on the task in which the combination of the light S and the CCD S must be discriminated (Fig. 27.3). RL with multiple actions using CMAC (5 inputs: 2 distance S, 2 light S, 1 data obtained from CCD S), and that using INGnet (10 inputs: 4 distance S, 4 light S, 2 data from CCD S) were compared with RL with simple actions. The modular RL using CMAC (3 control modules to process each sensor information and 1 selection module) and RL using INGnet (10 inputs) are used for RL with simple actions.

The RL with simple actions (modular RL using CMAC or RL using INGnet) learned the control of the task in which the correlation of light S and CCD S, AND (Fig. 27.3a), OR (Fig. 27.3b), and XOR (Fig. 27.3c), but the success rate of AND and XOR was low. RL with multiple actions using CMAC (Fig. 27.3 thin lines) and using INGnet (Fig. 27.3 thick lines) learned the tasks more efficiently than that with simple actions. The final success rate of RL using INGnet was similar to that using CMAC. The initial success rate of RL using INGnet was lower than that using CMAC, since INGnet had no effective processing unit in the initial condition. The network structure of CMAC is required to be initially determined. Since INGnet set processing units at the necessary position in the learning, INGnet is considered to have the ability to apply to general control tasks.

In front of objects or walls, appropriate multiple actions of rotating were selected (Fig. 27.4). In the neighborhood of the target, "approach" multiple actions were selected.

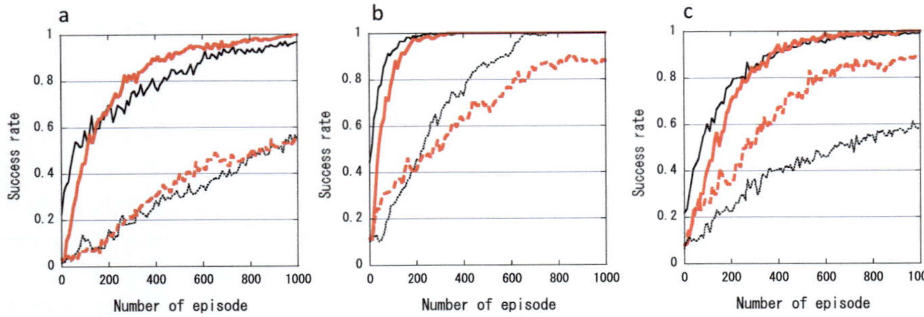

Fig. 27.3 Success rate on the learning tasks in which AND (**a**), OR (**b**), and XOR (**c**) between light S and CCD S. *Dotted line*: modular RL with simple actions using CMAC, broken line: RL with simple actions using INGnet, *thin line*: RL with multiple actions using CMAC, *thick line*: RL with multiple actions using INGnet

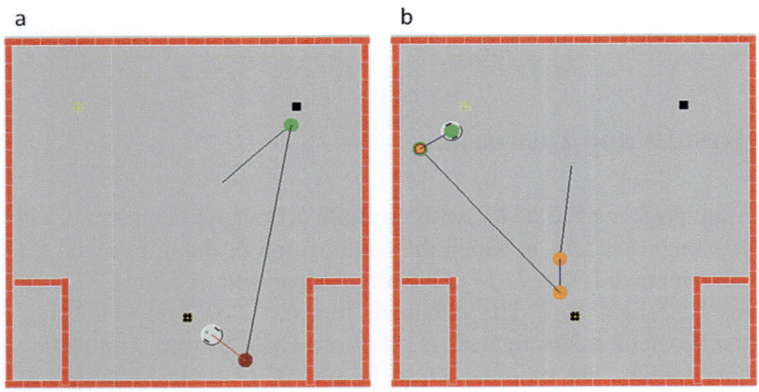

Fig. 27.4 Trajectory of robots controlled by trained networks, which were trained by RL with multiple actions using INGnet, for the task of AND (**a**) and XOR (**b**). *Several lines* denote the trajectories performed by GoStraight, ApproachBar, and ApproachLamp. *Circles* denote the selection of the other multiple actions

In the first 7 multiple actions, there are not behaviors which "approach lamp" or "search bar". To search and approach the lamp or the bar, two multiple actions are added (multiple action 8, 9). The success rate of RL with 9 multiple actions was not different from that with 8 or 7 for the task of AND (Fig. 27.5a) and OR (Fig. 27.5b), but was slightly larger than that with 8 or 7 for the task of XOR (Fig. 27.5c). The difference in XOR task is considered to depend on the number of the success to approach the lamp or the bar, where two additional multiple actions play important roles to approach them.

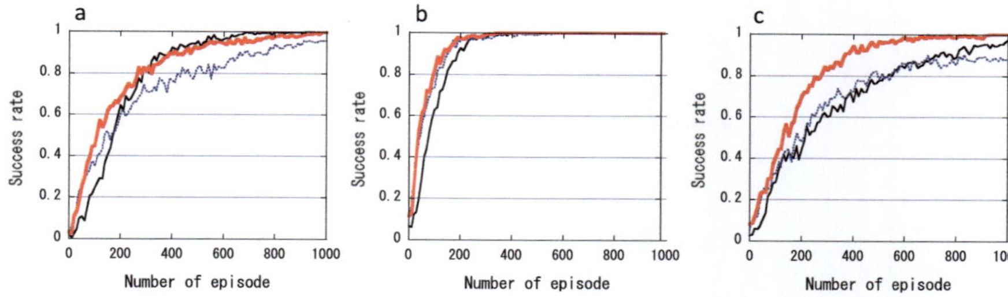

Fig. 27.5 Effect of the number of multiple actions on the success rate of RL with multiple actions using INGnet on the tasks in which AND (**a**), OR (**b**), and XOR (**c**) between light S and CCD S. *Dotted line*: the number of used multiple actions is 7 (1–7), *thin line*: 8 (1–8), *thick line*: 9

27.5 Conclusions

The reinforcement learning with multiple actions using INGnet showed the good learning performance on the task in which the combination of two types of sensor information must be discriminated. It is expected to be applicable to many practical tasks. The RL with multiple actions will be applied to more practical tasks.

Acknowledgement The authors would like to acknowledge financial support for this work provided by QOL (Quality of Life) Innovative Research (2012–2016) from the Ministry of Education, Culture, Sports, Science and Technology of Japan.

References

1. Barto, A.G., Sutton, R.S., Anderson, C.W.: Neuron like adaptive elements that can solve difficult learning control problems. IEEE Trans. Sys. Man Cybern. **SMC-13**, 834–846 (1983)
2. Sutton, R.S., Barto, A.G.: Reinforcement learning: an introduction. MIT, Cambridge (1998)
3. Stone, P., Sutton, R.S., Kuhlmann, G.: Reinforcement learning for RoboCup-Soccer keepaway. Adapt. Behav. **13**, 165–188 (2005)
4. Nakama, H., Yamada, S.: Modular reinforcement learning for control problems with multi sensors. Int. J. Adv. Mech. Syst. **3**, 251–258 (2011)
5. Asano, T., Yamada, S.: Modular reinforcement learning with adaptive networks. Int. J. Adv. Mech. Syst. **4**, 94–102 (2012)

Chapter 28
Wind Turbine Blade Load Alleviation Performance Employing Individual Pitch Control

Chin-Fan Chen, Chi-Jo Wang, Alireza Maheri, and Terrence Macquart

Abstract Nowadays, the usage of renewable energy has become popular. For the wind energy, as the wind turbine getting larger, more wind energy can be extracted. However, wind turbine components also became bulkier and hence increased the cost of installation. Furthermore, loads on the larger rotor also increased and made its life span shorter. As a result, the aim of the project is to investigate the implementation of individual pitch control (IPC) for alleviating wind turbine blade loads and then increase the cost effectiveness. This study covers the operation of variable wind speed with different controllers, namely pitch control, torque control and IPC. The result will corporate the implementation of proportional-integral-derivative controller (PID) into the torque and pitch control as well as the direct-quadrature (d-q) transformation for IPC.

28.1 Introduction

In the past decades the worldwide installed wind turbine capacity has significantly increased because of the development of technology. Large wind turbine made higher proportion of wind energy extraction. Because of the wind turbine rotor size increased, the "rotor disc" also increased which means more wind passed the rotor disc and generated more kinetic energy. As more energy can be generated,

C.-F. Chen (✉) • C.-J. Wang
Department of Electrical Engineering, Southern Taiwan University of Science and Technology, Tainan 71005, Taiwan
e-mail: jason780405@gmail.com

A. Maheri
Northumbria University, Newcastle Upon Tyne NE1 8ST, UK

T. Macquart
School of Computing, Engineering and Information Sciences, Northumbria University, Newcastle Upon Tyne NE1 8ST, UK

J. Juang (ed.), *Proceedings of the 3rd International Conference on Intelligent Technologies and Engineering Systems (ICITES2014)*, Lecture Notes in Electrical Engineering 345, DOI 10.1007/978-3-319-17314-6_28

the cost of that would become cheaper. When the size of rotor increases, the length of blade also should be adjusted. Therefore, the height of would be further from the ground. The size of rotor directly determines the quantity of power generation. This is because power generation is directly proportional to the rotor-swept area. At higher altitude, the wind turbine suffered stronger mean wind speed, according to the wind shear. Because the power generation is almost corresponding to the wind speed and the usage of wind energy would be more economical. Although the power growth is proportional to the square of radius, the mass will also increase proportionally to the cube of radius. Hence, larger wind turbine cannot simply consist of up-scaling of the previous generation. Wind turbines are frequently suffering from the fluctuating stress, which is caused by various factors, namely wind shear, tower shadow as well as turbulence and variable wind speed. These phenomena might cause more significant damage on components for taller wind turbine. Research in this field must be continuous until technology reaches maturity and the cost of energy will become difficult to further reduce. This project investigates how to decrease the load with active flow controller by using a MATLAB code, based on the NREL 5 MW wind turbine. In this study, an introduction of load, fatigue load in particular, will be explained. And then the type of wind turbine and its operating environment will be presented in the next section. After that the theory of different controllers used in this project will be demonstrated. All equations and simulation results will be compared in the last section. The sources of load taken into account might be briefly categorised as extreme loads and fatigue loads. Fatigue load is a key factor that should be considered for wind turbine design. Accordingly, wind turbine is a fatigue-critical machine [1, 2]. The design of many components is evaluated based on the consideration of fatigue load. In other words, fatigue is used to examine how wind turbine components are expected to withstand in a continuously changed wind condition over their life span. Fatigue load is caused by the cyclic stress or continuous fluctuation of press. Engineers have discovered that if the stress was applied and removed repeatedly from one component, the damage would accumulate and cause the crack after a number of cycles. Furthermore, loads also dramatically affect the life span of wind turbine and cost, so that the cost of installation and maintenance can be minimised.

28.2 Modelling and Methodology Analysis

The aim of the project is conducting an investigation, which focuses on the comparison of the load alleviation and performance between traditional torque, pitch controller and individual pitch controller.

28.2.1 Wind Turbine Model

The NREL 5 MW wind turbine is a three blades upwind, variable speed, collective pitch wind turbine, with 3 m hub diametre, 90 m hub height, and 20 m water depth [3].

28.2.2 Variable Wind-Speed Operation

For variable wind-speed operation, region I is not necessary because there is no any obvious aerodynamic behaviour. Region II is the section between cut in wind speed and rated wind speed and is a critical region for the power generation control [3]. The theoretical maximum power coefficient C_P for the ideal wind turbine is called Betz limit. In this region, the goal is to achieve the Betz limit as close as possible for maximum power generation. There is usually only torque control selected in this operating region. Pitch angle would be kept constant at its optimal value for maximising power extraction. The load is often much lower than the operation in region III, hence it is not essential to control the load. Region III is defined as the wind speed rises above the rated wind speed. The power generation would exceed the maximum requirement which wind turbine can afford and therefore lead to damage on the wind turbine structures. The variable wind turbine was designed to withstand in this condition. Pitch control is mainly used in this region to shed the power and alleviate the load in order to prevent the damage on components. Proportional-integral-derivative (PID) control is a popular tool and not only used to control the angular speed for torque control, it is also an effective method to control the pitch angle. For variable wind speed, it is possible to operate the wind turbine at its maximum power coefficient, as below:

$$P = \frac{\rho}{2} A C_P(\lambda) \tag{28.1}$$

$$\lambda = \omega \frac{R}{v} \tag{28.2}$$

in which, P is the power generation, ρ is the air density, A is the rotor disc area, and C_P is the power coefficient at relative-tip speed ratio. For tip speed ratio (λ), R is the rotor radius in metre, v is the actual wind velocity, is the rotor angular speed. When the wind turbine operating with the fixed speed rotor, power generation can only be maximised at specific wind speed. Since most of the factors are not controllable, it is not possible to store energy capture when the serious situations occurred, such as the turbulence or gust may affect the power generation. On the contrary, with variable speed rotor it is able to keep the wind turbine operating at its optimum power coefficient with relative-tip speed ratio (λ). Selecting different pitch angle can change the curve of the power coefficient and tip speed ratio. It is obvious that the pitch angle should be controlled to its ideal value in low-wind speed to allow the wind turbine operate at its optimal condition. In high-wind speed, the pitch angle is increased to spread some of the aerodynamic power. The detail of the theory of pitch control would be explained in the following section.

28.3 Methodology Analysis-Standard Controllers

In this section, torque control, pitch control and PID controller will be explained in detail, and the implementation and comparison with different factors.

28.3.1 *Proportional-Integral-Derivative Controller*

In this study, PID control would be adopted in the pitch controller and torque controller, the reference value of pitch angle and aerodynamic angular velocity, which is the term omega (ω), are process variables. On the other hand, the actual value of pitch angle and angular velocity are set points. The aim of using PID controller is to eliminate the difference between set point and process variable. The set point would be various in different circumstances. The set point usually represents an actual value, and the aim is trying to approach the reference target. PID controller consisted of three components: proportional control, integral control and derivative control [4]. The general equation of PID algorithm is described as below:

$$u(t) = K_P e(t) + \frac{K_I}{T_i}\int_0^t e(\tau)d\tau + K_d \frac{\mathrm{d}e(t)}{\mathrm{d}t} \tag{28.3}$$

In which, $u(t)$ is the system output. K_P, K_I and K_D are proportional gain, integral gain and derivative gain, respectively. $e(t)$ stands for the control error, which is referred to the difference between y_r and y. The term y is the measured process variable, and y_r denotes the reference variable, which is also called set point. In this study, y would be replaced by pitch angle and dynamic angular velocity.

28.3.2 *Torque Controller*

When the wind speed starts getting stronger, the lift force on the blade would become larger and might lead to the blade changing its shape if the lift force is too strong. All the lift forces on different blade points have accumulative effect on the tendency to bend, particularly on the furthest point. Therefore, it is obvious that the root of the blade must be thickest to stand against the stress. On the contrary, the blades need to have a thinner tip where the drag force has a crucial influence with respect to the stall for power regulation, because the lift force would place the load on the component. If the load is too large, the blade might be damaged permanently. This phenomenon is the so-called bending moment. Bending moment is an important consideration when designing the wind turbine.

Torque control is adopted to control the generator torque, by interposing a frequency converter between generator and network, and then the generator would be able to vary with time. The torque control allows the torque can be adjusted towards any desirable value, particularly in the region II when wind speed is below the rated speed in order to maximize the aerodynamic efficiency. When implementing PID controller, the following equation is the general rotor equation of the motion:

$$J\dot{\omega} = T_{aero} - K_t\omega_r - nT_{\text{gen}}. \tag{28.4}$$

$$T_{\text{gen}} = K_P e(t) + \frac{K_I}{T_i}\int_0^t e(\tau)d\tau + K_d\frac{\mathrm{d}e(t)}{\mathrm{d}t} \tag{28.5}$$

In which, J is the total inertia which is the combination of rotor inertia and generator inertia, ω is the speed of rotor, and T_{aero} and T_{gen} are aerodynamic torque and generator torque applied to the rotor, respectively. K_t is the total external damping, which is the summation of external generator damping and external rotor damping. n is representing the gearbox ratio, which is 97:1. The term T_{gen} is the control variable and would be replaced by the PID control equation. In this study, the inertia J is 38,759,766.166 kg × m2, $K_{\text{p\_torque}}$ is 10^6 and $K_{\text{I\_torque}}$ is 10^7. The result of adopting PID controller into torque control is shown in Fig. 28.1.

28.3.3 Pitch Controller

When the wind turbine works in the below-rated wind speed (region II), it aims to generate as much power as possible, increasing power coefficient (C_P) to its optimal value by setting the fixed pitch angle. Therefore, pitch control should be used to optimise the power output at low-wind speed and limit it at high-wind situation.

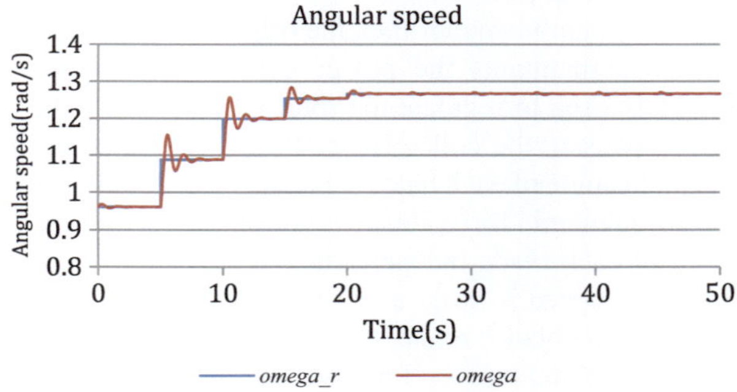

Fig. 28.1 The simulation result of angular speed against time

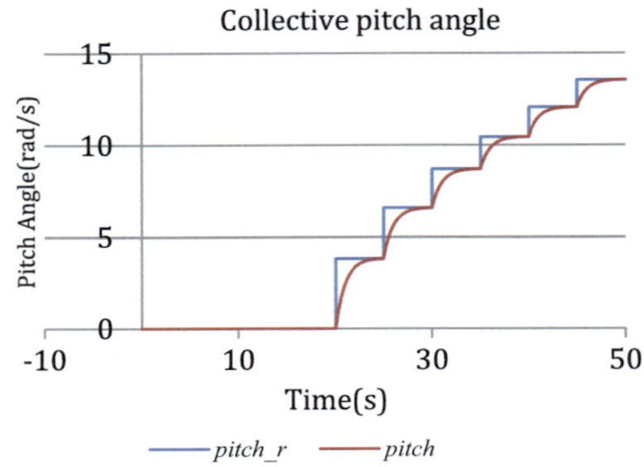

Fig. 28.2 The simulation result of pitch angle against time

A general equation for most control systems written by a second ordinary differential equation (ode) would be selected to evaluate and calculate the pitch angle (φ). PID controller is implemented as below:

$$m\ddot{\phi} + c\dot{\phi} + k = K_P e(t) + \frac{K_I}{T_i}\int_0^t e(\tau)d\tau + K_d\frac{\mathrm{d}e(t)}{\mathrm{d}t} \tag{28.6}$$

in which, m is the mass of the rotor, c is the linear damping constant, and k is the spring damping constant. Error here is the difference between reference pitch angle and actual pitch angle. $K_{\mathrm{P\_Pitch}}$ is 10^7, $K_{\mathrm{I\_Pitch}}$ is 10^9, mass is 28,600 kg and linear damping constant and spring damping are 206,000 (N × m/(rad/s)) and 971,350,000 (N × m/rad), respectively. The result of adding PID into pitch control is shown in Fig. 28.2. The pitch angle would rise rapidly to the reference ideal pitch angle. And then by increasing the integral gain, the error becomes smaller and the pitch angle is getting closer to the reference pitch angle.

28.3.4 Individual Pitch Control

The individual pitch control (IPC) is to minimise the load on each blade by making the load slightly smaller than the original value when blade at the highest position, and then maintains the power generation. Overall, collective pitch control is basically used to regulate the rotor speed, whilst IPC is designed as a supplement to compensate the collective pitch angle. IPC is an intelligent method to control the pitch angle of each blade separately in variable wind-speed conditions, and keep the aerodynamic force constant during the revolution. The reason for adjusting the pitch of each blade independently is particularly due to the effect of wind shear. The wind speed is faster at higher altitude and close to zero on the surface. As the wind turbine blades rotate, there is always one turbine blade suffered from stronger wind speed and stress. However, the blade should be adjusted in order to reduce load and damage. The collective pitch control can only fix all blades simultaneously. Therefore, the IPC has become increasingly important with respect to the alleviation of load. IPC can be used to alleviate the deterministic load caused by the wind shear or tower shadow [5]. Different factors resulting in blade loads at the rotational frequency (1P) or multiple frequencies are in consideration. The load reduction can be achieved by implementing the measurement of bending moment interns of yaw and tilt. By using the d-q transformation, the reliable simulation can be done and critically analyse the influence of the IPC. The result will be presented in the following sections.

28.3.5 Direct-Quadrature-Zero Transformation for IPC

The d-q-0 transformation control strategy is adopted in this study involving the IPC. The number of blades is presented in two ways of rotation transformation. One of the methods is to transform the blade root load measurement into non-rotating frame; another is a reverse transformation to transform the pitch demand back to the rotating frame. The d-q-0 transformation is easy to generate the solution for any number of blades. The forward equation is shown as below:

$$\begin{bmatrix} \beta_A \\ \beta_B \\ \beta_C \end{bmatrix} = \begin{bmatrix} \cos(\theta) & \sin(\theta) & 1 \\ \cos\left(\theta - \frac{2\pi}{3}\right) & \sin\left(\theta - \frac{2\pi}{3}\right) & 1 \\ \cos\left(\theta + \frac{2\pi}{3}\right) & \sin\left(\theta + \frac{2\pi}{3}\right) & 1 \end{bmatrix} \begin{bmatrix} \beta_D \\ \beta_Q \\ \beta_0 \end{bmatrix} \tag{28.7}$$

n which, β_i ($i = A, B, C$) are pitch demand for blade one to three, respectively. β_j ($j = D, Q, 0$) denotes the direct axis quantity, quadrature axis quantity and zero axis quantity, respectively. The individual pith angle would be the summation of the collective pitch angle and individual pitch angle demand increment. Thus, the pitch angle would be various on different blades that are suitable for the current wind conditions at unique height. θ is the rotor azimuth angle. On the other hand, the corresponding reverse transformation equation is shown as below:

$$\begin{bmatrix} \beta_d \\ \beta_q \\ \beta_0 \end{bmatrix} = \frac{2}{3} \begin{bmatrix} \cos(\theta) & \cos\left(\theta - \frac{2\pi}{3}\right) & \cos\left(\theta + \frac{2\pi}{3}\right) \\ \sin(\theta) & \sin\left(\theta - \frac{2\pi}{3}\right) & \sin\left(\theta - \frac{2\pi}{3}\right) \\ 1/2 & 1/2 & 1/2 \end{bmatrix} \tag{28.8}$$

In this equation, β_i ($i = a, b, c$) represent the blade root bending moment of each of the three blades. The matrix on the left hand side consists of transformed fixed axis loading (d and q) in the yaw and tilt moment, respectively. The zero axis loading is ignored because it only used to account the mean value and has a negligible impact

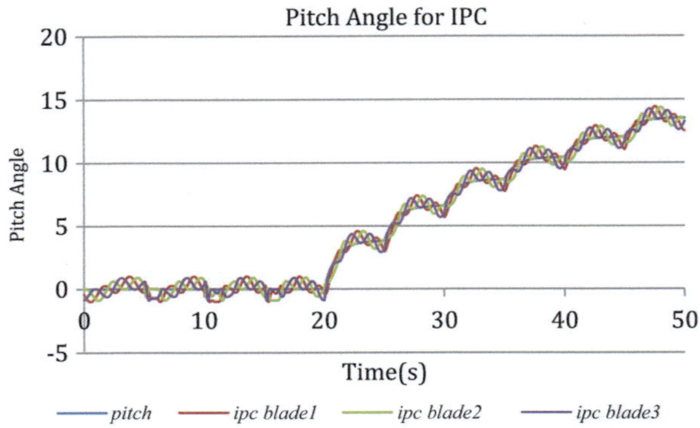

Fig. 28.3 Individual pitch angle during the stepwise change of wind speed

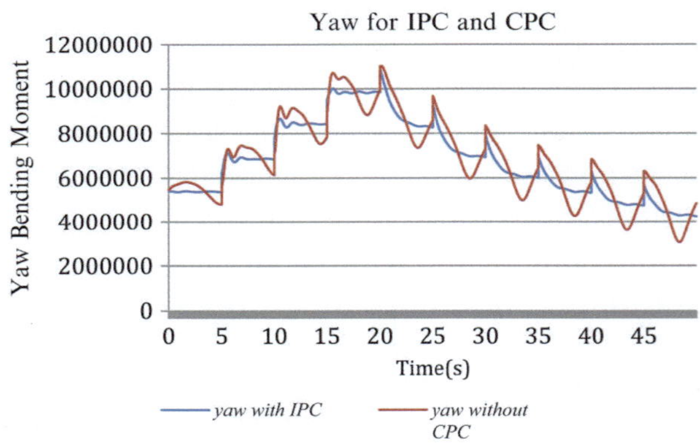

Fig. 28.4 The comparison of yaw bending moment between CPC and IPC

on the load reduction. Therefore, the zero sequence always omitted in IPC. This controller effectively reduced the load on the 1P loading. For the higher harmonic loading, altering the different factors might reduce the load, such that changing the azimuth angle for different number of blade wind turbine. The following results indicate the effect of pitch angle by using collective pitch controller and individual pitch controller. As we can see, the individual pitch angle is various around the collective pitch angle, as shown in Fig. 28.3. The pitch angle for each of the three blades is the result of the summation of collective pitch angle and individual pitch angle, which is calculated by d-q transformation. The pitch angle for each blade kept changing because of the rotation of the rotor. During the rotation, each suffered different wind speed caused by wind shear. Hence the pitch angle would be adjusted to adapt different conditions with a slight change of the collective pitch angle.

Figures 28.4 and 28.5 illustrate the results of implementing IPC into the wind turbine design. Figure 28.4 is the simulation of direct bending moment, and Fig. 28.5 is the quadrature bending moment. Direct and quadrature axis moment

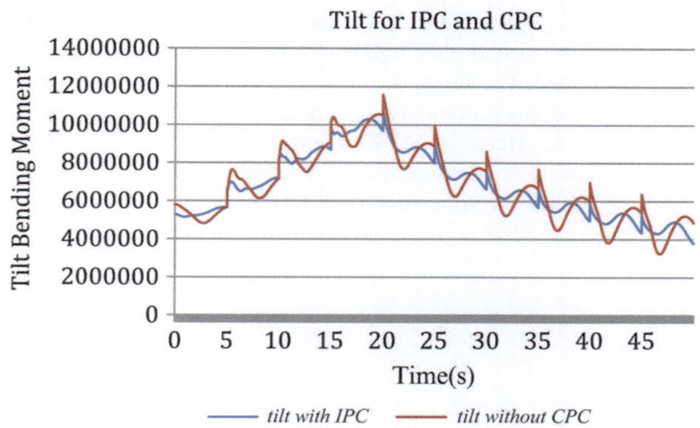

Fig. 28.5 The comparison of tilt bending moment between CPC and IPC

can be referred to the yaw moment and tilt moment. Compared with the collective pitch control, it is obvious that both direct and quadrature moment are becoming smaller when using IPC. By alleviating the yaw and tilt moment, the torque on the blade can also be adjusted in order to maintain the angular speed at its constant value. Hence, the load on each of the three blades has also been alleviated with a more intelligent way as the effect of moment reduced.

28.4 Conclusion

The analysis in this study demonstrates the basic idea for the collective pitch control and torque control, as well as the reduction in aerodynamic loading achieved by the IPC. In the beginning, the theory of collective pitch control and torque control is presented. The working region for each controller was explained. And then the reason using popular tool PID controller to improve and alleviate the error is also detailed in this study. After that, the mechanical method of bending moment transformation from rotational structure to the fixed part was calculated by using the d-q transformation. Finally, a list of simulation results is presented and analysed. As we can see, the IPC can effectively alleviate the load by turning the flap root bending moment into the direct and quadrature moment.

Reference

1. Burton, T., Jenkins, N., et al.: Wind energy handbook, 2nd edn, pp. 475–505. Wiley, Chichester (2011)
2. Barlas, T.: Smart rotor blades and rotor control for wind turbine. Delft University Wind Energy Research Institute (DUWIND), pp. 48–51. (2002–2006)

3. Jonkman, J., Butterfield, S., Musial, W., Scott, G.: Definition of a 5-MW reference wind turbine for offshore system development. National Renewable Energy Laboratory (2009)
4. Abdulhamed, H., Reza, K.: Wind turbine control using PI pitch angle controller. Industrial Control Centre, University of Strathclyde, Glasgow (2012)
5. Torben, J.L., Helge, A.M., Kenneth, T.: Active load reduction using individual pitch, based on local blade flow measurements, Wind Energy Department, Risø National Laboratory, PO Box 49, DK-4000 Roskilde, Denmark (2004)

Chapter 29
Planet Editing Method Using Leap Motions

Ji Won Kim, Phil Young Kim, and Yunsick Sung

Abstract Universal courseware using motion recognition devices is very useful for students to learn universal science because it enables interactive learning. However, given that students change the radius of planets by the distance between their hands, it is nearly impossible to denote larger radii with their hands within fixed-hand recognition area of the motion recognition devices by a single expression. This chapter proposes a method that controls planets using motion recognition devices for courseware contents. The distance between hands is expressed by the sum of the difference of each position of the user's hands. In the experiment, we validated the change of the radius of planets according to the difference of each recognized hand.

Keywords Leap motion • Universal courseware • Motion recognition

29.1 Introduction

Recent developments in courseware enable students to learn about phenomena such as the effect of a planet's radius on the force of gravity, which is too vast for a student to understand in a hands-on exercise. The term courseware is a combination of course and software and can be either online or offline presentations of content. Research supports a developmental approach of space science education by employing virtual reality courseware in a class setting [1]. Learner retention is much greater when students control the mouse and keyboard directly in planetary science lessons than in a one-sided lecture approach. The learning effect is increased even further when students can control the virtual planets and interact with them.

A variety of student interactions and full-body experiences enhance learning. Motion recognition devices enable physical interactions and full-body experiences to increase the effectiveness of space science education. However, they suffer from

J.W. Kim • P.Y. Kim • Y. Sung (✉)
Department of Game Mobile Contents, Keimyung University, Daegu, South Korea
e-mail: jiwon1219@kmu.ac.kr; kimpy1111@kmu.ac.kr; yunsick@kmu.ac.kr

J. Juang (ed.), *Proceedings of the 3rd International Conference on Intelligent Technologies and Engineering Systems (ICITES2014)*, Lecture Notes in Electrical Engineering 345, DOI 10.1007/978-3-319-17314-6_29

the following limitations. First, the recognition area of motion recognition devices is too limited to express the physical distances necessary to accommodate the size of planets. For example, within the recognition area, it is difficult to express the radius of a planet with a single motion using the distance between a student's hands. Next, given that a planet's size is incomparably bigger than the maximum distance of the recognition area, it is difficult to edit and control virtual planets elaborately.

This chapter proposes a method that controls planets using a motion recognition device for space science courseware. The distance between the user's hands is recognized by the motion recognition device and measured. The radius of the virtual planet is calculated based on that distance. The properties of the virtual planet are calculated by the radius, and then planets are rendered in real time. In addition, the effects on a planet's physical properties can be confirmed in real time as a student manipulates the radius of a virtual planet.

This chapter consists the following: Section 29.2 introduces related research. Section 29.3 explains a method that calculates the radius of planets using a motion recognition device. Section 29.4 introduces experiment results obtained using the proposed method. Finally, Sect. 29.5 summarizes our suggested method.

29.2 Related Work

Technology enables teachers to vary content and present the most current scientific concepts to students. This section introduces research about developing educational content.

Research proposes that virtual reality-based science education courseware enhances student retention [1]. This research provides functions that generate content about outer space. Users manipulate and interact with the content. Therefore, more effective learning can be expected through interacting with the content.

An app authoring tool is used to provide development systems and learning systems for school curriculum [2]. Students can select and learn the curriculum developed using the app authoring tool. It provides Q/A systems, learning progress, an evaluation system, and a wrong answer notation system.

There is a smart device authoring tool that generates content reflecting a teacher's own experiences [3]. This tool generates XML content that is platform-independent, so it can be used on any smart device.

This chapter proposes presentation of planetary science content through an interactive experience of using Leap Motion to increase learning effectiveness. A user can edit properties of planets with his or her hand motions and confirm the feedback of change in the properties in real time.

29.3 Method Explanation

29.3.1 Planet Control Process

In our method, planets are controlled by a motion recognition device and rendered as follows. First, the motion recognition device recognizes the user's hands and provides the distance d_t between the recognized hands in centimeters. Then the distance variation v_t of hands is calculated by the difference between the distance d_{t-1} and the distance d_t. When the hands are initially recognized or outside the recognition area, the distance variation becomes 0.

Next, the values of two factors are determined as shown in Fig. 29.1: one for a variable factor α and the other for constant factor β.

The variable factor α can be controlled by the user rotating his or her wrists, which determines the effect of variation v_t on radius r_t. The variable factor α is calculated as shown in (29.1).

$$\alpha = \frac{a_{l,t} + a_{r,t}}{2} \tag{29.1}$$

where factor $a_{l,t}$ is the value of the left wrist at time t and factor $a_{r,t}$ is that of the right wrist. Both factors range from 0 to 1. If both wrists face downwards, α becomes 0, which means variation v_t does not affect radius r_t at all. If both wrists face upwards, α becomes 1, which means 100 % dependence. The constant factor β is the fixed value set by the developers of the authoring tool and cannot be changed by users. This determines the maximum effective rate by variation v_t.

Next, radius r_t of a virtual planet is calculated by the distance variation v_t of the two factors. Therefore, radius r_t is calculated as shown in (29.2).

$$\begin{aligned} r_t\,\text{km} &= r\,\text{km} \times \sum_{t=1} (v_t \times a \times \beta + 1) \\ &= r\,\text{km} \times \sum_{t=1} \left(v_t \times \frac{a_{l,t} + a_{r,t}}{2} \times \beta + 1\right) \\ &= \frac{r}{2} \sum_{t=1} (v_t a_{l,t}\beta + v_t a_{r,t}\beta + 2)\,\text{km} \end{aligned} \tag{29.2}$$

where radius r is the radius of the real planet in kilometers.

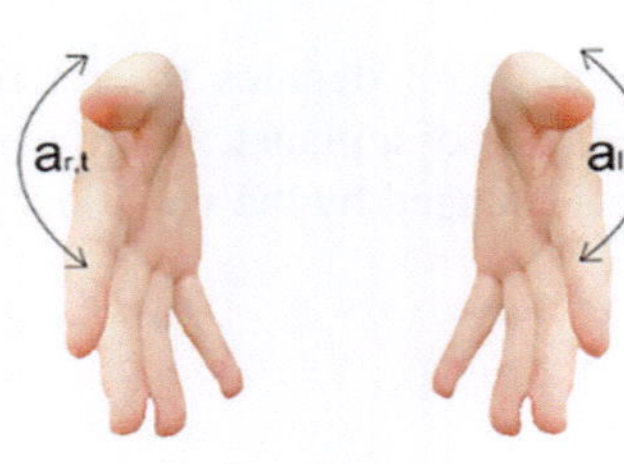

Fig. 29.1 Variable factor α

Next, the properties of a planet that are calculated by radius r_t are recalculated. Examples of properties dependent on a planet's radius are mass, density, acceleration of gravity, and so on.

Next, the recalculated properties of the planet are applied in real time for its rendering. Physical features are also reflected based on the changes in the properties.

29.4 Experiment

This chapter utilizes a motion recognition device, Leap Motion [4], to recognize detailed hand motions. In the experiment, after the distance variation between the hands was measured, we confirmed the changes in radius r_t of planets. In addition, the effect of radius r_t on gravity was analyzed.

In the experiment, the approach that increases radius r_t according to the distance between the user's hands was compared with the proposed method. The relationship between the recognized distance of the hands and radius r_t is shown in Fig. 29.2a. Depending on the distance between the hands, radius r_t was increased proportionally. However, only 0–12,756.2 km was possibly expressible within the hand recognition area.

The results of our proposed method are shown in Fig. 29.2b. Radius r_t was increased in proportion to the distance variation v_t. When the position of the hands was out of the device's recognition area, the distance variation v_t was measured by 0 and there were no changes in radius r_t. As a result, radius r_t was not proportional to the distance between the hands but proportional to the sum of the distance variation between the hands.

The mass of planets is defined as M. The acceleration of gravity of planets is defined as g. The universal gravitation G is 6.67289×10^{-11}. According to Newton's law, the gravity that affects the surface of the earth is obtained by multiplying mass M with acceleration of gravity g. In addition, gravity g is the value obtained by multiplying the mass of two objects and then dividing the square of the distance between the two objects. Based on these two equations, (29.3) is derived. Therefore, the acceleration of gravity g is proportionate to the square of the radius of planets.

$$g = \frac{{r_s}^2}{M} \times G \tag{29.3}$$

Figure 29.3 denotes the relationship between the acceleration of gravity g and radius r_t of a planet. When variation v_t was measured by a Leap Motion, radius r_t was changed by the variation, and the acceleration of gravity g was also changed.

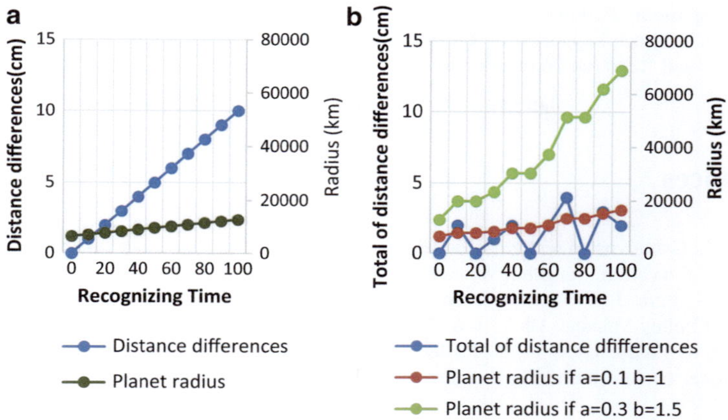

Fig. 29.2 Relationship between the hand distance difference. (**a**) The hand distance is measured at once. (**b**) The hand distance is calculated by the sum of the hand distance differences

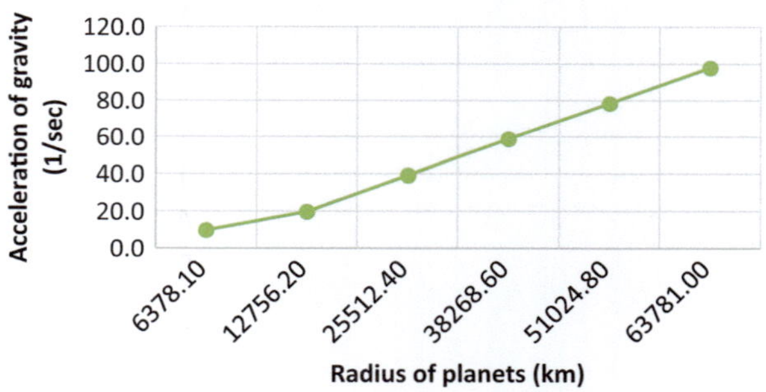

Fig. 29.3 The relationship between the radius and acceleration of gravity of planets. Increasing the radius of planets affects the acceleration of gravity and hence is increased proportionally

29.5 Conclusions

This chapter proposes a method that uses a motion recognition device to measure the distance between a user's hands and controls the properties of virtual planets using that distance. The distance variation between the user's hands was utilized to calculate the radius of planets, which modified the various properties of the planets.

In this experiment, the process to control planets by the distance variation of hands of users was validated. Depending on the distance between the hands, the calculated radius of the planets affected the acceleration of gravity of each planet. Therefore, within the limited motion recognition area, the process to calculate a distance greater than the maximum recognizable distance of the motion recognition device and express each planet with the calculated distance was also validated.

Acknowledgement Following are the results of a study on the "Leaders in Industry-university Cooperation" Project, supported by the Ministry of Education, Science & Technology (MEST) and the National Research Foundation of Korea (NRF).

References

1. Suk, J.-Y.: Using virtual reality the development of the science education courseware. Graduate School of Education, Sungshin Women's University, Seoul (2006)
2. Kim, E.-S., Park, J.-S.: A study on educational app development using the app authoring tool. J. Digital Policy Manage. **10**(5), 1–6 (2012)
3. Kwon, S.-O., Kim, J.-O., Ju, S.-Y., Jeong, J.-S., Yoo, K.-H.: Design and implementation of educational content authoring tool for smart devices. J. Korea Contents Assoc. **13**(12), 1–8 (2013)
4. Leap Motion: https://www.leapmotion.com

Chapter 30
Development of Flexible Haptic Robot Arm Using Flexible Pneumatic Cylinders with Backdrivability for Bilateral Control

Takafumi Morimoto, Mohd Aliff, Tetsuya Akagi, and Shujiro Dohta

Abstract In the remote controlled rehabilitation device, the physical therapists must recognize the situation of the patient. Therefore, a haptic device that the therapists can feel the reaction force is required. In our previous study, to realize the haptic device driven by pneumatic pressure, the flexible pneumatic cylinder with backdrivability was proposed and tested. In this chapter, the flexible haptic robot arm using the cylinders for human wrist rehabilitation is proposed and tested. The low-cost bilateral control system using two embedded controller and the tested cylinders for bilateral control is also proposed and tested. The bilateral control using the tested robot arm is carried out.

Keywords Flexible pneumatic cylinder with backdrivability • Flexible haptic robot arm • Rehabilitation device • Bilateral control

30.1 Introduction

The ratio of the elderly to young man is increasing in Japan. This causes a serious problem about lack of welfare caretakers for the elderly. The problem becomes worst when the society and the government failed to provide appropriate infrastructure to the aged and the disabled persons. The aged over 65 years old in Japan accounted for 24.1 % from the total of population in 2012 [1]. Therefore, in Japan, many studies have shown that the robot can be used in the medical field such as giving caregivers the extra strength which they need to lift patients [2] and also can be used in complex surgery [3]. In healthcare, robot can be used to assist in nursing care and also can support the activities of daily living for people with disabilities [4]. It must also have safety features such as not harming users and must have a flexible structure so as to be used in contact with human body [5]. Therefore,

T. Morimoto (✉) • M. Aliff • T. Akagi • S. Dohta
Department of Intelligent Mechanical Engineering, Okayama University of Science, Okayama 700-0005, Japan
e-mail: akagi@are.ous.ac.jp

J. Juang (ed.), *Proceedings of the 3rd International Conference on Intelligent Technologies and Engineering Systems (ICITES2014)*, Lecture Notes in Electrical Engineering 345, DOI 10.1007/978-3-319-17314-6_30

a novel flexible pneumatic actuator using low-cost embedded controller was proposed and tested in our previous study [6]. The flexible robot arm has also been proposed and tested for rehabilitation of human wrist [7]. In order to realize a haptic device driven by pneumatic pressure, the flexible pneumatic cylinder with backdrivability was proposed and tested. In this chapter, the flexible haptic robot arm using the flexible pneumatic cylinders with backdrivability for human wrist rehabilitation is proposed and tested. The low-cost bilateral control system using two embedded controller and the tested cylinders for bilateral control is also proposed and tested. The bilateral control is carried out by using the tested robot arm.

30.2 Flexible Pneumatic Cylinder with Backdrivability

In order to develop the haptic device driven by pneumatic pressure, it is necessary to construct the actuator to generate a force. However, the flexible robot arm can't be driven by human hand because of higher frictional force (about 10 N) in the slide stage of the flexible pneumatic cylinder. Therefore, the flexible pneumatic cylinder with backdrivability was proposed and tested. Figure 30.1 shows the construction and operating principle of the tested flexible pneumatic cylinder with backdrivability. The tested cylinder consists of two flexible tubes (SMC Co. Ltd., TUS1208) as a cylinder and gasket, two steel balls with the diameter of 9.5 mm as a cylinder head, and two slide stages connected with each other. Two tubes are set on parallel with a distance of 20 mm. Each steel ball is placed in each cylinder tube on the opposite side so that it can make the cylinder move upward or downward and also can clamp the slide stage during the "lock" mode. The size and the inner construction of each slide stage are same as previous one as shown in Fig. 30.1. In order to reduce the friction, it is coated with the grease in the tube.

Figure 30.1 on the right shows the operating principle of the tested cylinder for each motion. During the operation, the cylinder is driven while the slide stage is fixed. The tested cylinder has five motions such as "push," "pull," "lock," "reset,"

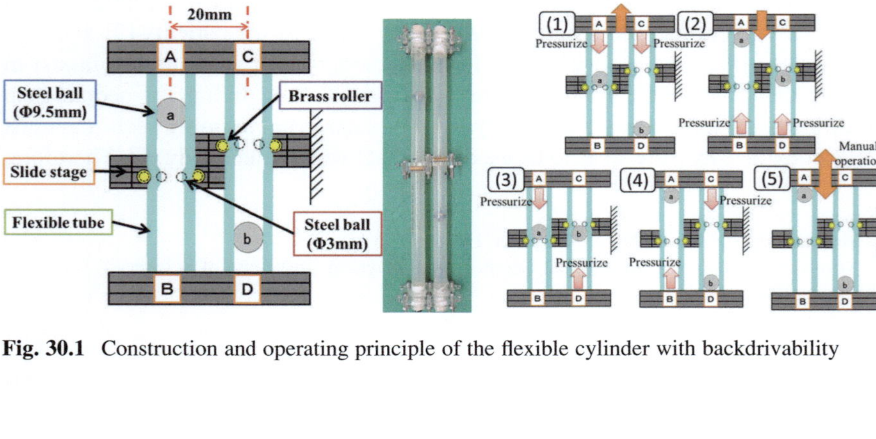

Fig. 30.1 Construction and operating principle of the flexible cylinder with backdrivability

and "free" modes. In the case of "push" mode as shown in Fig. 30.1 (1), the chambers A and C are pressurized. Then, the steel ball "a" and "b" are pushed downward. At the same time, the upper stage is pushed upward and the slide stage is pushed downward in left side tube. By using the generated force, two tubes move upward while the ball "a" deformed the tube. In the same manner of pushing, when the chambers B and D are pressurized, the two tubes move downward, that is "pull" mode as shown in Fig. 30.1 (2). In order to hold the tube, the chambers A and D are pressurized as shown in Fig. 30.1 (3). By pushing the slide stage from both sides, the tube can be held surely. This is a "lock" mode. In order to unlock it, the chambers B and C are pressurized. Then, each ball moves away from the slide stage as shown in Fig. 30.1 (4). This motion is called "reset" mode. In the case that balls "a" and "b" reached at each end of tubes, the tube can be moved freely by manual operation. This means that the cylinder has a backdrivability in this condition as shown in Fig. 30.1 (5).

30.3 Flexible Haptic Robot Arm Using the Cylinders

In order to develop the haptic device, the flexible haptic robot arm using three flexible pneumatic cylinders with backdrivability was developed. Figure 30.2 shows the construction and the view of bending motion of the tested robot arm. The robot arm consists of two round stages and three flexible pneumatic cylinders with a backdrivability described above. The outer diameter of the upper and lower stage is 100 mm. Each tested cylinder is also arranged so that the central angle of two adjacent slide stages becomes 120° on the point of 66 mm from the center of the stage. An end of each flexible cylinder is fixed to the upper stage. The size of the robot arm is 100 mm in outer diameter and 250 mm in height. The total mass of the robot arm is a little increased, that is 450 g.

Figure 30.2 on the right shows the motions of the robot arm when it extends and contracts vertically and it bends for every direction, respectively. In the case of bending, a cylinder which is on opposite side to be bent will be held by using "lock"

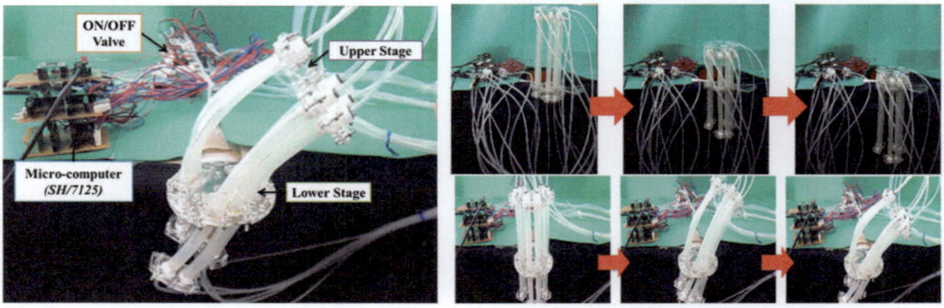

Fig. 30.2 Construction and view of movement of the tested robot arm

mode, and the other two cylinders will be extended or contacted by using of "push" or "pull" mode. The robot arm can bend in every direction by changing the cylinder with "lock" mode. In the case of extending, all three cylinders were used as "push" mode. In the experiment, the maximum bending angle of about 45° could be obtained. The bending angle depends on the locked point of a cylinder. It is also possible to increase the bending angle by using a longer flexible pneumatic cylinder.

30.4 Low-Cost Bilateral Attitude Control System

In order to execute the bilateral attitude control, the controller and multi-axis force sensor are needed. However, the typical controller such as personal computer that includes A/D and D/A converters is heavy, bulky, and expensive. Therefore, the small-sized and low-cost controller for bilateral attitude tracking control of tested robot arms was proposed. Figure 30.3 shows the construction and the schematic diagram of the whole control system of wrist rehabilitation device with bilateral control. The robot arm for patient (it is called "Slave arm" for short) consists of the flexible robot arm using three typical flexible pneumatic cylinders, an accelerometer, the embedded controller (Renesas Co. Ltd., SH7125), and six quasi-servo valves [8]. The flexible haptic robot arm for physical therapists (it is called "Master arm" for short) also consists of an accelerometer, the embedded controller (Renesas Co. Ltd., SH7125), and three flexible pneumatic cylinders with backdrivability mentioned above. Both robot arms have backbone flexible tubes connected with potentiometers to measure the displacement of the upper disk at center position. Figure 30.4 shows the block diagram of the bilateral attitude control system.

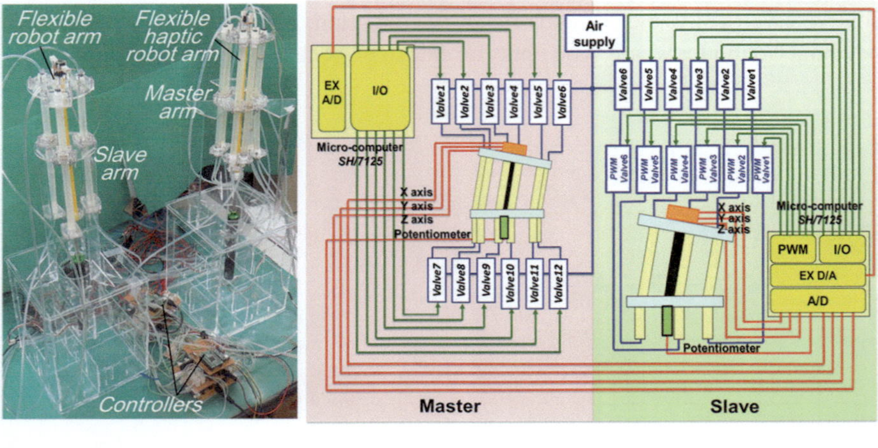

Fig. 30.3 Construction and schematic diagram of bilateral attitude control system

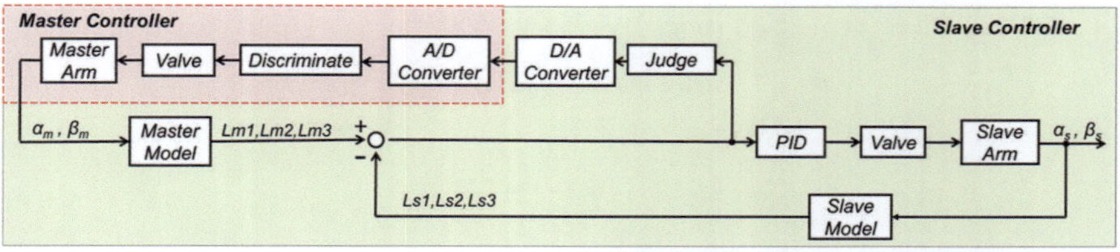

Fig. 30.4 Block diagram of the bilateral attitude control system

The bilateral attitude control of the robot is done as follows. In the attitude control of the slave arm, the embedded controller for the slave arm gets the output voltage from the potentiometer and *x*-, *y*-, *z*-axis output voltages from the accelerometer set on the master arm through the A/D converter as the desired value. From these values and the analytical model of the flexible robot arm ("Master Model" or "Slave Model" in Fig. 30.4) [9], each length of cylinders in master and slave arm is calculated. According to each displacement error of cylinders, the quasi-servo valves of the slave arm are always driven based on control scheme. As a control scheme for the slave arm, PID control scheme is used with the sampling period of about 3 ms. Then, in only case when the absolute value of the deviation of more than 5 mm for each cylinder length between the master and slave arm occurs, the master arm (flexible haptic robot arm) is also driven so as to decrease the deviation. The on/off control of the master arm is started according to the voltage signal from the slave arm's controller through the "Judge" in Fig. 30.4. The signal includes the information about which cylinder and direction are driven by using on/off valves, that is 27 patterns moving condition. In the case that the deviation is between −5 and 5 mm for each cylinder, the master arm's controller keeps the master arm "reset" mode. By this method, only when the specific condition is satisfied, the master arm is driven as a flexible haptic robot arm.

Figure 30.5 shows the transient view of movement of both robot arms in bilateral control mentioned above. It can be seen that the master arm was driven by the master arm's controller when the condition mentioned above is satisfied. In Fig. 30.5, blue and red arrows show the direction of moving force given by the operator and the master arm, respectively. From Fig. 30.5 (5), it can been seen that the steel balls in the cylinder are driven so as to pull the master arm toward the opposite direction to operator's desired direction when the slave arm is fixed by hand. It can be confirmed that the bilateral control in the tested system can be realized by using simple control method.

30.5 Conclusions

This study results aimed at developing the remote-controlled rehabilitation device for human wrist with bilateral control are summarized as follows.

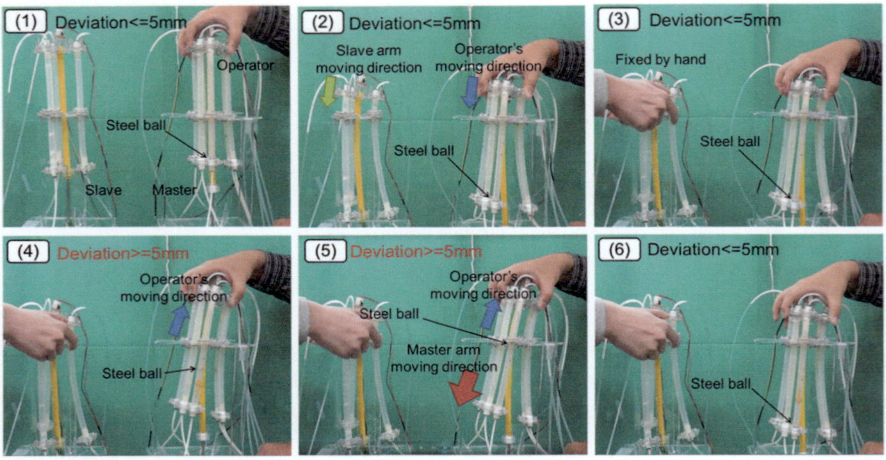

Fig. 30.5 Transient view of movement of both robot arms in bilateral control

1. The flexible haptic robot arm using the flexible pneumatic cylinders with backdrivability was proposed and tested. The performance of the robot arm was investigated. As a result, it was confirmed that the tested robot arm had five motions such as "push," "pull," "lock," "reset," and "free" modes. It was also confirmed that these motions were useful to construct the haptic robot arm.
2. The low-cost and small-sized bilateral attitude control system using the flexible robot arm, the flexible haptic robot arm, two inexpensive embedded controllers, and quasi-servo valves was proposed and tested. The simple bilateral attitude control method without expensive force sensors was proposed and tested. As a result of driving test of bilateral control, it is confirmed that the proposed control system is valid to apply to the remote-controlled rehabilitation device for human wrist with bilateral control.

Acknowledgments Finally, we express thanks that this work was supported in part by the Ministry of Education, Culture, Sports, Science and Technology of Japan through a QOL Innovative Research Program (2012-) and Grant-in-Aid for Scientific Research (C) (Subject No. 24560315).

References

1. Ministry of Internal Affairs and Communications, Statistics Bureau, Director-General for Policy Planning (Statistical Standards) & Statistical Research and Training Institute. Statistical handbook of Japan 2013. http://www.stat.go.jp/english/data/handbook/c0117.htm
2. Ishii, M., Yamamoto, K., Hyodo, K.: Stand-alone wearable power assist suit—development and availability. J. Robot. Mechatron. **17**(5), 575–583 (2005)
3. Piquion, J., Nayar, A., Ghazaryan, A., Papann, R., Klimek, W., Laroia, R.: Robot-assisted gynecological surgery in a community setting. J. Robot. Surg. **3**(2), 61–64 (2009)

4. Noritsugu, T., Takaiwa, M., Sasaki, D.: Development of power assist wear using pneumatic rubber artificial muscles. J. Robot. Mechatron. **21**(5), 607–613 (2009)
5. Kobayashi, H., Shiban, T., Ishida, Y.: Realization of all 7 motions for the upper limb by a muscle suit. J. Robot. Mechatron. **16**, 504–512 (2004)
6. Akagi, T., Dohta, S.: Development of a rodless type flexible pneumatic cylinder and its application. Trans. Robot. Automat. JSME (C) **73**(731), 2108–2114 (2007)
7. Fujikawa, T., Dohta, S., Akagi, T.: Development and attitude control of flexible robot arm with simple structure using flexible pneumatic cylinders. In: Proceedings of 4th Asia international symposium on mechatronics, Singapore, pp. 136–141 (2010)
8. Zhao, F., Dohta, S., Akagi, T.: Development analysis of small-sized quasi-servo valves for flexible bending actuator. Trans. Robot. Automat. JSME (C) **76**(772), 3665–3671 (2010)
9. Aliff, M., Dohta, S., Akagi, T., Li, H.: Development of a simple-structured pneumatic robot arm and its control using low-cost embedded controller. J. Procedia Eng. **41**, 134–142 (2012)

Chapter 31
Integration Method of Proxy and Producer–Consumer Patterns

Jeonghoon Kwak, Jaehak Uam, and Yunsick Sung

Abstract This chapter proposes a novel design pattern that combines proxy and producer–consumer patterns to solve problems of a scenario generator for verifications of sensors in smart space. In addition, how to apply the novel design pattern to scenario generator is introduced. Proxy pattern hides the complexity of the process of generating scenarios and solves the storage problem of a scenario generator. By applying the proposed Proxy-Producer pattern including multiple threads, the generation time of the huge number of scenarios was reduced.

Keywords Internet of things • Scenario generation • Bayesian probability • Design pattern • Proxy pattern • Producer–consumer pattern

31.1 Introduction

One method under research for automatic scenario generation of sensor verification is a process that regenerates scenarios based on sensory values obtained in smart space to reduce scenario generation time [1]. The obtained sensory values are divided into parts of scenarios and then the parts are combined to generate new scenarios, which in turn reduces the costs required to generate scenarios.

The problems of scenario regeneration are as follows: First, the storage problem. If all generated scenarios are stored, owing to a large number of scenarios, storage space shortage can be a problem. Second, it is difficult to predict the number of scenarios that will be generated. If the number of scenarios that are generated is more than the number that can be stored, then some of the scenarios will be lost. Third, there is the time problem associated with generating scenarios by combining parts of scenarios are continuously with each other to generate new scenarios. If a large number of parts of scenarios are utilized, scenario generation will require a long period of time because of the enormous number of combinations involved.

J. Kwak • J. Uam • Y. Sung (✉)
Department of Game Mobile Contents, Keimyung University, Daegu, South Korea
e-mail: jeonghoon@kmu.ac.kr; jaehak@kmu.ac.kr; yunsick@kmu.ac.kr

J. Juang (ed.), *Proceedings of the 3rd International Conference on Intelligent Technologies and Engineering Systems (ICITES2014)*, Lecture Notes in Electrical Engineering 345, DOI 10.1007/978-3-319-17314-6_31

Design patterns [2] can be applied to solve the structure problem of scenario generation. *Design pattern* is commonly utilized and proposed for solving the structure issues of various types of software. By applying the flexible structures of design patterns, the safety and flexibility of scenario generators can be improved.

This chapter proposes a combination method that applies the *Proxy* and *Producer–Consumer* patterns in order to solve the structure problem of scenario generators. The problems of storage according to the number of generated scenarios can be solved by *Producer–Consumer* patterns and then the complexity procedure required to generate scenarios can be hidden by the *Proxy* pattern. In the experiment, the *Producer–Consumer* pattern was utilized including multiple threads, which enabled the process to reduce generation time.

This chapter is organized as follows. Section 31.2 introduces scenario generation-related research and design pattern research. Section 31.3 proposes a scenario generating method applying the *Proxy* and *Producer–Consumer* patterns concurrently. Section 31.4 explains the scenario generation processes according to our proposed method. Finally, we conclude our proposal in Sect. 31.5.

31.2 Related Work

In order to solve the structural problems of complex systems, various design patterns can be applied. This section introduces a scenario generation method and related research that solves structure problems through design patterns.

There is research that applies Bayesian probability to generating scenarios [1]. Gathering sensor data in smart space in order to conduct activity recognition research incurs high costs. Therefore, the Bayesian probability is applied to the pregathered sensor data in order to generate new scenarios. The generated scenarios have probability features that are very similar to those of the measurement sensor data and are utilized for verification.

There is an example of applying aspect-oriented software development method (*Aspect-Oriented Programming*, AOP) and *Design patterns* in order to design the system structure of a secure entrance system [3]. The secure entrance system is a complex system that conducts a variety of functions. This system is designed by utilizing both *Producer–Consumer* and *Bridge* patterns, and also allows a complete segregation of business logic and security logic. In addition, flexible addition and deletion of new functions of a power plant is possible through customization, which does not require changes in the core or basic security modules.

There also exists an example of an application utilizing design patterns in mobile systems based on local-based system (LBS) [4]. Patterns such as *Abstract Factory*, *Prototype*, *Façade*, *Proxy*, *Command*, and *Mediator* patterns are utilized in order to produce a system design. An efficient system development is possible through redesigning by design patterns necessary for mobile systems based on LBS.

As stated above, various design patterns can be applied to solve different system problems. This chapter proposes a method of using integrated design patterns for improving the generation process of the scenario generator.

31.3 Proxy-Producer Pattern Definition

The proposed method solves the previous problems of a scenario generator by combining the *Proxy* and *Producer–Consumer* patterns as shown in Fig. 31.1. Each pattern can be applied separately to a scenario generator; however, unnecessary and redundant classes are invoked. Therefore, a combined pattern is introduced by reducing and removing the redundant and unnecessary classes.

Data is created by *RealProducer* and is delivered to *Client* through *Channel*, *MediationProducer*, and *ProxyProducer*. To describe the process in more detail; first, *RealProducer* generates *Data*. The created *Data* is delivered to *Channel* by calling the member function *put* of the *Channel*. The *Channel* is performed by executing exclusive controls between *RealProducer* and *MediationProducer* in order to maintain the safety and retention of the *Data*. When the *RealProducer* tries to send the *Data* to the *Channel*, if the *Channel* cannot receive the *Data*, the *Channel* waits until it can do so. The *RealProducer* then passes the data to the *Channel*. Therefore, by executing exclusive controls, when *Data* from multiple *RealProducers* is transferred, mutual interference does not occur. *MediationProducer* calls the member function *take* of the *Channel*, receives the *Data* created by the *RealProducer* and then delivers it when the member function *request* from *ProxyProducer*, a subclass of *Producer*, is called. The abstract class *Producer* provides the interfaces to unify the functions of the *RealProducer*, *MediationProducer*, and *ProxyProducer*.

Each class is created as shown in Fig. 31.2. The *Client* creates the *ProxyProducer*. The *ProxyProducer* creates the *MediationProducer* and the *MediationProducer* creates the *Channel* and *RealProducer* to receive Data. The *RealProducer* creates *Data* by calling the member function *run* itself, independent from calling the member function *take* of the *Channel*. The created *Data* is then pushed into the *Channel*.

The *Client* calls the member function *request* of the *ProxyProducer* to obtain the created *Data*. The *ProxyProducer* then calls the member function *request* of the *MediationProducer*. The *MediationProducer* calls the member function *take* of the *Channel* and then the *Channel* returns the created *Data*. When there is no

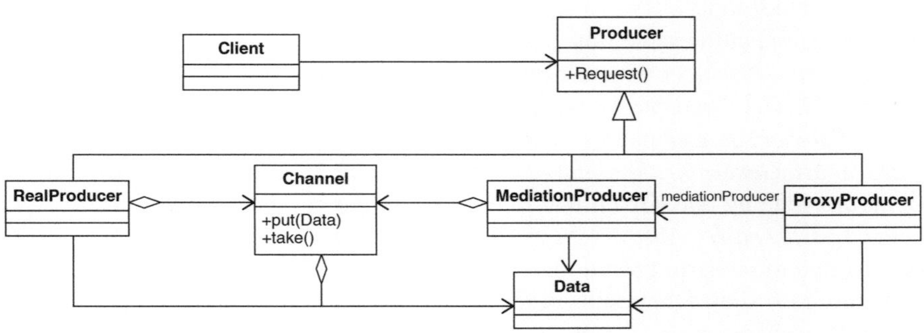

Fig. 31.1 Structure of proxy-producer pattern

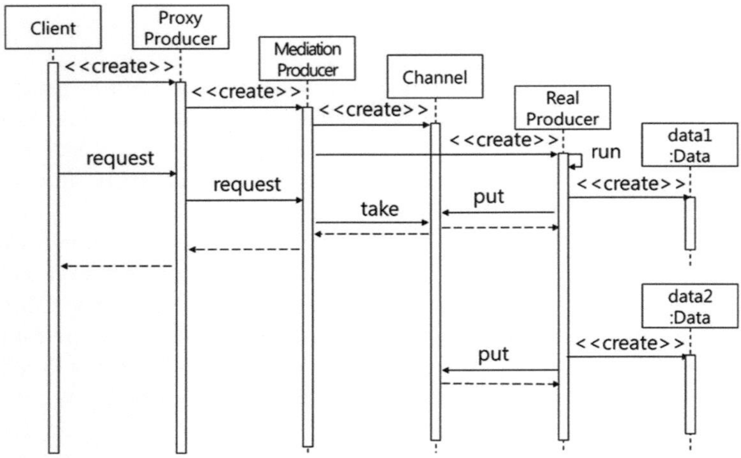

Fig. 31.2 Proxy-producer sequence diagram

created *Data* in the *Channel*, the *ProxyProducer* does not return any *Data* until new *Data* is created.

The advantages of the proposed pattern are as follows: First, during generating scenarios, the required space for scenario storage is reduced given that scenarios are generated when they are required. Second, generated scenarios are obtained by the *Client* without needing to know the complex process of scenario generation. Third, the time required to generate scenarios is reduced according to the number of threads.

31.4 Experiment

The proposed *Proxy-Producer* pattern was applied to a Bayesian probability-based scenario generator [1]. One scenario is composed of a plurality of consecutive sensory value tuple (SVT). SVT is defined by a tuple as sensory values (SV) measured at the same time.

The *Proxy-Produce*r pattern of the scenario generator was composed as shown in Fig. 31.3. The *Producer*, *ProxyProducer*, *MediationProducer*, and *RealProducer* of the *Proxy-Producer* pattern were matched to *AbstractOneScenarioGenerator*, *OneScenarioGenerator*, *MediationOneScenarioGenerator*, and *RealOneScenario-Generator* one by one. Given that the results of generation are scenarios, *Data* is defined as *Scenario*. *AbstractOneScenarioGenerator* contains the common interface related to scenario generation.

Figure 31.4 shows the results of the scenario generator obtained by applying the *Proxy-Producer* pattern. From 18 scenarios, 28,427 scenarios were generated. When only one thread was utilized for *RealOneScenarioGenerator*, it took 7 h

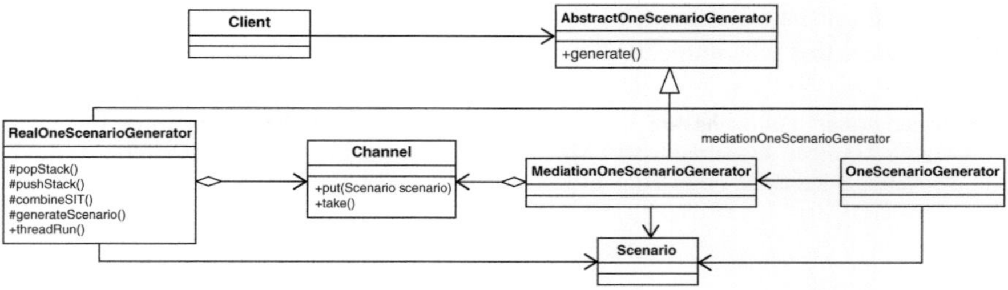

Fig. 31.3 Proxy-producer pattern using scenario generation

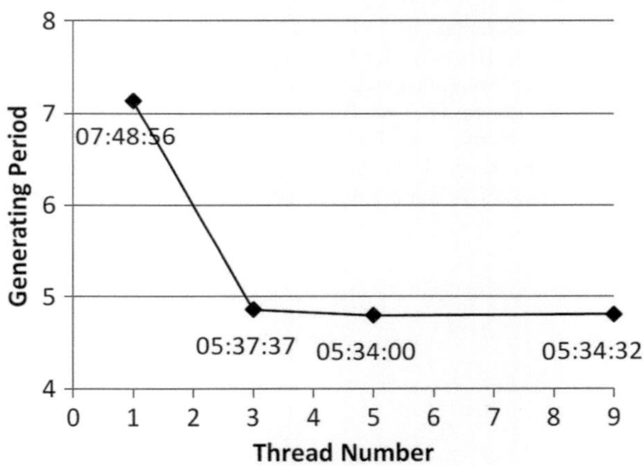

Fig. 31.4 Generating time according to the number of threads

48 min 56 s. In the case of three threads, the elapsed time was 5 h 37 min 37 s. Therefore, as the number of threads increased, when the same number of scenarios was generated, the elapsed time was further reduced. However, by five threads, the elapsed time was sparsely reduced because of the file-saving mechanism. Given that the time required to store scenarios is longer than the time required to generate scenarios, the time needed to generate scenarios is not reduced any further.

31.5 Conclusions

This chapter proposed a *Proxy-Producer* pattern. The structure of a scenario generator was improved by combining the advantages of *Proxy* pattern and that of *Producer–Consumer* pattern. Therefore, the complexity of the scenario generator was reduced more than in the case of each pattern being used independently. The problem of storing scenarios was solved by using the *Producer–Consumer* pattern. The complexity of the generation process of scenarios was overcome. Finally, by capitalizing on the advantages of the *Producer–Consumer* pattern, the generation time for scenarios was reduced.

Acknowledgment Following are results of a study on the "Leaders in Industry–University Cooperation" Project, supported by the Ministry of Education, Science and Technology (MEST) and the National Research Foundation of Korea (NRF).

References

1. Sung, Y., Helal, A., Lee, J.W., Cho, K.: Bayesian-based scenario generation method for human activities. 2013 ACM SIGSIM Conference on Principles of Advanced Discrete Simulation (PADS), Montreal, pp. 147–157, 19–20 May 2013
2. Gamma, E., Helm, R., Johnson, R., Vlissides, J.: Design patterns: elements of reusable object-oriented software. Pearson, Indianapolis (2007)
3. Kim, T., Cheon, H., Lee, H.: Development of secure entrance system using AOP and design pattern. J. Korea Acad. Ind. Coop. Soc. **11**(3), 943–950 (2010)
4. Lee, H.-R., Baek, J.-H., Mun, Y.-C.: Implementation of mobile system based on LBS using design pattern. J. Korean Assoc. Geogr. Inf. Stud. **12**(1), 26–35 (2009)

Chapter 32
Bayesian Probability and User Experience-Based Smart UI Design Method

Junhyuck Son and Yunsick Sung

Abstract This chapter proposes a method that controls User Interfaces by changing properties of UIs after analyzing the patters of events invoked by users. The events and criteria are defined and handled to determine the properties of UIs. In the experiment, our proposed method was applied to an education authoring tool. The changes of UIs after the analysis of user events were validated.

Keywords User interface • User experience • Bayesian probability

32.1 Introduction

The recent trends of functional diversity of smart phones have led to active UI design research to execute various functions within the limitation of small screens. For example, research is being conducted in the production of smartphone UI authoring tool through user needs analysis [1]. The smartphone UI is arranged in provided layout base, which leads to convenience UI arrangement and reduction of development time. However, this premanufactured UI displays versatility for convenient use by multiple users, and therefore does not highlight the UI preferred by a particular set of users.

It is necessary to emphasize frequently used or important UI by users through changing the properties of size, transparency, among other matters. For example, the more frequently used functions will be easier to find if the UI size is larger than other applications. In order to provide a user-customized UI, it is important to provide changes in UI base through the standards of analyzing user needs.

This chapter proposes a method that provides user-customized UIs through changing the UI properties based on an index of determining user needs. The user needs will be estimated through occurred events, calculating the index based on events and applying changes to UI.

J. Son • Y. Sung (✉)
Department of Game Mobile Contents, Keimyung University, Daegu, South Korea
e-mail: junhyuck12@kmu.ac.kr; yunsick@kmu.ac.kr

J. Juang (ed.), *Proceedings of the 3rd International Conference on Intelligent Technologies and Engineering Systems (ICITES2014)*, Lecture Notes in Electrical Engineering 345, DOI 10.1007/978-3-319-17314-6_32

This chapter is organized as follows: In Sect. 32.2, related work is described. Section 32.3 explains the method that analyzes the requirements of users through criteria and the configuration of our system. Section 32.3 proposes the approach. Finally, Sect. 32.4 shows the result of experiments and concludes our approach.

32.2 Related Work

An UI authoring tool for producing smartphone UI research has been conducted through analysis of user needs [1]. The UI layout must first be determined for arranging UI. The initial layout will be set after arranging UI, and the real-time preview of results may be examined, thus enhancing the speed of UI development.

There is also a research based on authoring tool for developing learning system and school applications [2]. The applications provide a base for registering various data including videos, images, and exams. The preinserted answer sheet leads to automatic grading and creation of incorrect answer note. The student users of this application may select registered teachers, and view the information and materials posted by the teacher. Since both the teacher and student interact through smart devices, they are undeterred by time and place.

A research on the utilization of WPF (Windows Presentation Foundation) and XML for designing and creating educational contents authoring tool for smart devices has been examined [3]. This research proposes a method of creating contents through WPF and saving them in XML format. The XML format allows for viewing in all smart devices, and the WPF format allows for resolution customization for each device.

There also exists a research related to a science education courseware development, where both courseware and virtual reality has been converged [4]. This courseware allows students to experience earth science subjects through virtual reality, providing an experience that is difficult in normal situations. Students are able to view planets in various aspects, and learn from the additional information stated for each subject. The students are in a way able to experience planets through interaction, which will ultimately lead to enhancement of academic achievement.

This chapter will focus on methods of providing user-customized UI for displaying various functions in the screen when developing courseware authoring tools, which will evidently lead to learning convenience.

32.3 User-Customized UI System

The smart UI control system proposed in this chapter provides user-customized UIs through the change of the UI properties based on the UI events conducted by a user. First, a user must first invoke one of predefined events. Then the event count e of the invoked event is increased by 1. The event handler decides the indices of criteria affected by the invoked events. In some cases, particular criteria may not be affected by the event. To handle such criteria, the predetermined event constant factor α is applied instead of the event count of the invoked event. The event constant factor α is from 0 or 1. To do that, influence factors are utilized. The influence factors determine whether each event is affected by one of the criteria or not. The influence factors are 0 or 1. Event Handler calculates criteria with event constant factor α, influence factors, and event counts.

$$c_k = \alpha\left(1 - f_{j,k}\right) + \frac{e_j - \gamma}{\beta - \gamma} f_{j,k} = \alpha - \alpha f_{j,k} + \frac{e_j - \gamma}{\beta - \gamma} f_{j,k} \tag{32.1}$$

The jth event count conducted by the user is defined as e_j. β is the maximum value of event counts, and γ is the minimum value. $f_{j,k}$ is the influence factor that determines influence of event count e_j to criterion c_k. If the event count e_j has influence on criterion c_k, a value of 1 is applied. If the event count e_j does not have influence on criterion c_k, a value of 0 is applied. In cases where event count e_j is not applied to criterion c_k, α is applied.

Next, the factor controller calculates UI factors based on the criteria. Each UI factor determines how much the corresponding UI property is affected. Each UI factor is calculated using (32.2). The weight $w_{i,k}$ is a predetermined value, as the degree of influence of criterion c_k on ith UI is different. The weight $w_{i,k}$ is assigned by 0 to 1. $\sum_{k=1}^{n} \sum_{j=1}^{m} w_{i,k}$ should be 1. In cases where there is no influence, the weight $w_{i,k}$ is assigned with a value of 0.

$$\begin{aligned} u_i = \sum_{k=1}^{n}\sum_{j=1}^{m}(w_{i,k}c_k) &= \sum_{k=1}^{n}\sum_{j=1}^{m}\left(w_{i,k}\left(\alpha\left(1 - f_{j,k}\right) + \frac{e_j - \gamma}{\beta - \gamma} f_{j,k}\right)\right) \\ &= \sum_{k=1}^{n}\sum_{j=1}^{m}\left(w_{i,k}\alpha - w_{i,k}\alpha f_{j,k} + w_{i,k} f_{j,k}\frac{e_j - \gamma}{\beta - \gamma}\right) \end{aligned} \tag{32.2}$$

Finally, the UI controller calculates UI properties based on the UI factors. Each property is calculated using (32.3). u_i is the amount of the effect to ith UI and should be adjusted with the predefined ranges. ζ is the maximum and ξ is the minimum of properties of UI provided by UI.

$$p_{i,y} = u_i\left(\zeta_y - \xi_y\right) + \xi_y \tag{32.3}$$

32.4 Experiment and Conclusion

In this experiment, our proposed method was applied to the authoring tool that utilizes motion recognition devices. Then, when events are invoked by users, the effect of UIs was analyzed. There are two types of events for each menu button: one for the event invoked by the button pressing motion from motion recognition devices and the other for mouse click event. Therefore, given that there are six menu buttons as UIs, 12 events are defined. There are Menu, Add, Planet, Orbit,

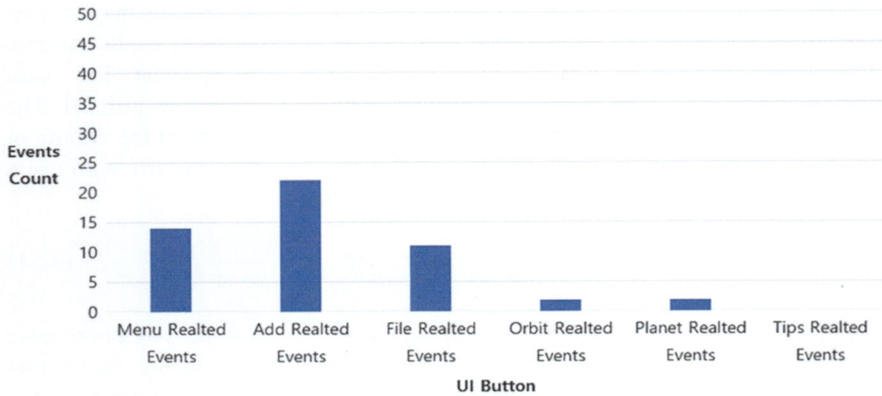

Fig. 32.1 The count of events by UI button clicks

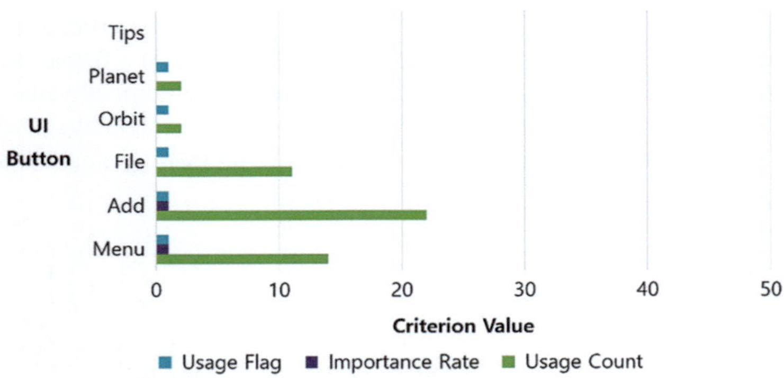

Fig. 32.2 The criteria according to invoked event by UI buttons

File, and Tips as UIs. In the experiment, the number of the invoked events is as shown in Fig. 32.1.

The criteria of each menu button are defined as usage count, importance rate, including function value, usage flag, and depth degree. The Including function value and depth degree are assigned in advance. When any event of two events has happened, the usage count is increased by 1. The usage count affects the importance rate and usage flag of the corresponding UIs. The importance rate and usage flag also affects other UIs that have the same depth degree. The criteria were changed according to the events of UI buttons as shown in Fig. 32.2. The sum of the criteria of each UI affects the UI Factors of the corresponding UI as shown in Fig. 32.3. Each menu button has two properties: scale and opacity. The changed properties based on the criteria are shown in Fig. 32.4.

In addition, the change of the UI is as shown in Fig. 32.5. The brightness of the menus, "Add," and "Planet," invoked more times than others was increased.

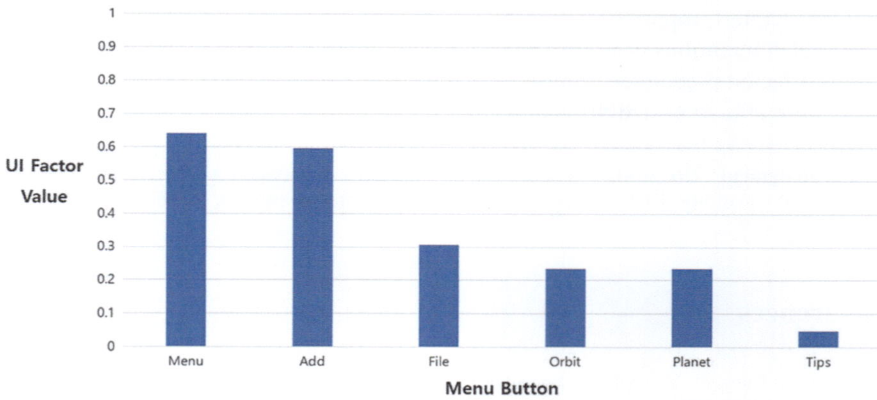

Fig. 32.3 The sum of the criteria of each UI affects the UI factors of the corresponding UI

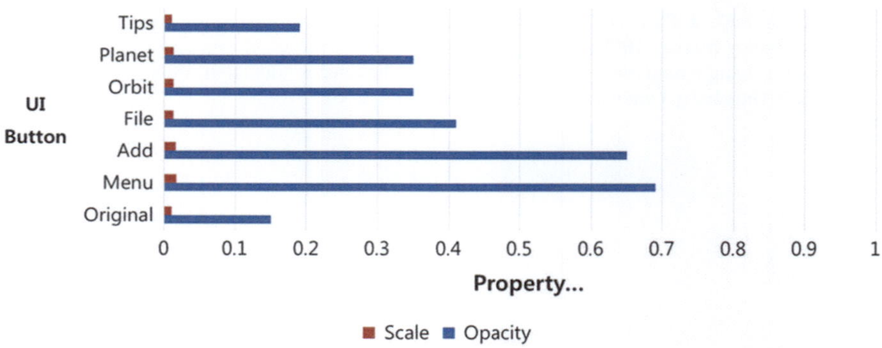

Fig. 32.4 The changes of properties according to criteria

Fig. 32.5 The comparison according to the change of properties

This chapter proposed a UI control method by changing the properties of UIs. The hidden intentions of users were predicted through invoked UI events. Invoked events affected predefined criteria, which were also applied to UI properties. Thus, users were able to utilize the changed UI.

Acknowledgment This work was supported by the Korea Foundation for the Advancement of Science and Creativity (KOFAC), and funded by the Korean Government (MOE).

References

1. Lee, J.-S., Rhee, D.-W.: A UI authoring tool user needs analysis through mobile game for smart devices. J. Korea Game Soc. **13**(1), 5–18 (2012)
2. Kwon, S.-O., Kim, J.-O., Ju, S.-Y., Jeong, J.-S., Yoo, K.-H.: Design and implementation of educational content authoring tool for smart devices. J. Korea Contents Assoc. **13**(12), 1–8 (2013)
3. Kim, E.-S., Park, J.-S.: A study on educational app development using the app authoring tool. J. Digit. Policy Manag. **10**(5), 1–6 (2012)
4. Suk, J.-Y.: Using virtual reality, the development of the science education courseware. Sungshin Women's University Graduate School of Education (2006)

Chapter 33
Bayesian Probability-Based Hand Property Control Method

Phil Young Kim, Ji Won Kim, and Yunsick Sung

Abstract Deficiencies in low-priced motion recognition devices lead to diverse kinds of errors in recognizing palms and hands. To utilize lower-priced devices better, the recognition rate of the properties of hands should be improved. This chapter proposes a method that revises recognition errors in properties of hands. By calculating the Bayesian probability of the directions of a recognized palm, the directions were revised.

Keywords Leap motion • Motion recognition device • Bayesian probability

33.1 Introduction

Low-priced motion recognition devices are being released commercially due to which various approaches can be taken to develop applications using them. For example, there are studies on using hand motion recognition devices to detect hand motions [1] and to input Korean words with Unistroke software [2]. Given that the Leap motion-based content makes difficult subjects easier to grasp, the experience is more effective for students trying to learn difficult subjects. Teachers can generate contents more easily and students can study difficult subjects more quickly if education-authoring tools with Leap motions are provided.

However, one such motion recognition device, Leap Motion [3], features a reduced number of sensors as a tradeoff for a reduction in its production cost. For example, there are only two infrared sensors to read a user's hands. Therefore, when the hands are far away from the Leap Motion, its recognition rate for those hands is decreased significantly. In addition, each user has differently sized hands, with regards to thickness and length of fingers. Because of these problems, Leap Motion sometimes recognizes a right hand as a left hand or vice versa, and interprets the direction of palms incorrectly. Therefore, the recognition rate of hands and palms should be improved.

P.Y. Kim • J.W. Kim • Y. Sung (✉)
Department of Game Mobile Contents, Keimyung University, Daegu, South Korea
e-mail: kimpy1111@kmu.ac.kr; jiwon1219@kmu.ac.kr; yunsick@kmu.ac.kr

J. Juang (ed.), *Proceedings of the 3rd International Conference on Intelligent Technologies and Engineering Systems (ICITES2014)*, Lecture Notes in Electrical Engineering 345, DOI 10.1007/978-3-319-17314-6_33

This chapter proposes a method based on Bayesian probability that revises the recognition errors of properties of hands. When users utilize Leap Motions, their experiences are recorded and then utilized to calculate Bayesian probability. When any properties of hands are recognized incorrectly, the errors are revised based on the Bayesian probability.

This chapter is organized as follows. Section 33.2 explains related work. Section 33.3 describes a process that revises recognition errors. In Sect. 33.4, the experiment that validated the revising processes is described and evaluated. Finally, Sect. 33.5 concludes and summarizes.

33.2 Related Work

There are diverse methods and approaches to teaching subjects that are difficult to learn through experience. This chapter introduces authoring tools to be used in the construction of coursewares.

Research has been conducted on earth science education coursewares with virtual reality applications [4]. Students gain experience with the subject of planetary science by experiencing it in virtual reality. Teachers can generate virtual solar systems, and students study the generated systems and the information that their teachers entered. Planets can be observed from various angles and controlled freely.

There is also research into the development of apps for school lectures, which pertain to lecture management applications that convey lecture content to students using smart devices [5]. Teachers upload lecture content using smart devices rather than just showing movies and pictures. Students study the content that their teachers upload and can pose questions to teachers through the apps.

There is also research into the development of educational content authoring tools using smart devices [6]. This research provides the environment to create learning content using interfaces such as Excel. Teachers generate learning content based on Windows Presentation Foundation (WPF). The generated content is stored after being converted to XML. The XML-based content can be used on any smart device.

To improve the effectiveness of learning in virtual reality, the utilization of motion recognition devices is required. To utilize low-priced motion recognition devices, it is necessary to increase the recognition rate of hands. This chapter proposes a revision method for hand recognition errors.

33.3 Recognition Error Detection and Revision

Our proposed method revises the hand recognition errors. The properties of recognized hands are acquired by the Leap Motion Library provided. Hand errors are revised as follows. First, if the recognition errors of right and left hands have occurred and are detected by Leap Motion, the recognition errors are revised. To provide information on the recognized hands to the Leap Motion Library, the Leap Motion provides Leftmost, Rightmost, and Frontmost fields as shown in Fig. 33.1.

Leftmost returns the recognized hand furthest to the left. Rightmost returns the recognized hand furthest to the right. However, the furthest left hand can be a person's left hand as shown in Fig. 33.2a or their right hand as shown in Fig. 33.2b. Because the recognition is carried out with only a few sensors, recognition errors are frequent.

In the proposed method, we assume that the situation as shown in Fig. 33.2b is not the case to increase recognition rates. Therefore, the leftmost hand always is treated as a left hand and the rightmost hand is treated as a right hand, as shown in Table 33.1. Each hand recognized at time t is defined as h_t.

Second, whether recognition errors of any properties of a hand have occurred or not is considered. For example, Leap Motion defines palm direction as $\pm\pi$, 0 indicates downside, "$-$"indicates left direction, and "+" indicates right direction. The proposed method distinguishes whether errors have occurred based on a Bayesian probability calculated by measured directions of palms. However, given that every direction expressed as from $-\pi$ to $+\pi$, the probability of each direction is low because of the diversity of the directions. Therefore, every measured direction is divided by γ and multiplied by γ to increase probability as shown in (33.1).

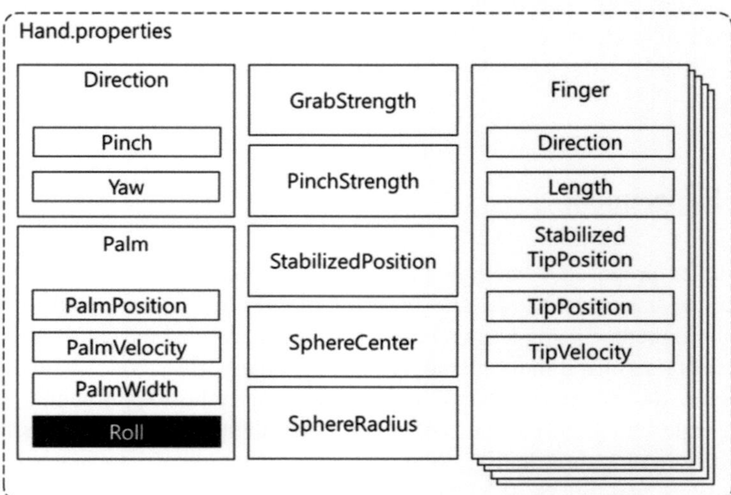

Fig. 33.1 Hand properties

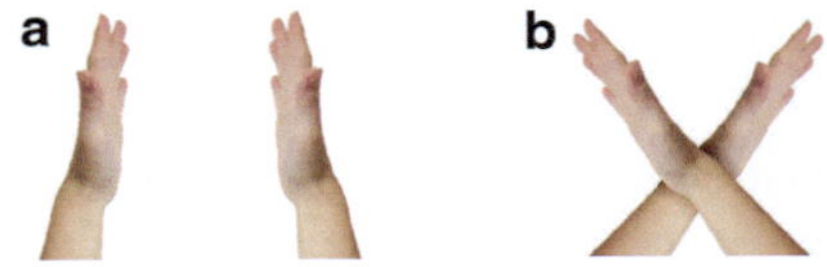

Fig. 33.2 Recognized hands in leap motion. (**a**) Straightforward direction of hands (**b**) Cross direction of hand recognition

Table 33.1 Revised hands

	Recognized hands	Revised hands
Leftmost	Left hand/right hand	LEFT
Rightmost	Left hand/right hand	RIGHT

$$h_t' = \lfloor \frac{h_t}{\gamma} \rfloor \times \gamma \tag{33.1}$$

Bayesian probability is calculated as shown in (33.2).

$$P\left(h_t'\right) = P\left(h_t' | h_{t-1}', h_{t-2}', \ldots\right) \tag{33.2}$$

If the probability $P(h_t')$' of a recognized hand is less than α, we assume that errors in recognition of that hand's direction have occurred.

Third, whether errors will be revised or not is determined. Even though errors of recognized hand directions may have occurred, the errors cannot be fixed if correct directions cannot be determined. The revision of hand errors is processed if $\max P\left(h' | h_{t-1}', h_{t-2}', \ldots\right)$ is greater than β. If not, no revision will occur.

Finally, the error hand is revised. The recognized hand h_t is allocated by h_t' adding half of γ. However, in the case when $\max P\left(h' | h_{t-1}', h_{t-2}', \ldots\right)$ is 0, h_{t-1} is utilized.

33.4 Experiment

This chapter proposed a method that recognizes and revises the errors of the properties of hands. Universe courseware authoring tool was utilized to validate the proposed method using Leap motion. Teachers generated the courseware-based virtual reality and hand motion recognition by the authoring tool. The direction of palms is one of properties of hands and was utilized to this experiment.

In the experiment, the constants of the proposed method were defined as follows. The direction of palms had the range from $-\pi$ to $+\pi$. To increase Bayesian probability, γ was set by 0.3. By setting α by 0.01, errors were arisen, when Bayesian probability was equal to and smaller than α (Fig. 33.3).

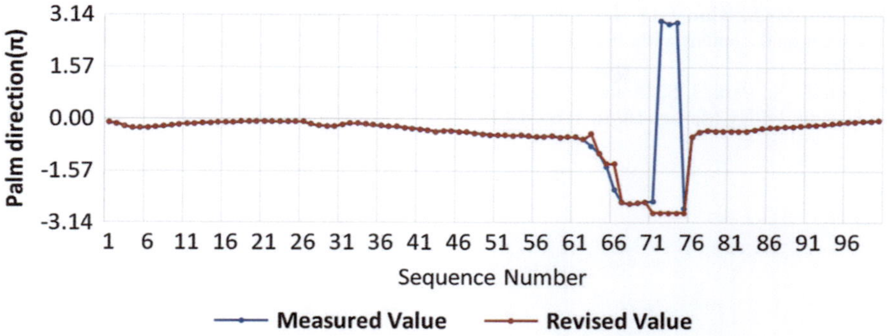

Fig. 33.3 The comparison between measured values and revised values

In the proposed method, the direction of a palm was revised as follows. −2.487496 measured at 71th of the measured values was changed to 2.981999 at 72th. The Bayesian probability at 72th was 0, which was smaller than α. Therefore, the average of the two maximum at 71th, −2.84, was assigned to the value at 72th instead of 2.981999.

33.5 Conclusions

In this chapter, Bayesian probability was calculated and applied to revise recognition errors in low-priced motion recognition devices. By utilizing the functions that are provided by Leap Motion, the direction values of hands were acquired. By analyzing the acquired values, whether the probability of the acquired value is less than α is determined. If the probability was erroneous, the value of the probability was substituted with the value of the maximum Bayesian probability.

Acknowledgments Following are results of a study on the "Leaders in Industry–University Cooperation" Project, supported by the Ministry of Education, Science, and Technology (MEST) and the National Research Foundation of Korea (NRF).

References

1. Nam. J.: A study on using leap motion and motion detection for interactive media projects. Dongguk University Graduate School of Digital Image and Contents (2014)
2. Kang, N.-H., Soe, J., Chae, S., Kim, D.-C., Han, T.-D.: LeapGraffiti: unistroke Korean character input method using leap motion. J. HCI Soc. Korea. (2014)
3. LeapMotion. http://leapmotion.com
4. Suk, J.-Y.: Using virtual reality the development of the science education courseware. Sungshin Women's University Graduate School of Education (2006)

5. Kwon, S.-O., Kim, J.-O., Ju, S.-Y., Jeong, J.-S., Yoo, K.-H.: Design and implementation of educational content authoring tool for smart devices. J. Korea Contents Assoc. **13**(12), 1–8 (2013)
6. Kim, E.-S., Park, J.-S.: A study on educational app development using the app authoring tool. J. Digit. Policy Manag. **10**(5), 1–6 (2012)

Chapter 34
Implementation of a Delta-Sigma Analog-to-Digital Converter

Chin-Fa Hsieh, Tsung-Han Tsai, Chun-Sheng Chen, and Yu-Hao Hsieh

Abstract The sigma–delta analog-to-digital converter (ADC) has less consumption of circuit power and can achieve higher resolution. In this chapter, a sigma–delta ADC which contains a second-order sigma-delta modulator is presented. The modulator architecture is first designed by using the behavioral simulation of MATLAB, and then the TSMC 0.18 μm single-poly six-metal process. Layout of each analog block has been shown. Simulation results show that, with an input of a −6 dB 1 kHz sine, the delta-sigma ADC can achieve an SNR of 87.2 dB. The core size is 0.6456 mm × 0.3340 mm. With a 16-bit resolution, it is suitable for audio applications.

Keywords ADC • Sigma–delta • Modulation

34.1 Introduction

Analog-to-digital converters (ADC) play a very important role in electronic systems. A direct-conversion ADC (flash ADC) converts an analog signal into a digital signal in parallel. But, it results in large die area, high input capacitance, and high-power dissipation. It is often used for video, wideband communications, or other fast signals processing device [1]. A Successive-Approximation-Register ADC (SAR ADC) uses a successive comparison operation [2] to successively narrow a range that contains the input voltage step-by-step to save power consumption and area, but its conversion rate is very slow. The Delta-Sigma (ΔΣ) modulators have the features of oversampling and noise-shaping. These can reduce the noise and move the in-band quantization error out of the band. It is widely used in the digital audio application. Because of these advantages, ΔΣ ADC is mainly used in the application in audio, measuring instruments, and so on [3–5].

C.-F. Hsieh (✉) • T.-H. Tsai
Department of Electrical Engineering, National Central University, Chung-Li 32001, Taiwan
e-mail: c0935@cc.cust.edu.tw

C.-S. Chen • Y.-H. Hsieh
Department of Electronic Engineering, China University of Science and Technology, Taipei 11581, Taiwan

J. Juang (ed.), *Proceedings of the 3rd International Conference on Intelligent Technologies and Engineering Systems (ICITES2014)*, Lecture Notes in Electrical Engineering 345, DOI 10.1007/978-3-319-17314-6_34

In this chapter, the implementation of a second-order ΔΣ modulator has been presented. This modulator is designed using 0.18 μm CMOS (Complementary Metal-Oxide-Semiconductor) technology. It is organized as follows. Section 34.2 describes the principal of ΔΣ ADC. The circuit is presented in Sect. 34.3. Simulation results and layout are presented in Sect. 34.4. Finally in Sect. 34.5, a conclusion of this article is presented.

34.2 Principal of ΔΣ

The ΔΣ ADC convert a continuous analog signal into a high-speed discrete digital signal with high frequency noise. After noise filtering and down sampling, these digital signals can be processed in a digital system. The diagram of the second-order ΔΣ modulator which mainly consists of integrators and a quantizer is shown in Fig. 34.1. It includes two integrators. The modulator processes the signal generated by sample and hold circuit. The feature of oversampling technique used in ΔΣ ADC can enhance the resolution.

34.3 Circuit Design

The architecture of the ADC can be designed as shown in Fig. 34.2. The main component is the operational amplifier, the common-mode feedback (CMFB) module, the quantizer module, and switches. The function of each part is described as follows:

The operational amplifier is composed of the main amplifier, CMFB, and bias circuit as shown in Fig. 34.3. The fully differential operational amplifier is applied for the input signal because it has a better performance than the single-end operational amplifier. To minimize the flicker noise, a PMOS input pair is used. The CMFB can stabilize at the common-mode level to avoid the signal distortion. The bias circuit provides bias voltage for the amplifier.

The quantizer is composed of comparator, dynamic latch, and static latch as shown in Fig. 34.4. The regenerative comparator is used. The RS latch circuit is to maintain the previous state and to result in a correct output.

34.4 Simulation Result and Layout

The circuit is verified by the simulations of a HSPICE. The frequency response of OP is shown in Fig. 34.5 and Table 34.1. The simulation result shows that the dc gain, unity-gain bandwidth, and phase margin are 120 dB, 60.5 MHz and 81, respectively. The simulation result of proposed ΔΣ ADC is shown in Fig. 34.6 and

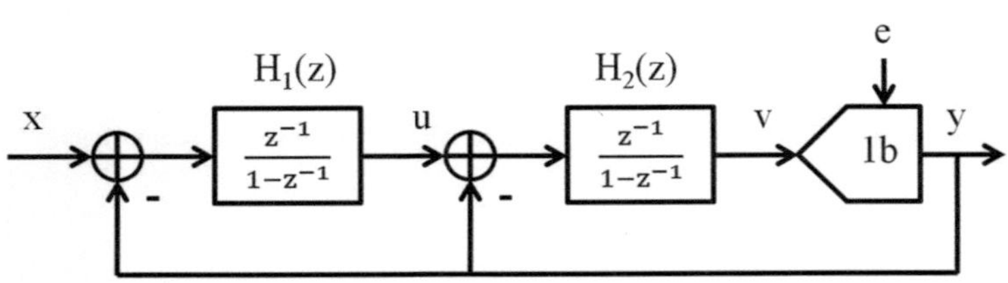

Fig. 34.1 Diagram of the ΔΣ modulator

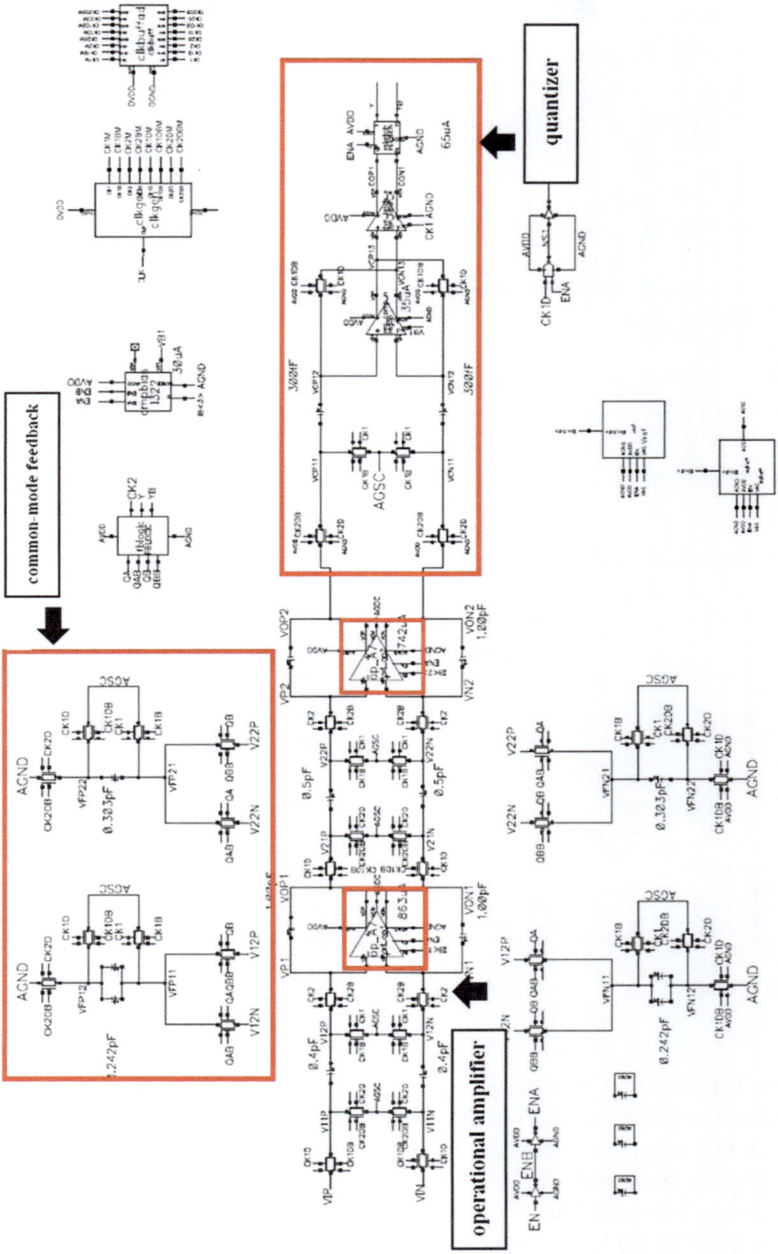

Fig. 34.2 Architecture of the ADC

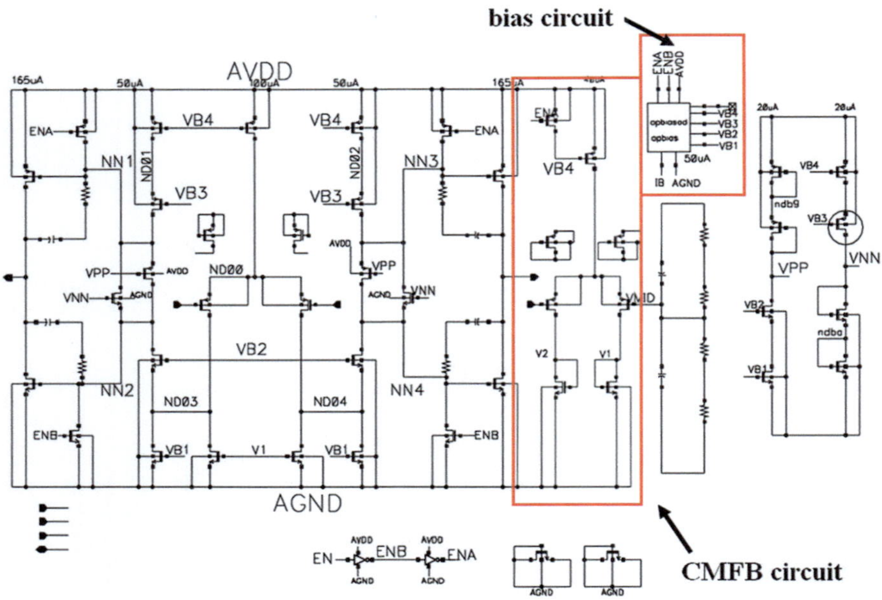

Fig. 34.3 Operational amplifier

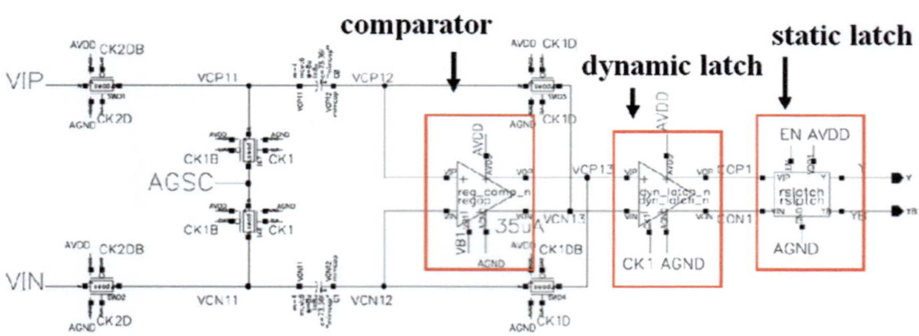

Fig. 34.4 Quantizer

Table 34.2. It indicates that the proposed modulator with input level −6 dB and −20 dB have an SNR of 87.2376 dB and 71.7615 dB, respectively. The effective number of bits is approximately 16. Figure 34.7 presents the whole chip layout. In order to verify the result, the LVS verification is applied. The schematics and layouts are in full compliance. The core size is 0.6456 mm × 0.3340 mm. The size included IO PAD is 0.860 mm × 0.733 mm.

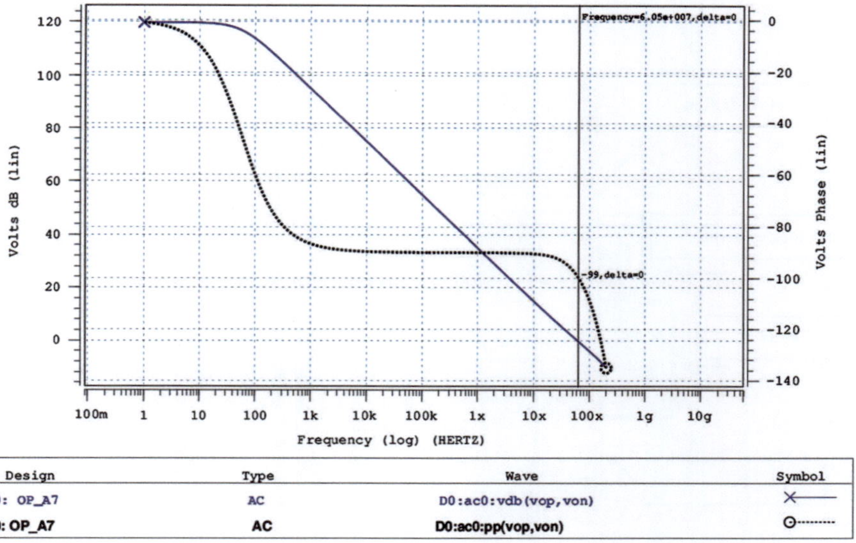

Fig. 34.5 Simulation result of OP

Table 34.1 Simulation results of OP

Parameter	Simulation result
Voltage	1.65 V
Gain	120 dB
UGBW	60.5 MHz
Phase margin	81
Load	4 pF

Fig. 34.6 Simulation result of ADC

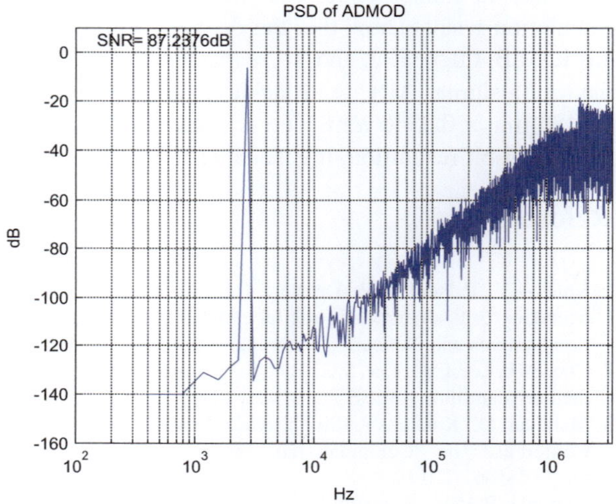

Table 34.2 Simulation results of modulator

Input level (dB)	SNR (dB)
−6	87.2376
−20	71.7615

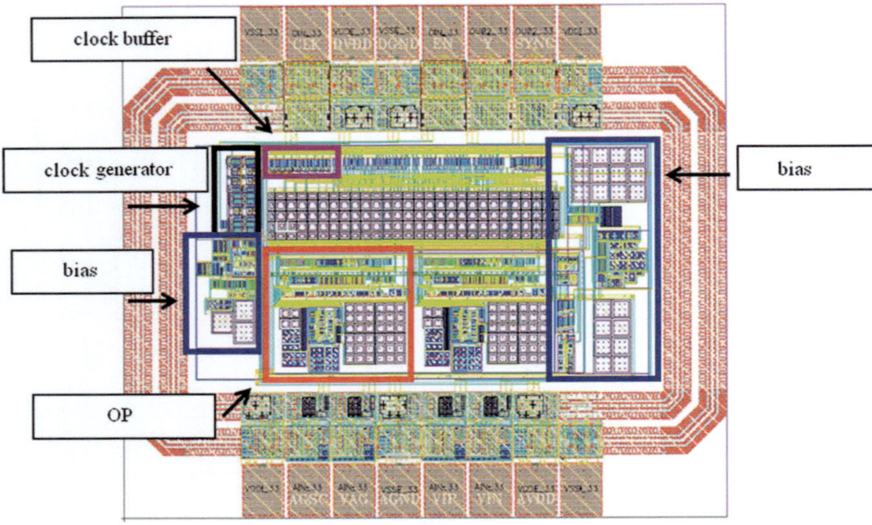

Fig. 34.7 Layout of ADC

34.5 Conclusions

In this chapter, a second-order ΔΣ modulator designed in TSMC 0.18 μm single-poly six-metal CMOS process is presented. The fully differential operational amplifier is applied to get a better performance. A PMOS input pair is used to minimize the flicker noise. The circuit is verified by the simulations of a HSPICE. Simulation results show that the dc gain, unity-gain bandwidth and phase margin are 120 dB, 60.5 MHz, and 81, respectively. With an input of a −6 dB 1 KHz sine, the delta-sigma ADC can achieve an SNR of 87.2 dB. The core size is 0.6456 mm × 0.3340 mm. The size-included IO PAD is 0.860 mm × 0.733 mm. With a 16-bit resolution, it is suitable for audio applications.

References

1. Kale, A.V., Palsodkar, P., Dakhole, P.K.: Comparative analysis of 6 bit thermometer-to-binary decoders for flash analog-to-digital converter: 2012 International Conference on Communication Systems and Network Technologies (CSNT), pp. 543–546. (2012)
2. Cho, S.H., Lee, C.K., Lee, S.G., Ryu, S.T.: A two-channel asynchronous SAR ADC with metastable-then-set algorithm. J. IEEE Trans. Very Large Scale Integr. Syst. **20**(4), 765–769 (2012)
3. Mahajan, D., Kakkar, V., Singh, A.K.: Analysis of delta sigma modulator: 2011 International Conference on Computational Intelligence and Communication Networks (CICN), pp. 182–186. (2011)
4. Koppula, R.M.R., Balagopal, S., Saxena, V.: Multi-bit continuous-time delta-sigma modulator for audio application: 2012 I.E. Workshop on Microelectronics and Electron Devices (WMED), pp. 1–5. (2012)
5. Kim, J.S., Kwon, T.I., Ahn, G.C., Kim, Y.G., Kwon, J.K.: A ΔΣ ADC using 4-bit SAR type quantizer for audio applications: 2011 I.E. International SoC Design Conference (ISOCC), pp. 73–75. (2011)

Chapter 35
An Experimental Investigation of Effect on Engine Performance by Controlling the Temperature of the Fuel

Ming-Hsien Hsueh

Abstract The advantages of controlling the fuel temperature in the engine system are presented in this chapter. By using the thermoelectric material, engine fuel can be cooled or heated to control the temperature of the inlet mixture. In this investigation, the preheating time of the engine can be reduced by heating the fuel. The fuel consumption, the concentration of oxides of nitrogen can be decreased by cooling the fuel. At the same time, the power of the engine also can be increased by cooling the fuel.

Keywords Fuel temperature • Thermoelectric material • Oxides of nitrogen • Power

35.1 Introduction

For more than a century, internal combustion engines have played an important role in the transportation sectors. Although the combustion engines can provide more power for vehicles to be driven faster, the reduction of petroleum resources and environmental problems are leading to an increasing the efficiency of the combustion. Kose and Ciniviz [1] discussed the results of engine torque, power, exhaust gases temperature by providing the additional hydrogen as an additional fuel in the diesel engine. Fang et al. [2] and Zheng and Mulenga [3] presented the effect of ethanol or biodiesel fuel on the combustion and emissions in premixed low-temperature combustion. Okude et al. [4] showed the results of the basic characteristics of premixed compression ignition combustion by using the single cylinder engine. Park et al. [5] investigated the fuel properties, droplet atomization, combustion performance, and exhaust emission characteristics of gasoline–diesel-blended fuels in diesel engine. Yun et al. [6–9] discussed the low-temperature

M.-H. Hsueh (✉)
Department of Industrial Engineering and Management, National Kaohsiung University of Applied Sciences, Kaohsiung 807, Taiwan
e-mail: mhhsueh@kuas.edu.tw

J. Juang (ed.), *Proceedings of the 3rd International Conference on Intelligent Technologies and Engineering Systems (ICITES2014)*, Lecture Notes in Electrical Engineering 345, DOI 10.1007/978-3-319-17314-6_35

combustion for the simultaneous reduction of NO_x and soot emissions. Dardiotis et al. [10] presented the gaseous emissions of gasoline and diesel vehicles measured in cold-start low-temperature test to generate data to contribute to the ongoing discussion on the revision of the low-temperature emission standards for Light-Duty vehicles. Weilenmann et al. [11] investigated the influence of ambient temperature on cold-start excess emission of both gasoline and diesel engine. Sajjad [12] discussed the temperature contour and the relationship between the equivalent fuel particles with the fuel injection pressure on diesel engine performance and emissions.

In order to reduce the more consumption of the fuel and air pollution by the internal combustion engines, the thermoelectric materials are applied in this chapter to control the temperature of fuel in the combustion engine. By using the device, the analysis of the fuel consumption, effective horsepower, gross torque, and exhaust emission can be found in this research. The comparison of the device either operated or not in the above performances will be discussed in the research to advantage the device in the internal combustion engine.

35.2 Thermoelectric Technology

Thermoelectric technology is widely applied in heat removal and temperature control of scientific research, military affairs, semiconductor industry, and so on. A thermoelectric material consists of equal number of P-type and N-type elements connected in series and parallel sandwiched in between two ceramic plates as shown in Fig. 35.1. When the electric power is transported across two metallization junctions (from P to N or N to P), the heat will be absorbed into the cold junction and dissipated from the hot junction, i.e., Peltier effect. The temperature of the cold junction and hot junction are defined as T_{cj} and T_{hj} respectively. The temperature of

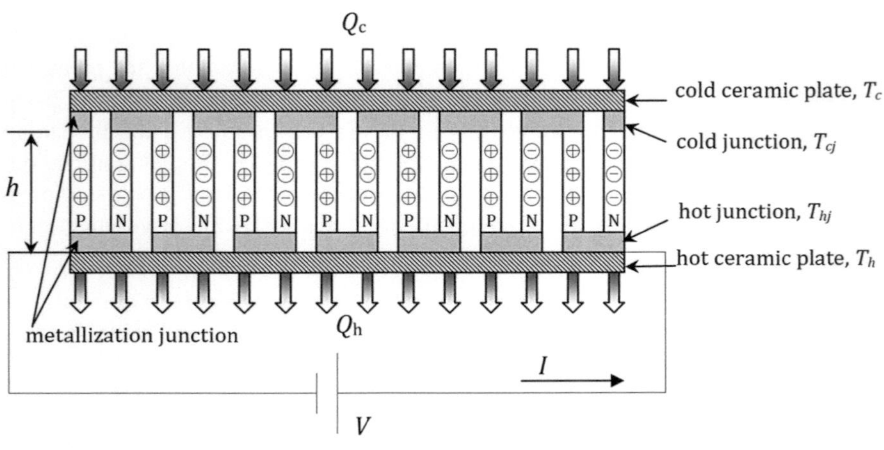

Fig. 35.1 Scheme of thermoelectric material construction

the cold ceramic plate connected cold junctions is defined as T_c, and the hot ceramic plate-connected hot junction is T_h.

The thermoelectric performance of the TEC is determined by three properties: the electrical resistivity (ρ), the Seebeck coefficient (α), and the thermal conductivity (k). These properties of the P and N arms are generally arbitrary functions of temperature. Because these parameters change slightly in the working temperature range for the commercial TEC chips, we assume that the three properties are independent of temperature in this chapter. The three parameters of TEC couple can be given as follows:

$$K = 2k \cdot \frac{S}{h} \tag{35.1}$$

$$R = 2\rho \frac{h}{S} \tag{35.2}$$

$$A = 2\alpha \tag{35.3}$$

where K, R, and A are represented as the thermal conductance, electrical resistance, and Seebeck coefficient of the TEC couple respectively, S and h are the cross-sectional area and the length of P- or N-type arms. The heat balance cooled or expelled by each TEC couple leads to the following equations:

$$Q_c = (T_c - T_{cj})K_c \tag{35.4}$$

$$Q_c = A \cdot I \cdot T_{cj} - \left[K(T_{hj} - T_{cj}) + \frac{1}{2}I^2R\right] \tag{35.5}$$

$$Q_h = A \cdot I \cdot T_{hj} - \left[K(T_{hj} - T_{cj}) - \frac{1}{2}I^2R\right] \tag{35.6}$$

$$Q_h = (T_{hj} - T_h)K_h \tag{35.7}$$

where Q_c represents the heat flow through the cold plate, Q_h the heat flow through the hot plate, K_c the thermal conductance of the cold ceramic plate, K_h the thermal conductance of the hot ceramic plate, and I the electric current that flows through TEC element. We assume that the thickness of the cold or hot ceramic plate is very thin, so the thermal conductance of the cold or hot ceramic plate can be limited to $K_c \to \infty$, $K_h \to \infty$. After being eliminated the T_{cj} and T_{hj} from (35.4) to (35.7) and, the heat balance equations of the TEC couple can be calculated by the following equations:

$$Q_c = A \cdot I \cdot T_c - \left[K(T_h - T_c) + \frac{1}{2}I^2R\right] \tag{35.8}$$

$$Q_h = A \cdot I \cdot T_h - \left[K(T_h - T_c) - \frac{1}{2}I^2R\right] \tag{35.9}$$

The voltage of the TEC chip can be given as:

$$V = I \cdot R + A \cdot (T_h - T_c) \tag{35.10}$$

The electric energy supplied into the TEC can be calculated as:

$$P = Q_h - Q_c = I^2 \cdot R + A \cdot I \cdot (T_h - T_c) \tag{35.11}$$

35.3 The Method of the Fuel Temperature Control Device

In the chapter, the thermoelectric materials are applied in the engine's fuel system which is illustrated in Fig. 35.2. When an electric current runs through the thermoelectric material, one surface of the thermoelectric material absorbs the heat energy and the other produces the heat energy. By applying this function, fuel can be cooled down together or warmed up together by the thermoelectric material. The purposes of the device applied in the engine are described in the following reasons:

1. During the first 2 min after starting the engine of a vehicle that has not been operated for several hours, the amount of emissions can be very high. Many of the motor parts inside the engine do not fit and work together well until they are at or near normal operating temperatures. Because the engine or the motor oil has the working performance after reaching the operating temperature, the driver needs to warm up the engine for a few minutes in order to preheat the parts of the engine. Even the catalytic converter needs the warm-up period to the working temperature and converts toxic pollutants in exhaust gas to less toxic pollutants. By the device, the fuel is heated by the thermoelectric material and increases the

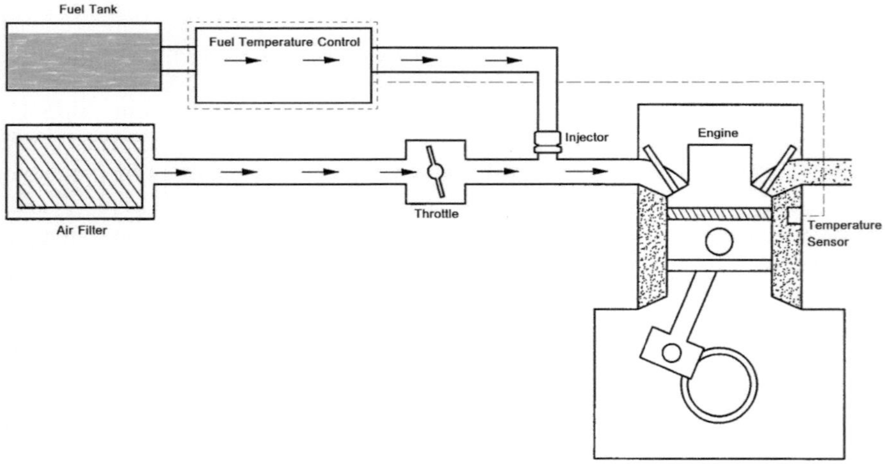

Fig. 35.2 Location of the fuel temperature device in an engine

temperature of the mixture. The prewarmed air can warm the engine's internal parts and the block, so the engine can take less time for the block and positions to expand to their optimum fit. Also, the warm-up time can be reduced which results in the temperature of the catalytic converter go up to the operating degree very quickly and decrease the amount of the toxic pollutants.

2. During the combustion in the engine, especially at high temperatures, the nitric oxide and nitrogen dioxide (NO_x) are produced from the reaction of nitrogen and oxygen gases. NO_x reacts with ammonia, moisture, and other compounds to form nitric acid vapor and related particles. Small particles can penetrate deeply into sensitive lung tissue and damage it, causing premature death in extreme cases. Inhalation of such particles may cause or worsen respiratory diseases, such as emphysema or bronchitis, or may also aggravate existing heart disease. By using this device, the fuel can be cooled by the thermoelectric material and causes the cooled mixture air to absorb the heat energy from the combustion chamber and lowers the adiabatic flame temperature. Therefore, NO_x can be reduced for cooling the fuel by the thermoelectric material, and makes exhaust gas to meet emission standards.
3. According to the physical property of thermal expansion and contraction, the more decrease in intake air temperature is, the more increase in intake air density produces. By increasing the intake air-charge density through isobaric (constant pressure) cooling can improve the volumetric efficiency in the internal combustion, the fuel can be cooled down by using the thermoelectric material and reduce the temperature of the mixture air into an engine's intake manifold and combustion chamber. The cooled mixture air increases the intake air-charge density through the thermoelectric material's cooling can promote more thorough combustion efficiency. The lowering of the intake charge air temperature also eliminates the danger of predetonation (knock) of the fuel/air charge prior to spark ignition. This preserves the benefits of more fuel/air burn per engine cycle, increasing the output of the engine.

35.4 The Result

35.4.1 The Design of the Fuel Temperature Device

The fuel temperature control device by using the thermoelectric material is illustrated in Fig. 35.3, which is mainly including the fuel temperature control block, the thermoelectric material, the water fan, radiator, and cooling fans. The inner side surface of the thermoelectric material is connected with the side surface of fuel temperature control block and the outer side surface is joined to the water block. The inlet fuel connector is joined to the fuel tank or fuel pump, and the outlet fuel connector is joined to the fuel pipe or fuel injection nozzle. The cooling fan is connected with the radiator by the fan bolts as shown in Fig. 35.3.

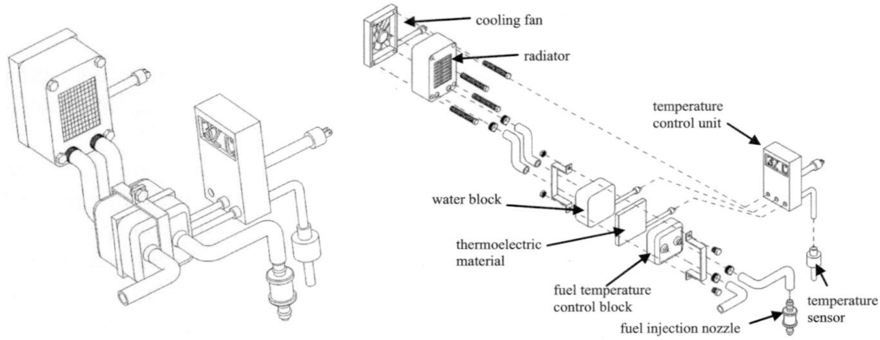

Fig. 35.3 The perspective view of the fuel temperature control device

35.4.2 The Experiment Results

Figure 35.4 shows the variation in the warm-up time of the engine before and after being applied the thermoelectric module chips to the fuel and intake air. Before warming the fuel and intake air by the thermoelectric module chips, it took 355 s to reach to the operating temperature of the engine (approximately 310 °C). By proheating the energy exchange manifold, the fuel and intake were warmed up and it took 165 s to reach to the operating temperature. The warm-up time of the engine heated by the device was smaller than that which did not be heated by the TEC module about 55 % at least. It can avoid the fuel vaporize incompletely and creating higher emissions of hydrocarbons, nitrogen oxides, and carbon monoxide. Figure 35.5 shows the variation in the concentration of oxides of nitrogen (NO_x) emissions in exhaust gas by using the thermoelectric module chips or not. From the results it is observed that the concentration of oxides of nitrogen is reduced substantially by the device for cooling the fuel and intake air. There is 76 % reduction of the oxides of nitrogen emissions than the original engine exhausted by cooling fuel and intake air at high engine speed which is shown that the cooled mixture air can massively absorb the heat from the combustion.

35.5 The Discussion

In this chapter, we invent a new method to absorb the heat from the fuel to increase the volumetric efficiency by using the thermoelectric material. The device effectively decreases the temperature of the mixture air to the engine and reduces the amount of NO_x emissions of the exhaust gas. Moreover, the device can decrease the warm-up time of the engine by heating the fuel by thermoelectric material to improve the emission of the exhaust gas from the engine. By applying the thermoelectric material in the device, we can promote the combustion efficiency of the internal combustion engine, increase the output of the engine and improve the environment of the earth.

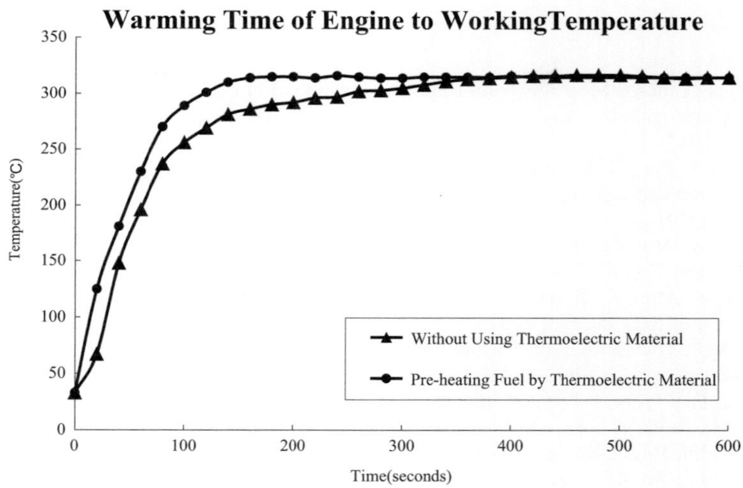

Fig. 35.4 Comparison warming time in the engine

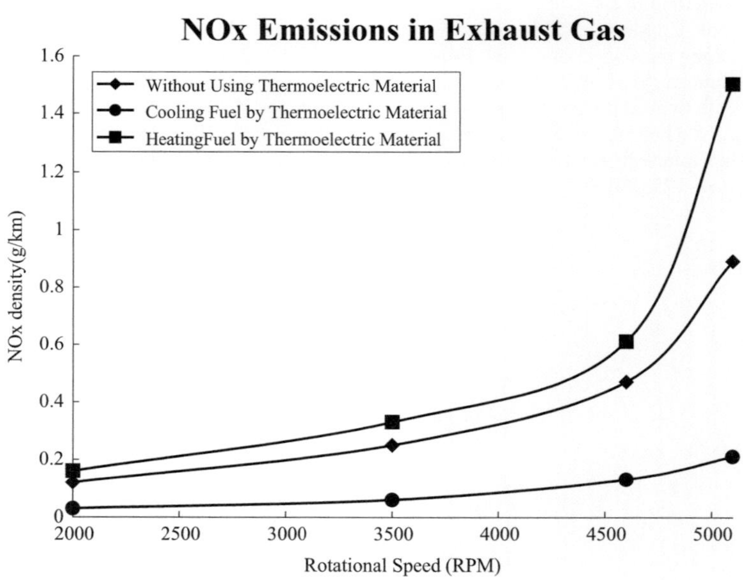

Fig. 35.5 The variation in the NO_x of the engine

References

1. Kose, H., Ciniviz, M.: An experimental investigation of effect on diesel engine performance and exhaust emission of addition at dual fuel mode of hydrogen. Fuel Process. Technol. **114**, 26–34 (2013)
2. Fang, Q.A., Fang, J.H., Zhuang, J.A., Huang, Z.: Effects of ethanol–diesel–biodiesel blends on combustion and emissions in premixed low temperature combustion. Appl. Therm. Eng. **54**, 541–548 (2013)
3. Zheng, M., Mulenga, G.T.: Biodiesel engine performance and emissions in low temperature combustion. Fuel **87**, 714–722 (2008)
4. Okude, K., Mori, K., Shiino, S.: Premixed compression ignition combustion for simultaneous reduction of NO_x and soot in diesel engine, SAE paper no. 2004-01-1907. (2004)
5. Park, S.H., Youn, I.M., Lim, Y.: Influence of the mixture of gasoline and diesel fuels on droplet atomization, combustion, and exhaust emission characteristics in a compression ignition engine. Fuel Process. Technol. **106**, 391–401 (2013)
6. Yun, H., Reitz, R.D.: Combustion optimization in the low-temperature diesel combustion regime. Int. J. Eng. Res. **6**, 513–524 (2005)
7. Kim, Y.J., Kim, K.B., Lee, K.H.: Effect of a 2-stage injection strategy on the combustion and flame characteristics in a PCCI engine. Int. J. Automot. Technol. **12**, 639–644 (2011)
8. Jung, S., Ishida, M., Yamamoto, S.: Enhancement of NOx-PM trade-off in a diesel engine adopting bio-ethanol and EGR. Int. J. Automot. Technol. **11**, 611–616 (2010)
9. Jacobs, T.J., Assanis, D.N.: The attainment of premixed compression ignition low-temperature combustion in a compression ignition direct injection engine. Proc. Combust. Inst. **31**, 2913–2920 (2007)
10. Dardiotis, C., Martini, G., Marotta, A.: Low-temperature cold-start gaseous emissions of late technology passenger cars. Appl. Energy **111**, 468–4780 (2013)
11. Weilenmann, M., Soltic, P., Saxer, C.: Regulated and nonregulated diesel and gasoline cold start emissions at different temperatures. Atoms Environ. **39**, 2433–2441 (2005)
12. Emami, S., Jafarmadar, S.: Multidimensional modeling of the effect of fuel injection pressure on temperature distribution in cylinder of a turbocharged DI diesel engine. Propul. Power Res. **2**(2), 162–175 (2013)

Chapter 36
Modeling of an AlGaAs-Based VCSEL with Bragg Mirrors

Shu-Hui Liao

Abstract This chapter discusses the methods of modeling Bragg mirrors for vertical cavity surface emitting lasers (VCSELs). The propagation matrix method was used to present the design and performance of a VCSEL with AlAs/AlGaAs Bragg mirrors for electromagnetic radiation centered at 980 nm. We adopted and compared with AlAs/AlGaAs periodic dielectric layer stacks consisting of 20 and 40 identical layer-pairs. The result indicates that the reflectivity of the Bragg mirror is close to unity over a wavelength band width centered at wavelength of 980 nm. It was found that the results achieved from the simulation showed a high correlation with predicted results.

Keywords VCSEL • Bragg mirror • Propagation matrix method

36.1 Introduction

Nowadays, the vertical cavity surface emitting lasers (VCSELs) provide a very exciting area of research. The optimization of the distributed Bragg reflector (DBR) structure is very important in increasing the performance of optical systems based on the VCSEL technology. Bragg mirrors are structures of high interest for a large set of applications that include the vertical cavity lasers and the upcoming range of devices which based on microcavity. In a surface emitting laser, the output light beam is perpendicular to the active layer and the surface of associated semiconductor [1]. The optical cavity is formed by two DBRs surrounding the active medium layer. Generally speaking, these DBRs have high reflectivity larger than 90 %. The high reflectivity is required since the optical gain per pass is small due to the small optical cavity compared to an edge emitting laser. Recently, the VCSEL usually has an active layer formed by multiple quantum wells. The benefits of a small optical cavity include low threshold current and single mode lasing since the mode separation is wide enough. Over the last 20 years, some attractive emission

S.-H. Liao (✉)
Department of Electronic Engineering, Chung Chou University of Science and Technology, Changhua County 51003, Taiwan
e-mail: liao@dragon.ccut.edu.tw

J. Juang (ed.), *Proceedings of the 3rd International Conference on Intelligent Technologies and Engineering Systems (ICITES2014)*, Lecture Notes in Electrical Engineering 345, DOI 10.1007/978-3-319-17314-6_36

properties of the VCSELs which compared to edge emitting semiconductor lasers have led to VCSELs' establishment as a powerful device within the semiconductor laser family. These properties include such as near-circular beam profiles, in contrast to astigmatic beams emitted by edge emitting lasers. A further property of VCSELs, resulting from their short cavities, is single longitudinal mode emission. Recently, there has been exciting research work in the area of microelectromechanical structures (MEMS) tunable VCSELs, with a special emphasis in increasing their wavelength tuning range [2–9]. However for conventional VCSELs, the DBR thickness imposes a significant design trade-off on the tuning speed, wavelength range, actuation voltage, as well as fabrication difficulties. Today although the compound semiconductors could perform very well in the optoelectronic applications, we can find that it is still not straightforward to realize Bragg mirrors based on these material systems due to the low optical index differences. Recently, it is found that novel Bragg reflecting structures which represent quite different from the conventional quarter-wavelength stacks used currently in the optoelectronics industry have been reported [10]. These structures can be used in designing high-reflectivity mirrors in strained material systems and mirrors that will provide high reflectivity. Moreover, such mirrors can be used in designing resonant- cavities for optical absorption and emission. In this chapter, the Bragg mirror and VCSEL with AlAs/AlGaAs Bragg mirror for electromagnetic radiation was investigated. The structures were modeled and optimized by the propagation matrix method [11].

36.2 Optical Performance of the VCSEL

The most general method of calculating the reflectance and the transmittance of a multilayer is based on a matrix formulation of the boundary conditions at the film surfaces derived from the Maxwell's equations. In this chapter, the Bragg mirrors are formed by alternating layers of AlAs and AlGaAs with quarter-wavelength thickness, and consist of a number of such pairs on both the n- and p-sides. Thus the device of VCSEL with Bragg mirrors is constructed by using the number of pairs of two times and making the thickness of the central GaAs layer at one wavelength wide such that there is half of layer-pair mirror above and half of layer-pair mirror below a central GaAs layer. The refractive index of the thin films are $n_{\mathrm{AlGaAs}} = 2.95$ for the refractive index of AlGaAs and $n_{\mathrm{AlAs}} = 3.52$ for that of AlAs. The two Bragg mirrors of the VCSEL structure consist of AlGaAs/AlAs alternating layers with refractive index difference of 0.57. The electromagnetic radiation was incident normally to the surface of a special AlGaAs/AlAs periodic dielectric layer stack which consists of corresponding identical layer pairs. In the analysis of the system, each individual dielectric layer which was used has a special thickness of $\lambda/4n$, where n is the refractive index of the dielectric material.

The propagation matrix method was used to investigate in designing high-reflectivity VCSEL with Bragg mirrors for electromagnetic radiation centered at

980 nm. The electromagnetic radiation was incident normally to the surface of a special AlGaAs/AlAs periodic dielectric layer stack which consists of some identical layer pairs.

36.3 Bragg Mirrors

The schematic VCSEL structure is shown in Fig. 36.1. The propagation matrix method was used to investigate a high-reflectivity VCSEL with Bragg mirror for electromagnetic radiation centered at 980 nm. We adopted and compared with three different conditions of layer-pairs as shown in Figs. 36.2, 36.3, 36.4, and 36.5. The result indicates that the reflectivity is very close to unity over a wavelength band width with a full-width half-maximum (FWHM). The FWHM decreases as the number of layer-pairs increases. Figure 36.6 shows the I–L and I–V characteristics of the VCSEL.

36.4 Summary

The VCSEL is capable of emitting light of wavelength dependent upon the design parameters. The reflectivity is close to unity over a wavelength band width and thus the addition of an optical band pass which is near unity transmission was established. The emission spectrum of the VCSEL with Bragg mirrors investigated in this work gives a full width of the image at half maximum value, or FWHM

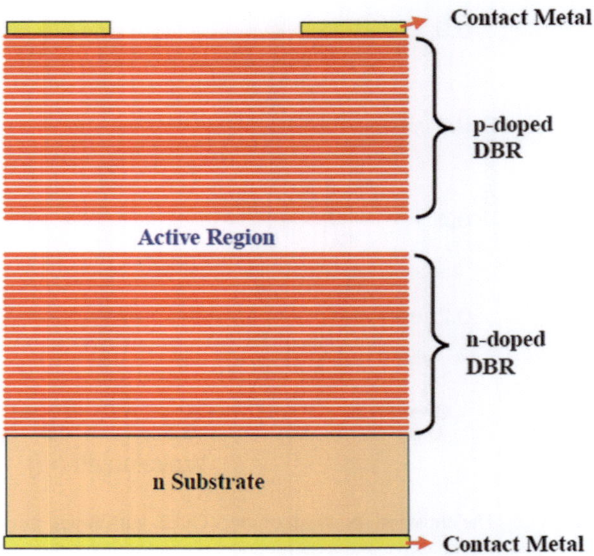

Fig. 36.1 Schematic layer structure of a VCSEL

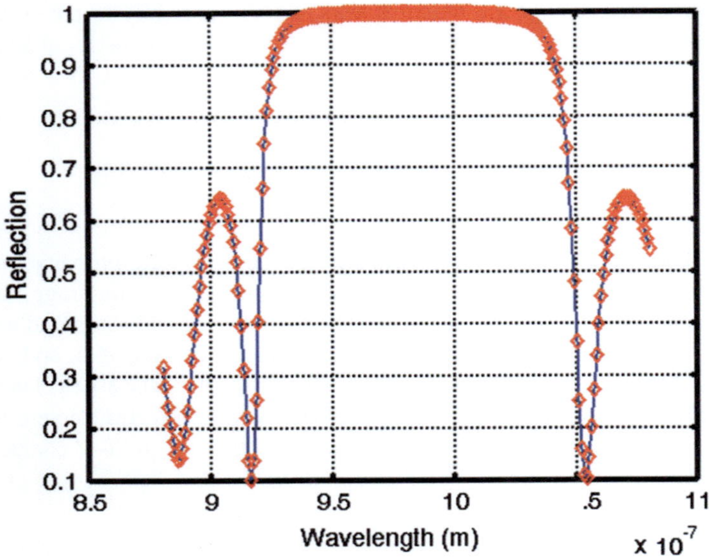

Fig. 36.2 The emission spectrum of a high-reflectivity Bragg mirror with 20 identical dielectric layer pairs

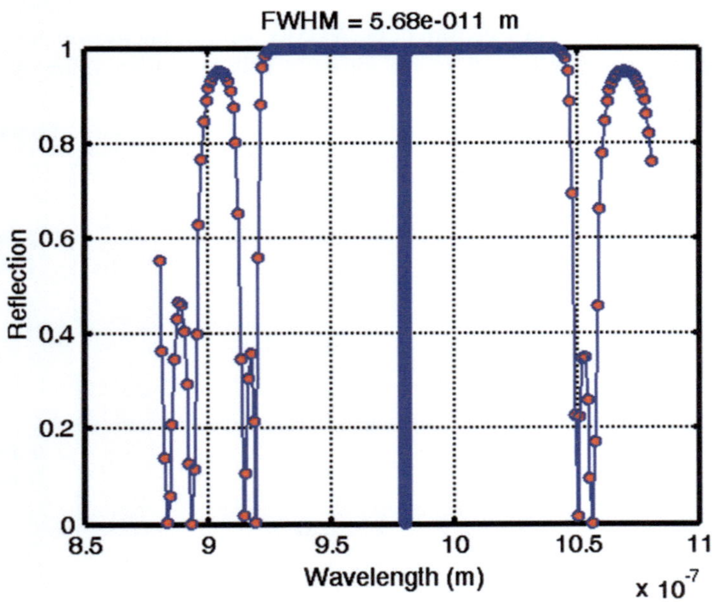

Fig. 36.3 The emission spectrum of the VCSEL with Bragg mirrors with a 20 identical dielectric layer-pair mirror above and a 20 identical dielectric layer-pair mirror below a central GaAs layer

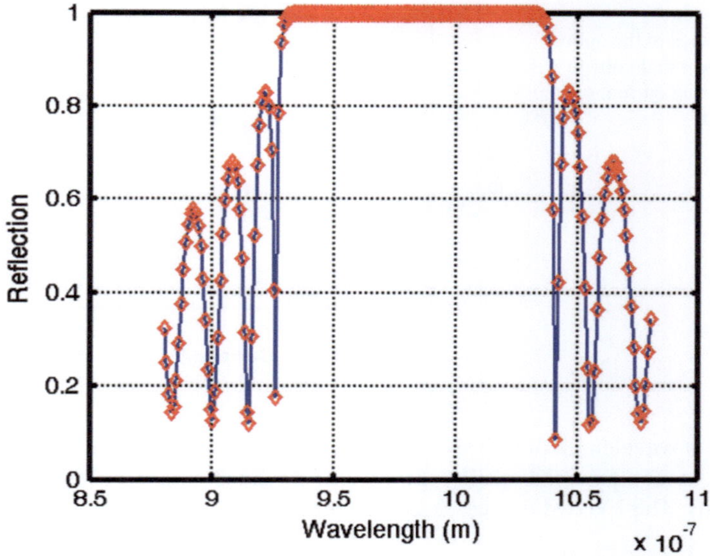

Fig. 36.4 The emission spectrum of a high-reflectivity Bragg mirror with 40 identical dielectric layer pairs

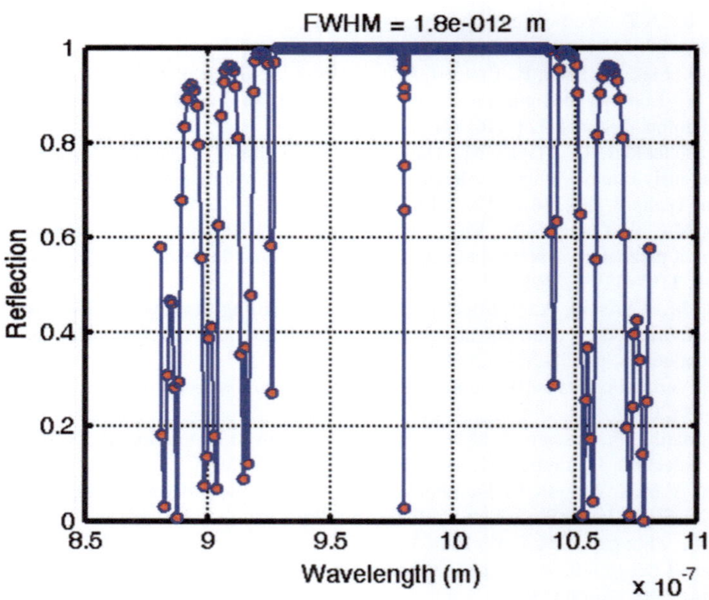

Fig. 36.5 The emission spectrum of the VCSEL with Bragg mirrors with a 40 identical dielectric layer-pair mirror above and a 40 identical dielectric layer-pair mirror below a central GaAs layer

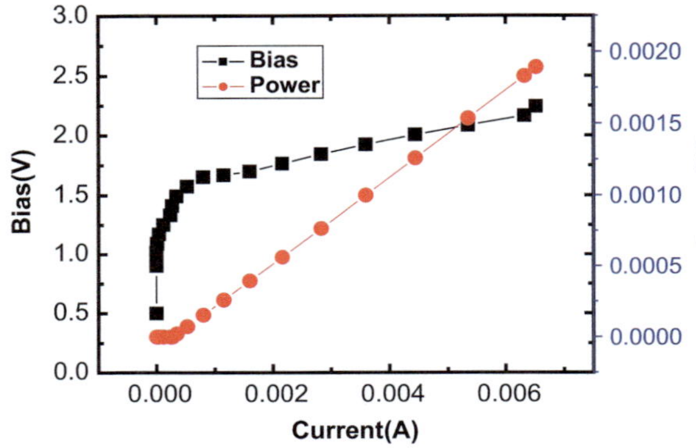

Fig. 36.6 The characteristics of the optical output power and voltage versus biasing current of the VCSEL

centered at wavelength of 980 nm. The FWHM is a well-defined number which can be used to compare the quality of images obtained under different observing conditions. The FWHM decreases as the number of layer-pairs increases.

References

1. Choquette, K.D.: Vertical cavity surface-emitting lasers: light for the information age. MRS Bull. **27**, 507–511 (2002)
2. Decai, S., Fan, W., Kner, P., Boucart, J., Kageyama, T., Dongxu, Z., Pathak, R., Nabiev, R.F., Yuen, W.: Long wavelength-tunable VCSELs with optimized MEMS bridge tuning structure. IEEE Photon. Technol. Lett. **16**, 714–716 (2004)
3. Riemenschneider, F., Maute, M., Halbritter, H., Boehm, G., Amann, M.C., Meissner, P.: Continuously tunable long-wavelength MEMS-VCSEL with over 40-nm tuning range. IEEE Photon. Technol. Lett. **16**, 2212–2214 (2004)
4. Huang, M.C.Y., Cheng, K.B., Ye, Z., Pesala, B., Chang-Hasnain, C.J., Pisano, A.P.: Demonstration of piezoelectric actuated GaAs-based MEMS tunable VCSEL. IEEE Photon. Technol. Lett. **18**, 1197–1199 (2006)
5. Linnik, M., Christou, A.: Vertical cavity surface emitting laser with AlGaInAs/InP Bragg mirrors fabricated for operation at 1.55 mm. In: IEEE International Symposium on Compound Semiconductors, pp. 38–388. (2000)
6. Iga, K.: Vertical-cavity surface-emitting laser: its conception and evolution. Jpn. J. Appl. Phys. **47**, 1–10 (2008)
7. De Leonardis, F., Passaro, V.M.N., Magno, F.: Improved simulation of VCSEL distributed Bragg reflectors. J. Comput. Electron. **6**, 289–292 (2007)
8. Passaro, V.M.N., Magno, F., De Leonardis, F.: Optimization of Bragg reflectors in AlGaAs/GaAs VCSELs. Laser Phys. Lett. **2**, 239–246 (2005)
9. Yana, C., Zhong, J., Zhao, Y., Hao, Y., Lib, T., Sun, Y.: Novel semiconductor/superlattice distributed Bragg reflector: experiment and simulation. J. Lumin. **122**, 838–840 (2007)
10. Murtaza, S.S., Anselm, K.A., Srinivasan, A., Streetman, B.G., Campbell, J.C., Bean, J.C., Peticolas, L.: High-reflectivity Bragg mirrors for optoelectronic applications. IEEE J. Quantum Electron. **31**, 1819–1825 (1995)
11. Levi, A.F.J.: Applied Quantum Mechanics, pp. 168–170. Cambridge University Press, Cambridge (2003)

Chapter 37
Risk Decision Method Based on Sensory Values of Smart Devices

Taejun Son, Jeonghoon Kwak, Suhyun Gong, and Yunsick Sung

Abstract Smartphone sensors can be useful in recognizing risk situations faced by users. However, the use of diverse kinds of sensors for risk estimation increases the complexity of such analyses. This chapter proposes a method to recognize the risk quotient in user situations by measuring and managing the data that smartphone sensors provide.

Keywords Risk recognition • Normalization • School zone • Internet of things

37.1 Introduction

As the daily commute to and from schools becomes riskier for children, our society is emphasizing on the importance of improving safety in and around schools. With this objective in mind, more research for the development of school-zone surveillance systems is being conducted [1].

The data from the smartphones can be useful in determining a child's location, and the riskiness of a particular situation he/she may face. By using more sensors, risky situations can be evaluated more precisely. However, the use of different types of sensors makes it more difficult to evaluate situations based on multiple sensors.

This chapter proposes a method for measuring values from various kinds of sensors and managing the measured values to recognize the level of riskiness of a situation faced by the users. The sensor values are normalized to reduce the difference in the values obtained from different users. This rest of the paper is organized as follows. Section 37.2 describes related work. Section 37.3 introduces our approach, and Sect. 37.4 explains the validation of this approach and concludes this paper.

T. Son • J. Kwak • Y. Sung (✉)
Department of Game Mobile Contents, Keimyung University, Daegu, South Korea
e-mail: jason.son@kmu.ac.kr; jeonghoon@kmu.ac.kr; yunsick@kmu.ac.kr

S. Gong
Department of Computer Engineering, Keimyung University, Daegu, South Korea
e-mail: suhyun.gong@kmu.ac.kr

J. Juang (ed.), *Proceedings of the 3rd International Conference on Intelligent Technologies and Engineering Systems (ICITES2014)*, Lecture Notes in Electrical Engineering 345, DOI 10.1007/978-3-319-17314-6_37

37.2 Related Work

Research has been conducted for developing child-safety systems in school zones based on context awareness [2]. By utilizing cutting edge IT devices and ITS technology, a child-safety system recognizes current situations. It monitors pedestrians, vehicles, and surroundings for surveillance to minimize the safety risks to children.

There is also research on realization of u-Care systems based on multisensors in u-Home environments [3]. These systems recognize risky situations using various sensors to monitor indoor situations and activities of elderly people. The results from these monitoring services are provided to families or medical institutions as a link service.

Research has also been conducted for designing care systems for the elderly based on OSGi by utilizing RFID and infrared sensors [4]. By calculating and comparing the positions of a ward with the amount of activity on a monthly and daily basis, the risk situations can be recognized and then provided to guardians through phone calls and email.

In existing research, various sensors like GPS, infrared sensors, and RFID are utilized to detect risk situations, which is not proper to outdoor environments because of huge space. This chapter proposes a method that utilizes various sensors in a smartphone to detect accurately the level of risks faced by users in outdoor environments.

37.3 Risk Recognition Processes

This chapter proposes a method for managing sensor values through a smartphone and recognizing risk situations step-by-step. The values of all the sensors in a smartphone are measured concurrently, and these values are utilized to detect risk levels of different situations. All six stages are handled in order as shown in Fig. 37.1.

In the SVE stage, the values of all sensors in a smartphone are obtained. The values of the sensors are divided into two groups. One group comprises the values that can be expressed and handled as numbers. The other group comprises the values such as voices, GPS information, etc., which cannot be expressed or handled as numbers. When there are $n+m$ sensors in a smartphone, $m-n$ sensors cannot be expressed or handled. These sensors are defined from sensor s_{n+1} to sensor s_{n+m}. The measured values, $x_{n+1,\,t}$ to $x_{n+m,\,t}$, of these sensors at time t, are delivered to the RLE stage directly. The n values of the sensor that can be expressed and handled as numbers defined by $x_{1,\,t}$ to $x_{n,\,t}$ from s_1 and s_n, are delivered to the NZE stage.

We assume that the frequency of each sensor value is relevant to the risk level of recognized risk situations. A higher frequency of the sensor values means that the recognized risk situations are safer. Conversely, a lower frequency of the sensor

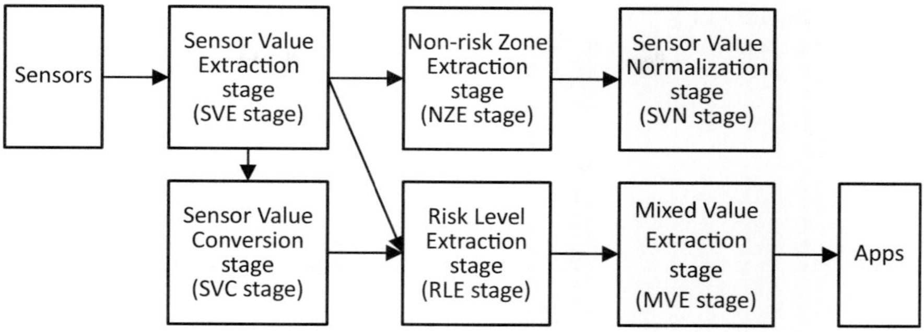

Fig. 37.1 Process for handling sensor values in a smartphone

values means that the recognized risk situations are more dangerous. Therefore, the NZE stage determines the non-risk zone as the range in which the sensor values have high frequencies. It makes this determination by defining two values: one for the starting point and the other for the ending point. In addition, there could be numerous possible reactions depending on the risk levels of the recognized risk situation. The NZE stage defines the levels of non-risk situations in detail, but does not handle the recognized risk situation in detail. This stage analyzes the premeasured values as per the below equation to decide a non-risk zone. Therefore, the values of sensors need to be collected for a certain period of time, 1 to ζ, in advance. First, the frequencies of the values of each sensor are calculated. Second, the frequencies are arranged in an ascending order according to their corresponding sensor values. Third, if $y_{i,\,j}$ is the jth value among the ordered values of sensor s_i, then the maximum frequency f_i of the values of its sensor can be obtained by using (37.1).

$$f_i = \text{Max}\left(f\left(y_{i,1}\right), f\left(y_{i,2}\right), \ldots\right) \tag{37.1}$$

where $f(y_{i,\,j})$ is the frequency of the value $y_{i,\,j}$. Fourth, the sensor value that is smaller than the sensor value of f_i, includes risk situation, and is the largest among the values smaller than the sensor value of f_i is defined as the minimum border value α_i, as shown in Fig. 37.2. If the sensor value that includes risk situation does not exist, then the smallest sensor value is defined as the minimum border value α_i. Fifth, the sensor value that is larger than the sensor value of f_i, includes risk situation, and is the smallest among the value larger than the sensor value of f_i is defined as the maximum border value β_i. A section between the minimum border value α_i and the maximum border value β_i is defined as a non-risk zone.

In the SVN stage, the premeasured values from α_i and β_i in the non-risk zone are normalized at time ζ once for revising the distribution of the values of each sensor. The average μ_i and standard deviation σ_i are calculated based on the measured values until time ζ.

In the SVC stage, the average μ_i and a standard deviation σ_i are utilized for extracting $Z_{i,t}$ using (37.2). The final output value $r_{i,\,t}$ of the SVC stage is assigned

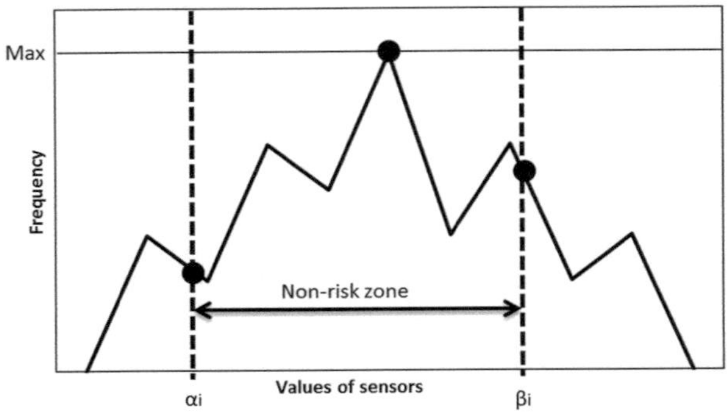

Fig. 37.2 Non-risk zone defined by two border values, α_i and β_i

by occurrence probability $P(x_{i,\ t})$ through $Z_{i,\ t}$. $P(x_{i,\ t})$ has a range from 0 to 1. The final output value $r_{i,\ t}$ affects the determination of the risk level, depending on the position of $P(x_{i,\ t})$.

$$Z_{i,t} = \frac{x_{i.t} - \mu_i}{\sigma_i} \tag{37.2}$$

In the RLE stage, the risk level $d_{i,\ t}$ is expressed as a value from 0 to γ, where 0 means a safe situation and γ means the most dangerous situation. The sensor values that cannot be expressed in numbers are determined within 0 to γ, considering the traits of its sensor. The sensor values that can be expressed as numbers are determined by (37.3).

$$d_{i,t} = \begin{cases} \text{if } x_{i,t} < \alpha \text{ or } x_{i,t} > \beta \\ \gamma \\ \text{else if } P(x_{i,t}) < 0.5 \\ \gamma - 1 - \left| \dfrac{\dfrac{P(x_{i,t}) - P(\alpha)}{0.5 - P(\alpha)}}{\gamma - 1} \right| \\ \text{else} \\ \gamma - 1 - \left| \dfrac{\dfrac{P(\beta) - P(x_{i,t})}{P(\beta) - 0.5}}{\gamma - 1} \right| \end{cases} \tag{37.3}$$

Each risk level is extracted from one value of one sensor. Therefore, all risk levels are multiplied at the same time, while considering the importance of all sensors using weights. These risk levels are allocated to a mixed value m_t by using (37.4).

$$m_t = w_1 \times d_{1,t} + w_2 \times d_{2,t} + \cdots + w_i \times d_{i,t} \tag{37.4}$$

where w_i is a weight of the ith sensor, which has a value between 0 and 1. The sum of all weights is 1. Then the calculated mixed value can be applied to smartphone apps.

37.4 Experiment and Conclusions

In this experiment, we compared risk levels calculated using the proposed method and those calculated using the values of sensors directly. The method of merging all sensor values is the same, which means that (37.4) is applied in both the approaches. In addition, we also applied the same weight for all sensors. All values of sensors were measured for 4 h through a user's smartphone. These values were utilized from the SVE stage to the SVN stage. Then, the smartphone sensor values were measured again for an hour the next day at the same place, including risk situations.

In this experiment, the risk levels of a speed sensor were compared. When the values of sensors were utilized directly, the risk level 1 is defined from 0 to 20, the risk level 2 is defined from 21 to 40, and so on after converting the values into percentages. γ is assigned the value 5. When the values of a speed sensor were processed in the SVN stage, the average μ_i was 1.21 and standard deviation σ_i was 0.439301 in Fig. 37.3. Both approaches showed a similar range of risk situations. Using our approach, level 4 and level 5 were obtained from time 1,435 to 1,474. With the direct approach, level 2 to level 5 were calculated from time 1,441 to 1,450. However, our approach could denote the difference between high- and low-risk situations.

Then the mixed values obtained by using both methods were compared as shown in Fig. 37.4. The values from 1 to 800 were calculated in a similar manner. However, given that the situations from time 2676 to time 2782 were risky, our approach could reduce the level of riskiness of situations during the period from time 800 to time 3500.

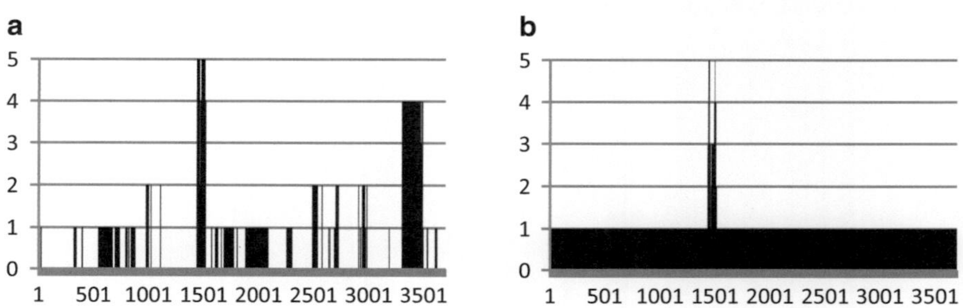

Fig. 37.3 Distributions of risk levels of a speed sensor, (**a**) By the proposed method, (**b**) By utilizing sensory values directly

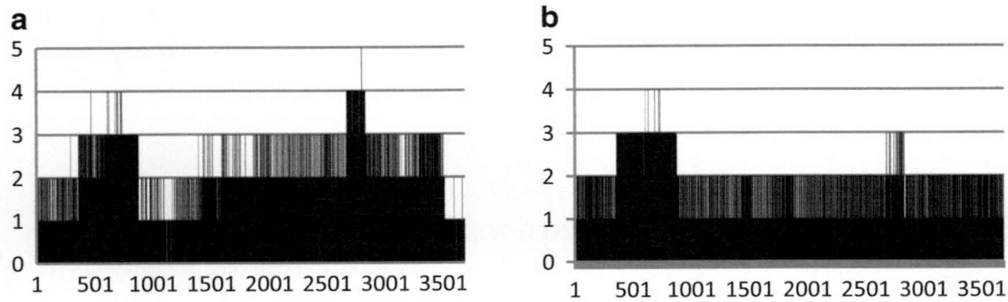

Fig. 37.4 Distributions of mixed values, (**a**) By the proposed method, (**b**) By utilizing sensory values directly

In this chapter, we proposed a method to use smartphone sensor values for extracting the level of riskiness of various situations, and also normalizing the sensor values because the distribution of these values is different, depending on individual users.

Acknowledgment Following are results of a study on the "Leaders in Industry–University Cooperation" Project, supported by the Ministry of Education, Science, and Technology (MEST) and the National Research Foundation of Korea (NRF).

References

1. Park, S.J., Lee, J.C., Jang, D.-S., Lee, G.S.: A study and design of monitoring module for schoolzone safety. Korea Acad. Ind. Cooper. Soc. **12**(4), 1940–1946 (2011)
2. Oh, S., Lee, D., Park, J., Lee, B., Kim, W.: A basic study on integrated safe conduct system based context awareness at school-zone. Korea Inst. Intell. Transp. Syst. **2012**(1), 237–243 (2012)
3. Lee, H.-J., Kang, S.-J., Jang, H.-G., Jeong, C.-W., Joo, S.-C.: Implementation of u-care system based on multi-sensor in u-home environment. Korean Soc. Internet Inf. **12**(2), 135–147 (2011)
4. Cha, S.-H., Kim, D.-Y., Choi, J.-H., Lee, J.-E., Kim, K.-H., Cho, K.-H.: Design and implementation of an OSGI-based old age patient care system in embedded programming on RFIDs and infrared sensors. J. Korea Inf. Commun. Soc. **33**(11), 1005–1012 (2008)

Chapter 38
A Flywheel Energy Storage System Suspended by Active Magnetic Bearings Using a Fuzzy Control with Radial Basis Function Neural Network

Van-Sum Nguyen, Hong-Van Tran, Lai-Khac Lai, Nguyen Thi-Hoai Nam, Tran Huu-Chau Giang, and Le-Phuong Truong

Abstract A flywheel energy storage system (FESS) is an effective energy-saving device. It works by accelerating a rotor flywheel disc at a very high speed and maintaining the energy in the system as rotational energy. Active magnetic bearings (AMBs) are ideally suited for use at high-speed and are so used in FESSs. This work develops a mathematical model of the levitation force and rotational torque of a flywheel. The system for controlling the position of the flywheel is designed based on a neural fuzzy controller. A mathematical model of an AMB system comprises identification followed by collection of information from this system. A fuzzy logic controller (FLC), the parameters of which are adjusted using a radial basis function neural network, is applied to the unbalanced vibration in FESS. The results obtained concerning the FESS indicate that the system exhibited satisfactory control performance including transient and steady-state responses under various operating conditions.

V.-S. Nguyen (✉)
Department of Electrical Engineering, Da-Yeh University, Changhua 51591, Taiwan
e-mail: nguyen.sum@gmail.com

H.-V. Tran
Department of Automation Technology, Thu Duc College of Technology, Ho Chi Minh, Vietnam

L.-K. Lai
Department of Science Technology and Environment, Thai Nguyen University, Thai Nguyen, Vietnam

N.T.-H. Nam
Department of Electrical Engineering, Hue Industrial College, Hue City, Vietnam

T.H.-C. Giang
Department of Training, Hue Industrial College, Hue City, Vietnam

L.-P. Truong
Department of Electromechanical and Electronics, LacHong University, Bien Hoa City, Vietnam

J. Juang (ed.), *Proceedings of the 3rd International Conference on Intelligent Technologies and Engineering Systems (ICITES2014)*, Lecture Notes in Electrical Engineering 345, DOI 10.1007/978-3-319-17314-6_38

Keywords Flywheel • Energy storage system • AMB • FLC • Neural network

38.1 Introduction

Saving energy has recently become very important globally, and technology for saving generated power that is surplus to requirements is required. The FESS is an effective energy-saving technology [1]. A FESS has various advantages over a traditional battery energy storage system, including a higher energy storage density, higher specific power and power density, lower risk of overcharging or overdischarging, a wide range of operating temperatures, a very long life cycle, and environmental friendliness. In FESS, not only is the speed of the rotor high, but also the position of the rotor must be controlled accurately to ensure that the vector of angular momentum is precisely and stably directed, so vibrations of the high speed rotor must be as small as possible. The AMB are ideally suited to high-speed and vacuum applications owing to their contact-free operation, low friction losses, adjustable damping, stiffness characteristics, and lack of a need for any lubricant [2, 3].

This chapter proposes a method for controlling the position of the rotor by using the neural fuzzy controller (NFC) approach. The method employs a fuzzy controller system with RBFNN rotation to identify the AMB system by Jacobian transformation [4]. The parameters of FLC can be optimally tuned to solve the problem of unbalanced vibration in the AMB in FESS by applying the gradient descent method [5] and the real-time values according to the AMB system information. The presented design uses a NFC to control AMB in a FESS and to reduce the vibration of the rotor in an AMB system.

38.2 Structure and Principles of Operation of a Flywheel

Figure 38.1 presents the proposed FESS. The developed system is composed of an upper radial AMB, a lower radial AMB, an axial AMB, a permanent magnet, a motor/generator, the flywheel's rim, hub and shaft, and other components. The axial AMB at the top of the FESS overcomes the weight of the components of the flywheel rotor, keeping it levitating. Two radial AMBs in the middle of the FESS keep the rotor suspended in the air. During charging mode, the flywheel's motor acts like a load and draws power from the grid to accelerate the rotor. During discharging, the motor is switched into generator mode, and generates electricity that is then sent back into the grid.

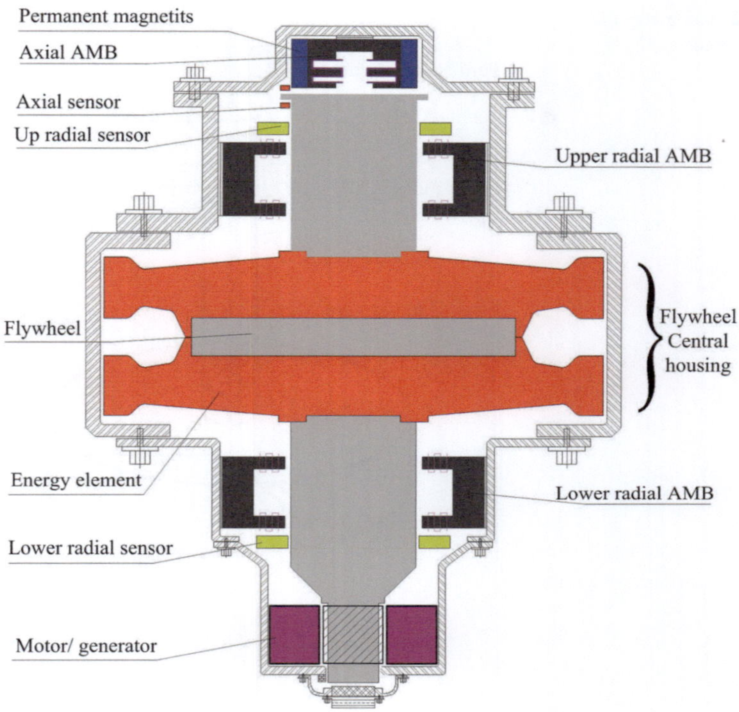

Fig. 38.1 Basic layout of a FESS

38.3 Dynamic Model of FESS

In this work, the rotor is assumed to be a rigid and symmetrical body with uniformly unbalanced mass. Figure 38.2 presents relationship between the center of gravity (CG: x_c, y_c, z_c) of the flywheel rotor and the five-DOF AMB system [6]. The dynamic equations that describe the motion of the rotor bearing system about the center of rotor are as follows.

$$\begin{cases} m\ddot{x}_c = F_{upx2} - F_{upx1} + F_{lox2} - F_{lox1} \\ J\ddot{\varphi}_x + J_z\Omega\dot{\varphi}_y = \left(F_{loy2} - F_{loy1}\right)l_{lo} - \left(F_{upy2} - F_{upy1}\right)l_{up} \\ m\ddot{y}_c = F_{upy2} - F_{upy1} + F_{loy2} - F_{loy1} \\ J\ddot{\varphi}_y - J_z\Omega\dot{\varphi}_x = \left(F_{upx2} - F_{upx1}\right)l_{up} - \left(F_{lox2} - F_{lox1}\right)l_{lo} \\ m\ddot{z}_c = F_{zAMB} - mg \end{cases} \tag{38.1}$$

where m denotes the mass of the rotor; g is the acceleration due to gravity; F_{upx1} and F_{upy1} represent the X and Y components of force in the negative direction on the upper magnetic bearing, respectively; F_{upx2} and F_{upy2} represent the X and Y components of force in the positive direction on the upper AMB. Similarly, F_{lox1} and F_{loy1} represent the corresponding forces on the lower AMB. F_{lox2} and

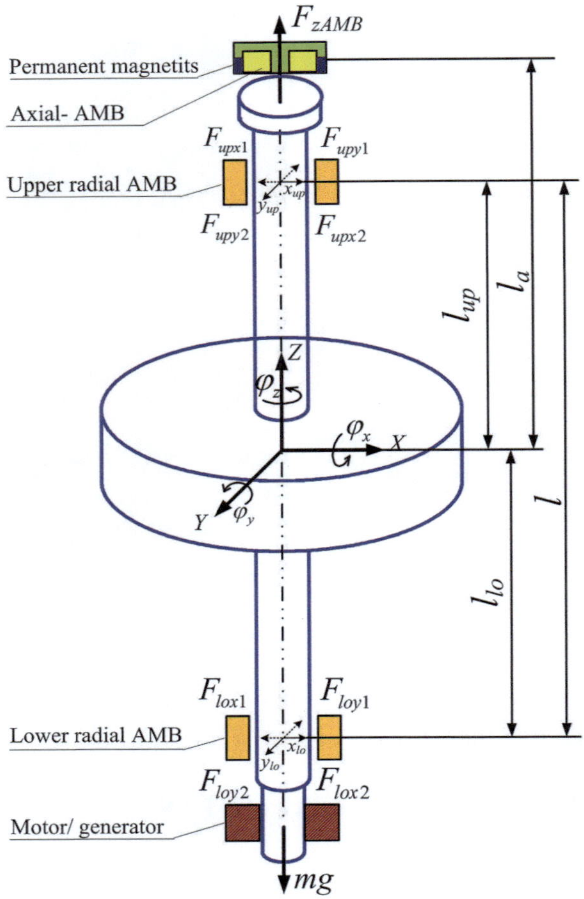

Fig. 38.2 Geometry of flywheel rotor

F_{loy2} are used to represent the X and Y components of force in the negative direction on the lower AMB; F_{zAMB} represents the magnetic suspension force on the axial AMB in the positive direction along the Z-axis; x_c, y_c, and z_c are the coordinates of the center of gravity (CG); φ_x, φ_y, and φ_z denote the angular displacement of pitch, yaw, and spin around the X, Y, and Z-axes of the rotor, respectively; J is the transverse moment of inertia of the rigid rotor around X-axis or Y-axis; J_z is the polar mass moments of inertia of the rotor; $-J_z\Omega\dot{\varphi}_x$ and $J_z\Omega\dot{\varphi}_y$ are the gyroscopic effects when rotor rotates at a speed of Ω around the Z-axis.

The rotor displacement from the desired position is assumed to be small, relationships between the shaft positions (x_up, x_lo, y_up, y_lo, z) and the CG (x_c, y_c, z_c) are expressed as follows.

$$x_c = \frac{l_{up}x_{lo} + l_{lo}x_{up}}{l}; \quad y_c = \frac{l_{up}y_{lo} + l_{lo}y_{up}}{l}; \quad z_c = z;$$

$$\varphi_x = \frac{y_{lo} - y_{up}}{l}; \quad \varphi_y = \frac{x_{up} - x_{lo}}{l} \tag{38.2}$$

where x_{up}, y_{up} are the displacements of the shaft at the radial upper AMB, and x_{lo}, y_{lo} are the displacements of the shaft at the lower radial AMB. The differential equation for a rigid rotor whose degrees of freedom transformed into bearing coordinates with displacement matrix output $X_c = \left[x_{up}x_{lo}y_{up}y_{lo}z\right]^T$ and input vector $u_c = \left[i_{xup}i_{xlo}i_{yup}i_{ylo}i_z\right]^T$. The dynamics of the AMB system as follows.

$$\mathbf{M_c}\ddot{X}_c + \mathbf{G_c}\dot{X}_c + \mathbf{K_{ds}}X_c = \mathbf{K_{is}}u_c + \mathbf{D}g \tag{38.3}$$

where $\mathbf{M_c}$ is the mass matrix; $\mathbf{G_c}$ is the gyroscopic matrix; $\mathbf{K_{ds}}$ is the matrix of force displacement factors; $\mathbf{K_{is}}$ is the matrix of force current factors; $\mathbf{D}$ is the gravity vector. The resulting continuous time state space model is given in bearing coordinates by

$$\dot{\mathbf{x}}=\mathbf{Ax}+\mathbf{Bu}; \quad \mathbf{y}=\mathbf{Cx} \tag{38.4}$$

with $\mathbf{x} = \left[X_c \quad \dot{X}_c\right]^T; \quad \mathbf{u} = \left[0 \quad u_c\right]^T;$

$$\mathbf{A} = \begin{bmatrix} 0 & \mathbf{I} \\ -\mathbf{M_c^{-1}K_{ds}} & -\mathbf{M_c^{-1}G_c} \end{bmatrix}; \quad \mathbf{B} = \begin{bmatrix} 0 \\ \mathbf{M_c^{-1}K_{is}} \end{bmatrix}; \quad \mathbf{C} = [\mathbf{I} \quad 0] \tag{38.5}$$

where $\mathbf{x}$ is the state vector, $\mathbf{u}$ is the control input signal, $\mathbf{y}$ is the output, $\mathbf{A}$ is the state matrix, $\mathbf{B}$ is the input matrix, and $\mathbf{C}$ is the output matrix.

38.4 NFC Design for FESS

The proposed NFC was designed to control an AMB in FESS, as shown in Fig. 38.3. It consists of a fuzzy controller, a reference model, an RBFNN, and an adjusting mechanism. The operating principle of this controller is based on AMB dynamics and control knowledge can be incorporated into a NFC design based on RBFNN identification.

where x^* and x_{rm} are a step reference and a reference model, respectively. x_{rbf} and u_{fn} are the outputs of neural and fuzzy neural controller, respectively. x_p is the response of displacement; u_f represents the output of the fuzzy controller. FPID is a fuzzy proportional integral derivative control for the current loop. F_d is the external torque from ventilator. FI and DFI are the fuzzification converted controller inputs (e and de), respectively.

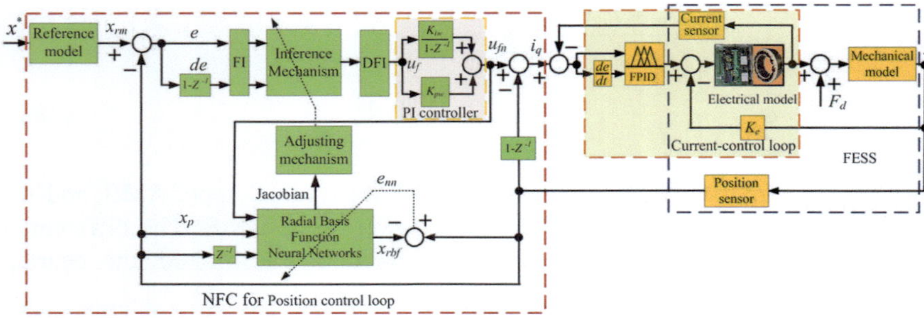

Fig. 38.3 Block diagram NFC for FESS

Based on Fig. 38.3, the tracking error e and the error change de are defined by the following equations:

$$e(k) = x_{\mathrm{rm}}(k) - x_{\mathrm{p}}(k); \quad \mathrm{de}(k) = e(k) - e(k-1) \tag{38.6}$$

The RBFNN adopted in this part of the study was a three-layer structure, as shown in Fig. 38.4, and comprised an input, a hidden, and an output layer. The RBFNN has three inputs by $u_{\mathrm{fn}}(k)$, $x_{\mathrm{p}}(k)$, $x_{\mathrm{p}}(k-1)$ and its vector form is represented by

$$X = \left[u_{\mathrm{fn}}(k), x_{\mathrm{p}}(k), x_{\mathrm{p}}(k-1)\right]^{\mathrm{T}} \tag{38.7}$$

The multivariate Gaussian function is used as the activated function in the hidden layer of the RBFNN, which is given by the following equations:

$$h_j(X) = \exp\left(\frac{1}{2}\frac{\left\|X - \mathbf{c_j}\right\|^2}{b_j^2}\right), \quad j = 1, 2, 3, \ldots, m \tag{38.8}$$

The network output in Fig. 38.4 is derived using the following equations,

$$x_{\mathrm{rbf}} = \sum_{j=1}^{m} w_j h_j \tag{38.9}$$

where w_j and h_j are the weights from the jth hidden layer neuron to the output layer neuron and the output of the jth node in the hidden layer, respectively.

The adjusting parameters of the fuzzy controller minimize the square error between the rotor displacement and the output of the reference model. The instantaneous cost function is defined as follows:

$$E_e = \frac{1}{2}e^2 = \frac{1}{2}\left(x_{\mathrm{rm}} - x_{\mathrm{p}}\right)^2 \tag{38.10}$$

The parameters of $c_{m,\,n}$ are adjusted according to

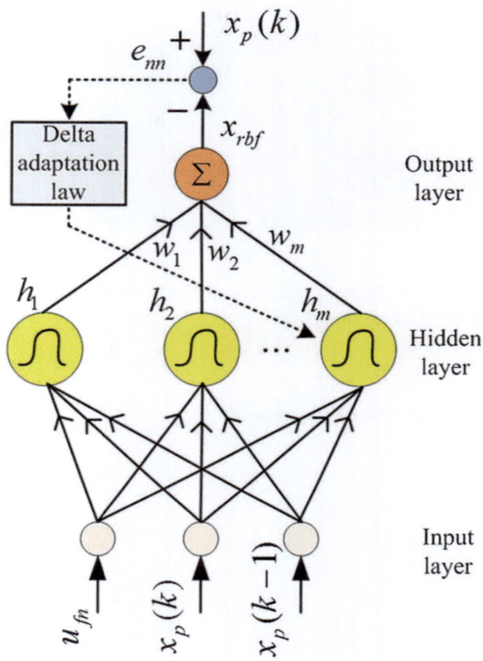

Fig. 38.4 The architecture of RBFNN

$$\Delta c_{m,n} = -\alpha(\partial E_e / \partial c_{m,n}) \tag{38.11}$$

where α represents the adaptive rate of the system. The chain rule is used, and the partial differential equation for E_e in (38.10) is written as follows:

$$\frac{\partial E_e}{\partial c_{m,n}} = -e \frac{\partial x_{\mathrm{p}}}{\partial u_{\mathrm{f}}} \frac{\partial u_{\mathrm{f}}}{\partial c_{m,n}} \tag{38.12}$$

To overcome this problem, and to increase the online learning rate of the connective weights, a delta adaptation law is proposed as follows:

$$\frac{\partial x_{\mathrm{p}}}{\partial u_{\mathrm{f}}} \approx \left(K_{\mathrm{pw}} + K_{\mathrm{iw}}\right) \frac{\partial x_{\mathrm{rbf}}}{\partial u_{\mathrm{fn}}} = \left(K_{\mathrm{pw}} + K_{\mathrm{iw}}\right) \sum_{j=1}^{m} w_j h_j \frac{c_{j1} - u_{\mathrm{fn}}(k)}{b_j^2} \tag{38.13}$$

38.5 Results and Discussions

Figure 38.5a presents the structural configuration of the AMB, including its electromagnetic coils. It is used in FESS. A current amplifier and current sensing circuit design for AMB are indicated in Fig. 38.5b. An experiment for the AMB in FESS has been verified by the current-control loop, using FPID control and current amplifier. The current response is very close to the reference signal and setting

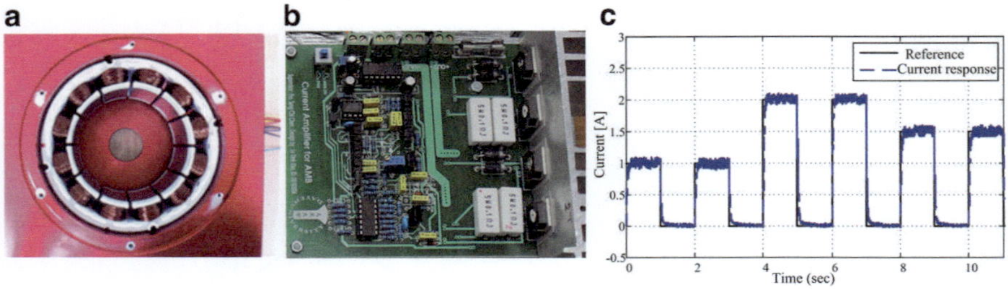

Fig. 38.5 Photograph of inside view of magnetic bearing (**a**); current amplifier (**b**); response of current loop control (**c**)

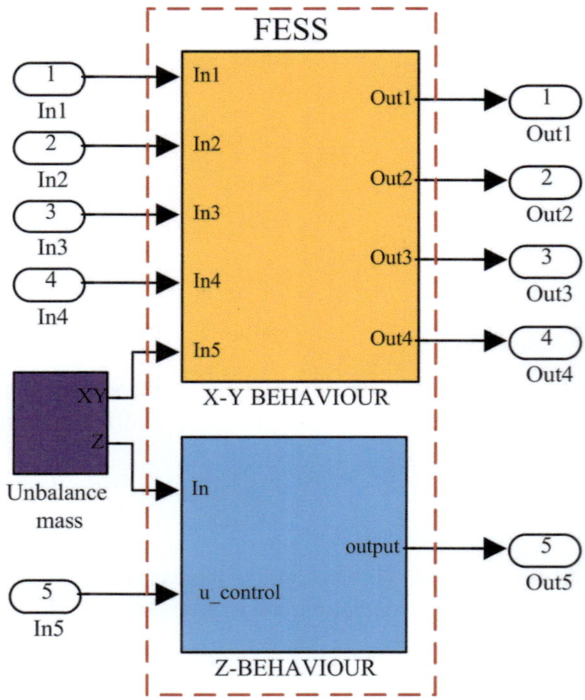

Fig. 38.6 Block diagram of a FESS

time about 0.01 s (Fig. 38.5c). The model of the FESS was used in simulations that involved a NFC method was using MATLAB-Simulink, as displayed in Fig. 38.6. Table 38.1 presents the parameters of the FESS.

Figures 38.7, 38.8, and 38.9 summarize those simulation results of FESS. The results shown in Figs. 38.7, 38.8, and 38.9a, b indicate that the operating envelope of the rotor displacements and orbits along the *X*-, *Y*-axes covered the entire feasible region (<0.5 mm). Figure 38.9a, b demonstrates that center of the rotor orbited the

Table 38.1 The parameters of the FESS

1	Mass of shaft (m)	70 kg
2	Displacement stiffness of lower/upper AMB-K_{ds}	973 kN/m
3	Current stiffness of lower/upper radial AMB-K_{is}	576 N/A
4	Displacement stiffness of axial AMB (K_{sa})	380 kN/m
5	Current stiffness of axial AMB (K_{ia})	520 N/A
6	Nominal length of air gap (s_0)	0.5 mm
7	Bias currents (i_{bias})	1.5 A
8	Transverse moment of inertia of rotor (J)	1.92 kg m^2
9	Polar moment of inertia of rotor (J_z)	1.64 kg m^2

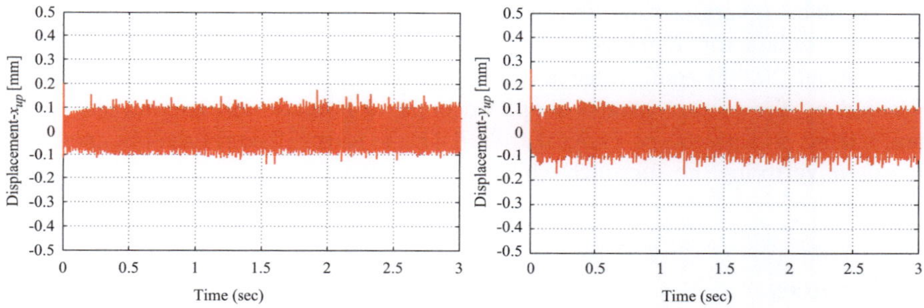

Fig. 38.7 Rotor displacement along X-, Y-axes in upper radial AMB

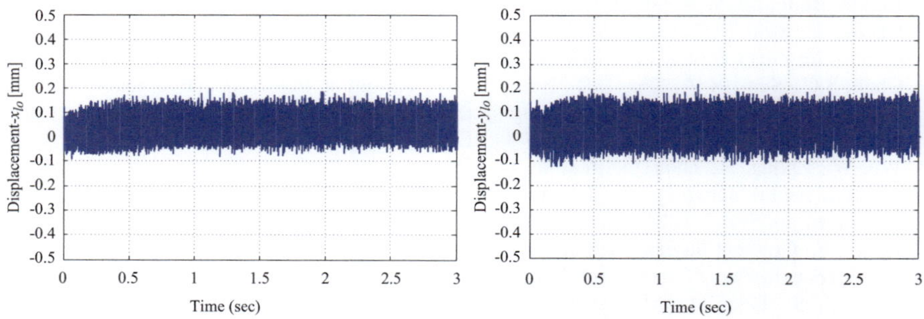

Fig. 38.8 Rotor displacement along X-, Y-axes in lower radial AMB

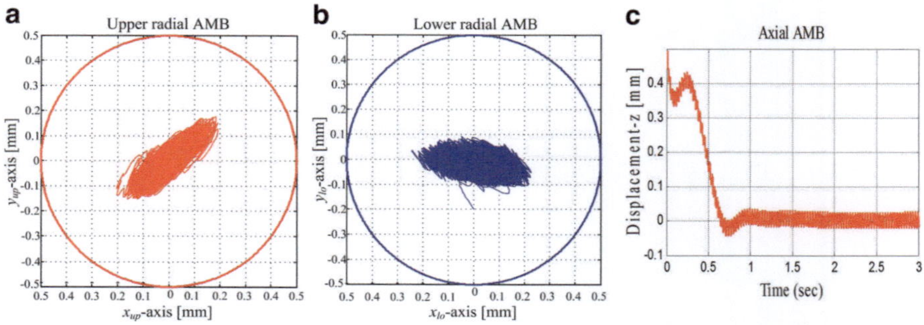

Fig. 38.9 Orbit of rotor in upper and lower radial AMB (**a**, **b**); rotor displacement along Z-axis (**c**)

X, Y axes rotated at speeds 20,000 rpm. The rotor displacement reduces about 0.15–0.23 mm. In Fig. 38.9c is present the rotor displacement along the Z-axis. The results further demonstrate that a short rise time implies a short settling time, low overshoot, and a small steady-state error.

38.6 Conclusions

This chapter develops an NFC algorithm to control for the position of AMB in a FESS. The results thus obtained indicate that the NFC enabled the FESS improved performance at various running rotor speeds. The results also reveal that the FESS exhibited satisfactory dynamic and steady-state responses at rotational speeds of the rotor about 20,000 rpm. Since an experimental for the AMB system has been verified by the current-control loop, we hope to prepare an experimental setup and testing on a prototype FESS with NFC in the future. The results obtained will facilitate verifying the approach which was developed in this study.

References

1. Liu, H., Jiang, J.: Flywheel energy storage-an upswing technology for energy sustainability. Energy Build. **39**, 599–604 (2007)
2. Chen, S.C., Sum, N.V., Kha, L.D., Hsu, M.M.: ANFIS controller for an active magnetic bearing system. In: Fuzzy Systems (FUZZ), IEEE, pp. 1–8, Indian (2013)
3. Chen, S.C., Sum, N.V., Kha, L.D., Nam, N.T.H.: An online trained adaptive neural network controller for an active magnetic bearing system. In: 2014 International Symposium on Computer, Consumer and Control, pp. 741–744. IEEE, Taiwan (2014)
4. Wu, W., Zhong, S., Zhou, G.: A study on PID intelligent optimization based on radial basis function neural networks. In: International Symposium Electronics, Communications and Networks, pp. 57–60. IEEE, Taiwan (2013)
5. Guely, F.: Gradient descent method for optimizing various fuzzy rule bases. In: International Conference on Fuzzy System, vol. 2, pp. 1241–1246. IEEE, USA (1993)
6. Gerhard, S., Eric, M.: Magnetic Bearings—Theory, Design and Application to Rotating Machinery. Springer, New York (2009)

Chapter 39
A Matrix Robot for Block Stacking

Yu-Ching Su, Ming-Chang Chen, and Wing-Kwong Wong

Abstract This research studies how to control a robot to stack blocks according to the block structure shown in a given image. In this project, a LabVIEW program is designed and run on a Lego Mindstorms NXT, which in turn controls a robot built with components of Matrix Robotics. There are two operation modes for the block stacking robot. In the first mode, a user uses an Android smartphone to control the robot with Bluetooth to move around, pick up blocks, and stack them. In the second mode, the robot identifies the block structure in a given image with OpenCV and picks up the needed blocks to stack them to build the block structure. The robot was tested with two images of simple block structures in the second mode. The success rates ranged from 7 to 80 %. Future work should further identify the weaknesses of the robot and improve its performance.

Keywords Android • LabVIEW • Matrix robotics • NXT • OpenCV

39.1 Introduction and Background

There are many possibilities of robotic applications at home. For example, six million units of Roomba, an automated vacuum cleaning robot, from the iRobot Corporation has been sold [1]. More personal and service robots are projected to be sold in a forecast by the International Federation of Robotics [2]. Suppose a robot is shown a picture of a simple structure consisting of wooden blocks. Then the robot figures out how to build the structure by getting and stacking wooden blocks. This might be an interesting toy robot for a child to observe and imitate. In this project, LabVIEW programs are used to control a robot to do block stacking.

There are some related research projects. For example, King used LabView and DaNI to make a set of experiments and explained how to use LabVIEW to control the DaNI robot [3]. Gómez-de-Gabriel used LEGO NXT mobile robots with LabVIEW for undergraduate courses on mechatronics [4, 5]. Oliver used the

Y.-C. Su • M.-C. Chen • W.-K. Wong (✉)
Department of Electronic Engineering, National Yunlin University of Science and Technology, Yunlin 64002, Taiwan
e-mail: wongwk@yuntech.edu.tw

J. Juang (ed.), *Proceedings of the 3rd International Conference on Intelligent Technologies and Engineering Systems (ICITES2014)*, Lecture Notes in Electrical Engineering 345, DOI 10.1007/978-3-319-17314-6_39

contact distributions to predict object interactions [6]. By observing the contacts between two objects, a robot learned to detect potential interactions between them. Thomas made an OpenCV demo application, and provided the information on a website [7, 8].

39.2 Method

Figure 39.1 is the overall architecture of this project. LabVIEW is a system-design platform and development environment from National Instruments, originally released in 1986 [9]. The programming language used in LabVIEW is a dataflow programming language, also known as G. Designer connects different modules by wires, resulting in a graphical block diagram that specifies how to execute the program.

Android [10] is an operating system based on Linux, and mainly used in mobile devices. Android is one of the most common mobile operating systems. Android SDK [10] is a software development kit. It provides API and tools for users. These tools include Android emulator and LogCat. And the SDK will be continuously updated with Internet.

OpenCV (Open Source Computer Vision) is a cross-platform library of computer vision [11]. It is free for use under the open-source BSD license. Image processing is done with an Android application and OpenCV library.

NXT is a programmable robotic microcontroller [12]. There are many sensors for NXT from various vendors. NXT, which is very popular in education, can control mechanical structures with motors and get readings from sensors driven by

Fig. 39.1 System architecture

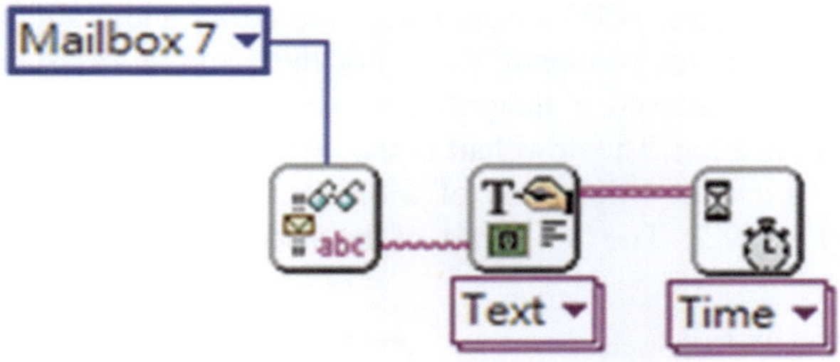

Fig. 39.2 LabVIEW program

Table 39.1 BT data format

Length LSB	Length MSB	Command type	Command	Inbox id

computer programs. Matrix Robotics [13] was created in 2011. Matrix can work with NXT and both are used to build the robot.

LabVIEW is used to control the NXT and Matrix after the installation of the LabVIEW for LEGO Mindstorm module and Matrix for LabVIEW. Then the OpenCV Library is added to the Android project with the installation of the OpenCV manager in a mobile device.

39.2.1 Send Messages to NXT via Bluetooth

In one operation mode of the robot, a user uses an Android phone to send commands via Bluetooth (BT) to control the robot. Figure 39.2 is a simple sample program in LabVIEW. It means BT messages will be read from the Mailbox7 and printed on screen. A message is sent to Mailbox7 with the BT data format (Table 39.1).

Message data is treated as a string. A Bluetooth message has two bytes in front of the telegram. The length of the message is divided into two bytes, the least significant byte followed by the most significant byte. Command type is commonly set to be 0x00 or 0x80. 0x00 means direct command telegram, response required; 0x80 means direct command telegram and no response is needed. The command is set to be write-message 0x09. And the fifth byte is the id of the mailbox.

39.2.2 Recognizing the Blocks in an Image

Blocks in an image are recognized with an OpenCV application modified from that used by Thomas [7]. The blocks in this project are all red, so color identification is simple with HSV [14]. HSV can transform RGB color mode into cylindrical coordinates. HSV means Hue, Saturation, and Value (brightness). The ranges of HSV values in OpenCV are not the same as the standard HSV. After the HSV ranges are set, a quadrilateral is drawn around each red area in the Android application. The flowchart is shown in Fig. 39.3 and the result is shown in Fig. 39.4.

First the orientation of a block must be decided to be horizontal or vertical (Fig. 39.5). The first vertex of the discovered quadrilateral of red area can be any of

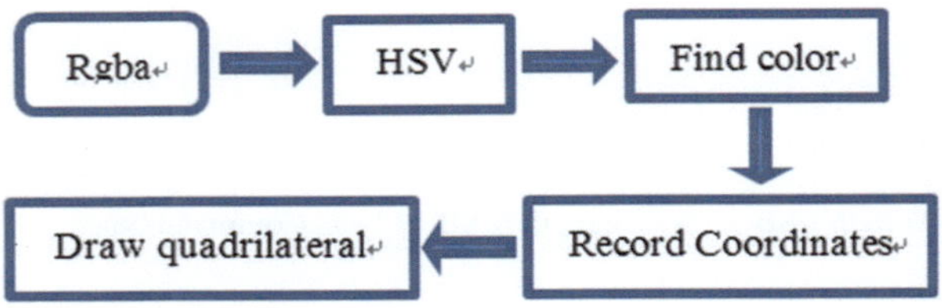

Fig. 39.3 Flowchart to identify a red area

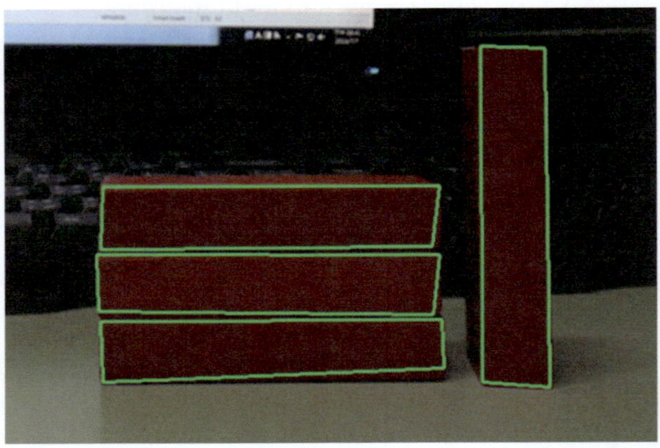

Fig. 39.4 Color identification

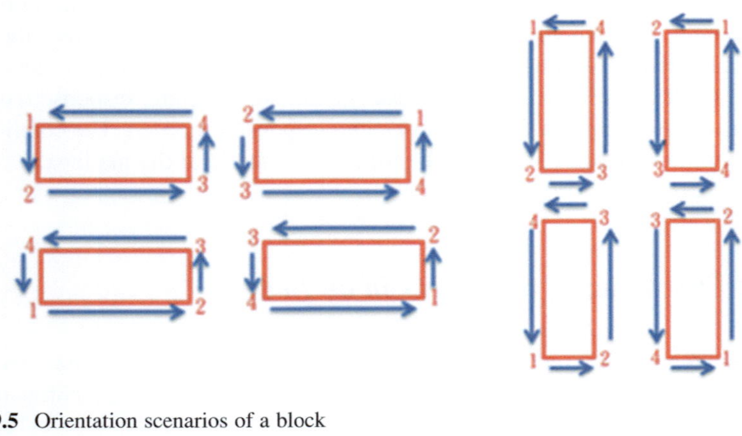

Fig. 39.5 Orientation scenarios of a block

the four vertices, and the vertices will be numbered in a counterclockwise way. Simply speaking, a block is horizontal if its horizontal length is longer than its vertical length. Otherwise, it is vertical.

39.3 Experimental Results

The robot was placed in an experimental environment, whose dimensions are 78.5 cm × 70.5 cm. There are four areas in Fig. 39.6. The target block structure is at the top right corner. The robot will pick up horizontal blocks from the bottom left area and vertical blocks from the bottom right area. The robot will build the target structure at the destination area.

At first, a researcher took a snapshot of the target block structure. An Android application then identified the blocks in the structure and commanded the robot to carry out a construction plan. The robot would pick up proper blocks from the horizontal or vertical area and stack them at the destination. After the robot had built the target structure, the task was done.

The black lines were used to align the robot with the horizontal and vertical areas, and the destination area. If the robot was not aligned precisely, it would fail to pick up a block or fail to stack one block on top of another. The robot used an NXT

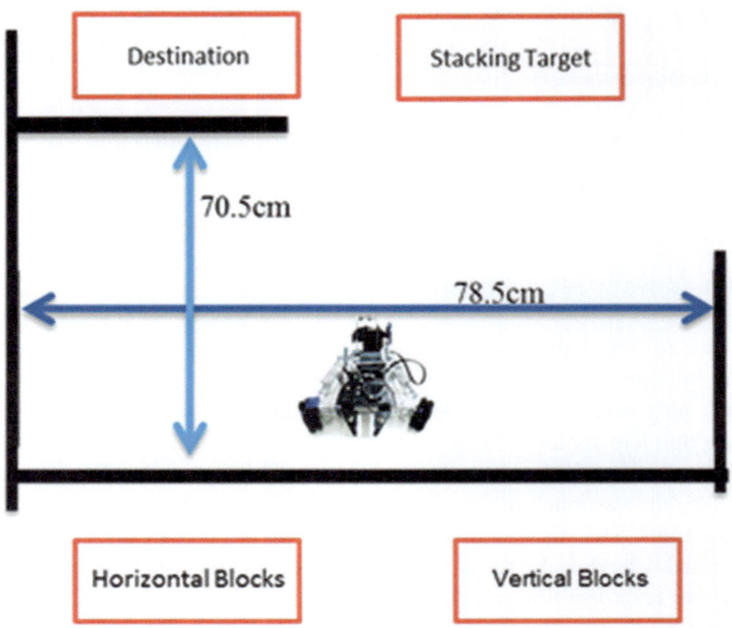

Fig. 39.6 Experimental environment

Table 39.2 Experiment success rates

Target	Result	Success rates
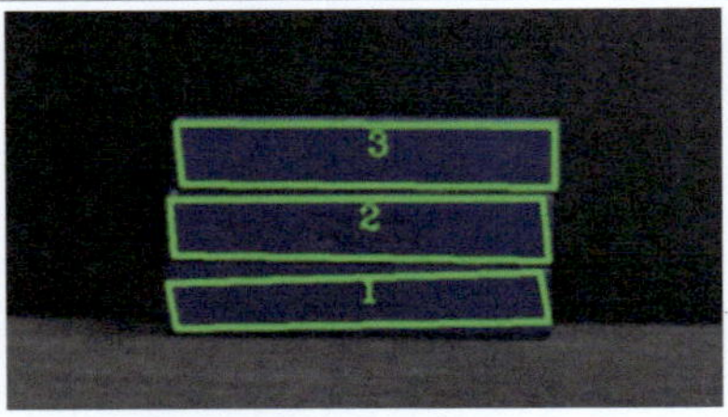		7/10
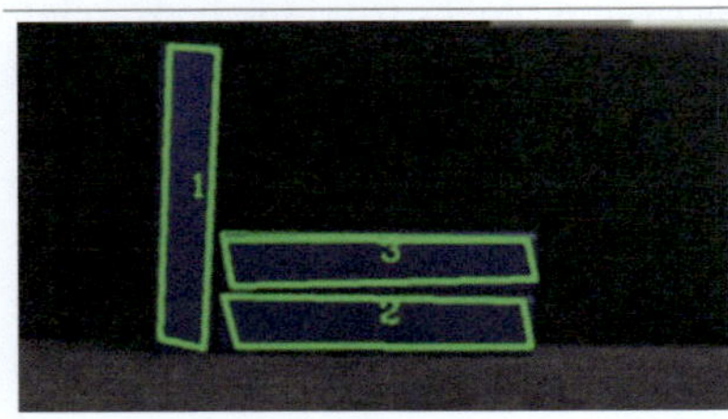	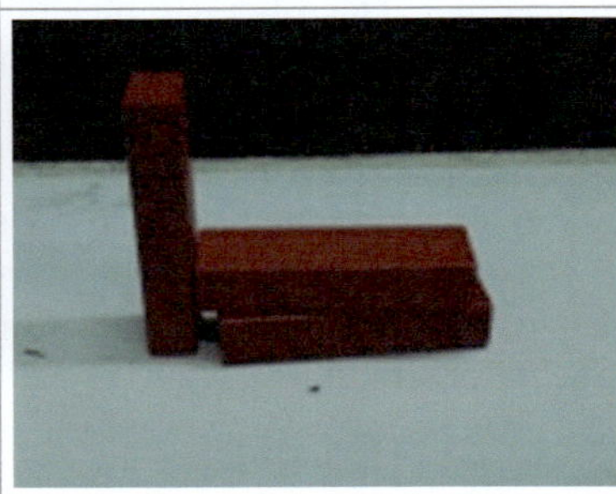	5/10

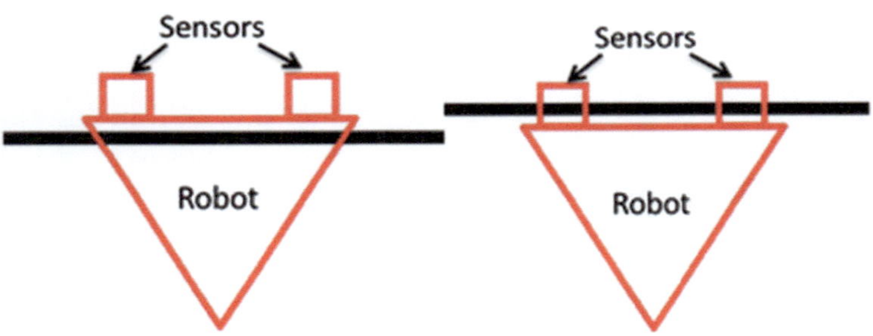

Fig. 39.7 *Left*: stop too late. *Right*: adjust to back up

light sensor to detect the black line. Robot will stop and adjust its position after detecting the line.

At first only one light sensor was used for this purpose, but the robot always failed to complete its task. Then two sensors were used and assembled near the motor. The reading of one sensor was used to control one motor. After this change, the success rate was much higher. The robot was tested with two images of simple block structures. The success rates ranged from 50 to 70 % (Table 39.2).

In its ideal behavior, the robot would stop immediately when detecting the black line. But the response time of the motors was too slow so that the robot always crossed the black line (the figure on the left of Fig. 39.7). In such case, the robot adjusted itself to back up to a position like the right figure of Fig. 39.7.

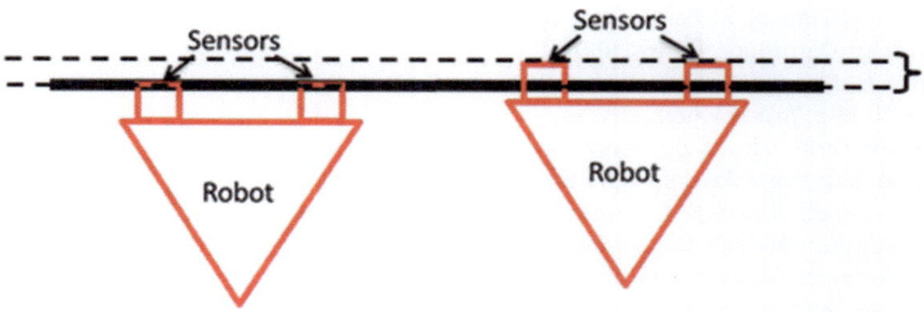

Fig. 39.8 Error status/correct status

When the robot turned 90° at a corner, e.g., the top left corner of Fig. 39.6, the robot would sometimes stop before reaching the black line (the figure on the left of Fig. 39.8). In this case, the robot would fail to complete its task. An adjustment in this case was to advance the robot until the light sensor had passed the black line (the figure on the left of Fig. 39.7). After such adjustment was implemented, the success rate increased from 70 % to 80 %.

39.4 Conclusions

In this study, a robot was designed to pick up blocks to build a target structure, which was given in a snapshot. The structure was identified in the snapshot with the OpenCV library. This robot can be a good toy for kids to build a structure and then ask the robot to build the same structure. Structure identification in an image was not an easy problem so only two simple structures were tried in this study. The success of building the target structures ranged from 70 % to 80 %.

This research has some limitations. The brightness of the environment will influence the result of image processing. The NXT memory may not be enough if the program gets longer. In future work, blocks of different colors should be used and more complicated structures should be built by the robot.

References

1. iRobot CEO discusses Q4 2010 results. http://seekingalpha.com/article/252090-irobot-ceo-discusses-q4-2010-results-earnings-call-transcript (2010)
2. International Federation of Robotics: World Robotics 2012 Service Robots. http://www.ifr.org/service-robots/statistics (2012)
3. King, R.: Mobile Robotics Experiments with DaNI. Colorado School of Mines, Golden (2011)
4. Gómez-de-Gabriel, J.M., Mandow, A., Fernández-Lozano, J., García-Cerezo, A.J.: Using LEGO NXT Mobile Robots with LabVIEW for Undergraduate Courses on Mechatronics. University of Malaga, Malaga (2011)
5. NXT. http://www.robotappstore.com/Knowledge-Base

6. Kroemer, O., Peters, J.: Predicting Object Interactions from Contact Distributions. Technische Universität Darmstadt, Darmstadt (2014)
7. Thomas B. http://www.barrythomas.co.uk/machinevision.html
8. OpenCV. http://opencv.org/
9. LabVIEW. http://zh.wikipedia.org/wiki/LabVIEW
10. Android. http://zh.wikipedia.org/wiki/Android
11. OpenCV. http://zh.wikipedia.org/wiki/OpenCV
12. NXT. http://en.wikipedia.org/wiki/Lego_Mindstorms_NXT
13. MatrixRobotics. http://matrixrobotics.com/
14. HSV. http://en.wikipedia.org/wiki/HSL_and_HSV

Chapter 40
A Study of Regular Transmission Delay in Bluetooth Communications

Komang Oka Saputra, Wei-Chung Teng, Pin-Yen Chou, and Tien-Ruey Hsiang

Abstract This chapter studies a special case of transmission delay when two devices communicate by Bluetooth technology. Transmission delays of packets are usually distributed randomly over some range, or the delay jitter, in wireless or wired communication. However, it is observed that under certain conditions, the transmission delays of consecutive packets may form into parallel dotted lines, and the intervals between a line and its next one are almost the same. The characteristics of the dotted-line delays, like the lifetime of one dotted line, are deduced to help develop an algorithm for detecting the period of this phenomenon. Experiments are further conducted to reveal how factors like operating system, packet sending period, and Bluetooth chips may affect the pattern of regular transmission delays.

Keywords Transmission delay • Bluetooth • Raining

40.1 Introduction

Transmission delays are essential, or at least useful, information to many applications including time synchronization, network traffic analysis, and device fingerprinting [1], to name but a few. Taking device fingerprinting as an example, it may be implemented by measuring the clock skews of the neighboring sensor motes in wireless sensor networks (WSN) [2], the clients device as a cloud service [3], or the other hosts inside the same wireless local area network (WLAN) [4]. A recent research even implemented clock skew-based Bluetooth device identification in personal area network (PAN) [5].

As Bluetooth is one of the most commonly used wireless communication technologies for many years, the communication performance and theoretical delay are studied [6, 7]. However, most of these researches studied Bluetooth's network behaviors on MAC layer instead of network layer. In the aforementioned

---

K.O. Saputra • W.-C. Teng (✉) • P.-Y. Chou • T.-R. Hsiang
Department of Computer Science and Information Engineering, National Taiwan University of Science and Technology, Taipe City 10607, Taiwan
e-mail: weichung@csie.ntust.edu.tw

J. Juang (ed.), *Proceedings of the 3rd International Conference on Intelligent Technologies and Engineering Systems (ICITES2014)*, Lecture Notes in Electrical Engineering 345, DOI 10.1007/978-3-319-17314-6_40

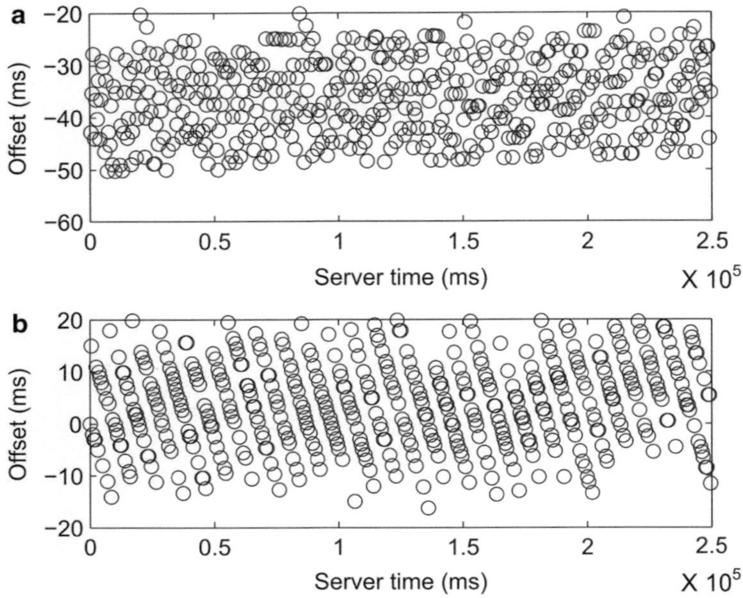

Fig. 40.1 Preliminary results. (**a**) Normal collection. (**b**) Collection with *dotted lines*

research [5], Huang et al. also implemented their BlueID technology based on the temporal feature of Bluetooth frequency hopping. This chapter presents a study of smart devices' time-stamping by transmitting the network layer timestamps to a measurer via Bluetooth communication. Nevertheless, during the experiments, it is found that there exist two types of communication delay patterns. In a normal collection, the offsets (difference between sending time of the smart device and the receiving time of the measurer) distributed randomly over some range, as shown in Fig. 40.1a. Yet in some conditions, the transmission delays of consecutive packets may form into parallel dotted lines, and the intervals between a line and its next one are almost the same (Fig. 40.1b). To the best of our knowledge, this *raining* phenomenon is not studied before.

To identify the presence of the dotted lines, a mechanism based on the patterns and the slopes of the dotted lines is developed. Furthermore, experiments covering variables of operating systems, server software languages, Bluetooth chips, and packet sending intervals are conducted to discover which combinations would the dotted lines would occur.

40.2 System Setup and Preliminary Results

The experiment system to perform device time-stamping collection is composed by several components:

Client: currently two mobile phones, HTC OneX and Samsung Galaxy Note3, are measured by their built-in Bluetooth chips.

Server: an ASUS K42JP notebook with 4GB RAM is used as server. The built-in AW-BT270 Bluetooth chip and an external BCM20702 Bluetooth chip are used to communicate with clients in the experiments.

Client application: an Android-based application which is developed to get the timestamps of the client device. The application reads the client's system time, and directly sends it to the server. The sending interval of the timestamps is adjustable from 0.5 s to few seconds.

Server application: an application that records the measurer's timestamp each time a packet contains the client's timestamp is received. For offsets like Fig. 40.1a, it is easy to estimate the clock skew of client device by using linear regression [8] or linear programming algorithm (LPA).

40.3 Dotted-Lines Detection Method

The preliminary experiments showed that the slopes of the dotted lines ranging from −1,850 to −1,350 ppm. Meanwhile, when the dotted line occurs, up to 90% of all offsets belong to the dotted lines. Due to these facts, a detection algorithm is developed as depicted in Fig. 40.2. It details how the detection starts from the first offset, the slope between offsets in the measured target are calculated. When the slope is detected to satisfy the threshold, the measured offset is then counted to be belonging to the dotted line. Otherwise, if the slope is out of range, then the offset is parked in a waiting area for later process. After the whole offsets have been processed, the dotted line is detected to happen if the number of offsets that belong to the dotted line is more that 90 % of all offsets.

40.4 Experiment Results

Four controlled variables are selected to study the behavior of the raining phenomenon. The values of these variables include: two types of server operating system, Ubuntu 14.0 (denoted as L) and Windows 7 Ultimate (W); three types of programming language, JAVA (J), C# (C), and Python (P); two types of Bluetooth chip from the onboard AW-BT270 chip (X) and the USB BCM20702 chip (Y); two devices: an HTC OneX (A) and a Samsung Galaxy Note3 (B); and three different clients sending intervals: 0.5, 1, and 2 s.

Table 40.1 summarizes if the dotted lines happen in the result offsets of all combination. As depicted in Table 40.1, the detection algorithm produced results that the dotted lines happen as long as the OS is Windows operating system. The statistic of the detection method is depicted in Table 40.2, which shows the average points in each dotted line for each combination and sending interval. The results

Fig. 40.2 *Dotted line* detection method

Table 40.1 The results if the raining phenomenon happens in various conditions

		W			L	
		C	J	P	J	P
X	A	Yes	No	Yes	No	No
	B	Yes	No	Yes	No	No
Y	A	Yes	No	Yes	No	No
	B	Yes	No	Yes	No	No

show negative correlation between the number of points and the sending interval, i.e., the bigger the interval, the less the number of points in each line.

To quantify the correctness of the detection method, we calculate the clock skew of all the combinations. As we did not know the real clock skews of the two devices,

Table 40.2 Relation between sending interval and the number of points in one dotted line

	AWPY	BWPY	AWPZ	BWPZ	AWCY	BWCY	AWCZ	BWCZ
0.5 s	17.85	18.51	16.56	18.34	18.50	18.50	19.03	18.20
1 s	8.94	9.42	8.82	8.02	9.65	9.48	9.52	9.42
2 s	4.77	4.79	4.69	4.64	4.83	4.75	4.81	4.82

Table 40.3 Clock skew results of all combinations (unit: ppm)

	Minimum	Maximum	Average	|Average − reference|
AWCY	−11.84	−8.48	−10.43	0.01
AWJY	−34.88	−31.14	−33.01	22.57
AWPY	−10.81	−8.78	−9.95	0.49
ALJY	−12.04	−9.60	−10.62	0.18
ALPY	−12.27	−10.31	−11.22	0.78
AWCZ	28.60	38.23	32.29	42.73
AWJZ	14.77	17.55	15.85	26.29
AWPZ	30.02	37.21	32.87	43.31
ALJZ	−11.96	−9.51	−10.73	0.29
ALPZ	−13.53	−9.54	−11.53	1.09
BWCY	12.01	15.35	13.82	0.46
BWJY	−15.91	−13.03	−14.55	28.83
BWPY	13.03	14.27	13.78	0.50
BLJY	12.02	15.97	14.48	0.20
BLPY	11.24	15.48	13.31	0.97
BWCZ	27.56	36.68	33.28	19.00
BWJZ	16.52	19.02	17.88	3.60
BWPZ	32.34	40.21	37.58	23.30
BLJZ	11.46	15.05	13.17	1.11
BLPZ	11.46	15.10	13.68	0.60

we used our previous research [9] which uses WiFi network to calculate the clock skews between the two devices and the server, and use both results as references for the clock skews in Bluetooth network.

As detailed in Table 40.3, all the experiments when using Linux as the server operating system show a relatively close clock skews when compared with the references. These results confirmed the correctness of the detection method's results in Table 40.1. For the Windows operating system, meanwhile, relatively far results of AWCZ, AWPZ, BWCZ, and BWPZ when compared with the references, confirmed that dotted line occurs in Windows when combined with C# and Python. However, we also discovered results which are inconsistent with the results in Table 40.1. AWJY, AWJZ, and BWJY, are three combinations that should produce a relatively close result to the reference, and BWPY which should produce a relatively far result. Since we use LPA as the estimator, LPA is known to find a slope of all observed data by creating a line that lies below all data (lower bound), there might be other factors that affect the lower bound of the observed device. For instance, even though AWJY is uncontaminated by the dotted line, the lower bound of AWJY's data could be not in a stable condition which caused the skew highly fluctuated.

40.5 Conclusions

This chapter verified the device time-stamping through Bluetooth network. A normal timestamp collection was obtained when using Linux as the server operating system combined with Java and Python, and when Windows combined with Java. The normal collection can be directly used to fabricate the clock skew, and the results were confirmed to be acceptable as they are close to clock skew references. For the dotted line, we detected its existence in Windows when combined with C# and Python. The clock skew results are relatively far when compared with the references, which shows the dotted line interferes with the clock skew measurement. We also observed that several clock skews are inconsistent with the results of the detection method, which might be caused by other factors that influence the lower bound of the collected data. As this work is a part of an ongoing research, our further work would include improving the precision of the Bluetooth device time-stamping and also to handle the problem of the lower bound instability.

Acknowledgments This study is partially supported by III Innovative and Prospective Technologies Project of the Institute for Information Industry.

References

1. Kohno, T., Broido, A., Claffy, K.: Remote physical device fingerprinting. IEEE Trans. Dependable Secure Comput. **2**, 93–108 (2005)
2. Rhee, I.-K., Lee, J., Kim, J., Serpedin, E., Wu, Y.-K.: Clock synchronization in wireless sensor networks: an overview. J. Sens. **9**, 56–85 (2009)
3. Huang, D.-J., Yang, K.-Y., Ni, C.-C., Teng, W.-C., Hsiang, T.-R., Lee, Y.-J.: Clock skew based client device identification in cloud environments. In: 26th IEEE International Conference on Advance Information Networking and Applications, pp. 52–533 (2012)
4. Jana, S., Kasera, S.K.: On fast and accurate detection of unauthorized wireless access points using clock skew. In: 14th ACM International Conference on Mobile Computing and Networking, pp. 10–115 (2008)
5. Huang, J., Albazrqaoe, W., Xing, G.: BlueID: a practical system for Bluetooth device identification. In: IEEE Conference on Computer Communications, pp. 284–2857 (2014)
6. Etxaniz, J., Alonso, A., Aranguren, G.: Influence of the distance between Bluetooth 2.0 nodes and their link mode with the communication delay. In: 3rd IEEE International Conference on Advances in Circuits, Electronics and Micro-Electronics, pp. 12–16. IEEE, New York (2010)
7. Morón, M.J., Luque, R., Casilari, E., Díaz-Estrella, A.: Analysis of Bluetooth transmission delay in personal area networks. In: 3rd IEEE International Symposium on Wireless Pervasive Computing, pp. 620–622 (2008)

8. Moon, S., Skelly, P., Towsley, D.: Estimation and removal of clock skew from network delay measurement. In: 18th Annual Joint Conference of the IEEE Computer and Communications Societies, pp. 22–234 (1999)
9. Cheng, L.-C.: Clock skew based identification technology for mobile devices. Master Thesis. Department of Computer Science and Information Engineering, National Taiwan University of Science and Technology (2013)

Chapter 41
IEEE 802.16j-Based Multihop Relays for Future Performance Enhancement of Aeronautical Mobile Airport Communications Systems (AeroMACS)

Behnam Kamali

Abstract The Aeronautical Mobile Airport Communication Systems (AeroMACS) has already been deployed in nine major US airports. This technology is used to support fixed and mobile ground-to-ground applications and services. In this article it is demonstrated that IEEE 802.16j-amendment-based WiMAX is most feasible for future enhancement of AeroMACS networks. Perhaps the most important benefit of application of multihop relays in AeroMACS is the flexible, power efficient, and cost-effective radio range extension into severely shadowed airport areas. Multihop relay operational modes are discussed. The key concept of "multihop gain," which explains how multihop relays enable performance enhancement in AeroMACS networks, is introduced. Under a reasonable set of assumptions, multihop gain is quantified in the form of an equation that provides a raw measure of this gain in Decibel.

Keywords AeroMACS • IEEE 802.16 • Multihop relay • WiMAX

41.1 Introduction

A joint FAA/EuroControl technology assessment, known as Future Communication Study, made specific recommendations on the development of data communications systems for various aeronautical applications over the legacy VHF band, the aeronautical L-band, and the newly available C-band (5,091–5,150 MHz). For AeroMACS a broadband wireless communication technology that enables rapid improvement was deemed necessary. The consensus was formed around a network-centric digital technology based on IEEE 802.16e standard [1]. To this end the 2007 ITU World Radiocommunication Conference took an action that allowed the development of a new international standard of aeronautical mobile route services

B. Kamali (✉)
Department of Electrical and Computer Engineering, Mercer University, Macon, GA 31207, USA
e-mail: Kamali_B@mercer.edu

J. Juang (ed.), *Proceedings of the 3rd International Conference on Intelligent Technologies and Engineering Systems (ICITES2014)*, Lecture Notes in Electrical Engineering 345, DOI 10.1007/978-3-319-17314-6_41

to support airport surface domain communications in the newly available C-band spectrum. The proposed standard will be used to support fixed and mobile ground-to-ground and ground-to-air data communications applications and services.

Owing to practical shortfalls arising from early implementation of IEEE 802.16e-based WiMAX networks, the need for some modification and amendment to the standard was recognized early on. To address this issue the IEEE802.16j Multihop Relay Task Group defined a new relay station (RS) that can be used as an extension to the base station (BS) and relay traffic between the BS and the subscriber station (SS). In this article it is proposed and demonstrates that IEEE 802.16j-amendment-based WiMAX is most feasible for AeroMACS applications.

41.2 WIMAX and AeroMACS

The IEEE 802.16 standard was originally developed to define air interface specifications for a fixed broadband wireless alternative to metropolitan area wired-access networks. The IEEE 802.16 standards exclusively standardize two aspects of air interface; the PHY layer and the MAC sublayer. The current version of IEEE 802.16 standard is IEEE 802.16-2012. The IEEE 802.16j represents a variation of the IEEE 802.16e standards in which relay-based multihop network capability has been incorporated. This amendment enables the operation of RSs over licensed bands without requiring any modifications in SSs.

Worldwide Interoperability for Microwave Access, WiMAX, is an IEEE 802.16 standard-based broadband cellular wireless solution in which OFDMA is the multiple access technique. WiMAX enables low-cost mobile access to the internet and provides integrated wireless fixed and mobile services using single air interface and network architecture. Some of the salient features of WiMAX networks, relevant to the AeroMACS application, are outlined below.

1. WiMAX applies Scalable OFDMA access technology; which enhances performance against frequency selective fading effect and enables bandwidth scalability over several spectral ranges.
2. WiMAX predominantly supports TDD, which supports the exchange of asymmetric traffic.
3. Adaptive modulation and coding (AMC) is another feature of WiMAX networks inherited from IEEE 802.16e standard.
4. Two levels of error control are provided in WiMAX standards.
5. In order to control subchannel interference, WiMAX applies a widely accepted method known as fractional frequency reuse. In this technique frequency reuse factor (FRF) is not a constant but rather adaptive.
6. WiMAX standard applies MIMO as a key technology.

A system profile is a subset of mandatory or optional PHY and MAC sublayer elements selected from IEEE 802.16 standards. A certification profile is established when the operating frequency, channel bandwidth, FFT size, and duplexing technique are defined for a system profile. WiMAX Forum System Profile Release 2.1 fully supports the operational and functional requirements of AeroMACS networks.

In a document entitled "AeroMACS system requirement document", EuroControl investigates the use of WiMAX Forum Mobile System Profile Released 1 Approved Specification against airport surface communication requirements. According to this document *no technical obstacles have been found which would make it impossible to apply this technology for AeroMACS* [2].

41.3 IEEE 802.16j-Based Multihop Relays

The IEEE 802.16j presents a variation of the IEEE 802.16e standards in which the relay-based multihop network capability has been incorporated. The main concept is to complement the base station (BS) nodes with relays instead of additional BSs in a wideband cellular network. A base station and its subordinate relay stations (RS) together are called multihop relay base station (MR-BS). MR-BS covers an extended area, beyond what the BS alone covers, which is denoted by multihop relay cell, MR-cell. MR-BS manages all communication resources within a MR-cell through a *centralized* or *decentralized* procedure. Resource management of subscriber stations (SS) may be carried out directly by the BS or via radio links through a relay station [3].

Traffic and signaling between BS and SS may be routed through *access RSs* or via a direct link between BS and the SS. Figure 41.1 shows a simple two-hop relay configuration. The physical channel between the RS-BS and a relay is called *relay link*, and the channel between an access relay and a SS is termed as *access link* as illustrated in Fig. 41.1.

A key feature of multihop relay architecture is the replacement of single-hop links over poor-quality channels with multihop links with greater efficiency, through which higher system capacity can be achieved. Moreover, multihop

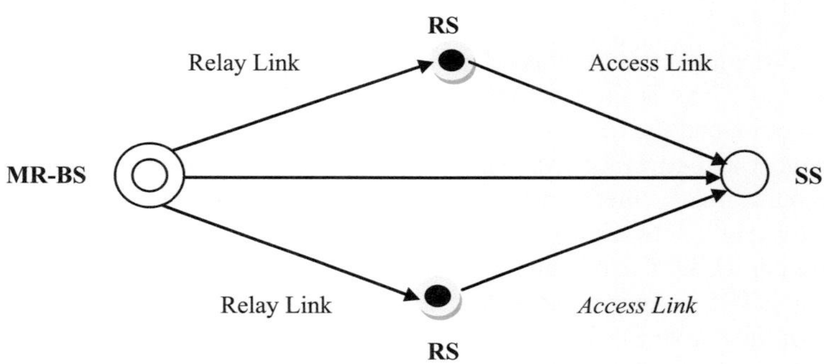

Fig. 41.1 A simple relay network consisting of two relay stations

communications can support spatial reuse, which can in turn boost the overall system capacity as well. The Standard specifies two modes of scheduling for controlling the allocation of bandwidth for RSs and SSs. In the *centralized scheduling* mode, the bandwidth allocation for a RS's subordinate stations is determined at the MR-BS, while in *distributed scheduling* they are carried out by the RS in coordination with MR-BS.

Relays may be classified according to their PHY layer and MAC layer functionalities. In terms of PHY layer processing, relay stations may be classified as Transparent Relays (TRS), and Nontransparent Relays (NTRS).

- A TRS essentially functions as a repeater which is transparent to the SS and bears no logical connection to it. As such TRS does not transmit preamble, nor does it broadcast control messages. The SS being served by a TRS receives traffic data from the TRS but control data comes from the corresponding MR-BS.
- A NTRS operates as a "mini BS" and thus is physically and logically connected to the SSs. The NTRS transmits preamble and broadcasts control messages, therefore the SSs served by a NTRS received both traffic and control data directly from the NTRS.

In so far as MAC functionalities are concerned, RSs can be characterized on the basis of their *scheduling arrangements* and *security capabilities*. In these respects the RS may operate in *centralized* or *distributed* modes. Distributed mode with respect to scheduling means that the RS is capable of scheduling network resources in coordination with MR-BS; otherwise the RS operates in centralized mode. Same can be said about security, i.e., the RS can be in distributed or centralized mode with respect to security arrangements. A TRS always operates in centralized mode. Hence bandwidth allocation and other scheduling procedures are carried out by the MR-BS. The main function of TRSs is network throughput enhancement. On the other hand a NTRS in distributed scheduling and security mode may provide radio outreach extension, higher bandwidth efficiency, as well as throughput enhancement in a WiMAX network. In centralized scheduling mode all information related to the access link; such as bandwidth request, is forwarded to the MS-BS. On the other hand a RS with distributed scheduling mode can process the information by itself.

Low cost and low complexity are important advantages of TRSs. Another advantage of TRSs is their ability to provide the lowest latency level, however they cannot extend the network radio outreach. NTRSs are used for both capacity enhancement and coverage extension. They offer several types of scheduling and radio resource management approaches. They enable a straightforward cell splitting scheme for quick network upgrade. The NTRS and its superordinate MR-BS may use different OFDM sub-channels, enabling control over network interference level. The NTRS is far more complex than the TRS to the extent that it can affect the cost of the network [4].

41.4 Multihop Relays for AeroMACS

This section is devoted to discussion of particular radio coverage situations in airport surface for which IEEE 802.16e-WiMAX system either fails to offer a viable solution, or the resolution it provides is inefficient, costly, or excessively power consuming. In all these cases, it is argued that 16j technology offers a much better alternative when multihop relays are applied [5].

1. When a portion of an airport is significantly shadowed by a new obstacle, such as a building constructed, a 16j-defined TRS or NTRS can be added to the airport network to provide higher capacity and acceptable QoS to the shadowed area. Adding a relay to an already established network does not require network reconfiguration and radio resource reallocation. The alternative that IEEE 802.16-2009 offers is the addition of another BS to the system which requires network redesign and entails reallocation of resources.
2. When a heavy load of traffic is expected in a certain location, or when an incident has occurred that requires a wireless link over a limited period of time, RSs may be deployed temporarily to provide coverage or additional capacity. In this case it may be possible to use a TRS that will add capacity by improving SS link throughput due to link improvement to and from the SSs.
3. If a station is outside of the airport area but needs connection to the AeroMACS network, an RS (as opposed to a BS) can be used to establish the connection.
4. Coverage to single point assets on the airport surface that are outside of the BSs coverage area can readily be rendered by an RS. This may be particularly suitable for airport security equipment such as cameras.
5. Relays may also be used to provide coverage to other airport assets such as lighting system, navigational aid, weather sensors and wake sleep sensor, etc.
6. Cooperative relaying enables improvement in spectrum efficiency and enhancement in reliability and can substantially increase the throughput and provide *cooperative diversity*. Multihop relays and MR-BS can function together to provide cooperative diversity. The IEEE 802.16j amendment allows for generation of cooperative diversity through several mechanisms [6].

41.5 Multihop Gain Analysis

The application of multihop relay configuration enables a reduction in path loss, and therefore a link budget "gain" is resulted. The author has designated this gain as "*multihop gain*". The multihop gain can then be translated into either radio outreach extension, or network capacity improvement, or reduction in transmit power, or a combination thereof [7].

Under the following assumptions, a simple analysis can provide a raw measure for the multihop gain measured in Decibel.

- RS and SS receivers have the same sensitivity shown by S_p.
- Propagation path loss between the MR-BS and RS is represented by L_{BR} dB.
- Propagation path loss between the RS and SS is represented by L_{RS} dB.
- Direct propagation path loss between the MR-BS and SS is represented by L_{BSS} dB.

The minimum required transmit power at the RS, P_{RS}, is then given by:

$$P_{RS} = S_p(10)^{L_{RS}/10} \quad (41.1)$$

Similarly, the minimum required transmit power from MR-BS, for BS to RS transmission, P_{BR}, is:

$$P_{BR} = S_p(10)^{L_{BR}/10} \quad (41.2)$$

The minimum required power for signal transmission from the MR-BS to the RS and then to the SS, P_{BRS} is the sum of the powers given in (41.1) and (41.2):

$$P_{BRS} = P_{BR} + P_{RS} = S_p\left[(10)^{L_{BR}/10}(10)^{L_{RS}/10}\right] \quad (41.3)$$

The minimum required transmit power for direct transmission of signal from BS to SS, P_{BSS}, is determined by:

$$P_{BSS} = S_p(10)^{L_{BS}/10} \quad (41.4)$$

One can define multihop gain as the ratio of (41.4) to (41.3). This gain, G_{MH}, in dB, can be therefore calculated by (41.5).

$$G_{MH} = L_{BSS} - 10\log\left[10^{L_{BR}/10} + 10^{L_{RS}/10}\right] \quad (41.5)$$

This is intuitively satisfying and can readily be generalized for the case in which RS and SS have unequal sensitivity values. Therefore in the general case the multihop gain is a function of receiver sensitivities of SSs as well. Equation 41.5 demonstrates that multihop gain depends on the propagation path loss between various stations in the network; in other words it varies from one propagation environment to the other. The final conclusion is that the multihop gain is directly affected by the following factors in the system:

- Relay stations positioning in the network
- Propagation characteristics of the terrain through which signal travels
- Transmit power setting and distribution

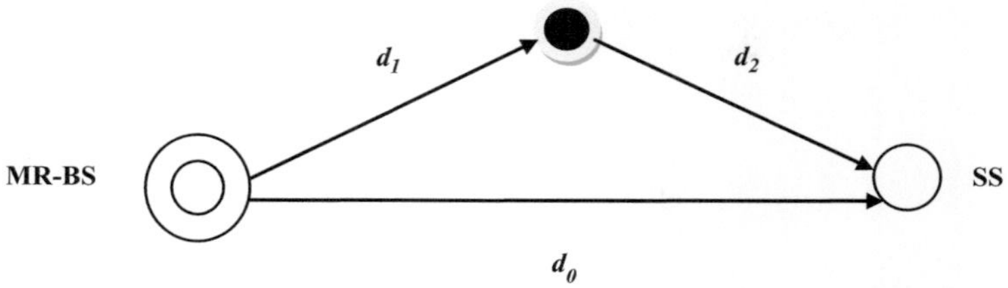

$$d_0 \leq d_1 + d_2 \rightarrow \; d_0 = \alpha(d_1 + d_2) \qquad \alpha \leq 1$$

Fig. 41.2 A multihop gain scenario

Path loss graphs can be plotted for various propagation loss models, such as Matolak model for airport surface case. One of the variables that can be taken as a parameter, is the factor α shown in Fig. 41.2.

41.6 Concluding Remarks

In this article we have discussed the benefits and challenges of employing multihop relays in AeroMACS networks. Furthermore, we have explored various classifications of multihop relays supported by IEEE 802.16j standards with an emphasis on their suitability for AeroMACS applications. Finally we have introduced the key concept of "multihop gain" which quantifies the benefits achieved when multihop relays are employed in AeroMACS systems.

References

1. Kerczewski, R., & Dyer, G.: EuroControl/FAA future communications study-phase II technology assessment. In: Proceedings of IEEE Aerospace Conference, Big Sky, Montana (2007)
2. EuroControl.: AeroMACS system requirement document. (2010)
3. IEEE Std. 802.16j™-2009, Part 16: Interface for broadband wireless access systems, Amendment1: Multihop Relay Specification (2009)
4. Gene, V., Murphy, S., Yu, Y., et al.: IEEE 802.16j relay-based wireless access network: an overview. IEEE Wirel. Commun. Mag **15**(5), 56–63 (2008)
5. Kamali, B., Wilson, J., & Kerczewski, R.: Multihop relay technology for enhancement of AeroMACS, NASA Technical Memorandum; NASA/TM-2012-217658 (2012). In: Proceedings of IEEE ICNS Conference, Herndon (2012)
6. Peters, S., Heath, R.: The future of WiMAX: multihop relaying with IEEE 802.16j. IEEE Commun. Mag **47**(1), 104–111 (2009)
7. Kerczewski, R., Kamali, B., Apaza, D., Wilson, J., & Diamond, R.: Considerations for improving the capacity and performance of AeroMACS. In: Proceedings of IEEE Aerospace Conference 2014, Big Sky, Montana (2014)

Chapter 42
Explore the Influential Factors on Maritime Accidents by Regression Approach

Chien-Chang Chou, Cheng-Yi Chen, Jeng-Ming Yih, Kur-Eng Chang, and Chung-Ping Wu

Abstract This chapter investigates the important factors on the maritime accidents in the harbors and waters surrounding Taiwan. Based on the collected data including wind, wave, tide, current, and maritime accidents, a regression model is used to discover the influential factors on the maritime accidents. It is found that various ports in Taiwan have different influential factors on the maritime accidents. Finally, some useful suggestions are given to the managers of harbors and the decision makers of governmental maritime departments to increase the navigation safety of ships in the harbors and waters surrounding Taiwan.

Keywords Regression analysis • Navigation safety • Safety management

42.1 Introduction

Marine accident analysis and navigation safety in most busy seaports are important issues in the water area within the port. Analyzing the causes of marine accidents is useful to prevent such marine accidents in the future. In the past, although many articles investigated the influential factors on navigation accidents within the water

C.-C. Chou (✉)
Department of Shipping Technology, National Kaohsiung Marine University, Kaohsiung, Taiwan

Chou's Science Research Center, Kaohsiung, Taiwan
e-mail: ccchou@mail.nkmu.edu.tw; chiench1@ms58.hinet.net

C.-Y. Chen
Department of Electrical Engineering, Cheng Shiu University, Kaohsiung, Taiwan

J.-M. Yih
Center of General Education, Min-Hwei College of Health Care Management, Tainan, Taiwan

K.-E. Chang
Department of Marine Information and Technology, National Kaohsiung Marine University, Kaohsiung, Taiwan

C.-P. Wu
Department of Shipping Technology, National Kaohsiung Marine University, Kaohsiung, Taiwan

J. Juang (ed.), *Proceedings of the 3rd International Conference on Intelligent Technologies and Engineering Systems (ICITES2014)*, Lecture Notes in Electrical Engineering 345, DOI 10.1007/978-3-319-17314-6_42

area of harbor by using various methodologies, e.g., interview, questionnaire, experience in navigation, fuzzy theory, MCDM, AHP, and so on, few identify directly the influential factors on navigation accidents by the mathematical model, e.g., regression analysis. Thus, this chapter applies regression analysis method to identify directly the influential factors on the marine accidents. The influential environmental factors on navigation accidents include the wind, wave, tide, and current. This chapter uses regression analysis method to identify directly the relationship between these influential factors and the marine accidents. Section 42.2 is the literature review. Regression analysis results are shown in Sect. 42.3. Finally, some conclusions and suggestions are given in Sect. 42.4.

42.2 Literature Review

Akten [1] analyzed the shipping casualties in the Bosporus. Aydogdu et al. [2] explored the navigation risk in the Istanbul Strait by using questionnaire survey. Aydogdu et al. [3] studied on local traffic management to improve marine traffic safety in the Istanbul Strait. Birpinar et al. [4] explored the environmental effects of maritime traffic on the Istanbul Strait. Bulut et al. [5] applied fuzzy-AHP model to shipping asset management. Chou et al. [6] used the grid to analyze the influential factors on marine accidents. Ece [7] analyzed the accidents in the Istanbul Strait. Kum et al. [8] analyzed the maritime accidents by approaching method for minimizing human error. Or and Kahraman [9] simulated the accident risk in the Istanbul Channel. Ozbas et al. [10] analyzed the risk analysis of maritime transit traffic in the strait of Istanbul by simulation methodology.

42.3 Result Analysis

In this section, results obtained using regression models are given in (42.1)–(42.12). The results for the wind, the wave, the tide, and the current in the port of Keelung are shown in (42.1)–(42.4) and in Figs. 42.1, 42.2, 42.3, and 42.4.

$$\text{Wind}: y_{11} = -0.0015x_{11} + 4.6584 \tag{42.1}$$

$$\text{Wave}: y_{12} = 0.0046x_{12} + 1.1489 \tag{42.2}$$

$$\text{Tide}: y_{13} = 0.0003x_{13} + 1.0232 \tag{42.3}$$

$$\text{Current}: y_{14} = 0.0112x_{14} + 16.886 \tag{42.4}$$

The results for the wind, the wave, the tide, and the current in the port of Taichung are shown in (42.5)–(42.8) and in Figs. 42.5, 42.6, 42.7, and 42.8.

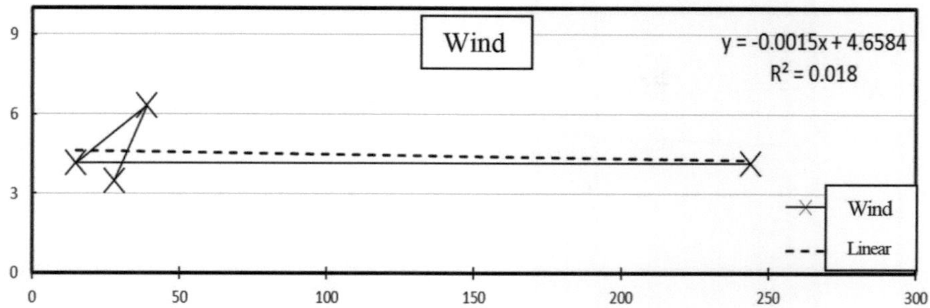

Fig. 42.1 The regression result for the wind in the port of Keelung

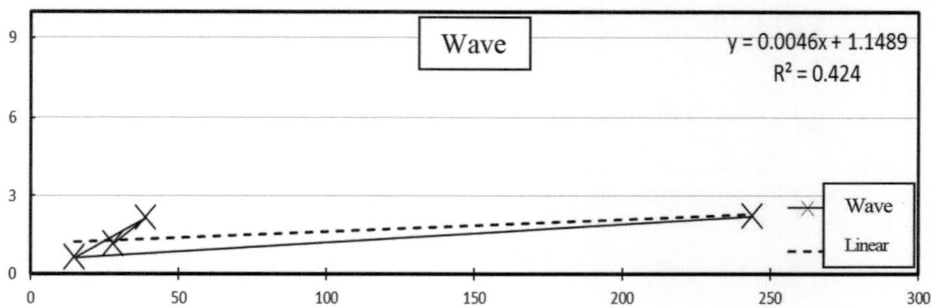

Fig. 42.2 The regression result for the wave in the port of Keelung

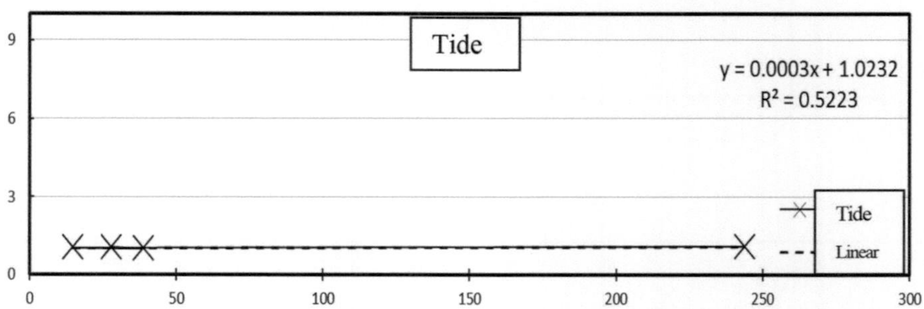

Fig. 42.3 The regression result for the tide in the port of Keelung

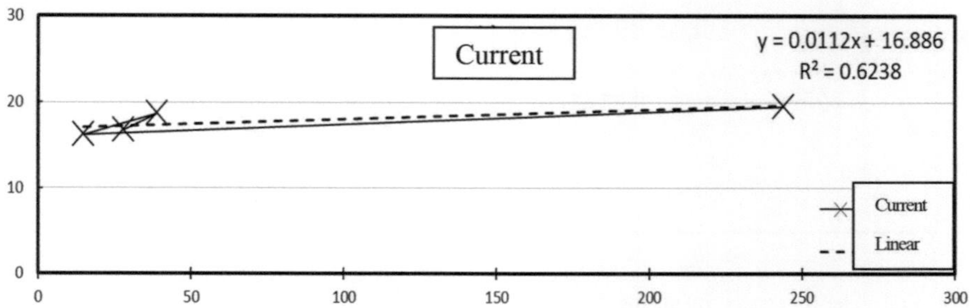

Fig. 42.4 The regression result for the current in the port of Keelung

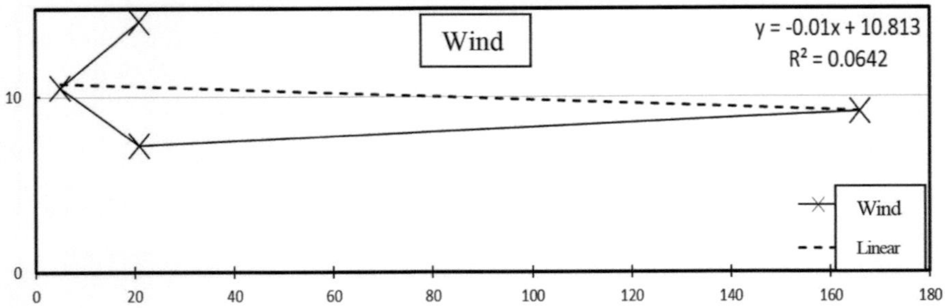

Fig. 42.5 The regression result for the wind in the port of Taichung

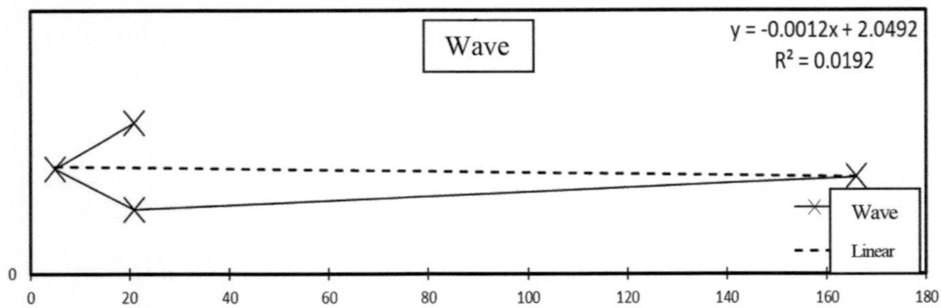

Fig. 42.6 The regression result for the wave in the port of Taichung

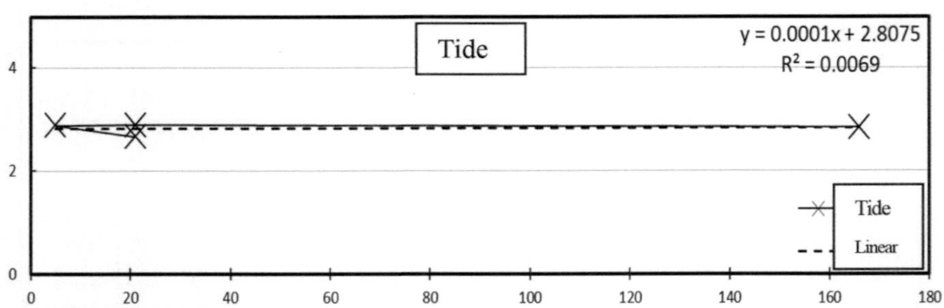

Fig. 42.7 The regression result for the tide in the port of Taichung

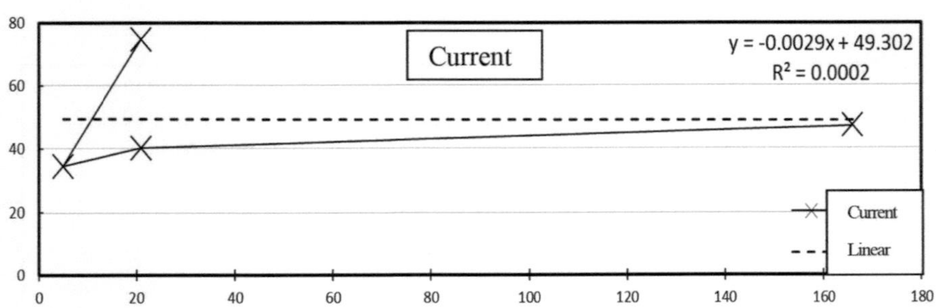

Fig. 42.8 The regression result for the current in the port of Taichung

$$\text{Wind}: y_{21} = -0.01x_{21} + 10.813 \quad (42.5)$$

$$\text{Wave}: y_{22} = -0.0012x_{22} + 2.0492 \quad (42.6)$$

$$\text{Tide}: y_{23} = 0.0001x_{23} + 2.8075 \quad (42.7)$$

$$\text{Current}: y_{24} = -0.0029x_{24} + 49.302 \quad (42.8)$$

The results for the wind, the wave, the tide, and the current in the port of Kaohsiung are shown in (42.9)–(42.12) and in Figs. 42.9, 42.10, 42.11, and 42.12.

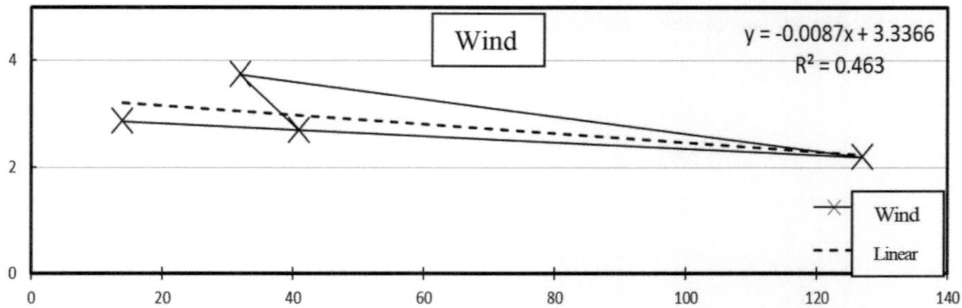

Fig. 42.9 The regression result for the wind in the port of Kaohsiung

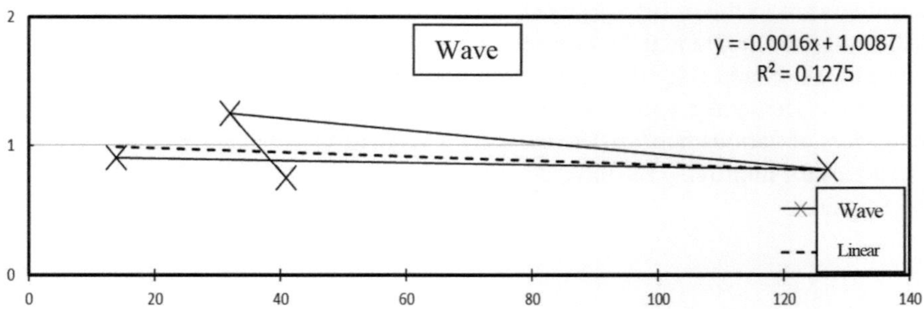

Fig. 42.10 The regression result for the wave in the port of Kaohsiung

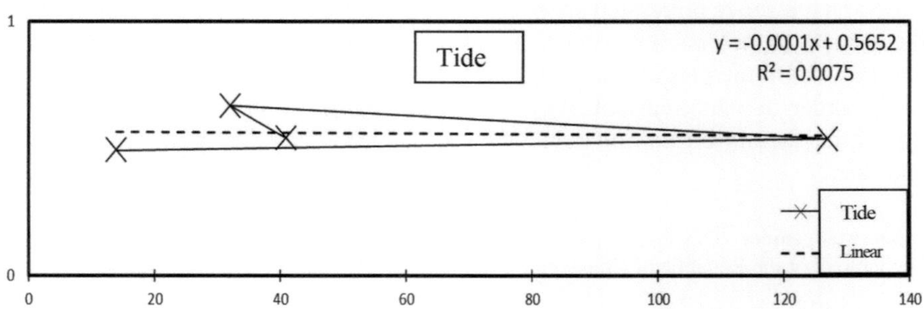

Fig. 42.11 The regression result for the tide in the port of Kaohsiung

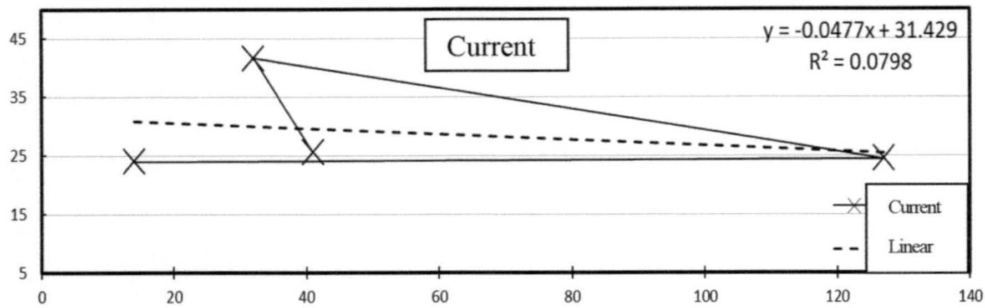

Fig. 42.12 The regression result for the current in the port of Kaohsiung

$$\text{Wind}: y_{31} = -0.0087x_{31} + 3.3366 \tag{42.9}$$

$$\text{Wave}: y_{32} = -0.0016x_{32} + 1.0087 \tag{42.10}$$

$$\text{Tide}: y_{33} = -0.0001x_{33} + 0.5652 \tag{42.11}$$

$$\text{Current}: y_{34} = -0.0477x_{34} + 31.429 \tag{42.12}$$

Finally, this chapter found that various ports in Taiwan have different influential factors on the marine accidents in the waters surrounding the harbors of Taiwan. For example, in terms of the port of Keelung, the influential factors on marine accidents are in the order of current, wave, wind, and tide. In terms of the port of Taichung, the influential factors on marine accidents are in the order of wind, current, wave, and tide. In terms of the port of Kaohsiung, the influential factors on marine accidents are in the order of current, wind, wave, and tide. Therefore, the managers of various ports in Taiwan should make different navigation policies and strategies for improving the navigation safety in the waters surrounding the harbors of Taiwan.

42.4 Conclusions

Based on the above regression analysis, this chapter finds (a) in the port of Keelung, the influential factors on marine accidents are in the order of current, wave, wind, and tide. (b) In the port of Taichung, the influential factors on marine accidents are in the order of wind, current, wave, and tide. (c) In the port of Kaohsiung, the influential factors on marine accidents are in the order of current, wind, wave, and tide.

Acknowledgement This research work was partially supported by the Ministry of Science and Technology of the Republic of China under Grant No. MOST 103-2410-H-022-002.

References

1. Akten, N.: Analysis of shipping casualties in the Bosporus. J. Navig. **57**, 345–356 (2004)
2. Aydogdu, Y.V., Yurtoren, C., Kum, S., Park, J.S., Park, Y.S.: Questionnaire survey on the risk perception in the Istanbul Strait. J. Navig. Port Res. **7**, 34 (2010)
3. Aydogdu, Y.V., Yurtoren, C., Kum, S., Park, J.S., Park, Y.S.: A study on local traffic management to improve marine traffic safety in the Istanbul Strait. J. Navig. **65**, 99–112 (2012)
4. Birpinar, M.E., Talu, G.F., Gonencgil, B.: Environmental effects of maritime traffic on the Istanbul Strait. Environ. Monit. Assess. **152**, 13–23 (2009)
5. Bulut, E., Duru, O., Keçeci, T., Yoshida, S.: Use of consistency index, expert prioritization and direct numerical inputs for generic fuzzy-AHP modeling: a process model for shipping asset management. Expert Syst. Appl. **39**, 1911–1923 (2012)
6. Chou, C.C., Chen, C.Y., Chang, P.C., Chang, T.M., Hsu, H.Z., Chang, K.E.: Use the grid to analyze the influential factors on marine accidents. Lect. Notes Electr. Eng. **293**, 867–874 (2014)
7. Ece, J.N.: The accident analysis of the Istanbul Strait from the points of safe navigation and environment and evaluation of innocent passage. Ph.D. Thesis, Institute of Science and Technology, Gazi University (2006)
8. Kum, S., Fuchi, M., & Furusho, M.: Analysing of maritime accidents by approaching method for minimizing human error. Proceedings of IAMU AGA-7, "Globalization and MET," Part 2, pp. 392–409 (2006).
9. Or, I., Kahraman, I.: A simulation study of the accident risk in the Istanbul Channel. Int. J. Emerg. Manag. **1**(2), 110–124 (2002)
10. Ozbas, B., Or, I., Uluscu, O.S., Altiok, T.: Simulation-based risk analysis of maritime transit traffic in the strait of Istanbul. Int. J. Mar. Navig. Saf. Sea Transport. **3**, 295–300 (2009)

Chapter 43
Analytical Model of Pipe Inspection Robot Using Flexible Pneumatic Cylinder

Haojun Qiu, Shujiro Dohta, Tetsuya Akagi, So Shimooka, and Shinsaku Fujimoto

Abstract A pipe inspection robot is useful to reduce the inspection cost. In our previous study, a novel pipe inspection robot using a flexible pneumatic cylinder that can be driven even if it bends has been proposed and tested. The built-in pneumatic driving system using a tiny embedded controller and small-sized valves have also been proposed and tested to decrease the mass of the robot for increasing inspection area. In this chapter, in order to find out the optimal driving pattern and length of the robot, an analytical model of the pipe inspection robot is proposed. The model consists of two on/off control valves and a sliding mechanism which is composed of a flexible pneumatic cylinder.

Keywords Pipe inspection robot • Flexible pneumatic cylinder • Analytical model

43.1 Introduction

From the Japan Water Works Association survey in 2012, it is found that the water supply pipe with length of about 30,800 km corresponding to 5 % of the total length had been already passed the statutory useful life of 40 years [1]. Water pipeline is very complex and includes enormous numerical corners and joints according to the development of human life. In the nuclear power plants, complex pipelines were also used. They require the pipe inspection to prevent serious accidents. The cost of the inspection will be reduced by using inspection robot. Under these backgrounds, many researchers study the inspection robots for pipe [2–5]. In the case of pipe inspection, the robot is required higher mobility. In order to get the higher mobility in a narrow and complex area, a variable structure that can change the body of the robot according to the pipe condition will be one of the solutions for this problem [4]. In ideal, it is more desirable that the shape of the robot changes naturally because of the lower energy consumption, decreasing the time for traveling. In our

H. Qiu (✉) • S. Dohta • T. Akagi • S. Shimooka • S. Fujimoto
Department of Intelligent Mechanical Engineering, Okayama University of Science, 1-1 Ridai-cho, Kita-ku, Okayama 700-0005, Japan
e-mail: t13rm02kk@ous.jp

J. Juang (ed.), *Proceedings of the 3rd International Conference on Intelligent Technologies and Engineering Systems (ICITES2014)*, Lecture Notes in Electrical Engineering 345, DOI 10.1007/978-3-319-17314-6_43

pervious study, a flexible sliding mechanism using the flexible pneumatic cylinders that can travel in the narrower space smoothly by changing the body according to the pipe had been proposed and tested [6–8]. In the next step, it is necessary to build an analytical model of the tested robot in order to predict the most appropriate driving pattern and optimal length of the robot. In this chapter, an analytical model of the robot is proposed. The analytical model consists of two on/off control valves and a sliding mechanism. The model of the on/off valve includes an electric circuit, equations of motion of an armature, and the fluid flow in the valve. The model of a sliding mechanism includes a flexible pneumatic cylinder and connecting pipes.

43.2 Flexible Pneumatic Cylinder

Figure 43.1 shows the construction of the flexible pneumatic cylinder developed before [6]. The cylinder consists of a flexible tube as a cylinder and gasket, two steel balls as a cylinder head, and a slide stage that can slide along the outside of the tube. The tube between two steel balls is pinched by the slide stage. The operating principle of the cylinder is as follows. When the supply pressure is applied to one side of the cylinder, the inner steel balls are pushed. At the same time, the steel balls push the brass rollers and then the slide stage moves while it deforms the tube. Table 43.1 shows the properties of the flexible pneumatic cylinder. The minimum radius of curvature of the cylinder, a maximum working pressure, and an allowable

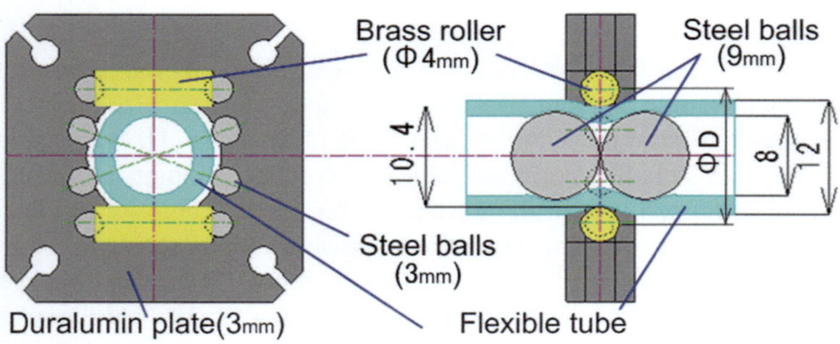

Fig. 43.1 Flexible pneumatic cylinder

Table 43.1 Properties of the cylinder

Min. driving pressure	130 kPa
Generated force	5 N (input: 500 kPa)
Max. moving speed	More than 1 m/s
Mass (stroke of 1 m)	Less than 0.1 kg
Min. radius of curvature	About 30 mm
Max. working pressure	600 kPa
Working temperature	From −20 to 60 °C
Movement	Push–pull actions

working temperature depend on the properties of the soft polyurethane tube (SMC Co. Ltd., TUS 1208). The minimum driving pressure of the cylinder is 130 kPa. The slide stage of the cylinder can move on the bending tube smoothly. The pushing motion of the cylinder can be also realized.

43.3 Pipe Inspection Robot and Control System

Figure 43.2 shows the pipe inspection robot using flexible pneumatic cylinder. The robot consists of the minicamera, the bending unit, the sliding mechanism, built-in valve unit, and built-in controller. The length of the robot is 1.65 m. The whole mass including the tested pneumatic driving system is 0.8 kg. The right upper photograph in Fig. 43.2 also shows the supply cable for air and electric signal. The line is the typical pneumatic supply tube coiled five-core flat cable. Five-core cable is used as a line for electric power supply (Vcc and GND), two serial communication lines (TXD and RXD), and a camera signal line. The mass of the line is 400 g with the length of 20 m. Figure 43.3 shows the relation between the pulling force of

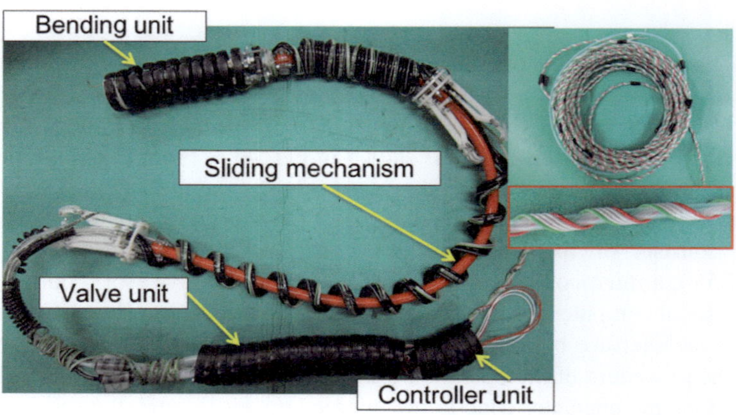

Fig. 43.2 Pipe inspection robot

Fig. 43.3 Investigation of searching area

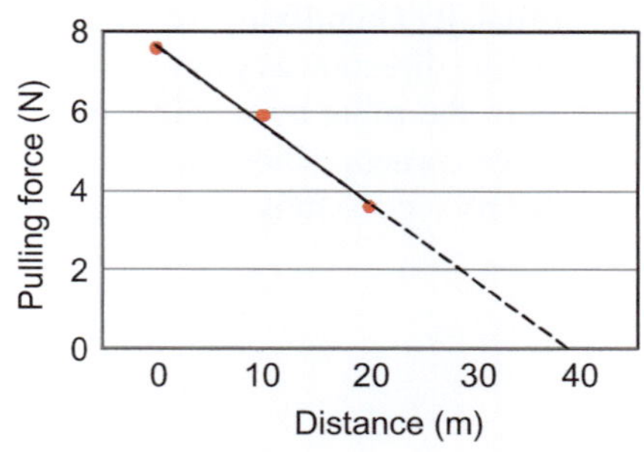

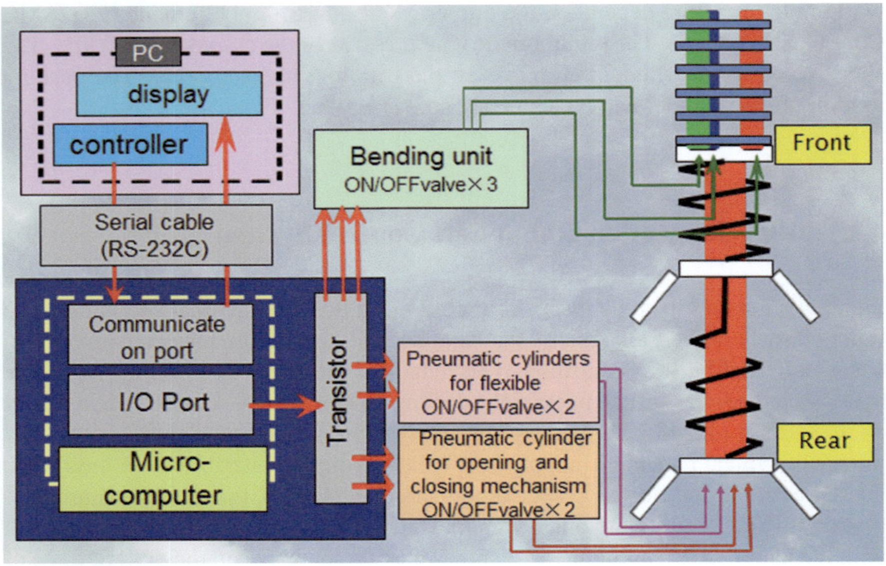

Fig. 43.4 Schematic diagram

the robot and measuring point on the supply cable from the robot end. In the experiment, the robot with the supply cable of 20 m was driven so as to move forward in the pipe with the inner diameter of 100 mm. The pulling force was measured by the force sensor and is shown in Fig. 43.3. From Fig. 43.3, it can be seen that the searching distance of the tested robot is about 38 m.

Figure 43.4 shows the schematic diagram of the control system of the robot. The system consists of the tested robot, seven on/off valves (SMC Co. Ltd., S070DC5V), a microcomputer (Renesas Co. Ltd. R8C/M12) as a robot controller and a personal computer (PC for short) for sending commands. The microcomputer works as an interface between the operator command given through PC and the robot. The procedure of the operation using the control system is as follows: First, the PC sends the character code to the microcomputer through the serial communication cable and port by the operator. The microcomputer selects to drive the valve based on the control algorithm for input code through the I/O port and the transistors. By simplifying the complex motion of the robot such as moving forward or changing direction as shown in Fig. 43.5, the controller can realize the sequential motion of the robot by sending only one character code. In the experiment, there were three corners (elbow and T-branch) in complex pipe, and the robot completed all motions successfully.

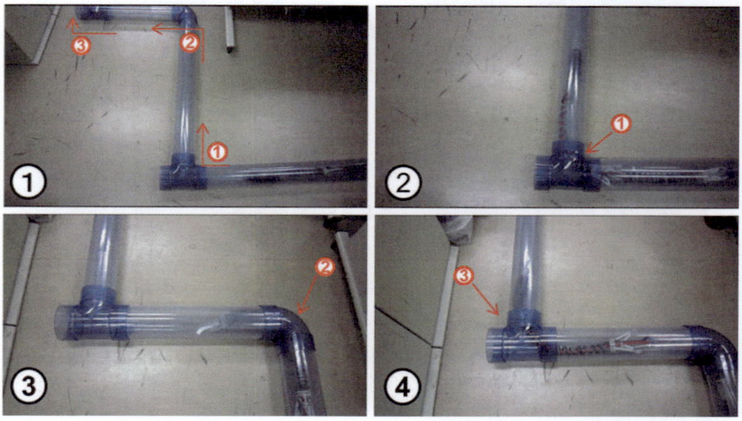

Fig. 43.5 Movement of inspection robot in complex pipe

Fig. 43.6 Analytical model of on/off valve

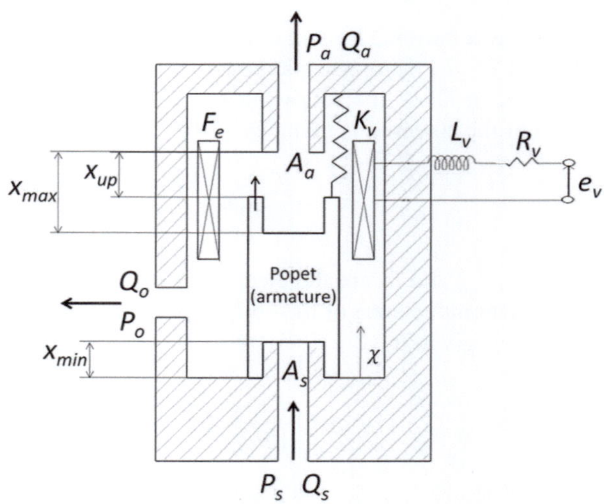

43.4 Analytical Model of Sliding Mechanism

Figure 43.6 shows the analytical model of an on/off valve. The size of the tested on/off valve is 12 × 33 × 7 mm, and the mass is only 6 g. The maximum output flow rate is 15 L/min at the supply pressure of 500 kPa. The valve consists of a solenoid, a poppet (armature with a spring), and 3 ports. Figure 43.7 shows the analytical model of the tested sliding mechanism which consists of two on/off valves and a flexible pneumatic cylinder. The generated electromagnetic force F_{ej} in Fig. 43.6 is given by

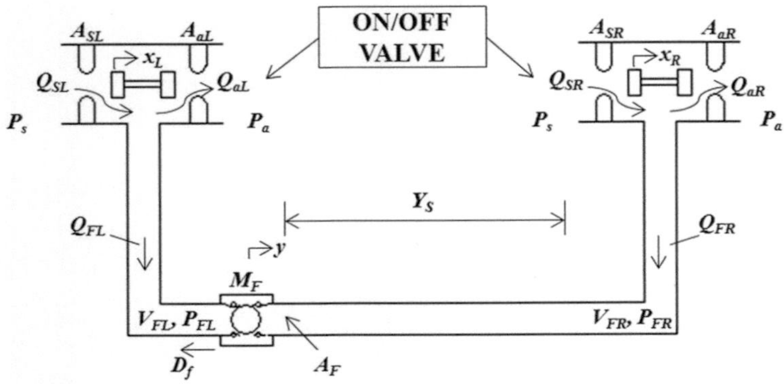

Fig. 43.7 Analytical model of the sliding mechanism connected with two on/off valves

$$F_{ej} = K_c i_j. \tag{43.1}$$

where i_j is a current, K_c is a conversion factor from the current to the force and subscript j (=L, R) represents the left-side valve (L) or the right-side valve (R) in Fig. 43.7. In all of the following equations, j represents L and R. The electrical circuit of the solenoid is given by

$$R_v i_j + L_v \frac{di_j}{dt} = e_{vj}. \tag{43.2}$$

where R_v, e_{vj}, and L_v are the electrical resistance, an input voltage of the solenoid coil, and an inductance of the coil, respectively. The equation of the motion of the armature is given by

$$M_v \frac{d^2 x_j}{dt^2} + C_v \frac{dx_j}{dt} + K_v x_j = F_{ej}. \tag{43.3}$$

where x_j, M_v, C_v, and K_v are displacements of the amature, mass of the armature, viscous damping coefficient, and spring constant, respectively.

The sectional area of the valve will change based on the displacement x_j. The sectional area of the supply port A_{sj} and the area of exhaust port A_{aj} start to change at the point of a certain displacement $x_{\min}$ (=0.27 mm) of the armature. Then both of the supply port and exhaust port open when $x_j < x_{\text{up.}}$ Where $x_{\text{up}} = x_{\max} - x_{\min}$. When $x_j = x_{\text{ up}}$, the exhaust port is closed and armature stops at the maximum displacement $x_{\max}$. Therefore, both sectional area A_{sj} and A_{aj} are given by (43.4a–43.4c). The sectional area must be chosen as the narrowest area of the flow field. Therefore, in the case if the cylindrical sectional area of the valve seat (orifice) is smaller than the sectional area of the orifice that is maximum value A_{aj0} or A_{sj0}, the effective sectional area is changed as the linear function of the displacement x.

$$A_{sj} = 0,\ \ A_{aj} = \sqrt{4\pi A_{a\,j0}}\left(x_{\mathrm{up}} - x\right)\left(0 < x_j < x_{\mathrm{min}}\right) \tag{43.4a}$$

$$\begin{aligned} A_{sj} &= \sqrt{4\pi A_{s\,j0}}\left(x_j - x_{\mathrm{min}}\right),\ \ A_{aj} \\ &= \sqrt{4\pi A_{a\,j0}}\left(x_{\mathrm{up}} - x_j\right)\left(x_{\mathrm{min}} < x_j < x_{\mathrm{up}}\right) \end{aligned} \tag{43.4b}$$

$$A_{sj} = \sqrt{4\pi A_{s\,j0}}\left(x_j - x_{\mathrm{min}}\right),\ \ A_{aj} = 0\left(x_{\mathrm{up}} < x_j < x_{\mathrm{max}}\right) \tag{43.4c}$$

Mass flow rates Q_{sj} and Q_{aj} shown in Fig. 43.7 are given as follows.

$$Q_{sj} = A_{sj}P_s\sqrt{\frac{2}{RT}}f(z),\ \ z = \frac{P_{Fj}}{P_s} \tag{43.5a}$$

$$Q_{aj} = A_{aj}P_{Fj}\sqrt{\frac{2}{RT}}f(z),\ \ z = \frac{P_a}{P_{Fj}} \tag{43.5b}$$

where, A_{sj}, A_{aj}, P_s, P_{Fj}, P_a, R, and T mean the sectional area of the supply orifice, the sectional area of the exhaust orifice, the supply pressure, pressure in the connected chamber, atmospheric pressure, gas constant, and absolute temperature, respectively. The function $f(z)$ that expresses the state of flow is given by

$$f(z) = \sqrt{\frac{\kappa}{\kappa - 1}\left(z^{\frac{2}{\kappa}} - z^{\frac{(\kappa+1)}{\kappa}}\right)}\,(0.528 < z \le 1) \tag{43.6a}$$

$$f(z) = \sqrt{\frac{\kappa}{\kappa + 1}\left(\frac{2}{\kappa + 1}\right)^{\frac{2}{(\kappa-1)}}}\,(0 \le z \le 0.528). \tag{43.6b}$$

where κ means a specific heat ratio (=1.4).

The output flow rate Q_{Fj} of the cylinder is given by

$$Q_{Fj} = Q_{sj} - Q_{aj}. \tag{43.7}$$

Then, the change of the pressure in the chamber of the flexible pneumatic cylinder is given by

$$\frac{dP_{Fj}}{dt} = \frac{\kappa RT}{V_{Fj}}Q_{Fj} - \frac{\kappa P_{Fj}}{V_{Fj}} \cdot \frac{dV_{Fj}}{dt}. \tag{43.8}$$

The volume in the chamber is given as follows.

$$V_{FR} = A_F \cdot (Y_{SR0} + Y_S) - A_F \cdot y + V_{0R} \tag{43.9a}$$

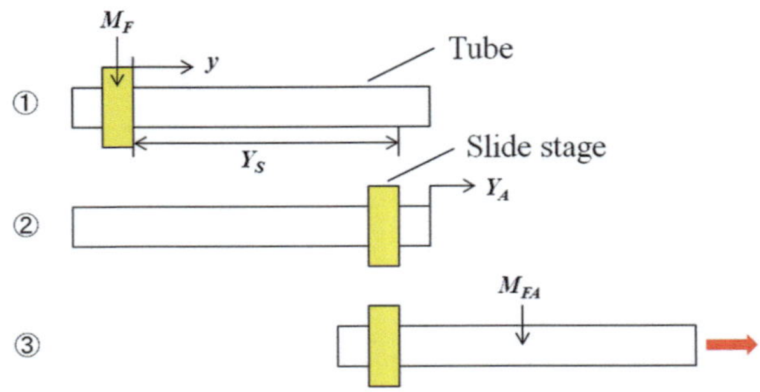

Fig. 43.8 Driving pattern of the sliding mechanism

$$V_{FL} = A_F \cdot Y_{SL0} + A_F \cdot y + V_{0L} \quad (43.9b)$$

where, A_F and Y_S are sectional area of the cylinder and stroke of the cylinder, respectively. Y_{SR0} and Y_{SL0} are minimum distance on the right side and on the left side, respectively. y shows the relative displacement of the slide stage.

Figure 43.8 shows the driving pattern of the sliding mechanism. The sliding mechanism is driven as follows. First, the tube is fixed and the left chamber is pressurized. Then the slide stage with a mass of M_F moves to the right with stroke Y_S (Pattern A shown in Fig. 43.8①–②). The equation of motion for this case can be expressed by

$$M_F \frac{dy^2}{dt^2} = A_F(P_{FL} - P_{FR}) - C_F \frac{dy}{dt} - D_f. \quad (43.10a)$$

Next, the slide stage is fixed and the tube with a mass of M_{FA} moves to the right (Pattern B, shown in Fig. 43.8③). The equation of motion for this case is given by

$$M_{FA} \frac{dy_A{}^2}{dt^2} = A_F(P_{FR} - P_{FL}) - C_{FA} \frac{dy_A}{dt} - D_{fA}. \quad (43.10b)$$

where, y_A is the relative displacement of the tube when the slide stage is fixed. C_F, C_{FA}, D_f, and D_{fA} mean viscous damping coefficients and frictional force, respectively. The total moving distance of the robot L_y is given by the following equation as a sum of y_A.

$$L_y = \sum y_A \quad (43.11a)$$

The relation between y and y_A is given by

$$y = -y_A. \quad (43.11b)$$

The nonlinear friction D_f can be expressed as follows.

When $\left|\frac{dy_A}{dt}\right| \le \varepsilon_v$,

$$D_f = A_F(P_{FL} - P_{FR}) \cdot \mathrm{sgn}\left(\frac{dy_A}{dt}\right) \quad (|A_F(P_{FL} - P_{FR})| \le F_s) \tag{43.12a}$$

$$D_f = F_s \cdot \mathrm{sgn}\left(\frac{dy_A}{dt}\right) \quad (|A_F(P_{FL} - P_{FR})| > F_s) \tag{43.12b}$$

When $\left|\frac{dy_A}{dt}\right| > \varepsilon_v$,

$$D_f = F_c \cdot \mathrm{sgn}\left(\frac{dy_A}{dt}\right). \tag{43.12c}$$

where F_s, F_c, ε_v mean static friction, Coulomb friction and small switching velocity between static and Coulomb friction. By using (43.1–43.12c), the behavior of the tested sliding mechanism can be predicted.

43.5 Conclusions

This study can be summarized as follows. The construction of the developed pipe inspection robot and its operating principle was described briefly. Then, the analytical model of the tested robot to estimate the performance of the robot theoretically and to carry out the optimal design of the robot was built and described. The optimal driving pattern and the optimal length of the robot can be predicted by using the proposed analytical model.

References

1. http://ja.wikipedia.org/wiki/%E6%B0%B4%E9%81%93%E7%AE%A1
2. Yuan, J., Wu, X., Huang, C.: Development of an inspection robot for long distance transmission pipeline on-site overhaul. Int. J. Ind. Robot **36**, 546–550 (2009)
3. Fukuda, T., Hosokai, H., Uemura, M.: Rubber gas actuator driven by hydrogen storage alloy for in-pipe inspection mobile robot with flexible structure. Proc. IEEE Int. Conf. Robot. Autom. **3**, 1847–1852 (1989)
4. Dertien, E., Stramigioli, S., Pulles, K.: Development of an inspection robot for small diameter gas distribution mains. Proc. IEEE Int. Conf. Robot. Autom. 281–294 (2011)
5. Nagatani, K., Kuze, M., Yoshida, K.: Development of a transformable mobile robot with a variable wheel diameter. J. Rob. Mechatronics **19**(3), 252–257 (2007)
6. Akagi, T., Dohta, S.: Development of wearable pneumatic actuator and multiport pressure control valve. J. Rob. Mechatronics **17**(5), 529–536 (2005)

7. Qiu, H., Dohta, S., Akagi, T., Wang, L.: Improvement of pipe inspection robot using flexible pneumatic cylinder. Proc. FLUCOME 2013, OS4-03-3.pdf, 1–10, (2013)
8. Akagi, T., Dohta, S., Wang, L., Kawasaki, T.: Development of sliding type inspection robot using flexible pneumatic cylinder. Int. J. Adv. Mechatron. Syst. **5**, 87–94 (2013)

Chapter 44
Thermal Resistance Characterization of SiGe-Based HBTs on Thick-Film and Thin-Film SOI

Shu-Hui Liao

Abstract The nonlinear behavior of thermal resistance (R_{Th}) of SiGe HBTs on both thick-film and thin-film SOI substrates were investigated at different levels of dissipated power densities by the ISE-TCAD simulator in this chapter. We examined both of the buried oxide thickness and the silicon thickness effects on the device characteristics for thermal resistance. DESSIS-ISE was used to compute the temperature distribution from the given power. Simulation results suggest that the amount of self-heating is strongly dependent on device structure. Owing to the thick-film SiGe on SOI works like a bulk SiGe HBT, the enhanced silicon thickness could degrade the thermal resistance. Thus it needs to be carefully considered in device design. The thermal resistance characteristics revealed for SiGe HBTs on thick-film and thin-film SOI may help us to establish more accurate thermal models for reliability of circuit design and device technology optimization.

Keywords SOI • HBT • SiGe • Thermal resistance

44.1 Introduction

SiGe BiCMOS using SOI technology is one of the preferred platforms for wireless and communication application. It is widely used for low-power, high-speed, and large-dynamic range RF application. Many research results have demonstrated that the Si/SiGe HBT on thin-film SOI substrate with a fully depleted collector can offer a higher base-collector breakdown voltage ($\mathrm{BV_{CEO}}$), higher early voltage (V_{A}), and a better $\mathrm{BV_{CEO}}$–f_{T} trade-off by experimental studies and simulations [1–6]. But, there are several advantages of fabricating the Si/SiGe HBTs on SOI substrate, including one key drawback. This drawback is the well-known effect called self-heating where the heat generated at the base/collector junction must be conducted

S.-H. Liao (✉)
Department of Electronic Engineering, Chung Chou University of Science and Technology, Changhua County 51003, Taiwan
e-mail: liao@dragon.ccut.edu.tw

J. Juang (ed.), *Proceedings of the 3rd International Conference on Intelligent Technologies and Engineering Systems (ICITES2014)*, Lecture Notes in Electrical Engineering 345, DOI 10.1007/978-3-319-17314-6_44

away from the active device area or else the device will be operated at some higher temperature that would reach an adverse circumstance in its performance as well as reliability.

In a bulk Si/SiGe HBT, heat can spread quickly through the sub-collector into the silicon substrate. For a Si/SiGe HBT on SOI substrate, however, the path of heat transfer to the Si substrate is effectively blocked by the buried oxide layer that has a much lower thermal conductivity. Usually, self-heating degrades not only the performance of the transistor but also potentially reduces its long-term reliability. In this work we consider Si/SiGe HBTs on thick-film and thin-film SOI substrates. Two-dimensional numerical simulation of heat transfer was verified by the numerical analysis performed by the semiconductor device simulator DESSIS-ISE [7].

44.2 Simulation Details

The schematically illustrated structure of the npn Si/SiGe HBT on thin-film (or -thick-film) SOI substrate discussed in this work is shown in Fig. 44.1. In the simulations, the thickness of the n^{-}-collector silicon layer T_{Si} and the buried oxide thickness T_{OX} of the devices are the only two variable parameters while all other parameters are kept in constant. The T_{Si} of the Si/SiGe HBT on thick-film SOI substrate is 1.5 μm, while it is 0.1 μm for the device on thin-film SOI substrate. For both type of devices, the distance from n^{+} reach-through to the edge of the intrinsic base L_{col} is 0.1 μm, and the buried oxide thickness T_{OX} is varied from 0.05 to 0.3 μm. The base width used is about 75 nm and the emitter size is 0.14×100 μm. Uniform doping concentrations were used in the emitter (1×10^{20} cm^{-3}), base (1×10^{18} cm^{-3}), collector (1×10^{17} cm^{-3}), and substrate (1×10^{15} cm^{-3}) of the device, respectively. In addition, the doping concentrations of p^{+}-extrinsic base region, n^{+}-reach-through region and n^{+}-sub-collector region are all 1×10^{20} cm^{-3}. For the sake of simplicity, only structures of single-emitter are adopted in this study and no metal depositions are considered. The substrate of the device is grounded in all our simulations.

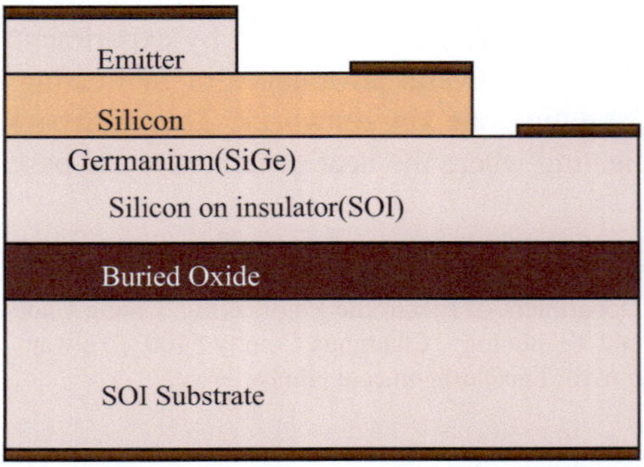

Fig. 44.1 Schematic cross-section of a SiGe HBT on thin-film (or thick-film) SOI substrate

All the simulations in this work are made by a two-dimensional electrical device simulator DESSIS-ISE. The system of differential equations comes from solving the Boltzmann transport equation and there are two main combinations of equations used. Drift-diffusion model is for isothermal simulations with stationary transport in low-power-density devices with long-active regions. Thermodynamic model extends the drift-diffusion approach to account for electrothermal effects, under the assumption that the charge carriers are in thermal equilibrium with the lattice. Thermodynamic model is taking into account self-heating effects in high power-density with long active regions. The drift-diffusion approximation and thermodynamic approximation models were used to solve the carrier continuity and Poisson's equation. The simulator allows for simultaneous solution of the Posson's equation, the continuity equation for electrons and holes and the heat flow equation.

To calculate the temperature distribution in the device due to self-heating, the following equation is solved [8]:

$$c\frac{\partial T}{\partial t} - \nabla \cdot \kappa \nabla T = -\nabla \cdot \left[(P_\text{n}T + \phi_\text{n})\vec{J}_\text{n} + (P_\text{n}T + \phi_\text{p})\vec{J}_\text{p}\right] - \left(E_\text{C} + \frac{3}{2}k_\text{B}T\right)\nabla \cdot \vec{J}_\text{n} - \left(E_\text{V} + \frac{3}{2}k_\text{B}T\right)\nabla \cdot \vec{J}_\text{p} + qR(E_\text{C} - E_\text{V} + 3k_\text{B}T) \tag{44.1}$$

where κ is the thermal conductivity, c is the lattice heat capacity, P_n and P_p are the absolute thermoelectric power, J_n is the electron current density, J_p is the hole current density, ϕ_n and ϕ_p are the electron and the hole quasi-Fermi potentials, E_C and E_V are the conduction and valence band energies, respectively, and R is the recombination rate.

Thermal resistance is a mathematical concept analogous to the electrical resistance. Several methods have been proposed and employed to characterize the thermal resistance R_Th. Thermal resistance is a measure of junction temperature T_junction rise relative to the ambient temperature due to certain power dissipation inside a transistor

$$T_\text{junction} = T_\text{ambient} + R_\text{Th} \times P_\text{diss} \tag{44.2}$$

where R_Th is the thermal resistance and P_diss is the power dissipation inside a transistor [9–12].

Bipolar transistors are often operated at high-power levels, thus are very sensitive to temperature variations. The amount of self-heating is strongly dependent on device structure. Self-heating is a known disadvantage in all SOI technology, especially for those using very thin SOI films [13]. In a bulk SiGe HBT, heat is quickly spread through the subcollector into the silicon substrate. For a SiGe HBT on SOI, however, this heat transfer path to the Si substrate is effectively blocked by the buried oxide layer, which has a much lower thermal conductivity. Self-heating degrades not only the performance of the transistor, but also potentially jeopardizes its long-term reliability [14]. The temperature increase due to power dissipation in the device self-heating effect affects both collector and base currents. The use of SOI is known to bring several benefits for advanced SiGe HBTs. The buried oxide significantly degrades thermal conduction from device to package. During the design of power transistors, simulation tools are needed to estimate the increase of junction temperature for different layouts on SOI substrate.

44.3 Results and Discussions

Figures 44.2a and 44.3a give the display of two-dimensional temperature distributions in the cases of SiGe HBTs on thick-film and thin-film SOI, respectively. Figures 44.2b and 44.3b are plots of the lattice temperature for the two structures with the vertical cutlines going through the centers of the emitters. It was found that in the device of SiGe HBT on thick-film SOI substrate, heat propagation occurred in all directions. Indeed, the behavior of the thick-film SOI device is similar to that of a conventional bulk device. On the other hand, in the case of SiGe HBT on thin-film SOI substrate heat first propagated laterally, then vertically.

The impact of self-heating effects SiGe HBTs on SOI substrates can be analyzed in several different areas. The first investigated is the effect on the plot of I_C as a function of V_{CE}, the well-known breakdown plot. The results shown in Fig. 44.4 indicate that in the higher voltage regimes the collector current will be higher when self-heating occurs, and this will give an effect on Early voltage V_A. In our simulation, the Early voltage of SiGe HBTs on thin-film SOI substrate device is slightly larger than that of on thick-film SOI substrate device.

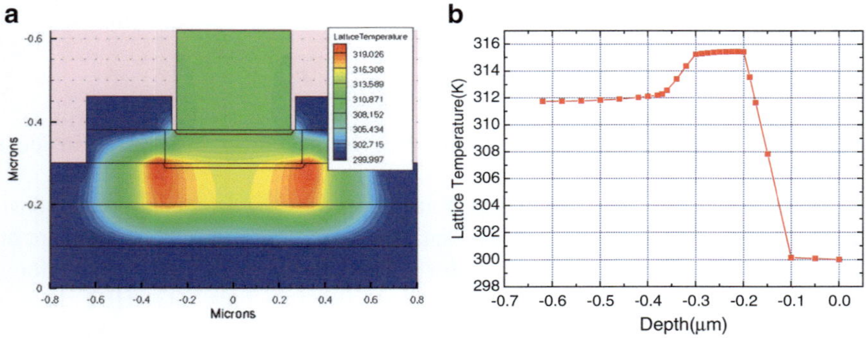

Fig. 44.2 (**a**) Two-dimensional temperature profile of Si/SiGe HBT on thin-film SOI substrate. (**b**) Simulated lattice temperature of Si/SiGe HBT on thin-film SOI substrate as a function of device depth. Cut is taken through the center of emitter

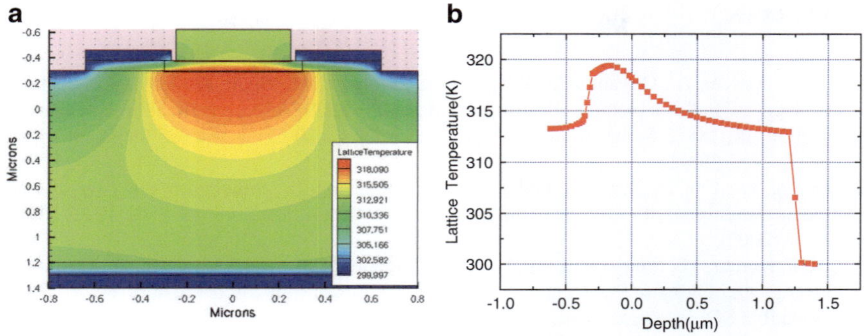

Fig. 44.3 (**a**) Two-dimensional temperature profile of Si/SiGe HBT on thick-film SOI substrate. (**b**) Simulated lattice temperature of Si/SiGe HBT on thick-film SOI substrate as a function of device depth. Cut is taken through the center of emitter

Fig. 44.4 Common-emitter output characteristics of the SiGe HBTs on thick-film and thin-film SOI substrates under various base current conditions. The collector current is higher in the higher voltage regimes. The Early voltage of SiGe HBTs on thin-film SOI substrate device is slightly larger than that of thick-film SOI device

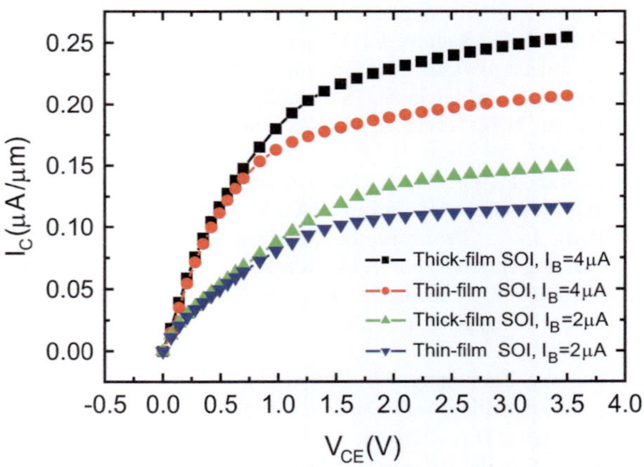

44.4 Conclusions

Simulation results indicate that the behavior of the SiGe HBT on thick-film SOI is similar to that of a bulk device. Heat propagation occurred in all directions. On the other hand, in the case of SiGe HBT on thin-film SOI substrate heat first propagated laterally, then vertically. Besides, in the condition of device on thin-film SOI substrate, self-heating effect is not serious under the situation of smaller oxide thickness. From the investigation of the Si/SiGe HBT on SOI substrate, the BOX thickness is of benefit to optimize and better understand the performance of the device. In addition, the enhancement of silicon thickness can degrade the thermal resistance and thus needs to be carefully considered in device design.

References

1. Cai, J., Ajmera, A., Ouyang, C., Oldiges, P., Steigerwalt, M., Stein, K., Jenkins, K., Shahidi, G., Ning, T.: Fully-depleted-collector polysilicon-emitter SiGe-base vertical bipolar transistor on SOI. In: 2002 Symposium on VLSI technology, 2002. Digest of technical papers, pp. 172–173 (2002)
2. Ouyang, Q., Cai, J., Ning, T., Oldiges, P., & Johnson, J.B.: Simulation study on thin SOI bipolar transistors with fully or partially depleted collector. Proceedings of the IEEE BCTM, 28–31 (2002)
3. Cai, J., Kumar, M., Steigerwalt, M., Ho, H., Schonenberg, K., Stein, K., Chen, H., Jenkins, K., Ouyang, Q., Oldiges, P., Ning, T.: Vertical SiGe-base bipolar transistors on a CMOS-compatible SOI substrate. IEEE BCTM Proceedings, pp. 215–218 (2003)
4. Avenier, G., Schwartzmann, T., Chevalier, P., Vandelle, B., Rubaldo, L., Dutartre, D., Boissonnet, L., Saguin, F., Pantel, R., Fregonese, S., Maneux, C., Zimmer, T., Chantre, A.: A self-aligned vertical HBT for thin SOI SiGe BiCMOS. Proceedings of the IEEE BCTM, pp. 128–131 (2005)
5. Chantre, A., Avenier, G., Chevalier, P., Vandelle, B., Saguin, F., Maneux, C., Dutartre, D., Zimmer, T.: SiGe HBT design for CMOS compatible SOI. In Proc. ISTDM, pp. 268–269 (2006)
6. Chevalier, P., Barbalat, B., Rubaldo, L., Vandelle, B., Dutartre, D., Bouillon, P., Jagueneau, T., Richard, C., Saguin, F., Margain, A., Chantre, A.: 300 GHz f_{max} self-aligned SiGeC HBT optimized towards CMOS compatibility. In Proc. IEEE BCTM, pp. 120–123 (2005)
7. DESSIS user's manual, ISE (2002)
8. Callen, H.B.: Thermodynamics and an Introduction to Thermostatistics. Wiley, New York (1985)
9. Palestri, P., Pacelli, A., Mastrapasqua, M.: Thermal resistance in $Si_{1-x}Ge_x$ HBTs on bulk-Si and SOI substrates. In Proc. BCTM, pp. 98–101 (2001)
10. Rieh, J.-S., Greenberg, D., Jagannathan, B., Freeman, G., Subbanna, S.: In Proc. Topical Meeting Silicon Monolithic Integrated Circuits in RF Systems, pp. 110–113 (2001)
11. Vanhoucke, T., Boots, H.M.J., Van Noot, W.D.: Revised method for extraction of the thermal resistance applied to bulk and SOI SiGe HBTs. IEEE Electron Device Lett. **25**, 150–152 (2004)
12. Hui, L., Zhenqiang, M., Pingxi, M., Marco, R.: Thermal resistance of SiGe HBTs at high power densities. Semicond. Sci. Technol. **22**, S68–71 (2007)
13. Dallmann, D.A., Shenai, K.: Scaling constraints imposed by self-heating in submicron SOI MOSFET. IEEE Trans. Electron Device **42**, 489–496 (1995)
14. Lim, H.T., Udrea, F., Garner, D.M., Milne, W.I.: Modelling of self-heating effect in thin SOI and partial SOI LDMOS power devices. Solid State Electron. **43**, 1267–1280 (1999)

Chapter 45
Improved Item Relational Structure Theory Based on Liu's Item Ordering Consistency Property

Hsiang-Chuan Liu and Jing-Ming Ju

Abstract The well-known Ordering Theory (OT) does not consider the non-independence property of ordered items. For improving this drawback, the Item Relational Structure Theory (IRST) and Liu's Ordering Theory (LOT) were proposed. Although IRST is more sensitive than LOT and OT, the IRST does not satisfy Liu's consistency property of ordering relationship as LOT. In this chapter, for overcoming the abovementioned drawback by mathematical analysis, an improved item relational structure theory (IIRST) was proposed, it satisfies item ordering consistency property, and it is more sensitive than LOT and OT.

Keywords Ordering theory • Item relational structure theory • Non-independence • Ordering consistency property

45.1 Introduction

To detect the item ordering relationships or directed structures of a group of students is an important issue in educational testing. Since if an easier item is a precondition of a more difficult one, then the joint probability of any student answers the easier one wrong and the more difficult one right is small. According to this reason, Airasian and Bart [1] and Bart and Krus [2] proposed the first item Ordering Theory (OT), but two items with ordering relation must not be independent, this is a drawback of it. Two improved ordering theories were proposed, one is Takeya's Item Relational Structure Theory (IRST) [3], and the other is Liu's Ordering Theory (LOT) proposed by the first author of this chapter [4].

H.-C. Liu
Department of Biomedical Informatics, Asia University, Taichung 41354, Taiwan
e-mail: lhc@asia.edu.tw

J.-M. Ju (✉)
Department of Early Childhood Education and Department of Psychology, Asia University, Taichung 41354, Taiwan
e-mail: jjm3222@gmail.com

J. Juang (ed.), *Proceedings of the 3rd International Conference on Intelligent Technologies and Engineering Systems (ICITES2014)*, Lecture Notes in Electrical Engineering 345, DOI 10.1007/978-3-319-17314-6_45

Another important property, item ordering consistency, was also proposed. It is that if two items has ordering relationship, then the easier one is a precondition of the more difficult one, and the ordering coefficient of the former to the latter is larger than that of the latter to the former, only OT and LOT satisfy this property. Furthermore, a counter example was presented that IRST does not satisfy this property; hence LOT is better than IRST.

In this chapter, the first author discovered that Takeya's ordering coefficient of IRST does not satisfy the item ordering consistency property, only if the association coefficient of two items is negative, hence IRST was revised and an improved IRST was proposed as follows. Since LOT is only dependent on the joint probability of the two items, and IIRS is not only dependent on the joint probability, but also dependent on the marginal probability of the two items, the same ordering coefficient of LOT about two different item pairs maybe have different ordering coefficients of IIRST about these two different item pairs, it means that IIRST is more sensitive than LOT, and IIRST is more reasonable and useful than before.

45.2 Ordering Theory (OT)

Bart and Krus [1] and Airasian and Bart [2] thought that for any two different items I_i and I_j in a test $I_T = (I_1, I_2, \ldots, I_n)$, if the easier item I_i is a precondition of the more difficult item I_j, then any examinee answers I_i wrong, always do so I_j. In other words, if I_i is a precondition of I_j, then the joint probability of I_i wrong and I_j correct is small. According to this consideration, they provided the first ordering relation theory of items, called the Ordering Theory (OT), for convenience of comparison, Liu [4] rewrote the definition of OT as follows;

Definition 1. Ordering Theory (OT) Let $\delta \in [0.96, 0.98]$, $P(I_i = 0, I_j = 1)$ represent the joint probability of students' answer item I_i wrong and item I_j correct

(1) The ordering coefficient from item I_i to item I_j, is defined as

$$\gamma_{ij}^{(\mathrm{OT})} = 1 - P(I_i = 0, I_j = 1) \tag{45.1}$$

(2) If $\gamma_{ij}^{(\mathrm{OT})} > \delta$, then we say that I_i is a precondition of I_j, denoted as $I_i \rightarrow_{\mathrm{OT}} I_j$, otherwise $I_i \nrightarrow_{\mathrm{OT}} I_j$. For convenience, let $\delta = 0.97$, in this chapter.

Remark *1* The ordering coefficient satisfies the logical property; the larger the value of the ordering coefficient is, the stronger the ordering relation is.

Example 1 Suppose that the joint and marginal probabilities of item I_i and I_j of a group of subjects are listed as in Table 45.1

Table 45.1 The joint probabilities of item I_i and I_j [4–6]

$P(I_i = x,\ I_j = y)$	$I_j = 1$	$I_j = 0$	$P(I_i = x)$
$I_i = 1$	0.48	0.48	0.96
$I_i = 0$	0.02	0.02	0.04
$P(I_j = y)$	0.50	0.50	1

From Table 45.1, it is known that

$$\gamma_{Ij}^{(OT)} = 1 - P(I_i = 0, I_j = 1) = 0.98 > \delta = 0.97 \Rightarrow I_i \rightarrow_{OT} I_j \tag{45.2}$$

$$\gamma_{ji}^{(OT)} = 1 - P(I_i = 1, I_j = 0) = 0.52 < \delta = 0.97 \Rightarrow I_j \not\rightarrow_{OT} I_i \tag{45.3}$$

Remark 2 In Example 1, from (45.3); we know that $I_i \rightarrow I_j$ but

$$P(I_i = 0, I_j = 1) = P(I_i = 0)P(I_j = 1) = 0.02 \rightarrow [I_i = 0] \perp\perp [I_j = 1] \tag{45.4}$$

Since if $I_i \rightarrow_{OT} I_j$, then I_i and I_j must not be independent, it leads to a contradiction, in other words, Ordering Theory has not considered that any two ordered items must be non-independent.

45.3 Item Relational Structure Theory

Since if item I_i is a precondition of item I_j, then I_i and I_j are non-independent. Takeya replaced the ordering coefficient of OT with the non-independent coefficient to propose his alternative item ordering structure theory [3]. For convenience of comparison, Liu [4] gave the modified definition of IRST as follows:

Definition 2. Item Relational Structure Theory

(1) The ordering coefficient from I_i to I_j is defined below:

$$\gamma_{ij}^{(IRST)} = 1 - \frac{P(I_i = 0, I_j = 1)}{P(I_i = 0)P(I_j = 1)} \tag{45.5}$$

(2) If $\gamma_{ij}^{(IRST)} > 0.5$, then I_i is a precondition of I_j, denoted as $I_i \rightarrow_{IRST} I_j$, otherwise $I_i \not\rightarrow_{IRST} I_j$.

Example 2 Data are the same as Example 1, from Table 45.1 [4–6]. Since I_i and I_j are independent, it is found

$$\gamma_{ij}^{(\mathrm{IRST})} = 1 - \frac{0.02}{0.5 \times 0.04} = \gamma_{ji}^{(\mathrm{IRST})} = 1 - \frac{0.48}{0.5 \times 0.96} = 0$$
$$\Rightarrow I_i \not\rightarrow_{\mathrm{IRST}} I_j, I_j \not\rightarrow_{\mathrm{IRST}} I_i \quad (45.6)$$

45.4 Liu's Ordering Theory

For improving OT, the first author of this chapter [4, 5, 6] replaced the independence degree with the association coefficient of two items, and then an improved ordering theory, called Liu's Ordering Theory (LOT) was proposed as follows.

Definition 3. Liu's Association Coefficient Liu's association coefficient of I_i and I_j (Lac_{ij}) is defined below

$$\begin{aligned} Lac_{ij} &= P(I_i = 0)P(I_j = 1) - P(I_i = 0, I_j = 1) \\ &= P(I_i = 1)P(I_j = 0) - P(I_i = 1, I_j = 0) \end{aligned} \quad (45.7)$$

Definition 4. Liu's Ordering Theory Let δ and $P(I_i = x, I_j = y)$ be defined as above,

(1) The ordering coefficient from I_i to I_j, is defined as

$$\gamma_{ij}^{(\mathrm{LOT})} = \begin{cases} 0 & \text{if} \quad Lac_{ij} = 0 \\ \gamma_{ij}^{(\mathrm{OT})} & \text{if} \quad Lac_{ij} \neq 0 \end{cases} \quad (45.8)$$

(2) If $\gamma_{ij}^{(\mathrm{LOT})} > \delta$, then I_i is a precondition of I_j, denoted as $I_i \rightarrow_{\mathrm{LOT}} I_j$, otherwise $I_i \not\rightarrow_{\mathrm{IRST}} I_j$. For convenience, let $\delta = 0.97$.

Theorem 1 Item ordering consistency property [4]

$$\text{If } Lac_{ij} \neq 0, \text{ then } P(I_i = 1) > P(I_j = 1) \Leftrightarrow \gamma_{ij}^{(\mathrm{LOT})} > \gamma_{ji}^{(\mathrm{LOT})} \quad (45.9)$$

Remark 3 Theorem 1 means that if two items are not independent, then the value of ordering coefficient of the easier one to the more difficult one is larger than that of the more difficult one to the easier one.

Example 3 Suppose that the joint and marginal probabilities of item I_i and I_j of a group of subjects are listed as in Table 45.2, [4–6].

Table 45.2 The joint probabilities of I_i and I_j

$P(I_i = x,\ I_j = y)$	$I_j = 1$	$I_j = 0$	$P(I_i = x)$
$I_i = 1$	0.08	0.82	0.9
$I_i = 0$	0.02	0.08	0.1
$P(I_j = y)$	0.1	0.9	1

From Table 45.2, it is known $Lac_{ij} \neq 0$ and

$$\begin{aligned} P(I_i = 1) = 0.9 \geq P(I_j = 1) = 0.1,\ \gamma_{ij}^{(LOT)} &= 0.98 > \gamma_{ji}^{(LOT)} \\ = 0.18,\ \gamma_{ij}^{(IRST)} = -1 < \gamma_{ji}^{(IRST)} &= -\frac{1}{81}, \end{aligned} \tag{45.10}$$

It means that LOT satisfies item ordering consistency property, but IRST does not.

45.5 Improved Item Relational Structure Theory

In this chapter, the first author proposed a modified Item ordering consistency property about IRST as Theorem 2. An improved IRST is as follows:

Theorem 2 Item ordering consistency property about IRST

$$\text{If } Lac_{ij} < 0, \text{ then } P(I_i = 1) > P(I_j = 1) \Leftrightarrow \gamma_{ij}^{(IRST)} < \gamma_{ji}^{(IRST)} \tag{45.11}$$

$$\text{If } Lac_{ij} \geq 0, \text{ then } P(I_i = 1) \geq P(I_j = 1) \Leftrightarrow \gamma_{ij}^{(IRST)} \geq \gamma_{ji}^{(IRST)} \tag{45.12}$$

Remark 4 Theorem 2 points that IRST does not satisfy the Item ordering consistency property, only if $Lac_{ij} < 0$.

Definition 5. Improved Item Relational Structure Theory

(1) The ordering coefficient from I_i to I_j, is defined as

$$\gamma_{ij}^{(IIRST)} = \begin{cases} 0 & \text{if } Lac_{ij} \leq 0 \\ \gamma_{ij}^{(IRST)} & \text{if } Lac_{ij} > 0 \end{cases} \tag{45.13}$$

(2) If $\gamma_{ij}^{(IIRST)} > 0.5$, then I_i is a precondition of I_j, denoted as $I_i \rightarrow_{IIRST} I_j$, otherwise $I_i \not\rightarrow_{IIRST} I_j$.

Theorem 4 Item ordering consistency property of IIRST

$$P(I_i = 1) \geq P(I_j = 1) \leftrightarrow \gamma_{ji}^{(\text{IIRST})} \geq \gamma_{ji}^{(\text{IIRST})} \tag{45.14}$$

Example 4 The data are the same as in Example 3, from Table 45.2, it was obtained that $P(I_i = 1) = 0.9 \geq P(I_j = 1) = 0.1$, $\gamma_{ij}^{(\text{IIRST})} = \gamma_{ji}^{(\text{IIRST})} = 0$. It shows that IIRST also satisfied the item ordering consistency property, but it is more sensitive than LOT, since LOT and OT are only dependent on the joint probability of the items, but IIRST is dependent not only on the joint probability but also on the marginal probabilities.

45.6 Conclusions

In this chapter, a modified Item Ordering Consistency Theorem was proposed, and an IIRST was also proposed. The IIRST satisfies the consistency property. Furthermore, it is more sensitive than LOT and OT, since it is based not only on the joint probability but also on the marginal probabilities of the items. LOT is an improvement from OT, as OT has not considered that any two ordered items must be non-independent. In the same way, the IIRST is an improvement from the IRST, as the latter lacks item ordering consistency property when the association coefficient between item I and item J is negative. IIRST should be considered when studying item relational structures.

Acknowledgments This work was supported in part by the National Science Council, Republic of China, under Grant NSC 100-2511-S-468-001.

References

1. Airasian, P.W., Bart, W.M.: Ordering theory: a new and useful measurement model. J. Educ. Technol. **5**, 56–60 (1973)
2. Bart, W.M., Krus, D.J.: An ordering theoretic method to determine hierarchies among Items. Educ. Psychol. Meas. **33**, 291–300 (1973)
3. Takeya, M.: Construction and utilization of item relational structure graphs for use in test analysis. J. Educ. Technol. **5**, 93–103 (1980)
4. Liu, H.-C.: Ideal Item Ordering Matrix and Liu's Ordering Theory. New Item Response Theories. Asia University, Tokyo (2012)
5. Liu, H.-C.: Liu's improved polytomous ordering theory based on extensional item ordering consistency theorem. In: International Conference on Information Management 2014, Taichung, Taiwan (2014)
6. Liu, H.-C., Shia, B.-C., & Cheng, D.-Y.: An improved Liu's ordering theory based on empirical distributive critical value. In: The 27th International Conference on Industrial, Engineering & Other Applications of Applied Intelligent Systems, Kaohsiung, Taiwan (2014)

Chapter 46
Enhancement of Digitized Old Film Using Inpainting-Based Method

Chung-Ming Li and Day-Fann Shen

Abstract In this chapter, our goal is to remove blotch noises as well as straight line scratches commonly seen in old film movies using digital image processing techniques. We proposed an adaptive blotch detection method based on our blotches characteristic analysis, where we calculate the size of each connected region using Connected Component methods to distinguish blotches and non-blotches, as a result, the proposed method can accurately mark the blotches noises. We also proposed a method to avoid misclassifying of small objects as blotches using and improving the accuracy of marked blotches. We adopt and improve Inpainting algorithm to repair the blotches noise. Inpainting algorithm is highly dependent on the order in which the filling proceeds; we also extend this approach to color images. In addition, we proposed criterion for performance evaluation of the proposed method.

Keywords Old film • Image restoration • Scratch • Blotch • Connected component method • Inpainting

46.1 Introduction

This chapter deals with digital image processing of old film movies, with focus on blotches detection and repair. A typical frame in digitized film (Southward Taiwan) is shown in Fig. 46.1.

In recent years, researches on old movie film restoration are briefly reviewed as follows: In 1996, Nadenau and Mitra [1] proposed an blotches restoration algorithm based on rank ordered differences (ROD), which are calculated from the data of the current image frame, and the preceding and the succeeding motion compensated frame. In 2011, Krishna et al. [2] proposed blotches detection algorithm. Their method combines the merits of image representation using Legendre moments, warped distance interpolation and adaptive particle swarm optimization. After

C.-M. Li • D.-F. Shen (✉)
Department of Computer, National Yunlin University of Science and Technology, Yunlin 64002, Taiwan
e-mail: shendf@yuntech.edu.tw

J. Juang (ed.), *Proceedings of the 3rd International Conference on Intelligent Technologies and Engineering Systems (ICITES2014)*, Lecture Notes in Electrical Engineering 345, DOI 10.1007/978-3-319-17314-6_46

Fig. 46.1 A typical frame in old film (Southward Taiwan)

detecting the blotches they apply the well-known Simplified Rank Order Detector (SROD). A warped distance approach is then used for better interpolation results. In 2012, Dangui et al. [3] proposed blotches detection method. First, an edge detection algorithm is adopted to obtain edge points of current frame image. Then, partial non-blotch edge points are omitted by proposed approach. Finally, for the remaining edge points as well as the surrounding pixels, an improved local motion compensation-based SROD is proposed to determine whether the point is a blotch or not, through the image-to-edge detection.

In 2013, Wun-Jheng Chen [4] proposed Enhancement of Digitized Old Film Using MCA-Based Method. Using the 1-D adaptive median filter to remove scratches. This chapter is the extending work of [4].

Chapter structure is as follows: Sect. 46.2 Adaptive scratch removal algorithm improvements, Sect. 46.3 Connected component method of blotch detection, Sect. 46.4 Inpainting algorithm, Sect. 46.5 Experimental results, Sect. 46.6 Conclusion.

46.2 Improvements on Adaptive Scratch Removal Algorithm

In this section, we present two improvements on the Adaptive 1-D median filter proposed in [4] for scratches detection: (1) the pillar or poles with sky background are misclassified as scratches.

Analysis shows that most frames (87 %) in the (Southward Taiwan) film movie are with skies. Based on this fact, we divide the upper 1/3 portion (160 lines) of the image into further two equal parts (80 liens each), with the assumption that the upmost 80 lines contain the skies. Detection method based on the above scheme has been proposed which can effectively avoid the false detections of poles (or pillars) as the straight line scratches (see Fig. 46.2).

Fig. 46.2 Divide the upper 160 lines into two equal parts

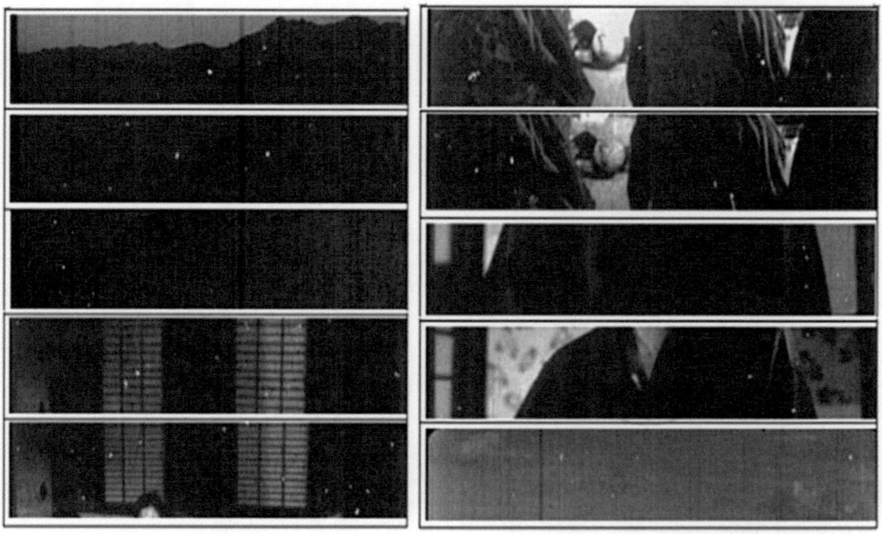

Fig. 46.3 Images with blotches

46.3 Connected Component Method for Blotch Detection

In this section, we proposed an adaptive blotch detection method based on blotches characteristic analysis.

46.3.1 Blotches Characteristic Analysis

We selected ten images containing irregular blotches (Fig. 46.3) from the Digitized old film (Southward Taiwan), which are referred to as blotches images. Blotches characteristic analysis results are shown in Fig. 46.4. Gray values distribution of

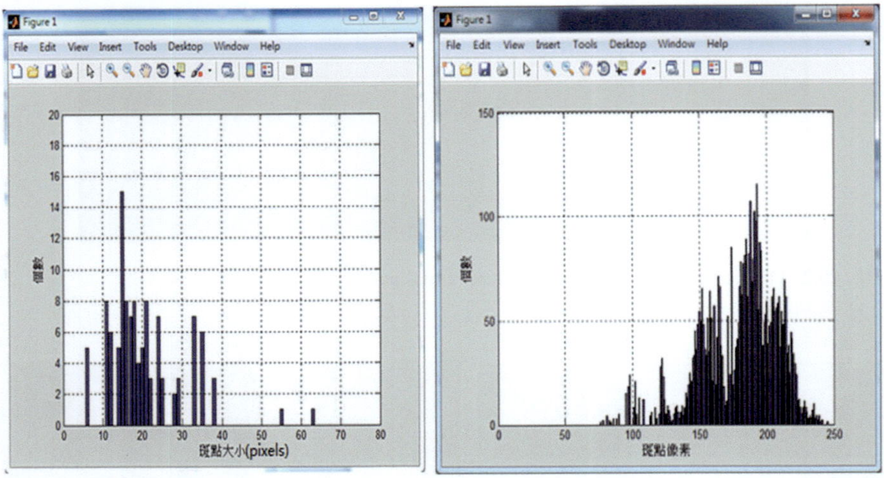

Fig. 46.4 Statistics of blotches analysis

blotches is between 150 and 230. These statistics are used for blotches detection and marking. The number of pixel threshold for blotch size is 40 pixels and below.

46.3.2 *Blotches Detection and Marking*

The blotches detection method is as follows:

Step 1: Apply bilateral low pass filter to the input frame I [5] to remove noises while keeping the edges. The results are the low-frequency image ILF.

Step 2: Apply canny edge detector [6] to ILF, the output is the binary image IC.

Step 3: Apply Morphological Dilation [7] on IC so that the edges in Ic are more complete. The output is IC-D.

Step 4: Apply the Connected Component method [8] to distinguish the blotches area and mark the blotches, threshold value is 40 (pixels), the output is the blotches marked frame J.

Figures 46.5a and 46.6a are the original blotch images, Figs. 46.5b and 46.6b are the corresponding blotch marked images.

46.3.3 *Blotches Marking: Using Temporal Hints*

In the blotches detection process, using the connected component method may cause errors: In Fig. 46.7, note that the cigarette in the upright man's mouth is

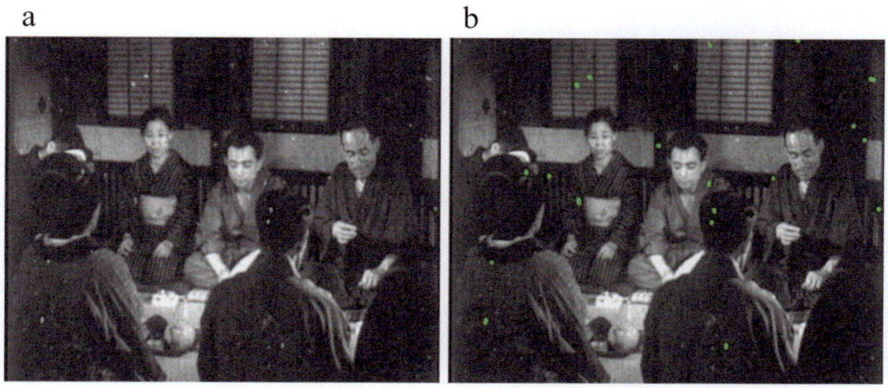

Fig. 46.5 (**a**) Blotches image 1, (**b**) Blotch marked results

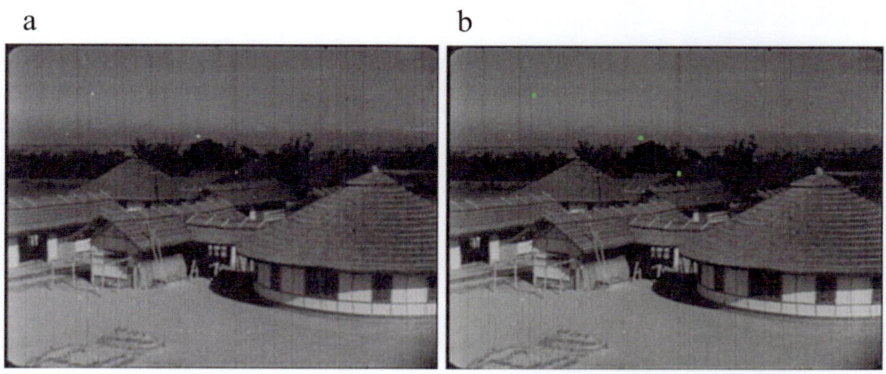

Fig. 46.6 (**a**) Blotches image 2 (**b**) Blotch marked results

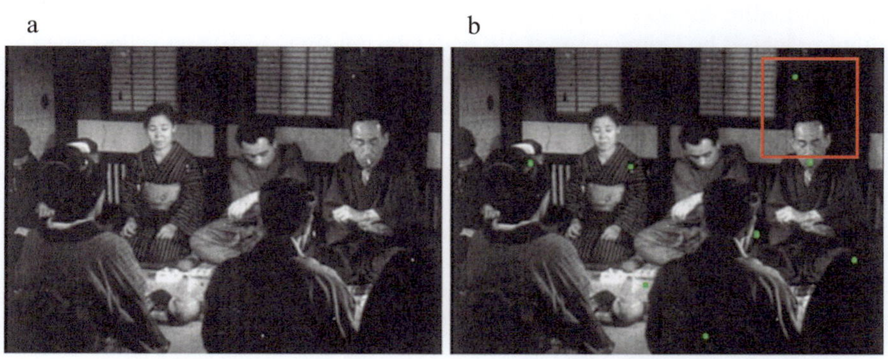

Fig. 46.7 (**a**) Blotches image 3 (**b**) Blotch marked results

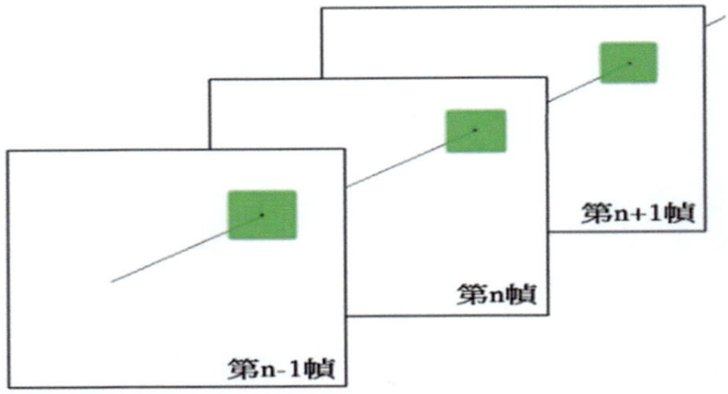

Fig. 46.8 Avoid false blotch detection using temporal hints

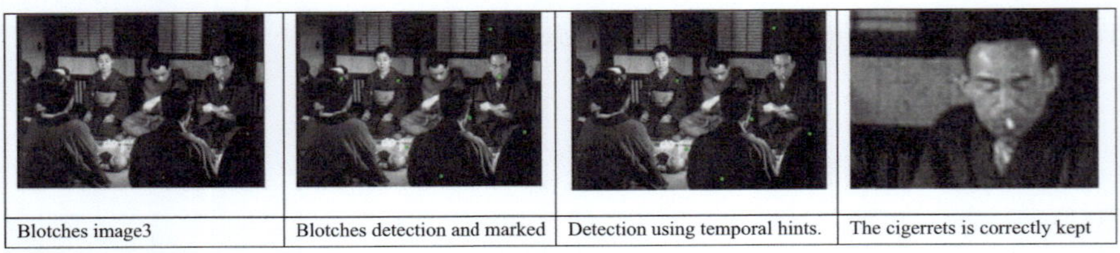

Fig. 46.9 Blotches image 3 detection with temporal hints

identified as a blotch. To avoid misdetection, if a detected blotch shows up in the same position for more than two frames, then it is classified as a small object rather than a blotch, as shown in Fig. 46.8. The cigarette is no longer to be marked as blotch as shown in Fig. 46.9.

46.4 Exemplar-Based Image Inpainting Algorithms

Once the blotches are marked, the Exemplar-Based Image Inpainting algorithm is then applied to fill-up the blotches. The Exemplar-Based Image Inpainting algorithm proposed by Criminisi et al. [9] is depicted in Fig. 46.10 and is described as follows: In order to extend the linear structure and propagate the textures properly for the region to be inpainted, it is necessary to determine the best (first priority) boundary pixel to start the inpainting process, determine the priority. In [9], for each boundary pixel p in a region to be inpainted, its confidence term $C(p)$ and data term $D(p)$ to determine its priority $P(p)$ value. The boundary pixel with the highest priority among all possible boundary pixels is chosen for inpainting.

where:

$$P(p) = C(p)^{*}D(p); \tag{46.1}$$

Fig. 46.10 Exemplar-based inpainting [9]

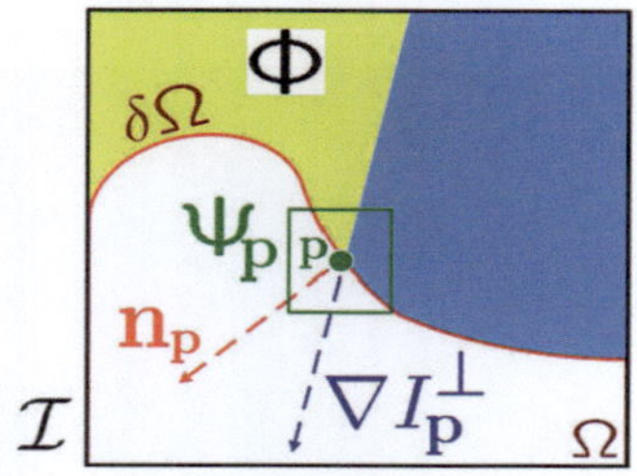

$$C(p) = \frac{\sum_{q \in \Psi p \cap (I - \varnothing)} C(p)}{|\Psi p|}; \tag{46.2}$$

$$D(p) = \frac{|\nabla I p \cdot n p|}{\alpha}. \tag{46.3}$$

For the best boundary pixel, search for the best matched patch by finding the minimal sum of squared differences of the Ψp and Ψq.

$$\Psi \hat{q} = \arg \min_{\Psi_q \in \Phi} d(\Psi \hat{p}, \Psi q) \tag{46.4}$$

46.4.1 Proposed an Improved Method Based on Exemplar-Based Image Inpainting

We make some improvement on the exemplar-based image inpainting: In [9], the target region Ω is filled with patches selected from the source region Φ. However, once the best matched patch Ψp is filled in the target region Ω, Ψp is then classified as the source region. Therefore, Ψp might be considered as the best patch to repair other areas which not only degradates the quality of inpainting but also increase the search areas. In our version of inpanting, the Ψp will not be in the source area Φ.

Summary of the proposed Inpainting algorithm:

Extract the manually selected initial front contour $\delta\Omega$.

Step 1: Compute priorities $P(p), \forall p \in \delta\Omega$.
Step 2: Find the patch $\Psi \hat{p}$ with the maximum priority, $\forall \boldsymbol{p} \in \delta\Omega$.
Step 3: Find the minimized $d(\Psi \hat{p}, \Psi \hat{q})$, $\Psi \hat{q} \in \Phi$.
Step 4: Copy image data from $\Psi \hat{q}$ to $\Psi \hat{p}$.
Step 5: Set $\Psi \hat{p}$ $C(p)$ close to 0.
Step 6: Update $C(p)$.
Step 7: Repeat until done.

a b

Fig. 46.11 (**a**) Blotches image 1. (**b**) Results of inpainting

a b

Fig. 46.12 (**a**) Blotches image 2. (**b**) Results of inpainting

46.5 Experimental Results

Figures 46.11a and 46.12a show the original blotch images and Figs. 46.11b and 46.12b are the repaired images.

We also compare the performance of the Inpainting algorithm proposed by Criminisi [9] and our version inpainting algorithm. Figure 46.13 shows the subjective inpainting results on Image Bungee by Criminisi [9] and our version respectively.

We also compare the performance using the objective SSIM (Structural Similarity Quality Index Measure) [10]. The results are shown in Figs. 46.14 and 46.15.

Original image (Bungee)	(1)Mark removal Region	(2) By Antonio Criminisi [9]	(3) By proposed method

Fig. 46.13 Subjective comparison of inpainting algorithms on image Bungee

Original image Lake	(1) Mark removal area	(2) By Antonio Criminisi [9]	(3) By Proposed methed

Fig. 46.14 Subjective comparison of inpainting algorithms on image lake

0.4967	0.4027	0.2614	0.2514	Avg. SSIM 0/3204 [9]
0.7097	0.5281	0.2862	0.2605	Avg. SSIM 0.5102 Propsed

Fig. 46.15 Objective SSIM comparison of inpainting algorithms on lake (marked areas)

46.6 Conclusion

In this chapter we have proposed several methods for repairing digitized old films with satisfactory results. We improved the Adaptive Scratch Removal Algorithm proposed in [4] to avoid misclassifying pole or pillar as the scratches. We also analyze the characteristics of blotches in the digitized old film and developed algorithm to detect and mark the blotches while avoiding misclassification of the small objects as blotches using the temporal hints. We then improve the Exemplar-Based Image Inpainting algorithm proposed by Criminisi et al. [9] for repairing the blotches. Experiments show that the improved version outperforms the original version [9] both subjectively and objectively in the SSIM sense on color images.

References

1. Nadenau, M.J., Mitra, S.K.: Blotch and scratch detection in image sequences based on rank ordered differences. In: 5th International Workshop on Time-Varying Image Processing and Moving Object Recognition, Florence, Italy (1997)
2. Krishna P.J., Santhosh Kumar, S.: Blotch removal for old movie restoration using legendre moment and particle swarm optimization. In: 2011 International Conference on Communications and Signal Processing (ICCSP), pp. 348–352 (2011)
3. Dangui, X., Junqing, L., Dong, R., Yong, L., Shuifa, S.: Fast and effective blotch identification algorithm in archive film restoration. In: World Automation Congress (WAC), pp. 17–21 (2012)
4. 沈岱**範**.陳文政, “應用信號型態成份分析法於數位化老舊膠捲影片**修**復的研究”, 2013 全國電信研討會, 台南大學, 11 (2013)
5. Tomasi, C., Manduchi, R.: Bilateral filtering for gray and color images. In: Sixth International Conference on Computer Vision, pp. 839–846 (1998)
6. 繆紹綱, Rafael C. Gonzalez, Richard E. Woods, Steven L. Eddins.: 原著, 『數位影像處理-運用MATLAB』 “Digital image processing using MATLAB”,台千,台灣培生教育、台灣東華書局合作出版,pp. 419–423 (2008)
7. 繆紹綱, Rafael C. Gonzalez, Richard E. Woods, Steven L. Eddins.: 原著, 『數位影像處理-運用MATLAB』 “Digital image processing using MATLAB”,台千,台灣培生教育、台灣東華書局合作出版,pp. 364–365 (2008)
8. 繆紹綱, Rafael C. Gonzalez, Richard E. Woods, Steven L. Eddins.: 原著, 『數位影像處理-運用MATLAB』 “Digital Image Processing Using MATLAB”,台千,台灣培生教育、台灣東華書局合作出版, pp. 386–389 (2008)
9. Criminisi, A., Perez, P., Toyama, K.: Region filling and object removal by exemplar-based image inpainting. IEEE Trans. Image Process. **13**(9), 1200–1212 (2004)
10. Wang, Z., Bovik, A.C., Sheikh, H.R., Simoncelli, E.P.: Image quality assessment: from error visibility to structural similarity. IEEE Trans. Image Process. **13**(4), 600–612 (2004)

Chapter 47
Study of Nickel Catalysts Deposited by Using the Electroless Plating Method and Growth of the Multiwall Carbon Nanotubes

Chih-Yi Lin, Jian-Liang Pan, Chia-Ching Wu, and Wen-Chung Chang

Abstract In this study, the electroless plating has been successfully applied for nickel catalyst layer and the multiwall carbon nanotubes (MWNTs) grown by chemical vapor deposition (CVD). Sulfuric acid solution was used as buffer to adjust and maintain pH value of electroless plating solution on 4.5. The structural, element, and quality of MWNTs were investigated with field emission scanning electron microscope (FE-SEM), X-ray diffraction patterns (XRD), energy dispersive spectrometer (EDS) and Raman spectrometer. From the FE-SEM image, it showed that the density of MWNTs increased as the deposition time of nickel catalyst layer increased. This result caused by the formation of nickel nucleation become rich as the immersion of the substrate in electroless plating solution was longer, and this benefited the growth of carbon nanotubes. The Raman analysis demonstrated that the I_D/I_G ratio of MWNTs decreases as the deposition times of nickel catalyst layer increases, indicating that more graphene MWNTs structures were formed.

Keywords Multiwall carbon nanotubes • Thermal chemical vapor deposition • Electroless plating method • Nickel catalyst

47.1 Introduction

The carbon nanotubes (CNTs) have been demonstrated to be excellent properties due to their large aspect ratio, high chemical stability, and small-end radius [1–3]. First noticeable discovery of carbon nanotubes was reported by Ilijima [4]

C.-Y. Lin • W.-C. Chang
Southern Taiwan University of Science and Technology, Tainan, Taiwan, R.O.C.

J.-L. Pan
Department of Applied Cosmetology, Kao Yuan University, Kaohsiung, Taiwan, R.O.C.

C.-C. Wu (✉)
Department of Electronic Engineering, Kao Yuan University, Kaohsiung, Taiwan, R.O.C.
e-mail: t10068@cc.kyu.edu.tw

J. Juang (ed.), *Proceedings of the 3rd International Conference on Intelligent Technologies and Engineering Systems (ICITES2014)*, Lecture Notes in Electrical Engineering 345, DOI 10.1007/978-3-319-17314-6_47

in 1991, when he found layers of carbon (graphene) rolled into tubular structure in the soot of arc-discharge method. The nanotubes consisted of up to several tens of graphitic shells with adjacent shell separation of 0.34 nm, diameters of 1 nm and high length/diameter ratio. Ilijima's discovery of carbon nanotubes in the insoluble material of arc-burned graphite rods created the buzz that greatly accelerated work on synthesis, production, and properties of carbon nanotubes. It took two more years for Iijima and Ichihashi at NEC [5], and Bethune et al. [6] at IBM to synthesize SWNT by addition of transition metal catalysts to carbon in an arc-discharge in 1993. There are two main types of carbon nanotubes [7] that can have high structural perfection. Single-walled nanotubes (SWNTs), these consist of a single graphite sheet seamlessly wrapped into a cylindrical tube. Multi-walled nanotubes (MWNTs), these comprise an array of nanotubes one concentrically placed inside another like rings of a tree trunk. Various synthesized methods of CNTs such as arc-discharge [8], pyrolysis [9], microwave plasma enhance chemical vapor deposition (MPCVD) [10] and thermal chemical vapor deposition (thermal CVD) [11] have been developed.

For synthesizing CNTs, typically, nanometer-size metal particles are required to enable hydrocarbon decomposition at a lower temperature than the spontaneous decomposition temperature of the hydrocarbon. Most commonly used metals are Fe, Co, and Ni, because of two main reasons: (1) high solubility of carbon in these metals at high temperatures; and (2) high carbon diffusion rate in these metals [12]. The catalyst layer can be deposition using variety of methods, such as sputtering [13, 14], evaporators [15, 16], sol-gel [17], and so on. However, these methods cost more deposition time and the capital of production. Here we demonstrate a simple and reliable method for deposition catalyst layer by electroless plating [18]. The purpose of maintaining low costs is to reach the expectation of mass-production. In this study, the nickel catalysts treated by electroless plating method and the MWNTs synthesized by using thermal chemical vapor deposition method.

47.2 Experimental Details

The multi-walled nanotubes (MWNTs) were synthesized on *n*-type (100) silicon substrate using a thermal chemical vapor deposition (Thermal CVD) system. Prior to MWNTs growth, different thickness of the nickel catalyst layer was deposited by electroless plating method. Sulfuric acid (H_2SO_4) solution was used as buffer to adjust and maintain pH value of electroless plating solution on 4.5. The procedure of electroless plating can be divided into three steps: activation, sensitization, and plating. Chemicals and their concentrations of the nickel plating solution were shown in Table 47.1. Subsequently, it used a thermal CVD growth the MWNTs. From the thermal CVD, the substrates were loaded on the aluminum oxide boat inside the thermal CVD quartz tube. Nitrogen (N_2) gas was flowed into the quartz reactor in order to prevent the oxidation of transition metal before up to growth

Table 47.1 Chemicals and concentrations of the electroless nickel plating solution

Chemical	Concentration (g/L)
$NiSO_4 \cdot 6H_2O$	87
$NaPH_2O_2 \cdot H_2O$	24
$C_4H_4Na_2O_4 \cdot 6H_2O$	4.1
$C_6H_8O_7 \cdot H_2O$	2
$C_4H_6O_4Pb \cdot 3H_2O$	0.0015
$CH_3COONa \cdot 3H_2O$	6
pH	4.5

temperature. The quartz tube was heated up to the growth temperature in the range of 800–900 °C with C_2H_2 gas of 15 sccm for the synthesis of MWNTs. The synthesize time of MWNTs is 30 min. Finally, the reactor was cooled slowly down to room temperature in N_2 ambient after the growth.

Following, the microstructure morphology of the MWNTs were measured by using the field emission scanning electron microscopy (FE-SEM). Energy dispersive spectrometer (EDS) attached to the FE-SEM and the elements of MWNTs were measured. The quality of MWNTs was identified by micro-Raman spectroscopy using the 532 nm line of a YAG laser and the laser beam was focused by a 50× objective onto the MWNTs surface with a beam size of approximately 10 nm in diameter.

47.3 Results and Discussions

The morphology of the nickel catalyst layer and the multi-walled nanotubes (MWNTs) were determined by FE-SEM. As seen in Fig. 47.1, the nickel catalyst layer was deposited on silicon substrate with different deposition times (10–300 s) by using electroless plating method. It is found that the surface roughness of nickel catalyst layer increased as the deposition time increased. This result is caused by the increase in thickness of nickel catalyst increased. The elements of nickel catalyst layer were measured by energy dispersive spectrometer (EDS). Figure 47.2 shows the EDS results of the nickel catalyst layer deposited by electroless nickel plating solution. It clearly shows that the elements of nickel catalyst layer contain Ni, S, P, Na, O, and Si. The oxygen signal might derive from air as the immersion of the substrate in electroless plating solution and dried out for 30 min in the air. The silicon signal might come from silicon substrate. It demonstrates that the nickel nucleation was formed as immersion in electroless plating nickel solution. In addition, it was found that the element of nickel increased as the deposition time of nickel catalyst layer increased. The detailed elements of nickel catalyst layer were shown in Table 47.2.

The cross-section FE-SEM image of the MWNTs with different deposition times of nickel catalyst layer is shown in Fig. 47.3. In Fig. 47.3a, the nickel catalyst layer was deposited at 10 s by electroless plating method and it found that the fewer

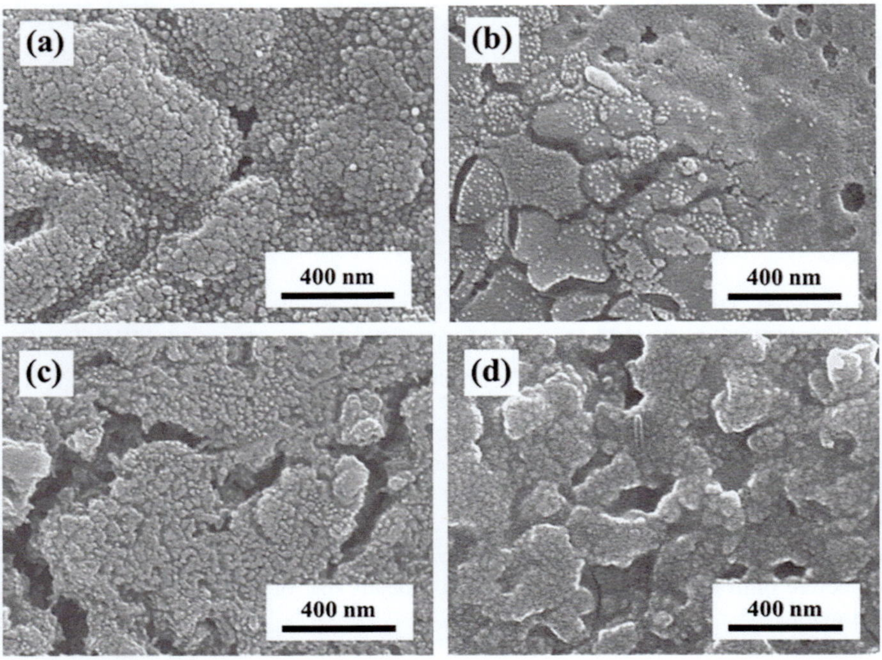

Fig. 47.1 FE-SEM image of nickel catalyst layer was deposited on silicon substrate at different deposition times. (**a**) 10 s, (**b**) 30 s, (**c**) 60 s, and (**d**) 300 s

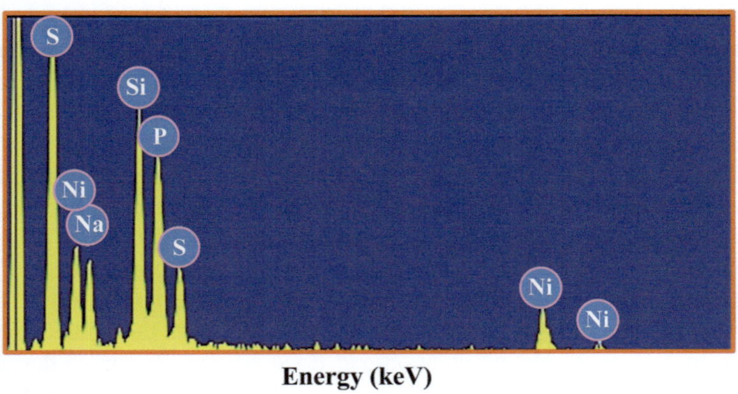

Fig. 47.2 The EDS analysis of the compositions with nickel catalyst layer was deposited by electroless plating method

MWNTs growth on the silicon substrate using the thermal CVD. This result caused by the less nickel nucleation on the silicon substrate as the nickel catalyst layer deposited at 10 s. As the deposition time of nickel catalyst layer increases, the nickel nucleation obtained on the silicon substrate become rich. The densities of the

Table 47.2 The elements of nickel layer were deposited at different times

Deposition times of Ni layer (s)	Element	Weight (%)	Atomic (%)
10	Ni	12.03	5.11
30	Ni	14.79	5.20
60	Ni	16.91	6.03
300	Ni	28.65	11.74

Fig. 47.3 The cross-section SEM images of MWNTs synthesized with different deposition times of nickel catalyst layer by thermal CVD. (**a**) 10 s and (**d**) 300 s

MWNTs with longer deposition time (300 s) of nickel catalyst layer are higher than the shorter deposition time (10 s), as shown in Fig. 47.3b. The surface FE-SEM image of the MWNTs with different deposition times of nickel catalyst layer is shown in Fig. 47.4. Compared with Fig. 47.4a, b, as the nickel catalyst layer deposited at 300 s, the high density MWNTs was obtained by using thermal CVD process. Inverse, in Fig. 47.4a, as the nickel catalyst layer deposited at 10 s, the density of the MWNTs is lower than Fig. 47.4b. The diameter of the MWNT with nickel catalyst layer deposited at 300 s is about 80–100 nm and the length of the MWNT is several micrometer, as shown in high magnification SEM images, as shown in Fig. 47.4c.

Figure 47.5 shows the Raman spectrum of MWNTs synthesized with different deposition times of nickel catalyst layer by thermal CVD. The characteristic peaks of the MWNTs are observed at 1,323 and 1,581 cm^{-1}. The G-band that originates in the graphitic sheet yields a peak at 1,583 cm^{-1} while the broad peak close to 1,323 cm^{-1} is associated with the D-band [19]. The D-band indicates the existence of a defective graphitic layer and/or some carbon particles, most of which are observed after the purification procedure [20]. The relative intensity ratios of the D-band to G-band peak (i.e., I_D/I_G) of the MWNTs were decreased from 1.42 to 1.24 as the deposition time of nickel catalyst layer increased. It was suggested that the MWNTs growth with 300 s deposition time of nickel catalyst layer could obtain more graphenes structure.

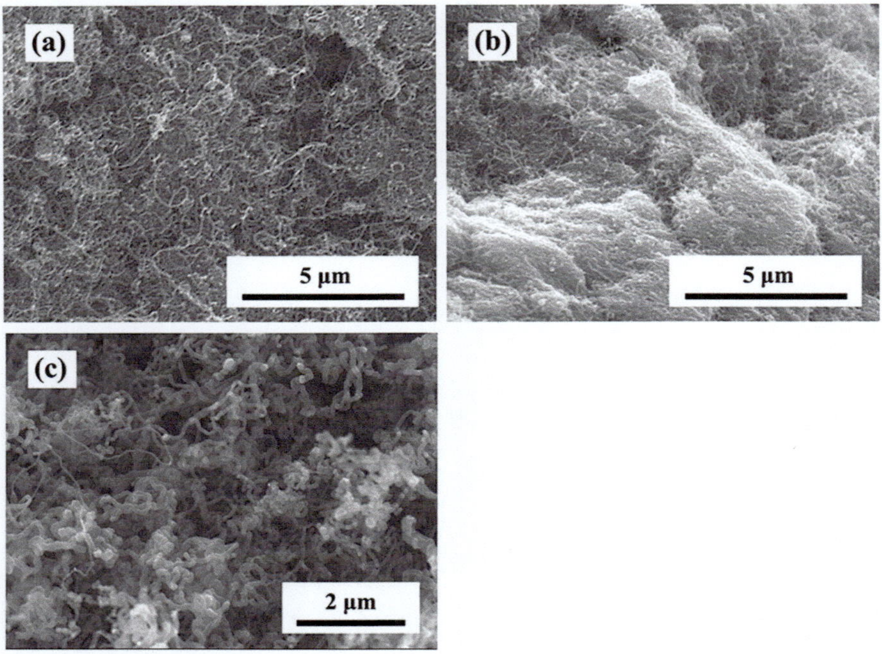

Fig. 47.4 The surface SEM images of MWNTs synthesized with different deposition times of nickel catalyst layer by thermal CVD. (**a**) 10 s (×10,000), (**b**) 300 s (×10,000), and (**c**) 300 s (×15,000)

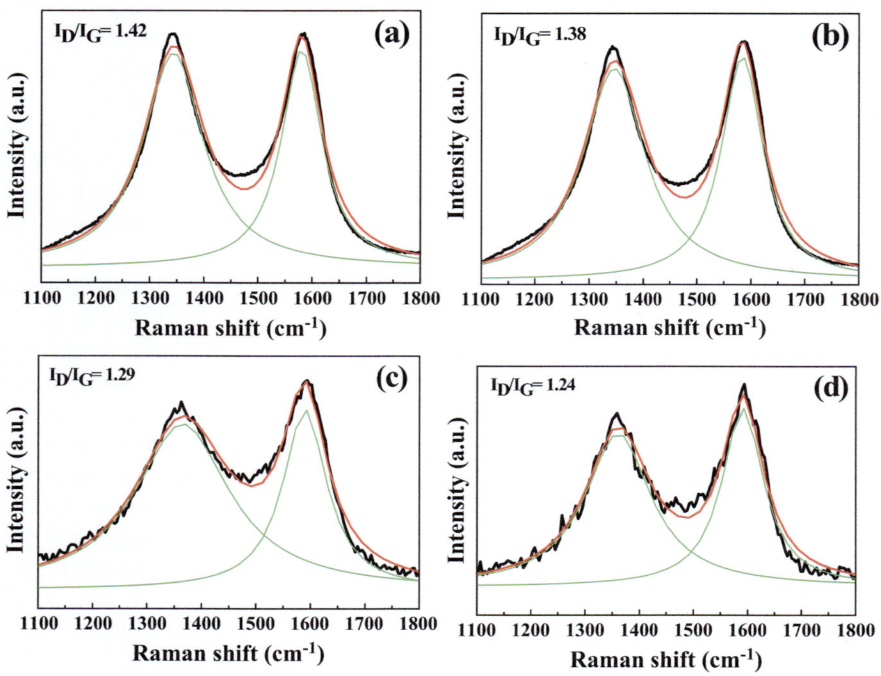

Fig. 47.5 Raman analysis of the MWNTs synthesized with different deposition times of nickel catalyst layer by thermal CVD. (**a**) 10 s, (**b**) 30 s, (**c**) 60 s, and (**d**) 300 s

47.4 Summary

The nickel catalyst layer was deposited by electroless plating method, and synthesized multi-walled carbon nanotubes (MWNTs) by thermal chemical vapor deposition (thermal CVD). From the SEM and EDS, it was found that the formation of nickel nucleation becomes rich as the immersion of the substrate in electroless plating solution was longer, and this benefited the growth of MWNTs. In addition, the Raman analysis demonstrated that the I_D/I_G ratio of MWNTs decreases as the deposition times of nickel catalyst layer increases, indicating that more graphene MWNTs structures were formed.

Acknowledgement The authors acknowledge the financial support of the Ministry of Science and Technology and National Science Council of MOST 103-2221-E-244-018 and 103-2514-S-218-001, NSC 102-2511-S-244-001 and NSC 102-2221-E-244-019.

References

1. Saito, Y., Hamaguchi, K., Hata, K., Uchida, K., Tasaka, Y., Ikazaki, F., Yumura, M., Kasuya, A., Nishina, Y.: Conical beams from open nanotubes. Nature **389**, 554–555 (1997)
2. Schmid, H., Fink, H.W.: Carbon nanotubes are coherent electron sources. Appl. Phys. Lett. **70**, 2679–2680 (1997)
3. Rafique, M.M.A., Iqbal, J.: Production of carbon nanotubes by different routes-a review. J. Encap. Adsorpt. Sci. **1**, 29–34 (2011)
4. Iijima, S.: Helical microtubules of graphitic carbon. Nature **354**, 56–58 (1991)
5. Iijima, S., Ichihashi, T.: Single-shell carbon nanotubes of 1-Nm diameter. Nature **363**, 603–605 (1993)
6. Bethune, D.S., Kiang, C.H., De Vries, M.S., Gorman, G., Savoy, R., Vazquez, J., Beyers, R.: Cobalt-catalysed growth of carbon nanotubes with single-atomic-layer walls. Nature **363**, 605–607 (1993)
7. Mcbride, W. S.: Synthesis of carbon nanotube by chemical vapor deposition. Undergraduate Degree Thesis, College of William and Marry in Virginia, Williamsburg (2001)
8. Mohammad, M.I., Moosa Ahmed, A., Potgieter, J.H., Ismael Mustafa, K.: Carbon nanotubes synthesis via arc discharge with a yttria catalyst. Nanomaterials **2013**, 1–7 (2013). Article ID 785160
9. Thess, A., Lee, R., Nikolaev, P., Dai, H., Petit, P., Robert, J., Xu, C., Lee, Y.H., Kim, S.G., Rinzler, A.G., Colbert, D.T., Scuseria, G.E., Tománek, D., Fischer, J.E., Smalle, R.E.: Crystalline ropes of metallic carbon nanotubes. Science **273**, 483–487 (1996)
10. Qin, L.C., Zhou, D., Krauss, A.R., Gruen, D.M.: Growing carbon nanotubes by microwave plasma-enhanced chemical vapor deposition. Appl. Phys. Lett. **72**, 3437–3439 (1998)
11. Li, W.Z., Xie, S.S., Qain, L.X., Chang, B.H., Zou, B.S., Zhou, W.Y., Zhao, R.A.: Large-scale synthesis of aligned carbon nanotubes. Science **274**, 1701–1703 (1996)
12. Kumar, M., Ando, Y.: Chemical vapor deposition of carbon nanotubes: a review on growth mechanism and mass production. J. Nanosci. Nanotechnol. **10**, 3739–3758 (2010)
13. Liu, C., Chen, Y.C., Tzeng, Y.: Effects of carbon content in iron catalyst coatings on the growth of vertically aligned carbon nanotubes on smooth silicon surfaces by thermal chemical vapor deposition. Diamond Relat. Mater. **13**, 1274–1280 (2004)

14. Jeong, H.J., Jeong, S.Y., Shin, Y.M., Han, J.H., Lim, S.C., Eum, S.J., Yang, C.W., Kim, N.G., Park, C.Y., Lee, Y.H.: Dual-catalyst growth of vertically aligned carbon nanotubes at low temperature in thermal chemical vapor deposition. Chem. Phys. Lett. **361**, 189–195 (2002)
15. Juang, Z.Y., Chien, I.P., Lai, J.F., Lai, T.S., Tsai, C.H.: The effects of ammonia on the growth of large-scale patterned aligned carbon nanotubes using thermal chemical vapor deposition method. Diamond Relat. Mater. **13**, 1203–1209 (2004)
16. Shyu, Y.M., Hong, F.C.N.: The effects of pre-treatment and catalyst composition on growth of carbon nanofibers at low temperature. Diamond Relat. Mater. **10**, 1241–1245 (2001)
17. Chaisitsak, S., Yamada, A., Konagai, M.: Hot filament enhanced CVD synthesis of carbon nanotubes by using a carbon filament. Diamond Relat. Mater. **13**, 438–444 (2004)
18. Kong, F.Z., Zhang, X.B., Xiong, W.Q., Liu, F., Huang, W.Z., Sun, Y.L., Tu, J.P., Chen, X.W.: Ni-layer on multiwall carbon nanotubes by an electroless plating method. Surf. Coat. Technol. **155**, 33–36 (2002)
19. Saito, R., Dresselhaus, G., Dresselhaus, M.S.: Physical Properties of CNTs. Imperial College Press, London (1999)
20. Lee, C.J., Park, Y.S., Kim, W.S., Lee, N.S., Kim, J.M., Choi, Y.G.: Synthesis of uniformly distributed carbon nanotubes on a large area of Si substrates by thermal chemical vapor deposition. Appl. Phys. Lett. **75**, 1721–1723 (1999)

Chapter 48
Photovoltaic Power Generation System Modeling Using an Artificial Neural Network

Cheng-Ting Hsu, Roman Korimara, Lian-Jou Tsai, and Tsun-Jen Cheng

Abstract This chapter presents a solar power modeling method using an application of the Levenberg–Marquardt (L–M) algorithm. This L–M algorithm has been adopted and incorporated into back propagation learning algorithm for training a feed-forward neural network. With this model, the photovoltaic power generation can be approximated. Meteorological data and the historical output power data of the Taiwan Chimei Island photovoltaic plant were selected for this study. The proposed model is evaluated by comparing the simulated results with the actual measured values and are found to be in good agreement.

Keywords Artificial neural networks • PV model • Chimei Island

48.1 Introduction

Photovoltaic power is a significant alternative source of energy in times of energy crises. Human kind is using fossil fuels as the primary energy source but nonpolluting sources of energy are preferable. Sunlight is a renewable source of energy that can be used to produce electricity, referred to as solar power. Advances in photovoltaic panel technology and the identification of rich solar energy resources in Taiwan improve prospects for the photovoltaic power industry. This motivates researchers to develop analysis and prediction methods for photovoltaic power output [1].

Prediction of power generation from solar energy is gaining considerable attention [2]. There are several modeling techniques that combine meteorological and historical generation data [2, 3]. In order to achieve the highest possible prediction accuracy, the methods should consider all appropriate parameters and data that may indicate future trends. In literature, artificial intelligence techniques such as artificial neural network (ANN), fuzzy logic, etc., are found to be more accurate than

---

C.-T. Hsu • R. Korimara • L.-J. Tsai • T.-J. Cheng (✉)
Department of Electrical Engineering, Southern Taiwan University
of Science and Technology, Tainan 71005, Taiwan
e-mail: chengtj@mail.stust.edu.tw

J. Juang (ed.), *Proceedings of the 3rd International Conference on Intelligent Technologies and Engineering Systems (ICITES2014)*, Lecture Notes in Electrical Engineering 345, DOI 10.1007/978-3-319-17314-6_48

the traditional statistical models [4]. In general, neural networks are universal approximators [5]. Therefore in this study, the ANN is used for the prediction of the photovoltaic power output.

Solar energy has become a technoeconomically viable source of energy in the renewable power sectors of Taiwan for sustainable power generation. For the present study, data collected from the Chimei Island photovoltaic system for a period of 3 months from July to Sept. 2013 were used. The ANN model constructed using these data shows good agreement between the estimated and the measured values.

48.2 ANN for Photovoltaic Power Output Prediction

Among different forms of ANN, the feed-forward neural network is proposed in this study. In order to train the feed-forward neural network (FFNN), the Levenberg-Marquardt was adopted into the back propagation algorithm. A multi-layer feed-forward network with the back propagation algorithm has often been applied for the ANN training. Fig. 48.1 shows a two-layer feed-forward network. The input values are multiplied by their weight to the hidden layer until it reaches the final output. Back propagation is assumed to be the learning technique in the training process. It includes the following steps: firstly, the calculation of the error with the initialized weight for the neurons multiplied by the desired value and actual value combined; secondly, with the changed weight based on the error, so that error is propagated backward to update the weights of the previous layer.

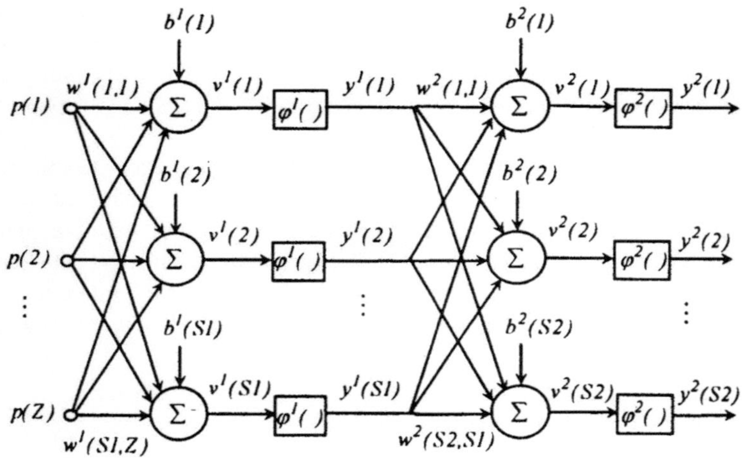

Fig. 48.1 A two-layer feed-forward network

48.3 Methodology

For this modeling approach, the data were collected from two different sources through the Penghu meteorological department and the SCADA reading system installed at the PV system. Meteorological and SCADA data readings are the inputs and outputs of the proposed ANN model. The data were collected from July to Sept 2013. All data were collected on the island for modeling the PV output power for the ANN architecture as shown in Fig. 48.2. The collected data were the hourly weather input data and the corresponding PV output power. In this analysis, the adopted neural network has four inputs and one output. The inputs are the wind speed, humidity, surrounding temperature, and solar irradiation; and the output is the solar power generation. The feed-forward back propagation is used as the neural network model for this study. It is further described in the following steps: (1) read training data; (2) read testing data; (3) use neural network training plus testing data; (4) get the prediction result; (5) plot the prediction result; and (6) plot the regression prediction results. For simplification, the above steps can be clarified into three steps:

Step 1: To prepare the input data for ANN, the meteorological and SCADA reading data have to be stored. The data of the selected input variables and the corresponding output are then divided into two data sets for training and testing, respectively. Before the training and testing process, all data sets should be normalized to the same range of values using (48.1).

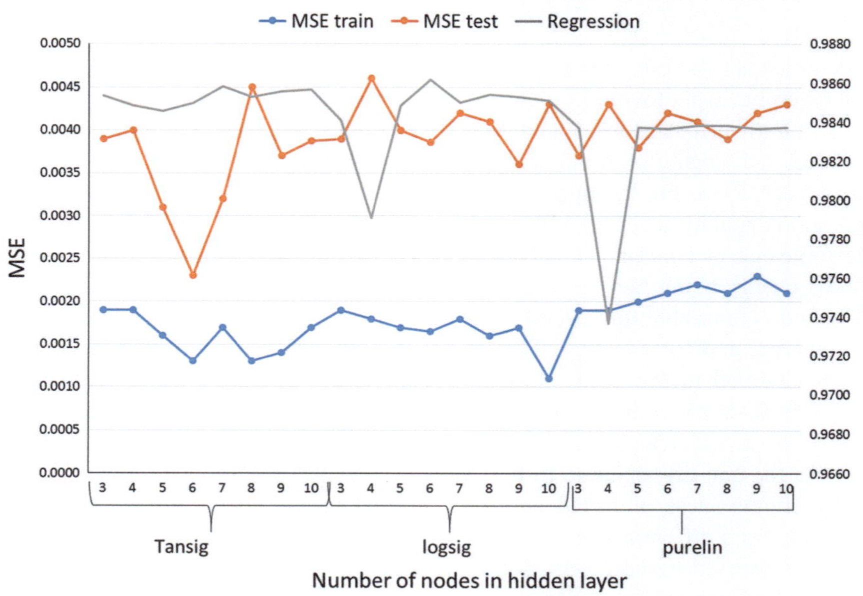

Fig. 48.2 MSE of three different types of ANN with the number of nodes in hidden layer

$$a = \frac{R_{\max}}{b} \tag{48.1}$$

where a is the normalized value, b is the non-normalized value, and $R_{\max}$ is the maximum number in a row vector.

Step 2: With the training and testing of inputs using different ANN architectures, the statistical error criteria for the MSE test, MSE test, and Regression for each ANN model must be observed. The error is recorded for each observation until the difference between the output of the network and the desired output is equal to a predefined threshold error. The input variable is redefined if the convergence of the ANN model cannot be obtained.

$$\mathrm{MSE} = \frac{1}{N}\sum_{i=1}^{N} \left(Z_i - \overline{Z}_i\right)^2 \tag{48.2}$$

where $\overline{Z}_i$ is a vector of the N prediction and Z_i is the vector of the real values.

Step 3: If the proposed ANN model converges to get the result, then the values must be normalized by using (48.3).

$$b = a \times R_{\max} \tag{48.3}$$

48.4 Results and Discussion

To demonstrate the effectiveness of the ANN method to predict the photovoltaic power output, a photovoltaic plant in Chimei Island was selected for computer simulation to verify the agreement between estimated and measured value.

The global generation trend of the solar energy increased to top 35 GW by the end of year 2013. Furthermore, Taiwan has a commitment to increase its solar generation capacity to 4,500 MW by the year 2020. Thus, the first Taiwan model renewable energy community in Penghu Island was adopted with about 56 % of all island's energy being met by renewable energy [6]. The geographical location of this island is favorable for several types of renewable energy such as tidal energy, wind energy, and solar energy. The electric power supply is connected to the PV system modeled in this study through 3.3 kV lines. The rated generator output voltage is 0.48 kV with an operating frequency of 60 Hz. And the study PV generation of 155 kWp is available to offset the customer demand as further shown in the one-line diagram in Fig. 48.3.

The effectiveness of FFBP has been verified in the results provided by the simulation for a different ANN architecture. After the simulation, the performance has to be assessed based on the criteria used in this paper which are: MSE train, MSE test and regression. ANN architecture includes: Logistics Sigmoid transfer

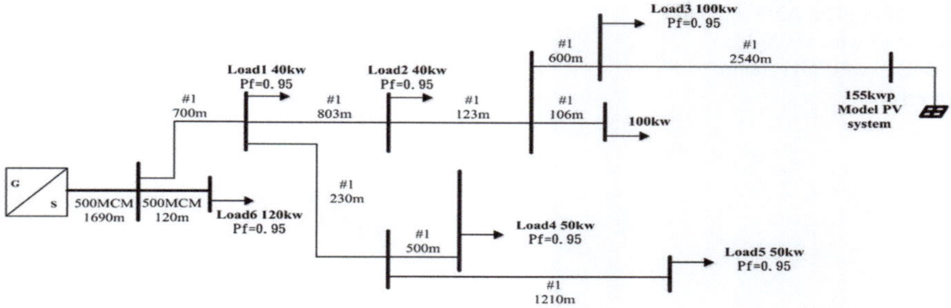

Fig. 48.3 Single-line diagram of the Chimei distribution network with the study PV system

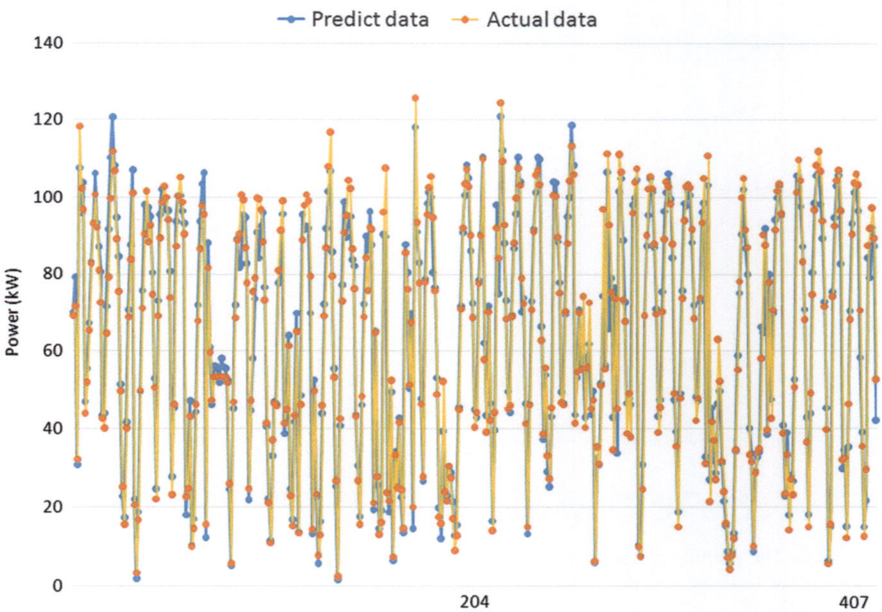

Fig. 48.4 Results for real data and ANN model (tan-sig) for 3 months

function (log-sig), tangent sigmoid transfer function (tran-sig) and linear transfer function (pure-lin) used in the hidden layers and output layers, respectively. The training MSE, testing MSE, and regression values for different numbers of hidden nodes for FFBP are shown in Fig. 48.2. For different input architectures, it is found that the best performance is obtained by the FFBP with the application of tangent sigmoid with six hidden neurons for the proposed criteria. Figure 48.4 shows the prediction level of accuracy is high when the network output is close to the value of the desired output with 6 hidden neurons and 12 epochs. Figure 48.5 presents the regression of calculated data and the real data. As shown in this figure, the measured

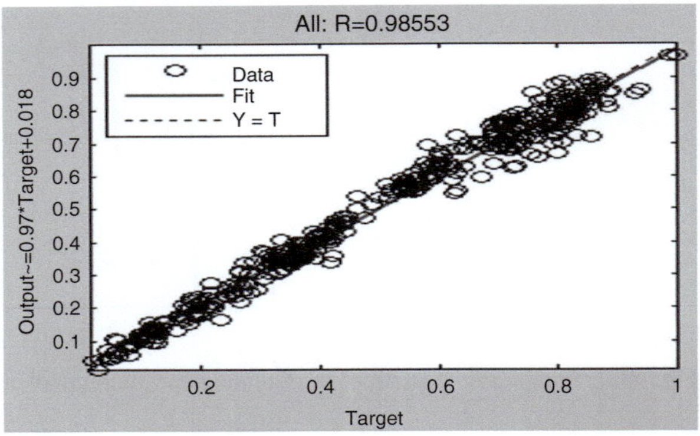

Fig. 48.5 The ANN model (tan-sig) with six hidden neurons with regression of error

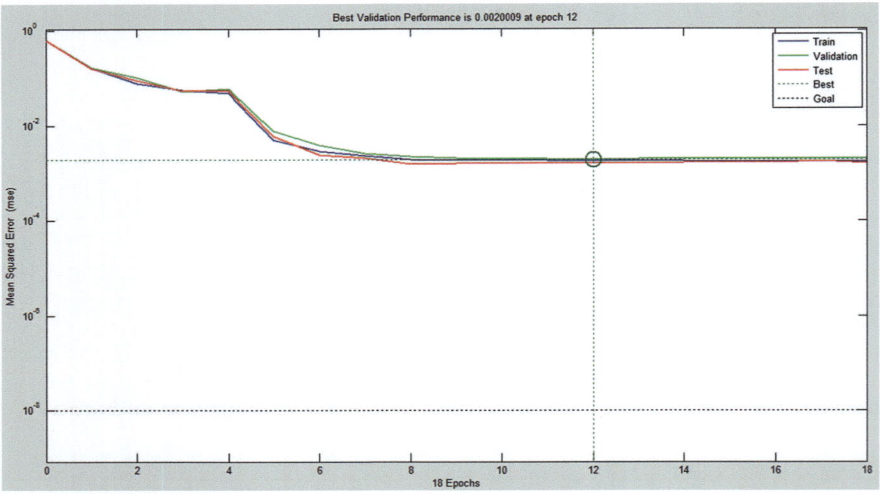

Fig. 48.6 The training performance for model ANN (tan-sig) with six neurons

error is very low. According to Fig. 48.6, the best validation performance is 0.0020009 at epoch 12. The performance error was 0.0013 for the MSE train, 0.0010 when using MSE test, and 0.985 for regression criterion.

The above figures show that the applied meteorological and climatological parameters have an influence on the estimation of the PV output power. This ANN simulation results proves that the wind speed, humidity, surrounding temperature, and solar irradiation all have an influence on the estimation of PV output power. Effectiveness of the ANN model has been shown in the high prediction accuracy of the simulation results. However, it was discovered that different inputs and learning rates as well as model structures, directly influence the prediction accuracy of ANN.

48.5 Conclusion

Predicting the future value of PV output power at an isolated site is challenging. This chapter has developed the prediction of PV output power using the ANN model with an LMBP algorithm for Chimei Island. The developed neural network model offers a reliable indication of the PV output power by using the input parameters like wind speed, surrounding temperature, humidity, and solar irradiation. The predicted PV output power for the Chimei Island PV plant using the ANN model shows good agreement with the actual value. This model is helpful for energy planners on the island for future planning and execution.

Acknowledgement This work was supported by the MOST of Taiwan (MOST 103-3113-E-214-002 and MOST 103-2221-E-218-016).

References

1. Hsu, C.T., Roman, K., Cheng, T.J.: Cost-effectiveness analysis of a PVGS on the electrical power supply of a small island. Int. J. Photoenergy **2014**, 1–9 (2014). Article ID 264802
2. Luis, M., Luis, F.Z., Jesús, P., Ana, N., Ruth, M., Marco, C.: Prediction of global solar irradiance based on time series analysis: application to solar thermal power plants energy production planning. Sol. Energy **84**(10), 1772–1781 (2010)
3. Almonacid, F., Rus, C., Pérez, P.J., Hontoria, L.: Estimation of the energy of a PV generator using artificial neural network. Renew. Energy **34**(12), 2743–2750 (2009)
4. Kelouwani, S., Agbossou, K.: Non-linear model identification of wind turbine with a neural network. IEEE Trans. Energy Convers. **19**(3), 608 (2004)
5. Kurt, H., Maxwell, S., Halbert, W.: Multilayer feedforward networks universal approximators. Neural Netw. **2**(5), 359–366 (1989)
6. Information on: Taiwan Bureau of Energy, http://web3.moeaboe.gov.tw

Chapter 49
Thermal Stress Analysis of Layer-wise Functionally Graded Material Beam Considering Neutral Plane

Young-Hoon Lee and Ji-Hwan Kim

Abstract Stress analysis of layer-wise functionally graded materials (FGMs) beam is investigated in this work. Layer-wise theory is a distributed model instead of the continuous model for the FGMs the material properties of FGMs are not isotropic. Therefore, neutral plane of FGMs is different from the isotropic material. And the surface is determined by using the first moment with respect to temperature-dependent Young's modulus. In the numerical results, axial and shear stress distribution are calculated by considering neutral plane according to the number of layers of FGMs.

Keywords Layer-wise • Functionally graded materials • Neutral plane

49.1 Introduction

Functionally graded materials (FGMs) are a kind of composite materials which have a continuous variation of material properties based on various directions. Layer-wise FGM are generated by numerous homogeneous and different isotropic material layers. Advantage is that manufacture is easier than the other technique. In recent years, the static and dynamic analysis of structural model with FGMs are considerable importance both in research as well as in industrial fields.

Recently, numerous research works have been studied widely for various fields. Alexraj et al. [1] analyzed static behavior of FGM Timoshenko beam model using ANSYS software. Simsek [2] studied static analysis FG beam subjected

---

Y.-H. Lee (✉)
Department of Mechanical and Aerospace Engineering, College of Engineering,
Seoul National University, Seoul 151-744, Republic of Korea
e-mail: jwhkim@Snu.ac.kr

J.-H. Kim
Department of Mechanical and Aerospace Engineering, Institute of Advanced
Aerospace Technology, College of Engineering, Seoul National University,
Seoul 151-744, Republic of Korea
e-mail: jwhkim@snu.ac.kr

J. Juang (ed.), *Proceedings of the 3rd International Conference on Intelligent Technologies and Engineering Systems (ICITES2014)*, Lecture Notes in Electrical Engineering 345, DOI 10.1007/978-3-319-17314-6_49

to a uniformly distributed load, and has been investigated by using Ritz method. Snakar [3] performed elasticity analysis of FGM beam subjected to transverse loads. Sanjay Anandrao et al. [4] researched free vibration of uniform FGM beam using classical Euler-Bernoulli beam theory and Timoshenko beam theory. Pradhan et al. [5] studied modeling of free vibration problem and the solution methodology. The focus of this chapter is to introduce the neutral surface and layer-wise FGM model. In this work, the neutral plane of FGM beam are chosen as a reference plane, and study to analyze static analysis of temperature dependent material property.

49.2 Formulation

49.2.1 Layer-Wise FGM Beam

The general layer-wise FGM beam is formulated by introducing piecewise approximations through the thickness for each state variable to achieve the desired level of detail in the analysis. Figure 49.1 shows a layer-wise FGM simply supported beam model subjected to a uniformly distributed load q.

Volume fraction of the material is defined as

$$V_m(z) = \left(\frac{z}{h} + \frac{1}{2}\right)^k, \quad V_c(z) + V_m(z) = 1 \tag{49.1}$$

where V, z, and h superscript k, c, and m represent the volume fraction, thickness direction and thickness, the volume fraction index, ceramic and metal, respectively.

The material properties P can be written as in [6]

$$P = P_0\left(P_{-1}T^{-1} + T^0 + P_1 T + P_2 T^2 + P_3 T^3\right) \tag{49.2}$$

where T is temperature and coefficients represent the cubic fit of the material properties. Using the linear Voigt rule of mixture, the effective material properties P_{eff} is

$$P_{\text{eff}}(z,T) = P_m(T)V_m(z) + P_c(T)V_c(z) = (P_m(T) - P_c(T))\left(\frac{z}{h} + \frac{1}{2}\right)^k + P_C(T) \tag{49.3}$$

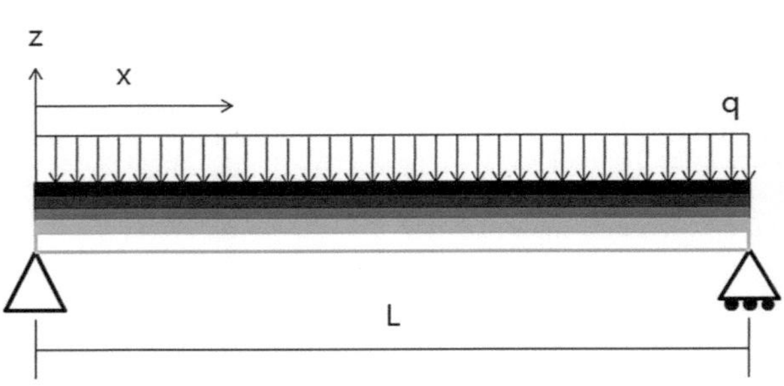

Fig. 49.1 A layer-wise FGM beam

Young's modulus E, Poisson's ratio υ, and shear modulus G apply on above (50.3) by temperature-dependent properties in the thickness direction.

49.2.2 Physical Neutral Plane

Neutral plane of FGM model is defined as the unstretched plane. The location of the neutral plane of the FGMs beam is determined to satisfy the first moment with respect to temperature-dependent Young's modulus equals to zero as in [5].

$$\int_{-h/2}^{h/2} E(z,T)(z - z_0(T))dz = 0 \tag{49.4}$$

Therefore, the position of neutral plane z_0 can be obtained as

$$z_0(T) = \frac{\int_{-h/2}^{h/2} zE(z,T)dz}{\int_{-h/2}^{h/2} E(z,T)dz} \tag{49.5}$$

Substituting (49.3) into (49.5) and integrating gives

$$z_0(T) = h\frac{k(E_m(T) - E_c(T))}{2(k+2)(kE_c(T) + E_m(T))} \tag{49.6}$$

49.2.3 Governing Equation

Assume the deformation of FGMs beam in the x–z plane and denote the displacement component along x, z directions by u_x, u_z, respectively. Based on a Higher-order shear deformation theory, the axial displacement (u_x) and the transverse displacement (u_z) of any point of the beam are given as in [2]

$$u_x(x,z) = u_0(x) + (z - z_0)\varphi(x) - \alpha(z - z_0)^3\left(\frac{\partial w_0}{\partial x} + \varphi(x)\right) \tag{49.7}$$

$$u_z(x,z) = w_0(x) \tag{49.8}$$

where $u_0(x)$, $w_0(x)$, $\varphi(x)$ are generalized displacements at the middle plane of beams ($z = 0$) and $\alpha = 4/(3h^2)$. Timoshenko beam theory is special case of higher order shear deformation theory and equation of equilibrium for Timoshenko beam theory are obtained by taking $\alpha = 0$.

Force and moment resultant as well as shear force are calculated using

$$\begin{bmatrix} N \\ M \\ Q \end{bmatrix} = \begin{bmatrix} A_{11} & B_{11} & 0 \\ B_{11} & D_{11} & 0 \\ 0 & 0 & A_{55} \end{bmatrix} \begin{bmatrix} \varepsilon_0 \\ \kappa \\ \gamma \end{bmatrix} - \begin{bmatrix} N_{\Delta T} \\ M_{\Delta T} \\ 0 \end{bmatrix} \tag{49.9}$$

where ε_0 is middle plane strain, κ is curvature strain, γ is the shear strain and φ is the rotation of a line element perpendicular to the original direction. N, M, Q are force, moment and shear resultant, respectively. A_{11}, B_{11}, D_{11} and A_{55} in (49.9) are extensional, coupling, bending and transverse shear rigidities, respectively. Meanwhile, $N_{\Delta T}$ and $M_{\Delta T}$ are the thermal inplane force resultant and thermal moment resultant vector, and defined as in [6]

$$[A_{11}, B_{11}, D_{11}] = \int_{-h/2}^{h/2} \frac{E(z,T)}{1 - \upsilon(z,T)^2} \left[1,\ z - z_0,\ (z - z_0)^2\right] dz \tag{49.10}$$

$$A_{55} = G(z,\ T)dz \tag{49.11}$$

$$[N_{\Delta T}, M_{\Delta T}] = \int_{-h/2}^{h/2} (1, z - z_0) E \begin{Bmatrix} \alpha(z) \\ \alpha(z) \\ 0 \end{Bmatrix} \Delta T(z) dz \tag{49.12}$$

The material constituent of FGMs beams to obey the generalized Hooke's law the state of resultant.

49.3 Numerical Results

In this part, the numerical results for stress components are compared for nondimensional data of layer-wise FGMs beam based on mid- and neutral surface. The axial normal stresses, σ_{xx}, are evaluated at the midpoint of the beam ($x = L/2$), and the shear stresses, τ_{xz}, are calculated at the left support of the beam ($x = 0$), and the normal and the shear stresses are nondimensionalized by $(\overline{\sigma}_{xx}, \overline{\tau}_{xz}) = \left(\frac{\sigma_{xx}A}{qL},\ \frac{\tau_{xz}A}{qL}\right)$ as in [2].

Figure 49.2 represents axial stresses for mid plane as in [2] and compared with two cases of reference plane at volume fraction $k = 5$ and number of layers $n = 10$. The magnitude of tensile stresses is larger than compressive stresses in layer-wise FG beam. Tensile stresses considering neutral plane are smaller than tensile stresses considering mid plane, however, the compressive stresses seem to have opposite trend. Nondimensional axial stresses are not zero at the neutral plane of the FG

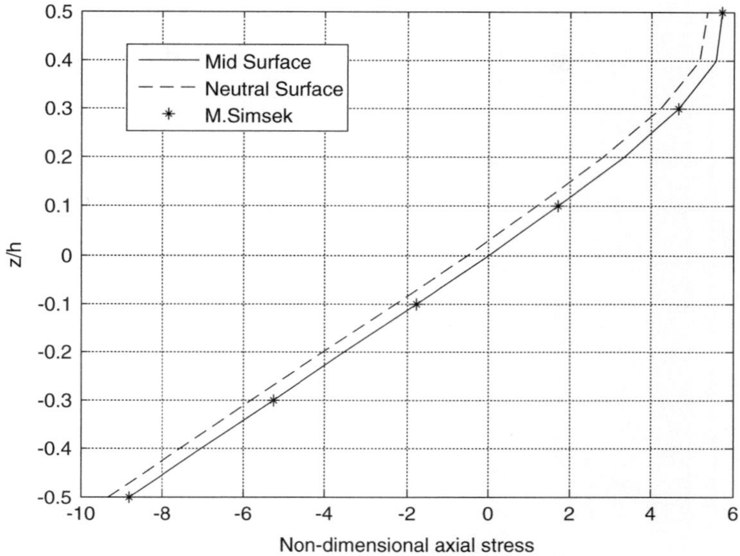

Fig. 49.2 Axial stresses distribution of the beam with mid & neutral plane

beam because neutral plane of the beam moves to upper side. This difference is due to change of modulus of elasticity.

Figure 49.3 shows the shear stress distribution along the thickness for various number of layers. All of shear stresses are symmetric about mid plane of the layer-wise beam due to volume fraction of FG beam is zero. Increasing the number of layers can increase the value of the shear stress and approach the exact value.

Figure 49.4 shows the thermal stresses considering various room temperatures with respect to neutral plane. When room temperature is increased, the thermal stresses are increased. Temperature effect on the metal parts is greater than ceramic part due to thermal expansion coefficient of metal is larger than ceramic.

49.4 Conclusions

Thermal stress analysis of layer-wise Functionally Graded Materials (FGMs)beam is investigated using neutral plane according to number of layers and various volume fractions. Numerical results indicate that considering appropriate reference plane is important for stress analysis, and the stresses are changed because the reference plane influence modulus of elasticity of the material. Furthermore, layer-wise model verifies the result that the number of layers plays a major role on the evaluation of stresses.

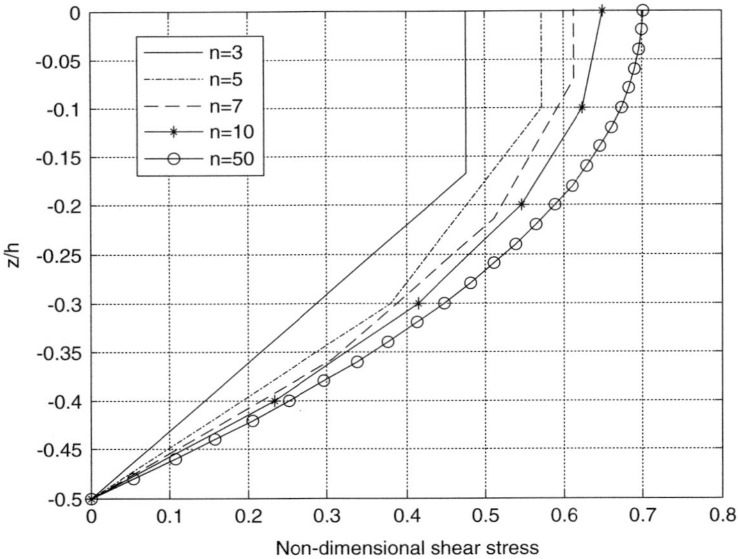

Fig. 49.3 Shear stress distribution according to the number of layers

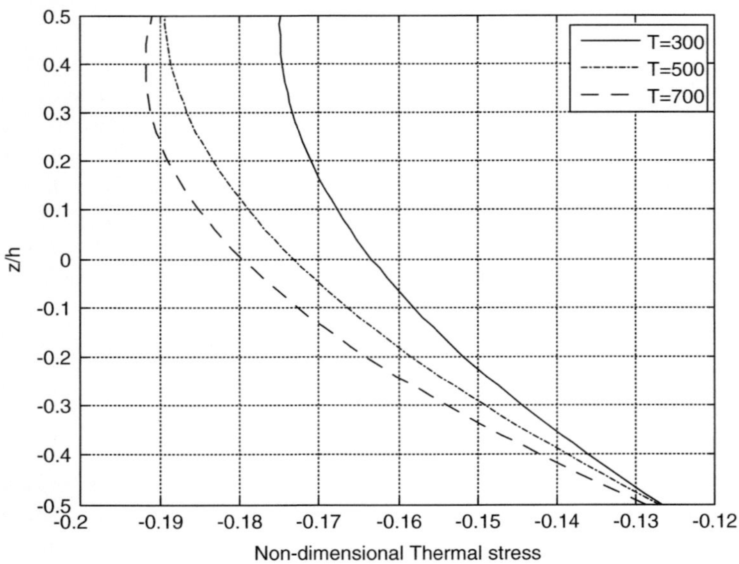

Fig. 49.4 Thermal stress distributions for various temperature based on neutral plane

In future works, various heat conductions are to be considered, and then analyze post-buckling behavior et al. Also, material properties and deflections are varies due to the temperature change.

Acknowledgement This work is financially supported by Ministry of Land, Infrastructure, and Transport as "NALDA Project".

References

1. Alexraj, S., Vasiraja N., Nagaraj, P.: Static behavior of functionally graded material beam using finite element method. International Conference on Energy Efficient Technologies for Sustainability (ICEETS), pp. 267–273 (2013)
2. Simsek, M.: Static analysis of a functionally graded beam under a uniformly distributed load by Ritz method. Int. J. Eng. Appl. Sci. **1**(3), 1–11 (2009)
3. SanjayAnandrao, K., Gupta, R.K., Ramachandram, P., Rao, G.V.: Free vibration analysis of functionally graded beams. Def. Sci. J. **62**(3), 139–146 (2012)
4. Pradhan, K.K., Chakraverty, S.: Free vibration of Euler and Timoshenko functionally graded beams by Rayleigh-Ritz method. Compos. Part B **51**, 175–184 (2013)
5. Zhang, D.G., Zhou, Y.H.: A theoretical analysis of FGM thin plates based on physical neutral surface. Comput. Mater. Sci. **44**, 716–720 (2008)
6. Reddy, J.N., Chin, C.D.: Thermomechanical analysis of functionally graded cylinders and plates. J. Therm. Stresses **21**(6), 593–626 (1998)

Chapter 50
A Smart Home Automation System

Chien-Yuan Liu

Abstract Internet of things enabled by the information and communication technologies is driving the promotion of smart living applications around the world. Convenient control with energy-saving consideration is one of the most important functions for smart homes. This chapter presents the research work for a smart-home automation system linked via wireless communications. The system is composed of a smart phone running an APP for control operations, a communication gateway, and multiple remote control devices. The functions of the system are correctly performed as the expected requirements of power socket control and environmental weather detection.

Keywords Smart APP • Automation system • Smart home

50.1 Introduction

The basic building blocks of Internet of things (IoT) includes microcontrollers, embedded memory, power chips, microsensors, and various wireless connectivities. The scale of IoT is tens or hundreds of that of Internet. Therefore, IoT is the biggest matter in the next decade seems the common consensus of information and communication technology (ICT) industry [1–3]. For instance, famous Acer and Asus, the leaders of Taiwan domestic companies, engage aggressively to build cloud services for compelling enterprises and smart homes [4, 5]. Furthermore, for driving the trend of IoT, Google Inc. acquired Nest Inc. in 2013 to involve in wireless sensors network and home automation [6, 7].

According to the above description, it is understood that IoT is the most critical section in ICT domain for the next decade. Meanwhile, Smart home is one of the most important applications of IoT. Smart home [8] contains widespread technologies, e.g., audio and video entertainment, safety monitor, home care, energy

C.-Y. Liu (✉)
Department of Computer Science and Information Engineering,
Cheng-Shiu University, Kaohsiung 83347, Taiwan
e-mail: chien.yuan.liu@gmail.com

J. Juang (ed.), *Proceedings of the 3rd International Conference on Intelligent Technologies and Engineering Systems (ICITES2014)*, Lecture Notes in Electrical Engineering 345, DOI 10.1007/978-3-319-17314-6_50

saving and control, green building, and wind and solar energy, etc. Among them, due to fossil energy exhaustion, the energy saving and control is one of the most imperative requirements of smart homes. A smart home automation system is required to realize the energy saving and control function. Hence, the research work is to design and construct the system which consisted of a smart phone acted as a control master, multiple controlled power sockets and sensor devices performed as slaves, and a gateway behaved as the bridge for the connection of these master and slaves.

The goal of the prototype system is to validate the feasibilities to design a smart APP on the control master, to make embedded modules for the control slaves, and to construct the gateway by proper and various wireless standards to connect the smart phone and the embedded slave devices. The APP supports a householder operation via touch sense interface. The embedded devices provide ON/OFF control capability through smart power sockets and measure ambient temperature and humidity through smart thermostats, respectively.

The following content of this chapter is organized as follows. Section 50.2 describes the scenario of smart home applications and depicts the requirement of the automation system, enclosing system architecture, wireless communication requirement, and basic functions of the gateway unit and the smart control devices. The results of the system integration and verification are demonstrated in Sect. 50.3. Finally, Sect. 50.4 presents the conclusions of the research work.

50.2 System Analysis and Design

The smart home gateway typically needs to furnish various communication interfaces for various network standards. The specifications of versatile connections generally are classified into two categories, e.g., wired and wireless communications. The IEEE 802.3, xDSL, RS-232, RS-485, and power line communication (PLC), etc. are the common specifications of wired communications. Whereas the Bluetooth, ZigBee, WiFi, GSM, 3G, and 4G, etc. are the pervasive specifications of wireless ones. The essential function of the gateway proposed in this chapter is focused on the energy saving and control for smart homes. For the sake of lower deployment cost and convenient, the wireless Bluetooth and ZigBee communications are elected for the smart home applications.

50.2.1 System Architecture

Figure 50.1 illustrates the architecture of the smart home automation system. The smart home gateway is depicted in the center of Fig. 50.1 to cascade the control commands issued from the smart APP of a householder at left side and to forward the commands to the smart controlled devices at the right side. On the other

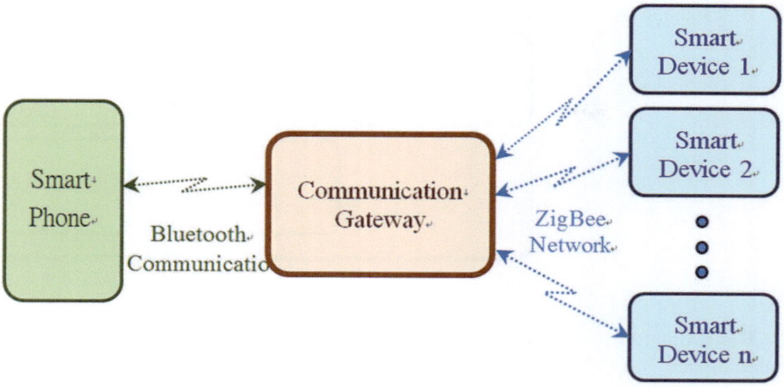

Fig. 50.1 Application scenario and system architecture

hand, the control states and ambient sensor data are transmitted from the devices at the right side to the smart phone at the left side via the middle communication gateway.

50.2.2 Wireless Communication Interfaces

Given the same sending power, ZigBee generally transmits a distance longer than that of WiFi. Meanwhile, the coverage of WiFi is larger than that of Bluetooth. In case of data rate, WiFi has the highest one among all. Bluetooth data rate is higher than that of ZigBee. As for the deploying popularity, Bluetooth is the highest, WiFi is the second, and ZigBee is the last. In the viewpoint of topology, the flexibility of ZigBee is the best, next WiFi, and then Bluetooth.

Under the application scenario about the energy saving and control of a smart home, the necessity of network topology is more crucial than that of data rate. In addition, lower transmit power always causes longer battery operation cycle. Therefore, to choose ZigBee as the wireless communication interface between the gateway and the controlled devices is more proper than others. As for the wireless communication interface between the gateway and the smart phone, it is able to utilize Bluetooth or WiFi. This time, Bluetooth is adopted for this project.

50.2.3 Gateway and Controlled Devices

Although the capability of a smart home gateway may be versatile, the more functions, the higher cost. In order to embody and verify the power saving and control goal of this project, the performance of a general microcontroller is enough for the gateway. Figure 50.2 illustrates the block diagram of the gateway. In middle

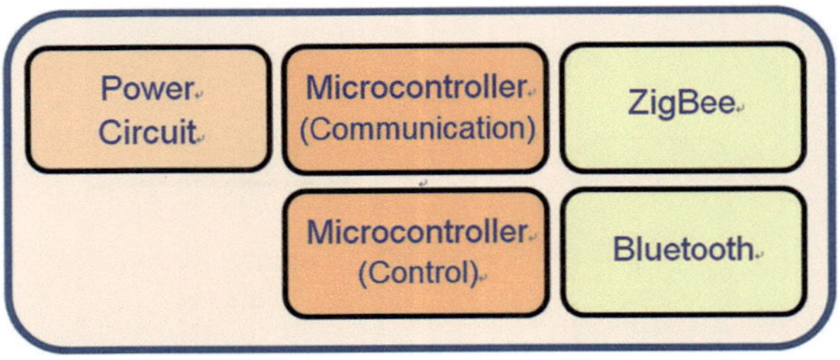

Fig. 50.2 Block diagram of gateway

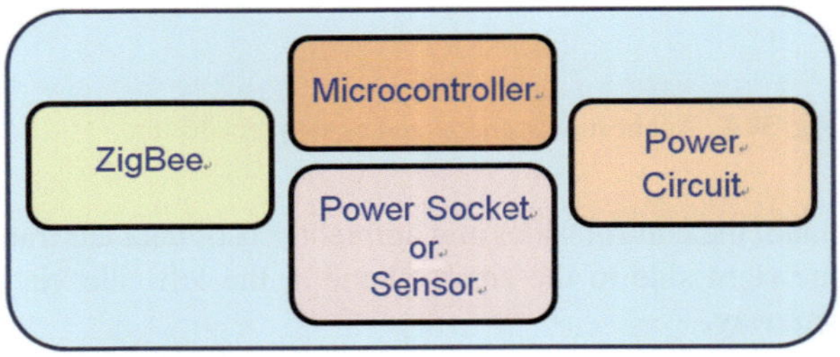

Fig. 50.3 Block diagram of controlled device

column of Fig. 50.2, there are two microcontrollers. One is used for communications, the other one is for control computation. ZigBee and Bluetooth are located in the right column of Fig. 50.2. The left block in Fig. 50.2 is the power module which converts AC to DC to supply power to the microcontrollers, the communication modules, and related electronic components.

Figure 50.3 demonstrates the block diagram of a smart controlled device. The upper middle block in Fig. 50.3 is the microcontroller for control processing or ambient data measuring. The lower middle block is the controlled power socket or the measuring sensor. The right block is the power circuit for the device. The left block is the ZigBee chip.

50.3 System Integration and Verification

50.3.1 Gateway Board

The embedded system of gateway and controlled devices are developed by the open-source Arduino [9]. The RISC chip of Atmel ATmega 328P is adopted for the microcontroller. It has 20 MIPS performance. It provides 6 AIO pins, 14 DIO pins. The Arduino IDE enables efficient coding and testing for the firmware. All Hardware circuits and layouts are sketched by the open-source Fritzing [10]. For instance, the schematic diagram of the gateway is demonstrated in Fig. 50.4.

The gateway module with the manufactured PCB, soldered chips and components is pictured in Fig. 50.5. Two ATmega 328P are located on the upper half and

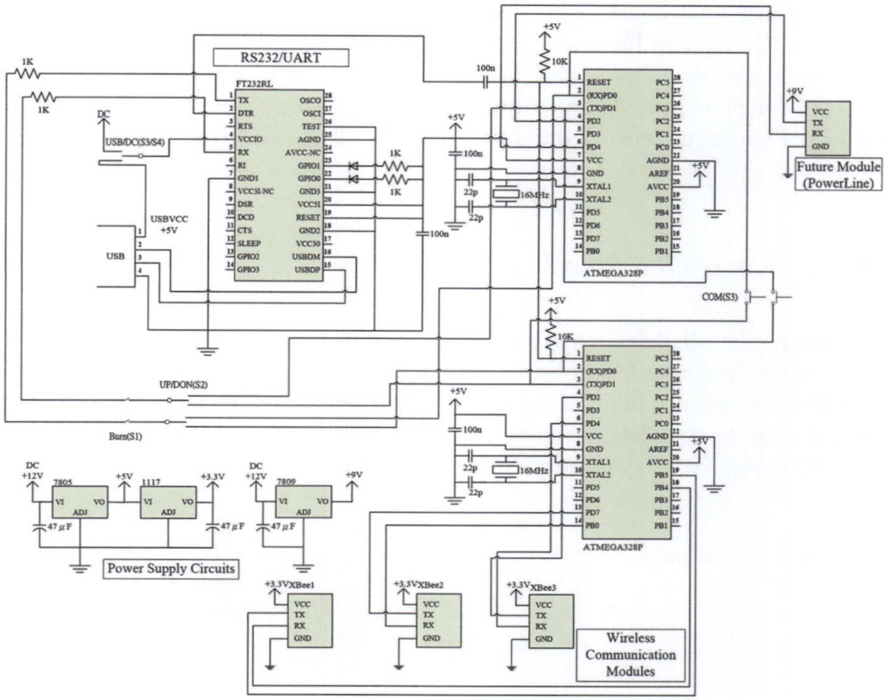

Fig. 50.4 Schematic diagram of gateway

Fig. 50.5 Picture of accomplished gateway

the lower half of the left column of the gateway board. At the upper middle, there is a USB to TTL chip for firmware uploading. The lower right has three XBee IO-pins sockets. They are designed for Bluetooth, ZigBee, and WiFi (WiFi socket is reserved for future utilization) chips. Above these three sockets, there is a larger IO socket for future network interface extension, such as Ethernet or PLC.

50.3.2 Controlled Board

Two smart controlled devices, power socket, and ambient sensor, are accomplished in this work. Figure 50.6 shows the smart power socket board. It is deployed to control an electrical appliance. Besides ON/OFF control, it also collects voltage and current data to calculate the power consumption of the appliance. The smart ambient sensor board is pictured in Fig. 50.7. It may measure indoor or outdoor

Fig. 50.6 Picture of smart power socket

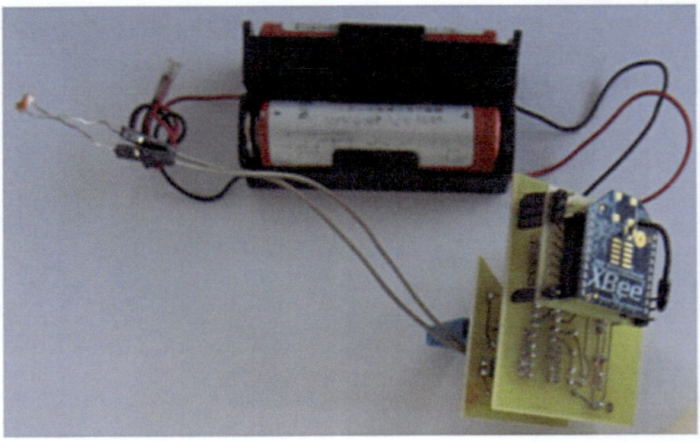

Fig. 50.7 Picture of ambient sensor board

temperature and humidity for wearing suggestion to householders. Both devices apply ZigBee to link to the gateway and via this gateway to communicate with the smart phone.

50.3.3 Smart APP

An APP may be developed by native, web-based, or hybrid coding methods [11–13]. The native coding manner is to develop an APP by Java or C# for Android or by object-C or Swift for iOS. In this work, the Android platform was chosen for the pilot development. Traditional way to develop an APP for Android platform is to utilize the java language combined with Android SDK. This way is named as native development. Another web-based manner is to execute APP within a browser. Recent popular method is hybrid model, in which an APP is developed by HTML5, CSS3, Javascript, jQuery Mobile, and Cordova (PhoneGap). At final step, the APP could be built for Android, iOS, Windows Phone, etc. by the Cordova. For the sake of the feature, it is easy to develop an APP for various platforms with same source codes.

In this work, the APP for the smart home automation system is developed by the hybrid model. The main menu screen of the accomplished APP is shown in Fig. 50.8a. In Fig. 50.8a, there are two option buttons, one links to the screen of smart power sockets, another links to the screen of smart sensor devices. Currently, each one instance is manufactured for a smart power socket and a smart sensor device, respectively. Figure 50.8b shows the data of power consumption by an electrical fan. The data of ambient weather is shown in Fig. 50.8c. On header division of Fig. 50.8b, c, there are two link buttons. The top-left button links to the

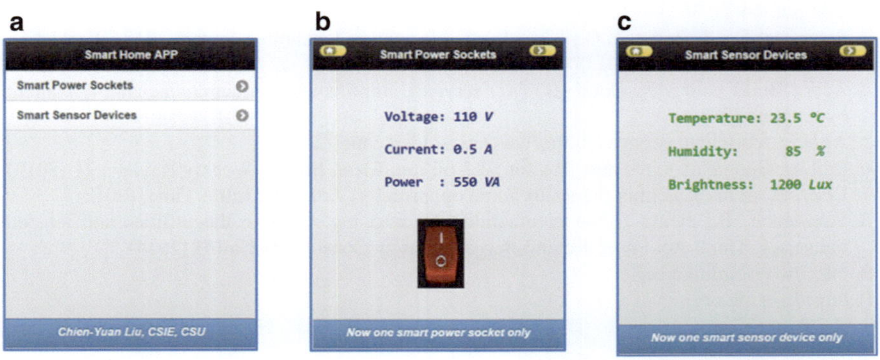

Fig. 50.8 (**a**) APP screen of main menu. (**b**) APP screen of smart power socket. (**c**) APP screen of smart sensor device

main menu screen. The top-right button links to the other screen of the Fig. 50.8c, b, separately. For instance, the current screen is Fig. 50.8b, and then next screen pointed by the top-right button in Fig. 50.8b would be Fig. 50.8c, vice versa.

50.4 Conclusions and Future Work

This chapter presents the design of a smart home automation system. The accomplished system is composed of a gateway to relay all devices, a smart phone APP for the operation of a householder, a smart power socket for controlling and monitoring an electrical fan, and a smart sensor device for measuring ambient weather condition. The hardware and software of the system are integrated and verified by the preset requirements and goals. The result conforms to the expectation. Certain improvements are expected in future, for instance, to utilize WiFi as the interface between the smart phone and the gateway because of WiFi has longer transmission distance than that of Bluetooth, to apply ARM processor instead of ATmega due to ARM provides higher performance to manipulate audio and video, to link to power line by PLC module for various accessibility of electrical appliances.

Acknowledgment The author appreciates the Ministry of Science and Technology, Taiwan, R.O.C., for the support of the project no. NSC 101-2632-E-230-001-MY3.

References

1. Vermesan, O., Friess, P.: Internet of Things: Converging Technologies for Smart Environments and Integrated Ecosystems. River, Aalborg (2013)
2. Kortuem, G., et al.: Smart objects as building blocks for the internet of things. IEEE Internet Comput. **14**(1), 30–37 (2010)
3. Soumitra, D., & Benat, B.O.: The global information technology report 2012. World Economic Forum, Geneva (2012)
4. Acer.: Build your won cloud (BYOC) services. http://www.acer.com.tw/ac/zh/TW/content/byoc-home (2014)
5. ASUS.: Asuscloud services. http://www.asuscloud.com/ (2014)
6. Editor.: Google acquires nest labs for $3.2 billion. Clean Energy Weekly Review. 21 (2013)
7. Levy, D., et al.: Catching the smart home opportunity. Arthur D. Little Time. (2012)
8. Schwenker, B., et al.: Home automation: the next big move in the utilities and telecom industries. Think Act Press, Roland Berger Strategy Consultants GmbH (2014)
9. http://www.arduino.cc/
10. http://fritzing.org/
11. Rainbow Riders.: How to choose the right technologies? Native, HTML5, or Hybrid, whitepaper (2014)
12. Perry, R.: Developing downloadable mobile apps using HTML5 and Apache Callback, Worklight (2014)
13. http://en.wikipedia.org/wiki/HTML5_in_mobile_devices

Chapter 51
Material Structure Selection of Solution Blue OLEDs Using a Design of Experiment

Yu-Sheng Tsai, Apisit Chittawanij, Lin-Ann Hong, Fuh-Shyang Juang, and Van-Huong Tran

Abstract A blue small-molecular organic light-emitting diode (SM-OLEDs) based on a solution-process is investigated in this study. Design of experiment (DOE) with response surface methodology (RSM) was applied to optimize the driving voltage and current efficiency of blue SM-OLED devices. The spin-coating speed of the PEDOT: PSS as hole injection layer and the 26DCzPPy: FIrpic as emitting layer were chosen as two main process input factors. Analysis of variance (ANOVA) was adopted to identify significant factors before regression models were obtained. The optimal material structure was determined by minimizing and maximizing a desirability function relating to selected critical quality characteristics including the driving voltage and current efficiency, respectively.

Keywords Solution-process • Blue OLEDs • Design of experiments • Stamping process

51.1 Introduction

The solution-based small molecule organic light-emitting diodes (SM-OLEDs) manufacturing process is ideal for large scale, low-cost mass production [1, 3]. However, when fabricating multilayer structures the solutions can easily attack layers other than the targeting layer [1–4]. Therefore, SM-OLEDs devices manufactured using solution-based processing can only have one layer [2, 3]. The spin-coating process is still required to build high efficiency of SM-OLEDs.

In the past, the tuning of material structure is mainly done via a trial-and-error [5–7] approach, which involves selecting various combinations of different

Y.-S. Tsai • A. Chittawanij (✉) • L.-A. Hong • F.-S. Juang
Institute of Electro-Optical and Materials Science, National Formosa University, Huwei, Yunlin 63208, Taiwan
e-mail: abcd_omo@hotmail.com

V.-H. Tran
Department of Power Mechanical Engineering, National Formosa University, Huwei, Yunlin 63208, Taiwan

J. Juang (ed.), *Proceedings of the 3rd International Conference on Intelligent Technologies and Engineering Systems (ICITES2014)*, Lecture Notes in Electrical Engineering 345, DOI 10.1007/978-3-319-17314-6_51

variables heavily depending on prior experience. Consequently, it is not cost-effective since the materials are very expensive. Moreover, there is no guarantee that the chosen combination is optimal. The objective of this study is to search for an optimal spin-coating speed of solution-process to improve SM-OLED performance. The spin-coating speed in selecting an optimal material structure includes the hole injection layer (HIL) and an emitting layer (EML), based on the driving voltage and Yield quality characteristics. To resolve the multilayer structure spin-coating speed problem, a response surface design of experiments (DOE) method is proposed [8]. Analysis of variance (ANOVA) was utilized to identify the influential factors affecting the selected critical quality characteristics.

51.2 Experimental

51.2.1 Fabrication of Blue OLED Device and Measurement

The device structure and materials used are shown in Fig. 51.1. The fabrication procedure of the blue SM-OLEDs is as follows. First, an indium–tin oxide (ITO) glass substrate was cleaned in ultrasonic baths of detergent, deionized water, acetone, and isopropanol, sequentially. Then, the ITO glass substrate was treated with O_2 plasma for 100 s. A HIL, poly(3,4-ethylenedioxy thiophene): poly(styrenesulfonate) (PEDOT:PSS), was spin-coated on the ITO glass substrate followed by baking for 10 min to remove the remaining solvent. Subsequently, EML which consists of host material 2,6-bis(3-(9H-carbazol-9-yl)phenyl) pyridine (26DCzPPy) and blue dopant bis[(3,5-difluoro-2(2-pyridyl) phenyl-(2-carboxypyridyl)iridium(III)] (FIrpic) are mixed and dissolved in chlorobenzene, which is then spin-coated onto HIL to form an EML, after spin-coating is over, the sample is directly baked for 15 min to solve the residual solution and to improve the properties of organic film. The 1,3,5-tris (*N*-phenylbenzimidazol-2-yl) benzene (TPBi) as ETL is subsequently deposited

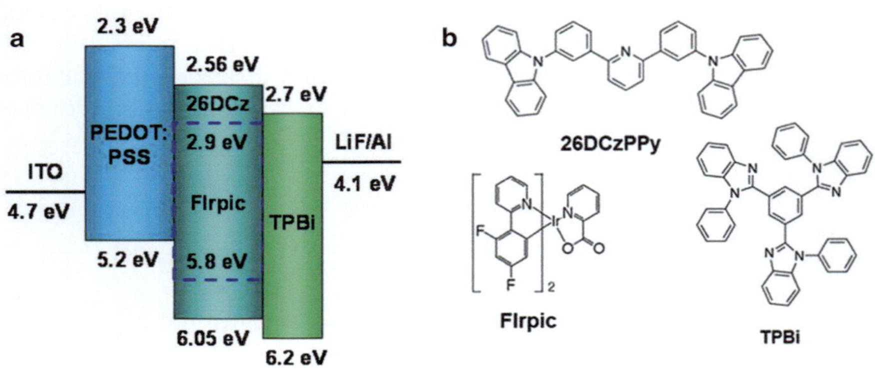

Fig. 51.1 (**a**) Energy band diagrams and (**b**) molecular structures of organic materials used

Table 51.1 Gathering experimental runs and results at 100 cd/m^2 and 20 mA/cm^2

Run	Factor 1 A:PEDOT (rpm)	Factor 2 B:26DCzPPy:FIrpic (rpm)	Driving voltage 100 cd/m^2	Yield (cd/A) 20 mA/cm^2
1	3,000.00	4,000.00	7.6	6.5
2	3,207.11	3,000.00	8.1	6.2
3	2,000.00	4,000.00	9.2	7.8
4	2,500.00	4,414.21	9.0	6.1
5	2,500.00	3,000.00	6.1	8.7
6	2,500.00	3,000.00	6.2	8.4
7	2,500.00	1,585.79	7.6	5.7
8	2,000.00	2,000.00	8.6	6.0
9	2,500.00	3,000.00	6.6	8.4
10	2,500.00	3,000.00	7.5	8.0
11	3,000.00	2,000.00	9.2	6.1
12	1,792.89	3,000.00	8.2	6.2
13	2,500.00	3,000.00	6.1	8.0

evaporated onto the EML surface. At the end of the device process, LiF and Al are thermally evaporated as cathode electrode.

The current, voltage, and luminance characteristics were measured simultaneously with a programmable Keithley 2,400 voltage–current and Spectra Scan PR650 sources. All measurements were carried out at room temperature under ambient conditions.

51.2.2 Design of Experimental Parameters

In this study, design of experiment (DOE) with response surface methodology (RSM) was applied to optimize the performance of OLED devices with two responses, yield at current density of 20 mA/cm^2 and driving voltage at luminance of 100 cd/m^2. The spin-coating speed of the HIL and EML are chosen as two main input factors. The ranges of spin-coating speed from 2,000 to 3,000 rpm and 2,000 to 4,000 rpm were selected, respectively. The OLED devices are setting follow-up design expert software version V8.0.6. Table 51.1 shows the experimental results conducted in a random order of 13 runs.

51.3 Methods and Results

51.3.1 Definition of an Objective Function for Optimization

Response surface method is a collection of math and statistical techniques useful for the modeling and analysis of problems where a response of interest is influenced by several variables and the objective is to optimize this response [9]. Suppose that to find the levels of spin-coating PEDOT:PSS (x_1) and 26DCzPPy:FIrpic (x_2) that maximize the yield (y) of an OLED device. The yield (y) is a function of the levels of spin-coating speeds PEDOT:PSS and 26DCzPPy:FIrpic, say

$$y = f(x_1, x_2) + \varepsilon$$

where ε represents the noise or error observed in the response y. Denoting the expected response by $E(y) = f(x_1, x_2) = \eta$, then the surface represented by

$$\eta = f(x_1, x_2)$$

is called a response surface. In this study, the parameters and analyzing simulations set up by using Design Expert software version V8.0.6

51.3.2 ANOVA Results

The ANOVA is conducted to identify the significant factors affecting the driving voltage and yield.

Table 51.2 shows the ANOVA result of the driving voltage, a "Model *F* value" of 6.13 with a "Model *P* value" less than 0.0170 implies that the selected model is significant and there is less than a 1.7 % chance that the "Model *F* value" could occur due to the noise. The "*P* value" for the model terms "A^2" and "B^2", are less than 0.05, indicating that the model terms are significant on the voltage. According to the hierarchy principle in model-building [10], the model terms "A"(HIL, PEDOT:PSS), "B"(EML, 26DCzPPy:FIrpic), and two-factor interaction terms, "AB" shall also be included in the regression model even though the "*P* value" of the model terms "A", "B," and "AB" are more than 0.05. The ANOVA result of the yield can be seen in Table 51.3. A "Model *F* value" of 12.30 with a "Model *P* value" less than 0.0023 implies that the selected model is significant and there is less than a 0.2 % chance that the "Model *F* value" could occur due to the noise.

Table 51.2 ANOVA results of the driving voltage (V)

Source	Sum of square	Degree of freedom	Mean square	*F* value	*P* value
Model	13.07	5	2.61	6.13	<0.0170
A	0.16	1	0.16	0.38	0.5561
B	0.12	1	0.12	0.28	0.6121
AB	1.21	1	1.21	2.84	0.1359
A^2	6.03	1	6.03	14.16	0.0071
B^2	7.04	1	7.04	16.53	0.0048
Residual	2.98	7	0.43	–	–
Cor. Total	16.05	12	–	–	–

Table 51.3 ANOVA results of the yield (cd/A)

Source	Sum of square	Degree of freedom	Mean square	*F* value	*P* value
Model	13.86	5	2.74	12.30	<0.0023
A	0.23	1	0.23	1.06	0.3384
B	0.96	1	0.96	4.30	0.0769
AB	0.49	1	0.49	2.20	0.1814
A^2	5.66	1	5.66	25.43	0.0015
B^2	7.88	1	7.88	35.42	0.0006
Residual	1.56	7	0.22	–	–
Cor. Total	15.24	12	–	–	–

The "*P* value" for the model terms "A^2" and "B^2" is far less than 0.05, indicating that the model terms are significant on the yield.

51.3.3 Regression Models for Critical Quality Characteristics Versus the Material Structure

Considering the significant parameters in Table 51.2 to 51.3 and the test results from Table 51.1, regression models can be derived for the driving voltage and the current efficiency, respectively. The mathematical models are shown as follows:

$$\begin{aligned}\text{Yield(cd/A)} = & -29.247 + 0.0198A + 0.009B - 7 \times 10^{-7}AB \\ & - 3.6 \times 10^{-6}A^2 - 1.05 \times 10^{-6}B^2\end{aligned} \tag{51.1}$$

$$\begin{aligned}\text{Driving voltage (V)} = & \; 30.933 - 0.0156A - 3.165B - 1.1 \times 10^{-6}AB \\ & + 3.725 \times 10^{-6}A^2 + 1.006 \times 10^{-6}B^2\end{aligned} \tag{51.2}$$

where *A* is rpm of PEDOT:PSS and *B* is rpm of 26DCzPPy:FIrpic

According to the diagnostics case statistics report of yield (cd/A) and driving voltage (V), it is anticipated that with 2,500 rpm of spin-coating speed for PEDOT:PSS, and 3,000 rpm for 26DCzPPy:FIrpic, the best performance of OLED devices in this study can reach to 8.31 cd/A at 20 mA/cm^2 and the lowest driving voltage of 6.5 V at 100 cd/m^2.

51.3.4 Confirmation Runs

To verify and consolidate the result of ANOVA from design expert software, two repeated experiments with the spin-coating speeds of 2,500 and 3,000 rpm of PEDOT:PSS and 26DCzPPy:FIrpic, respectively were conducted. After conducting

Table 51.4 Confirmation runs and an optimal material layer structure

	Factor 1	Factor 2	100 cd/m^2		20 mA/cm^2	
Runs	A:PEDOT rpm	B:26DCzPPy rpm	Driving voltage (V) Pred./Exp.	Error (%)	Yield (cd/A) Pred./Exp.	Error (%)
1	2,500	3,000	6.5/6.5	0	8.31/8.6	3.3
2	2,500	3,000	6.5/6.2	4.8	8.31/8.5	2.2
Average			–	4.8	–	5.5

two validation experiments, the results are close to the diagnostics from software. Table 51.4 shows the confirmation runs and an optimal materials layer structure. The first device of validation experiment showed the yield and driving voltage of 8.6 cd/A at 20 mA/cm^2 (error 3.3 %) and 6.5 V at 100 cd/m^2 (error 0 %), respectively. The second device of validation experiment showed the yield and driving voltage of 8.5 cd/A at 20 mA/cm^2 (error 2.2 %) and 6.2 V at 100 cd/m^2 (error 4.8 %), respectively.

51.4 Conclusions

This study used design of experiment (DOE) with response surface methodology (RSM) to design the blue SM-OLED material layer structure. An optimal material layer structure is selected of PEDOT:PSS (2,500 rpm) and 26DCz PPy:FIrpic (3,000 rpm), respectively. The yield and driving voltage based on this layer structure are 8.5 cd/A at 20 mA/cm^2 and 6.2 V at 100 cd/m^2, respectively. The proposed method provides a systematic approach to optimize the material layer structure of blue SM-OLEDs; thus, efficiency of searching for an optimal spin-coated has been improved. This approach can be easily extended to other material selection of OLEDs.

Acknowledgement The authors would like to acknowledge the financial support of the National Science Council of the Republic of China, project NSC-102-2221-E-150-066-MY2 and NSC-103-2622-E-150-018-CC3.

References

1. Choi, J.H., Kim, K.H., Choi, S.J., Lee, H.H.: Whole device printing for full colour displays with organic light emitting diodes. Nanotechnology **17**, 2246–2249 (2006)
2. Huang, J.H., Ho, Z.Y., Kuo, T.H., Kekuda, D., Chu, C.W., Ho, K.C.: Fabrication of multilayer organic solar cells through stamping technique. J. Mater. Chem. **19**, 4077–4080 (2009)
3. Kim, K.H., Huh, S.Y., Seo, S.M., Lee, H.H.: Solution-based formation of multilayers of small molecules for organic light emitting diodes. Appl. Phys. Lett. **92**, 093307 (2008)

4. Kim, C., Cao, Y., Soboyejo, W.O., Forrest, S.R.: Patterning of active organic materials by direct transfer for organic electronic devices. J. Appl. Phys. **97**, 113512 (2005)
5. Jeon, S.O., Lee, J.Y.: Red phosphorescent organic light-emitting diodes using pyridine based electron transport type triplet host materials. Mater. Chem. Phys. **127**, 300–304 (2011)
6. Haq, K.U., Khan, M.A., Zhang, X.W., Zhang, L., Jiang, X.Y., Zhang, Z.L.: Blue organic light-emitting diodes with low driving voltage and enhanced power efficiency based on MoO_3 as hole injection layer and optimized charge balance. J. Non-Crys. Solids **356**, 1012–1015 (2010)
7. Jeon, Y.M., Kim, J.W., Lee, C.W., Gong, M.S.: Blue organic light-emitting diodes using novel spiro[fluorene-benzofluorene]-type host materials. Dyes Pigments **83**, 66–71 (2009)
8. Douglas, C.M.: Design and Analysis of Experiments, 4th edn. Wiley, New York (1997)
9. Liu, J.M., Lu, P.Y., Weng, W.K.: Studies on modifications of ITO surfaces in OLED devices by Taguchi methods. Mater. Sci. Eng. **85**, 209–211 (2011)
10. Myers, R.H., Montgomery, D.C.: Response Surface Methodology, 2nd edn. Wiley, New York (2002)

Chapter 52
Development of Software Inspection Tools for Defect Examination in Machine Vision Applications

Jong-Hann Jean

Abstract In this chapter we developed a machine vision system, mainly focusing on implementation of software inspection tools for visual measurement and analysis. With reference to the support functions of several main brands of machine vision systems, we have proposed a visual inspection procedure for defect examination, consisting of five steps, namely, image capture and pre-processing, edge and feature extraction, image segmentation and positioning, template matching, and BLOB analysis. To fulfill the proposed inspection procedure, we implemented several software inspection tools. For each tool, we demonstrated its application example for defect examination in machine vision applications. All these tools can be further arranged in a proper sequence to realize a complicated defect inspection procedure. We have applied the proposed system to defect examination of mobile phone keypads to validate its effectiveness and obtained satisfactory results.

Keywords Visual measurement • Machine vision • Defect examination

52.1 Introduction

The machine vision technique, also called the automatic optical inspection (AOI) technique, has been widely used in the production line to examine the qualities of parts. It has the advantage of reducing the cost of workers, promoting the qualities of products and the producing capacity. It also helps workers avoid the wrong determination problem induced by overwork, fatigue, long hours and changeless work patterns.

Since the semiconductor, the communications, and the electronics industries have been booming these years, the demand of the machine vision system has been extensively expanded [1]. As the part size getting tinier, the necessary resolution of

J.-H. Jean (✉)
Department of Electrical Engineering, St. John's University, Taipei, Taiwan
e-mail: jhjean@mail.sju.edu.tw

J. Juang (ed.), *Proceedings of the 3rd International Conference on Intelligent Technologies and Engineering Systems (ICITES2014)*, Lecture Notes in Electrical Engineering 345, DOI 10.1007/978-3-319-17314-6_52

inspections has already reached the micrometer scale or even to the nanometer scale such that we cannot inspect the qualities of products by our own eyes. In addition, the trend of shortening product development life cycle and acceleration of globalization made manufactures realize that reducing the cost, increasing the production and its qualities are the key to survive. Consequently, the AOI systems have been considered as the indispensable facilities in production environment for quality management. These development trends drive the machine vision system toward the following five directions: high-speed inspection, sub-micro resolution, software intelligence, versatile applications, and system modularization.

Most machine vision systems are custom designed and each of them is only suitable for a specific application. In the literature, there have been several machine vision systems proposed for printed circuit board inspection [2–6]. Furthermore, Moganti et al. [7] provided a comprehensive review of the PCB inspection algorithms. Chen et al. [8] proposed a machine vision system for defect detection of analogical resistance type touch panel. Tsai et al. [9] developed machine vision techniques for micro-crack inspection of crystalline silicon solar cells. Xie [10] provided a systematic review of recent advances in surface defect detection algorithms based on visual inspection. There are also several commercial AOI products developed by companies such as BANNER, Euresys, Matrox, Narion Instruments, PPT Vision, DALSA, Cognex, and MVTec for diverse industrial applications.

This chapter presents our attempt to design and integrate a machine vision system which includes the image processing unit, the operational input and output interface, the illumination unit, the image measurement and analysis software. By investigating the current AOI products in the market, it shows that the main cost comes from the image processing software for visual measurements and analysis. Accordingly, this chapter is emphasizing on implementation of software inspection tools for visual measurement and analysis utilities.

With reference to the support functions of several main brands of the AOI products, we first summarized a general procedure for defect examination tasks in machine vision applications. Accordingly, by using the Windows-based programming software, Borland C++ Builder, we implemented nine software inspection tools, namely, the average grey level tool, the thresholding tool, the positioning tool, the edge detection tool, the binary large object (BLOB) tool, the template building tool, the smart matching tool, the inspection sequence tool, and the platform operation tool. All these tools can be used in an inspection task with only single operation and can also be arranged in a proper sequence of operations to fulfill a complicated inspection task.

As for the hardware components of the proposed system, we designed a dedicated inspection platform and an LED illumination unit with the backlighting configuration. The inspection platform is custom-tailored to reach the specifications of defect examination according to the demanding precision and the processing speed. To validate the performance of the proposed system, we applied the system to inspection of several defective samples of mobile phone keypads provided by the supplier and all obtain satisfactory results.

The remainder of the chapter is organized as follows. The system architecture is introduced in Sect. 52.2. The implementation steps and results are illustrated in Sect. 52.3. Application examples of the system to defect examination of mobile phone keypads are provided in Sect. 52.4. Conclusions are given in Sect. 52.5.

52.2 System Architecture

As shown in Fig. 52.1, the system architecture of the integrated machine vision system includes an inspection platform, a CCD camera, a frame-grabber, and an industrial computer.

In the following each part in the system architecture is further illustrated:

A. Frame-grabber: The system uses the ADLINK PCI-RTV24 frame-grabber which can retrieve four channels of color or grey level images.
B. CCD Camera: The system uses the SONY XC-75 interlaced camera with 768 (H) $\times$ 494(V) effective pixels, where one can program the trigger mode, the shutter speed, and the exposure time through RS232 interface.
C. Inspection Platform: The dedicated inspection platform includes a 3D XYZ mechanism surrounded by an open-up isolation box and a self-designed LED illumination unit with the backlighting configuration, as shown in Fig. 52.2. The camera is mounted on the XYZ mechanism where the parts together with the illumination unit are carried by a conveyor track driven by a DC motor to move along the X axis. Two additional shafts are used to adjust the Y, Z position of the camera for capturing a suitable image in size.
D. Industrial Computer: The system uses a PC-based industrial computer with Intel (R) Core™2 Quad CPU Q6600 @2.40GH processor to execute the software inspection tools for image measurement and analysis.

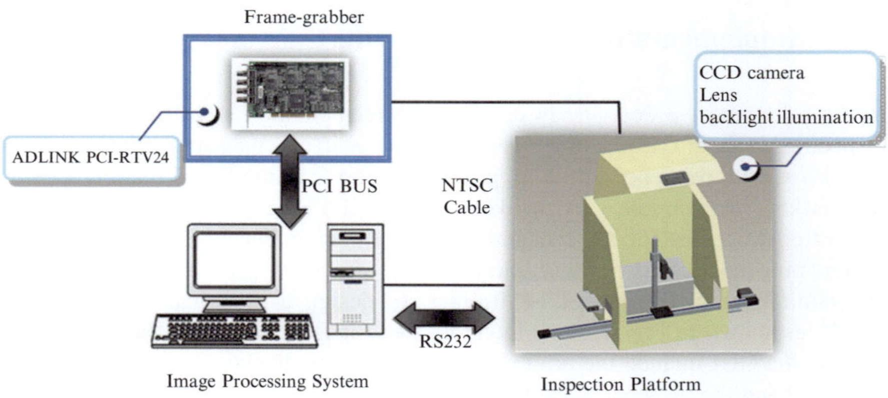

Fig. 52.1 The system architecture of the developed machine vision system

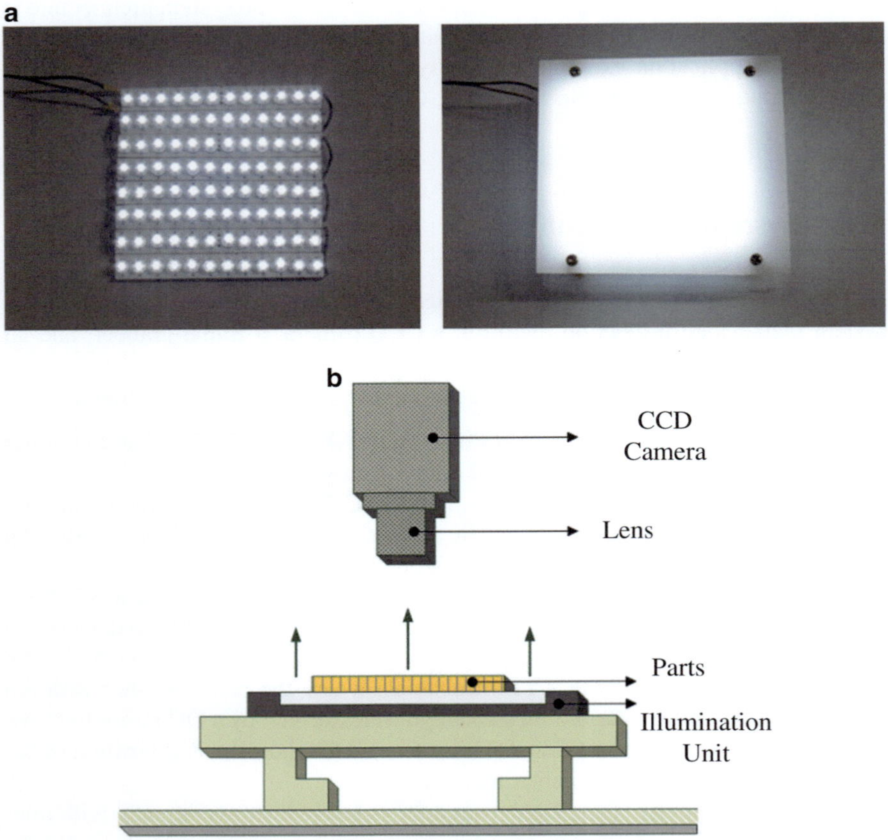

Fig. 52.2 (**a**) The self-designed backlight illumination unit. (**b**) The backlighting configuration adopted in the chapter

52.3 Implementation Steps and Results

In the following we describe in detail the implementation steps of the software inspection tools of our machine vision system for image measurements and analysis. First, with reference to the supporting functions from several main brands in the market, for example, NAIS, SHARP, and KEYENCE, we propose a visual inspection procedure for defect examination, as shown in Fig. 52.3, and accordingly design nine inspection tools to carry out the proposed procedure. We use the programming tool, Borland C++ Builder, to design a graphical user interface (GUI), as shown in Fig. 52.4, to integrate these inspection tools. The three main parts in the GUI include the original/processed image display area, the inspection tool area, and the inspection result area.

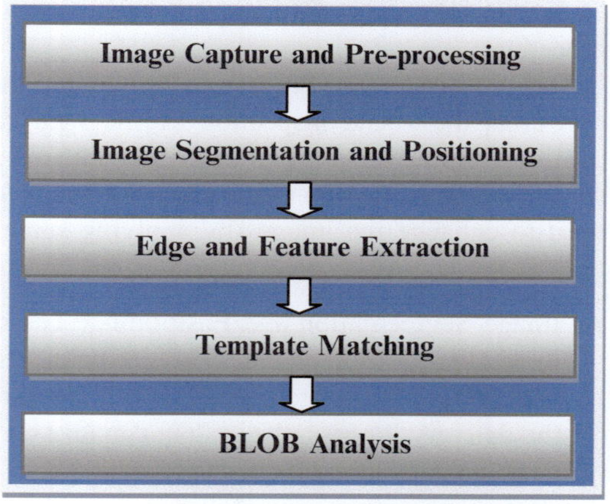

Fig. 52.3 The proposed visual inspection procedure for defect examination

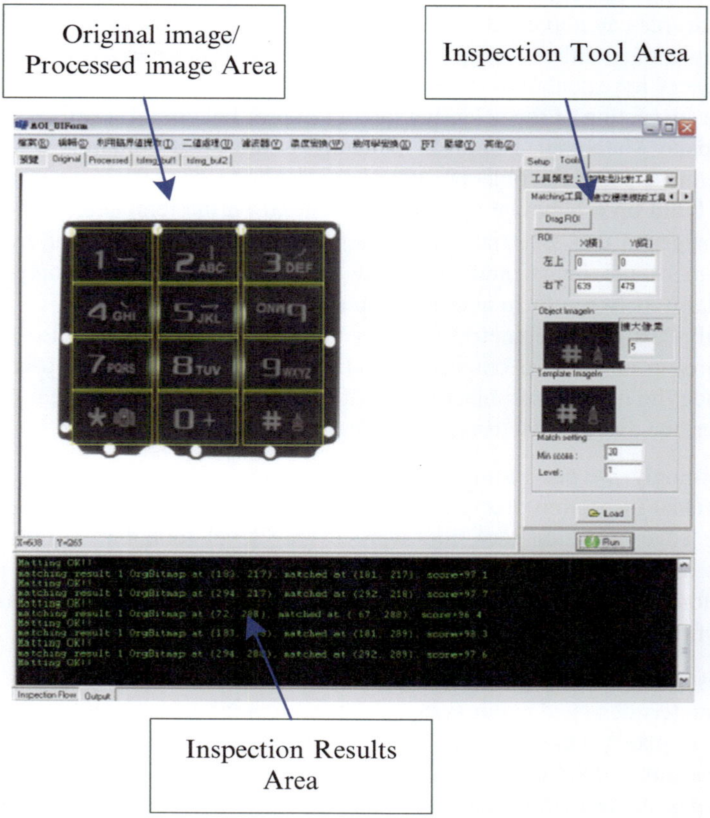

Fig. 52.4 The three main parts in the GUI of the developed inspection software

In the following we illustrate each step in the proposed visual inspection procedure.

Image Capture and Pre-processing: While the object is delivered to the inspection area, the visual inspection begins with capturing a sample image of good quality with the aid of the illumination unit and proper selection of lens and shutter speed. When the quality of the captured image is not good enough, the system may need to enhance the images by applying some image pre-processing techniques, such as median filter, averaging filter, histogram equalization, and so on.

A. Image Segmentation and Positioning: The next step of image inspection is to locate the position of the object in the captured image and separate it from the background by image segmentation. Because the object may be moved during the delivery process, the system cannot assume its position and orientation and needs to detect the image displacement and lean angle of the object. A common strategy of image segmentation is to process the grey level image by two-valued thresholding and follow up with morphological dilation and erosion to filter out oversized or tiny non-objects.
B. Edge and Feature Extraction: For most machine vision tasks, object edges and features play an important role in identification and classification of defects. For example, the number of edges and their separate distance in a chip are common evidences for calculation of the pin-to-pin distance and pin numbers in the chip or for inspection of the soldering quality in printing circuit boards.
C. Template Matching: Template matching is the most common way used in searching a particular object in an image. The first step of the template matching is to build up a template of the object. Based on the template, we can search subimages of the same size with the template from left to right and from top to bottom of the image and determine the object position by comparing the similarity between them and the template.
D. BLOB Analysis: Geometric parameters of object regions are also important features in image inspection for detection of defects. Commonly used features include the number of object regions in the image, the region area, the region center, the average intensity, the circumference, etc.

Accordingly, we design nine inspection tools to fulfill the supporting functions for the visual inspection procedure, namely, the thresholding tool, the positioning tool, the edge detection tool, the average gray level tool, the BLOB tool, the template building tool, the smart matching tool, the inspection sequence tool, and the platform operation tool. In the following we illustrate the implementation results of these inspection tools.

A. Thresholding tool: Figure 52.5a presents the thresholding tool and its relevant parameters designed in our system. There is a prefiltering checkbox for filtering out the pixels which are too bright or too dark in the thresholding process. Furthermore, there are several options for the thresholding tool to specify the region of interest (ROI), the selection (fixed or adaptive) of threshold value, and the modes of thresholding (active white or active black). Figure 52.6a is a

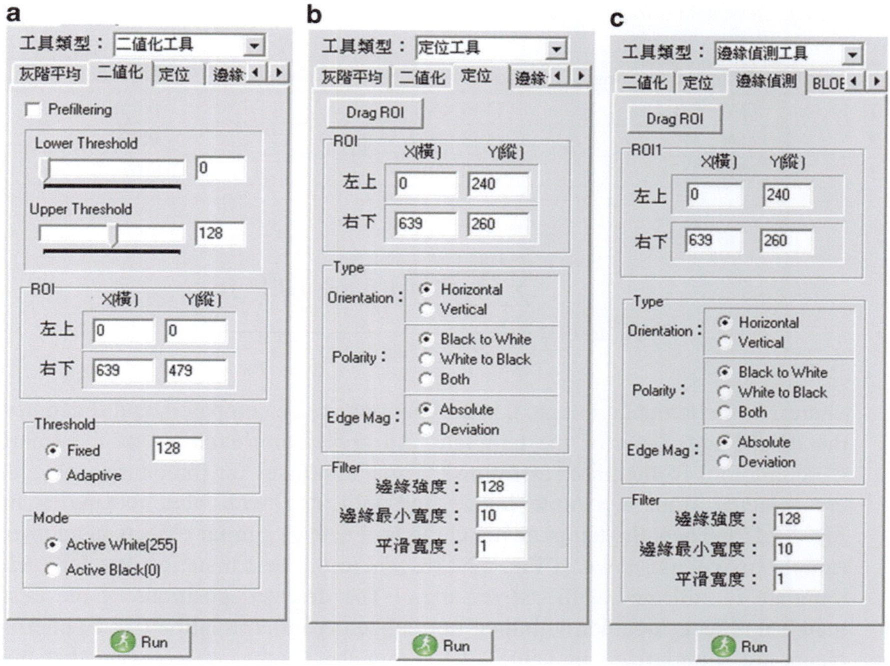

Fig. 52.5 (**a**) The thresholding tool, (**b**) the positioning tool, (**c**) the edge detection tool and the relevant parameters

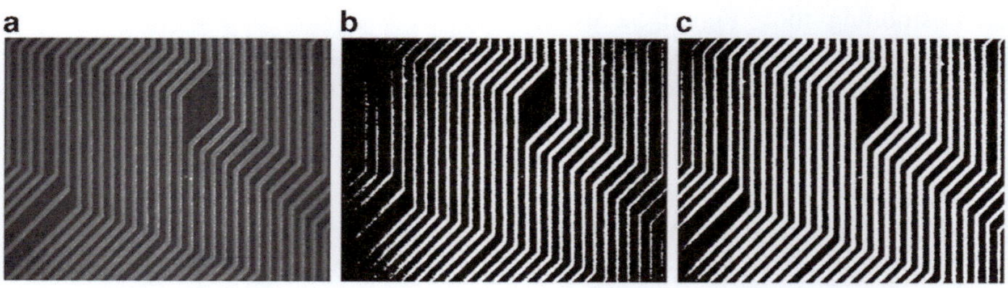

Fig. 52.6 (**a**) The sample image of TAB surface circuits to be inspected by the thresholding tool. (b) Processed result using a fixed threshold, 128. (**c**) Processed result using the adaptive threshold

sample image of the TAB surface circuits to be inspected. Application result of the thresholding tool to the sample image using a fixed threshold, 128, is indicated in Fig. 52.6b and another processed result using the adaptive threshold is shown in Fig. 52.6c. We can see that using the adaptive threshold value results in better performance in separating the object from the background image.

Table 52.1 Detected positions of the best matched objects shown in Fig. 52.11b and their NCC coefficients

No.	Template position	Position of matched objects	NCC coefficient
1	(132, 60)	(140, 61)	0.666
2	(140, 312)	(148, 313)	0.994
3	(312, 213)	(320, 214)	0.67

$$\mathrm{NCC}(u,v) \equiv \frac{\sum_{j=0}^{B-1}\sum_{i=0}^{B-1} I_t(u+i, v+j)T(i,j)}{\|I_t\| \cdot \|T\|} \tag{52.1}$$

where T is the template image, I_t is the subimage to be compared, and B denotes the height and width of the template. An application example of the smart matching tool is illustrated as follows. Three templates for three target objects are selected from a sample image and indicated with blue rectangles in Fig. 52.11a. Then, the template matching tool detects similar objects in another sample image, as shown in Fig. 52.11b, where the best matched one for each template is indicated with red rectangle. The detected positions of the best matched objects together with their normalized cross-correlation (NCC) coefficients are summarized in Table 52.1. From the results, we can find that the second object is successfully matched with a high NCC coefficient. However, the matched results for the first and third objects are not satisfactory because there are stains on the two objects and the NCC coefficients are only 0.666 and 0.67, respectively.

B. Positioning tool: Figure 52.5b shows the positioning tool and its relevant parameters, where the tool can scan the edge in horizontal or vertical direction to determine the x or y coordinates of the object position. The parameters for the positioning tool include the ROI, threshold of the edge strength, the polarities of the edge (black to white or white to black), and the minimal edge width and the smoothness depth. An application example of the positioning tool to a PCB sample image is illustrated in Fig. 52.7, where a horizontal scan is applied with the ROI set as (0, 240) ~ (639, 260), the threshold of edge strength 45, and the minimal edge width 10. Then, the detected x coordinate of the PCB is 97.

C. Edge detection tool: Figure 52.5c shows the edge detection tool and its relevant parameters which are quite similar with those of the positioning tool. The difference between the two tools lies in that the positioning tool outputs the coordinate of the first detected edge, but the edge detection tool will output coordinates of all edges scanned in the ROI. An application example of the edge detection tool to a sample image is illustrated in Fig. 52.8. By setting the ROI for horizontal scan as (0, 240) ~ (639, 260), selecting the black-to-white edge type, setting the edge strength as 45 and the minimal edge width as 15, the edge detection tool output the coordinates of all detected edges as 103, 129, 142, 155, 167, 180, 193, 206, 219, 231, 244, 257, 270, 283, 295, 308, 321, 334, 347, 359,

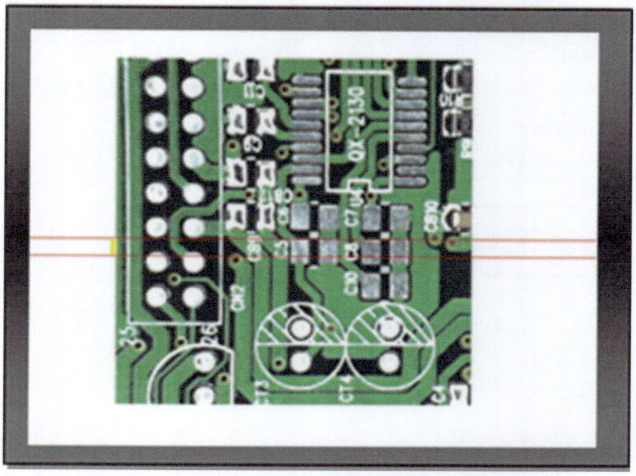

Fig. 52.7 Application result of the positioning tool for detecting the *x* coordinate of the PCB in the image

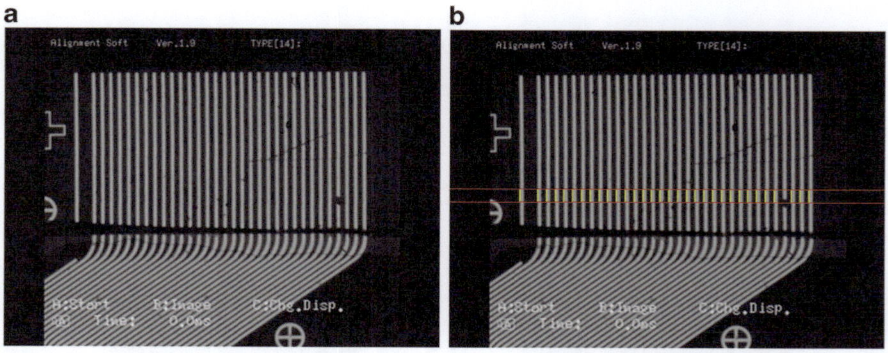

Fig. 52.8 (**a**) A sample image, (**b**) application result of the edge detection tool to the sample image for detecting the edges in the ROI

372, 385, 398, 411, 424, 436, 449, 462, 475, 500, 513, and 526. From the result, we can observe that the edge distance is 13 pixels in average and there is a broken edge in the coordinate range 475 ~ 500.

D. Average gray level tool: As mentioned, the average gray level of an object in ROI can be regarded as a simple feature. Figure 52.9a shows the average gray level tool and its relevant parameters. An application example of the tool for inspection of a ball grid array chip is illustrated in Fig. 52.10. The average gray level of a sample image depicted in Fig. 52.10a is computed as 101. However, the average gray level of the ROI for another sample shown in Fig. 52.10b is calculated as 107 and may be accordingly classified as a defective sample.

E. Smart matching tool: Figure 52.9b shows the smart matching tool and its relevant parameters. After specifying an object template, the tool can scan

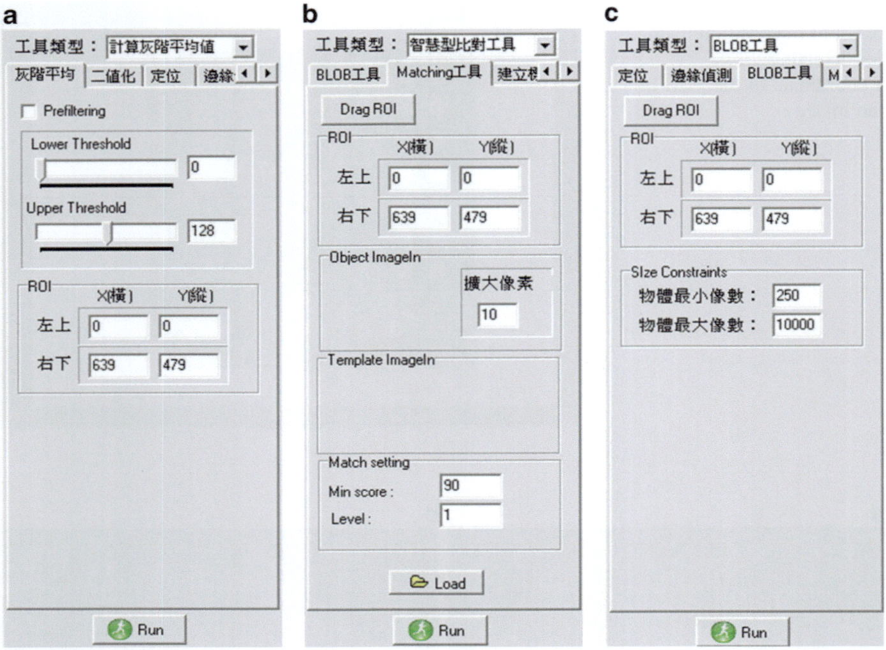

Fig. 52.9 (**a**) The average gray level tool, (**b**) the smart matching tool, (**c**) the BLOB tool and the relevant parameters

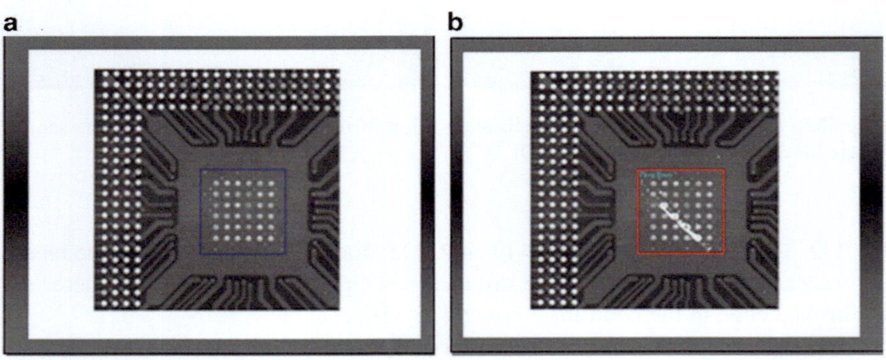

Fig. 52.10 (**a**) A normal sample image, (**b**) application result of the average gray level tool to another sample image for detecting the defect in the ROI

each subimage having the same size with the template in a search area and compare the similarity using the NCC method. The computation of the NCC coefficient can be expressed as the following equation.

F. BLOB tool: Figure 52.9c shows the BLOB tool and its relevant parameters. The BLOB tool requires a binary image with white pixels indicating object regions

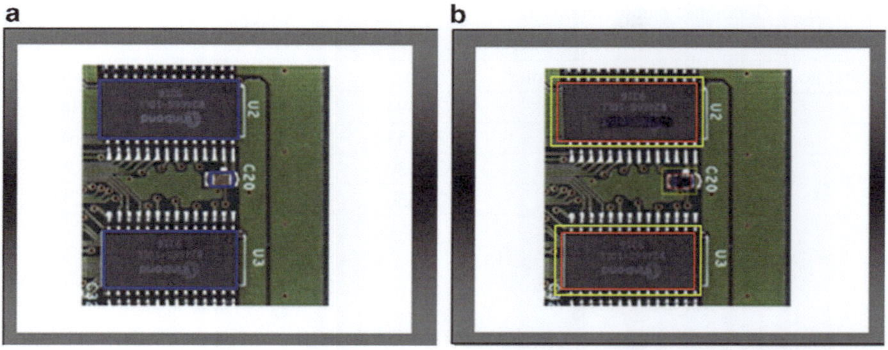

Fig. 52.11 (**a**) Three templates selected from a normal sample image, (**b**) application result of the template matching tool to search similar objects in another sample

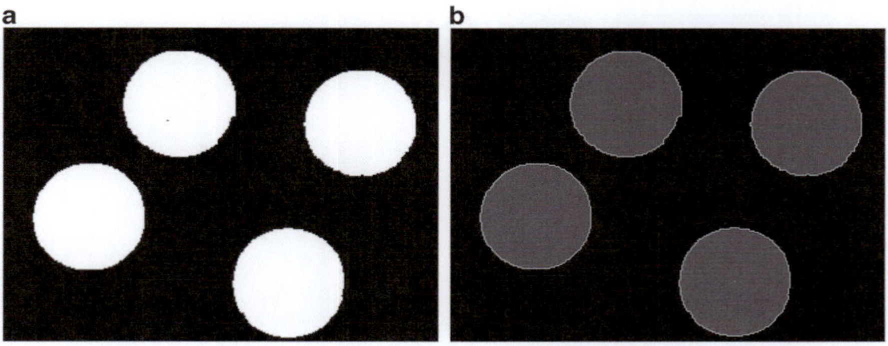

Fig. 52.12 (**a**) A sample image, (**b**) application result of the BLOB tool to the sample image for labeling the objects and calculating their geometric parameters

and black pixels indicating background region. By using the component labeling algorithm, the tool can indicate all objects in the image with different labels. Accordingly, the tool can calculate geometric properties of each object region, for example, region area and center of mass. It can further extract the contour of each object by using the boundary tracking algorithm. Figure 52.12 shows a test example for validating the BLOB tool. There are mainly four large regions indicating objects as shown in Fig. 52.12a. By applying the BLOB tool to the image, we have detected four regions as depicted in Fig. 52.12b and the geometric parameters of these regions are calculated and listed in Table 52.2.

G. Template building tool: Figure 52.13a shows the template building tool and its relevant parameters, which is a helping tool to facilitate choosing templates on a sample image for subsequent template matching process.
H. Inspection sequence tool: Figure 52.13b shows the inspection sequence tool and its relevant parameters. The tool can help arrange previous tools in a proper sequence of operations to fulfill a complicated inspection procedure.

Table 52.2 Geometric parameters of the four object regions detected by applying the BLOB tool to the image in Fig. 52.12a

Label no.	Position of region center	Area	Circumference	Compactness
0	(259,117)	20,594	594.551147	0.732103
1	(522,146)	20,553	595.379578	0.728614
2	(128,288)	20,530	585.622314	0.752252
3	(418,385)	20,630	597.036377	0.727290

Fig. 52.13 (**a**) The template building tool, (**b**) the inspection sequence tool, (**c**) the platform operation tool and the relevant parameters

I. Platform operation tool: Figure 52.13c shows the platform operation tool and its relevant parameters. The tool can be used to manually move the position of the inspection platform by controlling the motion of the conveyor track. It can also be used to automatically move the inspection platform according the inspection results of an inspection sequence and a specified criterion.

52.4 Application to Defect Inspection of Mobile Phone Keypads

To validate the performance of the developed inspection tools, we apply the proposed system to defect inspection of mobile phone keypads. Figure 52.14a shows a normal sample image of a keypad captured by our system using backlighting configuration.

The goal of inspection is to examine if there are defects on the 12 key buttons of the keypad. Accordingly, the inspection procedure is designed as follows. First, since the keypad may be moved during the delivery process, the positioning tool is used to determine the position of the keypad. Then, the template building tool is applied to choosing a template for each key button, respectively. The 12 templates built are indicated with blue rectangles as shown in Fig. 52.14b. In addition, the smart matching tool is used to examine the similarity between the button sample images and the templates and compute the NCC coefficients. Finally, if one of the NCC coefficient is smaller than 0.9, the system determines there is a defect occurred on the corresponding button of the keypad. Then, the platform operation tool is utilized to move the defective keypad to the NG area.

An application result of the inspection task is illustrated in Fig. 52.15, where the best matched subimages for the 12 templates are indicated with green rectangles and their correspondent NCC coefficients are listed in Table 52.3. From the results, we can find that there is a defect on the sixth button of the keypad because the NCC coefficient for that button is only 0.214. The defect is due to inverse mounting of the button.

Another inspection example is shown in Fig. 52.16, where we can observe there are some stains on button 1 and button 4 of the keypad. Similarly, the best matched subimages for the 12 templates are indicated with green rectangles and their correspondent NCC coefficients are listed in Table 52.4. The proposed system again successfully detects the defects according to the computed NCC values.

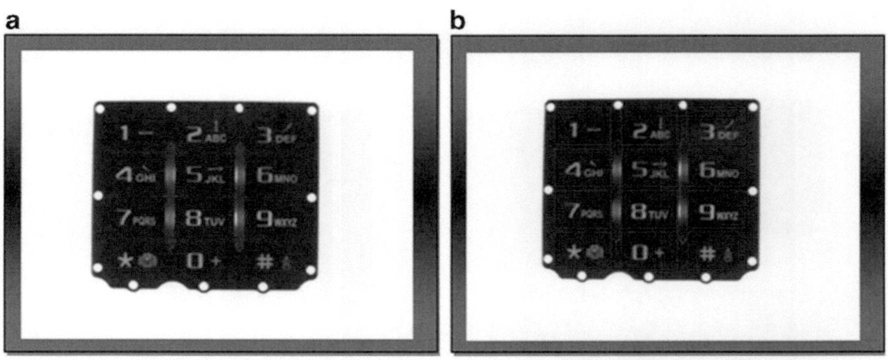

Fig. 52.14 (**a**) A normal sample image of a mobile phone keypad captured on our designed platform, (**b**) the 12 templates enclosed by *blue rectangles* for defect examination

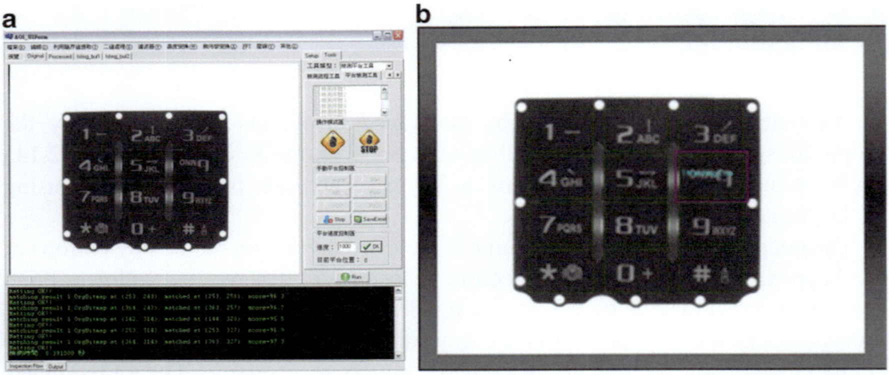

Fig. 52.15 (**a**) Application example for examination of a mobile phone keypad, (**b**) the template matching tool successfully detects a defect (inverse mounting) on *button 6* of the keypad

Table 52.3 Positions of matched buttons shown in Fig. 52.15b and their NCC coefficients

No.	Template position	Position of matched buttons	NCC coefficient
1	(143, 119)	(143, 116)	0.989
2	(253, 119)	(253, 114)	0.980
3	(363, 119)	(363, 116)	0.977
4	(143, 189)	(145, 186)	0.969
5	(253, 189)	(254, 185)	0.959
6	(363, 189)	(364, 187)	0.214
7	(143, 259)	(145, 257)	0.962
8	(253, 259)	(254, 256)	0.976
9	(363, 259)	(363, 255)	0.988
10	(143, 329)	(146, 325)	0.989
11	(253, 329)	(254, 324)	0.973
12	(363, 329)	(363, 326)	0.976

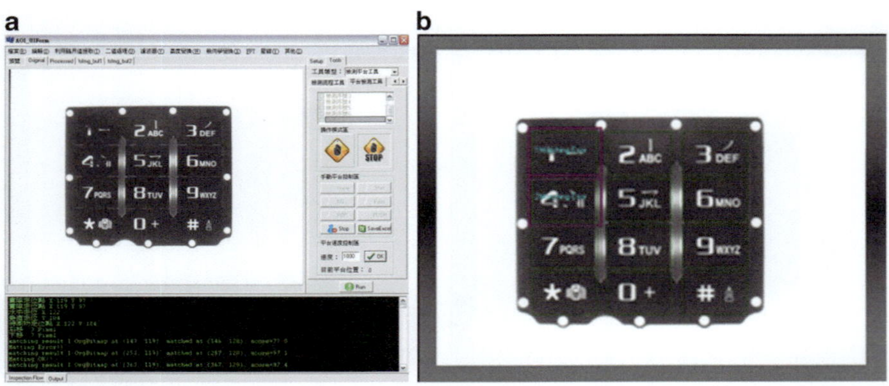

Fig. 52.16 (**a**) Another inspection example of keypad defects, (**b**) the template matching tool successfully finds there are defects (*stains*) on *button 1* and *button 4* of the keypad

Table 52.4 Positions of matched buttons shown in Fig. 52.16b and their NCC coefficients

No.	Template position	Position of matched buttons	NCC coefficient
1	(143, 119)	(146, 128)	0.77
2	(253, 119)	(257, 128)	0.971
3	(363, 119)	(367, 129)	0.974
4	(143, 189)	(148, 197)	0.865
5	(253, 189)	(257, 198)	0.976
6	(363, 189)	(366, 199)	0.975
7	(143, 259)	(148, 268)	0.985
8	(253, 259)	(257, 268)	0.962
9	(363, 259)	(365, 269)	0.98
10	(143, 329)	(147, 339)	0.925
11	(253, 329)	(259, 339)	0.97
12	(363, 329)	(365, 339)	0.983

52.5 Conclusions

In this chapter we have successfully integrated a machine vision system, emphasizing on implementation of software inspection tools for visual measurement and analysis. As for the hardware equipments, we designed an LED illumination unit and the custom-tailored machinery. With reference to the support functions of several main brands of the optical inspection machine, we propose an optical inspecting procedure and used the C++ Builder package to implement nine inspection tools, namely, the average gray level tool, the thresholding tool, the positioning tool, the edge detection tool, the BLOB tool, the template building tool, the smart matching tool, the inspection sequence tool, and the platform operation tool. All these tools can be used in an inspection with single operation and can also be arranged in a proper sequence of operations to fulfill a complicated inspection procedure.

Through the demonstrations of various application examples, we have validated the effectiveness of these proposed inspection tools. We have shown application examples of the proposed system to examine defects on mobile phone keypads and proven our system can be successfully used in the initial phase of an AOI task. In the future, we will continually develop additional inspection tools and improve the architecture of the proposed system to make it more flexible and automatic.

Acknowledgments This chapter was partially funded by NSC, Taiwan under grant NSC 101-2221-E-129-008 and partially supported under CTSP HTED program with the project number 302208201.

References

1. Fan, K. C., Hsu, C.: Strategic planning of developing automatic optical inspection (AOI) technologies in Taiwan. Proceedings of the 7th Conference on International Symposium on Measurement Technology and Intelligent Instruments, J. Phys. Conf. Ser. 13, 394–397 (2005)
2. Heriansyah, R., Abu-Bakar, S. A. R.: Defects classification on bare PCB using multiple learning vector quantization neural network paradigm. International Conference on Computer Graphics, Imaging, and Visualization, Penang, pp. 50–53 (2004)
3. Ibrahim, I., Ibrahim, Z., Khalil, K., Mokji, M.M., Abu-Bakar, S. A. R.: An algorithm for classification of five types of defects on bare printed circuit board. Int. J. Comput. Sci. Eng Syst. **5** (3), (2011)
4. Ibrahim, I., Bakar, S., Mokji, M., Mukred, J., Yusof, Z., Ibrahim, Z., Khalil, K., Mohamad, M.: A printed circuit board inspection system with defect classification capability. Int. J. Innov. Manag. Inf. Prod. **3**(1), 82–87 (2012)
5. Rau, H., Wu, C.H.: Automatic optical inspection for detecting defects on printed circuit board inner layers. Int. J. Adv. Manuf. Technol. **25**(9–10), 940–946 (2005)
6. Wu, W.Y., Wang, M.J., Liu, C.M.: Automated inspection of printed circuit boards through machine vision. Comput. Ind. **28**(2), 103–111 (1996)
7. Moganti, M., Ercal, F., Dagli, C.H., Tsunekawa, S.: Automatic PCB inspection algorithms: a survey. Comput. Vis. Image Underst. **63**(2), 287–313 (1996)
8. Chen, Y.C., Yu, J.H., Xie, M.C., Shiou, F.J.: Automated optical inspection system for analogical resistance type touch panel. Int. J. Phys. Sci. **6**(22), 5141–5152 (2011)
9. Tsai, D., Chang, C., Chao, S.: Micro-crack inspection in heterogeneously textured solar wafers using anisotropic diffusion. Image Vis. Comput. **28**, 491–501 (2009)
10. Xie, X.: A review of recent advances in surface defect detection using texture analysis techniques. Electron. Lett. Comp. Vis. Image Anal. **7**(3), 1–22 (2008)

Chapter 53
Design of a Light Guide for Highly Uniform Illumination

Kun-Yi Lee, Chun-Han Chou, Wei-Sung Weng, Wei-Ching Chuang, Wei-Yu Lee, and Allen Jong-Woei Whang

Abstract In this article, we propose a design for a light guide for highly uniform illumination. The design is based on a refractive equation and the law of conservation of energy. We applied these principles to design the secondary optics lens. The methods used can be utilized to quickly design a lens with high uniformity. Finally, we obtain a Fresnel lens for illumination with a uniformity of around 92 %.

Keywords Secondary optics lens • Fresnel lens

53.1 Introduction

Environmental pollution will grow increasingly serious in the coming years. Thus, many companies and researchers have become interested in sources of renewable energy, such as solar energy, wind energy, hydroelectric power, ocean wave energy, geothermal energy, and energy from biomass. Among sources of renewable energy, solar energy has many advantages – it is clean, abundant, and widely distributed – so it is desirable to use solar energy to reduce electricity consumption.

In residential buildings, lighting and air-conditioning account for most of the energy consumption. In commercial buildings, lighting accounts for a total of 30–40 % of the energy consumed. And in some countries, lighting accounts for

K.-Y. Lee (✉) • W.-Y. Lee
Department of Electrical Engineering, China University of Science and Technology, Taipei City 115, Taiwan
e-mail: kelvin119@gmail.com

C.-H. Chou • A.J.-W. Whang
Department of Electronic Engineering, National Taiwan University of Science and Technology, Taipei City 106, Taiwan

W.-S. Weng • W.-C. Chuang
Department of Electro-Optical Engineering, National Formosa University, Yunlin County 632, Taiwan

J. Juang (ed.), *Proceedings of the 3rd International Conference on Intelligent Technologies and Engineering Systems (ICITES2014)*, Lecture Notes in Electrical Engineering 345, DOI 10.1007/978-3-319-17314-6_53

15–20 % of the total power consumed annually. Illumination is a main factor in the consumption of electricity, and we have developed a system called the Natural Light Illumination System (NLIS). The NLIS can efficiently transmit natural light indoors where the light source is too collimated and, hence, unsuitable for illumination. For this reason, we used a novel method to design a lens that enabled us to illuminate a wide area with high uniformity. Finally, we obtained a Fresnel lens that allowed for illumination with a uniformity of close to 92 % [1, 2].

53.2 Natural Light Illumination System

We designed a daytime lighting system that emits a pleasant light. It can be placed on the floor, the balcony of a building, or on a wall. The system transmits sunlight from outdoors to indoors and illuminates rooms by sunlight. In the future, NLIS will be able to be used in healthcare facilities. The device is comprised of three main structures: a light collection system, an optical transmission system, and an illumination system. Sunlight is collected by the collection system and coupled to the transmission system indoors; finally, the light is emitted uniformly by the illumination system (Fig. 53.1).

The light collection system is a static lighting collection concentrator. The structure creates multiple isosceles right-angle prisms. The concentrators collect sunlight and convert parallax beams to focus the light. The optical transmission system transmits light in a highly efficient way through plastic optical fibers (12Φ), as shown in Fig. 53.2.

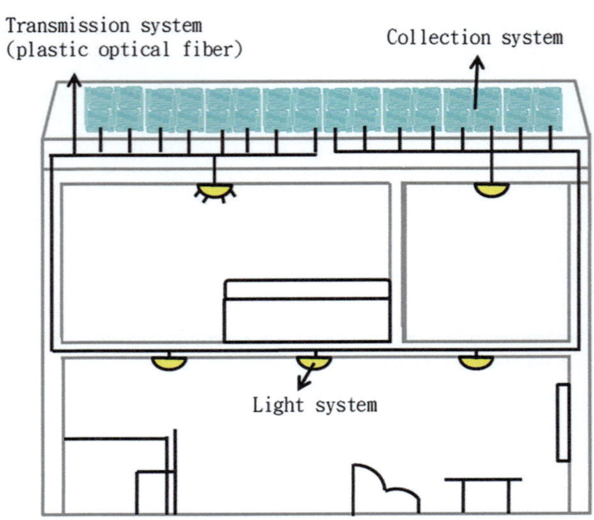

Fig. 53.1 Natural light illumination system

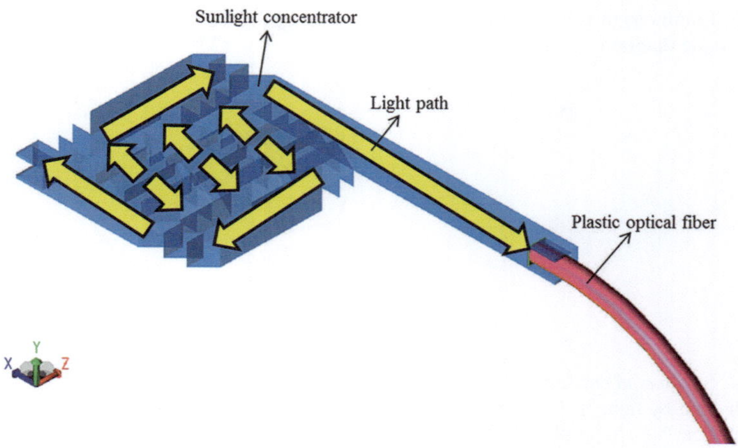

Fig. 53.2 Concentrator combined with plastic optical fiber

Finally, a high-density light is transmitted via a plastic optical fiber (12Φ) for indoor illumination, but the fiber output light distribution is not appropriate for indoor lighting [1, 3, 4].

53.2.1 Design Method of Freeform Lens for Illumination

The design method includes three parts: establishing a light energy mapping relationship between source and target, constructing a lens, and validating the lens design by numerical simulation. We used polymethyl methacrylate (PMMA, n=1.4935) to build the freeform lens.

53.2.2 Establishing Light Energy Mapping Relationship Between Source and Target

First we analyzed the surface light source intensity and flux distribution. Because our light source was a surface light source, we divided the light source surface into an $M \times N$ matrix. Following these steps, we establish the light source intensity and flux distribution (Fig. 53.3):

$$\Phi_{\mathrm{ab}} = \Delta C_{\mathrm{ab}} \times I_{\Delta \mathrm{ab}}, \tag{53.1}$$

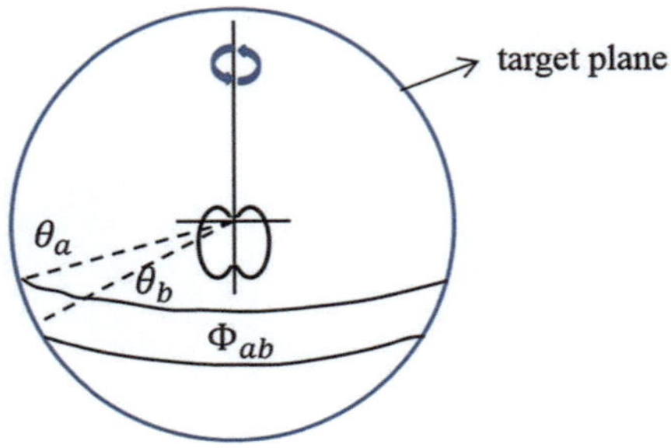

Fig. 53.3 Luminous flux of ring schematic diagram

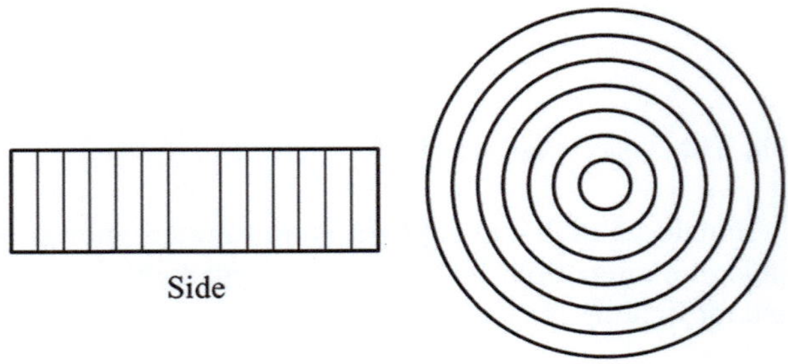

Fig. 53.4 Seven *concentric circles* divided by light source surface

$$\Delta C_{ab} = 2\pi(\cos\theta_a - \cos\theta_b). \tag{53.2}$$

Φ_{ab}: Total luminous flux of the ring from θ_a to θ_b degree.
ΔC_{ab}: Solid angle in steradians coefficient is equal to the area of the segment of a unit sphere, centered at the angle's vertex.
$I_{\Delta ab}$: Average luminous intensity from θ_a to θ_b degrees.

When we determine the luminous flux(Φ_{ab}) and solid angle(ΔC_{ab}), we could derive the light intensity distribution($I_{\Delta ab}$). After establishing the light source intensity, we divided the light source into seven concentric circles (Fig. 53.4) and analyzed each circle's light intensity. Then we designed the freeform surface [5–7].

53.2.3 Construction of Freeform Fresnel Lens

The illumination uniformity of indoor lighting is very important. If the uniformity of illumination varies too much, it will cause one's eyes to become fatigued and cause one's eyesight to deteriorate. For this reason, it is desirable to design a high-uniformity light. The average flux should approximate the following light intensity distribution equation (Fig. 53.5) [8–10]:

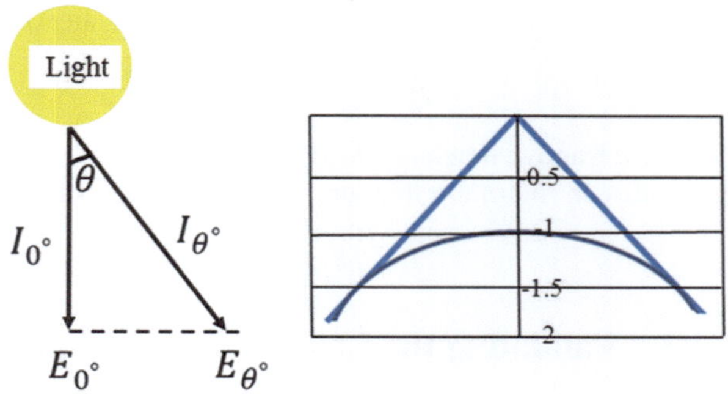

Fig. 53.5 Average light intensity distribution

Table 53.1 Uniformity flux of each angle's light intensity distribution

Emitting angle	0°	5°	10°	15°	20°	25°	30°	35°	40°
Ideal value $\left(\frac{I_{\theta°}}{I_{0°}}\right)$	1	1.01	1.04	1.10	1.20	1.34	1.53	1.81	2.23

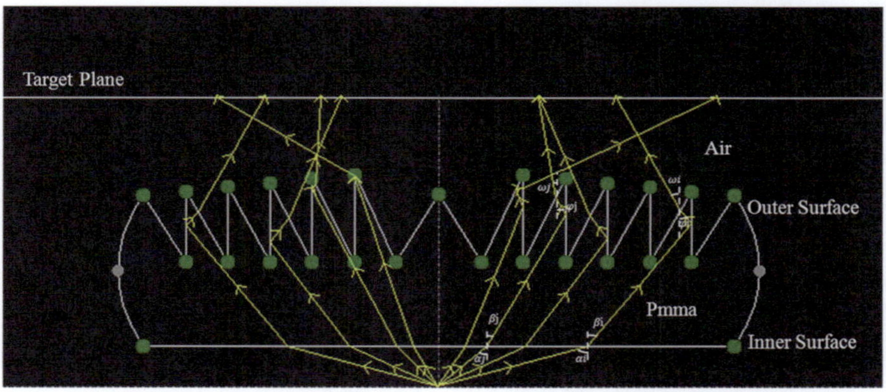

Fig. 53.6 Freeform Fresnel lens design schematic diagram

$$E_\theta = \left(I_\theta / h^2\right) \cos^3\theta. \tag{53.3}$$

$10If the working surface has flux uniformity, $E_{\theta°}$ should be equal to $E_{0°}$

$$I_{\theta°}/I_{0°} = 1/\cos^3\theta. \tag{53.4}$$

Using the preceding equation we can calculate each angle's light intensity distribution (Table 53.1).

We calculate the light intensity of each angle and use Snell's law to design each lens surface to construct the lens (Fig. 53.6). We fix every distance D_0 and adjust the distance D_1 to the other side to change the light intensity distribution so that it more closely approximates flux uniformity. We use (53.5) and (53.6) to calculate the light deflection angle:

$$n_{\mathrm{PMMA}} \sin \phi_j = n_{\mathrm{air}} \sin \omega_j, \tag{53.5}$$

$$n_{\mathrm{air}} \sin \alpha_j = n_{\mathrm{PMMA}} \sin \beta_j. \tag{53.6}$$

The refractive index of PMMA is 1.4935, which is higher than air. Incident light generates an internal reflection. The internal reflection is given by (53.7):

$$\phi_j = \arcsin(n_{\mathrm{air}}/n_{\mathrm{PMMA}}). \tag{53.7}$$

53.3 Validating the Lens Design by Numerical Simulation

In the optical design, the output light intensity distribution curve is a very important parameter. The optical fiber output is the size of a 12 × 12 mm cylinder. The output light distribution curves of the solar concentrator and plastic optical fiber are shown in Fig. 53.7. We found that the light intensity distribution didn't appropriate for indoor illumination, so we designed a freeform Fresnel lens to convert the focus light intensity distribution into widely light intensity distribution using the preceding equation.

We know the basic distribution of light, and so we were able to design a freeform Fresnel lens. The lens was designed starting from the outer regions and moving toward inner regions. We set the 0° energy reflection to 40° and used the curve of the lens edge so that the lens more closely approximated a light distribution of more uniform flux. Using the preceding equation, we construct the architecture of the

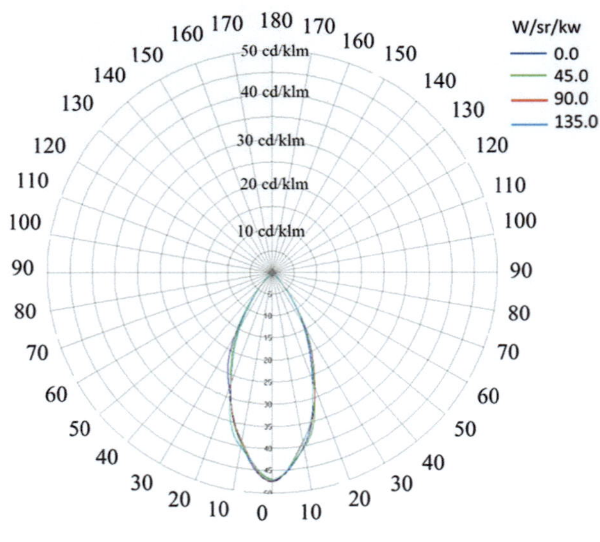

Fig. 53.7 General plastic optical fiber output of light distribution

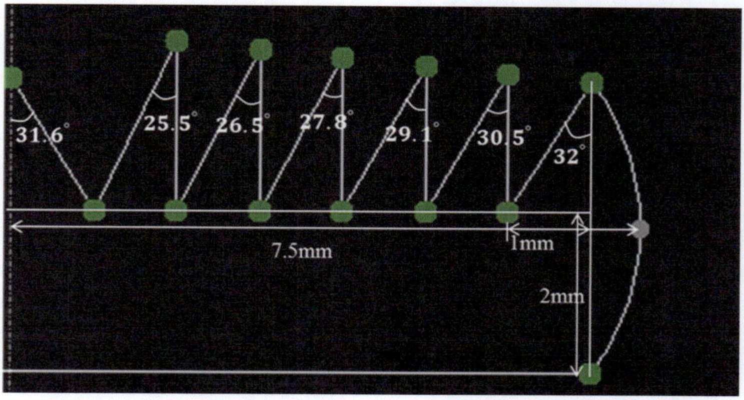

Fig. 53.8 Freeform Fresnel lens plane design

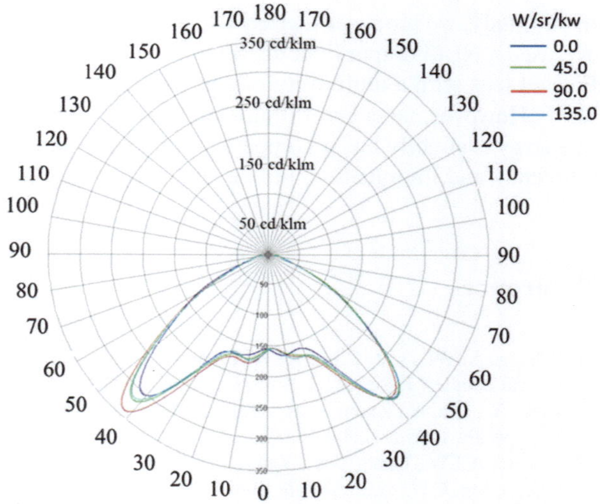

Fig. 53.9 Postdesign light distribution of plastic optical fiber output

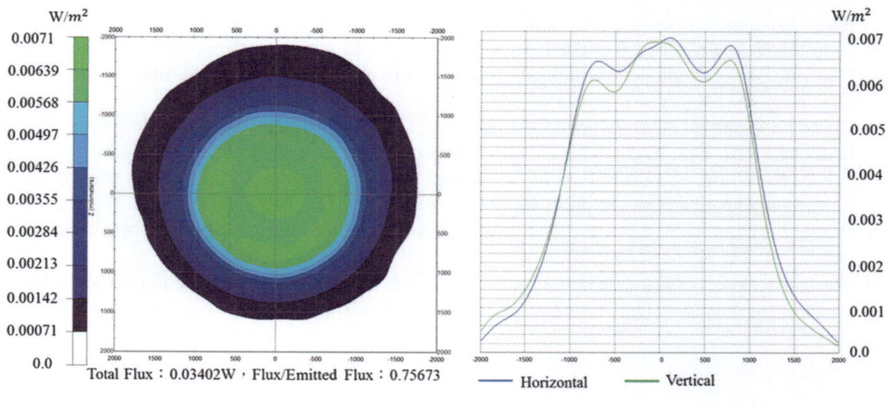

Fig. 53.10 Ray-tracing result of irradiance map

lens, as shown in Fig. 53.8. The general light distribution was converted by the lens to one of high uniformity (Fig. 53.9).

The results of ray tracing in the irradiance map are shown in Fig. 53.10. The vertical axis plots the energy (W/m^2) and the horizontal axis means illumination area coordinate. From Fig. 53.10 we see that the uniformity across the center is around 92 %. The lens output energy efficiency is approximately 75 %. But there is still some defect of the illumination area. The central area of the energy is slightly lower compared with its surroundings. In the future, the output efficiency and illumination will be enhanced by optimization.

53.4 Conclusion

In this study, we proposed an easy and efficient method for designing a Fresnel lens. We used a NLIS output light distribution to design the lens. Finally, we produced a Fresnel lens with a uniformity of approximately 92 % and an efficiency of around 75 %. However, the area of illumination area remains imperfect. The central point of energy is slightly lower compared with its surroundings. In the future, the output efficiency and illumination will be enhanced by optimization.

References

1. Whang, A.J.W., Lin, C.M., Ku, N.L., Wang, M.C.: A new static lighting concentrator with optical coupler. J. Renew. Sustain. Energy **4**, 063118 (2012)
2. Ding, Y., Liu, X., Zheng, Z.R., Gu, P.F.: Freeform LED lens for uniform illumination. Opt. Express **16**, 12958–12966 (2008)
3. Whang, A.J.W., Chen, Y.Y., Yang, S.H., Pan, P.H., Chou, K.H., Lee, Y.C., Lee, Z.Y., Chen, C.A., Chen, C.N.: Natural light illumination system. Appl. Opt. **49**, 6789–6801 (2010)
4. Whang, A.J.W., Wang, C.C., Chen, Y.Y.: Design of cascadable optical unit to compress light for light transmission used for indoor illumination. Renew. Energy **34**, 2280–2295 (2009)
5. Wang, K., Chen, F., Liu, Z.Y., Luo, X.B., Liu, S.: Design of compact freeform lens for application specific light-emitting diode packaging. Opt. Express **18**, 413–425 (2010)
6. Benı´tez, P., Min˜ano, J.C., ´Blen, J., Mohedano, R., Chaves, J., Dross, O., Herna´ndez, M., Falicoff, W.: Simultaneous multiple surface optical design method in three dimensions. Opt. Eng. **43**, 1489–1502 (2004)
7. Chen, F., Wang, K., Qin, Z., Wu, D., Luo, X.B., Liu, S.: Design method of high-efficient LED headlamp lens. Opt. Express **18**, 20926–20938 (2010)
8. Wang, S., Wang, K., Chen, F., Liu, S.: Design of primary optics for LED chip array in road lighting application. Opt. Express **19**, A716–A724 (2011)
9. Wang, L., Qian, K.Y., Luo, Y.: Discontinuous free-form lens design for prescribed irradiance. Appl. Opt. **46**, 3716–3723 (2007)
10. Wang, K., Wu, D., Qin, Z., Chen, F., Luo, X.B., Liu, S.: New reversing design method for LED uniform illumination. Opt. Express **19**, A830–A840 (2011)

Chapter 54
A Study on the Performance Characteristics of a Synchronous Elastic FIR Filter

Ren-Der Chen and Sheng-Yu Kao

Abstract Elastic systems provide tolerance to the unpredictable timing variations in computation and communication delays. For a synchronous elastic circuit, the handshaking mechanism to synchronize the data communication between sender and receiver is produced at the level of cycle in which the events are synchronized with the clock. In this chapter, the performance characteristics, i.e., area, delay, and power, of a synchronous elastic circuit are studied by the implementation of a finite impulse response (FIR) filter. The filter is designed as a two-stage pipeline, and to be compared with its non-elastic counterpart, both ASIC and FPGA implementations have been made for various orders of the filter.

Keywords Elastic • FIR • Filter

54.1 Introduction

Elastic systems provide tolerance to the unpredictable timing variations in computation and communication delays [1, 2]. This tolerance to delay variability requires the incorporation of handshake signals to synchronize the data communication between sender and receiver [3]. Synchronous elastic circuits are a form of discretized asynchronous circuits. The handshaking is produced at the level of cycle in which the events are synchronized with the clock. The implementation of elasticity relies on the clock-based synchronous handshake protocol to control the latching of data items. Through this synchronous elasticity, there is no need to fix the communication delays as used in traditional designs.

To make a synchronous circuit implemented in an elastic way, however, new data latching schemes and additional controllers have to be adopted. This may affect the performance characteristics, i.e., area, delay, and power, of the original circuit. To investigate the overhead resulting from elasticity, the concept of synchronous elasticity is incorporated into the design of an finite impulse response (FIR) filter.

R.-D. Chen • S.-Y. Kao (✉)
Department of Computer Science and Information Engineering, National Changhua University of Education, Changhua 500, Taiwan
e-mail: my1932@hotmail.com

J. Juang (ed.), *Proceedings of the 3rd International Conference on Intelligent Technologies and Engineering Systems (ICITES2014)*, Lecture Notes in Electrical Engineering 345, DOI 10.1007/978-3-319-17314-6_54

FIR filters are important building blocks for various digital signal processing (DSP) applications [4–6]. The synchronous elastic FIR filter is designed as a two-stage pipeline, and its non-elastic counterpart has also been implemented for comparison. The performance characteristics are compared in this chapter based on the Application-Specific Integrated Circuit (ASIC) and Field-Programmable Gate Array (FPGA) implementations for various orders of the filter.

This chapter is organized as follows. Section 2 introduces the concept of elastic circuits. Section 3 gives the architecture of the proposed synchronous elastic FIR filter. Section 4 includes the experimental results and discussions, and finally Sect. 5 concludes the chapter.

54.2 Elastic Circuits

54.2.1 Elastic Buffers

Synchronous elastic circuits use handshake signals to synchronize the data communication between sender and receiver. The handshaking of events is synchronized with the clock, and a pair of control signals, valid and stop, are used to control the flow of data by obeying a protocol called SELF [3]. This protocol consists of three basic actions: transfer, retry, and idle. Transfer occurs when valid data flow is not stopped. Retry occurs when valid data get stopped and need to be held until the stop is removed. When data are not valid, it is in the idle state.

Figure 54.1a shows one stage of data storage in a synchronous elastic pipeline, i.e., an elastic buffer [3]. It consists of two level-sensitive data latches (L), an elastic

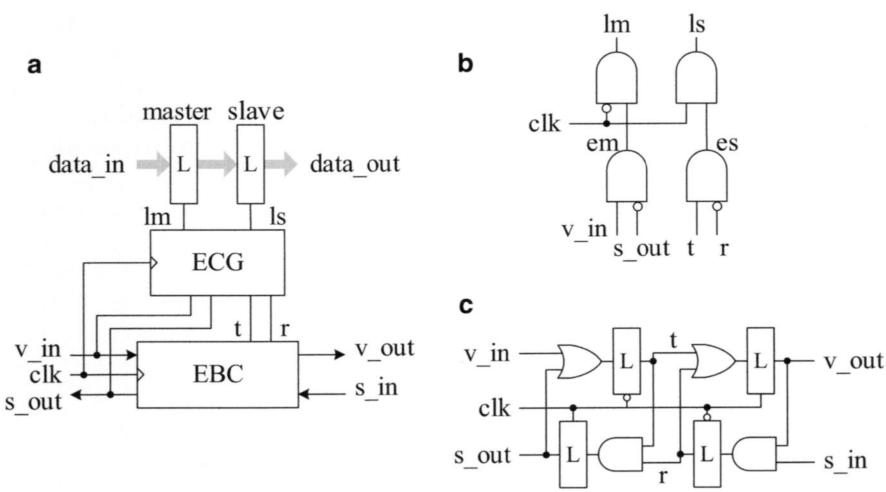

Fig. 54.1 (**a**) Elastic buffer. (**b**) Elastic clock generation (ECG) unit. (**c**) Elastic buffer control (EBC) unit

clock generation (ECG) unit, and an elastic buffer control (EBC) unit. The pair of latches work as an edge-triggered register when data are not stopped or as two separately controlled memory elements when any incoming stop occurs. The ECG unit is responsible for generating the two data latching signals *lm* and *ls* for the two master and slave latches, respectively. Figure 54.1b depicts an implementation of the ECG unit. When there is no valid data, signaled by the EBC unit, the two latching signals are gated in a logic low state to make the data latches opaque. When a valid data item enters the elastic stage, the EBC unit will switch to a clocked mode. A positive pulse will be generated on *lm* and then, after half a clock cycle, another positive pulse will also be generated on *ls* to latch the input data. An implementation of the EBC unit is given in Fig. 54.1c.

54.2.2 Join/Fork Elements

For an elastic block with multiple inputs, a Join element is needed to detect the validity of each input [3]. It asserts a valid output only when all the inputs are valid. Otherwise, the incoming valid inputs will be stopped. Figure 54.2a depicts the implementation of a two-input Join element. A multiple-input Join element can also be constructed similarly by composing multiple two-input Join elements into a tree-like structure. On the other hand, a Fork element is also needed for an elastic block with multiple outputs [3]. It sends a valid signal to all of its outputs if all the receivers are ready to accept it. If one or more receivers are not ready, the Fork element will keep the input signal valid for them. Figure 54.2b depicts the implementation of a Lazy fork element. It waits for both receivers to be ready before sending out the valid signal. The Eager fork element in Fig. 54.2c is a more efficient structure. It can send the valid signal to each receiver independently as long as the corresponding receiver is ready to accept it. This structure offers performance advantages when the two receivers have different back-pressures.

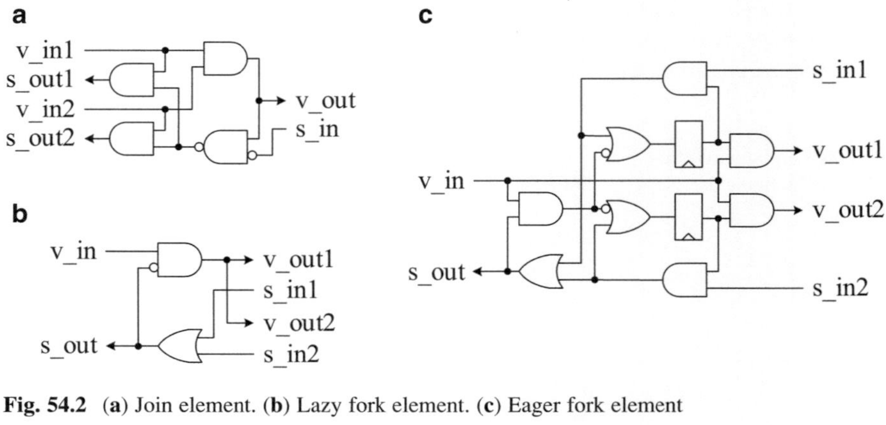

Fig. 54.2 (**a**) Join element. (**b**) Lazy fork element. (**c**) Eager fork element

54.3 Synchronous Elastic FIR Filter

54.3.1 FIR Filter

The output signal $y(n)$ of an FIR filter with order N is given by (54.1), where $x(n-k)$ and $h(k)$ denote the input samples and the filter coefficients, respectively, for $k=0$, 1, . . ., $N-1$.

$$y(n) = \sum_{k=0}^{N-1} h(k) \cdot x(n-k). \tag{54.1}$$

Shown in Fig. 54.3 is the general architecture of an N-tap FIR filter. It seeks to align the products $h(k) \cdot x(n-k)$ in time before being accumulated. This aligning of the input samples $x(n-k)$ is performed by the chained delay elements before parallel multiplication and accumulation.

54.3.2 Synchronous Elastic FIR Filter Architecture

Figure 54.4 illustrates the architecture of the proposed synchronous elastic four-tap FIR filter. It is designed as a two-stage pipeline. The chain of delay elements in the general FIR filter in Fig. 54.3 serve as the input pipeline registers. Each delay element in the chain is implemented by a pair of data latches associated with an *elastic latch control* (ELC) unit for achieving synchronous elasticity. The ELC element is composed of an ECG unit and an EBC unit as depicted in Fig. 54.1a, responsible for generating the two latching signals *lm* and *ls* for the pair of data latches. To achieve pipelining, another four pair of elastic latches are added to the architecture to serve as the internal pipeline registers. For the output signal $y(n)$, a pair of elastic latches are used as the output pipeline register.

The parallel multiplications in the general FIR filter are performed in the first pipeline stage by four multipliers, and three adders are used to perform accumulation in the second pipeline stage. For each input sample $x(n-k)$ that enters one of the input

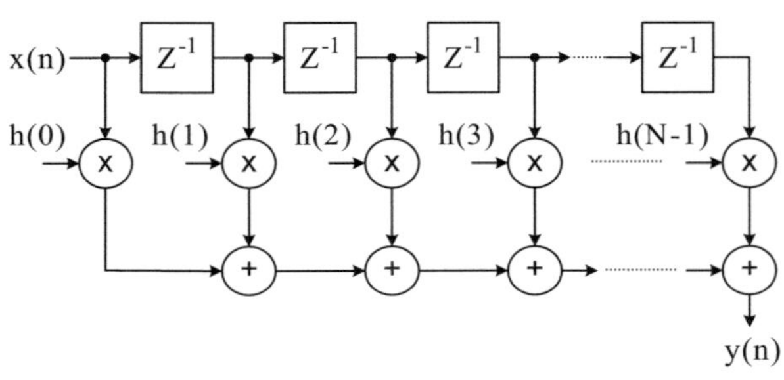

Fig. 54.3 A general N-tap FIR filter architecture

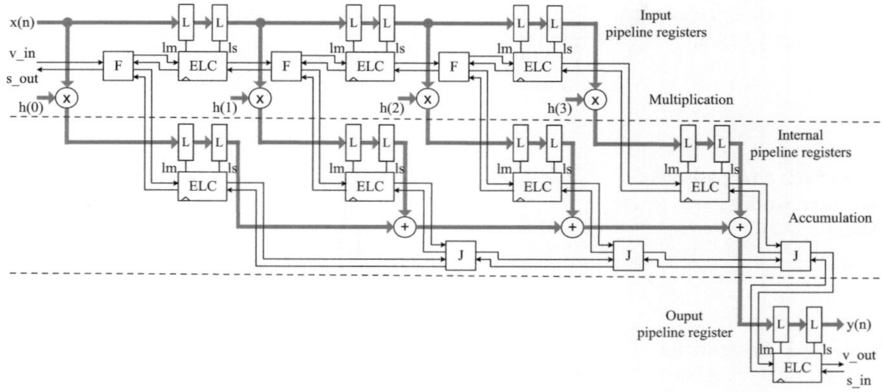

Fig. 54.4 Synchronous elastic four-tap FIR filter architecture

pipeline registers and, after multiplication, enters the corresponding internal pipeline register, a Fork element is used to fork the valid signal. Likewise, for the multiplier outputs to be accumulated in the second pipeline stage, three Join elements are needed to join all the valid signals coming from the internal pipeline registers.

54.4 Experimental Results

The synchronous elastic FIR filter was designed as a two-stage pipeline, receiving 16-bit wide input sample and multiplied by 8-bit wide coefficients. Both elastic and non-elastic (traditional) versions were implemented for comparison. All the designs were coded by the Verilog HDL for filters of order 4, 8, 16, 24, and 32. The performance characteristics compared here are area and maximum working frequency, obtained from synthesis. The other one is power-delay product, obtained from timing simulation results. Both ASIC and FPGA implementations have been made for each design.

54.4.1 ASIC Implementation

For ASIC implementation, the designs were synthesized by Synopsys Design Vision with the TSMC 180-nm cell library. The total power consumption was measured by Synopsys PrimeTime PX using 10,000 input samples. The comparisons of area, maximum working frequency, and power-delay product are listed in Tables 54.1, 54.2, and 54.3, respectively. Compared with the traditional design, the elastic design, on average, uses 1.13 times more area and consumes 1.67 times more energy.

Table 54.1 Comparison of area (gates) in ASIC

Design	4-Tap	8-Tap	16-Tap	24-Tap	32-Tap
Traditional	6,943	14,769	26,417	35,090	48,250
Elastic	7,821	16,764	30,377	39,122	54,961

Table 54.2 Comparison of maximum working frequency (MHz) in ASIC

Design	4-Tap	8-Tap	16-Tap	24-Tap	32-Tap
Traditional	220	177	141	126	122
Elastic	216	170	139	125	119

Table 54.3 Comparison of power-delay product (mW-ns) in ASIC

Design	4-Tap	8-Tap	16-Tap	24-Tap	32-Tap
Traditional	43	88	199	241	389
Elastic	78	160	297	404	595

Table 54.4 Comparison of area (LEs) in FPGA

Design	4-Tap	8-Tap	16-Tap	24-Tap	32-Tap
Traditional	124	291	628	1,249	1,680
Elastic	539	1,178	2,193	3,079	4,221

Table 54.5 Comparison of maximum working frequency (MHz) in FPGA

Design	4-Tap	8-Tap	16-Tap	24-Tap	32-Tap
Traditional	145	112	77	56	57
Elastic	66	45	34	30	28

Table 54.6 Comparison of power-delay product (mW-ns) in FPGA

Design	4-Tap	8-Tap	16-Tap	24-Tap	32-Tap
Traditional	710	733	1,152	1,824	1,944
Elastic	1,221	1,265	1,720	2,731	2,901

54.4.2 FPGA Implementation

For FPGA implementation, the Altera Quartus II 9.1 was used as the synthesis tool and an Altera Cyclone IVE FPGA (EP4CE40F29I8L) was used as the target device. The power estimations were obtained from the PowerPlay Power Analyzer Tool using 10,000 input samples. Tables 54.4, 54.5, and 54.6 show the comparisons of FPGA area, maximum working frequency, and power-delay product, respectively. The elastic design, on average, uses 3.37 times more FPGA area and consumes 1.58 times more energy, when compared with the traditional design.

54.5 Conclusions

In this chapter, the concept of synchronous elasticity has been incorporated into the design of a synchronous elastic FIR filter. The filter adopts a two-stage pipeline architecture and both elastic and non-elastic versions have been designed for comparison. The performance characteristics such as area, maximum working frequency, and power-delay product, obtained from both ASIC and FPGA implementations, have all been compared here for various orders of an FIR filter. Since the critical path in the filter architecture is dominated by a multiplication and a series of addition operations, determining an appropriate number of stages for pipelining will be the future work.

References

1. Carmona, J., Cortadella, J., Kishinevsky, M., Taubin, A.: Elastic circuits. IEEE Trans. Comput. Aided Des. Integr. Circuits Syst **28**(10), 1437–1455 (2009)
2. Casu, M.R.: Half-buffer retiming and token cages for synchronous elastic circuits. IET Comput. Digit. Tech. **5**(4), 318–330 (2011)
3. Cortadella, J., Galceran-Oms, M., Kishinevsky, M.: Elastic systems. In: 8th IEEE/ACM International Conference on Formal Methods and Models for Codesign (MEMOCODE), Grenoble, pp. 149–158 (2010)
4. Chang, T.-S., Jen, C.-W.: Hardware-efficient pipelined programmable FIR filter design. IEE Comput. Digit. Tech. **148**(6), 227–232 (2001)
5. Deepak, G., Meher, P. K., Sluzek, A.: Performance characteristics of parallel and pipelined implementation of FIR filters in FPGA platform. In: International Symposium on Signals, Circuits and Systems (ISSCS), pp. 1–4 (2007)
6. Singh, M., Tierno, J.A., Rylyakov, A., Rylov, S., Nowick, S.M.: An adaptively pipelined mixed synchronous-asynchronous digital FIR filter chip operating at 1.3 Gigahertz. IEEE Trans. VLSI Syst. **18**(7), 1043–1056 (2010)

Chapter 55
A Design of Cavity Filters Based on Photonic Crystal Slab Waveguide with Liquid Crystal

Wei-Sung Weng, Hui-Chun Lin, Kun-Yi Lee, Li-Ling Chu, Hsin-Jung Lee, and Wei-Ching Chuang

Abstract In this chapter, an optical filter based on photonic crystal structure with liquid crystal is proposed. Tunable spectral characteristics of the filters are demonstrated by changing the indexes of refraction of liquid via the external electric fields. The simulation results show the filters are promising in the optical communications.

Keywords Photonic crystals • Optical filters • Liquid crystals

55.1 Introduction

Photonic crystal cavities have received great attention over recent years due to the strong potential they provide for applications in the fields of quantum dot lasers [1], optical switches [2, 3], and optical filters [3, 4]. Most applications require several tuning of the resonant frequencies. The tuning mechanisms consist of changing the effective refractive index via temperature or inserting scanning probe tips inside the air holes [5]. Another approach is to change the refractive index of the cavity environment by infiltrating liquids. Liquid crystals have attracted much attention due to the ability to change the effective refractive index of the cavity by temperature or electric field. The effects for two-dimensional photonic crystal structures with infiltrated liquid crystals have been widely studied [6]. However, few of these studies have demonstrated the birefringent effect of the LC on the cavity resonances [7].

In this chapter, an optical filter based on photonic crystal structure with liquid crystal is proposed. The photonic crystal waveguide cavity filters with liquid crystal defects are designed and analyzed. Because of the birefringence property of the anisotropic liquid crystal, the filtering characteristics via changing the external electric

W.-S. Weng • H.-C. Lin • L.-L. Chu • W.-C. Chuang
Department of Electro-Optical Engineering, National Formosa University, Yunlin County 632, Taiwan

K.-Y. Lee (✉) • H.-J. Lee
Department of Electrical Engineering, China University of Science and Technology, Taipei City 115, Taiwan
e-mail: KELVIN119@gmail.com

J. Juang (ed.), *Proceedings of the 3rd International Conference on Intelligent Technologies and Engineering Systems (ICITES2014)*, Lecture Notes in Electrical Engineering 345, DOI 10.1007/978-3-319-17314-6_55

field to adjust the refractive index of the liquid crystal under different structure are analyzed. Tunable photonic crystal filters are demonstrated by employed electro-optic effect of liquid crystals. The optimal conditions obtained manifest the potential of applications in dense wavelength division multiplexing (DWDM) systems.

55.2 Theory and Methodologies

In photonic crystal devices, photonic band gap, being the wavelength window in which optical-wave penetration through the crystal is prohibited, is the most important parameter. The calculations of the photonic band gaps and optical transmission characteristics of the proposed structures are based on plane wave expansion (PWE) method and finite-difference time-domain (FDTD) method, respectively. In this chapter, a computer simulation software (CrystalWave, Photon Design, UK) is used to analyze the optical characteristics of the proposed devices.

The schematic of the proposed structure is shown as follows:

The filter is formed as a rhombus ring cavity with liquid crystal-filled nanorods distributed in the oscillation path. The basic structure of the photonic crystal is an air-rod triangular lattice, formed by using etching processes on the top silicon layer of a silicon-on-insulator (SOI) substrate, whose thicknesses of bottom silicon substrate, SiO_2 and top silicon layers are 1-, 0.5-, and 0.5 μm, respectively. The defects in the photonic crystals are introduced to form input and output waveguides beside the ring cavity, shown in Fig. 55.1. The incident light is introduced into the input channel and output 1 and 2 channels are thru- and drop ports, respectively.

The double refractive indexes of the liquid crystal depend on the propagation direction and the polarization states of the light and those for the extraordinary rays could be characterized by the following equation:

$$n_{\mathrm{eff}}(\theta) = \frac{n_{\mathrm{o}} n_{\mathrm{e}}}{\sqrt{n_{\mathrm{e}}^2 \cos^2\theta + n_{\mathrm{o}}^2 \sin^2\theta}} \tag{55.1}$$

where n_e, n_o, and θ are extraordinary, ordinary refractive indexes of the liquid

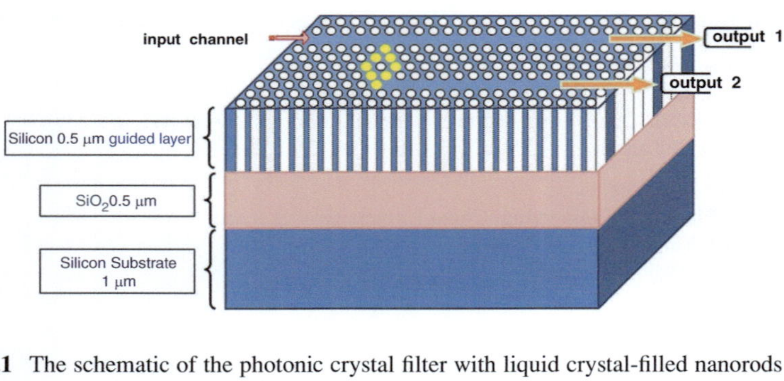

Fig. 55.1 The schematic of the photonic crystal filter with liquid crystal-filled nanorods

crystal and light propagation angle with respect to optical axis of the liquid crystal, respectively.

55.3 Simulation Results

The top views of the proposed rhombus ring filters are shown in Fig. 55.2a, b, schematically. In Fig. 55.2a, b, five and six liquid crystal nanorods are inserted into the defect regions of each side of the rhombus ring, respectively. Liquid crystal E7 is adopted in the devices and some important parameters of it are listed in Table 55.1. The diameter of the liquid crystal nanorods is 0.3 μm. To analyze the optical characteristics for the rhombus ring cavity, the waveguide modes in the photonic crystal are calculated first. The lattice constant, a, of the photonic crystal is 0.6 μm, and the diameter, r, of the air nanorod is 0.5 μm. By using the PWE method, the band diagrams for TE and TM wave can be firstly calculated and are as shown in Fig. 55.3a, b, respectively.

As shown in Fig. 55.3, the band gaps of the photonic crystal range from 1.48 to 2.52 μm and 2.28 to 2.70 μm for TE and TM modes, respectively. For the DWDM applications in optical fiber communications, of which the operating wavelength is

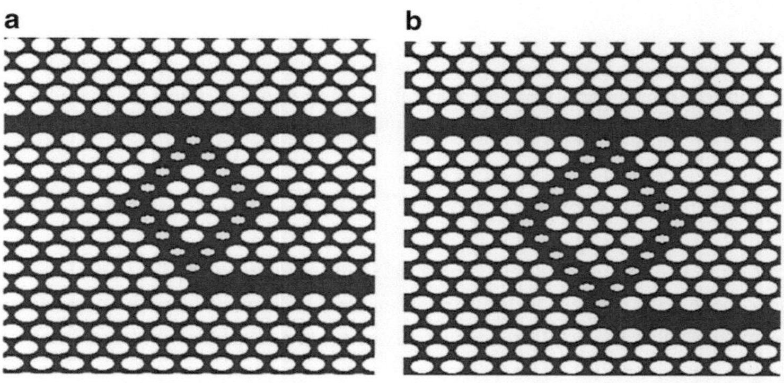

Fig. 55.2 The schematic diagrams of rhombus ring filters (**a**) five and (**b**) six liquid crystal nanorods (*yellow rods*) locate in each side of the rhombus rings

Table 55.1 Physical properties of liquid crystal E7

Property		
Type	Nematic	
Optical axis	Positive uniaxial material	
Index of refraction	Ordinary (n_o)	1.522
	Extraordinary(n_e)	1.747

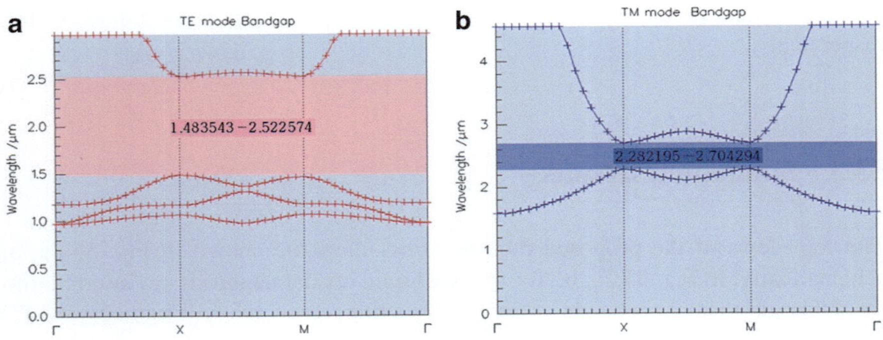

Fig. 55.3 The band diagrams for (**a**) TE and (**b**) TM optical modes

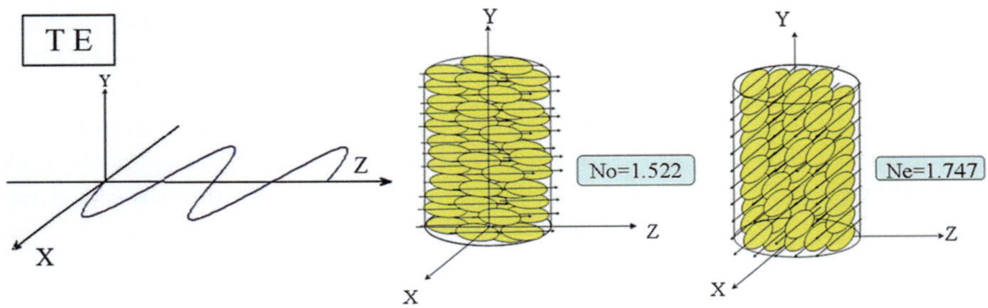

Fig. 55.4 The refractive index variation with respect to the orientation of molecular axis for TE polarized light (**a**) molecular align with *z* direction (the refractive index is equal to n_e), and (**b**) molecular align with *x* direction (the refractive index is equal to n_o)

around 1,550 nm, it is observed only the TE mode could be guided in the photonic crystal.

Since the refractive index of the liquid crystal vary with the orientations of the molecular axes with respect to the polarization directions of the propagation light, it could be modulated by applying an external electric field to change the molecular axes of the liquid crystal. The refractive index of liquid crystal for TE polarized wave is schematically described in Fig. 55.4.

A pulsed light with wavelength $\lambda = 1{,}550$ nm is engaged in input channel of the filter and the output lights are emerged at output channels 1 and 2. The power distributions for rhombus ring with six liquid crystal nanorods in each side are shown in Fig. 55.5a, b. In Fig. 55.5a, the refractive index of the liquid crystal is $n_o = 1.522$, and as well as in Fig. 55.5b, it is $n_e = 1.741$.

The performance of the PhC devices could be examined by calculating the spectra of the output port 1 (thru port) and 2 (drop port). The transmission spectra of the rhombus ring filters are depicted in Figs. 55.6, 55.7, 55.8, and 55.9 for TE and TM modes. In Figs. 55.6 and 55.7, the filter is of five liquid crystal nanorods and in Figs. 55.8 and 55.9, of six ones as well. The unguided phenomena for TM modes

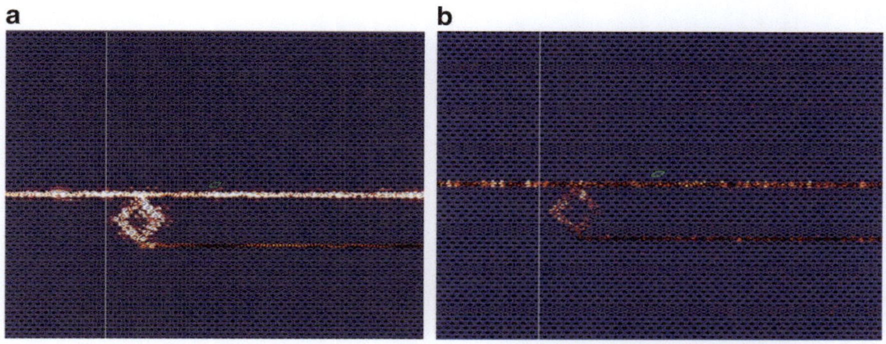

Fig. 55.5 The power distributions of the rhombus ring filter with refractive index of liquid crystal is (a) $n_o = 1.522$, and (b) $n_e = 1.741$

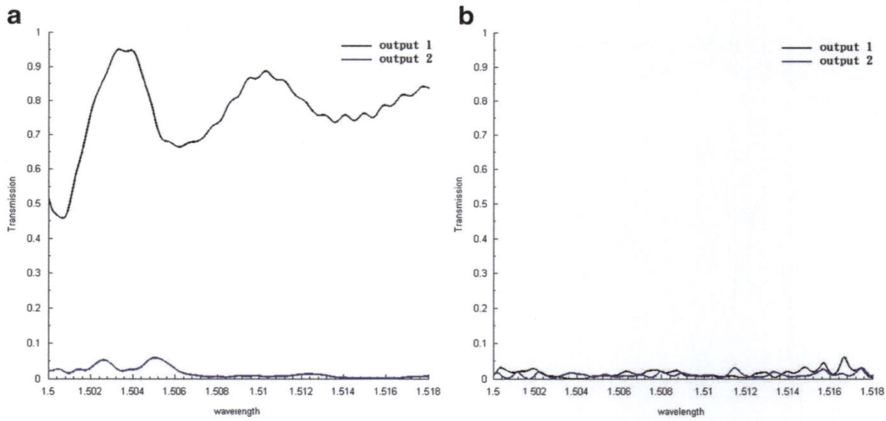

Fig. 55.6 The transmission spectra of the rhombus ring filter with inserted liquid crystal of refractive index $n_o = 1.522$ for (a) TE mode and (b) TM mode (five liquid crystal nanorods in each side)

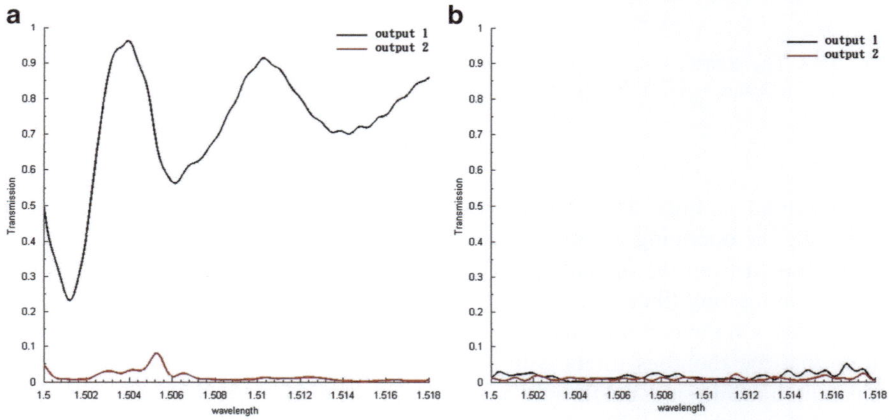

Fig. 55.7 The transmission spectra of the rhombus ring filter with inserted liquid crystal of refractive index $n_e = 1.741$ for (a) TE mode and (b) TM mode (five liquid crystal nanorods in each side)

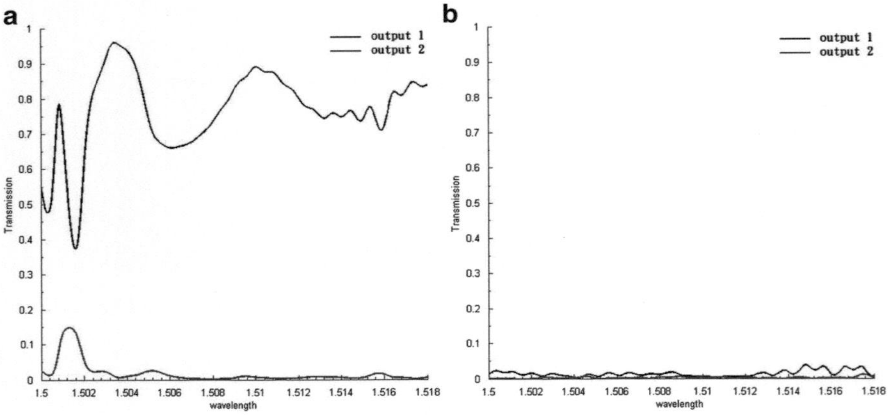

Fig. 55.8 The transmission spectra of the rhombus ring filter with inserted liquid crystal of refractive index $n_o = 1.522$ for (**a**) TE mode and (**b**) TM mode (six liquid crystal nanorods in each side)

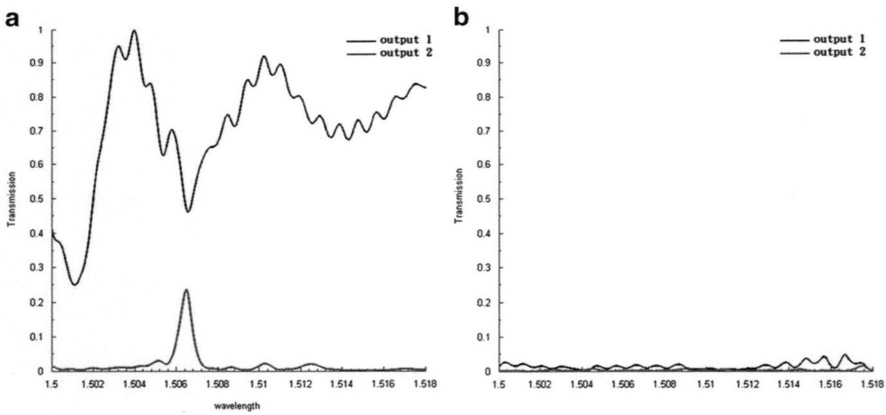

Fig. 55.9 The transmission spectra of the rhombus ring filter with inserted liquid crystal of refractive index $n_e = 1.741$ for (**a**) TE mode and (**b**) TM mode (six liquid crystal nanorods in each side)

are observed in Figs. 55.6, 55.7, 55.8, and 55.9b due to their forbidden bands excluding the operating wavelength range 1,500–1,518 nm.

It is noted from the optical spectra of output port 2 that the drop power is not obvious in the ring filter with five liquid crystal nanorods in each side (Figs. 55.6a and 55.7a). On the contrary, the results presented in Figs. 55.7a and 55.8a demonstrate that the rhombus filters with six liquid crystal nanorods in each side have apparent transmission notch and drop spectra. The filter with liquid crystal of refractive index $n_o = 1.522$ (Fig. 55.7a) has a transmission notch at wavelength

$\lambda = 1{,}501.4$ nm and drop peak at wavelength $\lambda = 1{,}501.2$ nm, whose 3 dB bandwidth is about 1.6 nm. Alternatively, the filter with liquid crystal of refractive index $n_e = 1.741$, a transmission notch locates at wavelength $\lambda = 1{,}506.4$ nm and drop peak at the same wavelength, whose 3 dB bandwidth is about 0.8 nm, complying with the channel width of 100 GHz in DWDM systems.

55.4 Conclusions

A rhombus ring filter with liquid crystal nanorods has been demonstrated. The simulation results reveal that filters with six liquid crystal nanorods on each side of the rhombus cavity have dramatic filtering characteristics, especially, with liquid crystal of refractive index $n_e = 1.741$, a drop channel at wavelength $\lambda = 1{,}506.4$ nm with 3 dB bandwidth of 0.8 nm are obtained. These characteristics are applicable in DWDM optical communication systems.

References

1. Yasutomo, O., Katsuyuki, W., Satoshi, I., Yasuhiko, A.: Self-frequency summing in quantum dot photonic crystal nanocavity lasers. Appl. Phys. Lett. **103**, 243115 (2013)
2. Nozaki, K., Kuramochi, E., Shinya, A., Notomi, M.: 25-Channel all-optical gate switches realized by integrating silicon photonic crystal nanocavities. Opt. Express **22**, 14263–14274 (2014)
3. Nozaki, K., Shinya, A., Matsuo, S., Sato, T., Kuramochi, E., Notomi, M.: Ultralow-energy and high-contrast all-optical switch involving Fano resonance based on coupled photonic crystal nanocavities. Opt. Express **21**, 11877–11888 (2013)
4. Faraon, A., Englund, D., Fushman, I., Vučković, J., Stoltz, N., Petroff, P.: Local quantum dot tuning on photonic crystal chips. Appl. Phys. Lett. **90**, 213110 (2007)
5. Rezaei, B., Khalkhali, T.F., Kalafi, M.: Tunable out-of-plane band gap of two-dimensional anisotropic photonic crystals infiltrated with liquid crystals. Opt. Commun. **284**, 813–817 (2011)
6. Rezaei, B., Kalafi, M.: Tunable full band gap in two dimensional anisotropic photonic crystals infiltrated with liquid crystals. Opt. Commun. **282**, 1584–1588 (2009)
7. Dündar, M.A., Kicken, H.H.J.E., Silov, A.Y., Nötzel, R., Karouta, F., Salemink, H.W.M., Heijden, R.W.: Birefringence-induced mode-dependent tuning of liquid crystal infiltrated InGaAsP photonic crystal nanocavities. Appl. Phys. Lett. **95**, 181111 (2009)

Chapter 56
A Self-Focus High Precision Scheme Applied to HCPV Solar Center Tracker

Yiing-Yuh Lin and Fu-Mao Jhuang

Abstract An image processor with a self-focus scheme to estimate the center of the solar disc is presented in this chapter. It can be utilized in the automatic tracking module on a HCPV power generation unit to closely follow the solar center with high precision under various weather conditions. To start the proposed procedures, a photo containing the solar image is taken first by a digital sensing device. Then, it is sent to the algorithm and goes through a series of steps to identify and to isolate the solar image. Finally, the solar center is located by the three-point center method. The proposed scheme, based on a weighted brightness factor of the image, evaluates the image center by comparing the estimated solar diameter to the reference one and iteratively modifies the thresholds in the image isolation step. The results show total processing time can be a fraction of a second and the estimation accuracy reaches to less than an arc second, depending on the pixels in the taken photo.

Keywords Solar tracking • Image processing • Threshold • Solar center • HCPV

56.1 Introduction

Since the last decade, direct conversion of solar radiation into electric power through photovoltaic materials has been actively studied as one of the important future energy resources [1]. Its continual challenge for the researchers is to increase the efficiency of the solar cell with low cost. Among various types of technology, the high concentration photovoltaic (HCPV) system using III–V GaAs solar cell is most likely to be used for building effective solar-electric power plant that can produce large power output [2]. To obtain the best possible photovoltaic conversion rate, putting the Fresnel lens on top of the particular type of solar cell is required for high sunlight concentration [3]. It can be achieved especially when sunlight impinges vertically on to the lens [4]. As such, pointing accurately to the center

Y.-Y. Lin (✉) • F.-M. Jhuang
Department of Aeronautics and Astronautics, National Cheng Kung University, Tainan 70101, Taiwan
e-mail: yinlin@mail.ncku.edu.tw

J. Juang (ed.), *Proceedings of the 3rd International Conference on Intelligent Technologies and Engineering Systems (ICITES2014)*, Lecture Notes in Electrical Engineering 345, DOI 10.1007/978-3-319-17314-6_56

of the sun is vital for the HCPV unit to gain an optimum electric power conversion from the sunlight.

A reliable approach is to integrate the theoretical sun trajectory with a real-time center positioning control, allowing the HCPV unit to follow the sun closely and aiming at the solar center precisely to receive the sunlight vertically [5, 6]. It usually relies on image processing technique to determine solar center with very high precision and known estimation error [7]. Also, when sunlight is blocked by dense clouds, the computed sun trajectory is utilized as a reference to rotate the HCPV unit to track the sun with open-loop control.

Using digital charge coupled device (CCD) to photograph the sun for real-time tracking purpose was proposed as early as in [8]. The difficulties involved are twofold: one is to assemble a set of digital sensing device that can take clear solar image and protect itself from over exposing to the sun; the other lies in the accurate estimation of the solar disc center. Furthermore, the related studies seldom consider the estimation under varieties of weather conditions. In this chapter, a digital image processor with a scheme is proposed which can be incorporated onto a HCPV unit for real-time pointing to the solar center. The scheme, by iterative self-focus method, shall be able to reduce the weather effects, which can blur or cover the solar image, and to enhance the estimation.

56.2 Image Processing Procedures

In this application, the objectives of the proposed scheme are to distinguish the solar image from its surrounding using its color and shape to sufficiently eliminate any possible noise on the acquired digital figure, effectively isolate the solar image, and accurately find the center of the sun. The scheme procedures are divided into the following steps [9, 10].

Image Acquisition. Obtaining a clear photo with large enough solar image is the first and determining step to accurately estimate the solar center. If the sun taken in the photo is too small, it may introduce uncharacteristic noise and errors to the image due to the digitizing effects and is difficult to filter out. The following units are assembled:

(a) Solar filter: Thousand oaks® optical type 2+, Dia. 82 mm, and transparent ratio 10^{-5}
(b) Telescope, model: Microtech® F90TR, diameter 90 mm, and magnifying ratio 50
(c) Digital Camera, model: Canon® A3300 IS, 16-million pixels, lens angle 55°, and magnifying ratio 1.645

In item a, the filter is for blocking most of the sunlight from damaging the camera. The telescope in item b is to magnify the solar image to have a better processing result. The high resolution digital camera in item c is used for taking solar image.

A full color solar image taken by the setup is sent through RS232 port and received by the image processor in a monitoring PC with 2.8 GHz Intel® Pentium 4 CPU. Custom-made instrument dedicated for the work shall be designed and manufactured after the test parameters are assured.

RGB to HSL. The sun's color doesn't change much but its brightness varies significantly under different weather conditions. RGB is transformed into HSL format to improve the identification of the solar image in a photo.

Binary Representation. Solar image usually is much brighter than its background. The photo is transformed into a binary format to separate the sun from its background with the following setting

$$g(x,y) = \begin{cases} 1, & f_{\min} \le f(x,y) \le f_{\max} \\ 0, & \text{otherwise} \end{cases}. \tag{56.1}$$

where function $g(x, y)$ stores the binary result, 1 as white and 0 as black, and $f(x, y)$ keeps the brightness info in HSL format at (x, y) photo coordinates. Also, $f_{\min}$ and $f_{\max}$ are preset lower and upper bounds of the binary conversion, respectively.

Image Filter. After the binary procedure, the image may not be clear and isolated. To improve the situation, the erosion and the distension are applied to remove the debris and the rough edges.

Boundary Detection. The boundary detection is required to identify the target edge after the filtering procedure and the Sobel method [11] is used.

Center Location. The three-point circle method is chosen which finds the center of a circle at the intersection of the lines that are perpendicular to the straight line segments connecting any three noncollinear points on the circle boundary.

56.3 Improvement of Threshold

Proper $f_{\min}$ is necessary to accurately determine the boundary of the binary image. An iterative correction scheme is implemented to improve $f_{\min}$ selection.

Reference solar diameter [12] D_{ref} relating to the solar size in a taken digital photo is also crucial to find correct solar center and it can be obtained in terms of pixel by:

$$D_{\text{ref}} = \theta \times M/\alpha \times P, \tag{56.2}$$

where θ in degree is the average solar view angle in the sky, M is the telescope magnifying ratio, α in degree is the camera view angle, and P is the photo pixel number along horizontal or vertical photo frame. In the study, the value is given as

Horizontal: $D_{\text{ref}} = 0.53 \times 50/38.90 \times 640 = 436$; Vertical: $D_{\text{ref}} = 0.53 \times 50/29.17 \times 480 = 436$.

To quantify the accuracy of the solar center, an estimated error $e(N)$ is defined by:

$$e(N) = \frac{1}{B(N)}\sum_{i=1}^{B(N)} \left| 2\sqrt{[X_i(N) - X_c(N)]^2 + [Y_i(N) - Y_c(N)]^2} - D_{\text{ref}} \right|, \quad (56.3)$$

where N is the iteration number, $B(N)$ is the number of points selected from the image boundary, $[X_i(N), Y_i(N)]$ is a boundary point and $[X_c(N), Y_c(N)]$ is the center.

The key factor to choose $f_{\min}$ is the brightness of solar image in contrast to its background in a photo. Weighted average value of brightness (WAVB) is defined to quantify the relation and is expressed as

$$\text{WAVB} = \sum_{i=0}^{255} i \times N(i) / \sum_{i=0}^{255} N(i) , \quad (56.4)$$

where i and $N(i)$ represent the brightness level of a pixel specified between 0 and 255 and the number of pixels with the same brightness in the HSL matrices, respectively.

In this study, three representative samples were chosen to form the equation, which are sample A: Sunny, sample B: partial cloudy, and sample C: very cloudy. For each sample, WAVB value and the corresponding $f_{\min}$ in the binary procedure were carefully evaluated and the pair that produces the smallest $e(N)$ was recorded. Then, an approximate linear equation was determined using the data obtained from the three samples and the well-known least square error method

$$f_{\min} = \alpha \times \text{WAVB} + b, \quad (56.5)$$

where α is 0.581 and b is -0.951. Initial $f_{\min}$ and $f_{\max}$ for the binary procedure are obtained from the HSL photo using (56.4) and (56.5) and the brightest point, respectively. To modify $f_{\min}$ from (56.5), WAVB computed from the binary figure and the bisection method are used. The iteration procedure stops when the estimated error $e(N)$ in N iteration is smaller than or equal to lower bound 0.5 pixels in the study.

56.4 Cases Study

Three solar images under different cloud conditions as shown below were taken for examining and comparing the proposed approach.

Case A, Case B, and Case C in Fig. 56.1 represent the sun in clear, light cloud, and heavy cloud sky conditions, respectively.

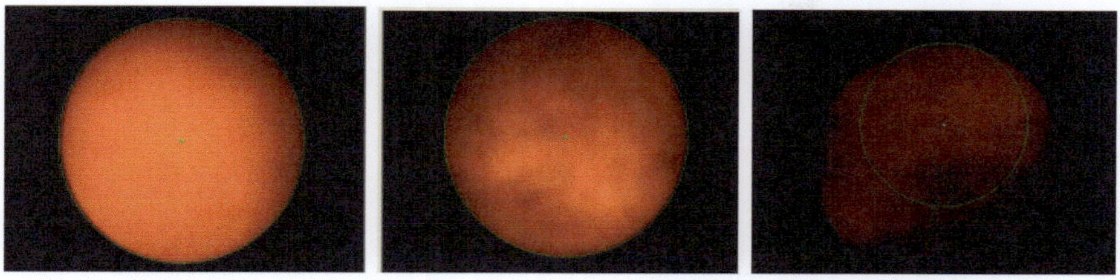

Case A Case B Case C

Fig. 56.1 Solar images from different sky conditions

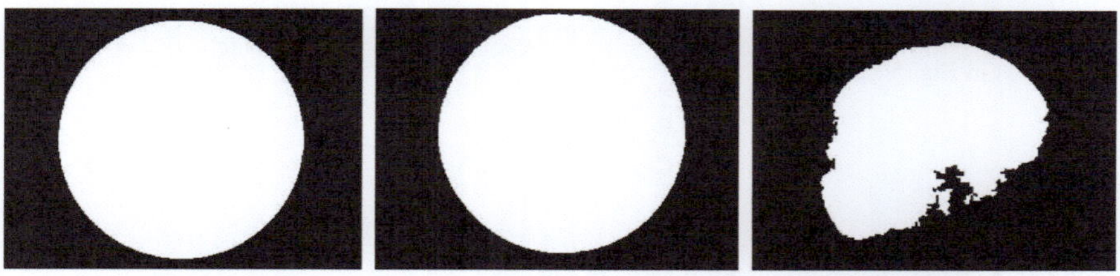

Case A_1 Case B_1 Case C_1

Fig. 56.2 Comparison of binary results from fixed $f_{\min}$

Table 56.1 Test results from cases in Fig. 56.2

Image	Center (x, y)	Diameter	$e(N)$ Error
Case A_1	313, 238	436	0.49
Case B_1	326, 225	435	1.48
Case C_1	332, 203	291	144.9

56.4.1 Test Using Fixed f_{min}

Lower bound $f_{\min}$ in binary representation is set at 20 for all three cases and results obtained after the process are shown below

where Case A_1, B_1, and C_1 in Fig. 56.2 are the binary solar images of Case A, B, and C, respectively, and their data are shown in Table 56.1.

In Table 56.1, each image is processed only once and the process times for each case are 0.56, 0.61, and 0.5 s, respectively, due to different point number selection.

From Fig. 56.2 and Table 56.1, the solar center can be effectively estimated in Case A_1 when the sun is very clear. But the estimation error function $e(N)$ of solar diameter for Case B_1 and C_1 are more than 0.5 pixels. Because the photos were taken in different time, their centers are not the same and the diameter is the only reference to check the estimation integrity. Also, the error is quite large for C_1 with heavy cloud coverage.

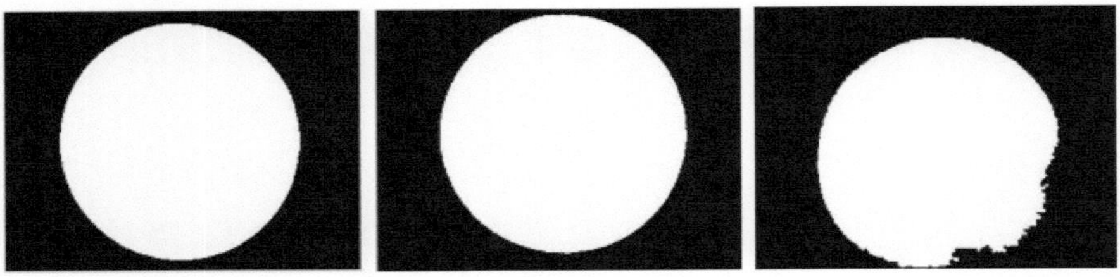

Fig. 56.3 Comparison of binary images from iterative f_{min}

Table 56.2 Iterative results from cases in Fig. 56.2

Image	Center (x, y)	Diameter	$e(N)$ Error
Case A_2	313, 238	436	0.44
Case B_2	326, 225	436	0.41
Case C_2	327, 273	436	0.15

Table 56.3 Iteration conditions from cases in Fig. 56.2

Image	Initial f_{min}	Final f_{min}	Iteration no.	Process time
Case A_2	29	29	1	0.48
Case B_2	21	21	1	0.47
Case C_2	6	2	5	0.59

56.4.2 Test Cases with Iterative f_{min}

The same cases in Fig. 56.1 are tested again with the proposed iterative scheme and the results and estimation data are shown in the following figure and tables.

Although the image differences between Case A_1 and A_2 or B_1 and B_2 are barely noticeable, images for Case C_1 and C_2 are quite different in Figs. 56.2 and 56.3.

Also shown in Tables 56.2 and 56.3, the scheme provides good initial f_{min} to obtain much accurate center point. For Case C_2 in Table 56.3, f_{min} was modified five times to obtain a very small error center position and the process time required is only 0.59 s.

56.5 Conclusions

The proposed scheme has effectively identified the solar center even with cloudy solar images. Through the digital image processing techniques and a simple three-point circle method, the scheme can efficiently determine the solar center under various degrees of dense cloud on the solar image. Feedback selection from the linear equation that approximates the relation between f_{min} and WAVB proves to be very useful for determining the boundary of a solar image under cloudy weather. Also, the known solar angular diameter taken and updated repetitively in clear days provides a vital reference in determining the error in estimating the solar center.

Analyses of the results from outdoor test cases also indicate that reducing the center estimation error requires a good size solar image, which can be taken from an image sensing device with proper lens angle and a large and compatible number of pixels. However, the computation number in each step of the scheme is very large and increases rapidly with more pixels in the photo. Further investigation on the solar image acquisition hardware, the algorithm, and the computing power are necessary to improve and to satisfy the requirements of the tracking module on a HCPV unit.

Acknowledgments The research presented in this chapter was supported by the Institute of Nuclear Energy Research, Atomic Energy Council, ROC, under the contract 1002001INER044 and 1012001INER026 and is gratefully acknowledged.

References

1. Razykov, T.M., Ferekides, C.S., Morel, D., Stefanakos, E., Ullal, H.S., Upadhyaya, H.M.: Solar photovoltaic electricity: Current status and future prospects. Sol. Energy **85**, 1580–1608 (2011)
2. Lung, I.T., Kou, C.T., Shin, H.Y., Hong, H.F., Lee, C.D., Lin, T.T.: Establishment of one MW HCPV system at Taiwan. ISESCO Sci. Technol. Vis. **6**(9), 50–53 (2010)
3. Sierra, C., Vazquez, A.J.: High solar energy concentration with a Fresnel lens. J. Mater. Sci. **40**, 1339–1343 (2005)
4. Kou, C.T., Shin, H.Y., Hong, H.F., Wu, C.H., Lee, C.D., Lung, I.T., Shu, Y.T.: Development of the high concentration III–V photovoltaic system at INER, Taiwan. Renew Energy **34**(8), 1932–1937 (2009)
5. Kribus, A., Vishnevetsky, I., Yogev, A., Rubinov, T.: Closed loop control of heliostats. Energy **29**, 905–913 (2004)
6. Rubio, F.R., Ortega, M.G., Gordillo, F., Lopez-Martinez, M.: Application of new control strategy for sun tracking. Energy Convers. Manag. **48**, 2174–2184 (2007)
7. Chang, Y.K., Lee, B.H.: Development of high-accuracy image centroiding algorithm for CMOS-based digital sun Sensors. Sensors Actuators **A9**, 7849–7865 (2009)
8. Bolle, G., Ciccarella, G., Marietti, P.: Automatic aiming and tracking of the solar disc using photosensitive charge coupled devices (CCD). Appl. Energy **19**, 197–208 (1985)
9. Gonzalez, R.C., Woods, R.E., Eddins, S.L.: Digital image processing using Matlab®, 2nd edn. Gatesmark Publishing, LLC (2009)
10. Pattanasethanon, S.: The solar tracking system by using digital solar position sensor. Am. J. Eng. Appl. Sci. **3**(4), 678–682 (2010)
11. Vincent, O.R., Folorunso, O.: A descriptive algorithm for Sobel image edge detection. Proceedings of Informing Science & IT Education Conference (InSITE) (2009)
12. Taiwan amateur astrophotography enthusiasts' forum. http://astro.freebbs.tw/viewthread.php/?tid=508\&extra=page\%3D22

Chapter 57
Controlling Flock Through Normalized Radial Basis Function Interpolation

Mankyu Sung

Abstract This chapter introduces a controllable real-time flocking simulation framework through a vector field based on normalized radial basis function. During the design process, the framework subdivides the entire simulating environment into small cells, a so-called grid structure, and then assigns a vector per each cell, which represents a 2D vector field. The vectors of the field are automatically calculated by specifying a set of *control vectors* which are used for interpolating all vectors on the field. The interpolation scheme is based on normalized radial basis function. Once the construction of vector field is done, at the low level, flocks are simulated by following the vector field in the grid structure. Throughout this process, the position of individual agents is updated and collisions between the flock and the static obstacles are avoided by emitting a repulsive vector around the obstacles on the field. Interindividual collisions are also handled through fast lattice-bin method which can minimize the number of comparisons for detecting collisions.

Keywords Flocking • Computer animation • Radial basis functions

57.1 Introduction

Flocking animation has been widely used for spectacle scenes in featured films or animations. Most likely, those scenes are shown at the most important time for the entire story [1]. Therefore, many Hollywood film makers use those scenes for advertising and showing off their technical capabilities. However, simulating those flocking scenes poses a lot of technical challenges [2]. First, because flocks consist of hundreds and thousands of characters, it is impossible to control each individual manually. Second, because the aggregate behaviors of a flock are more important than individual movements, there must be a mathematical control model to manipulate the overall shape or flow of the flock movement. Third, in the process

M. Sung (✉)
Department of Game and Mobile Contents, Keimyung University, Daegu, South Korea
e-mail: mksung@kmu.ac.kr

J. Juang (ed.), *Proceedings of the 3rd International Conference on Intelligent Technologies and Engineering Systems (ICITES2014)*, Lecture Notes in Electrical Engineering 345, DOI 10.1007/978-3-319-17314-6_57

of simulating flocks, collisions need to be avoided among individuals as well as between individuals and static obstacles.

In this chapter, a new control scheme for flock animation in real time is presented. It is based on a vector field which is embedded in the simulating environment. Depending on the environment, either 2D or 3D vector field is set. The only difference between 2D and 3D vector field is whether or not the vector has 2D or 3D coordinates. Given *m* by *n* vector field, a vector is defined at the center of each cell of the vector field. The vectors are used for controlling the flow of the flock. That is, if an individual enters one of the cells of the vector field, then it is influenced by the force which corresponds to the vector defined inside. Therefore, it is possible to cast the flock control problem into a vector field design problem. The proposed scheme automatically sets the vectors in the vector field from a set of *control vectors*. The control vectors are the user-specified vectors on the environment for controlling the movement of flow. From the control vectors, by applying radial basis function interpolation technique [3], it is possible to calculate all vectors in the vector field automatically. Since the interpolation can be done in real-time, users are allowed to add or delete control vectors at runtime, which makes it possible to change the movement of flock in real time. In this process, collisions between the flock and static obstacles are prevented by emitting repulsive force around the obstacles. Those repulsive forces are added to the existing vector field as another layer. Collisions between individuals are then avoided by using simple lattice-bin method which is another grid container structure for reducing the number of times neighboring individuals check the proximity. Several experiments have verified that this proposed algorithm is able to control the flock movement by setting control vectors on the environment without causing any collisions.

57.2 Related Works

Flocking simulation can be categorized into three different approaches, which are a flow-based, an entity-based, and an agent-based approach. The flow-based approach models a flock as a continuum fluid [4]. Like the fluid flows down on the ground, the flow-based approach is able to control the flow of crowd easily. This approach is good for specific scenes of sports events where the crowd enters or exits a big stadium. However, it is not easy to change the flow at runtime. Chenny et al. proposed a tile-based flow control algorithm for improving controllability [5]. This algorithm controls flow by laying out the predefined flow-tile-blocks on the environment. These two approaches are similar because both try to control the flock using flow tiles. However, this proposed approach uses a vector field instead of tile blocks. Also, it can arbitrarily change the flow at runtime by just specifying control vectors, which has benefits in the controllability over the flock. The entity-based approach considers the flock as a set of synchronous entities with same properties. The particle-based method is one of the popular algorithms of entity-based approaches. Helbing proposed a social force model for simulating crowds using a particle system [6]. This algorithm controls the crowd through social and psychological properties of crowds. Since those properties must be set before simulation, it is not easy to find exact values of those properties.

The agent-based approach considers each individual as an independent and autonomous robot. Each agent perceives the changes happening in the environment and makes a decision for the next action [7]. Since this approach considers each individual as an intelligent entity, it does not need any centralized control mechanism. However, it has a disadvantage for simulating aggregative behaviors such as formation control. In addition, when the total number of individuals increases, it is not easy to achieve a real-time performance because each one needs to be updated one by one.

57.3 Algorithms

In this proposed method, the environment is assumed to have w by h size. On the top of the environment, a vector field with n by m size was built. As a result, each entry of vector field takes a region of w/n by h/m size in the environment. The method then assigns a vector in the center position of each region. In the algorithm, the vectors are used as *force* vectors to drive flocks to move in a particular direction. For assigning all vectors in the vector field, this scheme uses a radial basis function interpolation technique over a few user-specified *control vectors* on the environment.

The radial basis function has been used for interpolating scattered data. Many methods exist for solving this problem. The problem is denoted as follows: for a given set of n data points, $\{x_j\}_{j=1}^{n}$ with actual values, $\{f_j\}_{j=1}^{n}$, the radial basis functions are defined as $\{\psi_j(x)_{j=1}^{n}\}$ so that the linear combination of these functions is able to interpolate the data [3]. Mathematically, a function $s(x)$ can be computed as in the following (57.1).

$$s(x) = \sum_{j=1}^{k} \lambda_j \psi_j(x) \tag{57.1}$$

where $s(x) = f_j$ for $j = 1, \ldots, k$ and λ_j are weights.

In this approach, $s(x)$ corresponds the unit vectors on the vector field at the position x and ψ_j are user-specified k control vectors. The control vectors can be defined in the 2D or 3D space. Even though this chapter explains mainly about 2D control vector, the approach itself can be easily extended to a 3D one just by adding z coordinate. Each 2D control vector ψ_j has a starting position $\psi_j(x_s)$ and an ending position $\psi_j(x_e)$ where both x_s and $x_e \in \mathbb{R}^2$. From these vectors, the λ_j can be computed using the only $\psi_j(x_s)$ as shown in (57.2).

$$\lambda_j = \sum_{j=1}^{k} \left\{ \frac{1}{\left\| x - \psi_j(x) \right\|} \right\}^2 \tag{57.2}$$

The λ_j then becomes a normalized radial basis function $\hat{\lambda}_j$ by normalization. Equation 57.3 explains the normalization step.

$$\hat{\lambda}_j = \frac{\lambda_j}{\sum_{j=1}^{k} \lambda_j} \tag{57.3}$$

From $\hat{\lambda}_j$, the $s(x)$, which is the vector in a vector field at x, can be calculated on the vector field by using $\psi_j(x_e)$ in (57.4).

$$s(x) = s(x) + \sum_{j=1}^{k} \hat{\lambda}_j \psi_j(x_e) \tag{57.4}$$

After that, $s(x)$ is normalized to ensure the unit size.

Once the vector field is constructed, a set of flocks is put on the environment for simulation. The initial positions of flocks are randomly set, and then they are updated according to the external forces existing on the environment. The proposed method adopts three different forces to control the flocks. Those forces are *flow force* and two *collision avoidance forces*. The flow force, denoted as F_f, is derived from the vector field explained above, and collision avoidances forces, which are subdivided into two different forces, F_o and F_i, are calculated separately. The F_o is for collision avoidance of static obstacles and F_i is for interindividual collision avoidance. All forces are added up with weighting values for each individual at every time step t. Equation 57.5 illustrates the external force which consists of F_f, F_o and F_i.

$$F = \alpha F_f + \beta F_o + \gamma F_i \tag{57.5}$$

where α, β and γ are all weighting values for flow force and two collision avoidance forces respectively. If the α is bigger than β or γ, then it means that the simulation puts more emphasis on making flock follow the vector field than on collision avoidance. Otherwise, it rather considers the collision avoidance than makes flock follow the vector field.

First, the F_f is calculated from the vector field $s(x)$. Let v be the current velocity of an individual. Then, F_f can be set by subtracting the current velocity from the vector $s(x)$ obtained from the vector field.

$$F_f = s(x) - v \tag{57.6}$$

The F_f is then truncated by the predefined maximum force F_{max}, if the value is bigger than F_{max}.

Second, in order to calculate the F_o, the proposed method embeds the obstacles into bounding circles and makes bounding circles emit repulsive vectors around them. These vectors are then added to the existing velocity field. Mathematically, let c be the center position of bounding circle, then x is the center position of an element of vector field located inside the bounding circle. Then, the F_o is calculated by (57.7).

$$F_o = x - c/\|x - c\| \tag{57.7}$$

Third, the collisions between individuals are avoided through the intercollision avoidance force F_i. The F_i is calculated by finding an *average* velocity vector that does not cause any collision with all neighbors. Let $\{A\}$ be all the neighbors around a particular individual of the flock. Then the individual calculates the distance vectors d_i between its current position and all neighbors' positions. The d_i is then normalized to $\hat{d}_i$ and divided by the length of the vector. The division by the vector length is for giving different weights on the F_i depending on the distance from the corresponding individual. Those weighted d_i are summed up for all neighbors and divided by the number of neighbors for calculating an average. Finally, the vector is subtracted by current velocity v. Mathematically, (57.8) explains how to compute F_i.

$$F_i = \frac{\sum_{j=0}^{m} \frac{\hat{d}_j}{\|d_j\|}}{m} - v \tag{57.8}$$

where m is the number of neighbors and v is the current velocity.

One important point to achieve real-time performance in simulation is that it needs a smart way to reduce the number of neighbors $\{A\}$ for F_i. The proposed algorithm uses a simple lattice-bin method to reduce the neighbors. The lattice-bin method divides the entire environment into a grid structure and makes each slot of the grid maintain a dynamic list structure for containing all individuals located within the slot. Because the computation of F_i is done only for all neighbors in the list, it can reduce the number of neighbors significantly.

After computing F in (57.5), the individual calculates its acceleration by dividing F by its mass. Once the individual knows its acceleration, it can update its positions through the simple Euler integration method.

57.4 Experiments

The proposed algorithm was implemented using *Processing* 2.2.1 programming language for experiments. Figure 57.1a shows the vector field, designed by the radial basis function interpolation. Note that red lines in the figure correspond to the control vectors that are user-specified. Thanks to the radial basis function interpolation, the entire vector field is automatically calculated by using only three or four control vectors. The two gray circles in the middle are bounding circles for static

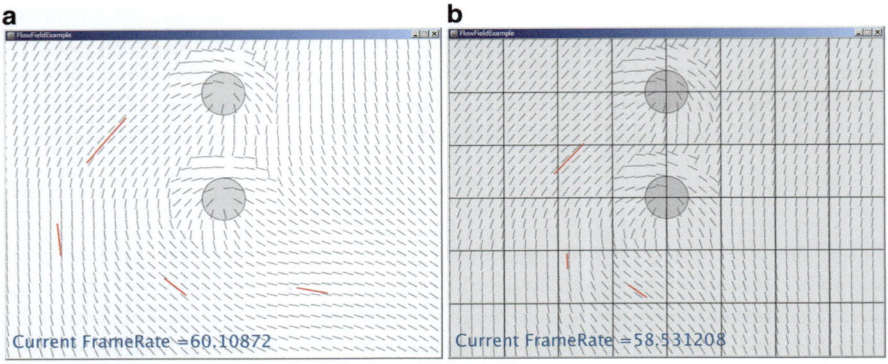

Fig. 57.1 (**a**) A vector field designed by radial basis function interpolation. (**b**) The lattice-bin for fast collision-checking between individuals

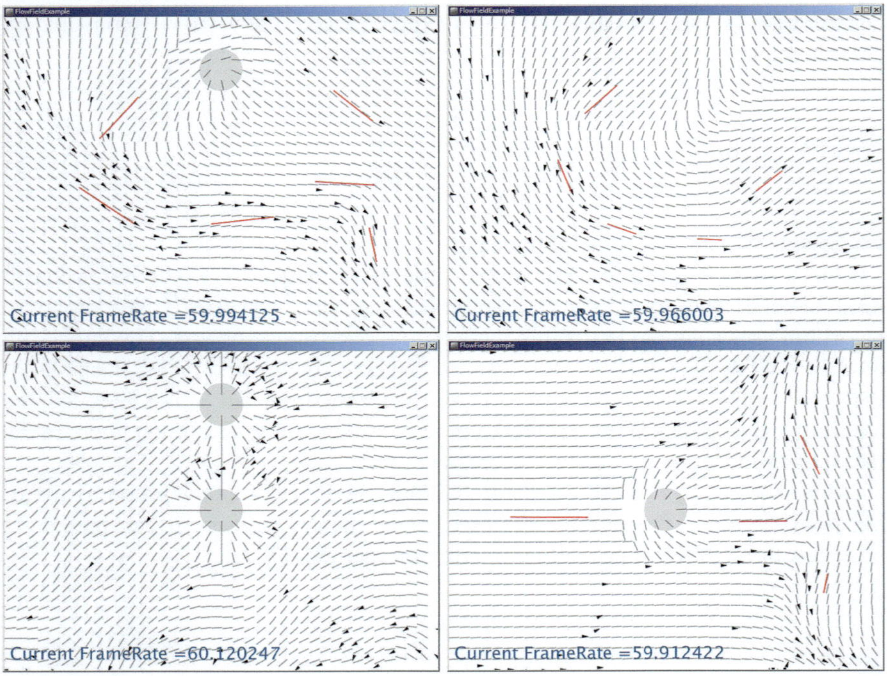

Fig. 57.2 Screenshots of real-time flock simulation using the proposed algorithm

obstacles. The vectors around the obstacles are different from other areas because of repulsive forces generated from the obstacles. Figure 57.1b shows the lattice-bin grid structure for fast collision detection between individuals.

Figure 57.2 shows several screen shots of flock simulation using the proposed algorithm. The flocks are represented as black triangles. In this simulation, users are allowed to add more and more control vectors on the environment to change the direction of flocks at runtime. The bottom of the figures indicates the current frame rate, which is around 60 frame/s. That guarantees that the proposed algorithm has the real-time performance.

57.5 Conclusion

This chapter introduces a novel control scheme for controlling flocks. For driving flocks to particular direction in the environment, the scheme uses a vector field that can be easily designed by the radial basis function interpolation technique based on only a few user-specified control vectors. On top of that, the algorithm combines collision avoidance forces for making individuals not bump into obstacles as well as other individuals. Through several experiments, the proposed algorithm verifies that it can run real-time performance. For future work, I would like to mix this algorithm with other control algorithms such as reciprocal velocity obstacle method (RVO) [8].

References

1. Greatest Crowd Film Scenes. http://www.filmsite.org/filmcrowds1.html
2. Zhou, S., Chen, D., Cai, W., Luo, L., Low, M.Y.-H., Tian, F., Tay, V.S.-H., Ong, D.W.S., Hamilton, B.D.: Crowd modeling and simulation technologies. ACM Trans. Model. Comput. Simul. **20**(4), 20 (2010)
3. Buhmann, M.: Radial Basis Functions: Theory and Implementations. Cambridge University Press, Cambridge (2003)
4. Hughes, R.L.: A continuum theory for the flow of pedestrian. Transp. Res. B **36**, 6 (2002)
5. Chenney S.: Flow Tiles. In: Eurographics/ACM SIGGRAPH Symposium on Computer Animation, pp. 233–242. University of Wisconsin, Madison (2004)
6. Helbing, D., Molnar, P.: Social force model for pedestrian dynamics. Phys. Rev. E **51**, 4282–4286 (1995)
7. Li, T.T., Chou, H.C.: Motion planning for a crowd of robot. In: IEEE International Conference on Robotics and Automation (2003)
8. Curtis, S., Manocha, D.: Pedestrian simulation using geometric reasoning in velocity space. In: Weidmann, U., Kirsch, U., Schreckenberg, M. (eds.) Pedestrian and Evacuation Dynamics, pp. 875–890. Springer, Heidelberg (2012)

Chapter 58
Novel AF Relay Design for Optimizing a MIMO Relay Network Under Backward Non-flat-Fading Channels

Chun-Hsien Wu

Abstract This chapter presents a novel AF relay design for a MIMO relay network optimization under received power constraint. With no CSI at the source, the optimum multiple relay precoders of a MIMO relay network are devised for combating the backward non-flat-fading channels. Simulation cases in terms of BER performance validate the proposed relay design and justify the proposed approach in its ability to pursue an optimum delay transmitted block for detection at the receiver.

Keywords Amplify-and-forward • AF • Oblique projection • Relay precoder

58.1 Introduction

Cooperative multiple-input multiple-output (MIMO) relaying techniques have evoked a lot of research interest because the spatial diversity of the network can be exploited and the communication range can be extended [1]. In such a network, the received signal superimposed by the emitted signals from multiple MIMO relays may achieve a much high power level, and thus, degrades the performance of neighbor receiver [2, 3]. This study extends the work of [3] to investigate the scenario where the multiple relays, suffered backward non-flat-fading channels, are selected for amplify-and-forward (AF) relaying. Particularly, the problem of optimizing MIMO relays is addressed for combating the backward interblock interference (IBI) effect, subject to the maximum-received power constraint at the receiver. To this end, all the proposed relays are devised according to the idea of recently built oblique-projection-based linear receiver with block delay detection [4, 5]. With no channel state information (CSI) at the source, the optimum relay precoders offered by the proposed multiple

C.-H. Wu (✉)
Department of Photonics and Communication Engineering, Asia University, Taichung 41354, Taiwan
e-mail: chwu@asia.edu.tw

J. Juang (ed.), *Proceedings of the 3rd International Conference on Intelligent Technologies and Engineering Systems (ICITES2014)*, Lecture Notes in Electrical Engineering 345, DOI 10.1007/978-3-319-17314-6_58

relays and the optimum delay transmitted block (for detection at the destination) are jointly pursued by the proposed approach so that the desired system criterion is achieved.

Notation: The superscripts $(\cdot)^H$, $(\cdot)^{\mathrm{T}}$, and $(\cdot)^{-1}$ are the Hermitian, transpose, and matrix inverse operators, respectively; $E[\cdot]$ is the statistical expectation; tr(**A**) is the trace of a matrix **A**; vec(**A**) is the operation that stacks the columns of **A** into a single column vector; $(\mathbf{M}_0 \oplus \cdots \oplus \mathbf{M}_{P-1})$ denotes a general block diagonal matrix whose block elements are given by $\mathbf{M}_i$ for $i = 0, 1, \ldots, P-1$.

58.2 Proposed Amplify-and-Forward Relay Design

Consider a MIMO relay network in which a single source and a single destination are consisted of M transmit antennas and M receive antennas, respectively. The K cooperative AF relays, each with N ($>M$) transmit/receive antennas, are used in this network. During data transmissions, the forward flat-fading FIR channels and backward non-flat-fading FIR channels are assumed linear and time-invariant (LTI). Additionally, there exists no direct line of sight channel between the source and the destination.

58.2.1 Received Signal Model of Each Relay

Let $\mathbf{s}_n = [s_n(0), s_n(1), \ldots, s_n(M-1)]^{\mathrm{T}}$ denote the nth block of M uncorrelated transmitted symbols on the source antennas. The nth block of N received symbols and the N zero-mean complex additive white Gaussian noise (AWGN) samples on the antennas of ith ($i = 0, 1, \ldots, K-1$) relay are denoted as $\mathbf{r}_{n(i)} = [r_{n(i)}(0), r_{n(i)}(1), \ldots, r_{n(i)}(N-1)]^{\mathrm{T}}$ and $\mathbf{v}_{n(i)} = [v_{n(i)}(0), v_{n(i)}(1), \ldots, v_{n(i)}(N-1)]^{\mathrm{T}}$, respectively. Since the backward non-flat-fading FIR channels are considered, the received signal at the ith relay is given by

$$\mathbf{r}_{n(i)} = \sum_{m=0}^{L} \mathbf{H}_{\mathrm{SR}(i),m}\mathbf{s}_{n-m} + \mathbf{v}_{n(i)}, \tag{58.1}$$

where $\mathbf{H}_{\mathrm{SR}(i),\,m}$ is a $N \times M$ matrix of mth ($m = 0, 1, \ldots, L$) FIR channel coefficient matrix with order L between the source and the ith relay. Note that in this work all the backward FIR channels are assumed the same order for simplicity. From the point of forwarding the target block $\mathbf{s}_n$ in $\mathbf{r}_{n(i)}$ of (58.1) in previous studies, $\mathbf{s}_n$ suffered the IBI effect caused by blocks $\mathbf{s}_{n-m}$.

Consequently, this study proposes a novel AF relay design based on the idea of our oblique-projection-based linear receiver with block delay detection [4, 5]. Unlike the receiver designs in ([5] and therein), the symbol detection is not operated at each relay but at the destination. Before forwarding the received signal, the ith relay is employed with a proposed design of matrix $\mathcal{W}_{d(i)}$ for simultaneously combating the backward IBI and the subsequent forward channel effects. More precisely, the proposed $\mathcal{W}_{d(i)}$ in each relay is applied on the consecutive blocks of $\mathbf{r}_{n(i)}$ in (58.1) to perform equalization for eliminating IBI against a certain delay transmitted block $\mathbf{s}_{n-d}$, where d is an integer and represents the block delay ($d \geq 0$). Let $\overline{\mathbf{r}}_{n(i)} = \text{vec}\left(\left[\mathbf{r}_{n-Q+1(i)}, \ldots, \mathbf{r}_{n(i)}\right]\right)$ denote the stacked Q consecutive blocks of $\mathbf{r}_{n(i)}$ in (58.1). Since $\mathbf{H}_{\text{SR}(i),\, m}$ of (58.1) is of order L, $\overline{\mathbf{r}}_{n(i)}$ can be constructed as

$$\overline{\mathbf{r}}_{n(i)} = \mathcal{C}_{(i)}\overline{\mathbf{s}}_n + \overline{\mathbf{v}}_{n(i)}, \tag{58.2}$$

where $\mathcal{C}_{(i)}$ denotes the $QN \times (Q+L)M$ composite block Toeplitz channel matrix (between the source and the ith relay) whose first block of rows is $[\mathbf{H}_{\text{SR}(i),L}, \ldots, \mathbf{H}_{\text{SR}(i),0}, \mathbf{O}_{N\times(Q-1)M}]$; $\overline{\mathbf{s}}_n = \text{vec}\left(\left[\mathbf{s}_{n-Q-L+1}, \ldots, \mathbf{s}_n\right]\right)$ and $\overline{\mathbf{v}}_{n(i)} = \text{vec}\left(\left[\mathbf{v}_{n-Q+1(i)}, \ldots, \mathbf{v}_{n(i)}\right]\right)$ denote the stacked $(Q+L)$ blocks of $\mathbf{s}_n$, and Q blocks of $\mathbf{v}_{n(i)}$ for ith relay, respectively. Based on the described design for ith relay, the proposed $N \times QN$ matrix $\mathcal{W}_{d(i)}$ is applied to $\overline{\mathbf{r}}_{n(i)}$ of (58.2) to yield the nth block of forwarded signal for ith relay, $\widetilde{\mathbf{r}}_{n(i)} = \mathcal{W}_{d(i)}\overline{\mathbf{r}}_{n(i)}$.

Applying the works of the oblique-projection framework of ([5] and therein), the proposed $\mathcal{W}_{d(i)}$ for ith relay, is implemented with

$$\mathcal{W}_{d(i)} = \mathbf{F}_{(i)}\mathbf{S}_e\mathbf{E}_{\mathcal{U}d(i)\mathcal{B}d(i)}\left(\widetilde{\mathcal{M}}_{d(i)}\right)^H_{Q(N-M)-LM}, \tag{58.3}$$

where the offered $N \times M$ block-based matrix $\mathbf{F}_{(i)}$ is known as relay precoder herein for ith relay, and $\mathbf{S}_e = \left[\mathbf{O}_{M\times(Q+L-1)M}, \mathbf{I}_M\right]$ extracts the lower M-size block of the $\mathbf{E}_{\mathcal{U}d(i)\mathcal{B}d(i)}\left(\widetilde{\mathcal{M}}_{d(i)}\right)^H_{Q(N-M)-LM}\overline{\mathbf{r}}_{n(i)}$; The above designs of $\left(\widetilde{\mathcal{M}}_{d(i)}\right)^H_{Q(N-M)-LM}$ and oblique projection $\mathbf{E}_{\mathcal{U}d(i)\mathcal{B}d(i)}$ for ith relay are constructed similarly as the derivations in [5]. Hence, with (58.2) and (58.3), and applying the properties of oblique projection, i.e., $\mathbf{E}_{\mathcal{U}d(i)\mathcal{B}d(i)}\mathcal{U}_{d(i)} = \mathcal{U}_{d(i)}$ and $\mathbf{E}_{\mathcal{U}d(i)\mathcal{B}d(i)}\mathcal{B}_{d(i)} = \mathbf{O}$ can be simplified as

$$\widetilde{\mathbf{r}}_{n(i)} = \mathbf{F}_{(i)}\mathbf{H}_{d(i)}\mathbf{s}_{n-d} + \mathbf{F}_{(i)}\widetilde{\mathbf{v}}_{n(i)}, \tag{58.4}$$

where $\mathbf{H}_{d(i)} = \mathbf{S}_e\mathcal{U}_{d(i)}$ denotes the lower $M \times M$ block of $\mathcal{U}_{d(i)}$, and $\widetilde{\mathbf{v}}_{n(i)} = \mathbf{S}_e \mathbf{E}_{\mathcal{U}d(i)\mathcal{B}d(i)}\left(\widetilde{\mathcal{M}}_{d(i)}\right)^H_{Q(N-M)-LM}\overline{\mathbf{v}}_{n(i)}$ denotes the $M \times 1$ oblique-projected Gaussian noise vector that has a positive-definite covariance matrix [4].

58.2.2 Design Problem Formulation

The received signal $\mathbf{y}_n$ at the destination is obtained from K relays that transmit signals through forward flat-fading channels, denoted by $M \times N$ matrix $\mathbf{H}_{\text{RD}(i)}$ for $i = 0, 1, \ldots, K-1$. Let $\mathbf{w}_n$ denote the nth block of M zero-mean complex AWGN samples on the receiver antennas. Hence, with $\bar{\mathbf{r}}_{n(i)}$ in (58.4) for ith relay, $\mathbf{y}_n$ is formulated as

$$\mathbf{y}_n = \sum_{i=0}^{K-1} \mathbf{H}_{\mathrm{RD}(i)}\widetilde{\mathbf{r}}_{n(i)} = \mathbf{H}_{\mathrm{RD}}\mathbf{F}\mathbf{H}_{\mathrm{SR}}\mathbf{s}_{n-d} + \mathbf{z}_n, \tag{58.5}$$

where $\mathbf{z}_n = \mathbf{H}_{\mathrm{RD}}\mathbf{F}\widetilde{\mathbf{v}}_n + \mathbf{w}_n$, $\mathbf{H}_{\mathrm{RD}} = \left[\mathbf{H}_{\mathrm{RD}(0)}, \ldots, \mathbf{H}_{\mathrm{RD}(K-1)}\right]$ is the $M \times KN$ composite forward channel matrix, $\mathbf{H}_{\mathrm{SR}} = \left[\mathbf{H}_{d(0)}^H, \ldots, \mathbf{H}_{d(K-1)}^H\right]^H$ is the $KM \times M$ composite effective backward channel matrix, and $\widetilde{\mathbf{v}}_n = \left[\widetilde{\mathbf{v}}_{n(0)}^H, \ldots, \widetilde{\mathbf{v}}_{n(K-1)}^H\right]^H$ is the $KM \times 1$ composite noise vector of the relays; Moreover, $\mathbf{F} = \left(\mathbf{F}_{(0)} \oplus \cdots \oplus \mathbf{F}_{(K-1)}\right)$ is the $KN \times KM$ block diagonal matrix. This study applies a block-based $M \times M$ equalizer $\mathbf{G}$ to $\mathbf{y}_n$ of (58.6) for equalizing the intrablock ISI effect caused by $\mathbf{H}_{\mathrm{RD}}\mathbf{F}\mathbf{H}_{\mathrm{SR}}$ so that the error $\mathbf{e}_n = \hat{\mathbf{s}}_n - \mathbf{s}_{n-d}$, exhibited in front of the detector, achieves minimum-mean-squared-error (MMSE) criterion, i.e., $E[\|\mathbf{e}_n\|^2]$, where $\hat{\mathbf{s}}_n$ denotes the nth block of the detector input. That is, $\hat{\mathbf{s}}_n = \mathbf{G}\mathbf{y}_n = \mathbf{G}\mathbf{H}_{\mathrm{RD}}\mathbf{F}\mathbf{H}_{\mathrm{SR}}\mathbf{s}_{n-d} + \mathbf{G}\mathbf{z}_n$.

For a given $\mathbf{F}$ and a specific d, the MMSE design of $\mathbf{G}$ in (58.8) is given by [6]

$$\mathbf{G}_{\mathrm{MMSE}} = \left(\sigma_s^{-2}\mathbf{I}_M + \mathbf{H}_{\mathrm{SR}}^H\mathbf{F}^H\mathbf{H}_{\mathrm{RD}}^H\mathbf{R}_{\mathbf{z}}^{-1}\mathbf{H}_{\mathrm{RD}}\mathbf{F}\mathbf{H}_{\mathrm{SR}}\right)^{-1}\mathbf{H}_{\mathrm{SR}}^H\mathbf{F}^H\mathbf{H}_{\mathrm{RD}}^H\mathbf{R}_{\mathbf{z}}^{-1}, \tag{58.6}$$

where $\mathbf{R}_{\mathbf{z}} = E\left[\mathbf{z}_n\mathbf{z}_n^H\right] = \mathbf{H}_{\mathrm{RD}}\mathbf{F}\left(\mathbf{R}_{\widetilde{\mathbf{v}}(0)} \oplus \cdots \oplus \mathbf{R}_{\widetilde{\mathbf{v}}(K-1)}\right)\mathbf{F}^H\mathbf{H}_{\mathrm{RD}}^H + \sigma_v^2\mathbf{I}_M$, and $\mathbf{R}_{\widetilde{\mathbf{v}}(i)} = E\left[\widetilde{\mathbf{v}}n(i)\widetilde{\mathbf{v}}_{n(i)}^H\right]$ is defined with $\widetilde{\mathbf{V}}n(i)$ in (58.4) for $i = 0, 1, \ldots, K-1$. Moreover, the covariance matrix of error, denoted by $\mathbf{R}_{\mathbf{e}} = E\left[\mathbf{e}_n\mathbf{e}_n^H\right]$, for intrablock MMSE equalization criterion, is

$$\mathbf{R}_{\mathbf{e}(\mathrm{MMSE})} = \left(\sigma_s^{-2}\mathbf{I}_M + \mathbf{H}_{\mathrm{SR}}^H\mathbf{F}^H\mathbf{H}_{\mathrm{RD}}^H\mathbf{R}_{\mathbf{z}}^{-1}\mathbf{H}_{\mathrm{RD}}\mathbf{F}\mathbf{H}_{\mathrm{SR}}\right)^{-1}. \tag{58.7}$$

Accordingly, with no available CSIs at the source, the design goal of this work is to jointly devise optimum $\mathbf{F}$, and d for minimizing $\mathrm{tr}(\mathbf{R}_{\mathbf{e}(\mathrm{MMSE})})$, subject to a bound P_R on the received power. That is, the design problem is formulated as

$$\min_{\mathbf{F},\, d} \mathrm{tr}\left(\mathbf{R}_{\mathbf{e}(\mathrm{MMSE})}\right) \tag{58.8}$$

subject to $\mathrm{tr}\left(\sigma_s^2\mathbf{H}_{\mathrm{RD}}\mathbf{F}\mathbf{H}_{\mathrm{SR}}\mathbf{H}_{\mathrm{SR}}^H\mathbf{F}^H\mathbf{H}_{\mathrm{RD}}^H + \mathbf{H}_{\mathrm{RD}}\mathbf{F}\left(\mathbf{R}_{\widetilde{\mathbf{v}}(0)} \oplus \cdots \oplus \mathbf{R}_{\widetilde{\mathbf{v}}(K-1)}\right)\mathbf{F}^H\mathbf{H}_{\mathrm{RD}}^H\right) \le P_R$. Assume $C(i)$ of ith relay is full column rank, $0 \le d \le Q + L - 1$ is satisfied in (58.8).

58.2.3 *Proposed Approach for Relay Optimization*

In the following, an approach of optimization steps is proposed to pursue an optimum solution of $\mathbf{F}$ and d to the design problem of (58.8).

S0. For a given signal-to-noise ratio (i.e., $\text{SNR} = \sigma_s^2/\sigma_v^2$), initialize $d = 0$, and employ the proposed design $\mathcal{W}_{d(i)}$ in (58.3) to construct the signal model $\widetilde{\mathbf{r}}_{n(i)}$ of (58.4) for ith relay, $i = 0, 1, \ldots, K-1$.

S1. Based on the signal model $\hat{\mathbf{s}}_n$, and the problem of (58.8) for $d = 0$, apply the method in [3] to devise the optimum $\mathbf{F}$, subject to the maximum received power constraint. Then, the proposed optimum $\mathcal{W}_{d(i)}$ in (58.3) for ith relay is built, and the optimum block-based $\mathbf{G}_{\text{MMSE}}$ of (58.6) is obtained accordingly.

S2. Repeat the steps S0 and S1 by increasing d that satisfies $0 \le d \le Q + L - 1$.

S3. Find the optimum d within $0 \le d \le Q + L - 1$ and the corresponding optimum designs $\mathbf{F}$, $\mathcal{W}_{d(i)}$ and $\mathbf{G}_{\text{MMSE}}$ whose $\text{tr}(\mathbf{R}_{\text{e(MMSE)}})$ of (58.8) yields the minimum.

Note that the above steps are applied after a suitable size of Q is set so that $C_{(i)}$ of (58.2) is of full column rank for ith relay.

58.3 Computer Simulations

This section provides the bit error rate (BER) simulations of the optimized MIMO relay network to validate the proposed AF relay design and investigate the significance of designing an optimum block delay d. The cases of existing backward channel IBI are considered. For simplicity, the 4-ary QAM constellation signaling was independently adopted for each symbol on different antenna for a MIMO relay network with parameters $(M, K, N) = (2, 3, 3)$. Figure 58.1 exhibits the BER curves of the proposed designs with variable d. The backward channels were of order $L = 2$. By applying the steps S0 ~ S3 with $Q = 4$, the proposed design with $d = 1$ is obtained and performed optimally through all SNR range. Obviously, the BER performance of the optimized network varied with different d. This finding signifies the importance of pursuing an optimum d with the proposed AF relay design under the case of existing backward IBI effect. Furthermore, another case of $L = 1$ was used for the proposed design with variable Q. As indicated in Fig. 58.2, the proposed optimum design could have an improved BER performance by increasing Q from 2 to 3, and 4 (nearly 1 dB, and 1.6 dB SNR, respectively, were gained at a BER of 10^{-5}). Interestingly, the above result reveals a merit that the proposed optimum design can even obtain a lower BER by processing larger stacked received blocks, i.e., size-Q, in each relay.

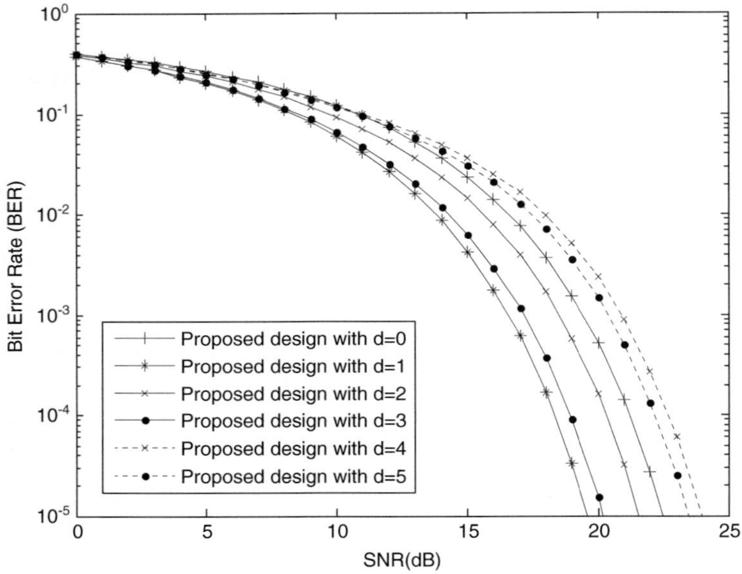

Fig. 58.1 BER curves of the proposed designs with variable d

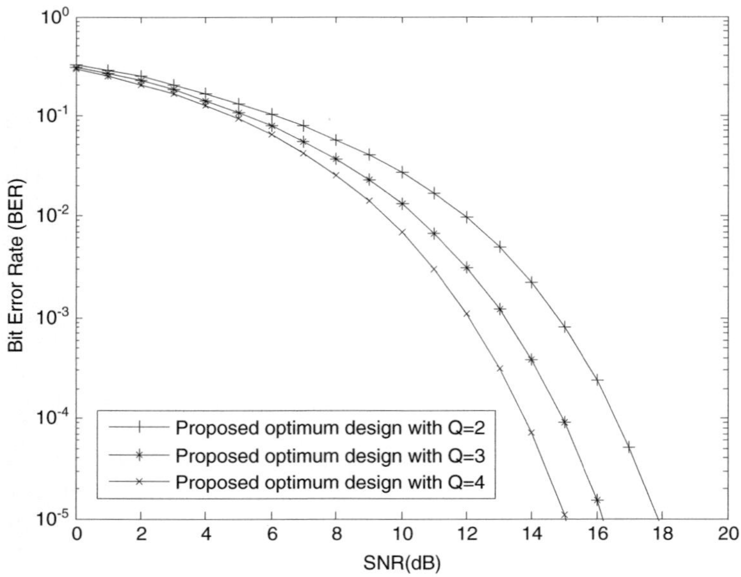

Fig. 58.2 BER curves of the proposed optimum designs with variable Q

58.4 Conclusions

This work has proposed a novel AF relay design for optimizing an MIMO relay network under the scenario of existing backward IBI effect. With no CSIs at the source, an approach of optimization steps was proposed to pursue optimum multiple relay precoders and the optimum delay transmitted block for detection at the receiver such that the intrablock MMSE equalization criterion is achieved, and the received power is constrained. Simulation cases have validated the proposed optimum AF relay design against the optimum delay transmitted bock and with larger stacked received blocks in each relay.

Acknowledgments The author would like to thank National Science Council (NSC) of R. O. C., Taiwan, for partly financially supporting this research under Contract Nos. NSC 101-2221-E-468-002, and 102-2221-E-468-002.

References

1. Scaglione, A., Goeckel, D.L., Laneman, J.N.: Cooperative communications in mobile ad hoc networks. IEEE Signal Process. Mag. **23**(5), 18–29 (2006)
2. Gastpar, M.: On capacity under receive and spatial spectrum-sharing constraints. IEEE Trans. Inf. Theory **53**(2), 471–787 (2007)
3. Behbahani, A.S., Merched, R., Eltawil, A.M.: Optimizations of a MIMO relay network. IEEE Trans. Signal Process. **56**(10), 5062–5073 (2008)
4. Wu, C.H.: The precoder design for intrablock MMSE equalization and block delay detection with a modified oblique projection framework. IEICE Trans. Fundam. **E94-A**(2), 829–832 (2011)
5. Wu, C.H., Tsai, C.Y.: Performance of joint linear transceiver design with BER-minimized block delay for detection. Lecture Notes in Electrical Engineering, vol. 293. Springer (2014)
6. Haykin, S.: Adaptive Filter Theory, 4th edn. Prentice-Hall, NJ (2002)

Chapter 59
Development of Wearable Power Assist Wear Using Pneumatic Actuators

Feifei Cho, Xiangpan Li, and Toshiro Noritsugu

Abstract This research focuses on developing a safe, lightweight, power assist device that can be worn by people during lifting or static holding tasks to prevent low back pain (LBP). In consideration of flexibility, light weight, and large force to weight ratio, two types of pneumatic actuators are employed in assisting low back movement for safety and comfort. The device can be worn directly on the body like normal clothing. Because there is no rigid exoskeleton frame structure, it is lightweight and user friendly.

Keywords Wearable • Assist wear • Pneumatic actuator

59.1 Introduction

Last year, the number of people in Japan aged 65 and over was 30.74 million, account for 24.1 % of the population. With the rapid arrival of an aging society, the demand for professional caregivers has increased drastically. It was said that 70 % of caregivers experience back pain reference. So the development of wearable power assist devices and nonwearable lifting assist devices are greatly needed in nursing care fields.

Wearable robots are mechatronic systems that are designed around the shape and function of the human body. Early designs were mainly exoskeleton types. These robots have multiple degrees of freedom (DOF), but are not suitable for use in small spaces and in daily life.

To meet the aging society's needs for power assist robotic technology, and devices that wear like normal clothing, this study proposes a new wearable power assist device for low back support using pneumatic actuators. First, the biomechanical model of the human spine was analyzed to better understand the reason for low

F. Cho (✉) • T. Noritsugu
National Institute of Technology, Tsuyama College, Tsuyama, Okayama 708-0825, Japan
e-mail: cho@tsuyama-ct.ac.jp

X. Li
Henan University of Science and Technology, Kaiyuan road, Luoyang, Henan 471023, China

J. Juang (ed.), *Proceedings of the 3rd International Conference on Intelligent Technologies and Engineering Systems (ICITES2014)*, Lecture Notes in Electrical Engineering 345, DOI 10.1007/978-3-319-17314-6_59

back pain (LBP). Based on this analysis, two types of pneumatic actuators were selected to support the human's back from two aspects: increased support force and related lever arm length. To measure the human's action, inertial sensors were selected by using sensor fusion technology. In order to provide efficient assisted force, control strategies were determined to minimize the interference of human actions and provide assistance effectiveness for the proposed device. The proposed design has been proven through experiments.

59.2 Low Back Pain

Since many health caregivers are troubled by LBP problems, the requirement for low back support is needed. It is critical to understand the cause of LBP when designing power assist wear for low back support. LBP is common in various occupations, its presence being related to activities requiring repetitive lifting, repeated activities in bending forward positions, and lifting high energetic loads. Related results also suggest that bending activities involving higher degrees of trunk flexion are associated with disabling types of LBP in certain working populations. Such work characteristics are common among nursing caregivers. The prevalence of LBP in nursing is high compared to other occupations and in relation to other types of work. Risk factors include physical work such as manual lifting and transferring of patients, working conditions such as working time and rest during the night shift, and the working environment. Among these factors, exposures to frequent manual lifting and transferring of patients are widely recognized factors.

59.3 Actuators of the System

59.3.1 Elongation-Type Pneumatic Rubber Artificial Muscle

Developing wearable power assist devices that wear like normal clothing requires the development of a new type of pneumatic rubber artificial muscle to meet the requirements of safety and user-friendliness. In this part, an elongation-type of pneumatic rubber artificial muscle referred as actuator A, is employed for the special requirement in assisting low back movement.

Figure 59.1 shows the structure of the elongation-type pneumatic rubber artificial muscle. It is composed of a rubber tube covered with a bellows sleeve and closed with ties at both ends. The outer diameter of the rubber tube is 12 mm; the inner diameter is 10 mm. The bellows sleeve is woven of twisted fiber cord, which has a maximum expansion diameter of 14 mm and a minimum contraction diameter of 4 mm. When first pressurized, the elongation-type pneumatic rubber artificial muscle expands slightly in the radial direction, putting the internal rubber tube into

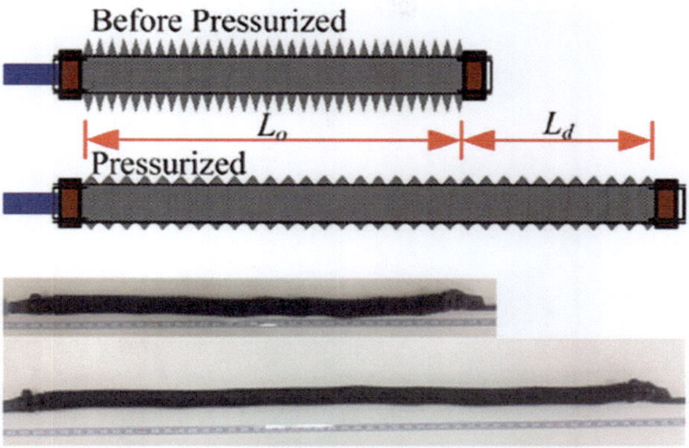

Fig. 59.1 Pneumatic rubber artificial muscle

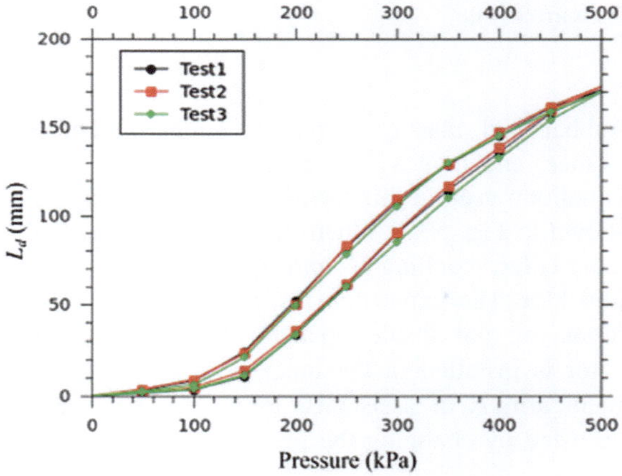

Fig. 59.2 Relation between supplied pressure and displacement

close contact with the external sleeve. As the pressure continues to rise, due to the fact that radial expansion is limited by the external sleeve, the artificial muscle expands in the axial direction, and it becomes longer. The higher the pressure, the longer it becomes. Figure 59.2 shows the relation between supplied pressure and Ld, related displacement of elongation-type pneumatic rubber artificial muscle with no external load connected. The maximum length is attained when it is pressurized; and the minimum length is attained when it is not. For this type of muscle (original length $L_o = 320$ mm), the total length can reach 490 mm ($L_d = 170$ mm) when pressurized to 500 kPa.

59.3.2 Layer-Type Pneumatic Actuator

The layer-type pneumatic actuator called actuator B, is composed of two TPU balloons. The TPU material is a composite material that combines the properties

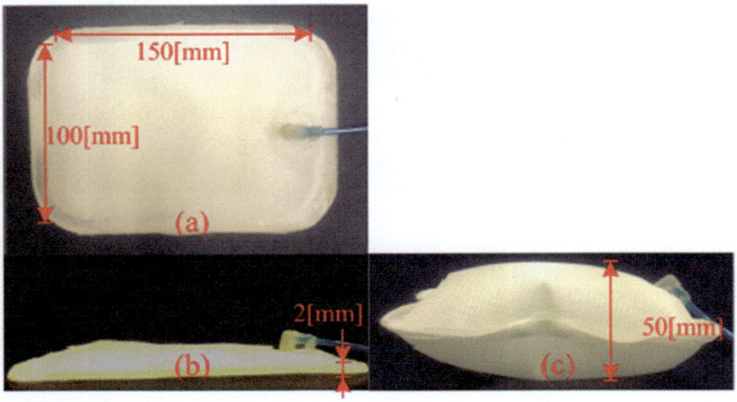

Fig. 59.3 Overview of TPU balloon. (**a**) Overview (top view); (**b**) Initial state (side view); (**c**) Pressurized state

of rubber and plastic. It has excellent weight bearing capacity and impact resistance, and it is widely used in producing massage chairs and airbags. The TPU balloon used in this device is 150 mm long, 100 mm wide, and 2 mm thick, as shown in Fig. 59.3. When the balloon is supplied with compressed air, it will become taller, reaching 50 mm. A TPU balloon can take a maximum air pressure of 250 kPa. The expansion force reaches 450 N at a pressure of 60 kPa. TPU balloons are put inside pockets made with nylon bands. In this device, the actuator is installed in the inner layer of the garment. In order to increase the moment arm A of assistance force, the expansion displacement in height can be adjusted by changing the pressure applied.

59.3.3 Bending Pneumatic Actuator

A wearable waist power assist device using a bending actuator as shown in Fig. 59.4. The device, composed of bending pneumatic rubber artificial muscles and a garment, can support waist movement by supplying the bending pneumatic rubber artificial muscles with the desired pressure.

The structure of the actuator is very simple; moreover it can be produced within 1 h. The operating principle is as follows: when the McKibben artificial muscle is pressurized, the muscle is expanded in the radial direction. The muscle is contracted in the axial direction for about 25 % in length. Then the actuator bends in the axial surface-anchoring direction.

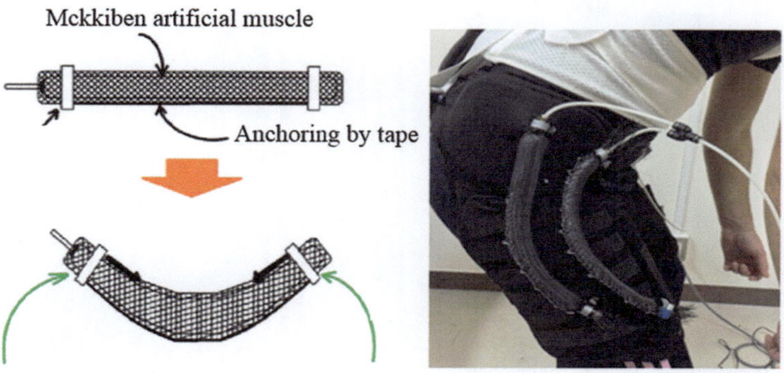

Fig. 59.4 Wearable waist power assist

59.4 Wearable Power Assist Wear Device

This research focuses on developing a safe, lightweight, power assist device that can be worn by people during lifting or static holding tasks to prevent LBP. In consideration of their flexibility, light weight, and large force to weight ratio, two types of pneumatic actuators were employed in assisting low back movement for their safety and comfort. The biomechanical model of the human spine was analyzed, and to understand the main causes of LBP: when the human is bending forward and lifting a load. The erector spinae muscles have a small lever arm, causing the spine to bear a large amount of force, several times the body weight. By taking into account the biomechanic structure of the human spine, this device can provide support in two ways. Actuator A acts as an external muscle power generator to reduce the force requirement for the erector spinae muscles. Actuator B acts as a moment arm for the contractile force generated by actuator A, and increases the effective torque. The device shown in Fig. 59.5 can be worn directly on the body like normal clothing. Because there is no rigid exoskeleton frame structure, it is lightweight and user friendly.

59.5 Static Holding Test

For the power assist wear proposed in this study, the elongation-type pneumatic rubber artificial muscle can be seen as a spring with a variable stiffness. Compared with exoskeleton-type power assist devices, this device has the advantage of maintaining a static posture during bending forward. During the experiment, the subject was told to stand on a force plate, bend forward, and maintain the static posture while holding a load (7 kg) in both hands. The subject maintained this position for 30 s. The force plate recorded the movement of the COG. The system

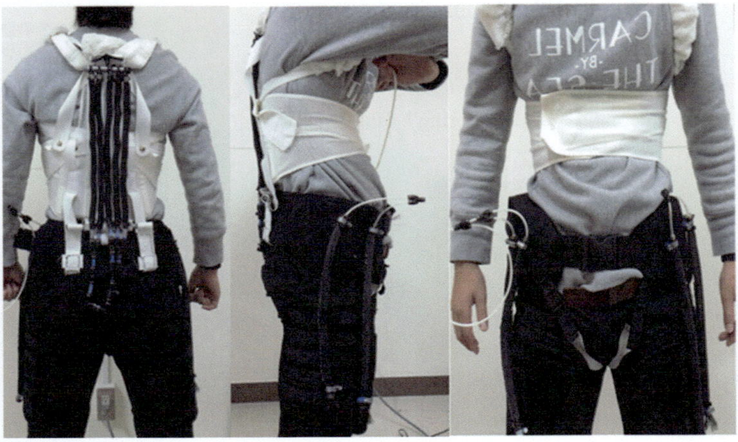

Fig. 59.5 Wearable power assist wear device

default sampling frequency was 60 Hz and 1,800 coordinate values in the X-axis and the Y-axis were obtained respectively during the experiment. In order to evaluate the movement of the COG, the collected data was analyzed according to the following equations:

$$\Delta L_i = \sqrt{(X_{i+1} - X_i)^2 + (Y_{i+1} - Y_i)^2} \tag{59.1}$$

$$\mathrm{LNG} = \sum_{i=1}^{n} \Delta L_i \tag{59.2}$$

$$\mathrm{LNG}_X = \sum_{i=1}^{n} \left|(X_{i+1} - X_i)\right| \tag{59.3}$$

$$\mathrm{LNG}_Y = \sum_{i=1}^{n} \left|(Y_{i+1} - Y_i)\right| \tag{59.4}$$

$$VX_{i+1} = (X_{i+1} - X_i)/T \tag{59.5}$$

$$VY_{i+1} = (Y_{i+1} - Y_i)/T \tag{59.6}$$

where the ΔL_i is the moving length of the COG in one sampling period, LNG is the length moved in 30 s, and n is the total number of samples ($n = 1{,}800$). LNG_X and LNG_Y are related to lengths moved in the direction of the X-axis and the Y-axis. VX_{i+1} and VY_{i+1} are the related moving velocities in the direction of the X-axis and the Y-axis. Here T is the sampling period ($T = 1/60$ s). The moving lengths of the

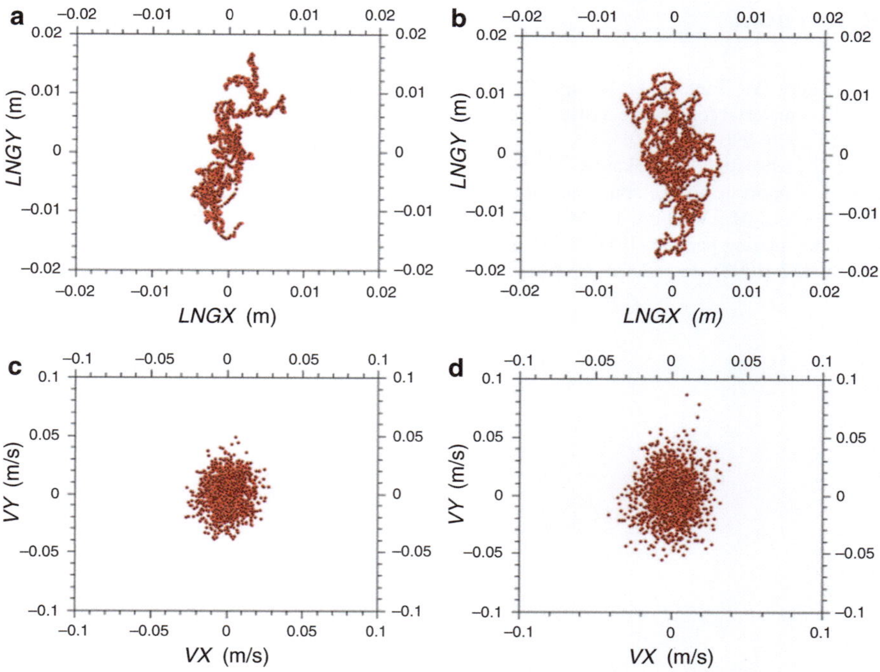

Fig. 59.6 The moving length and velocity distribution of the COG. (**a**) With assistance; (**b**) Without assistance; (**c**) With assistance; (**d**) Without assistance

COG were calculated and shown in Table 59.1. Related data are shown in Fig. 59.6, The top two figures show the moving trajectory of the COG and the last two figures show the velocity distribution of the COG.

59.6 Conclusion

In this study, A power assist wear for low back support, a device using new types of pneumatic actuators are proposed. Compared to that of McKibben-type pneumatic rubber artificial muscle, the contraction rate of elongation-type pneumatic rubber artificial muscle is larger. As it does not use an exoskeleton structure, the device can be worn on the human body just like normal clothing. It can provide assistance force for the low back, reducing the possibility of LBP, and the assistance power of the device can be adjusted by changing the pressure of the compressed air. The effectiveness of the device has been verified through experiments.

References

1. Imamura, Y., Tanaka, T., Suzuki, Y., Takizawa, K., Yamanaka, M.: Motion-based-design of elastic material for passive assistive device using musculoskeletal model. J. Rob. Mechatronics **23**(6), 978–990 (2011)
2. Li, X., Noritsugu, T., Takaiwa, M., Sasaki, D.: Design of wearable power assist wear for low back support using pneumatic actuators. Int. J. Automot. Technol. **7**(2), 228–236 (2013)
3. Josephson, M., Vingard, E.: Workplace factors and care seeking for low-back pain among female nursing personnel. MUSIC-Norrtalje Study Group. Scand. J. Work Environ. Health **24**, 465–472 (1998)

Chapter 60
Cost Model of Physical Activity Monitoring Systems

Jen-Liang Cheng, Chien-Chih Chen, and Ling-Hsuan Chen

Abstract A physical activity (PA) monitoring system comprises a number of wearable devices based on a certain measuring technique and a way to collect and record data on the devices. Recent studies showed that the system is an effective intervention tool to increase participants' PA level. Given the trend of physical inactivity among students, it is desired to widely deploy the system in schools to decelerate the trend. Based on the need of school, health educators must assess capability and cost of monitoring systems associated with various measuring techniques. The assessment, however, is difficult due to the lack of cost model for a PA monitoring system to be run in real setting. In this article, we propose the cost model so that cost of a PA monitoring system can be estimated according to the selected measuring technique. The model explains the infeasibility of pedometer- or accelerometer-based PA program in population level and implies the preference of an automated measurement system.

Keywords Physical activity (PA) promotion • School-based PA program • Pedometer • Objective measurement • Population-level setting

60.1 Introduction

Regular, brisk exercise reduces the risk of coronary heart disease, diabetes, hypertension, and bowel cancer. To assess the effectiveness of intervention strategy, many physical activity programs used activity monitor, e.g., pedometer or accelerometer, to objectively measure and monitor participants' physical activity. Recent researchers found that pedometer can increase individual's incentive of PA engagement. It is therefore strongly suggested as practical and accurate tools for measuring and motivating program participant's PA [1–4].

J.-L. Cheng (✉) • L.-H. Chen
Department of Medical Informatics, Tzu-Chi University, Hua-Lien City 97004, Taiwan
e-mail: quelay13@gmail.com

C.-C. Chen
Institute of Medical Sciences, Tzu-Chi University, Hua-Lien City 97004, Taiwan

J. Juang (ed.), *Proceedings of the 3rd International Conference on Intelligent Technologies and Engineering Systems (ICITES2014)*, Lecture Notes in Electrical Engineering 345, DOI 10.1007/978-3-319-17314-6_60

Define a PA monitoring system a system that comprises a number of wearable devices of a certain measuring technique and a way to collect and record activity data on the devices. The use of system is proven to be effective in PA promotion. For example, an investigation on 37 school-based projects summarized that: (1) pedometers can facilitate increased PA among children and adolescents. (2) interactive technologies can be particularly useful as an intervention tool because of their ability to provide personalized, tailored feedback. (3) pedometer provided an incentive to engage with other activities related to PA promotion; but little evidence can prove the sustainability of the effect. However, there are challenges in using the system, such as: (1) recording and collating the results, sorting out lost and forgotten pedometers and collecting them in, etc., are extremely time and energy consuming; (2) around 10–15 % of all pedometers were lost during the course of the month; (3) the reset button could be pressed by mistake easily, causing a reset to zero and de-motivation of the use of pedometers; (4) shaking pedometers to create false results is easy [5]. These challenges indicate that one has to pay the data collecting work, data checking works, and lost devices to ensure the operation of the system. The costs of these works are often omitted by a short-term program due to that its task is only to prove the effectiveness of its intervening strategy.

Conventionally, a PA monitoring system is used as a measurement tool. Participants were divided into several groups so that PA level of a group is sampled at a time. The number of devices required need not to be the number of participants of the program. When acting as an intervention tool, the system requires a quantity of devices equal to the number of program participants. Furthermore, school-based PA program in real setting has to be lasted for a long term to affect new students and form active climate; the works to operate the system can become unmanageable. In other words, school needs not only the budget to purchase a number of devices but also the budget to operate the system. Estimating the budgets is difficult due to the lack of previous reports. The uncertainty discourages the adoption of a monitoring system.

The purpose of this article is to reveal cost to initiate and operate a PA monitoring system and to propose a cost model so that the cost of PA monitoring systems based on various measuring techniques can be estimated. Based on the model, health educators may figure out what type of system is appropriate.

60.2 The Cost Model

Define a PA session as the period an individual has engaged a certain type of PA, e.g., walking, sleeping, and studying, in a certain place. The PA session can be described by the PA type, the space in which the PA is engaged, the duration of the session, and the energy expended for the session. The energy expended is often quantized by the number of METs (metabolic equivalents). One MET, which is equal to 3.5 mL/kg per minute, is the resting oxygen consumption of a resting human being after dividing his body weight [6]. An instrument in the lab may directly measure oxygen consumption of a PA session in a defined space of specific PA type for a defined period to obtain accurate energy expenditure (EE). On the other hand, a pedometer or an accelerometer can respectively measure a person's step count or acceleration without knowing the space, duration, and PA type of the person's PA session. They not only give the convenience in measuring a person's daily activity but also introduce uncertainty in the measurement result. The accuracy of the measurement result is thus limited.

To objectively measure PA of N participants, a PA program needs to purchase N devices, to distribute the devices, and to periodically collect data on the devices. Roughly speaking, the tasks can be divided into two stages: the initial stage and the operating stage.

The cost of the system in the initial stage can be formulated as:

$$s + d \times N \tag{60.1}$$

The meaning of the parameters is listed as below:

s—cost for setting up a data server and for components other than the wearable devices;
d—cost of a wearable device, which can range from a few to a few hundred US dollars.

In the operating stage, participants' cooperation in reporting data, uploading data, replacing battery, and charging battery are often thought free. However, these trifling tasks may weaken the sustainability of the program. Whether the tasks are done by a program staff or by participants, they should be treated as cost of a monitoring system. Thus, if the measurement service is to be delivered for Y years, then the cost in operating stage, referred as operating cost, can be formulated as:

$$((c + r \times d + w + b) \times N + m + u + p) \times Y \tag{60.2}$$

where:

c—communication cost. This value is nonzero if a charged network is used to transfer data.
r—probability that a device is lost or damaged per year.
w—cost of program staff's and participants' works for data collection. The works are depicted by one of the rows in Table 60.1 according to the combination of the system.
b—cost to replace or charge battery of a device.
m—cost to maintain data server and system components other than the measuring devices.
u—cost of training course for data uploading.
P—cost to prepare new participants' database.

Table 60.1 Collecting data works in systems of various combinations

Name	Device	Other components	Program staff's works	Participants' works
C1	Pedometer	None	Recording the data	Bring devices to a program staff
C2	Pedometer	A data server	Reminding participants and checking data validity	Self-reporting the data to the server via the Internet
C3	Computer-linked pedometer	A data server	Reminding participants	Connecting the device to a computer and executing data uploading procedure
C4	Passive RFID tags	Some RFID readers and an embedded computer	None	Engaging in exercise in defined space

The value of w, m, u, and b depends on salary level and whether full-time or part-time program staffs are recruited. Often, these costs are ignored in short-term PA programs. Volunteering staffs and participant's cooperating works in data collection is often thought free. For a long-term program, they can become significant and affect the sustainability of the program. For example, lack of time is the major barrier for students to engage in exercise, data reporting or uploading works may affect students' volition to participate in the program. The works are thus treated as part of the operating cost. Formula (60.2) reveals the cost to sustain a system.

School's health promotion program is often characterized by limited budget, long period, and large population. The two expressions explain the infeasibility of pedometer-based program in schools. Only a monitoring system of zeroed w and b, which are multiplied by $(Y \times N)$, can be sustained in population-level setting.

60.3 A Cost Comparison Among PA Monitoring Systems

In this section, monitoring system using Omron HJ-323U, Nike + FuelBand, dumb pedometer, and passive RFID tags, are compared. Table 60.2 lists s, d, m, l, and b, of the systems. Values of t and p are same among various systems and are not listed in Table 60.2. Parameter c is nonzero when a mobile phone network is used. The cost d of various devices is obtained from the Internet. Often, the device of mobile communication capability is unaffordable for a population-level program.

Obtaining accurate operating cost, which consists of data collecting cost and battery handling cost, can be complicated especially when the tasks are done by volunteering participants and volunteering teachers. Most tasks in operating stage can be eliminated by automatic system using passive RFID tags [7]. In Table 60.2, the setup and maintenance cost of an automated system called AXMS is calculated

Table 60.2 A cost comparison among monitoring systems of different measuring devices

Product's name	Dumb pedometer	*Omron HJ-323U*	Nike + FuelBand	AXMS
Technology	Dumb pedometer	Computer-linked pedometer	Wireless activity tracker	Passive RFID
s	(server)	(server)	(server)	8,000
d	10	25	75	5
m	(server)	(server)	(server)	1,000
l[a]	C1 or C2	C3	C3	C4
b	Replacing	Replacing/ charging	Charging	X
Space defined	Not in water or a vehicle	Not in water or a vehicle	Not in water	Track or pool
Types of a PA session to be measured	Body motion (walk is preferred)	Body motion (walk is preferred)	Body acceleration (walk is preferred)	Run, walk, and swim
Result	Step count/EE	Step count/EE	EE	Time/type/EE
Accuracy	Acceptable	Acceptable	Fair	Fair in walk and run

[a]Combination name in Table 60.1 is used. Value of *l* is calculated according to the works of the corresponding combination

based on our implementation experience. Though the system has higher initiating cost, its operating cost is irrelevant to *N*. Thus using it in a population-level setting can be feasible. However, the exercise type and space must be defined.

60.4 Conclusion

Role of activity trackers is shifting from a pure measurement device to an intervention tool. To leverage it, school has to build a monitoring system consists of them and a way to collect data. In a population-level setting, the cost is likely to become unmanageable. Our contribution is to propose a cost model of the system so that health educators may estimate the budget to build a system. The model explains the difficulty in using autonomous measurement devices in population-level setting. It also implies that a system for a population-level program is sustainable only when it can automatically measure and record participants' activity and uses batteryless devices.

There is no perfect solution to PA measurement. It depends on the property of the setting and the goal of the program that a health professional can choose a cost-effective monitoring system based on the model.

Acknowledgment The authors thank the National Science Department for its support on this study Grant NSC99-2410-H-320-019-MY2.

References

1. Fuller, N.R., Williams, K., Shrestha, R., Ahern, A.L., Holzapfel, C., Hauner, H., Jebb, S.A., Caterson, I.D.: Changes in physical activity during a weight loss intervention and follow-up: a randomized controlled trial. Clin. Obes. **4**(3), 127–135 (2014)
2. Laurson, K.R., Welk, G., Eisenmann, J.C.: Estimating physical activity in children: impact of pedometer wear time and metric. J. Phys. Act. Health **12**(1), 124–131 (2014)
3. Lubans, D.R., Plotnikoff, R.C., Miller, A., Scott, J.J., Thompson, D., Tudor-Locke, C.: Using pedometers for measuring and increasing physical activity in children and adolescents: the next step. Am. J. Lifestyle Med. (2014) doi: 1559827614537774
4. Tudor-Locke, C., Swift, D.L., Schuna, J.M., Dragg, A.T., Davis, A.B., Martin, C.K., Johnson, W.D., Church, T.S.: WalkMore: a randomized controlled trial of pedometer-based interventions differing on intensity messages. BMC Public Health **14**(1), 168 (2014)
5. Inchley, J., Cuthbert, L., Grimes, M.: An investigation of the use of pedometers to promote physical activity, and particularly walking, among school-aged children: review of evidence and scoping study. Child and Adolescent Health Research Unit, Edinburgh (2007)
6. American College of Sports Medicine: ACSM's Guidelines for Exercise Testing and Prescription. Lippincott Williams & Wilkins, Baltimore (2005)
7. Cheng, J.-L., Chen C.C, Li, Z.L., Chen, L.H.: AXMS: optimal tool for intramural exercise monitoring submitted to IJSK. Int. J. Res. Med. Health Sci. **4**(6) (2014)

Chapter 61
The BCI Control Applied to the Interactive Autonomous Robot with the Function of Meal Assistance

Shih-Chung Chen, Chih-Hung Hsu, Hsuan-Chia Kuo, and Ilham A.E. Zaeni

Abstract A brain–computer interface (BCI) system is proposed to control an interactive autonomous robot with a function to assist with feeding meals. The subject's electroencephalogram (EEG), regarded as the control command, can be utilized to combine with system integration technologies to establish a BCI control robot system with an automatic feeding function. At present, the integrated technologies of the automatic feeding robot encompasses image recognition, voice recognition, the robot's mechanism design, the gripper, tactile sensor design, etc. The automatic feeding robot can be controlled by steady state visual evoked potential (SSVEP)-based BCI to use the gripper grasping a utensil to ladle food to the subject's mouth successfully. The signal processing algorithm adopted for the SSVEP-based BCI is magnitude squared coherence (MSC). Ten subjects participated in the BCI test for choosing the food on the plate. The average of MSC values for different visual stimulation frequencies were calculated and compared.

Keywords Brain–computer interface (BCI) • Electroencephalogram (EEG) • Steady state visual evoked potential (SSVEP) • Magnitude squared coherence (MSC)

61.1 Introduction

Due to the aging population in the world and the other complicated problems such as the regional war, political conflict, economical decay, and many kinds of accidents, the number of disabled people is increasing. More and more impaired people will result in medical resource problems for the country and economic problems for the family. Many disabled may lose their life goals, self respect, and will to live because of losing self-reliance. The important motivation for this

S.-C. Chen (✉) • C.-H. Hsu • H.-C. Kuo • I.A.E. Zaeni
Department of Electrical Engineering, Southern Taiwan University of Science and Technology, Tainan 71005, Taiwan
e-mail: chung@mail.stust.edu.tw

J. Juang (ed.), *Proceedings of the 3rd International Conference on Intelligent Technologies and Engineering Systems (ICITES2014)*, Lecture Notes in Electrical Engineering 345, DOI 10.1007/978-3-319-17314-6_61

research is to enable the disabled to help themselves have new life goals and improve their life quality by means of a BCI control robot system.

In recent years, many laboratories in different countries have begun to develop brain–computer interface (BCI) systems that provide communication and control capabilities to people with severe motor disabilities. There have been many methods of BCI proposed by researchers, including p 300, motor imagery (event-related (de) synchronization) and SSVEP. The advantages of SSVEP include high information transfer rate and little user training. However these kinds of BCI's usually bring about annoying visual stimuli [1–3]. Recent SSVEP-based spelling systems use a moving cursor to select the target letter which leads to the target character [4]. Details on previous applications such as multiple movement commands include cursor movement via SSVEP were reviewed by Cecotti [5]. The use of LED and SSVEP was implemented in several ways, such as moving a cursor location or selecting a certain target. In larger cases such as a QWERTY keyboard spelling system, SSVEP requires the number of stimulating frequencies to be larger than that of the displayed characters [6].

The topic of this research is related to the development of the steady state visual evoked potential-based (SSVEP-based) BCI control automatic feeding robot. The BCI control structure of the robot with the function of meal assistance is shown in Fig. 61.1. Figure 61.2 shows a picture of the subject trying to choose the food on the plate by SSVEP-based BCI. The SSVEP-based BCI is the most popular issue in brain research [8]. It was used to control an automatic feeding robot to help the severe disabled or people with nerve degeneration have meals by themselves.

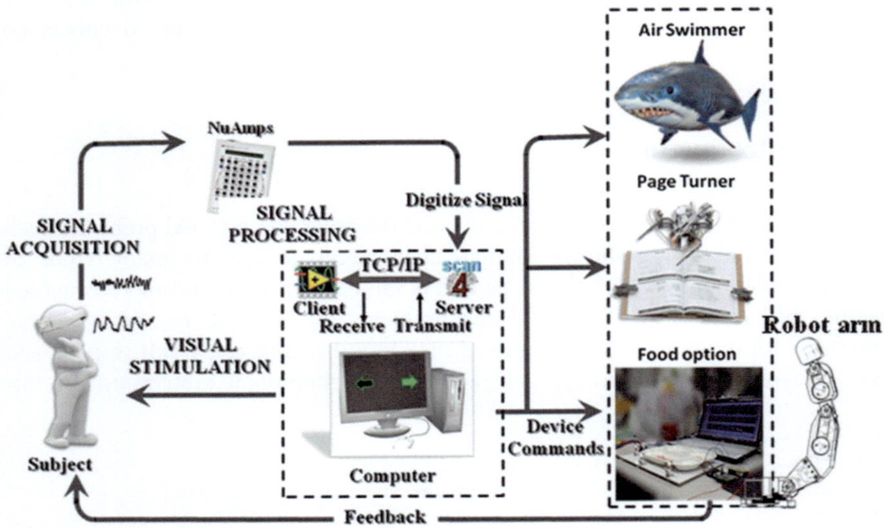

Fig. 61.1 BCI control structure of the robot with the function of meal assistance

Fig. 61.2 The subject was trying to choose the food on the plate by SSVEP-based BCI

BCI allows people to control the hardware operation by their own EEG acquired by the EEG measuring instrument. There are different stimulation methods that can be adopted in BCI applications and visual stimulation is the most popular making the visual SSVEP the choice for this research.

61.2 Materials and Methods

61.2.1 Data Acquisition

Regarding the EEG acquisition, the amplifier NuAmps™ manufactured by Neuroscan company adopted for the research includes 40 channels, 1,000 Hz sampling rate, and 22 bit A/D resolution as shown in Fig. 61.3a. The allocation of electrode follows the international rule of 10–20 electrode placements as shown in Fig. 61.3b. The channel chosen is at the Oz position near the occipital lobe. The electrode for detecting EEG is made of AgCl. The electrode must be injected into conductive gel to increase conductivity before data acquisition procedure. The reference and ground positions are at A1 and A2 near the backbones of two ears, respectively.

61.2.2 Signal Processing

In general, the frequency range of EEG is low frequency, below 30 Hz. The stimulation frequency adopted in the BCI control is between 6 and 10 Hz [9–10]. Therefore, the filtering bandwidth of the EEG amplifier NuAmp is set between 5 and 30 Hz to filter the noise or artefacts from the frequency range between 5 and

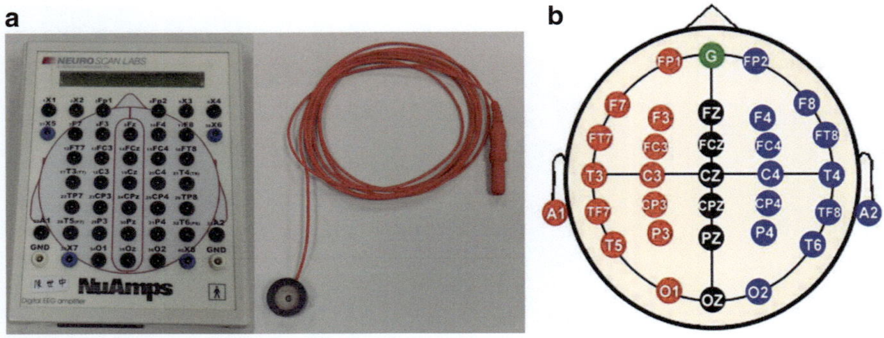

Fig. 61.3 EEG measurement instrument and electrode placements. (**a**) NuAmp EEG amplifier; (**b**) 10–20 electrode placements

30 Hz. The signal is refined again by a software filter programmed in LabVIEW language after the NuAmp hardware's filtering. The LabVIEW software filter is a second-order band pass filter for passing the signal whose frequency is between 5 and 11 Hz. It was used to reduce the effect of the noise or non-stimulation signal.

61.2.3 Feature Extraction

In order to understand the frequency response of EEG, the EEG signal processing is started by calculating the EEG signal's power spectrum using the fast Fourier Transform (FFT) [7] formula as shown in (61.1). Here $\overline{g_n}$ is the EEG after filtering and Ψ_d is the frequency response of the EEG after the FFT calculation in (61.1).

$$\Psi_d = \frac{1}{N}\sum_{n=0}^{N-1}\overline{g_n}\exp^{j\frac{2\pi}{N}nd}, d \in \{0, 1, \ldots, N-1\} \tag{61.1}$$

Next, substitute Ψ_d into (61.2) to calculate the power spectrum E

$$E = \lim_{T\to\infty}\frac{1}{2T}\int_{-T}^{T}\Psi(t)^2\mathrm{d}t \tag{61.2}$$

Five different flickering block stimulations were chosen whose frequencies are 6, 7, 8, 9, and 10 Hz as visual stimulations to help the disabled subject select the different foods on the plate. We can extract the frequency response features by observing the power spectrum of EEG after the identification of the decision model.

61.2.4 Decision Model

There are two kinds of methods for identifying the threshold value after the characteristic value is determined from the calculation result of the power spectrum of EEG. One is the magnitude squared coherence (MSC) method and the other is artificial threshold identification. The MSC method is used to calculate the coherence of two different signals. The MSC method was used to evaluate the coherence of the frequency response of the subject's EEG and the SSVEP model.

$$C_{xy}(f) = \frac{|P_{xy}(f)|^2}{P_{xy}(f)P_{xy}(f)} \tag{61.3}$$

where $P_{xy}(f)$ is the cross-spectral density between x and y, and $P_{xx}(f)$ and $P_{yy}(f)$ are the auto-spectral densities of two signals x and y at frequency f. The value of $C_{xy}(f)$ represents the degree of correlation between x[n] and y[n] at a given frequency f.

Finally, the threshold value is defined according to the coherence calculation result. The other artificial threshold identification means that the EEG feature recognition result will be identified as on or off according to (61.4). For an input EEG signal, a power spectrum, E, is estimated and used to decide a selection, $\hat{w}$, from the options of multilevel selection interface. The decision model is based on posterior probability and defined as

$$\hat{w} = \arg_{w_i}\max \; P(w_i|E) \tag{61.4}$$

If the amplitude of EEG frequency response exceeds the threshold value, then the EEG recognition result should be regarded as "on" otherwise should be "off".

61.3 Results

Four visual stimulations are offered with different flickering frequencies on the screen of user interface to match four kinds of food on the plate. The subject can look at one of four pictures or icons corresponding to four different options for four different kinds of food on the plate to choose the food he/she likes. The subject's EEG can be recognized by BCI transfered to the command of the automatic feeding robot after the procedure of signal processing, feature extraction, and decision model in the experiment.

As shown in Figs. 61.4, 61.5, and 61.6, the food type desired by the subject can be chosen through the procedure of EEG acquisition, power spectrum analysis, and MSC analysis. Ten subjects participated in the BCI tests for the food choice experiments. The result of the MSC analysis is shown in Table 61.1 and Fig. 61.7.

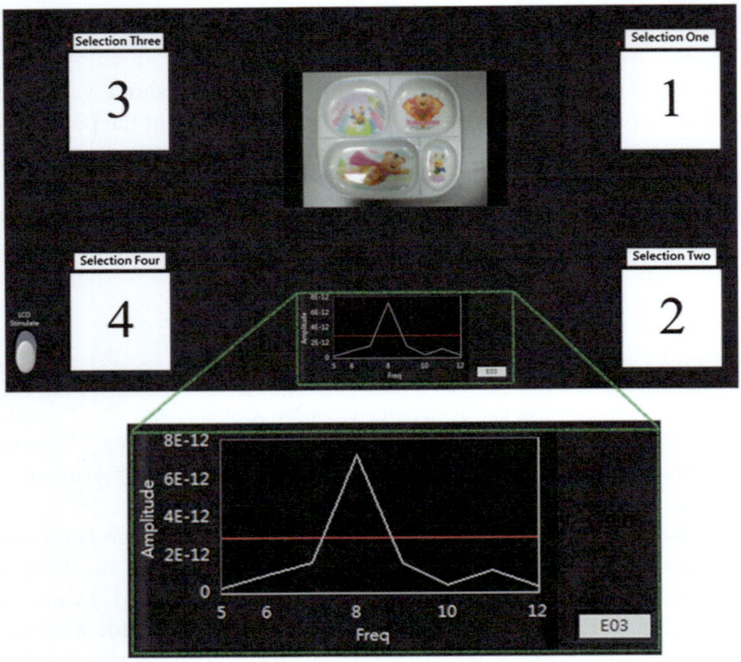

Fig. 61.4 Visual stimulation

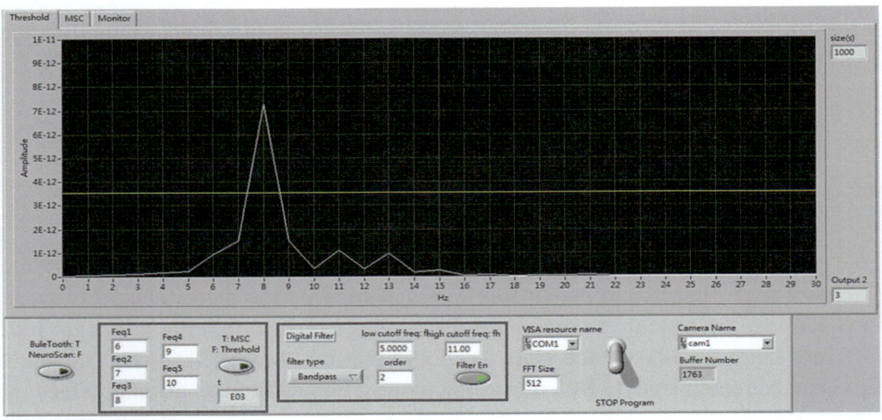

Fig. 61.5 Threshold decision and frequency response of EEG

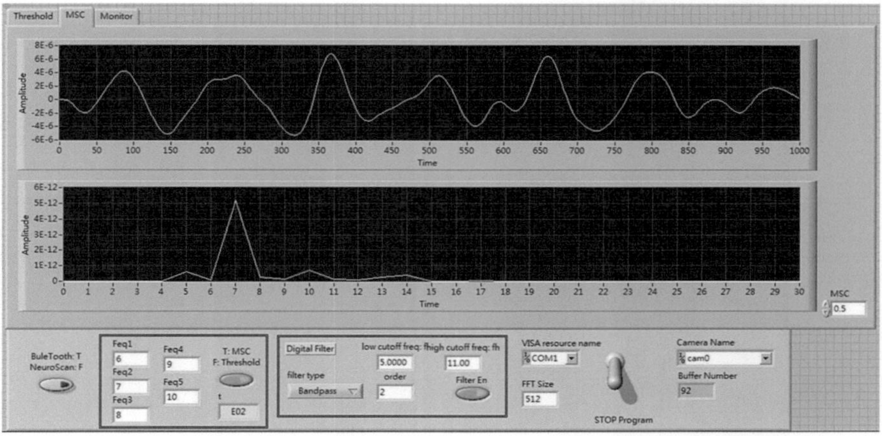

Fig. 61.6 MSC value calculation

Table 61.1 The MSC values for ten normal subjects

	MSC				
Subject	6 Hz	7 Hz	8 Hz	9 Hz	10 Hz
1	0.392	0.386	0.375	0.297	0.295
2	0.338	0.345	0.443	0.346	0.329
3	0.398	0.417	0.412	0.321	0.303
4	0.347	0.414	0.365	0.336	0.272
5	0.361	0.394	0.407	0.324	0.314
6	0.341	0.374	0.361	0.274	0.323
7	0.403	0.441	0.558	0.362	0.359
8	0.392	0.412	0.453	0.393	0.372
9	0.349	0.366	0.341	0.283	0.277
10	0.291	0.387	0.358	0.312	0.294

61.4 Conclusion

The detailed data in Table 61.1 and Fig. 61.7 can be used to further analyze the MSC results of the ten subjects' EEG (30 times each subject). The MSC values of all five different flickering pictures with five different frequencies are distributed among the range between 0.272 and 0.558. The averages of the MSC values stimulated by the 6, 7, and 8 Hz flickering frequencies are bigger than the average of MSC values stimulated by the 9 and 10 Hz flickering frequencies. Therefore, the three frequencies of 6, 7, and 8 Hz should be the better choices to act as the visual stimulations for the SSVEP-based BCI control automatic feeding robot. In the future, the information transfer rate (ITR) will be further studied for the SSVEP-based BCI control robot with the function of meal assistance.

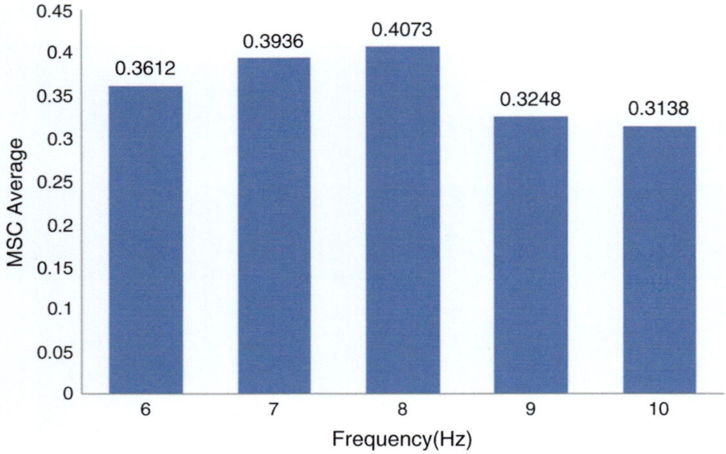

Fig. 61.7 Average of MSC values corresponding to different stimulation frequencies

Acknowledgements We would like to thank the National Science Council and Ministry of Science and Technology for funding this project under grant number NSC-100-2632-E-218-001-MY3 and MOST 103-2218-E-218-002, respectively.

References

1. Cecotti, H.: A self-paced and calibration-less SSVEP-based brain–computer interface speller. IEEE Trans. Neural Syst. Rehabil. Eng. **18**(2), 127–133 (2010)
2. Perego, P., Alamia, A., Maggi, L., Andreoni, G.: BCI keyboards: towards mind writing. Tools for brain–computer interaction (TOBI) workshop, p. 44 (2010)
3. Muller, S.M.T., Celeste, W.C., Bastos, T.F., Sarcinelli, M.: Brain-computer interface based on visual evoked potentials to command autonomous robotic wheelchair. J. Med. Biol. Eng. **30**, 407–415 (2010)
4. Lin, C.T., Lin, F.C., Chen, S.A., Lu, S.W., Chen, T.C., Ko, L.-W.: EEG-based brain-computer Interface for smart living environment auto-adjustment. J. Med. Biol. Eng. **30**, 237–245 (2010)
5. Cecotti, H.: Spelling with non-invasive brain-computer interfaces—current and future trends. J. Physiol. Paris **105**(1–3), 106–114 (2011)
6. Hwang, H.J., Lim, J.H., Jung, Y.J., Choi, H., Lee, S.W., Im, C.H.: Development of an SSVEP-based BCI spelling system adopting a QWERTY-style LED keyboard. J. Neurosci. Methods **208**(1), 59–65 (2012)
7. National Instruments: The fundamentals of FFT-based signal analysis and measurement in LabVIEW and LabWindows/CVI. http://www.ni.com/white-paper/4278/en/ (2012). Accessed 3 Sept 2013
8. Vialatte, F.-B., et al.: Steady-state visually evoked potentials: focus on essential paradigms and future perspectives. Prog. Neurobiol. **90**(4), 418–438 (2010)

9. Chen, S.-C., See, A.R., Chen, Y.-J., et al.: The use of a brain computer interface remote control to navigate a recreational device. Math. Probl. Eng. **2013**, 1 (2013)
10. Chen, S.-C., See, A.R., Yeng, C.-H., et al.: Recreational devices controlled using an ssvep-based brain computer interface (BCI). In: Meen, T.-H., Prior, S.D., Lam, A.D.K.-T. (eds.) Innovation, Communication and Engineering, pp. 175–178. CRC Press, Boca Rotan (2013)

Chapter 62
Design of a Transparent Pipeline-Based Multiplier

Ren-Der Chen and Xiang-Chih Kuo

Abstract This chapter implements an 8×8 multiplier based on the transparent pipeline architecture. A transparent pipeline can lower the power consumption by reducing the number of clock pulses required for data latch controlling. The efficiency of power saving is evaluated here by applying the multiplier to the multiplication of two sparse matrices. It can be seen from the experimental results that, when compared with the traditional synchronous multiplier using flip-flops as storage elements, the improvement in power consumption is obvious only when the sparsity of the matrix reaches a certain amount.

Keywords Transparent pipeline • Multiplier • Sparse matrix

62.1 Introduction

In modern VLSI design, clock power plays an important role in power dissipation. Traditional clock-gating method is an efficient technique widely used in low-power designs [1, 7]. Transparent pipelining is also a clock-gating solution for reducing dynamic power dissipation by reducing the amount of clock pulses which are redundant to the correct operation of a pipelined design [2–4]. In a traditional pipeline, the latches of a pipeline are assumed to be opaque by default to avoid data races between pipeline stages. But in a transparent pipeline, the internal latches are kept transparent by default to allow data items that are sufficiently separated to propagate through the pipeline without generating any clock pulses. Such separation often occurs in pipeline stalls of instruction execution caused by data dependencies. Since data races between latch stages can be avoided by properly separating data items in a transparent pipeline, a transparent latch needs to be clocked only when a true data race occurs. The number of required clock pulses can then be reduced by proper control of the gated-clock signals. Therefore, the

R.-D. Chen • X.-C. Kuo (✉)
Department of Computer Science and Information Engineering, National Changhua University of Education, Changhua 500, Taiwan
e-mail: gi5936128@gmail.com

J. Juang (ed.), *Proceedings of the 3rd International Conference on Intelligent Technologies and Engineering Systems (ICITES2014)*, Lecture Notes in Electrical Engineering 345, DOI 10.1007/978-3-319-17314-6_62

clock power can be significantly reduced by relaxing the clocking requirements of the latches in a pipeline.

A sparse matrix is a matrix in which most of the elements are zero. The fraction of zero elements in a matrix is called the sparsity of the matrix. Large sparse matrices often appear in scientific or engineering applications [5, 6]. The multiplication of sparse matrix can be processed in a specific way to avoid the huge number of trivial operations with zero elements. This chapter incorporates the concept of a transparent pipeline into the design of a multiplier for the multiplication of sparse matrices.

This chapter is organized as follows. Section 62.2 introduces the concept of a transparent pipeline. Section 62.3 gives the multiplier architecture. Section 62.4 includes the experimental results and discussions, and finally Sect. 62.5 concludes the chapter.

62.2 Transparent Pipelines

62.2.1 Behavior of a Transparent Pipeline

A transparent pipeline keeps its first and last latch stages opaque and its internal latch stages transparent by default. Data races of two items propagating concurrently through the transparent pipeline can be avoided by forcing a latch stage between them to enter an opaque mode [4]. However, data items that are sufficiently separated in time can propagate through the pipeline without requiring the latch stages to be clocked. By reducing the number of clock pulses generated in the pipeline, saving of clock power consumption can be achieved.

Table 62.1 illustrates the behavior of a five-stage transparent pipeline with three valid data items entering it separated by two clock cycles. The first stage $S1$ and last stage $S5$, operating in the traditional opaque mode by default, form the input and output environment of the pipeline, respectively. The internal three stages $S2$, $S3$, and $S4$ are the transparent stages. It can be seen from Table 62.1 that only a total of eight clock pulses are required for the three data items A, B, and C to propagate through the whole pipeline, shown as the eight shadowed data items in Table 62.1. Since the architecture is a five-stage pipeline, the latency for each data item is still five clock cycles. Due to the transparent property, there is no need to latch the data entering stages $S2$ and $S3$ for all the clock cycles. For example, when the valid data item A is latched in the first cycle in stage $S1$, it will also propagate through stages $S2$, $S3$, and $S4$ since these three stages are transparent by default. In the second cycle, since there is no valid data item entering $S1$, the value of A can still be kept in $S1$. Stage $S2$, which is transparent at this time, can get the valid data item A supported by $S1$. Therefore, there is no need for $S2$ to latch data item A in this cycle, i.e., no additional clock pulse is needed. For the traditional synchronous counterpart, however, it takes a total of 15 clock pulses (three for each stage) to

Table 62.1 An example illustrating the behavior of a five-stage transparent pipeline

Cycle	Input (Data, Valid)	Pipeline stages				
		Opaque	Transparent			Opaque
		S1	S2	S3	S4	S5
0	(A, 1)	(-, 0)	(-, 0)	(-, 0)	(-, 0)	(-, 0)
1	(-, 0)	(A, 1)	(A, 0)	(A, 0)	(A, 0)	(-, 0)
2	(-, 0)	(A, 0)	(A, 1)	(A, 0)	(A, 0)	(-, 0)
3	(B, 1)	(A, 0)	(A, 0)	(A, 1)	(A, 0)	(-, 0)
4	(-, 0)	(B, 1)	(B, 0)	(B, 0)	(A, 1)	(-, 0)
5	(-, 0)	(B, 0)	(B, 1)	(B, 0)	(B, 0)	(A, 1)
6	(C, 1)	(B, 0)	(B, 0)	(B, 1)	(B, 0)	(A, 0)
7	(-, 0)	(C, 1)	(C, 0)	(C, 0)	(B, 1)	(A, 0)
8	(-, 0)	(C, 0)	(C, 1)	(C, 0)	(C, 0)	(B, 1)
9	(-, 0)	(C, 0)	(C, 0)	(C, 1)	(C, 0)	(B, 0)
10	(-, 0)	(C, 0)	(C, 0)	(C, 0)	(C, 1)	(B, 0)
11	(-, 0)	(C, 0)	(C, 0)	(C, 0)	(C, 0)	(C, 1)

propagate the same number of data items through the pipeline. The number of clock pulses for data latching can hence be reduced from 15 to 8 so that power saving can be achieved by making a pipeline transparent.

62.2.2 Transparent Pipeline Architecture

Figure 62.1 illustrates the architecture of an eight-stage transparent pipeline with six transparent stages. A transparent stage is always clock-gated in transparent mode unless it has to separate two data items propagating concurrently through the transparent segment of the pipeline. In other words, a transparent stage has to switch from transparent mode to clocked mode if there is valid data at its input and there is also valid data either at the input of any upstream transparent stage or at the input of the outside environment. The detection of valid data for mode switching can be implemented by observing both the valid bit feeding into the current stage and the valid bit coming from the upstream pipeline stages by a look-behind function.

Figure 62.2a illustrates a transparent clock generation (TCG) element that supports transparent mode clock-gating. It contains one master and one slave latch internally for generating the clock-gated signals *lm* and *ls*. When the transparent stage is in transparent mode ($gate\_trans = 0$), both *lm* and *ls* are gated in a logic-low state to make the corresponding data latches transparent. When the transparent stage switches to clocked mode, a positive pulse will be generated on *lm* and then, after half a clock cycle, another positive pulse will also be generated on *ls* to latch data travelling through the transparent stages. Likewise, shown in

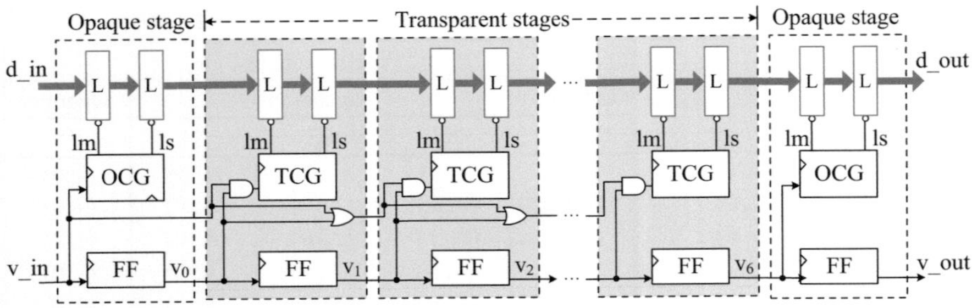

Fig. 62.1 Eight-stage transparent pipeline

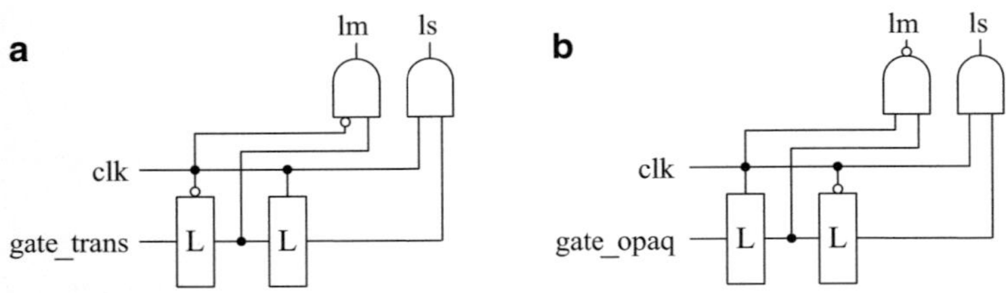

Fig. 62.2 (**a**) Transparent clock generation (TCG) unit. (**b**) Opaque clock generation (OCG) unit

Fig. 62.2b is an opaque clock generation (OCG) element that supports opaque mode clock-gating. When the opaque stage is in opaque mode, *lm* and *ls* are gated in logic-high and -low states, respectively. When it switches to clocked mode, a negative pulse will be generated on *lm* and, after half a clock cycle, another negative pulse will also be generated on *ls* to latch the incoming data item.

62.3 Multiplier Architecture

Shown in Fig. 62.3 is a traditional multiplier architecture using flip-flops as storage elements, i.e., A_0–A_8, P_1–P_8, and B_0–B_7. It is implemented as an eight-stage pipeline with a product generated at each cycle. For a multiplicand a_i to be multiplied by a multiplier b_i, registers A_i and B_i are responsible for storing the values of a_i and b_i, respectively, along the pipeline. At each pipeline stage, a corresponding bit of b_i will determine the shift amount for a_i, and the shifted a_i or zero is then added to the partial product, stored in register P_i. The Acc unit accumulates the partial products generated at each cycle and keeps the outcome of each multiplication.

To convert the traditional multiplier into a transparent pipeline-based architecture, registers from A_1 to A_7 in Fig. 62.3 are replaced by the transparent stages in Fig. 62.1 and the other registers A_0, A_8, P_1–P_8, and B_0–B_7 are replaced by the

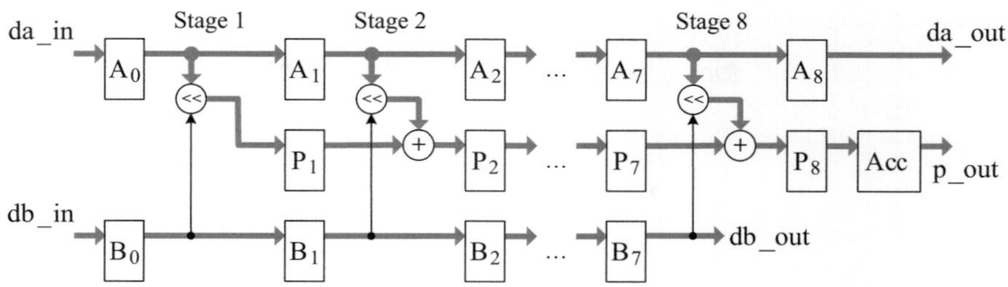

Fig. 62.3 Traditional multiplier architecture using flip-flops as storage elements

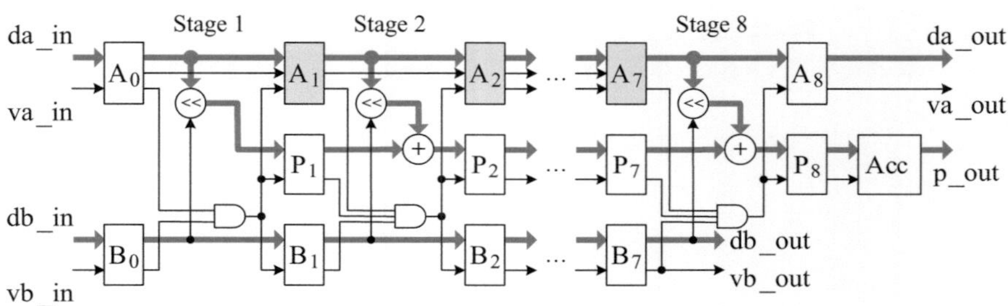

Fig. 62.4 Multiplier architecture based on transparent pipeline

opaque stages. The corresponding eight-stage multiplier based on the transparent pipeline is given in Fig. 62.4. The main difference between a traditional architecture in Fig. 62.3 and a transparent pipeline-based architecture in Fig. 62.4 is that each data item has to be associated with a valid signal to indicate the validity of the corresponding data. Besides that, each transparent stage uses an extra valid signal to indicate the validity of data items in its upstream stages through a chain of OR gate inside the transparent stages. Since there are two data items (A_0 and B_0) entering the first pipeline stage, and three data items (A_i, P_i, and B_i, for $i = 1$–7) entering the other seven pipeline stages, a two or three-input AND logic element is used to join the valid signals of these data.

62.4 Experimental Results

The transparent pipeline-based multiplier for sparse matrix multiplication was coded by the Verilog HDL and synthesized by Synopsys Design Vision with the TSMC 180-nm cell library. A traditional pipeline-based design using ordinary flip-flops as storage elements has also been implemented for comparison. Shown in Table 62.2 is the comparison of circuit area and maximum working frequency, obtained from synthesis. Due to the additional area overhead for control, the

Table 62.2 Comparisons of circuit area and maximum working frequency

Design	Area (#gate count)	Max. working freq. (MHz)
Traditional pipeline-based (FFs)	3,749	261.8
Transparent pipeline-based (latches)	4,922	240.4

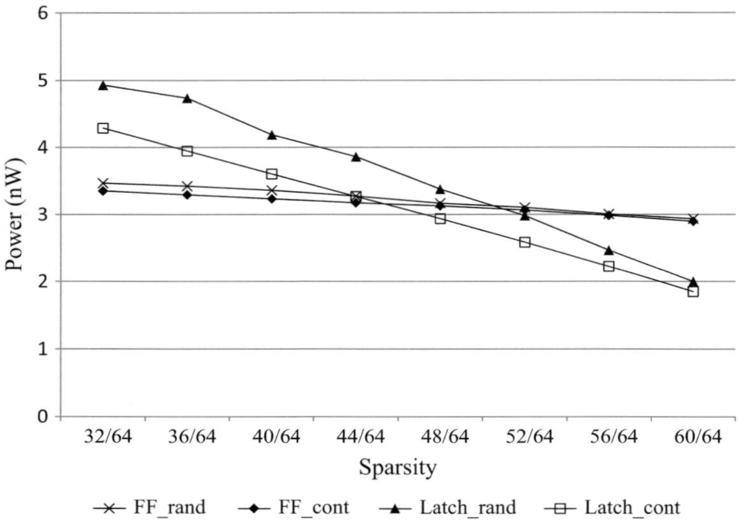

Fig. 62.5 Comparison of power consumption

transparent pipeline-based multiplier uses 1.31 times more area than the traditional pipeline-based design.

The total power consumption was measured by Synopsys PrimeTime PX. Since the efficiency of power saving for sparse matrix multiplication depends majorly on the sparsity of the matrix, various matrices of different sparsity have been evaluated for power comparison. For 8×8 matrix multiplication, since there are 64 elements in a matrix, eight matrices with sparsity $N/64$ will be compared here, where $N = 32$, 36, 40, 44, 48, 52, 56, and 60. Another important factor that determines how power consumption can be improved for a transparent-pipeline-based multiplier is the continuity of zeros in a matrix. For a specific sparsity, two types of matrices with different continuities will be compared. The first one is the best case where all the zeros are continuous in the matrix, called the continuous zero matrix. The other one is a matrix whose zeros are randomly distributed. In this case, ten randomly generated matrix multiplications are adopted to find the average power consumption.

Figure 62.5 shows the comparison of power consumption measured at 100 MHz simulation frequency. In Fig. 62.5, latch_cont and latch_rand refer to the results of the transparent pipeline-based multiplier evaluated by continuous zero and randomly generated matrices, respectively. Likewise, the results of the traditional pipeline-based multiplier evaluated by continuous zero and randomly generated matrices are denoted by FF_cont and FF_rand, respectively. In the continuous zero case, the power consumption of the transparent architecture will be lower than that of the traditional one only when the sparsity of the matrix is equal to or higher than 48/64. In the randomly distributed case, only when the sparsity is equal to or higher than 52/64 will the power consumption of the transparent architecture be lower than that of the traditional one. It can be seen that, the higher sparsity and the more continuous zeros in a matrix, the more power can be saved by using the transparent pipeline architecture.

62.5 Conclusions

In this chapter, a multiplier suitable for sparse matrix multiplication has been implemented based on the concept of a transparent pipeline. In a transparent pipeline architecture, the power consumption can be lowered by reducing the amount of clock pulses required for data latching. For sparse matrix multiplication, the efficiency of power saving is dependent on the matrix sparsity and how the zero elements are distributed in the matrix. The experimental result has shown that this efficiency is obvious only when the sparsity of the matrix reaches a certain amount. How to improve the power-saving efficiency for matrices with lower sparsity will be the future work.

References

1. Benini, L., De Micheli, G.: Automatic synthesis of low-power gated-clock finite-state machines. IEEE Trans. Comput. Aided Design Integr. Circuits Syst. **15**(6), 630–643 (1996)
2. Choi, J.-H., Kim, B.-G., Dasgupta, A., Roy, K.: Improved clock-gating control scheme for transparent pipeline. In: 15th Asia and South Pacific Design Automation Conference (ASP-DAC), Taipei, pp. 401–406 (2010)
3. Hill, E.L., Lipasti, M.H.: Transparent mode flip-flops for collapsible pipelines. In: 25th International Conference on Computer Design (ICCD), Lake Tahoe, pp. 553–560 (2007)
4. Jacobson, H.M.: Improved clock-gating through transparent pipelining. In: International Symposium on Low Power Electronics and Design (ISLPED), Newport Beach, pp. 26–31 (2004)
5. Kuzmanov, G., Taouil, M.: Reconfigurable sparse/dense matrix-vector multiplier. In: International Conference on Field-Programmable Technology (FPT), Sydney, pp. 483–488 (2009)
6. Matam, K., Indarapu, S.R.K.B., Kothapalli, K.: Sparse matrix-matrix multiplication on modern architectures. In: 19th International Conference on High Performance Computing (HiPC), Hyderabad, pp. 1–10. (2012)
7. Wu, Q., Pedram, M., Wu, X.: Clock-gating and its application to low power design of sequential circuits. IEEE Trans. Circuits Syst. I Fundam. Theory Appl. **47**(103), 415–420 (2000)

Chapter 63
An IP-Based Design to Achieve Power Reduction

Chin-Fa Hsieh, Tsung-Han Tsai, and Chih-Hung Lai

Abstract In this chapter, an IP-based design for power reduction on a one--dimension, lifting-based discrete wavelet transform (DWT) is presented. The prototype architecture is coded in VerilogHDL and simulated using Quartus-II to verify the function. Based on this prototype architecture, a low-power operator (adder and subtractor) IP, which is designed based on a full-custom design methodology, plays a role in replacing the main operations. The simulation result shows that power consumption can be reduced by 16.31 %. The architecture can be used as an independent IP core of a wavelet-based application.

Keywords IP • Discrete wavelet transform • VerilogHDL

63.1 Introduction

A wavelet-based video coding technique has recently gained much attention. The wavelet transform can decompose the signals into different subbands preserving both time and frequency properties and make it suitable for the analysis of signals. Based on this technique, discrete wavelet transform (DWT) has been widely applied in many different fields of audio and video signal processing, such as medical images denoising [1], EMG signal denoising [2], electrocardiogram (ECG) signal denoising [3], pattern recognition [4], and image coding [5]. In general, DWT can be implemented by direct convolution and several VLSI DWT architectures applying filtering and convolution technique have been proposed. Up to now, several VLSI implementations have been proposed based on the lifting scheme. An efficient power reduction solution has been proposed in [6–11]. Although these solutions reduce power consumption, it is only addressed on the architecture level to slow the working frequency. That work did not advance power reduction on the transistor level.

C.-F. Hsieh • T.-H. Tsai • C.-H. Lai (✉)
Department of Electrical Engineering, National Central University, Chung-Li 32001, Taiwan
e-mail: ch_lai@dsp.ee.ncu.edu.tw

J. Juang (ed.), *Proceedings of the 3rd International Conference on Intelligent Technologies and Engineering Systems (ICITES2014)*, Lecture Notes in Electrical Engineering 345, DOI 10.1007/978-3-319-17314-6_63

This chapter proposes an efficient VLSI architecture of 1D DWT that integrates a low-power operator IP. The architecture using the low-power operator IP is designed in TSMC 0.18 μm technology.

63.2 Principal of Discrete Wavelet Transform

DWT can decompose the input samples in multiresolution. As shown in Fig. 63.1, the implementation of the DWT is based on the filter banks, where G and H denote a high-pass filter and a low-pass-filter, respectively. After each filtering, the number of output samples is reduced by a factor of 2. The samples generated by the high-pass filters are completely decomposed; meanwhile, the other samples generated by the low-pass filters are applied to the next-level computation for further decomposition. The lifting scheme is a new algorithm proposed for the implementation of the wavelet transform. It can reduce the computational complexity of the DWT involved with the convolution implementation. Furthermore, the extra memory required to store the results of the convolution can also be reduced by in-place computation of the wavelet coefficient with the lifting scheme.

The lifting scheme consists of three steps to decompose the samples: splitting, predicting, and updating. Figure 63.2 illustrates the three steps associated with the

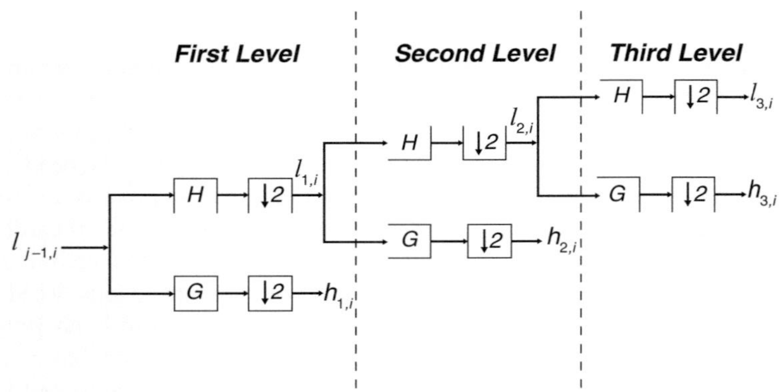

Fig. 63.1 Three-level analysis DWT

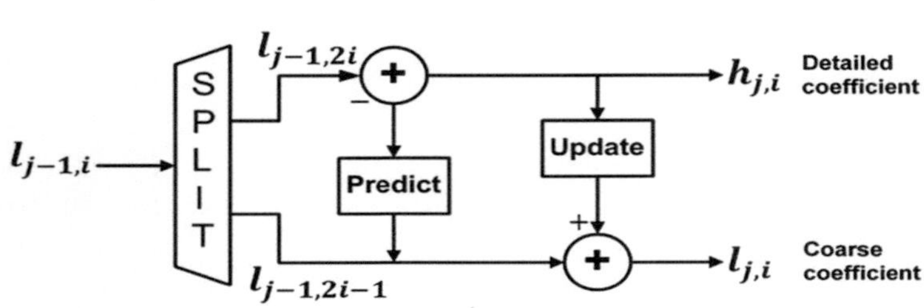

Fig. 63.2 Lifting DWT module

lifting scheme-based DWT for the one-dimensional signal. In the split step (S), the input samples, *l*, are split into even and odd samples. In the predict (P), the even samples are multiplied by the predict factor and then the results are added to the odd samples to generate the detailed coefficients. In the update step (U), the detailed coefficients computed by the predict step are multiplied by the update factors and then the results are added to the even samples to get the coarse coefficients.

63.3 Prototype Architecture

The DWT architecture using the Debuchies (5, 3) filter ((5, 3) DWT) is presented as the design example. The main module of the proposed architecture for the (5, 3) DWT is the predict module and the update module. The function of each part is described as follows: The predict module is shown in Fig. 63.3. As mentioned in Sect. 63.2, the predict module is used to compute the detailed coefficients. The update module is shown in Fig. 63.4. As mentioned in Sect. 63.2, the update module is used to compute the coarse coefficients.

The 1D1L prototype architecture is shown as Fig. 63.5. The main function of the registers (D) is to temporarily hold the data to meet the timing plan. DFF2 and

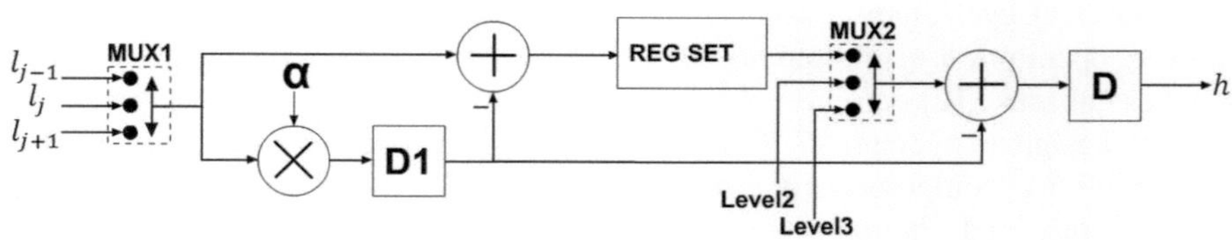

Fig. 63.3 Predict module of (5, 3) filter

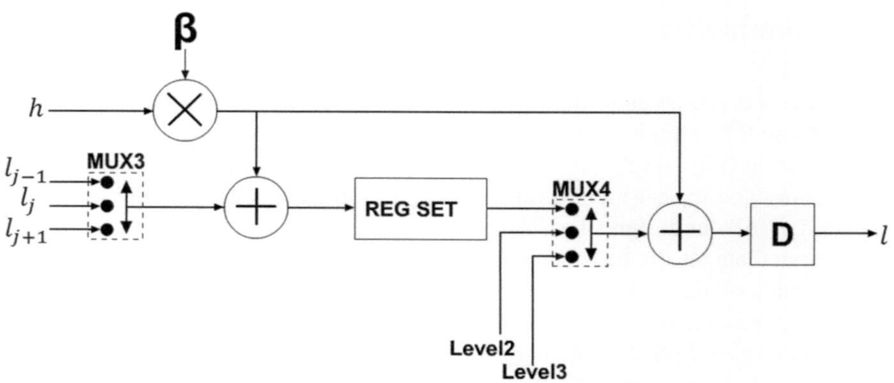

Fig. 63.4 Update module of (5, 3) filter

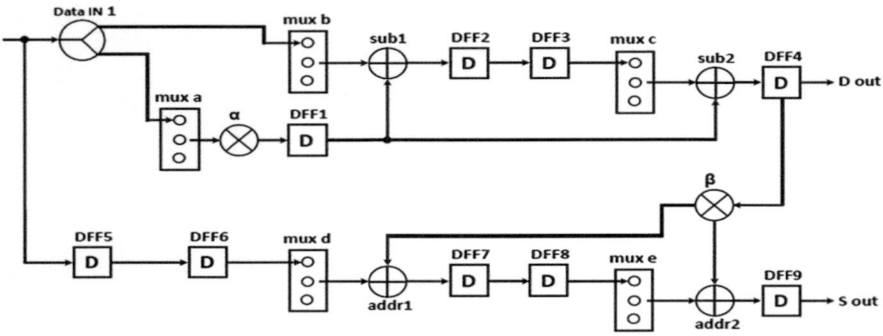

Fig. 63.5 1D1L prototype architecture

DFF3 are used to temporarily store the data for computing the detailed coefficients. DFF7 and DFF8 are used to temporarily store the data for computing the coarse coefficients. Multiplexers can ensure that the correct sample joins the associated computation according to the timing plan. In the architecture proposed in this article, there are five multiplexes denoted as muxa, muxb, muxc, muxd, and muxe. Under the timing control, the multiplexer will guide the samples to the correct path to join the proper computation to get the high-passed or low-passed coefficients. We use VerilogHDL to verify the function. Since the operators operate at every clock cycle, they consume a significant amount of power. In order to reduce power, operators are selected to be replaced by a low-power operator IP. The Chinese abacus [12] can meet the goal of this work. So a full-custom method is used to design an operator IP. The proposed 8-bit abacus adder architecture shown in Fig. 63.6 is composed of the binary-to-abacus (BA) module, parallel addition (PA) module, and Thermometric-to-Binary (TB) module.

63.4 Simulation Result

The proposed prototyping architecture has been designed using VerilogHDL to complete an RTL circuit and is implemented in the FPGA on the real-time platform, as shown in Fig. 63.7. Table 63.1 shows the performance comparisons of prototyping and former architectures reviewed in the literature. Compared to that reported in [6], our design also has the merits of not using external memory to store the intermediate results and of just using a global clock to control the registers. The critical path of our design only requires a single Ta, not two as described in [6]. Because external memory is not used in the present work as it was in [7], the extra memory controller for memory access is not required. As shown in Table 63.2, the proposed architecture in this work can reduce power consumption by 16.31 %.

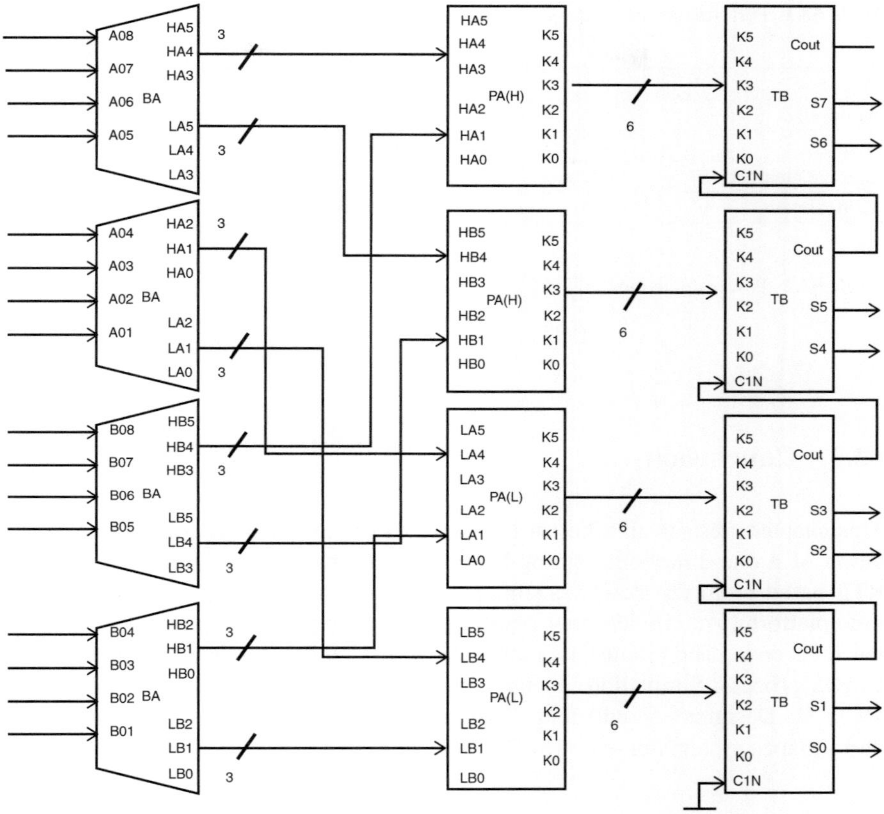

Fig. 63.6 Block diagram of the proposed 8-bit Chinese abacus adder

Fig. 63.7 Real-time platform

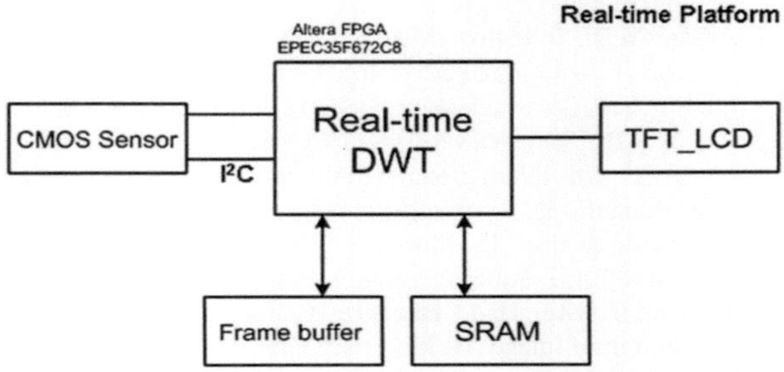

Table 63.1 Performance comparison of RTL architecture

	[6]	[7]	Proposed
Adders	4	4	4
Multipliers	2	2	2
Switch	4	Memory controller	4
Storage element	Register	Memory register	Register
Critical path	2Ta	Ta/memory access time	Ta

Table 63.2 Power reduction

Cell-based (mW)	IP-based (mW)	Power reduction (%)
4.66	3.90	16.31

63.5 Conclusion

This chapter presents an efficient low-power VLSI architecture for the implementation of a one-dimension, lifting-based discrete wavelet transform (DWT). The RTL-based prototype design is simulated using Quartus-II and verified by a real-time platform. An efficient low-power operator IP replaces the main operations in the RTL code. The circuit is simulated in TSMC 0.18 μm single-poly, six-metal CMOS process. Simulation results show a reduced power consumption of about 16.31 %. Designers should find it helpful to use IP blocks to implement a high performance systems-on-chip (SoC).

References

1. Raj, V.N.P., Venkateswarlu, T.: Denoising of medical images using undecimated wavelet transform comparative analysis of 6 bit thermometer-to-binary decoders for flash analog-to-digital converter. In: IEEE International Conference on Recent Advances in Intelligent Computational Systems (RAICS), Trivandrum, pp. 483–488 (2011)
2. Ahsan, M.R., Ibrahimy, M.I., Khalifa, O.O.: VHDL modelling of fixed-point DWT for the purpose of EMG signal denoising. In: IEEE Third International Conference on Computational Intelligence, Communication Systems and Networks (CICSyN), Bali, pp. 236–241 (2011)
3. Raj, V.N.P., Venkateswarlu, T.: ECG signal denoising using undecimated wavelet transform. In: IEEE 3rd International Conference on Electronics Computer Technology (ICECT), Kanyakumari, pp. 94–98 (2011)
4. Kronland-Martinet, R., Morlet, J., Grossman, A.: Analysis of sound patterns through wavelet transform. Int. J. Pattern Recognit. Artif. Intell. **1**(2), 273–302 (1987)
5. Sodagar, I., Lee, H.-J., Hatrack, P., Zhang, Y.-Q.: Scalable wavelet coding for synthetic/natural hybrid images. IEEE Trans. Circuits Syst. Video Technol. **9**, 244–254 (1999)
6. Chen, P.Y.: VLSI implementation for one-dimensional multilevel lifting-based wavelet transform. IEEE Trans. Comput. **53**(4), 386–398 (2004)
7. Andra, K., Chakrabati, C., Acharya, T.: A VLSI architecture for lifting-based forward and inverse wavelet transform. IEEE Trans. Signal Process. **50**(4), 966–977 (2002)

8. Dachasilaruk, S.: Efficient VLSI architectures of lifting-based discrete wavelet transform by systematic design method. In: Proceedings of IEEE International Symposium on Circuits and System, National Taiwan University, Taipei, vol. 5, pp. 565–568 (2002)
9. Xiong, C.-Y., Tian, J.-W., Jian, L.: Efficient high-speed/low-power line-based architecture for two-dimensional discrete wavelet transform using lifting scheme. IEEE Trans. Circuits Syst. Video Technol. **16**(2), 309–316 (2006)
10. Marino, F.: Efficient high-speed/low-power pipelined architecture for the direct 2-D discrete wavelet transform. IEEE Trans. Circuits Syst. II Analog Digit. Signal Process. **47**(12), 1476–1491 (2000)
11. Mohanty, B.K., Meher, P.K.: VLSI architecture for high-speed/low-power implementation of multilevel lifting DWT. In: IEEE Asia Pacific Conference on Circuits and Systems, Singapore, pp. 458–461 (2006)
12. Maloberti, F., Chen, G.: Performing arithmetic functions with the Chinese abacus approach. IEEE Trans. Circuits Syst. II Analog Digit. Signal Process. **46**(12), 1512–1515 (1999)

Chapter 64
Crystalline Indium-Doped Zinc Oxide Thin Films Prepared by RF Magnetron Reactive Sputtering

Chien-Chen Diao, Chia-Ching Wu, and Cheng-Fu Yang

Abstract The characteristic of indium-doped zinc oxide (IZO) thin films are closely related with the composition of the target, the deposition technique, and various process parameters such as the substrate temperature, the deposition pressure, the distance from target to substrate, and so on. In this study, IZO thin films have been deposited onto glass substrates with different deposition pressures by using the radio frequency magnetron reactive sputtering method. The structural, optical, and resistivity properties of IZO thin films were investigated using a field emission scanning electron microscope (FE-SEM), X-ray diffraction patterns (XRD), UV-visible spectroscopy, and Hall-effect analysis. XRD analysis on IZO thin films showed that only the (002) diffraction peak was observable, indicating that the IZO films showed a good *c*-axis orientation perpendicular to the glass substrates. As the deposition pressure of IZO thin films change from 5×10^{-3} to 5×10^{-2} Torr, the thickness decreased from 220 to 72 nm, the grain size increased from 74 to 23 nm, and the resistivity increased from 2.03×10^{-3} to 1.65×10^{-1} Ω cm. The lowest resistivity value of 2.03×10^{-3} Ω m was obtained for a deposition pressure of 5×10^{-3} Torr.

Keywords Transparent conductive oxide • Indium-doped zinc oxide • RF magnetron sputtering • Pressure

C.-C. Diao • C.-C. Wu
Department of Electronic Engineering, Kao Yuan University, Kaohsiung, Taiwan, R.O.C.
e-mail: t10025@cc.kyu.edu.tw

C.-F. Yang (✉)
Department of Chemical and Materials Engineering, National University of Kaohsiung, Kaohsiung, Taiwan, R.O.C.
e-mail: cfyang@nuk.edu.tw

J. Juang (ed.), *Proceedings of the 3rd International Conference on Intelligent Technologies and Engineering Systems (ICITES2014)*, Lecture Notes in Electrical Engineering 345, DOI 10.1007/978-3-319-17314-6_64

64.1 Introduction

Transparent conductive oxide (TCO) thin films are of great importance due to their applications in various electronic and optoelectronic devices, such as solar cells, gas sensors, varistors, and diodes [1–5]. Good TCOs should have a wide optical band gap (>3.5 eV), low electrical resistivity ($<10^{-3}$ Ω cm), high optical transmittance (>80 % in the visible region), and good etching properties. Impurity-doped indium oxides, tin oxides, and zinc oxides are known to satisfy these conditions well [6, 7]. Especially, impurity-doped indium oxide systems, such as tin-doped indium oxide (ITO), have been widely used for numerous optoelectronic applications. However, in order to have high electrical conductivity and high transmittance, ITO thin films must be deposited and then annealed at high temperature (>300 °C). This high temperature makes the ITO films rough due to the crystallization, which leads to significant deterioration of the device reliability [8, 9].

Zinc oxide (ZnO) is an n-type semiconductor with a large binding energy and a wide bandgap. Doped ZnO thin films are promising alternatives to replace ITO thin films as TCOs due to the former's stable electrical and optical properties. The low resistivity of ZnO-based thin films arises from the presence of oxygen vacancies and zinc interstitials [10]. These properties can be improved by doping with appropriate elements. For example, Group III elements such as In^{3+}, Al^{3+}, and Ga^{3+} are used to improve and/or control the electrical conductivity [11–13].

Numerous thin film deposition techniques have been employed by different workers e.g., chemical vapor deposition (CVD) [14], radio frequency magnetron sputtering [15], pulsed laser deposition [16], and spray pyrolysis [17] for the preparation of undoped and doped ZnO films. Indium-doped zinc oxide (IZO) thin films can be deposited by using the RF magnetron sputtering technique which has been widely used due to its advantageous features including the simple apparatus, high deposition rates, and low deposition temperature.

In this work, IZO thin films of a fixed composition (ZnO = 98 mol% and In_2O_2 = 2 mol%) were deposited by radio frequency magnetron sputtering from an oxide target mixed with ZnO and In_2O_3, the composition for which the lowest resistivity was observed when processed at room temperature. The structural, morphological, optical, and electrical characteristics of IZO thin films were characterized by field emission scanning electron microscope (FE-SEM) images, X-ray diffraction (XRD) patterns, UV-visible spectroscopy, and Hall measurements.

64.2 Experimental Procedures

Raw materials (ZnO and In_2O_2, 99.99 % in purity) were weighed according to the composition formula ZnO = 98 mol% and In_2O_2 = 2 mol% (IZO), and ball-milled with deionized water for 1 h. After being dried and ground, the powder was calcined at 800 °C for 2 h and ground again. The calcined powder was uniaxially pressed into a 2-in. plate in a steel die, and sintering was carried out at 1,250 to 1,450 °C in air for 2 h. Substrates were 20 mm × 20 mm × 1 mm Corning 1,737 glass. The substrates were cleaned with acetone, isopropyl alcohol, deionized (D.I.) water, and dried under blown nitrogen gas, before IZO films were deposited on the glass substrates. The working distance between the substrate and target was fixed at 10 cm. The base pressure was 5×10^{-6} Torr and the working pressure was maintained at 5×10^{-3} Torr. The deposition temperature of IZO films was kept at room temperature, the RF power was 100 W, and the deposition times varied from 30 to 90 min. The crystalline structures of IZO films were determined with an X-ray diffractometer using Cu$K\alpha$ radiation ($K = 1.5418$ Å). The surface morphology and thickness of IZO films were measured using the FE-SEM. The mobility, carrier concentration, and resistivity were obtained from Hall-effect measurements using the Van der Pauw method. Optical transmittance was measured by a UV spectrophotometer in the wavelength range of 200–1,100 nm.

64.3 Results and Discussion

The surface field emission scanning electronic micrograph (FE-SEM) images of the IZO ceramics were investigated as a function of sintering temperature, and the results are shown in Fig. 64.1. The grain growths were observed in the IZO ceramics as the sintered temperature increased from 1,250 to 1,450 °C. As the sintered temperature reached 1,450 °C, the micrograph shows a smooth surface on IZO ceramic. It appears that the sintered temperature at 1,450 °C of IZO ceramic is above the ideal values.

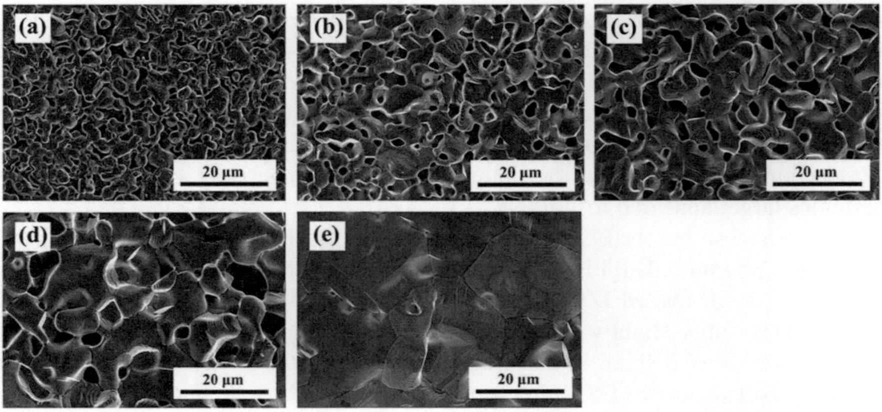

Fig. 64.1 Surface SEM images of the IZO ceramics as a function of sintered temperature. (**a**) 1,250 °C, (**b**) 1,300 °C, (**c**) 1,350 °C, (**d**) 1,400 °C, and (**e**) 1,450 °C

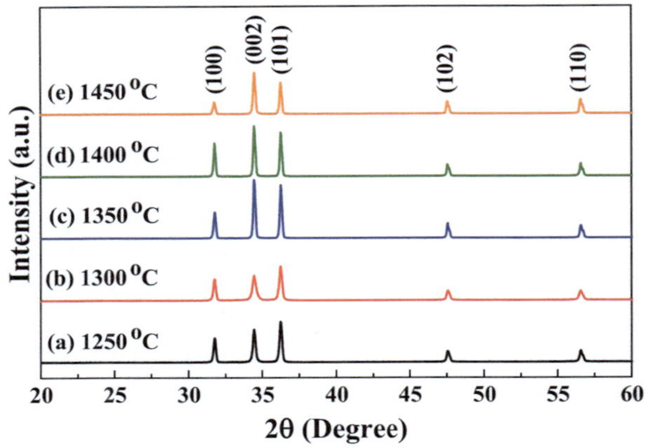

Fig. 64.2 XRD results of the IZO ceramics as a function of sintered temperature. (**a**) 1,250 °C, (**b**) 1,300 °C, (**c**) 1,350 °C, (**d**) 140 °C, and (**e**) 1,450 °C

Fig. 64.3 Surface SEM images of the IZO thin films as a function of deposition pressure (**a**) 5×10^{-3} Torr, (**b**) 1×10^{-2} Torr, and (**c**) 5×10^{-2} Torr

Figure 64.2 shows the XRD patterns of the IZO ceramics sintered at different temperatures. The XRD patterns of the IZO ceramic show the characteristic peaks of the ZnO (JCPDS no. 36-1451) ceramic. Only the hexagonal wurtzite structure is found, no secondary or unknown phases are detectable. The crystalline intensities of the characteristic peaks of IZO at 2θ values increase with the increasing sintered temperature from 1,250 to 1,400 °C. The crystalline intensities of the characteristic peaks of IZO slightly decrease as the sintered temperature at 1,450 °C. This result caused by the sintered temperature at 1,450 °C is higher than the melt point of the IZO ceramic. In this study, the optimum sintered temperature of the 2 in. IZO ceramics target appeared to 1,400 °C.

Figure 64.3 shows the FE-SEM surface images of IZO thin films as a function of deposition pressure. The FE-SEM surface images of the 5×10^{-3}, 1×10^{-2}, and 5×10^{-2} Torr-deposited IZO thin films had similar morphologies. But the grain size of IZO films slightly decreased as the deposition pressure changes from 5×10^{-3} to 5×10^{-2} Torr. The average grain sizes of the 5×10^{-3}, 1×10^{-2}, and 5×10^{-2} Torr-deposited IZO thin films were about 74, 42, and 23 nm, respectively. The cross-section FE-SEM images of IZO thin films with different deposition pressure are shown in inset Fig. 64.4. The thickness of IZO thin films decreased from 220 to 72 nm as the deposition pressure change from 5×10^{-3} to

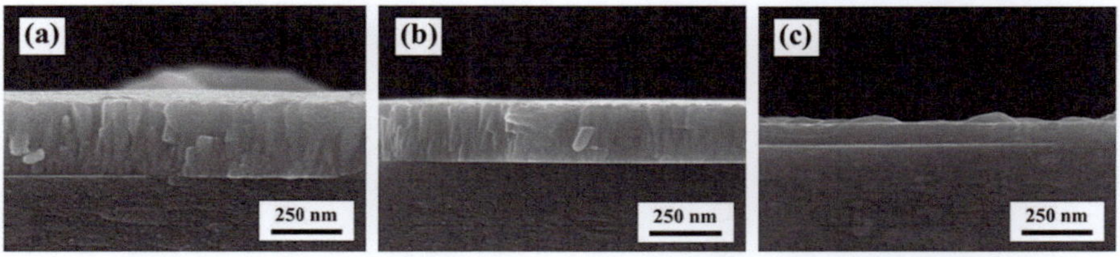

Fig. 64.4 Cross-section SEM images of the IZO thin films as a function of deposition pressure (**a**) 5×10^{-3} Torr, (**b**) 1×10^{-2} Torr, and (**c**) 5×10^{-2} Torr

Fig. 64.5 XRD patterns of IZO thin films as a function of deposition pressure: (*a*) 5×10^{-3} Torr, (*b*) 1×10^{-2} Torr, and (*c*) 5×10^{-2} Torr

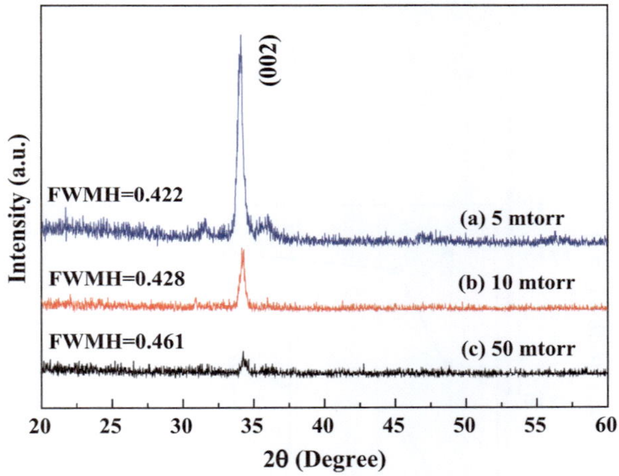

5×10^{-2} Torr. This result was caused by the mean free patch (MFP) becoming shorter as the deposition pressure changed from 5×10^{-3} to 5×10^{-2} Torr. The shorter MFP leads to fewer sputtered particles arriving at the substrate surface. In addition, the crystallization IZO thin films reveal a preferential orientation growth with the columnar structure, as shown in Fig. 64.4a, b.

XRD patterns of IZO thin films with different deposition pressures are shown in Fig. 64.5. All patterns exhibit the (002) peaks of IZO crystallization preferential orientation along the *c*-axis at diffraction angles (2θ) near 34.1°, with a hexagonal structure; no characteristic peak of In_2O_3 phase was found. The intensity of the characteristic peak of IZO thin films decreases as the deposition pressure changes from 5×10^{-3} to 5×10^{-2} Torr.

In the Fig. 64.6, as deposition pressure changes from 5×10^{-2} to 5×10^{-3} Torr, the full width at half maximum (FWHM) values decreased from 0.461 to 0.422. The decrease in the FWHM value suggests the crystallization of IZO thin films increases with deposition pressure change from 5×10^{-2} to 5×10^{-3} Torr. Comparing the diffraction spectra shows that the 2θ value of the (002) peak shifted from 34.25° to 34.11° as the deposition pressure of IZO films changed from 5×10^{-2} to

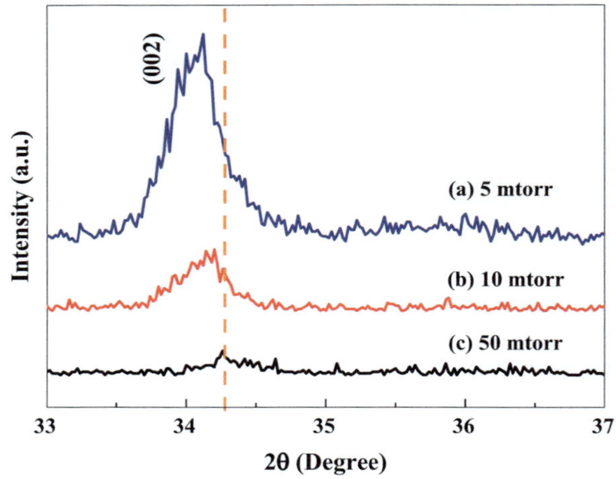

Fig. 64.6 XRD patterns of IZO thin films as a function of deposition pressure: (*a*) 5×10^{-3} Torr, (*b*) 1×10^{-2} Torr, and (*c*) 5×10^{-2} Torr

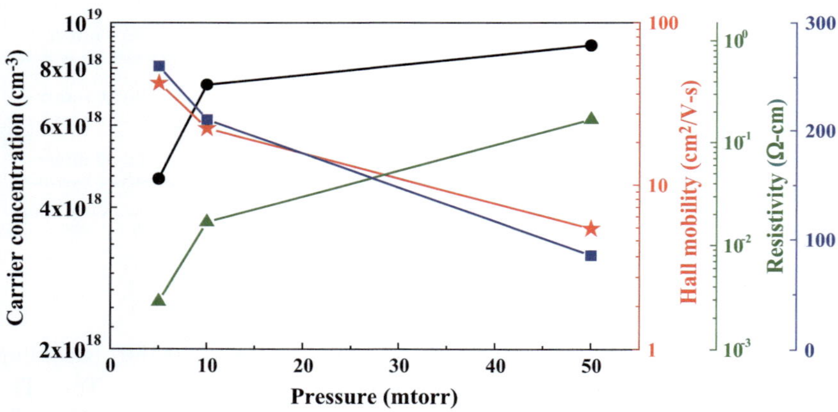

Fig. 64.7 Carrier concentration, resistivity, hall mobility, and thickness of the IZO thin films as a function of deposition pressure

5×10^{-3} Torr. This means that the lattice constant c increased with deposition pressure change from 5×10^{-2} to 5×10^{-3} Torr. This result is caused by that the ionic radius of In^{3+} (80 pm) is being larger than that of Zn^{2+} (74 pm), and the 2θ value of the (002) peak is expected to shift downwards.

Figure 64.7 shows the mobility, carrier concentration, and resistivity of IZO thin films as a function of deposition pressure. As the deposition pressure changes from 5×10^{-3} to 5×10^{-2} Torr, the mobility of IZO thin films decreases from 7.42 to 3.61 cm^2/V s. This result is caused by the larger grain sizes and better crystal quality of IZO thin films at the deposition pressure of 5×10^{-3} Torr. Therefore, a reduced grain boundary barrier is obtained, leading to an increase in carrier mobility. The carrier concentration of IZO thin films slightly increased from 4.82×10^{18} to

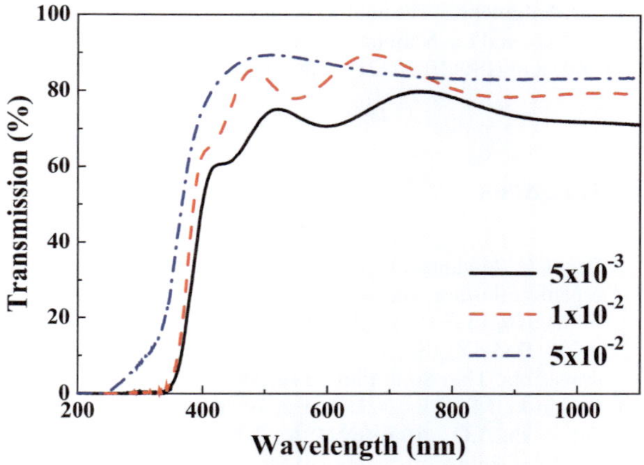

Fig. 64.8 Transmittance spectra of the IZO thin films as a function of deposition pressure

8.97×10^{18} cm^{-3} as the deposition pressure changed from 5×10^{-3} to 5×10^{-2} Torr. The resistivity of IZO thin films increased from 2.03×10^{-3} to 1.65×10^{-1} Ω cm when the deposition pressure changed from 5×10^{-3} to 5×10^{-2} Torr.

The optical transmittance spectra of IZO thin films in the wavelength range of 200–1,100 nm are shown in Fig. 64.8. The average transmittance rate of the IZO thin film is about 84 % in the 400–700 nm range for a deposition pressure of 5×10^{-2} Torr, and the average transparency of IZO thin films slightly decreased as the deposition pressure change to 5×10^{-3} Torr. In the ultraviolet range, all the IZO thin films showed a sharp absorption edge and exhibited a stronger shift phenomenon as the deposition pressure changed.

64.4 Conclusions

In this study, crystalline indium–zinc oxide (IZO) thin films were deposited on glass substrates using the RF magnetron sputtering method. The (200) preferred orientation of IZO thin films increased as the deposition pressure changed from 5×10^{-2} to 5×10^{-3} Torr, which would cause better conductive properties in IZO thin films. The resistivity of IZO thin films increased from 2.03×10^{-3} to 1.65×10^{-1} Ω cm as the deposition pressure changes from 5×10^{-2} to 5×10^{-3} Torr. The average transmittance rates of IZO thin films in the wavelength range from 400 to 800 nm were 84, 82, and 78 % as the deposition pressure changed from 5×10^{-2} to 5×10^{-3} Torr. These results suggest that the deposition pressure of IZO thin films at 5×10^{-3} Torr is better.

Acknowledgments The authors acknowledge the financial support of the Ministry of Science and Technology and the National Science Council of MOST 103-2221-E-244-018, NSC 102-2511-S-244-001, and NSC 102-2221-E-244-019.

References

1. Orgiu, E., Manunza, I., Sanna, M., Cosseddu, P., Bonfiglio, A.: Transparent dielectric films for organic thin-film transistors: a perspective for low cost, low size technologies. Thin Solid Films **516**, 1533–1537 (2008)
2. Paine, D.C., Yaglioglu, B., Beiley, Z., Lee, S.: Amorphous IZO-based transparent thin film transistors. Thin Solid Films **516**, 5894–5898 (2008)
3. Berginski, M., Hüpkes, J., Reetz, W., Rech, B., Wuttig, M.: Thin solid films (2007, in press)
4. Dikovska, A.O., Atanasov, P.A., Tonchev, S., Ferreira, J., Escoubas, L.: Periodically structured ZnO thin films for optical gas sensor application. Sensors Actuators A **140**, 19–23 (2007)
5. Suvaci, E., Ozer, I.O.: Processing of textured zinc oxide varistors via templated grain growth. J. Eur. Ceram. Soc. **25**, 1663–1673 (2005)
6. Haacke, G.: Transparent conducting coatings. Annu. Rev. Mater. Sci. **7**, 73–93 (1977)
7. Chena, K.J., Fang, T.H., Hung, F.Y., Ji, L.W., Chang, S.J., Young, S.J., Hsiao, Y.J.: The crystallization and physical properties of Al-doped ZnO nanoparticles. Appl. Surf. Sci. **254**, 5791–5795 (2008)
8. Taga, N., Maekawa, M., Shigesato, Y., Yasui, I., Kamei, M., Haynes, T.E.: Deposition of heteroepitaxial In_2O_3 thin films by molecular beam epitaxy. Jpn. J. Appl. Phys. **37**, 6524–6529 (1998)
9. Tuna, O., Selamet, Y., Aygun, G.: High quality ITO thin films grown by dc and RF sputtering without oxygen. J. Phys. D Appl. Phys. **43**, 055402–055409 (2010)
10. Chen, X.L., Geng, X.H., Xue, J.M., Li, L.N.: Two-step growth of ZnO films with high conductivity and high roughness. J. Cryst. Growth **299**, 77–81 (2007)
11. Chang, H.P., Wang, F.H., Wu, J.Y., Kung, C.Y., Liu, H.W.: Enhanced conductivity of aluminum doped ZnO films by hydrogen plasma treatment. Thin Solid Films **518**, 7445–7449 (2010)
12. Le, H.Q., Lim, S.K., Goh, G.K.L., Chua, S.J., Ong, J.X.: Optical and electrical properties of Ga-doped ZnO single crystalline films grown on $MgAl_2O_4$ (111) by low temperature hydrothermal synthesis semiconductor devices, materials, and processing. J. Electrochem. Soc. **157**, H796–H800 (2010)
13. Huang, C.C., Wang, F.H., Wu, C.C., Huang, H.H., Yang, C.F.: Developing high-transmittance heterojunction diodes based on NiO/TZO bilayer thin films. Nanoscale Res. Lett. **8**(1), 206 (2013)
14. Wang, A.C., Dai, J.Y., Cheng, J.Z., Chudzik, M.P., Marks, T.J.: Charge transport, optical transparency, microstructure, and processing relationships in transparent conductive indium-zinc oxide films grown by low-pressure metal-organic chemical vapor deposition. Appl. Phys. Lett. **73**, 327–329 (1998)
15. Minami, T., Yamamoto, T., Toda, Y.: Transparent conducting zinc-co-doped ITO films prepared by magnetron sputtering. Thin Solid Films **373**, 189–194 (2000)
16. Naghavi, N., Marcel, C., Dupont, L., Rougier, A., Leriche, J.B., Guery, C.: Structure and physical characterization of transparent conducting pulsed laser deposited In_2O_3-ZnO thin films. J. Mater. Chem. **10**, 2315–2319 (2000)
17. Pasquarelli, R., Hest, M., Miedaner, A., Curtis, C., Perkins, J., Berry, J., O'Hayre, R., Ginley, D.: Solution deposition of amorphous IZO films by ultrasonic spray pyrolysis. In: Conference Record of the IEEE Photovoltaic Specialists Conference, Colorado School of Mines, Golden (2009)

Chapter 65
The Construction and Efficiency Validation of an Educational Module of Pneumatics Robotics Arm Controlled by Programmable Logic Controller (PLC)

Pornpirom Fak-orn, Utjanapol Yomkurd, Natawat Punfong, and Mana Thanaon

Abstract This chapter presents construction and efficiency validation of an educational module of pneumatics robotics arm controlled by programmable logic controller (PLC). This module was a student project in bachelor of science in technical education in electrical engineering, that also expected to be used in a subject, PLC, of high vocational certificate or diploma (electrical power) curriculum at Rajamangala University of Technology Lanna (RMUTL). The module consists of three main systems, pneumatics, robotic arm, and PLC, integrated in as a laboratory station. PLC Omron CPM2A model was selected to play as controller here but not limited. Atleast there lab sheets were set and tested. The verification and validation have been done in three topics, body, usage, and experiment through 5 experts and 20 students in the second year of diploma. The reports from all experts and all topics show the average *t*-score at 4.52 and the average score from students at 4.535.

Keywords Pneumatics • Robotics arm • Programmable logic controller (PLC)

65.1 Introduction

To learn how does a programmable logic controller (PLC) work and make students deeply understand, it needs to demonstrate correctly and let those students reach to sufficient education material. Unfortunately, many vocational schools in Thailand still lacks some educational modules in skill training, such as hydrolic system, pneumatic system, and also PLC system because it is very expensive and inadequate for registered students. With the reasons, bachelor students in final year of

P. Fak-orn • U. Yomkurd • N. Punfong • M. Thanaon (✉)
Department of Electrical Engineering, Rajamangala University of Technology Lanna, Muang, Tak 63000, Thailand
e-mail: thanaon1983@gmail.com

J. Juang (ed.), *Proceedings of the 3rd International Conference on Intelligent Technologies and Engineering Systems (ICITES2014)*, Lecture Notes in Electrical Engineering 345, DOI 10.1007/978-3-319-17314-6_65

RMUTL Tak have an idea to propose their project to construct an educational module for diploma students studying in PLC subject applied to pneumatic robotics arm. After the proposed educational module is built, it will be treated for verification by some experts and registered diploma students.

This chapter consists of four topics, introduction, educational module, verification, and conclusion as follows.

65.2 Educational Module

The proposed educational module mounted on a table, has size 70 cm × 55 cm × 60 cm shown in Fig. 65.1. The top side is install robotic arm, including many sensors and actuators. Pneumatic panel is attached on the left hand side below and the other right hand side is installed with the Omron PLC and its peripherals, connectors and switches.

65.2.1 Robotic Arm

Robotic arm of SMC company [1] as shown in Fig. 65.2a stands on a sliding plate which can be moveable in horizontal range along pneumatic cylinder. It also can

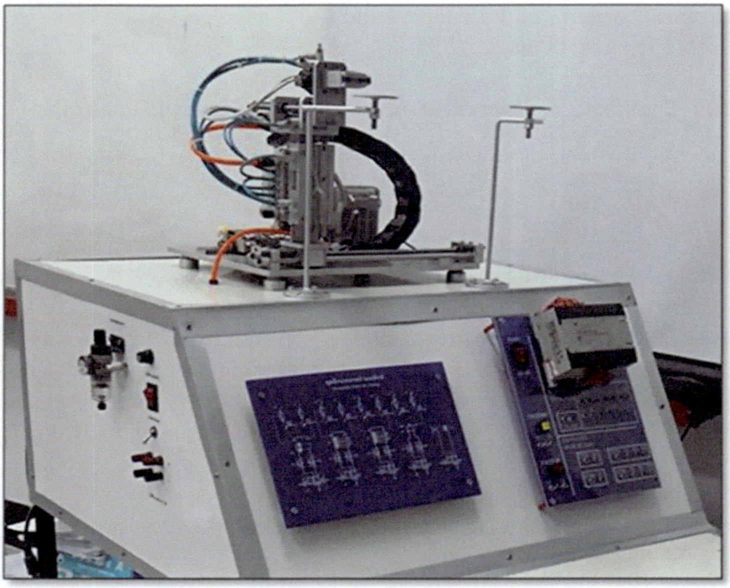

Fig. 65.1 Proposed educational module, robotic arm (*above*), pneumatic panel and PLC panel (*left* and *right*)

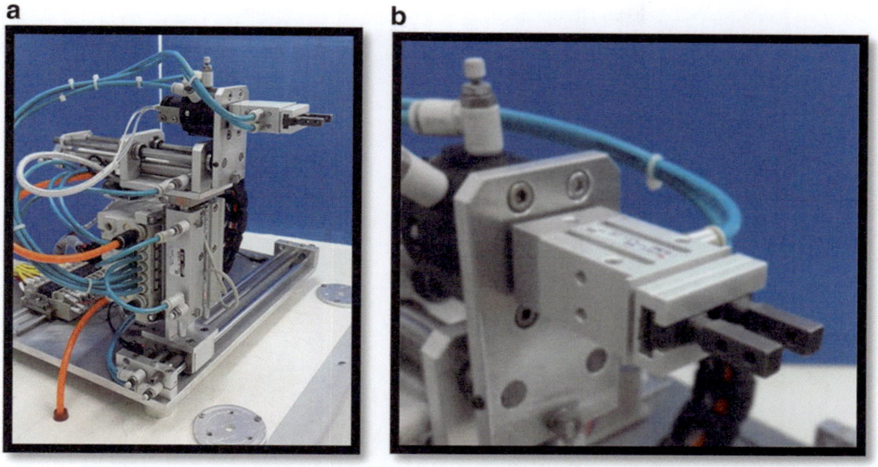

Fig. 65.2 Robotic arm (a), gripper (b)

move up and down along vertical axis. The end of the arm connected with a rotational gripper over 360° for picking an object, see detail of the gripper in Fig. 65.2b.

65.2.2 Pneumatic Panel

Pneumatic panel shown in Fig. 65.3 is designed for learning mechanics of pneumatic system installed here. First cylinder from left hand side is to control horizontal position (x-axis), second for vertical position (y-axis), third for z-axis, fourth for gripper rotation and the last on the right hand side for gripping task.

65.2.3 PLC Panel

Omron PLC CPM2A-30CDR-A model used here has 18 inputs, 12 output, and AC supply requirement. Software for PLC programming is CX-programmer [2].

65.2.4 Lab Sheets

Three laboratories were set for using the educational module, such as basic operation of pneumatic system, basic operation of robotic arm, and robotic arm control by PLC.

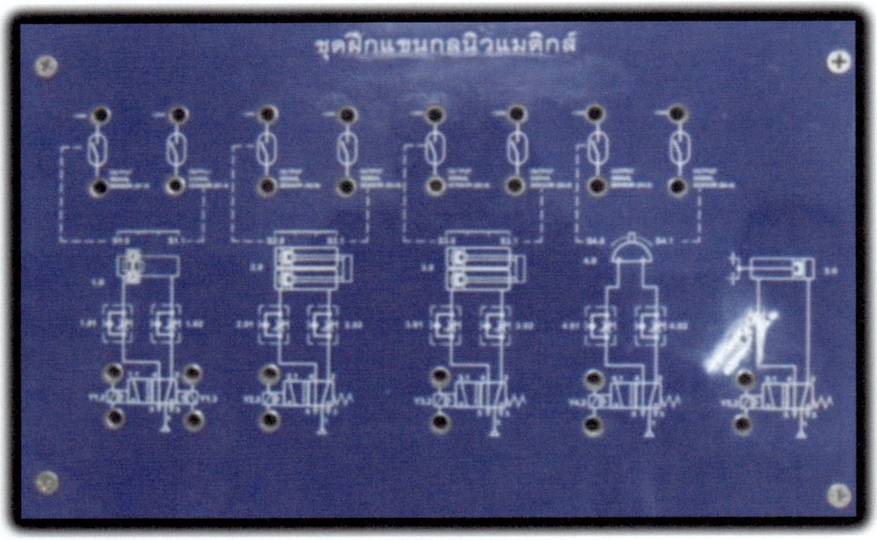

Fig. 65.3 Pneumatic panel

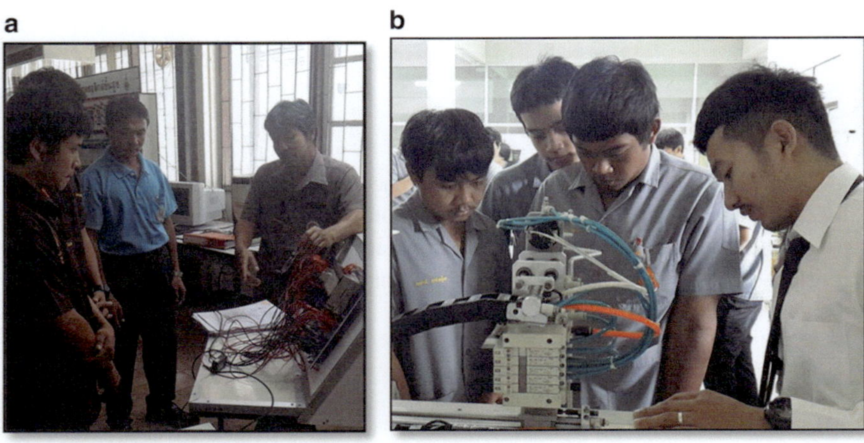

Fig. 65.4 Verification of the educational module by (**a**) experts and (**b**) students

65.3 Verification and Validation

Based on curriculum development and teaching strategy [3, 4], the verification and validation have been investigated in three items, body, usage, and experiment through 5 experts and 20 students in the second year of diploma shown in Fig. 65.4.

Table 65.1 Verification results from 5 experts via *t*-test

Items	Mean scores	Standard deviation	Meaning
Body	4.60	0.53	Excellent
Usage	4.60	0.40	Excellent
Experiment	4.35	0.79	Very good
Average	4.52	0.57	Excellent

Table 65.2 Verification results from 20 students via *t*-test

Items	Mean scores	Standard deviation	Meaning
Body	4.58	0.57	Excellent
Usage	4.78	0.47	Excellent
Average	4.68	0.52	Excellent

Five experts have been evaluated in the educational module and all results are summarized in Table 65.1, the mean score is 4.52. We can interpret that it is excellent. Range of score is illustrated in Appendix.

The students have been divided into two groups, treatment group and control group, for comparison purpose. The obtained results show in Table 65.2, the mean score of both items, body and usage, are over 4.5 satisfying excellent level.

65.4 Conclusion

This chapter presents construction and efficiency validation of an educational module of pneumatics robotics arm controlled by PLC. This module was a student project in bachelor of science in technical education in electrical engineering, that also expected to be used in a subject, PLC, of high vocational certificate. The module consists of three main systems, pneumatics, robotic arm and PLC, integrated in as a laboratory station. There lab sheets were set and tested. The verification and validation have been done in three topics, body, usage and experiment through 5 experts and 20 students in the second year of diploma. The results are ranked in excellent level and also meet the project objectives.

Acknowledgments The authors would like to thank all staff of department of electrical engineering for their kind help and Rajamangala University of Technology Lanna Tak for partial financial support.

Appendix: Range of Scores

Score 4.50–5.00 means "Excellent"
Score 3.50–4.49 means "Very good"
Score 2.50–3.49 means "Good"
Score 1.50–2.49 means "Fair"
Score 1.00–1.49 means "Poor"

References

1. SMC World. http://www.smcworld.com/products/en/
2. Omron Industrial Automation. http://industrial.omron.eu/en/products/
3. Wongyai, W.: Modern Techniques in Curriculum Development and Teaching Strategy. Odeon Store, Bangkok (1982) (in Thai)
4. Wongrat, T.: Application of SPSS 17 to Statistical Analysis. Se-Ed Book, Bangkok (2012) (in Thai)

Chapter 66
GPU Computations on Hadoop Clusters for Massive Data Processing

Wenbo Chen, Shungou Xu, Hai Jiang, Tien-Hsiung Weng, Mario Donato Marino, Yi-Siang Chen, and Kuan-Ching Li

Abstract Hadoop is a well-designed approach for handling massive amount of data. Comprised at the core of the Hadoop File System and MapReduce, it schedules the processing by orchestrating the distributed servers, providing redundancy and fault tolerance. In terms of performance, Hadoop is still behind high performance capacity due to CPUs' limited parallelism, though. GPU accelerated computing involves the use of a GPU together with a CPU to accelerate applications to data processing on GPU cluster toward higher efficiency. However, GPU cluster has low level data storage capacity. In this chapter, we exploit the hybrid model of GPU and Hadoop to make best use of both capabilities, and the design and implementation of application using Hadoop and CUDA is presented through two interfaces: Hadoop Streaming and Hadoop Pipes. Experimental results on *K*-means algorithm are presented as well as their performance results are discussed.

Keywords Hadoop • GPU • HPC • Massive data processing

W. Chen • S. Xu
School of Information and Technology of Lanzhou University, Lanzhou, China
e-mail: chenwb@lzu.edu.cn; xushg12@lzu.edu.cn

H. Jiang
Department of Computer Science, Arkansas State University, Jonesboro, AR, USA
e-mail: hjiang@astate.edu

T.-H. Weng (✉) • Y.-S. Chen • K.-C. Li
Department of Computer Science and Information Engineering, Providence University, Taichung, Taiwan
e-mail: thweng@pu.edu.tw; kuancli@pu.edu.tw

M.D. Marino
Piazzale Umbria 15, Sanfatucchio (PG) 06060, Italy

J. Juang (ed.), *Proceedings of the 3rd International Conference on Intelligent Technologies and Engineering Systems (ICITES2014)*, Lecture Notes in Electrical Engineering 345, DOI 10.1007/978-3-319-17314-6_66

66.1 Introduction

Hadoop is an outstanding open-source MapReduce framework for storing and processing big data in a distributed fashion on large clusters of commodity hardware. Hadoop splits a big task into small tasks and then distributed these small tasks to the whole cluster computing nodes. It has a high performance distributed file system with fault tolerance capabilities. But there is a kind of massive data processing that can hardly be processed by current cloud computing technology which has both data-intensive and computing-intensive traits since the computing capacity of each of the computing node cannot meet our needs. GPU has evolved into general-purpose computing [9]. Its computing capacity is far beyond CPU, this makes it more popular to be used in general-purpose acceleration tasks. Thus, it is necessary to migrate the intensive computing from CPU to GPU to bring up computational power.

This chapter takes advantage of Hadoop and CUDA for Massive Data Processing, and makes the following contributions:

1. Implementation of application using Hadoop and CUDA for heterogeneous computing is presented.
2. Two methods of Hadoop using CUDA were proposed.
3. The performance and effectiveness of the two methods are compared and analyzed via the experimental results.

The organization of this chapter is as follows. Section 66.2 reviews the related works. Hadoop is written in Java and GPU code is essentially written in C/C++, this brings us lots of difficulties porting it to GPU. To realize this, we proposed two methods and used *K*-Means clustering algorithm as an example in Sect. 66.3. In Sect. 66.4, we analyzed and compared our experimental results. Finally, the conclusion and future work are given in Sect. 66.5.

66.2 Related Work

These years, there are a lot of studies that have been done to port distributed MapReduce to CPU–GPU architectures. In Map phase of MapReduce, the computing is totally parallel among map functions. In Combine phase, there is also partial parallelism.

It is likely to think that running the Map and Combine phases on GPUs will improve the performances. MGMR [1, 2] is a version of MapReduce on multi-GPUs environment, and PMGMR is an updated and pipelined version of MGMR [3], which employs multiple GPUs and makes use of new features of GPU. Mars [4] is the first framework for distributed MapReduce on CUDA, but it is restricted to single node system.

Hadoop implementation is independent of CUDA. However, recent studies indicate that this framework is likely to incorporate CUDA interface elements to explore its performance parallelization benefits [10]. In [5], it presented a novel parallel implementation of RSA algorithm using JCUDA and Hadoop, where RSA algorithm is encapsulated in JNI, and used JCUDA to implement for parallelization. It is depicted in [6], a novel approach for processing very large files by using Hadoop in handling file splitting and merging, and CUDA for accelerating the computation of the FFT algorithm, in which JCUFFT and JTransforms were applied. In [7], streaming, JCuda, JNI were mentioned, and the authors employ JNI method at last. However, all previous works are specifically focused on image processing discussion; do not approach the integration between GPU CUDA interface and Hadoop.

66.3 Methods of Integrating GPU with Hadoop

In this section, we first introduce Streaming and Pipes in detail. The two methods are tested in our experiment in order to integrate GPU with Hadoop by using *K*-means algorithm. First, according to MapReduce model, we divided *K*-means into two parts which correspond to Map phase and Reduce phase respectively. Next, we accelerated one part on CUDA. Finally, we use the two methods to integrate CUDA code with Java code in Hadoop.

66.3.1 K-*Means in Hadoop*

In Hadoop, we divided *K*-means into two parts: (a) Find nearest cluster for each sample. This part corresponds to the Map stage of Hadoop. The output <key, value> pair of map is <nearest cluster, sample>. (b) Update its centroids for each cluster. This part is accelerated by GPU, and this part is corresponding to the Reduce stage of Hadoop. Figure 66.1 shows the flow chart of *K*-means in Hadoop.

66.3.2 Hadoop Interface Programming Methods

As known, the Hadoop framework is implemented in Java. While Hadoop provides two mechanisms to support the use of other languages: Streaming and Pipes. These two programming methods will be described in detail in this chapter later; they all provide C/C++ MapReduce programming interfaces. So utilizing CUDA in Hadoop is feasible. The Map and Reduce functions just act as proxies, the actual calculating work is done on GPU.

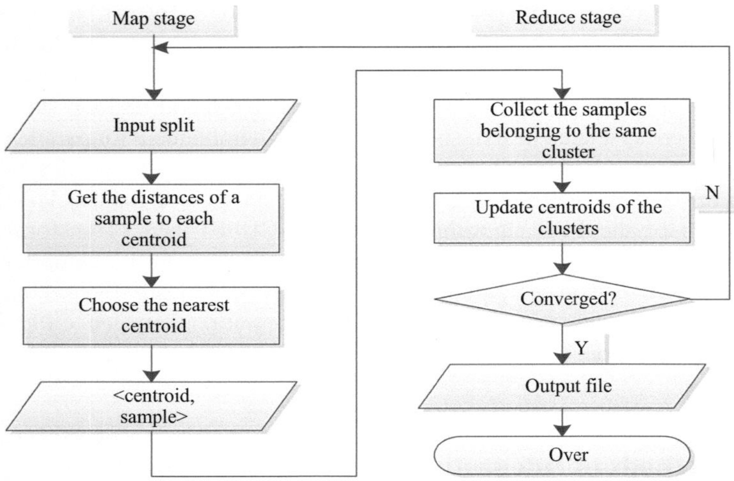

Fig. 66.1 *K*-means flowchart in Hadoop

66.3.2.1 Streaming

Hadoop Streaming is a utility that allows users to create and run Map/Reduce jobs with any executable or script as the mapper and/or the reducer. Mapper and reducer are executables that read the input from stdin (line by line) and emit the output to stdout [8]. Figure 66.2 is the flowchart of streaming MapReduce. When an executable is specified for mappers, each mapper task will launch the executable in a separate process. When the mapper task is running, its inputs are split into lines which are fed to the standard input of the process. At the same time, mapper collects the lines from the standard output of the process and converts them into <key, value> pairs as mappers' outputs.

66.3.2.2 Pipes

Hadoop Pipes provide a C/C++ library to support Hadoop programs in C/C++ only. C/C++ application is launched as a subprocess of the Java task. C/C++ code communicates with Java code through socket. In many ways, Hadoop Pipes are similar to Hadoop streaming, the differences between them are the methods of communication: one is standard input and output, the other is sockets. Hadoop Pipes allow users to define five basic components: mapper, reducer, partitioner, combiner, and recordReader. Figure 66.3 shows the design detail of mapper and reducer component of Hadoop Pipes mechanism.

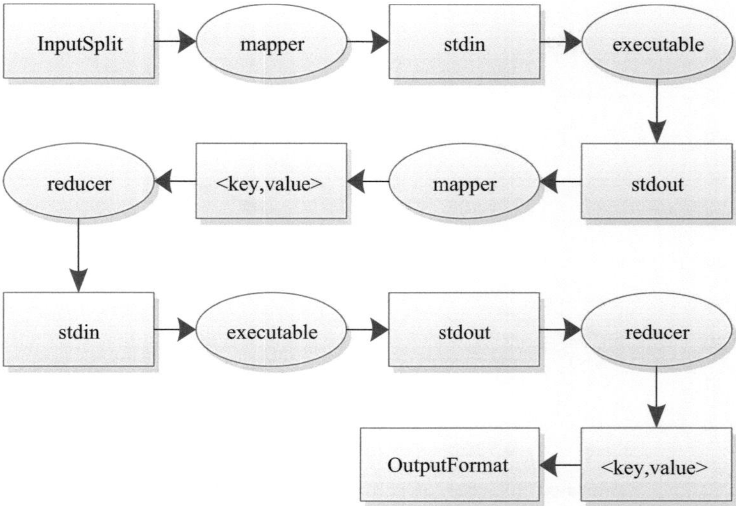

Fig. 66.2 Hadoop streaming MapReduce

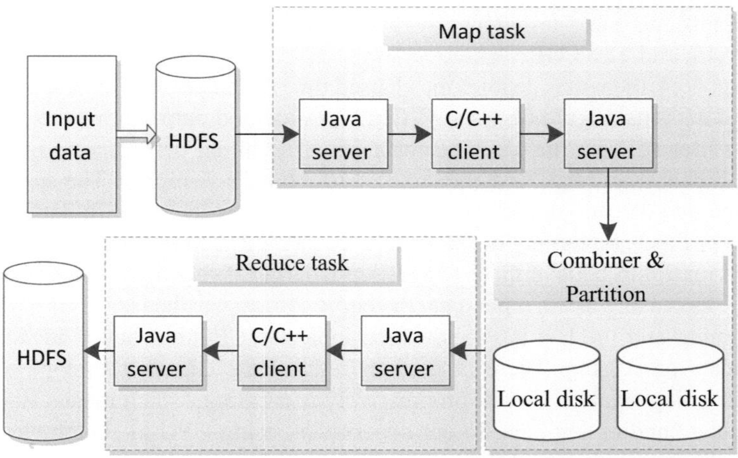

Fig. 66.3 Hadoop pipes details

66.4 Experimental Results

Experiments were performed on heterogeneous cluster of four servers (one master and three slaves), whereas each slave has a different type of GPU cards ranging from NVIDIA GeForce GTX 470, Tesla K20C, and Tesla K40C. All servers run 64-bit Ubuntu server 12.04 LTS, NVIDIA CUDA 6.0 and Hadoop-1.2.1. In our

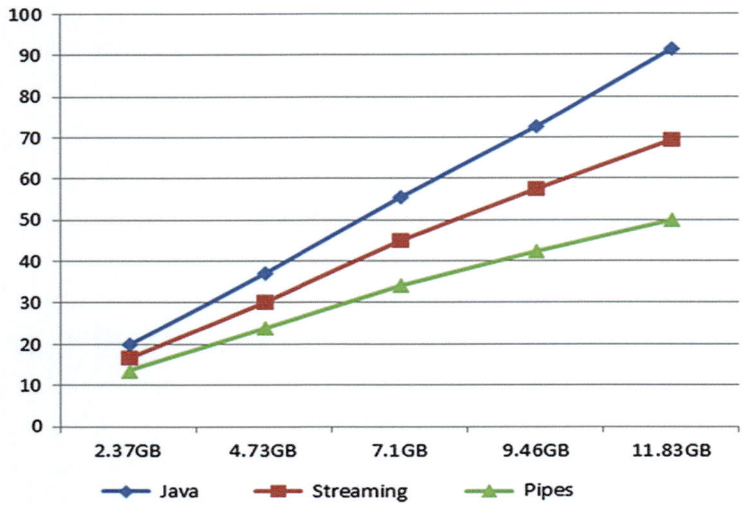

Fig. 66.4 The elapsed time

experiments, we have tested five datasets whose sizes range from 2.365, 4.73, 7.095, 9.46, and 11.83 GB.

In Fig. 66.4, the y-axis denotes the elapsed time (minutes), and the x-axis denotes the input datasets. Figure 66.4 shows that Pipes method outperforms the other two whereas Java delivers the worst performance. The initial performance of Pipes is 48.4 % higher than Java. And Streaming's is about 20 % higher. The speedups of both Pipe and Streaming increase as the input datasets increase. In dataset 5, the speedup of Pipes, Streaming is nearly 83 % higher, 32 % individually higher. Thus, we have reason to believe that the speedup is larger if the input dataset is bigger. With the increasing of the input data, the curve of Java has the fastest riser, the next is Streaming and the last is Pipes, which indicates GPU has better advantage in bigger data processing. There is another trend of the curves: the slope of these curves is going smaller, that is, progressively decreasing, which means Hadoop is adequate at handing big data. In our experiment, using GPU in Hadoop has some extra costs, however it's still worthy. Pipes' implementation is higher than the Streaming one, mainly due to the fact that Pipes' one skips the stdin/stdout buffer.

66.5 Conclusions and Future Work

With the recent advent of GPU computing, it has given a step forward in computing architecture by introducing a hybrid model where GPUs work in conjunction with CPUs. In this research, we investigated the K-means algorithm over different methods in cluster of heterogeneous CPU–GPU environments. We presented two ways to fuse GPU and Hadoop together to take the advantage of their merit fully.

Our preliminary results have shown good promises. Though, still more work to be done to improve the overall performance, for instance, redesign the scheduling algorithm of Hadoop, do performance tuning of Hadoop. There are two other ways to combine CUDA and Java: JNI and JCuda. As future work, we can consider to compare these two methods, as well as with other approaches.

References

1. Chen, Y., et al.: MGMR: Multi-GPU based MapReduce. In: Park, J.J., Arabnia, H.R., Kim, C., Shi, W., Gil, J.-M. (eds.) Grid and Pervasive Computing, pp. 433–442. Springer, Heidelberg (2013)
2. Jiang, H., et al.: Scaling up MapReduce-based big data processing on multi-GPU systems. Clust. Comput. **18**(1), 369–383 (2015)
3. Chen, Y. et al.: Pipelined multi-GPU MapReduce for big-data processing. In: Lee, R. (ed.) Computer and Information Science, pp. 231–246. Springer, New York (2013)
4. Fang, W., et al.: Mars: accelerating MapReduce with graphics processors. IEEE Trans. Parallel Distrib. Syst. **22**(4), 608–620 (2011)
5. Fan, W. et al.: Parallelization of RSA algorithm based on compute unified device architecture. In: Proceedings of the 9th International Conference on Grid and Cooperative Computing (GCC), IEEE, (2010)
6. Tsiomenko, R., Rees, B.S.: Accelerating Fast Fourier Transforms Using Hadoop and CUDA. (2013)
7. Zhu, J., et al.: Embedding GPU computations in Hadoop. Int. J. Netw. Distrib. Comput. **2**(4), 211–220 (2014)
8. Ding, M. et al.: More convenient more overhead: the performance evaluation of Hadoop streaming. In: Proceedings of the 2011 ACM Symposium on Research in Applied Computation. ACM, New York, pp. 307–313 (2011)
9. Kirk, D.: NVIDIA CUDA software and GPU parallel computing architecture. In: ISMM, vol. 7, pp. 103–104 (2007)
10. Jiang, H., et al.: Accelerating MapReduce framework on multi-GPU systems. Clust. Comput. **17**(2), 293–301 (2014)

Chapter 67
A HHT-Based Music Synthesizer

I-Hao Hsiao, Chun-Tang Chao, and Chi-Jo Wang

Abstract Synthesizing musical sound plays an important role in modern music composition. Composers nowadays can easily take advantage of powerful and user-friendly personal computers to produce the desired musical sound with a good music synthesis method. In this chapter, the Hilbert-Huang Transform (HHT) time-frequency analysis method is employed, in an attempt to implement a new efficient music synthesizer. By applying the HHT technique, the original varying-pitch music signals can be decomposed into several intrinsic mode functions (IMF) based on the empirical mode decomposition (EMD). The instantaneous amplitude and frequency of IMFs can be further obtained by Hilbert transform. By extracting the main spectrum coefficients of the instantaneous amplitude and frequency of the IMFs, the original musical signal can be reconstructed with little error. Experimental results indicate the feasibility of the proposed method.

Keywords Music synthesis • Hilbert-Huang transform (HHT) • Empirical mode decomposition (EMD) • Intrinsic mode function (IMF)

67.1 Introduction

For regular music synthesis methods, the two most popular methods may be the wavetable music synthesis [1] and FM synthesis [2]. A good music synthesis allows music creators to synthesize the sound signal accurately and quickly. However, the two methods have been unable to provide satisfactory quality for high performance applications.

In recent years, the trend for a musical-tone generator has been based on physical modeling of sound production mechanisms [3]. The digital waveguide filter [4, 5] can be applied to simulate a wide class of musical instruments. Figure 67.1 shows the nonlinear predictive model of an instrument. The excitation unit (Exciter) is the nonlinear part, responsible for generating an oscillatory signal source. And the

I.-H. Hsiao • C.-T. Chao (✉) • C.-J. Wang
Department of Electrical Engineering, Southern Taiwan University of Science and Technology, Tainan, Taiwan
e-mail: tang@mail.stust.edu.tw

J. Juang (ed.), *Proceedings of the 3rd International Conference on Intelligent Technologies and Engineering Systems (ICITES2014)*, Lecture Notes in Electrical Engineering 345, DOI 10.1007/978-3-319-17314-6_67

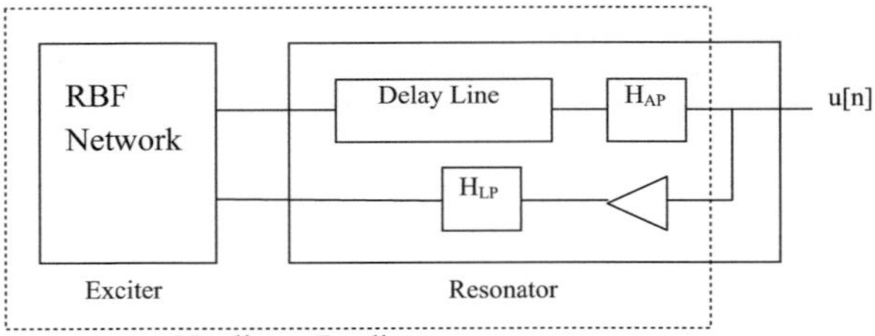

Fig. 67.1 The nonlinear predictive model of an instrument

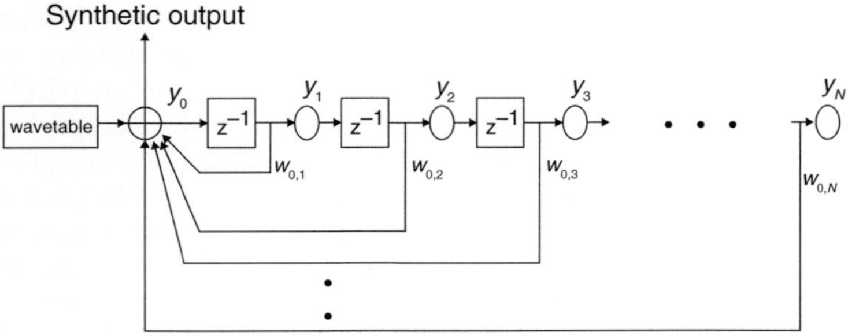

Fig. 67.2 The nonlinear predictive model of an instrument

resonance unit (Resonator) belongs to the linear filter part, responsible for modulating out the sound signal.

Figure 67.2 shows a simple model-based structure implemented by IIR (infinite impulse response) synthesis, consisting of a prediction filter and a delay line to synthesize tones produced by instruments [6]. The design of the coefficients for the IIR synthesizer is accomplished by using a neural network (NN)-based training algorithm. A recurrent NN (RNN) is applied for the prediction filter design. However, such kinds of design approaches can be time-consuming during the training process.

Huang et al. [7] in 1998 developed a new method called Hilbert-Huang Transform (HHT) for analyzing nonlinear and nonstationary data. The HHT should be more powerful and suitable in timbre analysis when compared with traditional Short-Time Fourier Transform (STFT). Through the understanding of the HHT method, this chapter proposes a more efficient HHT-Based Music Synthesizer.

67.2 The HHT and EMD

The HHT was pioneered by Huang et al., for adaptively representing nonstationary signals as sums of zero-mean amplitude modulation frequency modulation components. The Fourier Transform views the signal as a combination of many fixed-frequency and fixed-amplitude sinusoids. The HHT regards the signal as a combination of many intrinsic mode functions (IMF), which have time-varying frequency (instantaneous frequency) and time-varying amplitude (instantaneous amplitude) [8]. Thus, the HHT provides a more powerful analysis and synthesis tool for the pitch and timbre of a music sound. In this section, the HHT and EMD are briefly introduced.

There are two steps in the HHT: (1) For a given signal $x(t)$, extract the IMFs by means of empirical mode decomposition (EMD); and (2) apply the Hilbert Transform on each IMF to get the corresponding instantaneous frequency and amplitude. Step 1 is iteratively finished until the residue becomes a monotonic function or a function with only one cycle from which no more IMFs can be extracted. Equation (67.1) shows the decomposition of the $x(t)$ into N-empirical modes, where $c_j(t)$ is the jth IMF and $r_N(t)$ is the final residue.

$$x(t) = \sum_{j=1}^{N} c_j(t) + r_N(t) \tag{67.1}$$

In Step 2, the Hilbert Transform is utilized to obtain an analytic complex representation $z(t)$ for each IMF $c(t)$, as shown in (67.2), where $d(t)$ is the Hilbert Transform of $c(t)$. The instantaneous amplitude and instantaneous phase are denoted as $a(t)$ and $\theta(t)$, respectively.

$$z(t) = c(t) + id(t) = a(t)e^{i\theta(t)} \tag{67.2}$$

Then the original signal $x(t)$ can be represented as

$$x(t) = \mathrm{Re}\left\{ \sum_{j=1}^{N} a_j(t) e^{i\int \omega_j(\tau)d\tau} \right\} \tag{67.3}$$

where $\omega(t) = \frac{d\theta(t)}{dt}$ is the instantaneous frequency and Re denotes real part. Equation (67.3) shows the difference between the HHT and the Discrete Fourier Transform. In the HHT, each component is considered as time-varying amplitude and time-varying frequency sinusoid. For brevity, the instantaneous frequency and the instantaneous amplitude will be referred to as "IF" and "IA" in the following text. For each IMF, its corresponding "IF" and "IA" can be calculated. On the contrary, the IMF can be reconstructed by its corresponding "IF" and "IA."

In this chapter, the FFT will be applied to the "IF" and "IA," instead of directly applied to the IMF. Using this approach, less FFT coefficients are needed and will yield better synthesis performance.

67.3 Simulation Results

The simulation was implemented in the MATLAB environment. Different sound signals, including piano, trumpet, violin, and bird chirps, are provided. Figure 67.3 shows the EMD analysis of trumpet music (pitch A4 or A440), including the original signal, IMF1–IMF8, and the final residue.

For each IMF, the corresponding "IF" and "IA" can be obtained. Figure 67.4 shows the "IA" analysis for each IMF. In the proposed method, only the first four IMFs (IMF1–IMF4) are considered and the latter IMFs (IMF5–IMF11) are omitted. The FFT is applied to "IF" and "IA," and 128 main coefficients for "IF" and "IA," respectively, are selected. Thus for the first four IMFs, all the 1,024 FFT coefficients are stored. The proposed synthesis method is compared with the original sound with direct 1,024-point FFT analysis, to demonstrate its efficiency and feasibility. Table 67.1 shows the synthesis error comparison under the same coefficients number 1,024 for different instruments, where the error is measured by the Euclidean distance defined in (67.4).

$$d(x, y) = \sqrt{\sum_{i=1}^{n} (x_i - y_i)^2} \tag{67.4}$$

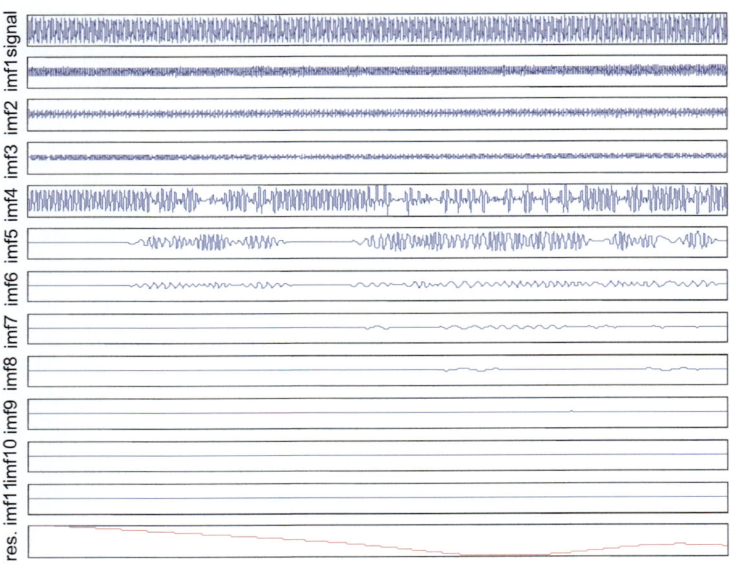

Fig. 67.3 EMD analysis of trumpet music (A4)

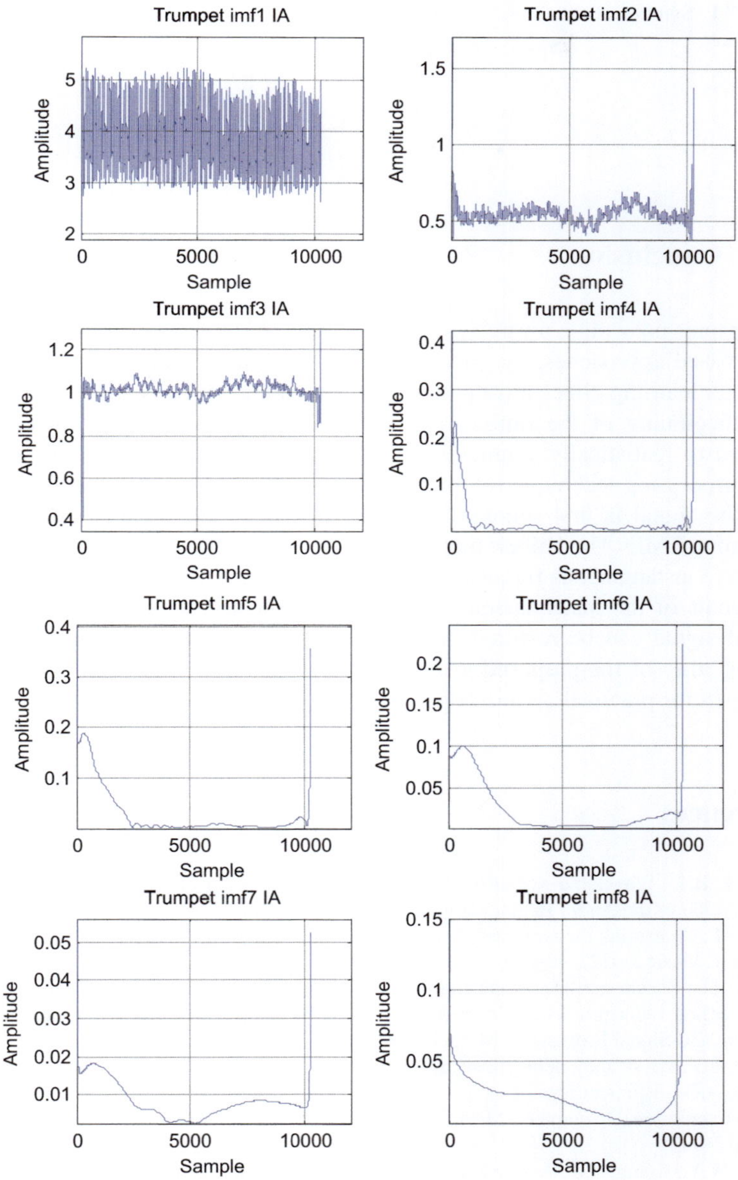

Fig. 67.4 Instantaneous amplitude (IA) analysis for each IMF

Table 67.1 Synthesis error comparison

Method / Sound	The proposed method	Direct FFT
Piano	0.3407	0.8267
Trumpet	0.5202	0.8229
Violin	0.2152	0.8219
Bird chirps	0.0596	0.2952

67.4 Conclusion

This chapter presents a music synthesizer based on the HHT. For some advanced model-based approaches, the procedure may be tedious and time-consuming for parameter learning. Since most practical music sounds are not stationary, especially in the beginning of the timbre, the conventional Fourier Transform cannot be expected to realistically synthesize the music sounds. The HHT is an advanced signal-processing technique for analyzing nonlinear and nonstationary time series data. The signal is first segregated into narrow band components, the IMFs, by performing EMD. The Hilbert transform is then applied on each mode to obtain the respective instantaneous frequency and the amplitude. By extracting the main FFT coefficients of the instantaneous frequency and the amplitude for each IMF, the original signal can be restored in a good performance. Simulation results show the feasibility of the proposed synthesis method. Further improvement should be developed for practical applications.

References

1. Robert, B.J.: Wavetable synthesis 101, a fundamental perspective. In: Proceedings 101st Convention of the Audio Engineering Society, Los Angeles (1996)
2. John, M.: Chowning: the synthesis of complex audio spectra by means of frequency modulation. Comput. Music J. **1**(2), 46–54 (1977)
3. Drioli, C., Rocchesso, D.: A generalized musical-tone generator with application to sound compression and synthesis. In: Proceeding of the IEEE International Conference on Acoustics, Speech, and Signal Processing, Munich, vol. 1, pp. 431–434 (1997)
4. Smith, J.O.: Physical modeling using digital waveguides. Comput. Music J. **16**(4), 74–87 (1992)
5. Smith, J.O.: Efficient synthesis of stringed musical instruments. In: Proceedings of the 1993 International Computer Music Conference, pp. 64–71, Computer Music Association, Tokyo (1993)
6. Su, A.W.Y., Liang, S.F.: A new automatic IIR analysis/synthesis technique for plucked-string instruments. IEEE Trans. Speech Audio Process. **9**(7), 747–754 (2001)
7. Huang, N.E., Shen, Z., Long, S.R., Wu, M.C., Shih, H.H., Zheng, Q., Yen, N.C., Tung, C.C., Liu, H.H.: The empirical mode decomposition and the Hilbert spectrum for nonlinear and non-stationary time series analysis. Proc. R. Soc. Lond. Ser. **454**(1971), 903–995 (1998)
8. Gloersen, P., Huang, N.E.: Comparison of interannual intrinsic modes in hemispheric sea ice covers and other geophysical parameters. IEEE Trans. Geosci. Remote Sens. **41**(5), 1062–1074 (2003)

Chapter 68
Decentralized Robust Estimation of Interconnected Systems with State-Dependent Impulse Disturbances

Cheng-Fa Cheng and Shih-Jyun Lin

Abstract The problem of local observer synthesis for uncertain interconnected systems subjected to nonlinear interaction functions, bounded uncertainties, and state-dependent impulse disturbances is investigated. The observation scheme is only based on certain functional properties of the uncertainty and interaction bounds and sufficient conditions will be provided such that the observation error of the uncertain state-jump interconnected system will be practically stable by the proposed local observer scheme. Furthermore, within the derived lower bound of the impulse interval, the boundedness of the observation errors of the uncertain interconnected system with the equidistant impulse disturbance will be guaranteed and the radius of the attraction ball can be measured. Finally, an example with simulations is given to illustrate the application of our results.

Keywords Interconnected systems • Robust • Local observer • State-dependent impulse disturbance • Practical stability

68.1 Introduction

The basic assumption in state feedback control is that all state variables are available. However, in practical situations, the system states are not completely available; therefore, it is necessary to estimate the system states from measurements of some of its outputs. Ever since the pioneering work of [1], several researchers have devoted to the observers design for reconstructing the system states in a decentralized framework for interconnected systems [2–4]. It is clear that the number of computations for an interconnected system will be prohibitive if the centralized method is adopted. And perhaps even more important, the structure of

C.-F. Cheng (✉) • S.-J. Lin
Department of Communications, Navigation and Control Engineering,
National Taiwan Ocean University, Keelung 20224, Taiwan
e-mail: cfcheng@ntou.edu.tw

J. Juang (ed.), *Proceedings of the 3rd International Conference on Intelligent Technologies and Engineering Systems (ICITES2014)*, Lecture Notes in Electrical Engineering 345, DOI 10.1007/978-3-319-17314-6_68

an interconnected system is most naturally represented in decentralized form. Thus a single observer is not feasible.

Impulse disturbance may arise from the sudden changes of the environment and causes the systems to undergo rapid changes in the states at certain instants. The problem of systems with impulse effect in the fields of the control theory, biology, and electronics has already received great attention [5–7]. Sudden changes of the environment make the systems suffer the impulse effect. How to overcome the impulse effect of the impulse disturbances is another important topic.

In this chapter, a systematic constructive procedure for designing local observers of uncertain interconnected jumping systems is attacked. Without precise knowledge of the plant, the observer is only based on certain functional properties of the uncertainty and interaction bounds and the practical stability of the observation errors of the whole systems will be guaranteed by the proposed local observer schemes.

68.2 Problem Statement and Definitions

Let us consider an interconnected system S composed of N subsystems S_i, $i = 1, 2, \ldots, N$. Each S_i is modeled by the equation

$$\dot{x}_i(t) = A_i x_i(t) + B_i(u_i(t) + \upsilon_i(t,\ u_i(t),\ \sigma_i(t), x_i(t))) + g_i(t, \sigma_i(t), x(t)),\ t \neq \tau_k \tag{68.1a}$$

$$x_i\left(t_k^+\right) = x_i(t_k) + w_i(t_k, x_i(t_k)),\ t = \tau_k \tag{68.1b}$$

$$y_i(t) = C_i x_i(t) \tag{68.1c}$$

where A_i, B_i and C_i denote the nominal system matrix, input connection matrix, and output matrix of appropriate dimensions, respectively; moreover A_i is not necessary a Hurwitz matrix. $g_i(t, \sigma_i, x)$ is the interaction function is represented as

$$g_i(t, \sigma_i, x) = B_i g_i^{'}(t, \sigma_i, x) + (D_i - B_i) g_i^{'}(t, \sigma_i, x) \tag{68.2}$$

where D_i is chosen by the designer, $\sigma_i(t)$ is the uncertain parameter, and $\left\| g_i^{'}(t, \sigma_i, x) \right\| \leq \sum_{j=1}^{N} \varpi_{ij} \left\| x_j(t) \right\|$. Whereas the function $\upsilon_i(t,\ u_i, \sigma_i, x_i)$ can be considered as input disturbances and satisfies

$$\|\upsilon_i(t, u_i, \sigma_i, x_i)\| \leq \rho_i(t,\ u_i, y_i) \tag{68.3}$$

Impulse disturbance $w_i(t_k, x_i(t_k))$ has the effect of suddenly changing the state of the system at the instant t_k and is bounded by

$$\|w_i(t_k, x_i(t_k))\| \leq \gamma_{ik} \|x_i(t_k)\| \tag{68.4}$$

with $\gamma_{ik} \geq 0$, where $t_0 < t_1 < t_2 < \cdots < t_k < \cdots$, $\lim_{k \to \infty} t_k = \infty$, $k \in \mathbb{N}$, and $\mathbb{N}$ denotes the set of natural numbers. Without loss of generality, assume both matrices B_i and C_i are of full rank and all pairs (A_i, C_i), $i = 1, 2, \ldots, N$, are observable.

In this chapter, the following nonlinear local observer scheme will be adopted

$$\begin{aligned}\dot{\hat{x}}_i(t) &= A_{oi}\hat{x}_i(t) + \overline{P}_i^{-1}C_i^T\overline{F}_i^T\overline{F}_iC_i\widetilde{x}_i\rho_i^2(t,\, u_i,\, y_i)/\left(\left\|\overline{F}_iC_i\widetilde{x}_i\right\|\rho_i(t,\, u_i,\, y_i) + \hat{\eta}_i\right)\\ &\quad - G_iy_i(t) + B_iu_i(t) + 1/2\sum_{j=1}^{N}\varpi_{ij}\left(\ell_{1i} + \xi_j\right)\overline{P}_i^{-1}C_i^T\overline{F}_i^T\overline{F}_iC_i\widetilde{x}_i(t)\end{aligned} \tag{68.5}$$

with $\overline{F}_i = B_i^T\overline{P}_iC_i^T\left(C_iC_i^T\right)^{-1}$, where $\overline{P}_i$ is a positive-definite symmetric matrix satisfying

$$A_{oi}^T\overline{P}_i + \overline{P}_iA_{oi} + \sum_{j=1}^{N}\left\|D_i - B_i\right\|\left(\ell_{2i} + \xi_j\right)\varpi_{ij}\overline{P}_i^2 + \sum_{j=1}^{N}\varpi_{ji}\left(1/\ell_{1j} + \left\|D_i - B_i\right\|/\ell_{2j}\right)I = -\overline{Q}_i \tag{68.6}$$

with $\overline{Q}_i = \overline{Q}_i^T > 0$, $\ell_{1i} > 0$, $\ell_{2i} > 0$ subject to $\left\{I - C_i^T\left(C_iC_i^T\right)^{-1}C_i\right\}\overline{P}_iB_i = 0$.

68.3 Robust Local Observer Synthesis

For derivation convenience, some notations are denoted as

$$\rho_{ik} = (1 + a_{ik}) + (1 + 1/a_{ik})(1 + b_{ik})\lambda_M\left(\overline{P}_i\right)/\lambda_m\left(\overline{P}_i\right)\gamma_{ik}^2;\ h_{\max} = \max\{t_k - t_{k-1}\}$$

where a_{ik} and b_{ik} are both positive constants for all $i = 1, 2, \ldots, N$ and $k \in \mathbb{N}$.

Theorem 1 For the uncertain interconnected system (68.1a)–(68.1c) with impulse disturbance (68.3), if there exists a positive symmetric matrix $\overline{P}_i$ such that (68.6) holds then, by local observer (68.5), the observation error $\widetilde{x}(t)$ will be practically stable with respect to $\left(\mu,\, \sqrt{\chi/\overline{\nu}_1}\right)$ for $\mu \geq \left\|\widetilde{x}(t_0)\right\|$ and

$$\begin{aligned}\chi &= \overline{\nu}_2\mu^2\prod_{i=1}^{\infty}\rho_i\mathrm{e}^{-\overline{\nu}(t-t_0)} + \overline{\nu}_4/\overline{\nu}\sum_{j=1}^{\infty}\prod_{i=j}^{\infty}\rho_i\left(\mathrm{e}^{-\overline{\nu}\left(t-t_j\right)} - \mathrm{e}^{-\overline{\nu}\left(t-t_{j-1}\right)}\right)\\ &\quad + \overline{\nu}_4/\overline{\nu}\left(1 - \mathrm{e}^{-\overline{\nu}\tau_{\max}}\right) + \sum_{j=1}^{\infty}\prod_{i=j}^{\infty}\rho_i/\rho_jf_j\mathrm{e}^{-\overline{\nu}\left(t-t_j\right)}\end{aligned} \tag{68.7}$$

where $\rho_k = \max_i\{\rho_{ik}\}$, $f_k = \sum_{i=1}^{N}(1 + 1/a_{ik})(1 + 1/b_{ik})\lambda_M\left(\overline{P}_i\right)\gamma_{ik}^2\left\|\hat{x}_i(t_k)\right\|^2$, $\overline{\nu}_1 = \min_i\left\{\lambda_m\left(\overline{P}_i\right)\right\}$, $\overline{\nu}_2 = \max_i\left\{\lambda_M\left(\overline{P}_i\right)\right\}$, $\overline{\nu}_3 := \min_i\left\{\lambda_m\left(\overline{Q}_i\right)\right\}$,

$\overline{\nu}_4 = \sum_{i=1}^{N}\{2\hat{\eta}_i + \sum_{j=1}^{N}(1 + \|D_i - B_i\|)\varpi_{ij}\xi_j\}$, and $\overline{\nu} = \overline{\nu}_3/\overline{\nu}_2$ for all $i = 1, 2, \ldots, N$ and.

Proof Define a Lyapunov function $V = \sum_{i=1}^{N}\widetilde{x}_i^T(t)P_i\widetilde{x}_i(t)$ with $\widetilde{x}_i(t) = x_i(t) - \hat{x}_i(t)$. The derivative of the Lyapunov function evaluated along the trajectories of the error equation yields

$$D^+V(t) = \sum_{i=1}^{N}\left\{\dot{\widetilde{x}}_i^T P_i\widetilde{x}_i + \widetilde{x}_i^T P_i\dot{\widetilde{x}}_i\right\} \le -\overline{\nu}_3\sum_{i=1}^{N}\left\|\widetilde{x}_i\right\|^2 + \overline{\nu}_4 \le -\overline{\nu}V(t) + \overline{\nu}_4 \quad (68.8a)$$

for $t \neq t_k$. For $t = t_k$, we have

$$\begin{aligned} V(t_k^+) &= \sum_{i=1}^{N}\widetilde{x}_i^T(t_k^+)\overline{P}_i\widetilde{x}_i(t_k^+) \\ &= \sum_{i=1}^{N}[\widetilde{x}_i(t_k) + w_i(t_k, x_i(t_k))]^T\,\overline{P}_i[\widetilde{x}_i(t_k) + w_i(t_k, x_i(t_k))] \\ &\le \rho_k V(t_k) + f_k \end{aligned} \quad (68.8b)$$

It follows from Theorem 3.1 of [8] with $g(t, m, m) = -\overline{\nu}m + \overline{\nu}_4$ and $\psi_k(m) = \rho_k m + f_k$ that

$$\begin{aligned} m(t) \le m(t_0)\prod_{i=1}^{k}\rho_i e^{-\delta_0(t-t_0)} + \overline{\nu}_4/\overline{\nu}\sum_{j=1}^{k}\prod_{i=j}^{k}\rho_i\left(e^{-\overline{\nu}(t-t_j)} - e^{-\overline{\nu}(t-t_{j-1})}\right) \\ + \overline{\nu}_4/\overline{\nu}\left(1 - e^{-\overline{\nu}\tau_{\max}}\right) + \sum_{j=1}^{k}\prod_{i=j}^{k}\rho_i/\rho_j f_j e^{-\overline{\nu}(t-t_j)} \end{aligned}$$

for $t_k < t \le t_{k+1}$. Since $\|\widetilde{x}(t_0)\| \le \mu$, we have

$$\begin{aligned} \mu\overline{\nu}_1\|\widetilde{x}(t)\|^2 \le \overline{\nu}_2\mu^2\prod_{i=1}^{k}\rho_i e^{-\overline{\nu}(t-t_0)} + \overline{\nu}_4/\overline{\nu}\sum_{j=1}^{k}\prod_{i=j}^{k}\rho_i\left(e^{-\overline{\nu}(t-t_j)} - e^{-\overline{\nu}(t-t_{j-1})}\right) \\ + \overline{\nu}_4/\overline{\nu}\left(1 - e^{-\overline{\nu}\tau_{\max}}\right) + \sum_{j=1}^{k}\prod_{i=j}^{k}\rho_i/\rho_j f_j e^{-\overline{\nu}(t-t_j)} \end{aligned} \quad (68.9)$$

and then (68.7) holds. The proof is completed.

Remark 1 Without impulse disturbance, it follows from [9] that $\widetilde{x}(t)$ converges exponentially with the degree $\overline{\nu}_3/2$ toward the finial attractor $B(0, \widetilde{r}) := \{\widetilde{x}(t) \in R^n : \|\widetilde{x}(t)\| \le \widetilde{r}\}$ with $\widetilde{r} > \widetilde{r}^* := \sqrt{\overline{\nu}_4/(\overline{\nu}_1\overline{\nu}_3)}$ and arbitrary initial conditions $\widetilde{x}(0)$.

Remark 2 For the uncertain interconnected system (68.1a)–(68.1c) with equidistant impulse interval h, if $\ln(\rho) < \overline{\upsilon} h$ then, by local observer (68.5), the observation error $\widetilde{x}(t)$ will attract into a ball $B(0, \widetilde{r})$ with

$$\overline{r} > \overline{r}^* := \sqrt{\frac{1}{\overline{\upsilon}_1}\left\{\frac{\overline{\upsilon}_4}{\overline{\upsilon}}(1 - \mathrm{e}^{-\overline{\upsilon}h})\frac{\rho^2 \mathrm{e}^{-\overline{\upsilon}h}}{1 - \rho \mathrm{e}^{-\overline{\upsilon}h}} + \frac{f}{1 - \rho \mathrm{e}^{-\overline{\upsilon}h}} + \frac{\overline{\upsilon}_4}{\overline{\upsilon}}(1 - \mathrm{e}^{-\overline{\upsilon}h})\right\}}, \quad \text{where}$$

$$\rho = \max_i \{\rho_i\}.$$

68.4 Illustrative Example

Consider the inverted pendulums which are coupled by the springs and subject to three distinct input moments $u_i(t)$ [10]

$$S_1 : \dot{x}_1(t) = \begin{bmatrix} 0 & 1 \\ g/l & 0 \end{bmatrix} x_1(t) + \begin{bmatrix} 0 \\ 1/(\tau_1 l^2) \end{bmatrix} (u_1(t) + \upsilon_1(t)) + \begin{bmatrix} 0 & 0 \\ -k_1 a^2/(\tau_1 l^2) & 0 \end{bmatrix} x_1(t) + \begin{bmatrix} 0 & 0 \\ -k_1 a^2/(\tau_1 l^2) & 0 \end{bmatrix} x_2(t) \tag{68.10a}$$

$$y_1(t) = [1 \quad 1] x_1(t) \tag{68.10b}$$

$$S_2 : \dot{x}_2(t) = \begin{bmatrix} 0 & 1 \\ g/l & 0 \end{bmatrix} x_2(t) + \begin{bmatrix} 0 \\ 1/(\tau_2 l^2) \end{bmatrix} (u_2(t) + \upsilon_2(t)) + \begin{bmatrix} 0 & 0 \\ -k_1 a^2/(\tau_1 l^2) & 0 \end{bmatrix} x_1(t) + \begin{bmatrix} 0 & 0 \\ -\frac{k_1 a^2}{\tau_1 l^2} - \frac{k_2 b^2}{\tau_1 l^2} & 0 \end{bmatrix} x_2(t) + \begin{bmatrix} 0 & 0 \\ -\frac{k_2 b^2}{\tau_1 l^2} & 0 \end{bmatrix} x_3(t) \tag{68.10c}$$

$$y_2(t) = [1 \quad 1] x_2(t) \tag{68.10d}$$

$$S_3 : \dot{x}_3(t) = \begin{bmatrix} 0 & 1 \\ \frac{g}{l} & 0 \end{bmatrix} x_3(t) + \begin{bmatrix} 0 \\ \frac{1}{\tau_3 l^2} \end{bmatrix} (u_3(t) + \upsilon_3(t)) + \begin{bmatrix} 0 & 0 \\ -\frac{k_2 b^2}{\tau_3 l^2} & 0 \end{bmatrix} x_3(t) + \begin{bmatrix} 0 & 0 \\ -\frac{k_2 b^2}{\tau_3 l^2} & 0 \end{bmatrix} x_2(t) \tag{68.10e}$$

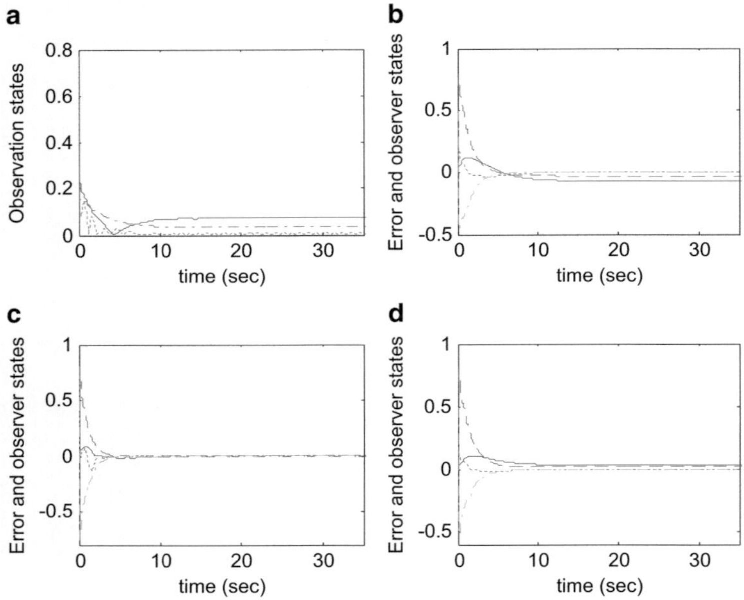

Fig. 68.1 The trajectories of the inverted pendulums system (10) without state jumps (**a**) $|C_i\hat{x}_i(t)|$; (**b**) subsystem 1; (**c**) subsystem 2; (**d**) subsystem 3

$$y_3(t) = [1 \quad 1]x_3(t) \tag{68.10f}$$

Letting $g = 9.8(\mathrm{ms}^{-2})$, $l = 1(\mathrm{m})$, $\tau_1 = 0.8(\mathrm{kg})$, $\tau_2 = 0.4(\mathrm{kg})$, $\tau_3 = 1.25(\mathrm{kg})$, $k_1 = 1(\mathrm{N\ m}^{-1}), k_2 = 1.5(\mathrm{N\ m}^{-1})$ and $\upsilon_1 = -0.5 + 0.3y_1, \upsilon_2 = 0.2\sin y_2 - 0.4y_2$, and $\upsilon_3 = 0.6 + y_3$, we have $\rho_1 = 0.5 + 0.3|y_1|$, $\rho_2 = 0.2 + 0.4|y_2|$, and $\rho_3 = 0.6 + |y_3|$. Choosing $\ell_{11} = \ell_{12} = \ell_{13} = 5$ and $G_1 = \begin{bmatrix} 0 \\ -22.5 \end{bmatrix}$, $G_2 = \begin{bmatrix} 0 \\ -45 \end{bmatrix}$, $G_3 = \begin{bmatrix} 0 \\ -22.5 \end{bmatrix}$, yields $\overline{F}_1 = 0.625$, $\overline{F}_2 = 1.25$, $\overline{F}_3 = 0.4$, and $\overline{P}_1 = \overline{P}_2 = \overline{P}_3 = \begin{bmatrix} 2.5 & 0.5 \\ 0.5 & 0.5 \end{bmatrix}$. It follows from Theorem 1 that the observation error states $\widetilde{x}(t)$ of the pendulum systems (68.10a)–(68.10f) will be bounded and the time responses of the observation error states will be plotted in Fig. 68.1 with $w_i(t_k, x_i(t_k)) = c_i(-1)^k(1 + e^{-k})\sin\left(\sqrt{\|x_i(t_k)\|}\right)x_i(t_k)$, $c_1 = 0.05$, $c_2 = 0.05$, and $c_3 = 0.06$ and the impulse instant t_k is chosen randomly between 0 and 0.5 s.

68.5 Conclusion

A design algorithm for local observers of uncertain interconnected systems with nonlinear interaction functions, bounded uncertainties, and impulse disturbances has been proposed. The observation scheme is only based on certain functional properties of the uncertainty and interaction bounds and will make the estimation errors of the considered systems be robustly practically stable. Furthermore, the relation between the robust local observer design and the admissible intervals of the impulse disturbances has been established. Finally, the observation error will attract into a ball with finite radius once the equidistant impulse interval is greater than the derived bound.

References

1. Hodzic, M., Siljak, D.D.: Decentralized estimation and control with overlapping information sets. IEEE Trans. Autom. Control **31**, 83–86 (1986)
2. Hou, M., Muller, P.C.: Design of linear state function observers. Automatica **30**, 1801–1805 (1994)
3. Jiang, L., Wu, Q.H., Wen, Y.J.: Decentralized nonlinear adaptive control for multimachine power systems via high-gain perturbation observer. IEEE Trans. Circuits Syst. **51**, 2052–2059 (2004)
4. Yang, P., Freeman, R.A., Lynch, K.M.: Multi-agent coordination by decentralized estimation and control. IEEE Trans. Autom. Control **53**, 2480–2496 (2008)
5. Bainov, D., Simeonov, P.S.: Systems with Impulse Effect: Stability, Theory and Applications. Halsted, New York (1989)
6. Lakshmikantham, V., Bainov, D.D., Simeonov, P.S.: Theory of Impulse Differential Equations. World Scientific, Singapore (1989)
7. Yang, T.: Impulsive Control Theory, vol. 272. Spinger, Berlin (2001)
8. Gong, Z., Wen, C., Mital, D.P.: Decentralized robust controller design for a class of interconnected uncertain systems: with unknown bound of uncertainty. IEEE Trans. Autom. Control **41**, 850–854 (1996)
9. Liu, X.Z.: Practical stabilization of control systems with impulse effects. J. Math. Anal. Appl. **116**, 563–576 (1992)
10. Han, M.C., Chen, Y.H.: Decentralized robust control of nonlinear systems with bounded time-varying uncertainties. Control Theory Adv. Technol. **7**, 609–628 (1991)

Chapter 69
A Lean Analysis to Automate the Draw Operations at a Tube Manufacturing Company

Jun-Ing Ker, Chandra Mani Shrestha, and Yichuan Wang

Abstract The objective of this research is to conduct a lean analysis of the Draw Cell operations at a manufacturer that fabricates stainless steel tubes and to implement Lean concepts for automation. Videotaping operations, interviewing operators, and studying the engineered layout of the operation were carried out to understand the existing operations. Industrial engineering knowledge of Lean, 8 wastes, Kaizen, Single Minute Exchange of Die, and Value Stream Map were applied to improve the setup of workstations and material handling procedures. These principles helped make the current draw cell to operate more effectively and economically without major renovations.

Keywords Tube drawing • Draw operation automation • Lean analysis • Kaizen • Single Minute Exchange of Die

69.1 Introduction

The tube manufacturing company under study is located in north Louisiana in the United States. The company is on the process to automate its Draw Cell. At present, the Draw operation is the bottleneck for the plant. The company currently has many manual machine systems where human workers operate machines to complete given tasks. In these systems, operators control machines and inspect the quality of products. Operators also setup the machines according to the customer requirements. Some machines wait for operators to start and finish the jobs, while in some cases operators wait for machines to finish their jobs. Idles of operators and machines are commonly seen in the plant.

J.-I. Ker (✉) • C.M. Shrestha
Industrial Engineering, Louisiana Tech University, Ruston, LA 71272, USA
e-mail: ker@latech.edu

Y. Wang
Department of Aviation and Supply Chain Management,
Auburn University, Auburn, AL 36830, USA

J. Juang (ed.), *Proceedings of the 3rd International Conference on Intelligent Technologies and Engineering Systems (ICITES2014)*, Lecture Notes in Electrical Engineering 345, DOI 10.1007/978-3-319-17314-6_69

69.1.1 Drawing of Tubes

The tube drawing operation is an essential step in tube manufacturing. The tube draw operation changes the tubes' inner diameter (IDs), outer diameter (ODs), wall thicknesses, and also improves the surface finish by refining the grain structure. Use of die to change the shape of a tube is a common procedure in the tube drawing process. Die is a sintered tungsten carbide insert encased in steel that contains cobalt. Higher cobalt content provides more shock resistance while lower content provides better wear resistance. There are various drawing methods like sinking, rod drawing, floating plug drawing, tethered plug drawing, and fixed plug drawing [1]. The company in this case study uses the rod drawing method to draw the tube. The rod drawing process draws the tube over a hardened steel rod or bar that passes through the die resulting in the reduction of the OD, ID, and the wall thickness. Then, the reeling process slightly expands the diameter of the tube to extract the bar.

69.1.2 Benefits of Automation

At present, the Draw Cell operation is not fully automated and most of the operation is controlled and carried out by the operators. Mostly the production and quality inspection are done by the decision of the operators. Only some portion of the transportation of tube from workstations to workstations is automated. Thus, the first step to initiate the automation of Draw Cell is to understand the detail operation of the process by the engineers, managers, and maintenance staffs with the help of operators in the Draw Cell.

As shown in Fig. 69.1, the operation of the Draw Cell requires four operators. Operator 1 is responsible for controlling box 1 and loading bar into the tube. Operator 2 performs the drawing of tube in die, reeling operation, and controlling boxes 2 and 3. Operator 3 is responsible for controlling box 4 and the sink process. Operator 4 controls control box 5 and cuts the tube end. As the tube passes each workstation and operator, there are many value-added and non-value-added tasks associated with the operations. A value-added task is an activity that adds value to the product as the product is moved a step further towards the finished stage [2]. For example: activities such as lubing the tube, running tube in the die, and obtaining OD and ID of the tube are considered value-added tasks. Whereas there are many necessary support tasks that must be performed in the process of operation but do not directly further the completion of the workpiece. These tasks do not make any money to the company and are known as non-value-added tasks [2]. For example, dropping tub and bars to the v-roll machine, changing die, and performing an inspection are considered non-value-added tasks. Any activity that does not add value to the product is a waste because it only adds time and cost.

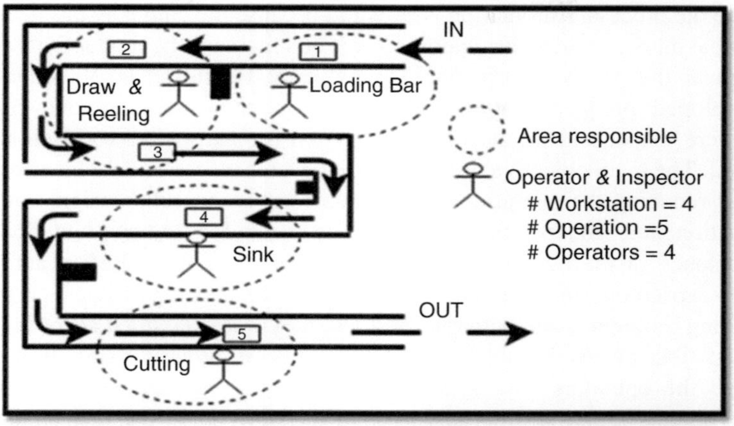

Fig. 69.1 Present state of the tube assembly line

At this study site, inventory is built in between the workstations. For example, Operator 3 has to wait for 20–25 tubes from the reeling operation to start the sink operation. This wait time maintains the heat of the tube. Similarly Operator 4 has to wait 15–20 tubes to start cutting the tube before the tube cools down to certain temperature. The operator decides the temperature of the tube by feeling with hands. Some of the tasks done by the operators can be automated easily, for example, cutting the tube by Operator 4 and loading the bar in the tube carried out by Operator 1. There are some complicated tasks like setup of the machines and inspections of the quality of tube are still done by operators manually. These tasks can be automated only after a detail study of the process is done that would require a joint effort of managers, engineers, and operators.

There are numerous tasks carried out by the operators in their day-to-day operations which need to be understood and addressed in order to automate the Draw Cell. For example, the temperature of the tube needs to be recorded before the tube proceeds to the sink or cut operation. Once the exact temperature is obtained, the process of automating the sink and cut operation would be easy. In this research, the breakdown of the entire Draw Cell and study of each operation was done in detail to better understand the existing process through Kaizen events.

69.1.3 Kaizen Events

Kaizen is one of the most successful Lean tools [3]. Kaizen is conducted to empower frontline staff so that they may utilize their knowledge to create more effective and efficient processes [1]. The operators in this plant have experience for more than 10 years and have most control over the operations. The operators prefer to work in the traditional manual operation mode, but the management wants to automate the process. Automation is essential because the company has to compete with global market, reduce the operation cost and increase the productivity. Kaizen events made the process transparent in many ways: The flowchart of the Draw Cell was developed; the present-state work steps of operations and the noncyclic operations were listed; and the need of automation for particular workstations were discussed. The main objective of these Kaizen events was to initiate the automation and start reducing the operation cost. It also focused on cells design, equipment modification, and sequence of operations. The events focused on yield, downtime, and efficiency of the Draw Cell before and after automation. Before automation could be carried out, the team decided to resolve the existing obstacles first. After the existing problems were better understood, they were classified according to the priorities. The events brought good breakdown of process using flow charts. Although the operators have great experience they lack the concepts of lean, resource management, and automation. On the other hand the management had little knowledge about numerous problems encountered daily by operators. The events helped to share, coordinate, and team up and acted as the first step towards automation of the Draw Cell.

69.2 Problems

At present, Workstation 1 frequently encountered machine failure and misplacement of the parts, making Operator 1 often to remain idle. If Operator 2 is busy or if draw machine is down for some reason, Operator 1 will stay idle. Operator 1 can pass a maximum of two bars and tubes from his workstation to the v-roll 2 and 3, then he has to wait for the machine or Operator 2 to complete the task. After finishing the first order, if the second order does not arrive on the tube table, Operator 2 would be idle. This is because Operator 2 has to wait for the bundle tag to start his setup for the next order as the information needed for the setup is recorded on the bundle tag on the tubes. Missing any parts like strippers, squeezer, and other tools also cause operators to stop the operation as they try to find the parts. All the manual works performed by the operator at this workstation need to be minutely studied before one can find proper ways to automate the process. Manual works need to be automated one after another. At this point it is very essential to separate the work done by the operator and by machine. There are 12 tasks associated at this workstation out of which 9 tasks are done manually, and only 3 are automated. Out of these 12 tasks, only 2 tasks are considered value-added. The remaining ten manual jobs need to be automated to eliminate non-value-added tasks.

At Workstation 2 the inspection task is conducted manually, and it is the bottleneck station. This is the most important operation, which determines the ID, OD, wall thickness, and length of the tube. In spite of two operations in one workstation, there is only one operator assigned to perform these operations due to a long setup time. The long setup time at Workstation 2 often causes Operator

1 and Operator 3 to remain idle. Operator 1 has to wait for Operator 2 to take the tube, whereas Operator 3 waits for Operator 2 to pass the tube to his workstation. So if we break down the system and look at the beginning of a new job when setup is done, Workstation 2 becomes the bottleneck station that prevents greater throughput. It is also the rate-limiting step that determines the capacity of the process. Identifying the bottleneck process allows one to improve the particular process and increase the flow of whole value stream [4]. The long setup time is due to the associated manual inspection like measuring OD, length, and wall thickness of the tube the operator has to conduct.

Automation of the draw and reeling operations will decrease the setup time and increase the tube flow through the process. There are 20 tasks associated with Workstation 2 out of which 11 are done manually and only 7 tasks are value-added. The entire inspection tasks are conducted manually in the draw and reeling operations, which is the root cause for the machine stoppage as he has to engage himself in the inspection task.

The main problem of Workstation 3 is the inventory build up for the cooling of tube. The operators have their own ways of estimation to determine the temperature of tubes to sink. There is no any specified temperature recorded for the sink operation, and the first tube that goes through the sink operation varies with size of the tube, OD, wall thickness, and length. From the Kaizen events, operators revealed that the use of fans under the incline tube holder 1 has shown better result, and the front part of the tube gets colder faster than the back part of the tubes. The operator pointed out that fans were installed only in the front part but not in the back part of the tube holder rack. The hot tubes yield the bad quality surface in the tube, which the operator has been inspecting with naked eyes. Another problem is as the tubes come out of the sink operation they bend for some unknown reason. Even the experienced operators are unable to figure out the root causes. Although there are devices installed to hold the tubes, still the problem exists. There are 15 tasks in this workstation, out of which 7 are automated and are considered value-added. Almost half of the tasks need automation at this workstation.

There are problems at Workstation 4, and automation is only possible once these problems are fixed. The tube has to cool down to certain temperature before it is loaded in the v-roll for cutting operation. Hot tube does not contribute to producing better edge quality of the tube. There are 13 tasks and only 4 tasks are automated and only 6 tasks are value-added. The company purchased a new automated saw, but there were some problems associated with the new saw. First, the operator has to depend on the maintenance crew to change the saw. Moreover, the manual saw operator can pass the end of first tube to the wash machine and begin cutting the front end of the next tube. In contrast, the automated saw has to wait till the entire tube passes through the tube wash in order for it to perform a cut at new tube. In addition, those bent tubes produced from the sink operation need special attention while cutting and passing through the tube wash. Sometimes the operator has to hold the tube straight by hand until the operation is completed.

69.3 Conclusions and Recommendations

The study aimed to automate the Draw operation to make it more effective and economic, without major renovation. The automation of the draw cell can be implemented in future by adding more modern machines. Changes in the tools used and modernizing the process will help the company increase the productivity. All the workstations in the Draw Cell involve the setup of the machine, counting of the tubes, receiving bundle tag, and recording data manually. To better automate the setup process, they need to apply the method of Single Minute Exchange of Die (SMED) to reduce die changeover time to less than 10 min. The SMED enables the ability of workstations to manufacture smaller lots, reduce the inventory, and improve the customer responsiveness [5]. The SMED can be performed by having an operator make external setup tasks while the tube is being processed in the machine. The internal setup needs to be simplified by eliminating nonessential operations.

The company needs to invest on developing a Visual Factory, which is also a tool of Lean, where the data can flow through the computer system instead of through the Bundle Tags. Once the order is received it has to be generated on the computer system, and order data has to be visible to all operators, managers, or any other staffs within the company. The system should also be able to track the present status and new updates of the order. This would improve the communication and eliminate the wait time operators are facing at present. It would also make the state and condition of manufacturing processes easily accessible and very clear to everyone.

For any major reoccurring problems, they need to perform root cause analyses, or 8D analyses. For example, at present the plant has many problems without solutions. Like, why the tube bends at sink operation? Why the surface of the tube does not meet quality if processed at certain temperature? Therefore, methods focusing on resolving the underlying problems instead of applying quick fixes which only treats immediate symptoms of the problem must be found and implemented.

Instead of developing full energy and resources to automation, the company can follow the principles of Jidoka to achieve its ultimate goal. Jidoka is the design of equipment to partially automate the manufacturing process. This is a much less expensive approach and can utilize existing equipment instead of investing on new costly equipments. After Jidoka, single operator can frequently monitor multiple stations, reducing the labor cost. And many quality issues can be detected immediately [5]. The future state Draw Cell needs to be developed so single operator can control it with one main control unit as shown in the Fig. 69.2. The automated system should be able to control all the operations and be able to inspect the quality of the tubes. If any failure occurs, the system should be able to shutdown itself and inform the operator about the issue.

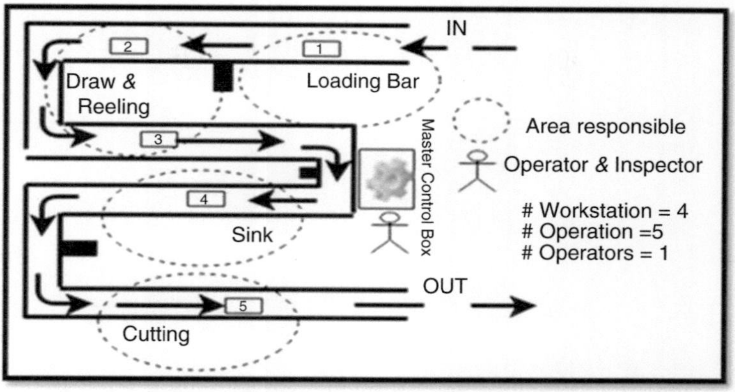

Fig. 69.2 Future state generic assembly line layout

References

1. Shaheen, L.: Tube Drawing Principles: Understanding Processes, Parameters Key to Quality. Publication of the Fabrications & Manufacturers Association International. http://www.thefabricator.com/article/tubepipeproduction/tube-drawing-principles
2. http://www.mmsonline.com/columns/applying-value-added-principles
3. Gene, S., Randolph, G., et al.: Applying lean principles and Kaizen rapid improvement events in public health practice. J. Public Health Manag. Pract. **18**(1), 52–54 (2012)
4. http://www.velaction.com/bottleneck/
5. http://www.leanproduction.com/top-25-lean-tools.html

Chapter 70
Classification of the ECG Signal Using Artificial Neural Network

Andrew Weems, Mike Harding, and Anthony Choi

Abstract Recording of electrocardiogram (ECG) signals and the correlation to cardiovascular diseases are a major problem in today's society. A common abnormality is arrhythmia, which is unexpected variation in cardiac rhythm. The goal of this study is to analyze these types of signals and find a more efficient way to classify these signals. Currently, medical devices for detecting ECG signals are at least 85 % accurate in analyzing the data. Neural networks have progressed quickly over the past few years, and have the capability of recognizing many types of variation in these signals. The pattern recognition power of Artificial Neural Networks (ANNs) is a valuable tool when classifying ECG signals in cardiac patients. Data obtained from the PhysioBank ATM was used to analyze the structure of an ANN and the effect that it has on pattern recognition. The results show that only one misclassification occurred resulting in an accuracy of 96 %.

Keywords Artificial neural network • ECG • MATLAB • Signal classification • Cardiac abnormalities • Cardiac arrhythmias

70.1 Introduction

Cardiovascular diseases are the number one killer worldwide of heart-related deaths, which encompass approximately one out of every three deaths. Most of these are sudden cardiac deaths after myocardial infarction and 90 % are due to cardiac arrhythmias. This type of effect is typically caused by the deficiency of oxygen transported to the coronary artery. Coronary artery disease is due to athermanous narrowing and occlusion of the coronary vessel. This, in turn, can manifest as angina, silent ischemia, unstable angina, myocardial infarction, heart failure, and sudden death. Some studies show that abnormal symptoms can arise

A. Weems • M. Harding
Department of Biomedical Engineering, Mercer University, Macon, GA 31207, USA

A. Choi (✉)
Department of Electrical and Computer Engineering, Mercer University,
Macon, GA 31207, USA
e-mail: choi_ta@mercer.edu

J. Juang (ed.), *Proceedings of the 3rd International Conference on Intelligent Technologies and Engineering Systems (ICITES2014)*, Lecture Notes in Electrical Engineering 345, DOI 10.1007/978-3-319-17314-6_70

Fig. 70.1 A typical ECG signal

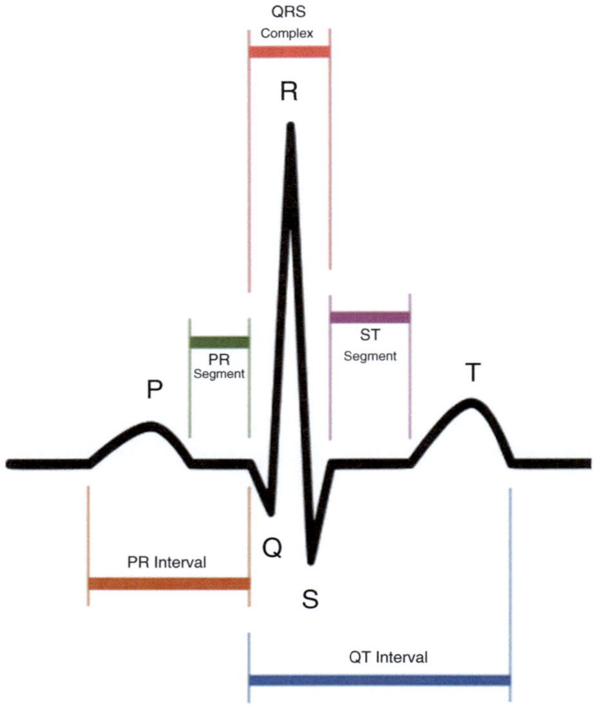

before the sudden occurrence of a heart attack. If these symptoms can be detected and diagnosed early, there would be time to prevent a heart attack or start treatment [1, 2].

An ECG signal is one of the most frequently taken measures from the human body. This simple and noninvasive measure of looking at the heart leads to the determination of irregular function. A typical ECG contains specific areas of interest that indicate each process occurring within the heart (Fig. 70.1). The P-wave represents the atrial contraction. The largest signal, the QRS-complex, is indicative of the ventricular contraction that sends blood throughout the body. The magnitude of this signal is larger due to the increased electrical signal that is provided to the larger muscular tissue. The T-wave represents the ventricular recovery time and subsequent refilling of the atrium. As many abnormalities are consistent with irregular ventricular function, the QRS-complex is arguably the most frequently analyzed signal of the ECG. For many cardiovascular problems, the duration and/or amplitude of this part of the signal reveals information for classifying a specific type of cardiac arrhythmia. It may be possible to decrease the occurrence frequency of severe cardiac conditions if monitoring systems were in place as detectors for abnormal rhythms [3]. These abnormal rhythms are referred to as cardiac arrhythmias and are caused by irregular firing patterns from the sinoatrial node (SA) or other areas of signal origin [4]. The accuracy of artificial neural networks (ANNs) for the classification of specific signals of ECG have been shown to be very accurate (within 99.0 % or greater) [3, 5, 6]. Just as a physician determines abnormalities through visual inspection of these signals, ANNs mimic this process through training of normal data and classifying areas of potential abnormality occurrence. In an attempt to collect biological signal data, artifacts may exist from sources unrelated to the area of interest. ECG artifacts that are present can change these signals and result in incorrect signal analysis. Therefore, it is important to identify potential sources of these artifacts and eliminate them as much as possible. Without this complete removal, misdiagnosis is highly probable [1–4, 7].

The heart is one of the most important organs in the body, as it is in control of the blood transport system that delivers oxygen and nutrients to the body. This process is controlled by the autonomous nervous system, which regulates the heart. If these communication signals are not propagated appropriately or if muscular contraction irregularities exist, it can have a devastating effect on the body [4]. The heart beat is generated by the movement of the heart walls. The atria chambers of the heart are the entry points for blood. These chambers prime the heart and then the ventricles provide the pulsations of the muscle walls, which in turn, produce the flow rate of the blood.

In order for sufficient cardiovascular function, the cardiovascular tissue must be able to receive a propagated signal in order for contraction to occur. At rest, the heart has a negative membrane potential, which will approach zero as ions cross-cellular membranes to reach ionic equilibrium. When the equilibrium is reached, muscular contraction occurs (depolarization). The entire region of cardiac muscle undergoes this process at an orderly rate. The signal originates in the SA-node and proceeds throughout length of the tissue. The signal is then propagated through the atrium and then to the ventricles. As the signal travels, this produces a voltage gradient that is detected between the electrodes (leads) that are attached to the chest of the patient. For the basic ECG detection, only three of these leads are needed. As more electrodes are added, many other types of cardiovascular signals can be obtained, yielding many potential electrical viewpoints for cardiac observation. As a result, diseases that affect only specific regions of the heart, such as myocardial infarction, can be better identified [8–11].

Of the signal abnormalities analyzed, the first one discussed is myocardial infarction. Myocardial infarction (or heart attack) is the interruption of blood flow to portions of the heart, resulting in oxygen deprivation and eventual tissue death. This normally occurs in the coronary arteries, which means that the blockage is occurring on the surface of the heart itself. Many heart attacks (approximately a quarter of patients) do not exhibit specific symptoms of a heart attack [8, 12–15]. Myocardial infarctions can result in cardiac failure, but do not necessary imply that the heart muscle has ceased to output blood and an electrical signal [16, 17]. This case can result in a continuation of the ECG signal from patients that are suffering from this condition.

Cardiomyopathy, known as heart muscle death, is defined as the deterioration of the heart muscle. This condition is a long-term onset condition, with the potential of slight changes in the ECG signal over time [16, 17]. This potential change occurs due to reduction in signal magnitudes resulting from muscular atrophy. In most cases, this disease classification is reserved for conditions that lead to heart failure [18].

Dysrhythmia is classified as any arrhythmia where there is abnormal electrical activity of the heart. Many signal indicators of dysrhythmias exist including irregular beat frequency, atrial fibrillation or flutter, and premature supraventricular or ventricular ectopic beats. These arrhythmias can be life-threatening, while others are only prevalent in palpitations [19].

Other signals that have been classified accurately with ANNs include arrhythmias, bundle branch block, and carditis [3, 5, 9–12, 16–18, 20, 21]. These signals are potential signs of future myocardial infarction. Bundle branch Block (BBB) is in a class of heart problems called Intraventricular Conduction Defects (IVCD). The two types of BBB are right and left. The right BBB carries nerve impulses that cause contraction of the right ventricle and vice versa. BBB consists of a slow or interrupted transmission of nerve impulses. Typically, the patient has little or no symptoms unless BBB is severe. Left BBB is usually caused by other diseases like myocardial infarction and arteriosclerosis, while right BBB usually occurs less often from underlying heart disease. BBB is usually diagnosed by the length of the QRS wave. This duration is typically greater than 110 ms.

Carditis, an inflammation of the heart or its surroundings, can be further classified by the area that is affected. The three areas that could be affected by this inflammation are the pericardium, cardiac muscle, and the endocardium. These problems are called pericarditis, myocarditis, and endocarditis, respectively. Symptoms of this irregularity in patients include night sweats, weight loss, fever, and chest pain. Carditis is symptomatically very similar to myocardial infarction pain; therefore, pericarditis can be misdiagnosed as an acute myocardial infarction solely based on the clinical data. As a result of this misdiagnosis, further analysis on the ECG signal is required by the diagnostician. A normal heart signal has a width of 40–100 ms for the QRS pulse [22, 23]. The usage of neural networks to detect these patterns in cardiac signals of diseased patients has been widely recorded [24–26].

A problem for researchers interested in the various classification techniques of ECG signals using ANNs presents itself in common academic literature searches. Many published articles present a variety of approaches and network geometries to obtain optimal pattern recognition results. The proposed network is to implement a MATLAB-based ANN that will resolve geometric and structural problems. The network is then trained using ECG signals that are unfiltered, to maximize and display the power of pattern recognition applications in artificial neural networks.

70.2 Methods and Materials

The training data was obtained from the PTB Electrocardiogram database and from the PhysioBank ATM database. A visual example of the data used can be seen in Fig. 70.2. The network analyzed these signals for the patient and the ANN was used through a variety of algorithms to train, validate, and test the validity of the chosen network. Pattern recognition was confirmed by analysis of the data points and similarities in other training data.

Network Structure: The neural network was assembled using the pattern recognition toolbox in MATLAB software. The network used was the multilayer feed-forward network with backpropagation, as shown in Fig. 70.3.

Preliminary research and testing showed that the optimal design for neural network structure in classifying ECG signals includes input and output layers with two internal hidden layers. The input layer consisted of an input neuron for each input of the corresponding ECG signals from the dataset. The number of neurons in the hidden layer was altered for optimization through trial-and-error. The output layer consisted of six output neurons that correspond to each of the specific conditions being examined. Previous literature research also recommended six neurons in the hidden layer [12, 16]. It was later determined that only a single input neuron would be used for network data acquisition.

Using the MATLAB toolbox nftool, the initial geometry was set to two hidden layers with an adjusted neuron count from five to nine. Initially, the default network conditions were used as to include the Tansig function for neuron function.

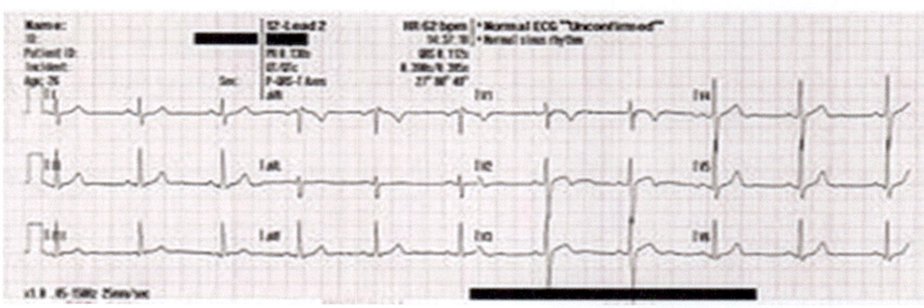

Fig. 70.2 Sample signal from the PTB database

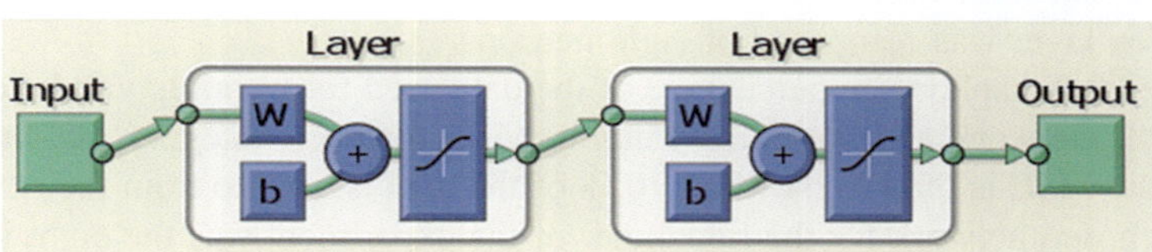

Fig. 70.3 The assembled neural network, recommended from literature searches

Table 70.1 Categorical distribution of data

Myocardial infarction	148
Cardiomyopathy/heart failure	18
Bundle branch block	15
Dysrhythmia	14
Carditis	14
Healthy controls	52

The chosen backpropagation learning algorithm featured the Levenberg-Marquadt algorithm [9].

PhysioBank has the capability of taking the recorded patient files and converting them into MATLAB compatible files. These files are read directly by the ANN in the training, validation, and testing phase of the pattern recognition network. The training data was compiled from PhysioBank's ATM, under the file of the PTB diagnostic ECG database and categorically classified to the six predetermined outputs of interest. These categories of interest include myocardial infarction, cardiomyopathy, bundle branch block, dysrhythmia, carditis, and normal healthy signals. Table 70.1 indicates the conditions present in the dataset used for training of the network and the number of signals for each desired output. This classification of data was supplemented using the associated header files that indicate physician diagnosis.

The majority of all of the signals are healthy, taken by a 14-lead ECG with a sampling frequency of 1,000 Hz. Splicing the healthy signals surrounding the disease markers in the ECG wave increased the number of healthy signals available for training the network.

Additional testing of the network was performed on the selected design. This data was taken from multiple sources, in order to check the network's accuracy of the signals. During this testing, it was observed that the symptoms of the abnormalities of interest persisted for a period of time prior to correction, either naturally or by physician intervention.

70.3 Results

The mean squared error (MSE) of the network geometry analysis (Fig. 70.4) corresponds to the number of neurons in the hidden layer for different network geometry that were tested. As shown in Fig. 70.4, the MSE was minimized when the hidden layer was composed of eight neurons.

Table 70.2 displays the percentage of the data used for training, validation, and testing the network and the corresponding MSE for each category. As the table shows, the MSE is the lowest when 70 % of the data is used to train the ANN.

The chosen network for the remainder of the experiment was the eight neuron/hidden layer, with a validation set of 70 %, 15 %, and 15 % for training, validation, and testing, respectively. The data was taken from the PhysioBank ATM, but not

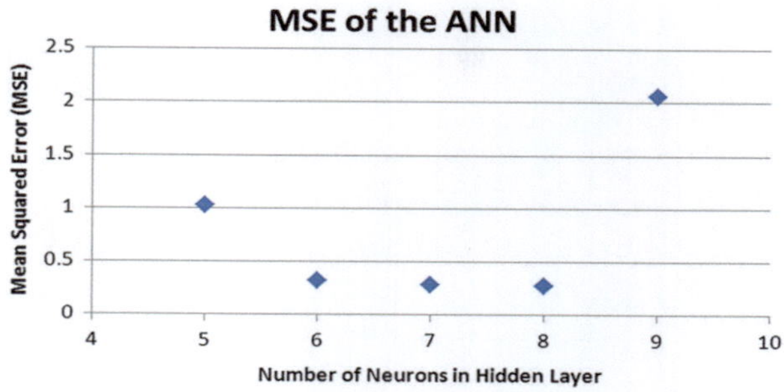

Fig. 70.4 The mean squared error of different numbers of neurons tested in the hidden layer

Table 70.2 Validation of the network with associated repetitions and data ratios

Network validation (eight neurons per hidden layer)			
Training repetitions	Validation	Testing	MSE
60	25 %	15 %	9.1757×10^{-1}
60	20 %	20 %	5.8033×10^{-1}
70	15	15	2.6969×10^{-1}
75	15	15	6.7451×10^{-1}
80	10 %	10 %	7.6969×10^{-1}
90	5	5	9.0573×10^{-1}

Table 70.3 Classification results

Type of signal	Number of cases	Accuracy
Healthy	52	99.1
Myocardial	148	86.4
Cardiomyopathy	18	90.1
Dysrhythmia	14	99.2
Bundle branch block	15	98.7
Carditis	14	98.9
Total	261	95.4

from the PTB diagnostic ECG compilation. Table 70.3 shows the total accuracy for each case using the chosen percentages for training, validation, and testing. The testing results of the dataset are shown in Table 70.3, indicative of each categorical type and associated accuracy for each type.

The network misclassified cases of myocardial infarction with cardiomyopathy. The error in analyzing this signal was 11 % according to the confusion matrix, shown in Fig. 70.5. Further tests and training were performed using additional data obtained from the PhysioBank ATM. The error found with classifying the myocardial infarction and cardiomyopathy signals was the only misclassification in the test confusion matrix. The other ECG signals were classified correctly without any problems.

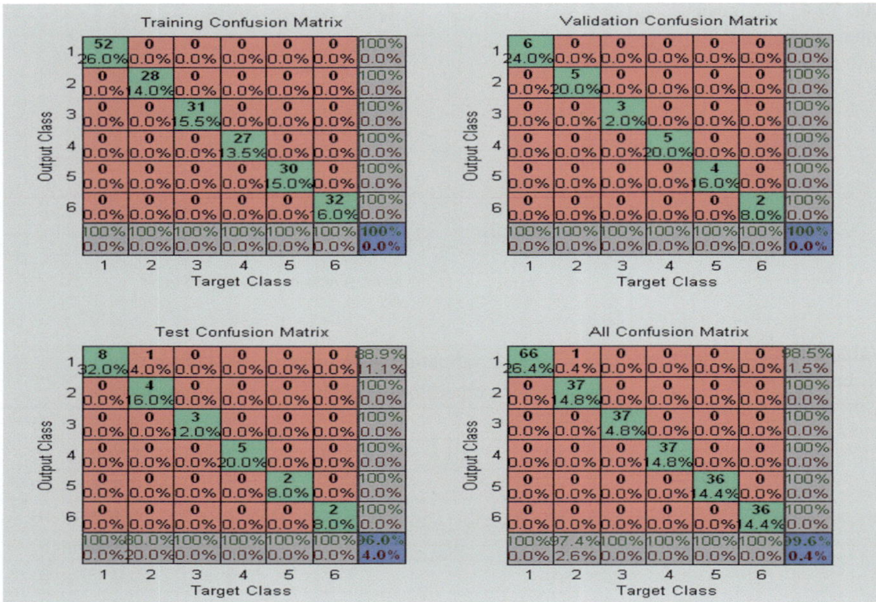

Fig. 70.5 Test results of the neural network

70.4 Discussion

The geometry of the neural network was optimized using trial-and-error for the testing data. This resulted in a four-layer neural network, with one input neuron, eight neurons in each of the two hidden layers, and six output neurons. The network functioned very well with the test data, as shown in the previous results. The overall accuracy of the eight neuron network was 95.4 %. This value was then compared to other training data approximations to ensure that the ideal network would be used in the classification of future data. The accuracy of this network is consistent with other networks recorded in literature of this type [3, 5, 9–12].

The issues with classification occurred most frequently for cardiomyopathy and myocardial infarction. This may be due to some similarities between the ECG waves of the two diseases. The T-wave of the cardiomyopathy patients is generally flattened slightly, which can appear as a decreased signal wave. Most analysis usually involves the QRS wave. When the ventricles are repolarizing, the signal is not as strong and as a result, the signal is weaker. Also, myocardial infarction is characterized by the decrease in the signal of T-wave. It may also slightly minimize the QRS portion of the wave, resulting in symptoms that are similar to patients with cardiomyopathy. In other cases, the waveform of the ECG may not be diseased at all for myocardial infarction patients, even in the midst of an attack. Abdominal ultrasound is then required to confirm that the output of the heart is not correct, despite the signal to the contrary. This indicates that future analysis to find a better system that can differentiate these two abnormalities should be completed. Although a limited dataset was used in the development of this network, only one major misclassification of data occurred. Future work will include analyzing more data from these two cardiac irregularities in order to better distinguish between the two and achieve higher accuracy.

One technique that was not used for this experiment was filtering the quality of the data that was presented to the neural network. In addition to not filtering the data, the raw amount of data was not filtered from what would be taken directly from a 14-lead ECG of a patient. Both of these factors may have added variation to the data sets that the network was trying to learn. Future studies could filter the data in order to decrease potential variation in the data to help the ANN achieve better results. For cardiomyopathy, the onset of heart muscle death is normally a gradual process. Therefore, the diagnosis of an oncoming myocardial infarction as cardiomyopathy could have disastrous results for the patient in the immediate future.

For trained physicians, time is needed to definitively differentiate between the two conditions. In future work, differing results could be obtained by lumping both of these conditions into single category. This category could be labeled as an immediate threat to patient condition. By providing a category with this label, additional recommendations could be attached to such an output from the network. An abdominal ultrasound, in conjunction with this NN output, could be used to determine between the two categories with a potential increase in the accuracy of the neural network. The other misclassifications of the neural network were significantly less frequent, and provided a similar ratio of correct diagnoses as physicians. This was characteristic of signals that are different in amplitude and frequency (T, QRS waves). Few minor changes could be made to optimize this network for future implementation in clinical settings.

70.5 Conclusion

Artificial neural networks are a powerful tool that can be used to diagnose cardiac patients with certain disorders. Signals that have distinct differences in the ECG waveform are simple for an ANN to classify correctly. Signals that have similar characteristics, or that do not display distinct characteristics, are more difficult for the network to classify accurately. The proposed artificial neural network currently exhibits minor confusion associated with the misclassification of cardiomyopathy and myocardial infarction. Both of these abnormalities exhibit very different symptoms and require different treatments, but have similar signals in the ECG waveform. The network successfully classified other abnormalities with high accuracy. With the increase of future research and optimization, neural networks will achieve higher accuracy and have the potential for integration of medical devices such as pacemakers.

Acknowledgments We would like to thank Dr. Edward O'Brien (professor/chair of Biomedical Engineering Department at Mercer University School of Engineering) for his knowledge of ECG and the electrical activity in the heart. Also, Roland Adams (Electrical and Computer Engineering Department at Mercer University School of Engineering) for his assistance with MATLAB and the initial problems encountered with the study.

References

1. Benchaib, Y., Chikh, M.: A specialized learning for neural classification of cardiac arrhythmias. J. Theor. Appl. Inf. Technol. **6**(1), 92–100 (2009)
2. Kavitha, K., Ramakrishnan, K., Singh, M.: Modeling and design of evolutionary neural network for heart disease detection. Int. J. Comput. Sci. Issues **7**(5), 272–283 (2010)
3. Ghorbanian, P., Ghaffari, A., Nataraj, C.: Heart arrhythmia detection through continuous wavelet transform and principal component analysis with neural network classifier. Comput. Cardiol. **37**, 669–672 (2010)
4. Vijaya, V., Rao, K., Rama, V.: Arrhythmia detection through ECG feature extraction using wavelet analysis. Eur. J. Sci. Res. **66**(3), 441–448 (2011)
5. Ozbay, Y., Karlik, B.: A recognition of ECG arrhythmias using artificial neural networks. In: Proceedings—23rd Annual Conference, 23, pp. 1–5 (2001)
6. Aprasit, W., Laosen, N., Chevakidagarn, S.: Data filtering technique for neural networks forecasting. WSEAS Int. Conf. Simul. Model. Optim **7**, 225–230 (2007)
7. Baranchuk, A., Shaw, C., Alanazi, H., Campbell, D., Bally, K., Redfearn, D., Simpson, C., Abdollah, H.: Electrocardiography pitfalls and artifacts: the 10 commandments. Crit. Care Nurse **29**(1), 67–73 (2009)
8. Ayub, S., Saini, J.P.: ECG classification and abnormality detection using cascade forward neural network. Int. J. Eng. Sci. Technol. **3**(3), 41–46 (2011)
9. Ball, R., Tissot, P.: Demonstration of artificial neural network in Matlab. Div. Nearhsore Res. **12**, 1–5 (2006)
10. Kannathal, N., Acharya, U., Lim, C., Sadasivan, P., Krishnan, S.: Classification of cardiac patient states using artificial neural networks. Intell. Inf. Syst. Conf. **8**(4), 206–211 (2003)
11. Nugent, C., Lopez, J., Smith, A., Black, N.: Prediction models in the design of neural network based ECG classifiers: a neural network & genetic programming approach. BMC Med. Inform. Decis. Mak. **2**(1), 1–6 (2002)
12. Hedén, B., Öhlin, H., Rittner, R., Edenbrandt, L.: Acute myocardial infarction detected in the 12-lead ECG by artificial neural network. Dep. Clin. Physiol. Cardiol. **96**, 1798–1802 (1997)
13. Silipo, R., Marchesi, C.: Artificial neural network for automatic ECG analysis. Trans. Signal Process. **46**(5), 1417–1425 (1998)
14. Maglaveras, N., Stamkopoulos, T., Diamantaras, K., Pappas, C., Strintzis, M.: ECG pattern recognition and classification using non-linear transformations and neural networks: a review. Int. J. Med. Inform. **52**(1), 191–208 (1998)
15. Goslar, T., Podbregar, M.: Acute ECG ST-segment elevation mimicking myocardial infarction in a patient with pulmonary embolism. Cardiovasc. Ultrasound **8950**, 1–7 (2010)
16. Stamkopoulos, T., Diamantaras, K., Maglaveras, N., Strintzis, M.: ECG analysis using non-linear pca neural networks for ischemia detection. Trans. Signal Process. **46**(11), 3058–3067 (1998)
17. Bortolan, G., Willems, J.: Diagnostic ECG classification based on neural networks. J. Electrocardiol. **75**(9), 1–6 (1993)
18. Suzuki, Y.: Self-organizing QRS-wave recognition ECG using neural networks. Trans. Neural Netw. **6**(6), 1469–1477 (1995)

19. Manchanda, S., Ehsanullah, M.: Suspected cardiac syncope in elderly patients: use of the 12-lead electrocardiogram to select patients for Holter monitoring. Gerontology **47**, 195–197 (2001)
20. Edenbrandt, L., Devine, B., Macfarlane, P.: Neural networks for classification of ECG ST-segments. Eur. Heart J. **14**(4), 464–468 (1992)
21. Bortolan, G., Degani, R., Willems, J.: ECG classification with neural networks and cluster analysis. Comput. Cardiol. **20**, 177–180 (1991)
22. Belhachat, F., Izeboudjen, N.: Conception of intelligent classifiers for cardiac arrhythmias detection. Centre de Développement des Technologies Avancées, pp. 1–10 (2010)
23. Ceylan, R., Özbay, Y.: Wavelet neural network for classification of bundle branch blocks. In: Proceedings of the World Congress on Engineering, vol. 2, pp. 1–5 (2011)
24. Al-Fahoum, A., Howitt, I.: Combined wavelet transformation and radial basis neural networks for classifying life-threatening cardiac arrhythmias. Med. Biol. Eng. Comput. **37**(5), 566–573 (1999)
25. Papaloukas, C., Fotiadis, D., Likas, A., Michalis, L.: An ischemia detection method based on artificial neural networks. Artif. Intell. Med. **24**, 167–178 (2002)
26. Haraldsson, H., Edenbrandt, L., Ohlsson, M.: Detecting acute myocardial infarction in the 12-lead ECG using Hermite expansions and neural networks. Artif. Intell. Med. **32**, 127–132 (2004)

Chapter 71
Integration Method of Composite Pattern for Solving Structure Problems of Visitor Pattern

Jaehak Uam, Jeonghoon Kwak, and Yunsick Sung

Abstract This paper proposes a method in which the Visitor pattern is used to apply the Composite pattern for solving the structure problems of the scenario generator, based on Bayesian probability. The traditional Visitor pattern is not ideal for adding new member functions due to its structure problems. By applying the Composite pattern, our approach provides flexibility to interfaces for adding and deleting new member functions.

Keywords Internet of things • Scenario generator • Design pattern • Visitor pattern • Composite pattern

71.1 Introduction

In the Internet of Things (IoT) environments, the abundance of sensors and devices necessitates diverse scenarios for the validation of sensors and devices. Preparing numerous scenarios for validation costs time and money. This can be resolved by generating scenarios automatically [1–7].

There exists a research that generates scenarios automatically utilizing the principles of Bayesian probability [8]. Through this approach, new scenarios are generated by calculating and utilizing the Bayesian probability of measured and predefined scenarios. The generation of new scenarios for diverse environments require a scenario generator possessing a flexible structure. This needs *Design pattern* [9] to be applied. The *Structure pattern* within the Design pattern will be better suited to solve the structure problem for the scenario generator.

This chapter proposes a method that combines *Visitor* and *Composite* patterns to solve the structure problem of Visitor pattern. The member functions of the Visitor patterns are divided and contained into the classes based on *Composite* pattern separately, which can increase the extension ability of the Visitor pattern.

J. Uam • J. Kwak • Y. Sung (✉)
Department of Game Mobile Contents, Keimyung University, Daegu, South Korea
e-mail: jaehak@kmu.ac.kr; jeonghoon@kmu.ac.kr; yunsick@kmu.ac.kr

J. Juang (ed.), *Proceedings of the 3rd International Conference on Intelligent Technologies and Engineering Systems (ICITES2014)*, Lecture Notes in Electrical Engineering 345, DOI 10.1007/978-3-319-17314-6_71

This chapter is structured as follows. Section 71.2 describes design pattern-related research. In this section, the integration method applying Composite pattern to solve the structure problem of Visitor Composite is proposed. Section 71.3 introduces the case study of the proposed method. Finally, Sect. 71.4 provides conclusions.

71.2 Integration Method

In the proposed method, *ConcreateLeafVisitors* are new classes defined by binding the related member functions instead of the *Visitor*. Thereafter, the new classes are managed hierarchically, which can be handled by applying the *Composite* pattern. The *Composite* pattern includes *Component* and *Composite* to manage multiple *Leaves* hierarchically, where *Composites* include other *Composites* and Leafs. By utilizing the *Composite* pattern, the *ConcreateLeafVisitor* can add and delete member functions in a flexible manner. Therefore, the modification problem of the *Visitor* can be solved whenever member functions are added. The proposed pattern is defined as *Composited Visitor* pattern. Figure 71.1 shows the class diagram of the proposed *Composited Visitor* pattern.

The *Composite* pattern is applied to the *Visitor* pattern as follows. First, the *LeafVisitors* are managed by the *Composite* pattern. The *Components* of the *Composite* pattern include *Composites* and *LeafVisitors* corresponding to the *Leaf* of the *Composite* pattern and the *Visitor* of *Visitor* pattern. The *Composite* is an abstract class and super class of the *LeafVisitor* and the *Composite* and is

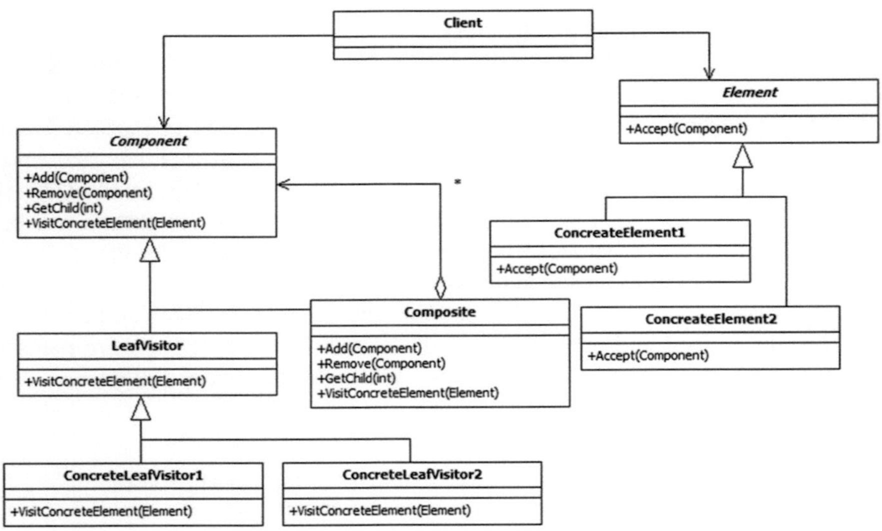

Fig. 71.1 Class diagram of *Composited Visitor* pattern

constructed hierarchically. Therefore, the *LeafVisitor* has the rolls of both the *Visitor* and *Leaf*. *VisitConcreteElement* implements an abstract function which contains a parameter as Element. Therefore, the subclasses derived from LeafVisitor should define the body of the abstract function.

Thereafter, the member functions of the *Visitor* are labeled as *Concrete-LeafVisitor*. The *ConcreteLeafVisitor* can be defined multiple times. In the case of the *Visitor* pattern, all member functions are included in one *Visitor*, which is tasked with the difficulty of expanding the *Visitor*. In the proposed method, because the member functions are divided into the *ConcreteLeafVisitors*, more member functions can be added and deleted without any modification of the *Concrete-LeafVisitor*. The *ConcreteLeafVisitor* is implemented by deriving it from *LeafVisitor* and overriding a *VisitConcreteElement* function.

Element is a super class of *ConcreteElement*. It defines an abstract function *Accept*, which receives a *Component* as a parameter. *ConcreateElement* should define the body of the abstract function *Accept*.

Finally, *ConcreteElement* is implemented by overriding the abstract function *Accept*, which calls the function *VisitorConcreteElement* referencing all *ConcreateLeafVisitors* registered through *Component* parameters.

Figure 71.2 shows the sequence diagram of creating *Composited Visitor* pattern. *Client* creates *Concreate Elements*, *Composite*, and *ConcreateLeafVisitors*.

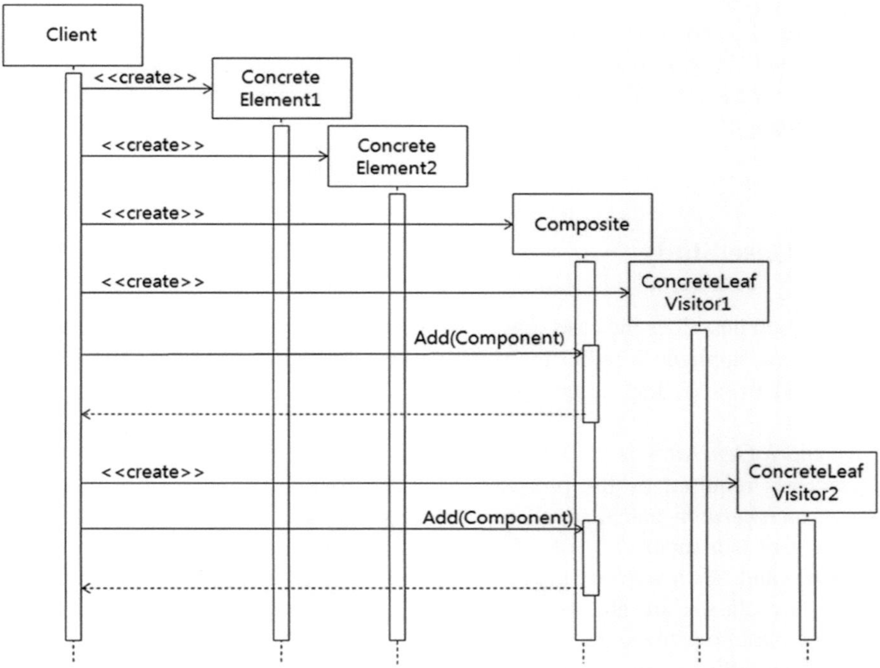

Fig. 71.2 Sequence diagram of creating *Composited Visitor* pattern

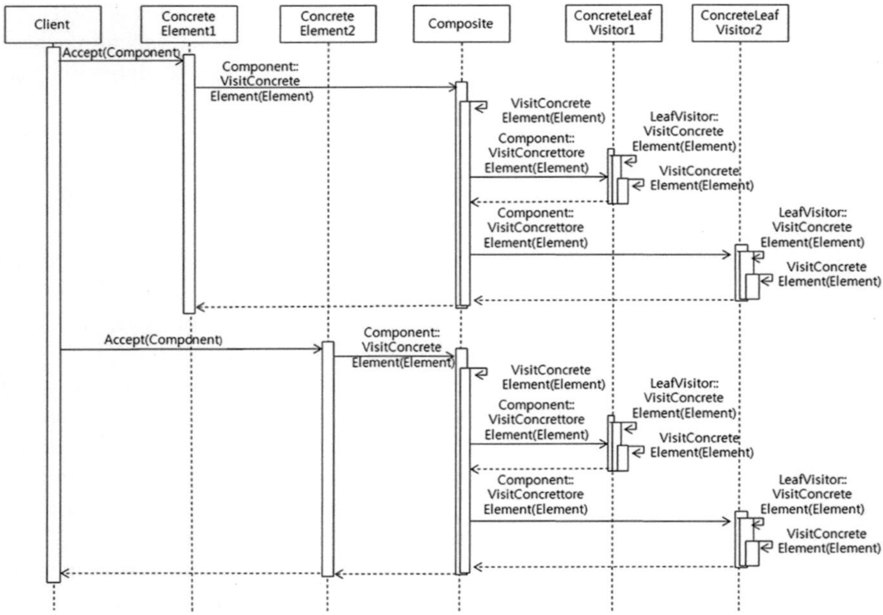

Fig. 71.3 Sequence diagram of Utilizing *Composited Visitor* pattern

Figure 71.3 shows the sequence of calling the member functions of the classes in *Composited Visitor* pattern. Whenever *Client* calls the function *Accept* of *ConcreteElement*, the *ConcreateElement* calls the function *VisitConcreteElement* hierarchically.

71.3 Case Study

This section describes the case study of a scenario generator to which the proposed method was applied. Whether the structure problem of the scenario generator can be solved by applying *Composited Visitor* pattern was validated, as shown in Fig. 71.4.

ScenarioGenerator is the main class of the scenario generator. After receiving parameters required by the processes of the scenario generator from users, the *ScenarioGenerator* managed all processes of the scenario generator. *AbstractComponent* is a super class of all *Handlers*, which manages events. *AbstractDataWriter* and *AbstractGenericDecorator* are derived from *AbstractComponent*. These two classes are abstract classes, which possess subclasses and define the abstract functions executed by the subclasses. The subclasses derived from *AbstractDataWriter* are in charge of the role of recording generated scenarios. The scenario generator includes one subclass of *AbstractDataWriter* and multiple

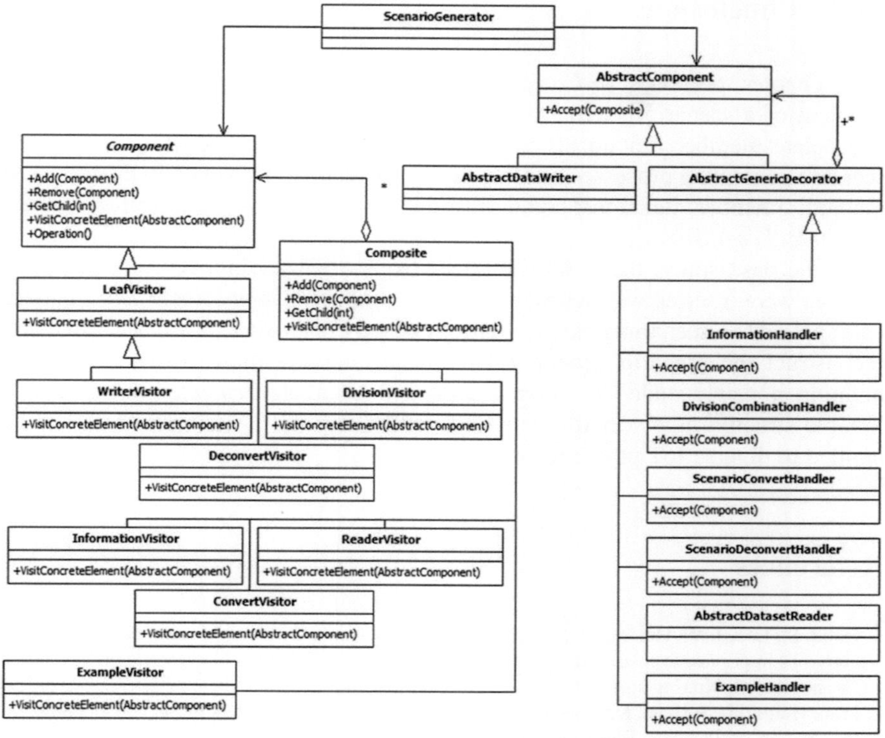

Fig. 71.4 The class diagram of *Composited Visitor* pattern with embedded classes

subclasses of *AbstractGenericDecorators*. Therefore, it can be concluded that the structure of the scenario generator was reconstructed by applying the *Decorator* pattern.

InformationHandler, *DivisionCombinationHandler*, *ScenarioConvertHandler*, *ScenarioDeconvertHandler*, and *AbstractDatasetReader* handle the generating roles of the scenario generator by dividing the entire process into multiple processes. These classes implement the function *Accept* after inheriting from *AbstractGenericDecorator*. The scenario generator defines *WriterVisitor*, *InformationVisitor*, *DivisionVisitor*, *ConvertVisitor*, *DeconvertVisitor*, and *ReaderVisitor* as Concrete-LeafVisitor. These classes implement VisitConcreteElement through overriding functions.

When the *ExampleHandler* is inserted into the scenario generator, no further modification to the *LeafVisitor* is invoked, except adding the new *ExampleVisitor* as *ConreteLeafVisitor*. As a result, when new member functions are added, no modification of *Visitor* is invoked. When some of the *ConreteLeafVisitors* are not needed, they can be deleted easily.

71.4 Conclusions

This chapter proposed a *Composited Visitor* pattern to solve the structure problem of a scenario generator. The previous scenario generator could not add and delete member functions flexibly because the configuration required by the scenario generating processes was handled by the *Visitor*. However, in the proposed method, member functions could be added without any modification of the *ConcretedLeafVisitors*.

In the case study, the modification invoked when additional *ConcretedLeaf-Visitors* were inserted was described after the proposed *Composited Visitor* pattern was applied to a scenario generator. The proposed *Composited Visitor* pattern could reconstruct the scenario generator without any modification to *LeafVisitors* by defining a newly added *ConcreteVisitor* and *ConcreteElement*. Owing to this flexible structure, the scenario generator with *Composited Visitor* pattern can be applied to diverse IoT environments.

References

1. Helal, S., Lee, J.W., Hossain, S., Kim, E., Hagras, H., Cook, D.: Persim—simulator for human activities in pervasive spaces. In: Proceedings of the 7th International Conference on Intelligent Environments, Nottingham, UK, pp. 192–199 (2011)
2. Pranevicius, H., Sutiene, K.: Scenario tree generation by clustering the simulated data paths. In: Proceeding of the 21st European Conference on Modelling and Simulation, Prague, Czech Republic, 4–6 June 2007
3. Pham-Nguyen, C., Garlatti, S.: Context-aware scenarios for pervasive long-life learning. IEEE/WIC/ACM International Conference on Web Intelligence and Intelligent Agent Technology, vol. 1, 9–12 Dec 2008, pp. 824–827, (2008)
4. Grois, E., Hsu, W.H., Voloshin, M., Wilkins, D.C.: Bayesian network models for generation of crisis management training scenarios. In: Proceedings of the Fifteenth National/Tenth Conference on Artificial Intelligence/Innovative Applications of Artificial Intelligence, AAAI 98/IAAI 98, pp. 1113–1120, July 26–;30, 1998, Madison, Wisconsin, USA
5. Kuchler, C., Vigerske, S.: Decomposition of multistage stochastic programs with recombining scenario trees, Stochastic Programming E-Print Series, 2007–09, 1–21 (2007)
6. Høyland, K., Kaut, M., Wallace, S.W.: Generating scenario trees for multistage decision problems. Manag Sci. **47**(2), 295–307 (2001)
7. Helal, A., Cho, K., Lee, W., Sung, Y., Lee, J.W., Kim, E.: 3D modeling and simulation of human activities in smart spaces. In: Proceeding of Ninth International Conference on Ubiquitous Intelligence and Computing and Ninth International Conference on Autonomic and Trusted Computing, Fukuoka, Japan, 4–7 Sept. 2012, pp. 112–119 (2012)
8. Sung, Y., Helal, A., Lee, J.W., Cho, K.: Bayesian-based scenario generation method for human activities. In: 2013 ACM SIGSIM Conference on Principles of Advanced Discrete Simulation (PADS), Montreal, QC, Canada, 19–20 May 2013, pp. 147–157 (2013)
9. Erich, G., Helm, R., Johnson, R., Vlissides, J.: Design Patterns: Elements of Reusable Object-Oriented Software. Pearson, Indianapolis (2007)

Chapter 72
A Circuit Design of Dimming T8 LED Tube for High Power Factor Correction

Manh Tran Van, Wei-Sung Weng, Chun-Shan Liu, Shen-Yuar Chen, and Wei-Ching Chuang

Abstract An advanced method to raise the power factor (PF) with fewer components than the traditional PFC circuits is proposed. It also provides an efficient solution to supply power for Light-Emitting Diodes (LEDs). The structure of driver and control circuit for T8-LED tube is designed and verified as well.

Keywords Pulse-width modulation (PWM) • Flyback converter • Power factor correction

72.1 Introduction

High-brightness light-emitting diodes (LED) have been regarded as the "green" light source for next generation due to high efficiency, environmental friendliness, long lifetime, etc. With the significant improvement in the performance of high-brightness LEDs, they hold promise towards many applications, and might even be considered for common use in daily lighting systems [1, 2]. However, how LEDs can be driven with high efficiency, long lifetime, and low cost is still the key challenge lying ahead.

Recently, control methods for dc–dc converters to drive LEDs efficiently with constant current have been proposed [3]. When the LED drivers utilize the commercial ac voltage source instead of a dc one, such as a lithium–ion battery, an ac–dc power supply is required to convert the ac voltage to a regulated dc one for the drivers. In daily lighting applications, the input power factor is an important index, which needs to be higher than 0.9 for most commercial luminaries as required by regulation standards such as the Energy Star [4]. So, the ac–dc power supply must be designed to achieve power factor correction (PFC).

M.T. Van • W.-S. Weng • S.-Y. Chen • W.-C. Chuang (✉)
Department of Electro-Optical Engineering, National Formosa University, Yunlin 632, Taiwan
e-mail: eocwc@nfu.edu.tw

C.-S. Liu
Department of Electrical Engineering, National Formosa University, Yunlin 632, Taiwan

J. Juang (ed.), *Proceedings of the 3rd International Conference on Intelligent Technologies and Engineering Systems (ICITES2014)*, Lecture Notes in Electrical Engineering 345, DOI 10.1007/978-3-319-17314-6_72

In this chapter, a two-stage configuration to control the brightness of LED is demonstrated. A completed prototype facilitates the measurement of some waveforms and necessary parameters of this circuit.

72.2 System Design and Experimental Implementations

The proposed LED dimming system for high PFC is depicted diagrammatically as in Fig. 72.1. In the first stage, the Flyback converter, performing as an active PFC circuit, is introduced to raise the power factor as high as possible by using the controller L6561. The second stage is the boost converter to tune the brightness of the T8 LED by applying the UC3845 controller with peak current mode control. The Buck converter, converting 17 V DC to 3.5 V DC, supplies power to the MUC as well.

The Flyback converter, Fig. 72.2, is based on Transition Mode (TM, i.e., on the boundary between continuous and discontinuous inductor current mode) operating at a frequency depending on both input voltage and output current. The L6561D controller is incorporated into the Flyback converter, as shown in Fig. 72.2.

It is quite different from the conventional Flyback converter. The input capacitance is so small that the input voltage is very close to a rectified sinusoid. The control loop has a narrow bandwidth so as to be little sensitive to the twice mains frequency ripple appearing at the output. Figure 72.3 shows the internal circuit block diagram of the L6561D.

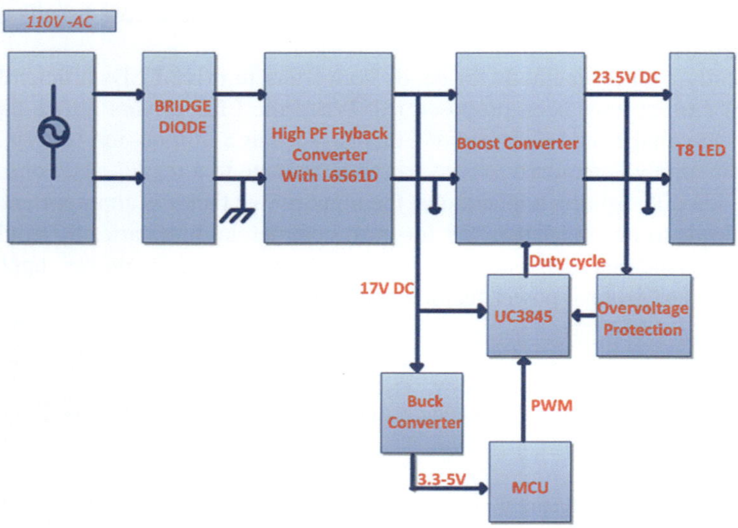

Fig. 72.1 The schematic of a LED dimming system

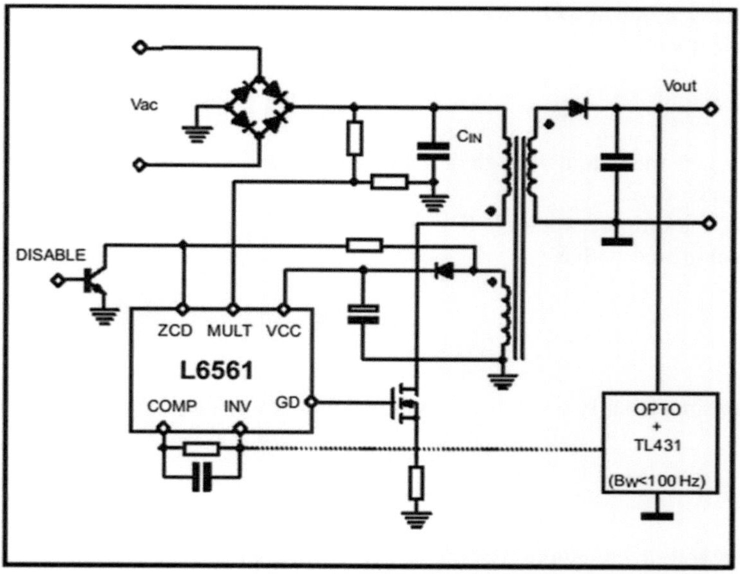

Fig. 72.2 Schematic of high power factor Flyback converter

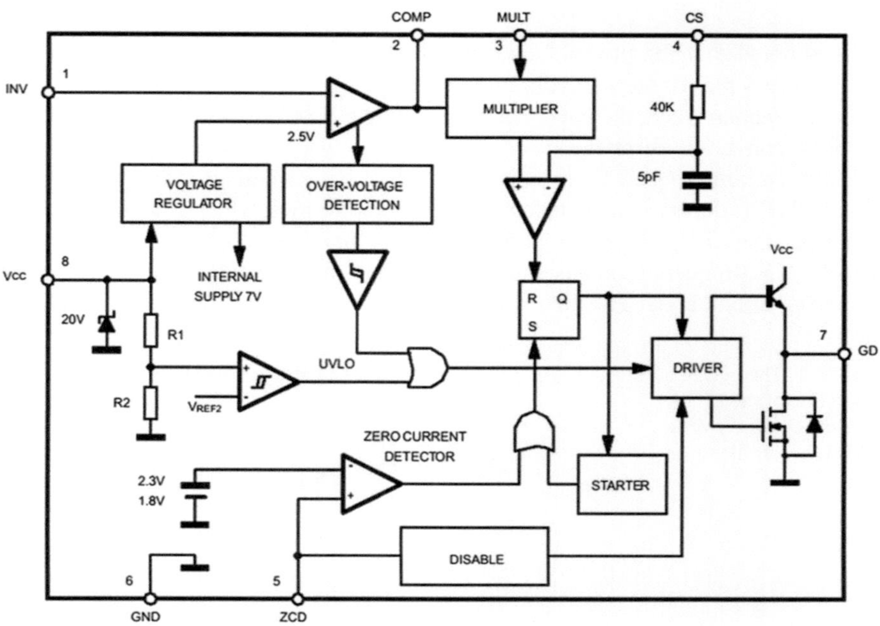

Fig. 72.3 Internal block diagram of L6561D [5]

To analyze how the Flyback converter with the L6561D raises the power factor, a specific quantity, called reflected voltage V_R, is defined by

$$V_R = n \cdot (V_{out} + V_f) \quad (72.1)$$

where V_{out} is the output voltage of the converter, and V_f is the forward drop on the output catch diode.

Then, the ratio between the line peak voltage V_{pk} and the reflected voltage V_R, will be indicated with K_V,

$$K_V = \frac{V_{pk}}{V_R} \quad (72.2)$$

If the sinusoidal line voltage is assumed, the power factor could be defined as:

$$\mathrm{PF} = \frac{\text{Real input power}}{\text{Apparent input power}} \quad (72.3)$$

The theoretical prediction of the power factor could be obtained, and it depends only on the K_V value. Their relation is depicted in Fig. 72.4. For practical application, PF could be approximated by

$$\mathrm{PF}(K_V) = 1 - 8.1 \times 10^{-3} K_V + 3.4 \times 10^{-4} K_V^2 \quad (72.4)$$

It is noted that the power factor approaches unity and much larger than 0.9, which is feasible for the LED dimming circuits.

In the second stage, the boost converter and the buck converter, that follows the Flyback converter, form the T8 LED dimming system. There are two popular methods for dimming LEDs in switched-mode driver circuits: Pulse-Width Modulation (PWM) dimming and analog dimming. The former, which is actual start and restart of the LED current for short periods of time, is adopted in the proposed system. The frequency of this start–restart cycle must be fast enough to avoid a flickering effect, about 200 Hz or faster is usually acceptable. MCU generates the

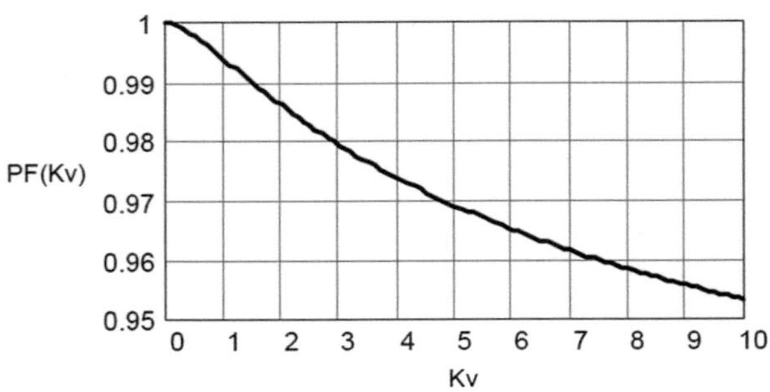

Fig. 72.4 Theoretical prediction of the PF for Flyback converter with L6561D controller

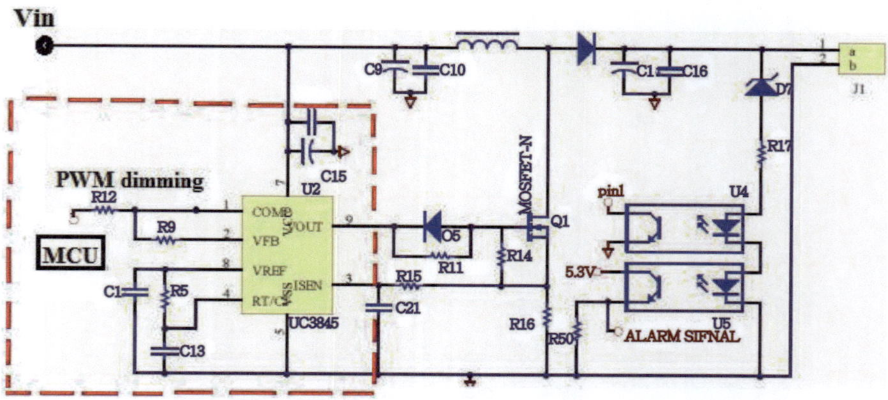

Fig. 72.5 The PWM dimming circuit

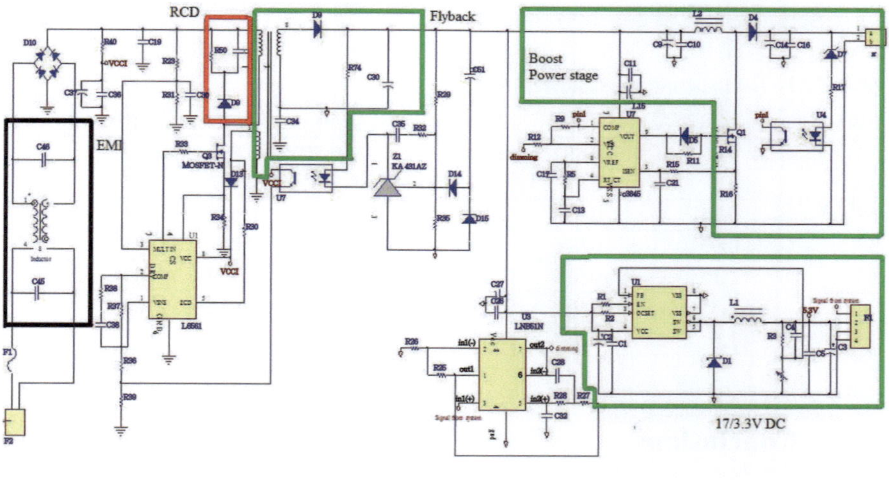

Fig. 72.6 Schematic of the complete LED dimming system

PWM signal which has a duty cycle change in the range of 0–100 %. The PWM signal is filtered and supplied to the UC3845 by R12 resistor. Figure 72.5 shows the actual circuit diagram of the PWM dimming circuit. The schematic of the complete circuit, generated from Altium software, is shown in Fig. 72.6.

A prototype circuit device that concludes the designed parameters and their components is shown in Fig. 72.7, and the experimental results, measured by using the power analyzer (PM 1000+, Agilent DSO—X 3024A) are listed in Table 72.1.

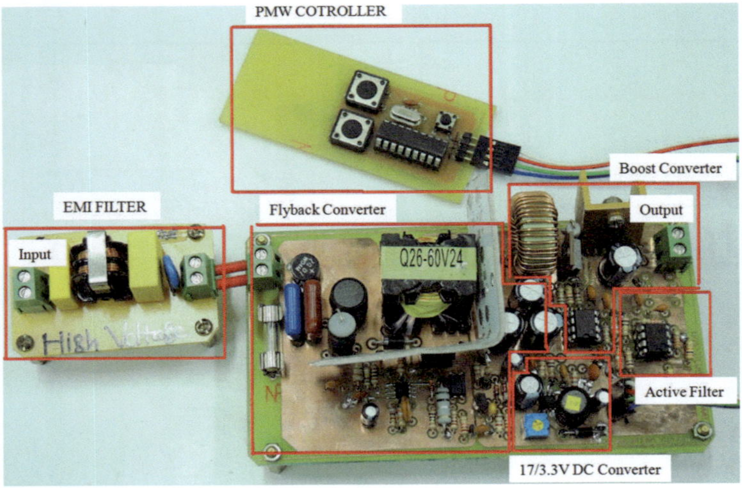

Fig. 72.7 The prototype of the T8 LED dimming circuit

Table 72.1 The experimental results

PWM (%)	I load (A)	PFC	P_{in} (W)	V_{out} (V)	P_{out} (W)	Efficiency (%)
20	0.3	0.892	9.46	20.99	6.297	67
30	0.5	0.92	15.7	21.5	10.75	68
40	0.75	0.974	21.4	22	16.5	77
50	0.95	0.976	26.2	22.3	21.85	81
60	1.15	0.985	32.32	22.78	26.197	81
70	1.35	0.989	37.6	22.9	30.915	82
100	1.6	0.99	43.57	23.316	37.3	85

72.3 Conclusions

A single phase power factor correction circuit with a wide output voltage and current range was demonstrated. The topologies of Flyback and Boost Converter are completely built in order to supply power for T8 Tube LED. The MCU dimming circuit helps to control the duty cycle of the UC3845. Therefore the LED's current is controlled as well. The experimental results show the PFCs of the prototype circuits reach 0.99 while the PWM is 100 %. High Power factor correction with the Flyback converter performed by a L6561 driver is an important requirement in some standards of the world. Active Power factor makes T8 Tube LED behave as a resistive load so that the input current waveform is in phase with input voltage waveform.

References

1. Wendt, M., Andriesse, J.W.: LEDs in real lighting applications: from niche markets to general lighting. In Proc. IEEE IAS, 2601–2603 (2006)
2. Tsao, J.Y.: Solid-state lighting: lamps, chips and materials for tomorrow. IEEE Circuits Devices Mag. **20**(3), 28–37 (2004)
3. Carraro, G.: Solving high-voltage off-line HB-LED constant-current control-circuit issues. In Proc. IEEE APEC, 2007, 1316–1318 (2007)
4. *ENERGY STAR*^{R} Program requirements for solid state lighting luminaires.eligibility criteria—version 1.1, USA (2008). http://www.energystar.gov/index.cfm?c=new_specs.ssl_luminaires
5. Design equations of high-power-factor Flyback converters based on the L6561 (AN1059)

Chapter 73
Anti-ESD Improvement by the Bulk-FOX Structure in HV nLDMOS Devices

Shen-Li Chen, Shawn Chang, Yu-Ting Huang, and Shun-Bao Chang

Abstract This work is focused on a 0.25 μm 60 V high-voltage nLDMOS devices which will be integrated with a FOX structure in the bulk region, and evaluate the impacts on its anti-ESD protection ability. It is found that as an FOX sturcture adding, and as the FOX area ratio is increased, I_{t2} value will be enhanced too. When the FOX area ratio is about 83.5 %, I_{t2} value has a maximum value ~6A. However, as the FOX area ratio is increased, the R_{on} value will be declined. From the experimental data, it is revealed that V_{t1} (V_h) value decreased more than 16.9 % (35.6 %), anti-ESD ability increased more than 170.2 %, and R_{on} decreased more than 81.2 % as compared with the Ref. DUT.

Keywords Electrostatic discharge (ESD) • Field oxide (FOX) • Holding voltage (V_h) • n-Channel lateral-diffused MOS (nLDMOS) • On-resistance (R_{on}) • Secondary breakdown current (I_{t2}) • Trigger voltage (V_{t1})

73.1 Introduction

High-voltage laterally diffused MOS transistors (LDMOS) are often used in a power I/O driver circuits or a high-voltage (HV) electrostatic discharge protection device, so how to design a high anti-ESD ability and high latch-up immunity are in need during a high-voltage operating environment. In order to achieve the required ESD ability, ESD protection devices are often designing with a large cell size area. Usually, a designing by a parallel multi-finger structure is used to reduce the cell across length. Thus, a typical multi-finger gate-grounded nMOS (GGnMOS) component is widely used in the ESD protection circuits. However, many studies have shown that a multi-finger structure of GGnMOS cannot uniformly turn-on in an ESD bombardment [1–4]. Therefore, for a large cell area GGnMOS, a sub-transistor of all the fingers uniformly turned-on is difficult especially in a

S.-L. Chen (✉) • S. Chang • Y.-T. Huang • S.-B. Chang
Department of Electronic Engineering, National United University, MiaoLi City 36003, Taiwan
e-mail: jackchen@nuu.edu.tw

J. Juang (ed.), *Proceedings of the 3rd International Conference on Intelligent Technologies and Engineering Systems (ICITES2014)*, Lecture Notes in Electrical Engineering 345, DOI 10.1007/978-3-319-17314-6_73

high-voltage process, thus causing high-voltage circuits cannot have an excellent anti-ESD capability.

Revealed from the past literatures, many nLDMOS techniques were to reorganize the structure of LDMOS device, such as the embedded-type SCR structure [5–9], or the source-side engineering of component studies [10, 11]. These efforts are shown that an nLDMOS device has a serious uneven turned-on problem and the holding voltage too low issues. Then, is there any other solutions? Therefore, a systematic changing and modulation of the cell layout and structure of an nLDMOS device will be proposed in this chapter. This method is focused on the anti-ESD capability and latch-up immunity considerations; it is hoping to effectively improve the high-voltage components reliability capabilities.

73.2 Device Structures of DUTs

Figures 73.1 and 73.2 are the cross-section and schematic layout of an HV nLDMOS device, respectively. The bulk side does not integrate with any FOX structure; we called it as the reference DUT. And, all the experimental test devices were fabricated by a TSMC 0.25 μm 60 V BCD process. The channel length (L) is kept to be 2 μm, channel width of each finger (W_f) is 100 μm, finger numbers $M = 6$, and the total channel width (W_{tot}) is kept at constancy, 600 μm. In this chapter, an increasing the R_{bulk} value of nLDMOS at the source-side to bulk region will be focused. Therefore, under the minimum design rule of an FOX region, the area of an FOX (0.58 μm × 5.22 μm) was fixed to insert into the bulk region. The number of these FOX blocks is from minimum one, then it is gradually increasing the number of inserting blocks as 2, 3, 4, 5 (see Fig. 73.3) and finally the number of blocks is 17. Here, with inserting the FOXs into bulk side is used to observe the impact on anti-ESD protection ability.

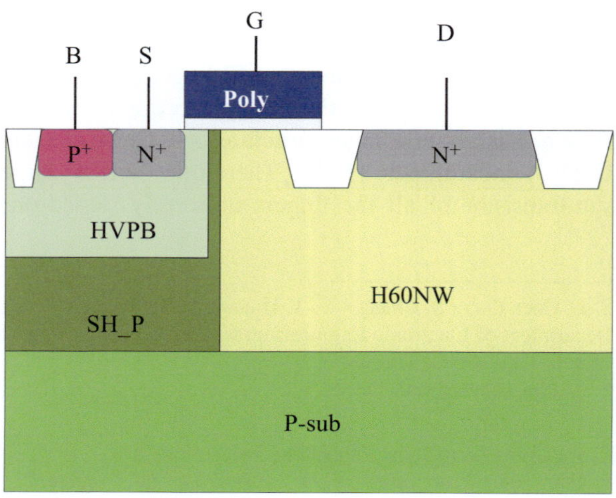

Fig. 73.1 Cross-section diagram of an HV nLDMOS device without adding any FOX structures in the bulk region (Ref. DUT)

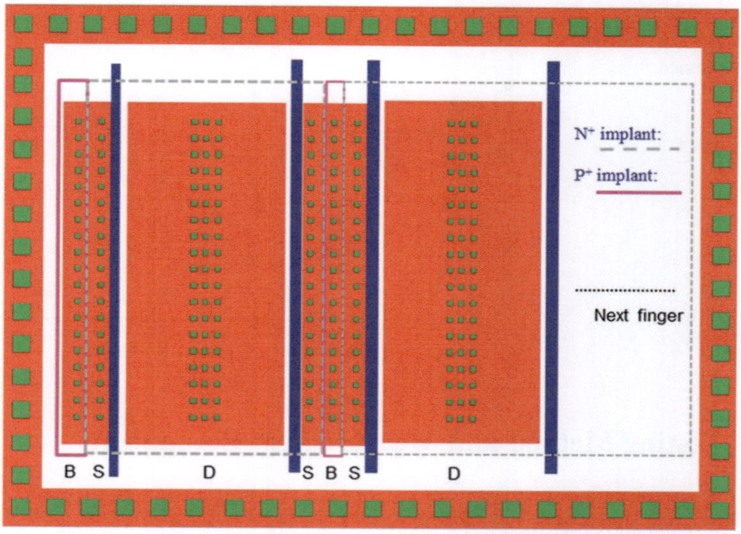

Fig. 73.2 Layout scheme of an HV nLDMOS device without adding any FOX structures in the bulk region (Ref. DUT)

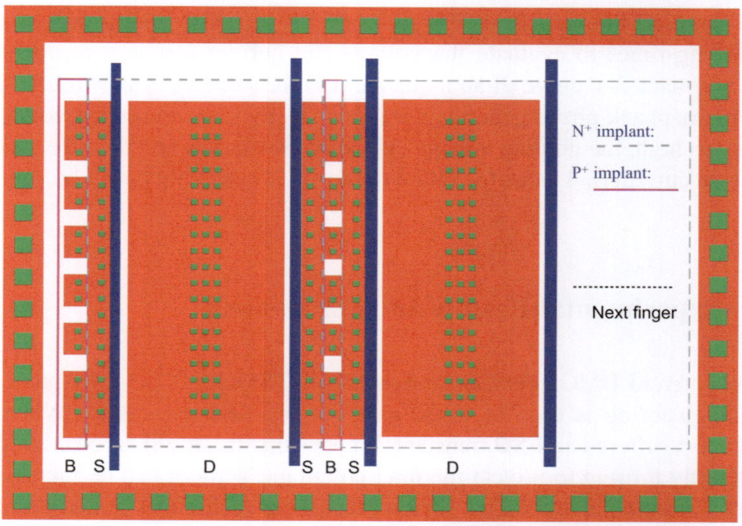

Fig. 73.3 Layout scheme of an HV nLDMOS as FOXs adding in the bulk region (# of bulk area replaced by FOX's in bulk region is 5)

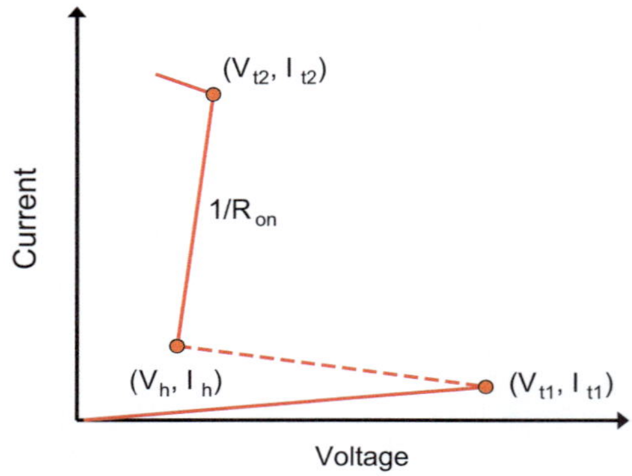

Fig. 73.4 Typical snapback *I–V* curve of an nLDMOS device

73.3 Testing Method and Measurement Equipment

A transmission-line-pulse (TLP) system for experimental testing is controlled by the LabVIEW software. It managed the subsystem electrical machine such as the ESD pulse generator, the high-frequency digital oscilloscope and the digital power electric meter instruments to achieve the automatic measurement. This machine can provide a continuous step-high square wave to device, and short raise time of the continuous square wave can also simulate transient noise of ESD. This HBM-like system has used the short square wave with 100 ns pulse widths and less 10 ns rising/falling times to evaluate the voltage and current response of device. The typical snapback *I–V* curve of an n-channel LDMOS device is shown in Fig. 73.4, due to its snapback phenomenon is very obvious, then the triggering voltage (V_{t1}) value is too high, the holding voltage (V_h) value is too low and then the secondary breakdown current (I_{t2}) value (or called the anti-ESD capability) will be very small.

73.4 Experimental Results and Discussions

In this chapter, nLDMOS devices fabricated by a TSMC 0.25 μm 60 V process with a bulk engineering is used to investigate the anti-ESD capability. When an IC external port suffered an ESD transient, it hopes that these ESD protection devices can promptly turn-on to protect the internal circuit, while it can be increasing the resistance of bulk-side-to-source-region to let the parasitic BJT quickly turned-on. Therefore, in this chapter, with increasing the source-side-to-bulk-side resistance by adding the FOX blocks in the bulk region will be proposed to evaluate the ESD protection response.

The testing results of device *I–V* snapback behaviors were shown in Fig. 73.5 and Table 73.1. After the bulk region adding FOX slot structures, V_{t1} and V_h values will decrease as the FOX ratio increased. Due to the R_{bulk} value will be increased as

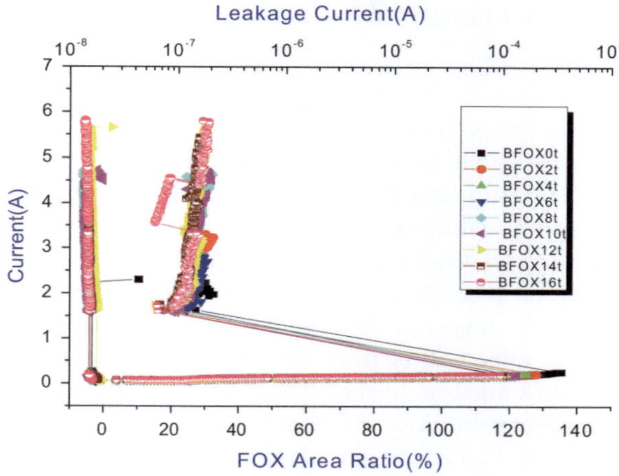

Fig. 73.5 Snapback *I–V* and leakage current curves of different FOX adding structures in the bulk region of HV nLDMOS devices

Table 73.1 Snapback key parameters of nLDMOS DUTs as FOXs adding in the bulk side (σ was averaged by three DUTs)

Sample type (stripe number of FOX adding)	V_{t1} (V)	V_h (V)	I_{t2} (A)	R_{on} (Ohm)	FOX area ratio (%)
0 (none adding)	133.684	27.632	2.208	11.42	0.0
1	125.745	21.561	3.479	5.99	5.2
2	124.91	20.689	4.070	5.39	10.4
3	123.684	20.399	4.139	4.57	15.7
4	122.32	20.252	4.392	5.63	20.9
5	122.49	19.852	4.395	3.44	26.1
6	121.865	20.245	4.550	3.94	31.3
7	122.384	19.647	4.625	3.47	36.5
8	122.177	21.659	4.789	3.09	41.8
9	120.665	19.413	5.457	2.98	47.0
10	121.018	20.845	4.945	3.21	52.2
11	120.497	19.109	5.008	3.06	57.4
12	118.438	18.774	5.445	2.63	62.6
13	119.427	19.125	5.159	2.45	67.9
14	118.69	18.788	5.140	2.53	73.1
15	119.247	19.724	4.921	3.25	78.3
16	117.472	18.746	5.966	2.15	83.5
17	116.747	17.778	4.889	3.24	88.7

the FOX ratio increasing, it makes the V_{t1} and V_h values decreased and shown in Table 73.1. Mysteriously, the I_{t2} value will increase as the FOX ratio increased and shown in Table 73.1. Moreover, the R_{on} value will decrease as the FOX ratio increased too. Due to easily happen the uniform turned-on behavior, overall the on-resistance will slowly decrease shown in Table 73.1.

73.5 Conclusions

This chapter focuses on the anti-ESD robustness enhancement of nLDMOS devices which fabricated by a 0.25 μm 60 V BCD process and added with FOX's blocks in the bulk region. By the bulk region adding FOX structures, and through the actual samples testing, the I_{t2} value will be increased as the ratio of FOX's increasing. For the 83.5 % adding ratio of FOX's structure, the I_{t2} has a highest value of about 6 A, it means that it has a very good anti-ESD capability. Moreover, the V_{t1} value will be decreased as the ratio of FOX's increasing too. As more FOX slots added in the bulk region, the higher an R_{bulk} value, which will make the V_{t1} decreased. Meanwhile, the R_{on} value will be decreased as the FOX ratio is increasing. It is resulting in a uniform turned-on phenomenon happened originally, then the value of R_{on} resistance decreased slowly. In this work, as the FOX's adding to the bulk region, which can let the V_{t1} (V_h) decreased more than 16.9 % (35.6 %), the capability of anti-ESD increased more than 170.2 %, and R_{on} decreased more than 81.2 % as compared with the Ref. DUT.

Acknowledgments In this work, authors would like to thank the National Chip Implementation Center in Taiwan for providing the process information and fabrication platform. And, authors would like to acknowledge the financial support of the Ministry of Science and Technology of Taiwan, through grant number MOST 103-2221-E-239 -014.

References

1. Pendharkar, S., Teggatz, R., Devore, J., Carpenter, J., Efland, T., Tsai, C.-Y.: A novel LDMOS device with ESD robustness. In: 12th International Symposium on Power Semiconductor Devices and ICs, 341–344 (2000)
2. Ker, M.-D., Chen, J.-H., Hsu, K.-C.: Self-substrate-triggered technique to enhance turn-on uniformity of multi-finger ESD protection devices. In: IEEE VLSI-TSA International Symposium on VLSI Technology, 17–18 (2005)
3. Ker, M.-D., Chen, J.-H.: Self-substrate-triggered technique to enhance turn-on uniformity of multi-finger ESD protection devices. IEEE J. Solid-State Circ. **41**(11), 2601–2609 (2006)
4. Chen, S.-L., Lee, M.-H., Lai, Y.-S., Lin, C.-J.: Influences of source pick-up and well engineering on the ESD robustness of LV process nMOSTs. In: IEEE International Future Energy Electronics Conference (IFEEC), 740–745 (2013)
5. Zhang, P., Wang, Y., Jia, S., Zhang, X.: Study of LDMOS-SCR: a high voltage ESD protection device. In: 10th IEEE International Conference on Solid-State and Integrated Circuit Technology, 1722–1724 (2010)
6. Liu, S.-Y., Sun, W.-F., Pan, H.-W., Wang, H., Qian, Q.-S.: A novel latch-up free SCR-LDMOS for power-rail ESD clamp in half-bridge driver IC. In: IEEE 11th International Conference on Solid-State and Integrated Circuit Technology, 1–3 (2012)
7. Huang, Y.-C., Dai, C.-T., Ker, M.-D.: Self-protected LDMOS output device with embedded SCR to improve ESD robustness in 0.25-μm 60-V BCD process, 116–119 (2013)
8. Ma, F., Zhang, B., Han, Y., Zheng, J., Song, B., Dong, S., Liang, H.: High holding voltage SCR-LDMOS stacking structure with ring-resistance-triggered technique. IEEE Electr. Device Lett. **34**(9), 1178–1180 (2013)

9. Altolaguirre, F.A., Ker, M.-D.: Power-rail ESD clamp circuit with embedded-trigger SCRdevice in a 65-nm CMOS process. In: IEEE 57th International Midwest Symposium on Circuits and Systems, 250–253 (2014)
10. Chen, W.-Y., Ker, M.-D., Jou, Y.-N., Huang, Y.-J., Lin, G.-L.: Source-side engineering to increase holding voltage of LDMOS in a 0.5-μm 16-V BCD technology to avoid latch-up failure. In: 16th IEEE International Symposium on the Physical and Failure Analysis of Integrated Circuits, 41–44 (2009)
11. Fujiwara, S., Nakaya, K., Hirano, T., Okuda, T., Watanabe, Y.: Source engineering for ESD robust NLDMOS. In: IEEE 33rd Electrical Overstress/Electrostatic Discharge Symposium (EOS/ESD), 1–6 (2011)

Chapter 74
N^+ Extended-Distribution Influences on Anti-ESD Ability in the 60-V pLDMOS-SCR (NPN-Arranged-Type)

Shen-Li Chen, Yu-Ting Huang, Shawn Chang, and Shun-Bao Chang

Abstract In order to effectively improve the reliability capability, a p-channel lateral-diffused MOSFETs with an embedded SCR which is formed by implanting N^+ doses in the both sides of the drain end, and this structure called as the *npn*-arranged-type of pLDMOS-SCR in this chapter (diffusion region of drain-side is N^+-P^+-N^+). Then, changing the layout manner of N^+ implants in both sides of a drain-side P^+ region is investigated in this chapter by a 0.25-μm 60-V BCD process. In this planning idea, the layout types of N^+ region are continuously extended into the drain-side. From the experimental results, due to all of their secondary breakdown current (I_{t2}) values are so good that reached above 7 A, it can be found that the layout manner of continuous extended types in the drain-side have a little impact on the ESD capability. However, the major repercussion is the V_h value will be decreased about 10.8–49.5 %.

Keywords Electrical overstress (EOS) • Electrostatic discharge (ESD) • Latchup (LU) • p-Channel lateral-diffused MOSFET (pLDMOS) • Secondary breakdown current (I_{t2}) • Transmission-line pulse (TLP) • Trigger voltage (V_{t1})

74.1 Introduction

Lateral-diffused MOSFET (LDMOS) devices are widely used in many applications nowadays, such as communication modules, power electronic components, power management circuits, automotive electronics, and LCD drivers [1–4], therefore its importance is increased in the HV applications. Due to its high operating voltage, it needs to have a good reliability. Then, expects extremely these devices have a good ability in the anti-ESD (electrostatic discharge) and/or anti-EOS (electrical overstress).

S.-L. Chen (✉) • Y.-T. Huang • S. Chang • S.-B. Chang
Department of Electronic Engineering, National United University, MiaoLi City 36003, Taiwan
e-mail: jackchen2100@hotmail.com

J. Juang (ed.), *Proceedings of the 3rd International Conference on Intelligent Technologies and Engineering Systems (ICITES2014)*, Lecture Notes in Electrical Engineering 345, DOI 10.1007/978-3-319-17314-6_74

Usually, an n-channel LDMOS device has a large cell area especially in the high-voltage driver application; therefore it can be served as an ESD protection component for its low on-resistance behavior. However, it has several obvious shortcomings which including the V_{t1} value is too high and the device with a multi-finger structure can't completely turn-on, then it will result in per unit length of the ESD capacity is too low. On the other hand, a conventional SCR is also used in the HV applications due to have a very strong ESD capacity per unit length, but it also has some disadvantages especially the V_h value is very low. Then, there are some studies brightening an idea to combine these two components [5–8]. An ESD protection circuit in the power port, the nLDMOS device is easily prone to the latchup happening, so using a pLDMOS will have some advantages; however the pLDMOS has a higher V_h value than an nLDMOS, and its anti-ESD efficiency is lower than an nLDMOS. In this chapter, the N^+ implants used in the drain-side of a pLDMOS to form a pLDMOS with an embedded SCR structure, so that these two elements can be corporated their characteristics to achieve a high anti-ESD and a high anti-latchup (LU) abilities. Meanwhile, the experimental test components of pLDMOS and pLDMOS-SCR were fabricated by a TSMC 0.25-μm 60-V BCD process. The channel length (L) is kept to be 2-μm, channel width of each finger (W_f) is 100-μm, finger numbers $M = 6$, and the total channel width (W_{tot}) is kept at constancy, 600-μm.

74.2 Layout Design of HV Test Devices

74.2.1 pLDMOS-SCR (npn-Type) with a Drain-Side N^+ Continuously Extended

Figures 74.1 and 74.2 are the cross-section and schematic layout of a pLDMOS-SCR, respectively. From this cross-section illustration, a pLDMOS is divided into three zones in the drain-side, and then the N^+ doses implanted into the first region and third region to form parasitic SCR structures, which called the pLDMOS-SCR npn-type structure. From Figs. 74.1 and 74.2, a reference DUT is defined as an original embedded striped-type SCR structure of the pLDMOS device which was not modulated in the drain-side. In the next step, the layout type of these N^+ regions from the original striped-type was changed to the N^+ region continuously extended toward the middle region of a P^+ diffusion region by some fixed spacing intervals. Then, the N^+ diffusion (parasitic SCR) area of a pLDMOS will be increased, which is used to verify whether its anti-ESD ability of a pLDMOS-SCR will be increased or not?

The experimental groups of DUTs are shown in Table 74.1, where there are five different planning parameters. The interval between two adjacent extended are kept two contact spacing and the segment lengths of every N^+ extension area inserting into the middle P^+ zone are kept to be 3, 11, 24, 63, 128 contact spacing.

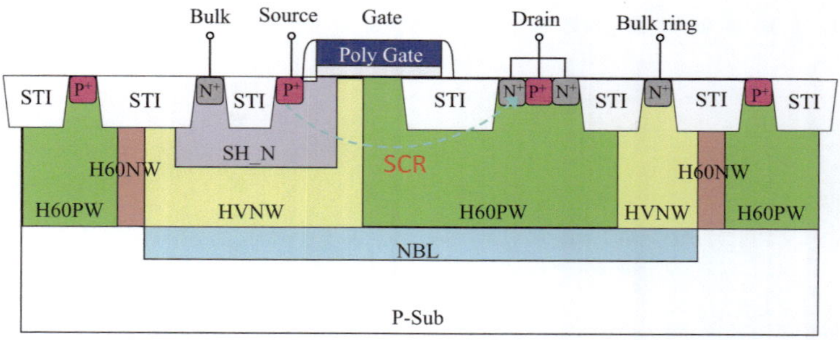

Fig. 74.1 Cross-section diagram of an HV pLDMOS-SCR with the npn-striped-type (Ref. DUT)

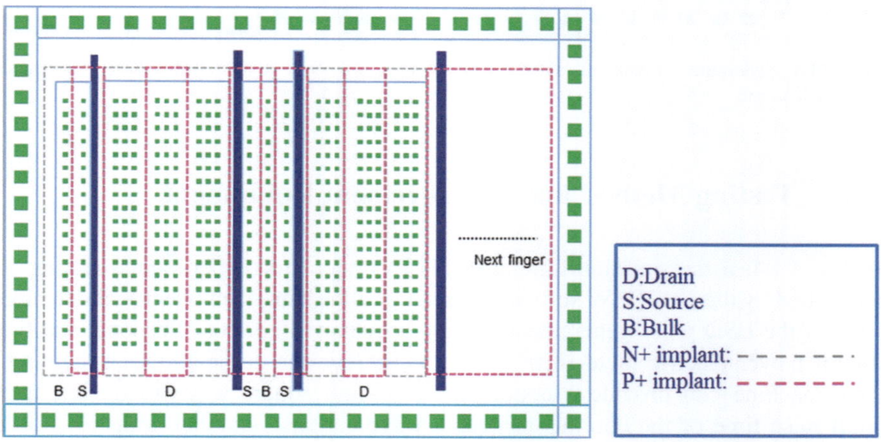

Fig. 74.2 Schematic layout of an HV pLDMOS-SCR with the npn-striped-type (Ref. DUT)

Table 74.1 Device lists of pLDMOS-SCR (npn) DUTs with an N^+ continuously extended type

pLD_npn_DE3_2co
pLD_npn_DE11_2co
pLD_npn_DE24_2co
pLD_npn_DE63_2co
pLD_npn_DE128_2co

Meanwhile, for example, a DE3_2co layout device of the high-voltage pLDMOS-SCR (npn-type) with an N^+ continuously extended manner in the drain-side is shown in Fig. 74.3. Therefore, the N^+ to P^+ area ratios of these samples are well adjusted. Therefore, these DUTs will be compared with the reference DUT (a pure pLDMOS-SCR (npn striped-type)). Eventually, from this planning, it hopes to find an optimum ratio value for the best anti-ESD ability.

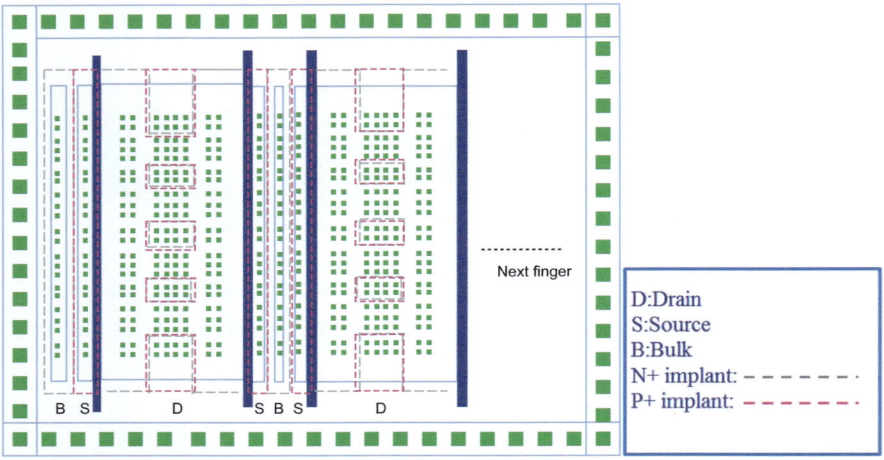

Fig. 74.3 Schematic layout of a pLDMOS-SCR DUT with a P^+ continuously extended (DE3_2co) type

74.3 Testing Method and Measurement Equipment

In Fig. 74.4, a transmission-line-pulse (TLP) system for experimental testing is controlled by the LabVIEW software. It managed the subsystem electrical machine such as the ESD pulse generator, the high-frequency digital oscilloscope, and the digital power electric meter instruments to achieve the automatic measurement. This machine can provide a continuous step-high square wave to device, and short raise time of the continuous square wave can also simulate transient noise of ESD. This HBM-like system has used the short square wave with 100 ns pulse widths and less 10 ns rising/falling times to evaluate the voltage and current response of device.

74.4 Experimental Results and Discussions

74.4.1 pLDMOS-SCR (npn-Type) with a Drain-Side N^+ Continuously Extended

Snapback *I–V* curves of HV pLDMOS-SCR DUTs (npn-arranged) with a drain-side N^+ continuously extended manner are shown in Fig. 74.5. It can be found that all the I_{t2} values of these DUTs are greater than 7 A (due to the electric power limitation of this TLP measurement system, a measurement will be stopped when the internal current of DUTs is more than 7 A). Meanwhile, the V_{t1}, V_h, and I_{t2} distributions are shown in

Fig. 74.4 A TLP measurement system

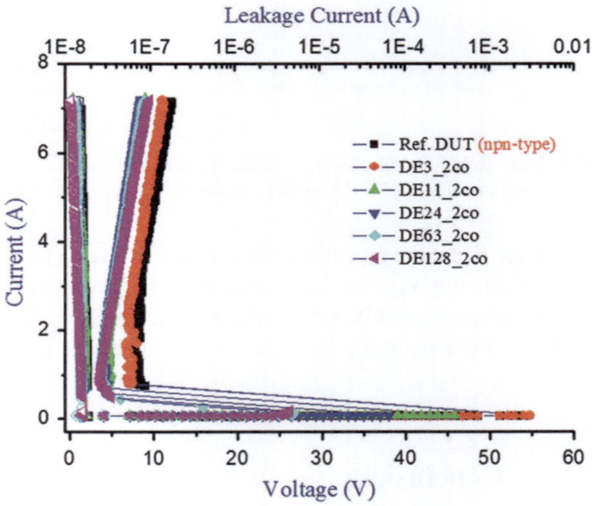

Fig. 74.5 Snapback *I–V* curves and leakage currents of pLDMOS-SCR DUTs (npn) N^+ drain-side continuously extended manner in the drain-side

Figs. 74.6 and 74.7a. And, the breakdown-voltage distribution by using a Keithley 2410 SMU is shown in Fig. 74.7b (under the Vg = Vs = VDD bias condition).

Here, the N^+ continuously extended type in drain-side makes the parasitic SCR (N^+) area expansion. In other words that strengthens the contribution of the parasitic SCR. So, the capability of anti-ESD which we originally predict will significantly

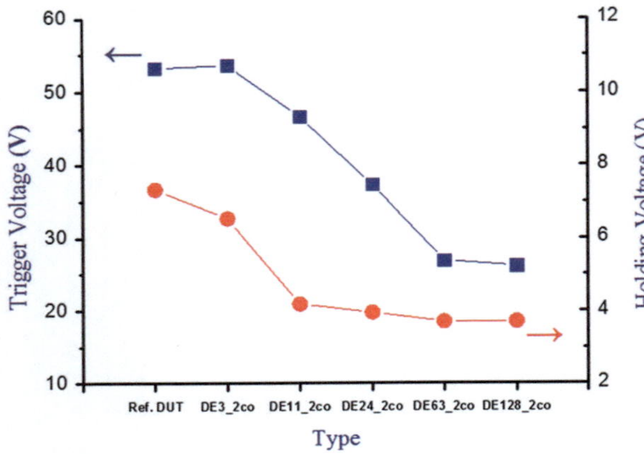

Fig. 74.6 Relationship diagrams of V_{t1} and V_h values of pLDMOS-SCR DUTs (npn) N^+ drain-side continuously extended manner in the drain-side

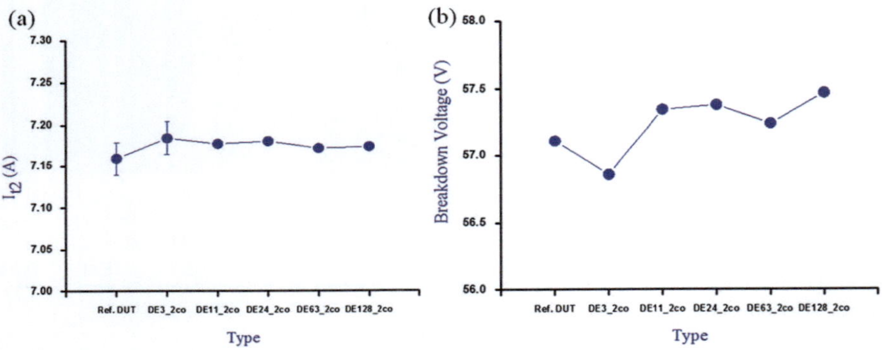

Fig. 74.7 Relationship diagrams of (**a**) I_{t2}, and (**b**) breakdown-voltage values of pLDMOS-SCR DUTs (npn-arranged) N^+ drain-side continuously extended manner in the drain-side

increase. Then, the I_{t2} values of these DUTs have a good capability that is more than 7 A. And, the V_{t1} and V_h values will be decreased. In the same meaning that performance is mainly based on the parasitic SCR. The testing results also showed that V_h values have the tendency decreased slightly.

74.5 Conclusions

From this chapter, it can be found that the cathode region of an embedded parasitic SCR (N^+) implanted location and ratio percentage of a pLDMOS in the drain-side by the npn-arranged-type doesn't have a strong impact on the anti-ESD ability, in other words, all of its I_{t2} values can be reached over 7 A. However, the V_{t1} and V_h values are decreased as the percentage of embedded SCRs in the drain-side increased. Eventually, the performance of a pLDMOS-SCR device is mainly based on the parasitic SCR if the area ratio of an embedded parasitic SCR (N^+ area) increased. Then, in the N^+ continuously extended manner in the drain-side of pLDMOS-SCR devices, the major impact is V_h values will be decreased about 10.8–49.5 %.

Acknowledgments In this work, authors would like to thank the National Chip Implementation Center in Taiwan for providing the process information and fabrication platform. And, authors would like to acknowledge the financial support of the Ministry of Science and Technology of Taiwan, through grant number NSC 102-2221-E-239-015.

References

1. Kurohane, K., Senjyu, T., Yona, A., Urasaki, N., Goya, T., Funabashi, T.: A hybrid smart AC/DC power system. IEEE Trans. Smart Grid **1**, 199–204 (2010)
2. Wang, S., Liu, J., Chen, J.-J., Liu, X.: Power sleep: a smart power-saving scheme with sleep for servers under response time constraint. IEEE Trans. Emerging and Selected Topics in Circuits and Systems **1**, 289–298 (2011)
3. Chang, H., Jang, J.-J., Kim, M.-H., Lee, E.-K., Jang, D.-E., Park, J.-S. Jung, J.-H., Yoon, C.-J., Bae, S.-R., Park, C.-H.: Advanced 0.13um smart power technology from 7V to 70V. In: 24th International Symposium on Power Semiconductor Devices and ICs. 217–220 (2012)
4. Stanzione, S., van Liempd, C., van Schaijk, R., Naito, Y., Yazicioglu, R.F., Van Hoof, C.,: A self-biased 5-to-60V input voltage and 25-to-1600μW integrated DC-DC buck converter with fully analog MPPT algorithm reaching up to 88% end-to-end efficiency. In: IEEE International Solid-State Circuits Conference, 74–75 (2013)
5. Zhang, P., Wang, Y., Jia, S., Zhang, X.: Study of LDMOS-SCR: a high voltage ESD protection device. In: 10th IEEE International Conference on Solid-State and Integrated Circuit Technology, 1722–1724 (2010)
6. Zhang, P., Wang, Y., Jia, S., Zhang, X.: Analysis of LDMOS-SCR ESD Protection Device for 60V SOI BCD Technology. In: IEEE International Conference on Electron Devices and Solid-State Circuits, 1–4 (2010)
7. Liu, S.-Y., Sun, W.-F., Pan, H.-W., Wang, H., Qian, Q.-S.: A novel latch-up free SCR-LDMOS for power-rail ESD clamp in half-bridge driver IC. In: IEEE 11th International Conference on Solid-State and Integrated Circuit Technology, 1–3 (2012)
8. Ma, F., Zhang, B., Han, Y., Zheng, J., Song, B., Dong, S., Liang, H.: High holding voltage SCR-LDMOS stacking structure with ring-resistance-triggered technique. IEEE Electr. Device Lett. **34**(9), 1178–1180 (2013)

Chapter 75
WristEye: An Elderly Computer Learning Assistant System with Wrist-Wearable Devices

Wan-Jung Chang, Yi-You Hou, Rung-Shiang Cheng, and Ming-Che Chen

Abstract This chapter develops a computer learning assistant system with wrist-wearable devices for the elderly, designated as WristEye, which can be used to analyze the computer learning attitudes, reactions, and behaviors of elderly individuals whilst in computer learning classes. WristEye is equipped with a kinematic sensor to effectively detect the changes in the orientation and vertical acceleration of the elderly wrist and to determine the corresponding operations in the learning computer, i.e., moving the mouse, hitting the keyboard, idle, and swinging the mouse. Furthermore, a remote backend server receives the detected signal from the wearable unit via a Wireless Sensor Network (WSN) and then identifies the corresponding computer learning effectiveness. The experimental results show that WristEye has a classification accuracy to recognize computer learning status of elderly individuals.

Keywords Wearable • Elderly • Learning

W.-J. Chang (✉)
Department of Electronics Engineering, Southern Taiwan University of Science and Technology, Tainan 71005, Taiwan (R.O.C.)
e-mail: allenchang@stust.edu.tw

Y.-Y. Hou
Department of Electrical Engineering, Far East University, Tainan 74448, Taiwan (R.O.C.)

R.-S. Cheng
Department of Computer and Communication, Kun Shan University, Tainan 71070, Taiwan (R.O.C.)

M.-C. Chen
Institute of Computer and Communication Engineering, National Cheng Kung University, Tainan 701, Taiwan (R.O.C.)

J. Juang (ed.), *Proceedings of the 3rd International Conference on Intelligent Technologies and Engineering Systems (ICITES2014)*, Lecture Notes in Electrical Engineering 345, DOI 10.1007/978-3-319-17314-6_75

75.1 Introduction

Due to rising life expectancies as a result of advancements in medical science and the continuing aging of the baby boomers born in the middle of the last century, the elderly populations of many countries around the world have increased significantly in recent years. For example in Japan, the percentage of the population greater than 65 years of age was 15.1 % in 1996, and increased to 21.8 % in 2008 [1, 2]. Moreover, many countries in the world are currently planning and developing various e-government projects (e.g., e-Japan in Japan, A Framework for Global EC in American, UK Online in E England, e-Korea in Korea, etc.) to turning them into the advanced Information Technology (IT) nationals [3]. These two worldwide demographic shifts have many significant social implications, not least of which includes an additional burden on the computer learning services for elderly populations.

The computer learning effects of the elderly is limited by some factors, such as educational background, the knowledge of digital information, small texts on the monitor, unacquainted with keyboard and mouse, web page operation, and so on [4, 5]. These factors always cause elderly people to feel barriers and be disappointed in learning how to operate computers. Such phenomenon of leading the elderly to decrease their computer learning interest and desire is called "Computer Phobia" [6].

To resolve the Computer Phobia issue, the current study develops a sophisticated elderly computer learning assistant system with wrist-wearable devices, designated as WristEye, which employs a kinematic sensor for the continuous monitoring and reporting of an individual's computer learning status over a Wireless Sensor Network (WSN). Utilizing an accelerometer and a gyroscope, the kinematic sensor measures the changes in the orientation and acceleration of the wrist and then communicates this data to a Micro Controller Unit (MCU), where the signal is processed by a wrist recognition algorithm to determine the corresponding activities in learning computer, i.e., moving the mouse, hitting the keyboard, idle, and swinging the mouse. Furthermore, a remote backend server receives the detected signal from the MCU via a Wireless Sensor Network (WSN) and then identifies the corresponding computer learning effectiveness. Significantly, WristEye systems not only detect each occurrence of the computer learning status, but also records the approximate amount of time spent in performing each computer learning activity, and therefore enables a rough estimate to be made of the computer learning effectiveness of the individual. Through the use of WristEye, abnormal computer learning activities of elderly individuals can be immediately detected and appropriate actions from the instructor are followed. As a result, the proposed WristEye system is expected to be of great use to the e-learning class infrastructure vendors in implementing an elderly e-learning system.

Section 75.2 introduces the proposed WristEye system comprising the sensor unit hardware, the signal processing scheme, and the wrist recognition algorithm.

Section 75.3 presents the results of a series of experimental trials designed to evaluate the performance of the WristEye. Finally, Sect. 75.4 draws brief conclusions.

75.2 WristEye System

75.2.1 *Overall WristEye Structure*

The objective in the present study is to realize the real-time recognition of computer learning activities utilizing a small, wireless, low-power kinematic device. The system hardware of WristEye system comprises just two parts, namely a wearable unit and the backend server. The wearable unit is attached to the subject's wrist and consists of a data acquisition board and a kinematic sensor. Whereas the ambulatory systems presented in [7–9] utilize a single triaxial accelerometer for motion sensing purposes, the kinematic sensor in WristEye makes use of information of two axes from an accelerometer and information of single axis from a gyroscope in order to increase the reliability of the movement information provided to the recognition algorithm. Figure 75.1a presents a schematic illustration of the system architecture. The wearable unit performs an initial preprocessing of the signals produced by the gyroscope and accelerometer to determine the wrist orientation and vertical acceleration, respectively, and then transmits this information to MCU which processes the information using a wrist recognition algorithm to classify the corresponding computer learning activities. The wearable unit transmits the recognition information via a wireless transceiver module to the backend server through the wireless base station attached to the server. The server then processes the information to classify the computer learning effectiveness of elderly individuals.

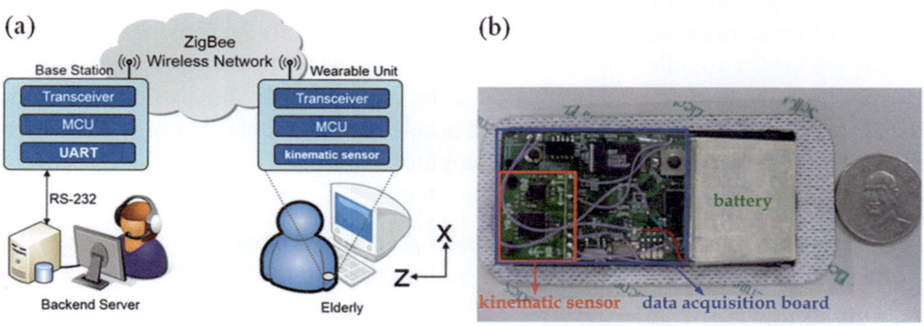

Fig. 75.1 (**a**) Schematic of system architecture. (**b**) Photograph showing full-size view of wireless transceiver and sensor device

75.2.2 Wearable Unit Hardware

Figure 75.1b presents a full-size photograph of the wearable unit within the WristEye system. The data acquisition board includes an Atmel Atmega 128 L microcontroller, 4 KB of RAM, various ICs, and a wireless transceiver module, while the kinematic sensor comprises an IDG300 gyroscope [8] and an ADXL330 accelerometer [9]. The sensor unit has two main functions, namely to detect changes in the posture and movements of the wearer, and to merge the data signals of the accelerometer and gyroscope, respectively, prior to their transmission to the base station. The sensor unit is equipped with a low-voltage, low-power consumption ZigBee-compliant transceiver module (CC2420, Chipcon, TI) to facilitate its WSN communication ability with the base station.

75.2.3 Wrist Recognition Algorithm

Upon receiving the wrist orientation and vertical acceleration from kinematic sensor, the MCU utilizes a wrist recognition algorithm to identify the corresponding computer learning activities. The algorithm commences by applying vertical wrist acceleration threshold value to determine whether the wrist posture of the subject falls into a "moving mouse" category or a "hitting keyboard" category. For example, when changing from a "moving mouse" to a "hitting keyboard" computer learning active, the vertical of the wrist posture exceeds the vertical threshold since the elderly invariably hit the keyboard in alphabets-by-alphabets manner, i.e., use only one hand to key-in a word in alphabets-by-alphabets. Consequently, the corresponding wrist motion is classified as a "hitting keyboard." Conversely, when moving the mouse, the wrist tends to be in a relatively horizontal position, and thus the vertical acceleration is less than the critical vertical value and the corresponding wrist motion is classified as a "moving mouse."

To further distinguish the wrist motion of the commands for moving the mouse, hitting the keyboard, idle, and swinging the mouse, the vertical wrist acceleration is closely examined. Figure 75.2a illustrates the measurement signals obtained from the kinematic sensor for the wrist vertical acceleration as a subject wares a unit (see Fig. 75.2b) and executes a simple computer learning course, namely *Google Search*, including the following execution steps:

- *Step 1*: *Move the mouse until the mouse pointer is located at the icon of web explore* (e.g., *Chrome*, *Explore*, etc.) *and then click the icon. Move the mouse until the mouse pointer is located at the search textbox and then click the mouse.*
- *Step 2*: *Hit the keyboard to insert the search key word in the search textbox in an alphabet-by-alphabet manner.*
- *Step 3*: *Move the mouse to the location of search target and then click the mouse.*

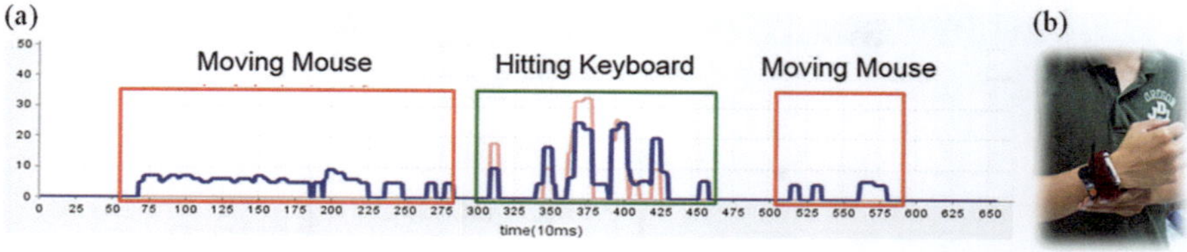

Fig. 75.2 (**a**) Wrist vertical acceleration when performing *Google Search* learning course. (**b**) Photograph showing wear of a wearable unit on the wrist

Fig. 75.3 Decision-tree manner

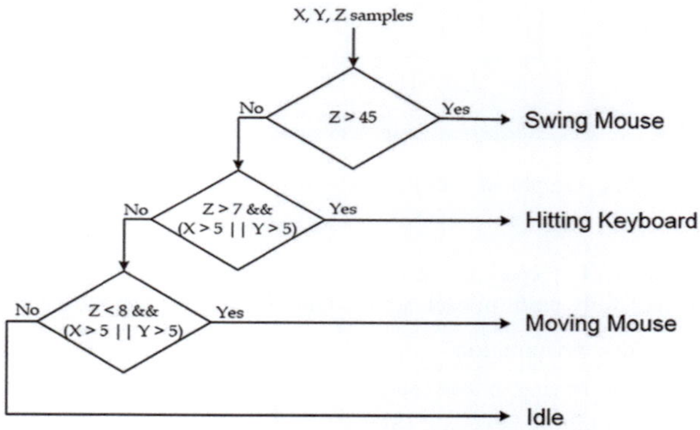

The acceleration profile associated with the moving mouse activity has a lower vertical acceleration than that of hitting the keyboard. Thus, the wrist recognition algorithm developed in this study exploits characteristics of the acceleration profiles to distinguish the four motions. Figure 75.3 illustrates the details of the posture recognition algorithm in flowchart form.

75.3 Experimental Results

The performance of the WristEye system was evaluated by performing a series of trials using a total of 30 elderly people with various computer abilities. During the trials, the 30 elderly people were further divided into ten groups and each group was organized by three different kinds of computer abilities of the elderly students:

- *Class A*: *The elderly student has various experiences in using computers.*
- *Class B*: *The elderly student only has little experiences in using computers.*
- *Class B*: *The elderly student has no experiences in using computers.*

Each trial of a group performed a *Google Search* computer learning course as described in Sect. 75.2.3. The performance of the proposed WristEye system was quantified by comparing the actual computer abilities of the elderly people with that

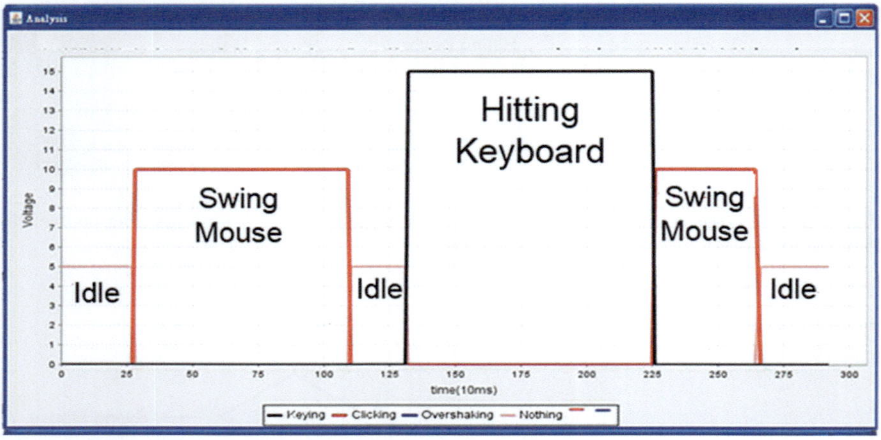

Fig. 75.4 Graphical user interface of PC-based monitoring system

Table 75.1 Experimental results of the average time(s) for each kind of elderly students who successfully perform each step of *Google Search* computer learning course

Abilities wrist motion	Class A (s)	Class B (s)	Class C (s)
Time of moving mouse (Step 1)	4.5 ± 0.23	4.3 ± 0.1	8.2 ± 0.25
Time of hitting keyboard (Step 2)	$6.2 \pm 0.3.1$	10.8 ± 0.2	18.3 ± 1
Time of moving mouse (Step 3)	2.7 ± 0.17	3.2 ± 0.11	7.8 ± 0.16

identified by the wrist recognition algorithm (see Fig. 75.4). Table 75.1 summarizes the overall experimental results of the average time for each kind of elderly student who successfully perform each step of the *Google Search* computer learning course. The recognition results show that Class A participants spend on average less time for successfully performing each step of the *Google Search* computer learning course than Class B and Class C participants. This result reflects the actual computer abilities of participants and therefore demonstrates that the proposed WristEye achieves recognition accuracies compared with the actual computer abilities of participants.

75.4 Conclusions

This study has developed a MEMS-based monitor, called WristEye, for identifying the wrist motions of elderly individuals while in computer learning classes. The proposed WristEye applies a kinematic sensor to derive more accurate vertical acceleration signals of the monitoring target for recognition of the wrist posture by employing a posture recognition algorithm. These recognition results are then forwarded wirelessly in real time to a backend server for further processing via a

Zigbee WSN. Upon receipt of the signals, the backend server extracts the corresponding wrist motion (i.e., "moving the mouse," "hitting the keyboard," "idle," and "swinging the mouse"). The experimental results have confirmed that WristEye system is capable of identifying a variety of common daily activities of computer learning for elderly individuals in real time with an excellent accuracy. Therefore, it may be considered as a suitable solution for the emerging elderly e-learning paradigm aimed at computer learning institutes located in communities.

References

1. Statistics Bureau and the Director-General for Policy Planning (Statistical Standards), Ministry of Internal Affairs and Communication Home. http://www.stat.go.jp/english/index.htm
2. Japan eyes robots to support aging population, *Boston Global* Home. http://www.boston.com/news/world/asia/articles/2007/09/18/japan_eyes_robots_to_support_aging_population/
3. Jupiter Communications (2000). http://www.jup.com
4. Pei-Yun Wang, Trends in Media Usage and Life Style of Mature Ages-A Case Study of 2004 Shih Hsin Communication Database, Master Thesis, Taipei, Taiwan (R.O.C) (2006)
5. Morrell, Q.W., Echt, K.V.: Designing Written Instructions for Older Adults Learning to Use Computers. In: Fisk, A.D., Rogers, W.A. (eds.) Handbook of Human Factors and the Older Adult. Academic, San Diego, CA (1996)
6. Jay, T.: Computerphobia: what to do about it. Educ. Comm. Tech. J. **21**, 47–48 (1981)
7. IDG 300 Gyroscope Home. http://www.invensense.com/products/idg_300.html
8. ADXL330 Accelerometer Home. http://www.analog.com/en/mems-and-sensors/imems-accelerometers/adxl330/products/product.html
9. Sun, S.-L., et al.: Multi-sensor optimal information fusion Kalman filter. Automatica **40**(6), 1017–1023 (2004)

Chapter 76
A Study of Ultrafast Laser-Based X-Ray Sources for In-Line Phase-Contrast Biomedical Imaging

Mario do Nascimento and Jengnan Juang

Abstract Ever since the Coolidge tube was created in 1913, the same design has been used for X-ray imaging devices. As modern medicine advances, the Coolidge tube is reaching its limited full potential. For this reason, this research analyzes two of the modern alternatives for X-ray imaging: laser-based X-ray sources and phase-contrast imaging. According to this study, lasers are able to provide significantly better focal spot sizes, much better temporal resolution, and a high repetition rate. As for integrating phase-contrast imaging, this research depicts how phase-contrast imaging is overwhelmingly more sensitive to elements found in soft tissue and refraction is far more likely to occur than absorption, which has the potential to decrease exposure time. Although these technologies have been found to be more advantageous than the current technologies, there are still essential improvements and research that need to be accomplished before clinical use.

Keywords Laser X-rays • Phase-contrast imaging • Biomedical imaging

76.1 Introduction

Ever since X-ray radiation was discovered by Wilhelm Roentgen in 1895 [1], X-ray imaging has increasingly become a major tool in medical diagnostics. As medicine progressed, so did X-ray imaging technology. However, as the present-day medicine converges with technology more than ever before, the conventional X-ray tube has reached its limited potential and new technologies have become necessary in order for X-ray imaging to follow modern medical advancements.

M. do Nascimento (✉)
Department of Electrical and Computer Engineering, Mercer University,
Macon, GA 31207, USA
e-mail: mario.kdn@outlook.com

J. Juang
Department of Electrical and Computer Engineering, School of Engineering,
Mercer University, Macon, GA, USA

J. Juang (ed.), *Proceedings of the 3rd International Conference on Intelligent Technologies and Engineering Systems (ICITES2014)*, Lecture Notes in Electrical Engineering 345, DOI 10.1007/978-3-319-17314-6_76

As research evolved, several technologies drew the attention of scientific communities as a viable modern alternative to the conventional X-ray tube. This study addresses the necessary background information in order to understand the technology that, according to this study, is the most feasible technology to be incorporated into medical clinics in the future. This technology is based on the modern category of lasers called ultrafast lasers, which combine the advantages of a coherent light source and an ultra short laser pulse. In addition to the X-ray source technology, this study also elaborates on a modern method of generating images from X-rays: phase-contrast imaging. This method uses the property of refraction from X-rays, instead of absorption, in order to create an image. Using this alternate property can provide a much better depiction of soft tissues in the image and also decrease radiation exposure [2].

76.2 Advancing Methods and Technologies

76.2.1 Laser-Based X-Ray Sources

When ultrafast laser pulses are focused onto a solid target made of a material with a high nuclear charge Z, intense X-ray radiation of picoseconds in duration and with keV of energy is emitted. An experimental setting of how this system would be organized is shown in Fig. 76.1 [3]. Collisionally pumped laser-based X-rays, which is what this study reviews, works by heating the target material very quickly with ultrafast lasers, which creates a thin line of plasma on the surface of the target. This extreme heating puts the atoms of the target material in the pumped level, which consequently releases electrons from the atoms in the target, which in turn

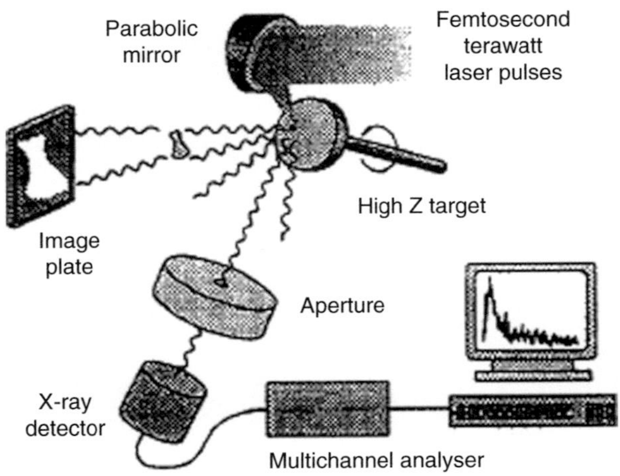

Fig. 76.1 Experimental setting for an ultrafast laser X-ray source using a rotating metal target [3]

collide with other ions on the target [4]. This collision takes the ions back to their rest level and releases X-ray radiation. This process is also called laser-produced plasma-based X-ray sources.

76.2.2 Phase-Contrast Imaging Through Laser-Based X-Ray Sources

76.2.2.1 Phase-Contrast Imaging

Phase-contrast is an imaging method that relies on the refraction that the X-ray beam suffers when it impinges on a material. This particular method is called the phase-contrast method because the angular deviation is directly proportional to the gradient of the phase of the refracted beam. There are three general categories through which phase-contrast imaging may be performed: free-space propagation, interferometry-based techniques, and analyzer systems. However, all of the different approaches have the same goal: determine the angular deviation of the X-ray beam [5].

Although there are three different types of techniques for acquiring the phase-contrast image, this study is going to focus on the free-space propagation types of imaging, also called in-line phase-contrast or in-line holography. Using the conventional absorption method, the image detector is placed immediately after the sample image, so the X-rays travel a minimal distance before reaching the detector, as shown in Fig. 76.2a. Using the phase-contrast method, however, X-rays are transmitted through the sample object at multiple angles and then propagated over a certain distance until the X-rays reach the detector, as shown in Fig. 76.2b [2].

Therefore, variations in the X-ray refractive index and the thickness within the sample object will change the X-ray wavefront once it passes through the sample object. This change of the X-ray wavefront provides the conventional attenuation image an edge enhancement at the interfaces within the sample object that have different refractive indices. Phase-contrast imaging can yield much more improved contrast than convention attenuation techniques. This reason can be explained by the refractive index that characterizes the optical properties of tissues, and it can be expressed by the complex equation

$$n = 1 - \delta - i\beta \qquad (76.1)$$

where δ is the decrement of the real part of the refractive index, which characterizes the phase-shift, and β is the imaginary part, which characterizes the absorption property for the tissue. In a conventional X-ray image, the real part of the refractive index is not commonly realized [2]. However, as shown in Fig. 76.3, at large photon energies [5] and for soft tissues composed of light elements, the real part δ plays a more significant role than the imaginary part [2].

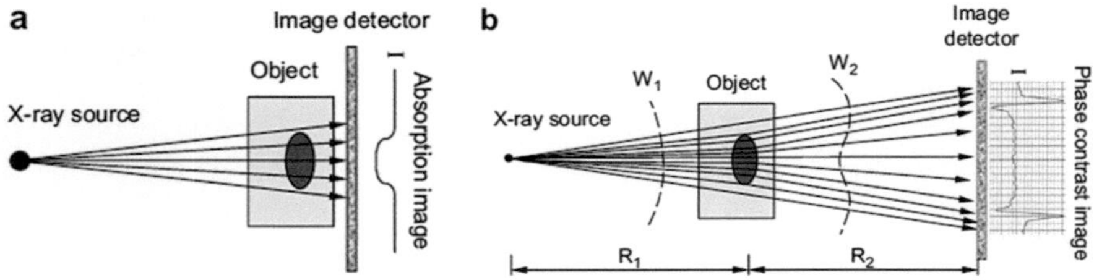

Fig. 76.2 It is shown the conventional attenuation-based X-ray imaging system (**a**); and the in-line phase-contrast imaging system (**b**), with the first wavefront (W_1), the second wavefront (W_2), the distance between the X-ray source (R_1), the sample object (*Object*), and the distance between the object and the detector (R_2) [2]

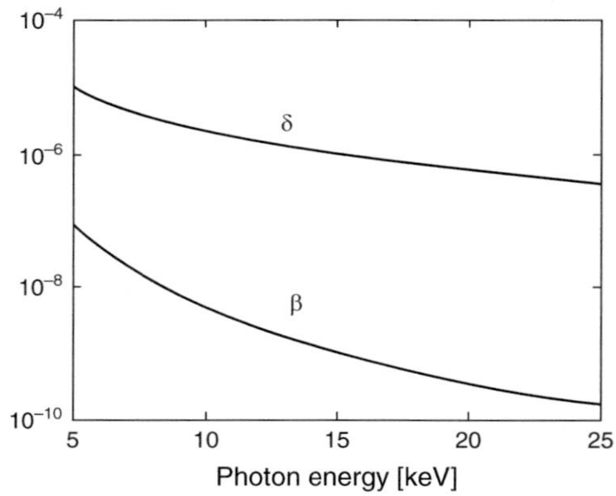

Fig. 76.3 The imaginary and real parts of the refractive index as a function of photon energy. As it can be clearly seen in the graph, the real part (δ) is theoretically always higher than the imaginary part (β) from 5 to 25 keV [5]

76.3 Results and Discussion

76.3.1 X-Ray Tubes and Their Limitations

Even over 100 years after its creation [1], the X-ray tube is still the common device to be utilized for X-ray imaging throughout the globe. However, the Coolidge tube design seems to have encountered roadblocks that it is not able to overcome, therefore hindering improvements in medical imaging.

In mammography, for instance, microfocal X-ray tubes have been experimented in order to effectively depict microcalcifications within breast tissue. Yet, these X-ray tubes have the limitation of focal spots of approximately 100 μm, which is not concentrated enough to clearly depict microcalcifications that are around 30 μm. This lack of resolution leads to longer exposure times and consequently possible motion blur. For this reason and for the limited heat load, microfocal X-ray tubes have been deemed as not successful in diagnostic mammography [6].

In dual-energy subtraction angiography (DESA), conventional X-ray tubes suffer from motion, loss of sharpness, and artifacts when the initial image (before the contrast) is subtracted from the final image. X-ray tubes also show low sensitivity to iodine, a commonly used contrast. Furthermore, conventional tubes have spatial resolutions that interfere with the quantization of stenosis, which makes it challenging to depict collateral vessels when diagnosing coronary artery disease [6].

76.3.2 *Advantages of Laser-Based Sources*

Coming from the invention of the Chirped Pulse Amplification (CPA) technology, ultrafast laser-produced-plasma (LPP) X-ray sources have rather small focal spot sizes due to the extremely short laser pulse duration which freezes hydrodynamic motion in the plasma. That means that, in a mammography using laser-based X-ray sources, the laser target surface is parallel to the tray and yields an effective focal spot size which will not change more than 5 % across the mammogram cassette. Consequently, every part of a breast, with any thickness, can be depicted with an appropriate spatial resolution [6].

In DESA, extremely short pulses of laser sources eliminate most of the motion-related loss of sharpness and artifacts. Additionally, it is significantly more sensitive to iodine, which in turn allows less invasive angiography that uses less volume of contrast agent, which may lead to a reduction of the risk of renal failure [6].

Two recent experimental research studies regarding laser-based X-ray imaging have been using ultrafast Ti-Sapphire laser sources with peak and average power on the terawatts and watts, respectively [7, 8]. The X-ray radiation is produced by Kα emission and bremsstrahlung emission using a variety of target materials (Cu, Mo, Ag, Sn, La, and W). It is deduced that, by integrating the spectrum lines shown in Fig. 76.4, around 50 % of the total X-ray emission is caused by Kα emissions. After several enhancement methods, the most recent experiment was able to yield an effective X-ray source with a spot size of 10 μm [7], which would be an excellent resolution for mammography, where microcalcifications are commonly approximately 30 μm in diameter [6].

As seen in Table 76.1, laser-based X-ray sources have several advantages over conventional, broadly used X-ray sources. Smaller focus spot size leads to reduction in patient exposure times and higher spatial resolution leads to better depiction of details, which is extremely significant for medical diagnosis. Furthermore, higher iodine sensitivity can improve the image contrast when using iodine-based contrast agents. The only disadvantage of laser-based sources nowadays is their current complexity, yet further research has the potential to simplify and miniaturize these systems.

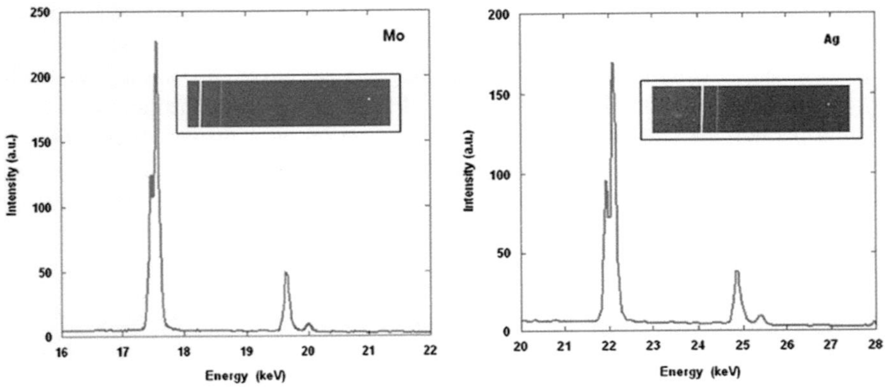

Fig. 76.4 The graph on the left depicts the laser intensity using a Mo target for a Ti:Sapphire laser-based X-ray. The graph on the right shows the laser intensity using an Ag target instead. The first peak in both graphs is the high-intensity Kα emission [7]

Table 76.1 Comparison between laser-based and conventional X-ray sources

	Focal spot size (μm)	Spatial resolution (lp/mm)	Iodine sensitivity	Soft tissue sensitivity	Complexity
Laser-based	≈30	≈20	High	Low	High
Conventional	>300	<13	Low	Low	Low

76.3.3 Advantages of Phase-Contrast Imaging

Several characteristics native to the phase-contrast imaging method have been found to be advantageous to medical imaging. The refraction properties of X-ray radiation are particularly sensitive to light elements such as oxygen, nitrogen, carbon, and hydrogen, which are found commonly in soft tissue. For that reason, this method is especially more beneficial for medical procedures that require high contrast for soft tissues, such as mammography and angiography, than conventional attenuation-based X-ray imaging. Also, X-ray phase shifting has a probability of occurring about 1,000 times greater than the probability for X-ray attenuation, which results in phase-contrast imaging being able to depict details that conventional X-ray methods would not [9]. Moreover, radiation exposure can be reduced since large levels of noise could then be tolerated without losing too much image detail [2].

A phase-contrast research [10] performed using conventional X-ray tubes has additionally shown that phase-contrast imaging has the potential to better depict pathologies, since it depicts tumor masses with better accuracy and detail. That ability, if in conjunction with the laser-based X-ray sources can improve the diagnosis of tumors and other pathologies with much greater accuracy and clarity.

The shorter exposure time and better contrast of laser-based sources, combined with the higher exposure of soft tissues of phase-contrast methods, have shown the potential to dramatically improve early cancer diagnostics.

76.4 Conclusions

Ultrafast laser sources have been found to be the most feasible and applicable replacement of conventional X-ray tubes. They provide enough average power for clinical applications, which can shorten the patient exposure time. They have extremely small focal sizes, which enable the imaging to depict even the smaller microcalcifications in mammography [6].

As for phase-contrast imaging, it uses a different X-ray property in order to generate an image [5]. This technique has been found to be more advantageous because it can provide information about soft tissue in a much higher degree, since X-rays can be refracted differently even when they go through distinct soft tissues [9].

Although major improvements have been made in the phase-contrast imaging techniques and in ultrafast laser technology, commercial and clinical uses are still not entirely practical. The research depicted in this chapter had the purpose of communicating the potential of such technologies, but with no intention to provide practical conclusions [7, 8].

References

1. Selman, J.: The fundamentals of X-rays and radium physics. Charles C Thomas, Springfield (1994)
2. Zhou, S.-A., Brahme, A.: Development of phase-contrast X-ray imaging techniques and potential medical applications. Phys. Med. **24**(3), 129–148 (2008)
3. Svanberg, S.: Some applications of ultrashort laser pulses in biology and medicine. Meas. Sci. Technol. **12**, 1777–1783 (2001)
4. Da Silva, L.B., Cauble, R.C., Libby, S.B.: X-ray Lasers and High-Density Plasma. E&TR, 9–19 (1995)
5. Als-Nielsen, J., McMorrow, D.: Elements of modern X-ray physics. Wiley, West Sussex (2011)
6. Keiffer, J.C., Krol, A., Jiang, Z., Chamberlain, C.C., Scalzetti, E., Ichalalene, Z.: Future of laser-based X-ray sources for medical imaging. Appl. Phys. B **74**(9), S75–S81 (2002)
7. Toth, R., Fourmaux, S., Ozaki, T., Servol, M., Kieffer, J., Kincaid, R.J., Krol, A.: Evaluation of ultrafast laser-based hard X-ray sources for phase-contrast imaging. Phys. Plasmas **14**, 0535061–0535068 (2007)
8. Toth, R., Keiffer, J.C., Fourmaux, S., Ozaki, T.: In-line phase-contrast imaging with a laser-based hard X-ray source. Rev. Sci. Instrum. **76**, 0837011–0837016 (2005)
9. Laperle, C.M., Wintermeyer, P., Wands, J.R., Shi, D., Anastasio, M.A., Li, X., Ahr, B., Diebold, G.J., Rose-Petruck, C.: Propagation based differential phase contrast imaging and tomography. Appl. Phys. Lett. **91**, 1739011–1739013 (2007)
10. Pfeiffer, F., Weitkamp, T., Bunk, O., David, C.: Phase retrieval and differential phase-contrast imaging with low-brilliance X-ray sources. Nat. Phys. **2**, 258–261 (2006)

Chapter 77
PC-USB-Based Real-Time Control Systems Using Quadratic Optimal Control Method

Dershyang Ker

Abstract The goal of this chapter is to implement the quadratic optimal control algorithm in PC-USB-Based real-time control systems by using mathematic model blocks in VisSim/USB software package environment and a USB-based module. PC is used as a controller to apply quadratic optimal control algorithm in PC-USB-Based real-time control systems. First, a design example based on the quadratic optimal control algorithm is given. The theoretical state feedback controller can be obtained. An alternative approach using VisSim's OptimizePro to automatically calculate optimal state feedback controller has been verified to be a very powerful method to avoid tedious theoretical design. Second, the model of the real control system, FB-33 control system, can be set up in VisSim. The quadratic optimal controller can be easily obtained without the theoretical design. Finally, the VisSim/USB is used to implement the quadratic optimal control and USB-Based module to control the FB-33 servo control system. The satisfied results are shown in this chapter.

Keywords PC-USB-based real-time control system • Quadratic optimal control • USB-based module

77.1 Introduction

Depending on the popularity and development of USB-Based Interface, the demand of market is rapidly growing. The PC system of popularity use is found that is lack of the mathematical calculation functions. This makes modern and constant progress control algorithms to be unable to really apply in industrial control. The related fields of domestic and abroad, implementing modern control algorithms in USB-Based real-time control systems, still have many problems to be needed to overcome.

D. Ker (✉)
Program of Game Design, China University of Science and Technology, Taipei 115, Taiwan
e-mail: kirk@cc.cust.edu.tw

J. Juang (ed.), *Proceedings of the 3rd International Conference on Intelligent Technologies and Engineering Systems (ICITES2014)*, Lecture Notes in Electrical Engineering 345, DOI 10.1007/978-3-319-17314-6_77

This chapter uses VisSim as a system developing environment to design and develop advanced modern control algorithms by mathematic model blocks. VisSim is a Windows-based program for the modeling, design, and simulation of complex control systems without writing a line of code [1–5]. It combines an intuitive drag-and-drop block diagram interface with a powerful mathematical engine. The visual block diagram interface offers a simple method for constructing, modifying, and maintaining complex control system models. Furthermore, VisSim offers unprecedented ease-of-use and consequently a shorter learning curve than competitive systems.

In designing optimal control systems, It is desired to select a control vector such that a given performance index is minimized [6–16]. It can be proved that a quadratic performance index will yield linear control law. Setting up a simulation in VisSim is simple. Connect the controller to the mathematical model and to a plot block. VisSim/Optimize can determine optimal values for design variables subject to the given performance indexes starting from initial guess values [2]. VisSim/Optimize can automatically calculate optimal control vector that gives minimal performance indexes. It allows user-specified cost functions that can consider controller behavior, such as steady-state error, overshoot, and rise and settling times. Finally, the optimal control vector and USB-Based module are used to control the FB-33 servo control system [13].

77.2 Quadratic Optimal Control

The control system is shown in Fig. 77.1. The optimal control problems [8] are based on quadratic performance indexes that, given the system equations

$$\dot{x} = Ax + Bu \tag{77.1}$$

In designing control systems, it is desired to determine the optimal feedback gain matrix K of the optimal control vector

$$u(t) = -Kx(t) \tag{77.2}$$

So as to minimize the performance index

$$J = \int_0^\infty L(x,u)\mathrm{d}t = \int_0^\infty (x^\mathrm{T}Qx + u^\mathrm{T}Ru)\mathrm{d}t = \int_0^\infty x^\mathrm{T}(Q + K^\mathrm{T}RK)x\ \mathrm{d}t \tag{77.3}$$

where Q is a positive-definite Hermitian or real symmetric matrix and R is also a positive-definite Hermitian or real symmetric matrix.

From Fig. 77.2, it can be found that the state equation for the plant is

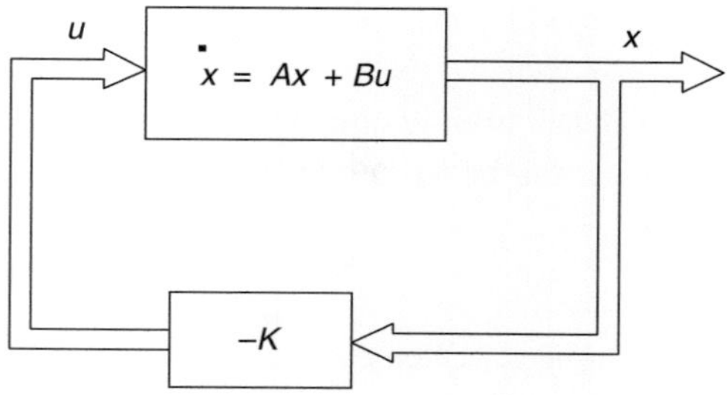

Fig. 77.1 Optimal control system

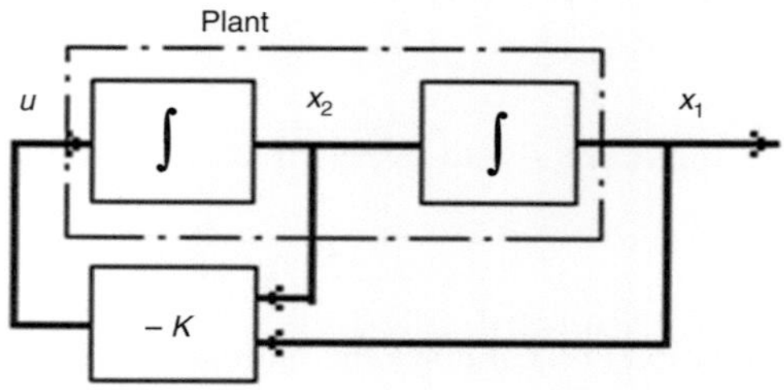

Fig. 77.2 Example control system

$$A = \begin{bmatrix} 0 & 1 \\ 0 & 0 \end{bmatrix} \quad B = \begin{bmatrix} 0 \\ 1 \end{bmatrix} \quad Q = \begin{bmatrix} 1 & 0 \\ 0 & \mu \end{bmatrix} \tag{77.4}$$

Let us assume that

$$x^{\mathrm{T}}(Q + K^{\mathrm{T}}RK)x = -\frac{d}{\mathrm{d}t}(x^{\mathrm{T}}Px) \tag{77.5}$$

where P is a positive-definite Hermitian or symmetric matrix. Therefore

$$x^{\mathrm{T}}(Q + K^{\mathrm{T}}RK)x = -x^{\mathrm{T}}[(A - BK)^{\mathrm{T}}P + P(A - BK)]x \tag{77.6}$$

Comparing both sides of this last equation, it can be obtained

$$(Q + K^{\mathrm{T}}RK) = -[(A - BK)^{\mathrm{T}}P + P(A - BK)] \tag{77.7}$$

Using the second method of Liapunov, if $A - BK$ is a stable matrix, there exists P that satisfies (77.7).

Hence to evaluate the performance index J as

$$J = \int_0^{\infty} x^{\mathrm{T}}(Q + K^{\mathrm{T}}RK)xdt = -x^{\mathrm{T}}Px\Big|_0^{\infty} = -x^{\mathrm{T}}(\infty)Px(\infty) + x^{\mathrm{T}}(0)Px(0) \tag{77.8}$$

It can be obtained that $X(\infty)$ tends to zero because all eigenvalues of $A - BK$ have negative real parts. Then, the value of J can be given by

$$J = x^{\mathrm{T}}(0+)Px(0+) \tag{77.9}$$

Since R has been assumed to be a positive-definite Hermitian or real symmetric matrix, it can be written to $R = T^{\mathrm{T}}T$. The (77.7) will become

$$(A^{\mathrm{T}} - K^{\mathrm{T}}B^{\mathrm{T}})P + P(A - BK) = -(Q + K^{\mathrm{T}}T^{\mathrm{T}}TK) \tag{77.10}$$

This can be rewritten as

$$A^{\mathrm{T}}P + PA + [TK - (T^{\mathrm{T}})^{-1}B^{\mathrm{T}}P]^{\mathrm{T}}[TK - (T^{\mathrm{T}})^{-1}B^{\mathrm{T}}P] - PBR^{-1}B^{\mathrm{T}}P + Q = 0 \tag{77.11}$$

To minimize J with respect to K, this yields

$$K = T^{-1}(T^{\mathrm{T}})^{-1}B^{\mathrm{T}}P = R^{-1}B^{\mathrm{T}}P \tag{77.12}$$

Equation (77.12) gives the optimal matrix K. The P matrix in (77.12) must satisfy the reduced matrix Ricatti equation as

$$A^{\mathrm{T}}P + PA - PBR^{-1}B^{\mathrm{T}}P + Q = 0 \tag{77.13}$$

Noting that the matrix A is real and matrix Q is real symmetric, matrix P is the real symmetric matrix in (77.13). Solving (77.13) for the P_{ij}, then obtains

$$P = \begin{bmatrix} p_{11} & p_{12} \\ p_{21} & p_{22} \end{bmatrix} = \begin{bmatrix} \sqrt{\mu+2} & 1 \\ 1 & \sqrt{\mu+2} \end{bmatrix} \tag{77.14}$$

Substituting (77.14) to (77.12), the optimal feedback gain matrix K can be written as

$$K = R^{-1}B^{\mathrm{T}}P = [1]\begin{bmatrix} 0 & 1 \end{bmatrix}\begin{bmatrix} p_{11} & p_{12} \\ p_{21} & p_{22} \end{bmatrix} = \begin{bmatrix} 1 & \sqrt{\mu+1} \end{bmatrix} \tag{77.15}$$

Then, the optimal control signal is

$$u(t) = -Kx(t) = -x_1 - \sqrt{\mu + 2x_2} \tag{77.16}$$

Therefore, the block diagram of this optimal control system is shown in Fig. 77.3.

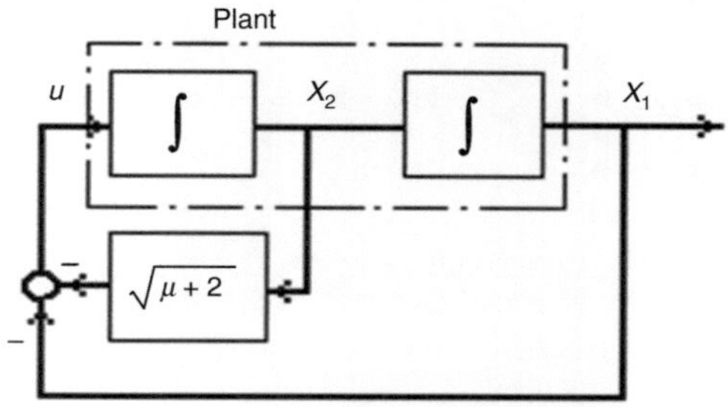

Fig. 77.3 Optimal control of the plant shown in Fig. 77.2

77.3 Verifying Quadratic Optimal Control with VisSim/Optimize

To verify the quadratic optimal control law described above, it can be easily done using VisSim/OptimizePro. It is convenient to use VisSim/OptimizePro for optimization problems on VisSim simulations and models because VisSim/Optimize was designed and implemented to take advantage of the VisSim environment. It is convenient to use all of the features of VisSim to create the cost function. VisSim dialog box makes it easy to set upper and lower bounds on design variables. The Optimization Properties dialog box is used to select an optimization run.

The control system shown in Fig. 77.2 has a feedback gain matrix with two parameters. Following the quadratic optimal control law, this feedback gain matrix K can be also optimized using VisSim/OptimizePro shown in Fig. 77.4. The state feedback controller has two variables k_1 and k_2. The cost function is computed by integrating the quadratic function $L(x,u)$ with $\mu = 1.5$ in (77.15). The plant is produced with two integrator blocks.

Once the initial conditions of the plant and initial values of k_1 and k_2 are set, a simulation run produces the results in Fig. 77.4. The final value of the cost function is 1.87092. The final values of k_1 and k_2 are 0.999954 and 1.87079, respectively. It can be seen that the errors between the simulation results and the theoretical values, $k_1 = 1$ and $k_2 = 1.870829$ with $\mu = 1.5$ in (77.15), are less than 0.0001 %.

77.4 PC-USB-Based Quadratic Optimal Control in Control Systems

Implementing Quadratic Optimal Control algorithm to the mathematical model of FB-33 system using VisSim/OptimizePro.

The FB-33 Feedback control system shown in Fig. 77.5, the apparatus of automatic control laboratory, is a typical second-order system [8, 9]. The objective of this system is to control the position of the mechanical load in accordance with the reference position. The open-loop transfer function of the FB-33 control system obtained by the author [15, 16] can be written as

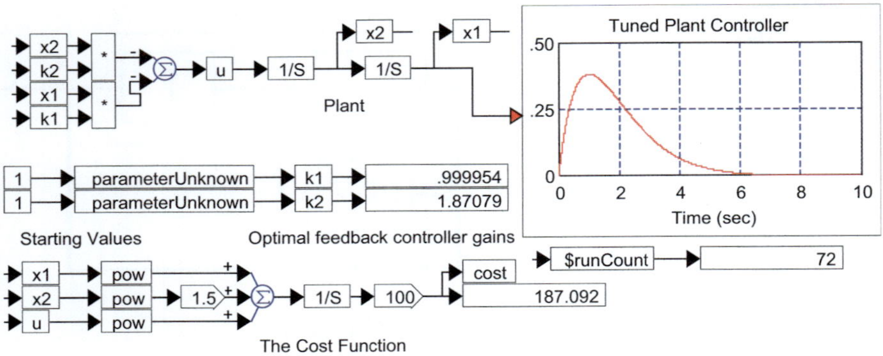

Fig. 77.4 The optimal feedback controller tuning diagram with VisSim/Optimize

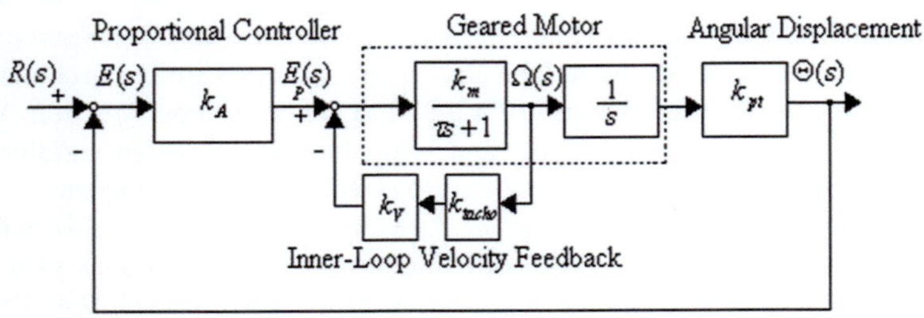

Fig. 77.5 Block diagram of FB-33 closed-loop control system with $k_A = 10$, $k_\theta = k_m k_{pt} = 4.36$, and $k_\omega = k_m k_{tacho} = 0.87$

$$\frac{\Theta(s)}{E(s)} = \frac{k_A k_\theta}{s(\tau s + 1)} = \frac{10 \times 4.36}{(1 + 0.5\ \ s)s} \tag{77.17}$$

where θ is the angular displacement of the motor shaft, ω is the angular velocity of the motor shaft, τ is the time constant of the motor, k_A is the gain of the amplifier, k_m is the gain of the motor, k_{pt} is the coefficient of the potentiometer, k_{tacho} is the coefficient of the tachometer, $k_\theta = k_m k_{pt} = 4.36$, and $k_\omega = k_m k_{tacho} = 0.87$.

After the mathematical model is obtained, it is easy to build the mathematical mode of the FB-33 control system by simply selecting and connecting predefined function blocks in VisSim. It is convenient to use VisSim/OptimizePro [2] for optimization problems on VisSim simulations. The quadratic optimal control simulation can be set up in VisSim as shown in Fig. 77.6. After 26 simulation runs, the final value of the cost function is almost zero. The final values of k_1 and k_2 are 1.52 and 4.47, respectively.

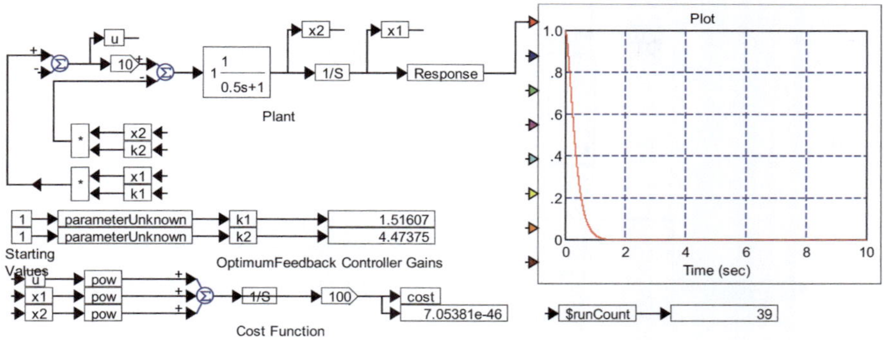

Fig. 77.6 The quadratic optimal controller tuning diagram of the FB-33 control system

77.4.1 Real-Time Hardware-in-the-loop

Once the quadratic optimal control design is complete, the controller performance with the FB-33 control system can be tested. VisSim/Real-Time is a VisSim option that provides the capability to couple a VisSim system model directly with a real-world hardware. Hardware-in-the-loop (HIL) systems can be configured and executed by interfacing VisSim controller mode with the FB-33 control system.

With VisSim/Real-Time, the PC can be used as online servo controller for the FB-33 control system through a high-speed data acquisition module (RT-DAC/USB). The A/D conversion time of the RT-DAC/USB board is equal to 5.4 μs. The RT-DAC/USB interface module connected with the PC offers a high sampling rate for A/D and D/A channels larger than the bandwidth of the FB-33 servomotor. Figure 77.7 shows a PC-based real-time control diagram and the actual step response of the FB-33 control system with quadratic optimal controller.

77.5 Concluding Remarks

1. This chapter has presented an approach using VisSim to design the quadratic optimal controller. This approach is a very powerful method to avoid tedious theoretical design.
2. VisSim is a Windows-based software package for the modeling, design, and simulation of complex control systems without writing a line of code. It offers unprecedented ease-of-use and consequently a short learning curve.
3. The PC-USB-based real-time control system using the quadratic optimal control algorithm and data acquisition system to control the real-world have been developed in this chapter.

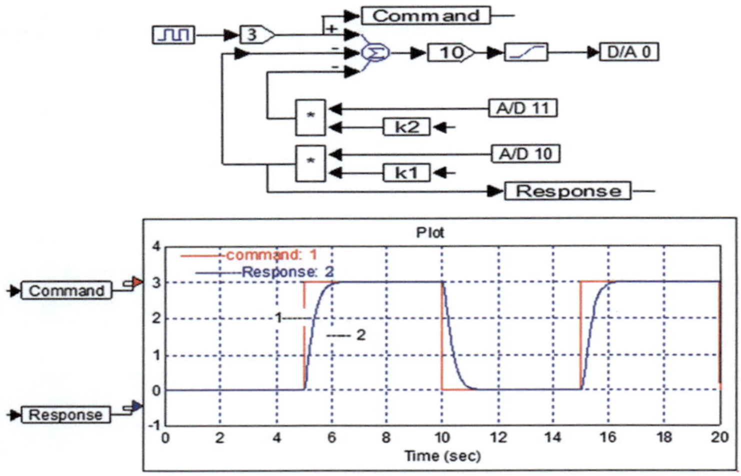

Fig. 77.7 PC-based real-time control diagram and the actual step response

4. Directions for future work include studying time-optimal controller, dead-beat controller, fuzzy controller, and adaptive control algorithm, and designing various interfacing capabilities for the design of USB-based in-the-loop control systems.

References

1. Visual Solutions: VisSim user's guide. Visual Solutions, Westford, MA (1995)
2. VisSim/OptimizePRO user's guide—version 4.5. Visual Solutions. (2000)
3. VisSim/Real-TimePRO user's guide—version 4.5. Visual Solutions Inc. (2000)
4. RT-DAC/USB user's manual. Intelligent Technology for Control Ltd. (2004)
5. RT-Motion-USB user's manual- Chinese version, Cho-Chieh Ltd. (2005)
6. Ogata, K.: Modern control engineering. Prentice Hall, Upper Saddle River (2002)
7. Kuo, B.C., Golnaraghi, F.: Automatic control systems, 8th edn. Wiley, New York (2002)
8. Bryson, A.E., Ho, Y.C.: Applied optimal control. Hemisphere Publishing Corporation, New York (1975)
9. Goodwin, G.C., Graebe, S.F., Salgado, M.E.: Control system design. Prentice Hall, Upper Saddle River (2001)
10. Analogue Servo Fundamentals Trainer 33-002. Feedback Instruments Ltd.
11. Ker, D.: The implementation of PC-based real-time control systems using Gain-Optimization Algorithm. Journal of China Institute of Technology **26**, 132–144 (2003)
12. Ker, D., Chang, H.W.: The implementation of PC-based real-time control systems using Quadratic Optimal Control. Journal of China Institute of Technology **27**, 439–447 (2003)
13. Ker, D.: Automatic control simulation analysis and experiments. Gau Lih Book, (2008)
14. Ker, D.: Application of virtual instrumentation in control system analysis and design. Journal of China Institute of Technology **25**, 64–74 (2002)

15. Ker, D.: The implementation of PC-based real-time control systems using Pole-Zero Cancellation Method. In: Proceedings of the 6th Conference on Integrated Opto-Mechatronic Technology and Intellectual Property Rights, December 20, 2011, pp. 19–29
16. Ker, D.: DSP-based real-time control systems using Pole-Placement Method. The 7th Conference on Integrated Opto-Mechatronic Technology and Intellectual Property Rights, May 14, 2013, pp. 12–19

Chapter 78
A Bibliometric Analysis on Data Mining Using Bradford's Law

Jiann-Min Yang, Shu-Feng Tseng, and Yu-Ling Won

Abstract The subject-specific Bibliometric Analysis on Data Mining complies with the Bradford's Law to show the $1{:}n{:}n^2$ relationship and the zone distribution of research journals. The overall count of articles in each zone is about the same. The top three author countries contribute to more than 50 % of total journal articles. However, big differences in productivity exist among them. Most of the high ranking author organizations are academic institutions and many of them belong to the top three author countries. The total article count grows year by year except a sudden drop in 2007 but grows again in 2008 and afterwards. The top three research areas are quite stable. There are some variations in other research areas. Some drop down the article counts and some others grow gradually.

Keywords Data mining • Bradford's law • Bibliometric analysis

78.1 Introduction

Data Mining is evolved from disciplines of Database, Statistics, and Artificial Intelligence, etc. [1]. Various data analysis technologies are integrated into this subject to support forecasting and decision making. Data Mining has been an important subject in either academic research or practical application. How is the distribution of the subject-specific journals and articles revealed in the literature? What are the most popular author countries, organizations, and research areas? These are research questions to be addressed in this chapter.

In Bibliometrics, Bradford's Law has been addressed the most and has far-reaching impact. Most bibliometric laws delineate some relationships among journals and articles in the literature. This chapter adopts Bradford's Law to examine the literature and observe the compliance of the addressed relationships

J.-M. Yang • S.-F. Tseng • Y.-L. Won (✉)
Department of MIS, National Chengchi University, Taipei 116, Taiwan
e-mail: tony@cc.cust.edu.tw

J. Juang (ed.), *Proceedings of the 3rd International Conference on Intelligent Technologies and Engineering Systems (ICITES2014)*, Lecture Notes in Electrical Engineering 345, DOI 10.1007/978-3-319-17314-6_78

as described in the Bradford's Law. The core journals are then identified for the Data Mining subject. Further analyses with respect to author affiliated country, organization, and research area in the recent 10 years, 2004–2013 are also presented.

78.2 Literature Review

78.2.1 Data Mining

Based on Technology Review published by MIT in 2001/1–2, Data Mining would be one of the top ten new technologies changing the world [2]. Data Mining models include predictive models and descriptive models [1]. Predictive models are used to predict future occurrences based on the pattern recognized from the past data. Descriptive models, although not predicting future occurrences, are used to exhibit the analysis patterns for users to make decision by referencing the recognized patterns..

78.2.2 Bradford's Law

With journals aligned by the sequence of decreasing article count for each, Bradford's Law [3] delineates the $1{:}n{:}n^2$ relationship and the zone distribution of research journals. The overall count of articles in each zone is about the same. The journals falling into the first zone is called core journals. Total number of articles in each zone is about the same. The figure of n is close to 5.

Bradford Distribution is a semi-log graph which visualizes data with an exponential relationship. This semi-log graph is a lin-log graph which uses a logarithmic scale on the x-axis and a linear scale on the y-axis. The later part of the distribution exhibits a close-linear relationship, as shown in Fig. 78.1.

Suppose $\overline{ox_1} = r$, $\overline{x_1x_2} = \overline{x_2x_3} = s$,
and log values of α, β, γ correspond to $\overline{ox_1}$, $\overline{ox_2}$, $\overline{ox_3}$,
then $\log\alpha = r$, $\log\beta = r + s$, $\log\gamma = r + 2\,s$.
Let $10^S = n$, then $\alpha{:}\beta{:}\gamma = 1{:}n{:}n^2$.

Tsay [4] describes the distribution as similar to a comet. The journals falling into $\overline{ox_1}$ zone belong to "core journals." The detritus part of the comet belongs to "border journals" containing small amount of subject-specific articles occasionally.

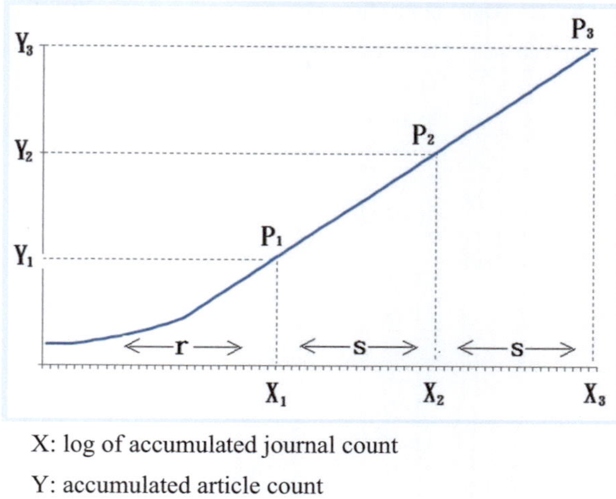

Fig. 78.1 Bradford distribution. X: log of accumulated journal count Y: accumulated article count [*Source*: Bradford, S. C.: Sources of Information on Specific Subjects. Engineering. 137, 85–86 (1934)]

78.3 Bibliometric Analysis

Articles with keyword of "Data Mining" or "Knowledge Discovery from SCI (Science Citation Index Expanded) and SSCI (Social Sciences Citation Index) and AHCI (Arts & Humanities Citation Index) [5] during 1983–2013 were collected for analysis. The total count of these subject-specific articles is 16,689. The subject is named as "Data Mining."

The collected literature data were analyzed based on Bradford's Law to observe the compliance the addressed distribution for the subject-specific journal and article counts. Furthermore, analyses of article counts by author country, organization, and research area were also conducted to observe various aspects of the literature distribution for the recent 10 years.

78.3.1 Initial Bibliometric Data

Based on the databases of SCI/SSCI/AHCI databases, the term of "Data Mining" is coined in 1983 by Lovell [6]. Only two articles mention this term in 1985 [7, 8]. More articles address related issues after 1991 [9–12]. After 1997, the article count grows faster to more than 100 per year. After 2005, the article count grows further up to more than 1,000 per year. The count comes down a little bit in 2007 and grows again afterwards.

Table 78.1 Data mining literature distribution (1983–2013)

Journal count	Article count	Accumulated journal count	Accumulated article count
1	968	1	968
1	884	2	1,852
1	752	3	2,604
1	393	4	2,997
12	101–300	16	4,753
9	71–100	25	5,514
7	61–70	32	5,973
8	51–60	40	6,407
13	41–50	53	6,987
24	31–40	77	7,803
46	21–30	123	8,959
157	11–20	280	11,243
3,271	1–10	3,551	16,689

Table 78.2 Distribution zone of data mining literature (1983–2013)

Journal count	Article count	Accumulated journal count	Journal article count	Accumulated article count	Log of accumulated journal count
25	71–968	25	5,514	5,514	1,398
255	11–70	280	5,729	11,243	2,447
3,271	1–10	3,551	5,446	16,689	3,550

78.3.2 Bradford's Law Analysis

The literature distribution in the subject of Data Mining is summarized in Table 78.1. Using the Bradford literal description, the article counts can be divided into three zones, as shown in Table 78.2. The first zone contains 25 journals with high article count ($\geq$71) for each. The second zone contains 255 journals with less article count (11–70) for each. The third zone contains 3,271 journals with low article count (1–10) for each.

As shown in Table 78.2, the overall article counts in three zones are about the same, near to 5,563 which is one-third of the accumulated total 16,689. The ratio of journal counts in three zones is 25:255:3,271, close to 22:268:3,274 or $1:12.2:12.2^2$. Therefore, $n = 12.2$. Although the figure of n is not exactly the same as 5 in Bradford's Law, the distribution of journal counts $1:n:n^2$ is quite similar to the literal description in Bradford's Law. Let's draw a lin-log graph using the Table 78.1 data, as shown in Fig. 78.2.

As shown in Fig. 78.2, the later part of the distribution exhibits a close-linear relationship, similar to Bradford's graphic description.

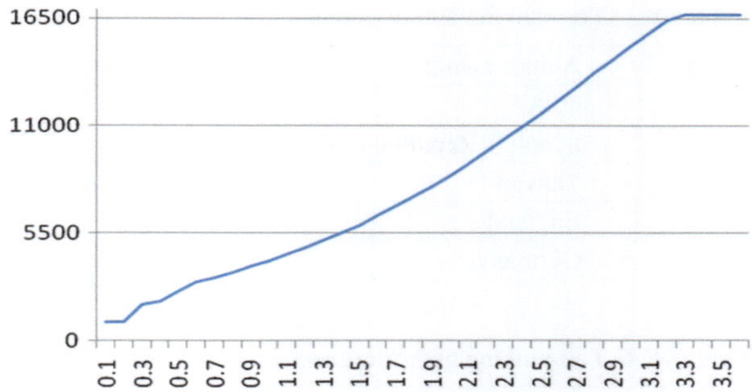

Fig. 78.2 Lin-log graph of data mining literature during 1983–2013

Corresponding to the Journal Article Counts in Table 78.2, $\overline{oy_1} = 5,514$, $\overline{y_1y_2} = 5,729$, and $\overline{y_2y_3} = 5,446$ are similar in three zones. Besides, $\alpha = 25$, $\beta = 280$, and $\gamma = 3{,}551$ correspond to the Accumulated Journal Counts in Table 78.2.

Suppose $\overline{ox_1} = r$, $\overline{x_1x_2} = s1$, $\overline{x_2x_3} = s2$,
and Log values of α, β, γ correspond to $\overline{ox_1}$, $\overline{ox_2}$, $\overline{ox_3}$,
then, $\log\alpha = r$, $\log\beta = r + s1$, $\log\gamma = r + s1 + s2$.
Therefore $r = 1.398$, $s1 = 2.447 - 1.398 = 1.049$, $s2 = 3.550 - 2.447 = 1.103$.
According to the Bradford's graphic description, $s1$ should be equal to $s2$.
In our data analysis, $s1 = 1.049$ is close to $s2 = 1.103$ with a bias of
$5\ \% = (1.103 - 1.049)/(0.5 \times (1.103 + 1.049))$.
Let $10^S = n$, then $\alpha{:}\beta{:}\gamma = 25{:}280{:}3{,}551$,
which is close to $25{:}300{:}3{,}600 = 1{:}12{:}12^2 = 1{:}n{:}n^2$.

This is similar to what described in Bradford's graphic description, $\alpha{:}\beta{:}\gamma = 1{:}n{:}n^2$ but $n = 12$ not 5. $\overline{ox_1}$ represents zone 1 containing the core journals. Therefore, the journals having subject-specific articles equal to or greater than 71 belong to the "core journals."

78.3.3 Aggregation Analysis

The Data Mining literature is analyzed by author country, organization, and research area for the recent 10 years, 2004–2013.

Table 78.3 shows the first five ranks with respect to Author Country. The overall article count of the top three ranks is 6,608, occupying more than half (53 %) of the total. There are big differences among the top three, compared to the small differences for the ranks 3, 4, and 5. Table 78.4 shows the top three ranks in Author Organization. Their article counts are 438 together, occupying only 3.51 % of the total.

As shown in Fig. 78.3, the total article count in Data Mining has been growing up till 2013. There is only a short fall down in 2007 and grows up again in the next

Table 78.3 Top ranking author countries for 2004–2013

Rank	Author country	Article count	Article percentage (%)
1	USA	3,959	31.71
2	People's Republic of China	1,672	13.39
3	Taiwan	977	7.83
4	England	785	6.29
5	Germany	687	5.50

Table 78.4 Top ranking author organizations for 2004–2013

Rank	Author organization	Article count	Article percentage (%)
1	Chinese Acad Sci	214	1.71
2	Univ Illinois	124	0.99
3	Natl Cent Univ	100	0.80
4	Natl Chiao Tung Univ	99	0.79
5	Univ Iowa	98	0.79

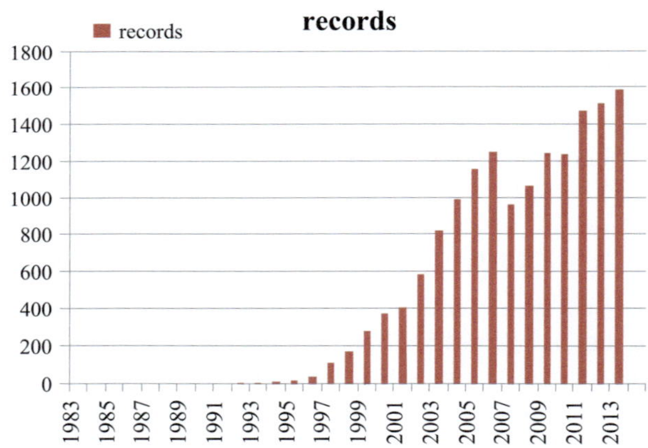

Fig. 78.3 Article count distribution by year

year. The reason may be due to the withdrawals of LECTURE NOTES IN COMPUTER SCIENCE (LNCS) and LECTURE NOTES IN ARTIFICIAL INTELLIGENCE (LNAI) from SCI. The article count reduces 300–400 every year. This may be the cause of drop in 2007. However, the continuous growth brings the count back to later increase again.

As shown in Fig. 78.4, the article counts of COMPUTER SCIENCE ARTIFICIAL INTELLIGENCE, COMPUTER SCIENCE INFORMATION SYSTEMS, and COMPUTER SCIENCE THEORY METHODS drop dramatically. They are the top three ranks before the sudden drop in 2007. Especially, COMPUTER SCIENCE THEORY METHODS drops to rank 6–9 and never comes back to the original higher rank again. There may be some development obstacles in these research areas. On the other hand, the article count of ENGINEERING

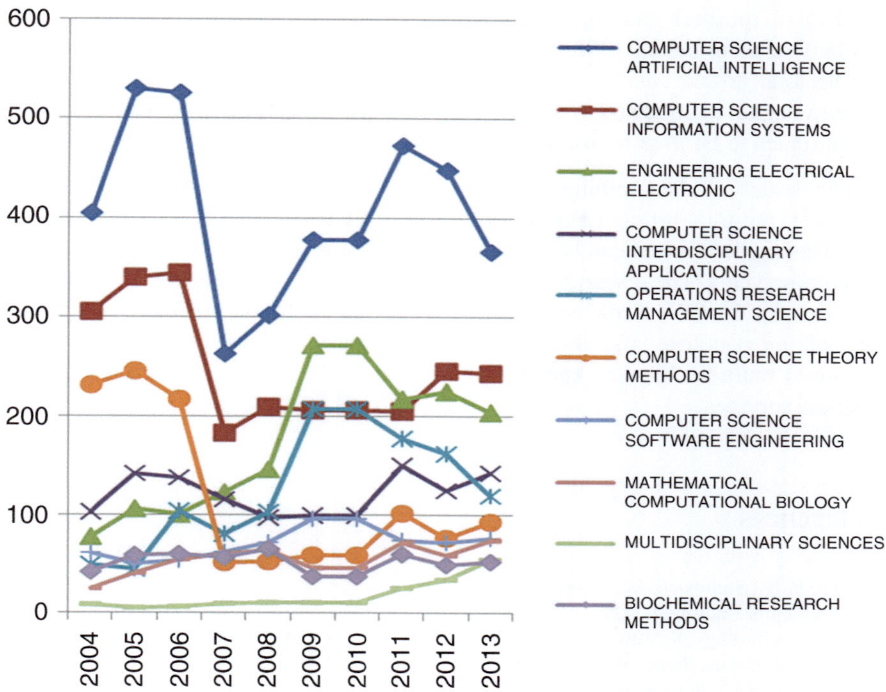

Fig. 78.4 Article count distribution by research area

ELECTRICAL ELECTRONIC has a big growth in 2009 and remains as high as rank three in 2013. Another worth noting journal is MULTIDISCIPLINARY SCIENCES. The rank is not high but has been growing up stably, indicating that Data Mining has been a growing- up subject in cross-domain research.

78.4 Conclusion and Future Research

This chapter analyzes the Data Mining literature during 1983–2013 in SCI/SSCI/AHCI databases to observe the compliance of the Bradford's Law. Further aggregation analyses are based on various aspects of author country, organization, and research area for the recent 10 years.

The research is concluded as follows.

1. The subject-specific Bibliometric Analysis in Data Mining complies with the Bradford's Law to show the $1{:}n{:}n^2$ relationship and the zone distribution of research journals. The overall count of articles in each zone is about the same.
2. The top three author countries contribute to more than 50 % of total journal articles. However, big differences in productivity exist among them.

3. Most of the high ranking author organizations are academic institutions and many of them belong to the top three author countries.
4. The total article count grows year by year except a sudden drop in 2007 but grows again in 2008 and afterwards. This reflects that the subject of Data Mining continues to be important and the article publication is not yet saturated.

The issues of data mining may evolve into issues of big data applications. However, the total article count of Big Data in SCI/SSCI/AHCI databases is still low. The big data issues can be anticipated to become important and grow faster in the coming years. The research approach presented in this chapter can be applied or extended for the "Big Data" subject to observe the evolution process. Also, in the top ranking countries, the abundant research expertise can be used to apply the important multidisciplinary knowledge to enhance the productivity and quality of business services.

References

1. Tan, P. N., Steinbach, M., Kumar, V.: Introduction to data mining. Addison-Wesley/Pearson International Edition (2006)
2. MIT Technology Review, http://www2.technologyreview.com/featured-story/400868/emerging-technologies-that-will-change-the-world/
3. Bradford, S.C.: Sources of information on specific subjects. Engineering **137**, 85–86 (1934)
4. Tsay, M.Y.: Informetrics and characteristics of literature. Hwa Tai Publishing, Taipei (2003)
5. Web of Science, https://app.webofknowledge.com
6. Lovell, M.C.: Data mining. Rev. Econ. Stat. **65**(1), 1–12 (1983)
7. Denton, F.T.: Data mining as an industry. Rev. Econ. Stat. **67**(1), 124–127 (1985)
8. Marquez, J., Shackmarquez, J., Wascher, W.L.: Statistical-inference, model selection and research experience—a multinomial model of data mining. Economics Lett. **18**(1), 39–44 (1985)
9. Delong, J.B., Lang, K.: Are all economic hypotheses false. J. Political Econ. **100**(6), 1257–1272 (1992)
10. Veall, M.R.: Bootstrapping the process of model selection—an econometric example. J. Appl. Econometrics **7**(1), 93–99 (1992)
11. Zivot, E., Andrews, D.W.: Further evidence on the great crash, the oil-price 14_Shock, and the unit-root hypothesis. J. Bus. Econ. Statistics **10**(3), 251–270 (1992)
12. Piatetskyshapiro, G.: Knowledge discovery in databases. IEEE Expert-intellegent Syst. Their Appl. **6**(5), 74–76 (1991)

MIX
Papier aus verantwortungsvollen Quellen
Paper from responsible sources
FSC® C105338

If you have any concerns about our products,
you can contact us on
ProductSafety@springernature.com

In case Publisher is established outside the EU,
the EU authorized representative is:
Springer Nature Customer Service Center GmbH
Europaplatz 3, 69115 Heidelberg, Germany

Printed by Libri Plureos GmbH
in Hamburg, Germany